NOVEL APPROACHES OF NANOTECHNOLOGY IN FOOD

NOVEL APPROACHES OF NANOTECHNOLOGY IN FOOD

Nanotechnology in the Agri-Food Industry, Volume 1

Edited by

ALEXANDRU MIHAI GRUMEZESCU

Department of Science and Engineering of Oxide Materials and Nanomaterials, Faculty of Applied Chemistry and Materials Science, University Politehnica of Bucharest, Bucharest, Romania

ELSEVIER

AMSTERDAM • BOSTON • HEIDELBERG • LONDON • NEW YORK • OXFORD
PARIS • SAN DIEGO • SAN FRANCISCO • SINGAPORE • SYDNEY • TOKYO
Academic Press is an imprint of Elsevier

Academic Press is an imprint of Elsevier
125, London Wall, London EC2Y 5AS, UK
525 B Street, Suite 1800, San Diego, CA 92101-4495, USA
50 Hampshire Street, 5th Floor, Cambridge, MA 02139, USA
The Boulevard, Langford Lane, Kidlington, Oxford OX5 1GB, UK

Notices

Knowledge and best practice in this field are constantly changing. As new research and experience broaden our understanding, changes in research methods, professional practices, or medical treatment may become necessary.

Practitioners and researchers must always rely on their own experience and knowledge in evaluating and using any information, methods, compounds, or experiments described herein. In using such information or methods they should be mindful of their own safety and the safety of others, including parties for whom they have a professional responsibility.

To the fullest extent of the law, neither the Publisher nor the authors, contributors, or editors, assume any liability for any injury and/or damage to persons or property as a matter of products liability, negligence or otherwise, or from any use or operation of any methods, products, instructions, or ideas contained in the material herein.

British Library Cataloguing-in-Publication Data
A catalogue record for this book is available from the British Library

Library of Congress Cataloging-in-Publication Data
A catalog record for this book is available from the Library of Congress

ISBN: 978-0-12-804308-0

Publisher: Nikki Levy
Acquisition Editor: Patricia Osborn
Editorial Project Manager: Karen Miller
Production Project Manager: Caroline Johnson
Designer: Mark Rogers

Typeset by Thomson Digital

CONTENTS

LIST OF CONTRIBUTORS

Ann Rose Abraham
Mahatma Gandhi University, International and Inter University Centre for Nanoscience and Nanotechnology; Mahatma Gandhi University, School of Pure and Applied Physics, Kottayam, Kerala, India

Cristóbal N. Aguilar-González
Autonomous University of Coahuila, Food Research Department, Chemistry Faculty, Col. Republic East, Saltillo, Coahuila, Mexico

Asad Mohammad Amini
Ferdowsi University of Mashhad (FUM), Food Hydrocolloids Research Centre, Department of Food Science and Technology, Mashhad, Iran

Morata Antonio
enotecUPM, Polytechnic University of Madrid, Chemistry and Food Technology Department, Higher Technical College of Agronomic Engineers, Avenida Complutense S/N, Madrid, Spain

Robin Augustine
Department of Materials Science and Engineering, Technion Israel Institute of Technology, Technion City, Haifa, Israel; Mahatma Gandhi University, International and Inter University Centre for Nanoscience and Nanotechnology, Kottayam, Kerala, India

Claudia Baumung
Chemisches und Veterinäruntersuchungsamt Karlsruhe, Karlsruhe, Baden-Württemberg, Germany

Arturo Armone Caruso
ENT Department, AIAS Structure of Afragola, Contrada Leutrec snc, Naples, Italy

Visith Chavasit
Mahidol University, Research and Education Unit on Sensory Science for Better Nutrition, Institute of Nutrition, Nakhonpathom, Thailand

Bing Huei Chen
Fu Jen University, Department of Food Science, Taipei, Taiwan

Amparo Chiralt
Universitat Politècnica de València, Research Institute of Food Engineering for Development, Valencia, Spain

Kai On Chu
The Chinese University of Hong Kong, Department of Ophthalmology and Visual Sciences, Hong Kong Eye Hospital, Kowloon, Hong Kong

Juan C. Contreras-Esquivel
Autonomous University of Coahuila, Food Research Department, Chemistry Faculty, Col. Republic East, Saltillo, Coahuila, Mexico

Gülşen Akın Evingür
Piri Reis University, Faculty of Engineering, Department of Electrical and Electronics Engineering, Tuzla- Istanbul, Turkey

Lydia Ferrara
University of Naples Federico II, Department of Pharmacy, Naples, Italy

Monica Gallo
University of Naples Federico II, Department of Molecular Medicine and Medical Biotechnology, Naples, Italy

Konstantinos G. Gatos
R&D Center, Megaplast S.A., Athens, Greece

Padmini S. Ghugre
SNDT Women's University, Department of Food Science and Nutrition, Mumbai, Maharashtra, India

Alexandru Mihai Grumezescu
University Politehnica of Bucharest, Department of Science and Engineering of Oxide Materials and Nanomaterials, Faculty of Applied Chemistry and Materials Science, Bucharest, Romania

Anna Ilyina
Autonomous University of Coahuila, Nanobioscience Group, Chemistry Faculty, Col. Republic East, Saltillo, Coahuila, Mexico

Baskaran Stephen Inbaraj
Fu Jen University, Department of Food Science, Taipei, Taiwan

Loira Iris
enotecUPM, Polytechnic University of Madrid, Chemistry and Food Technology Department, Higher Technical College of Agronomic Engineers, Avenida Complutense S/N, Madrid, Spain

Alberto Jiménez
Universitat Politècnica de València, Research Institute of Food Engineering for Development, Valencia, Spain

Suárez-Lepe José Antonio
enotecUPM, Polytechnic University of Madrid, Chemistry and Food Technology Department, Higher Technical College of Agronomic Engineers, Avenida Complutense S/N, Madrid, Spain

Nandakumar Kalarikkal
Mahatma Gandhi University, International and Inter University Centre for Nanoscience and Nanotechnology; Mahatma Gandhi University, School of Pure and Applied Physics, Kottayam, Kerala, India

Muhammad Issa Khan
University of Agriculture, National Institute of Food Science and Technology, Faculty of Food, Nutrition and Home Sciences, Faisalabad, Pakistan

Evamaria Kratz
Chemisches und Veterinäruntersuchungsamt Karlsruhe, Karlsruhe, Baden-Württemberg, Germany

Aditya S. Kulkarni
Aten Porus Lifesciences, Bengaluru, Karnataka, India

Dirk W. Lachenmeier
Chemisches und Veterinäruntersuchungsamt Karlsruhe, Karlsruhe, Baden-Württemberg, Germany

Virangkumar Nanubhai Lad
Sardar Vallabhbhai National Institute of Technology, Department of Chemical Engineering, Surat, Gujarat, India

Sophiya Devi Lamabam
Mizoram University, Department of Biotechnology, School of Life Sciences, Mizoram, India

Danielle Branta Lopes
University of Campinas (UNICAMP), Department of Food and Nutrition, Food Engineering Faculty, Campinas, São Paulo, Brazil

Gabriela Alves Macedo
University of Campinas (UNICAMP), Department of Food and Nutrition, Food Engineering Faculty, Campinas, São Paulo, Brazil

Gene Chi Wai Man
The Chinese University of Hong Kong, Department of Orthopaedics and Traumatology, Faculty of Medicine, Prince of Wales Hospital, Shatin, New Territories, Hong Kong

Santi M. Mandal
Vidyasagar University, Department of Microbiology, Midnapore, West Bengal, India

José L. Martínez-Hernández
Autonomous University of Coahuila, Nanobioscience Group, Chemistry Faculty, Col. Republic East, Saltillo, Coahuila, Mexico

Zagabathuni Venkata Panchakshari Murthy
Sardar Vallabhbhai National Institute of Technology, Department of Chemical Engineering, Surat, Gujarat, India

Daniele Naviglio
University of Naples Federico II, Department of Chemical Sciences, Naples, Italy

Sócrates A. Palacios-Ponce
Autonomous University of Coahuila, Nanobioscience Group, Chemistry Faculty, Col. Republic East, Saltillo, Coahuila, Mexico

Önder Pekcan
Kadir Has University, Faculty of Engineering and Natural Sciences, Department of Bioinformatics and Genetics, Cibali- Istanbul, Turkey

Tabea Pflaum
Chemisches und Veterinäruntersuchungsamt Karlsruhe, Karlsruhe, Baden-Württemberg, Germany

Ubaid ur Rahman
University of Agriculture, National Institute of Food Science and Technology, Faculty of Food, Nutrition and Home Sciences, Faisalabad, Pakistan

Rodolfo Ramos-González
Autonomous University of Coahuila, Nanobioscience Group, Chemistry Faculty, Col. Republic East, Saltillo, Coahuila, Mexico

Seyed Mohammad Ali Razavi
Ferdowsi University of Mashhad (FUM), Food Hydrocolloids Research Centre, Department of Food Science and Technology, Mashhad, Iran

Anupam Roy
Agricultural and Food Engineering Department, Indian Institute of Technology Kharagpur, West Bengal, India

Jaquelina Sánchez-Ramírez
Autonomous University of Coahuila, Nanobioscience Group, Chemistry Faculty, Col. Republic East, Saltillo, Coahuila, Mexico

Amna Sahar
University of Agriculture, Department of Food Engineering, Faculty of Agricultural Engineering and Technology, Faisalabad, Pakistan

Nobuyuki Sakai
Tohoku University, Department of Psychology, Graduate School of Arts and Letters, Sendai, Japan

Kerstin Schöberl
Chemisches und Veterinäruntersuchungsamt Karlsruhe, Karlsruhe, Baden-Württemberg, Germany

Elda P. Segura-Ceniceros
Autonomous University of Coahuila, Nanobioscience Group, Chemistry Faculty, Col. Republic East, Saltillo, Coahuila, Mexico

Shanker Lal Shrivastava
Agricultural and Food Engineering Department, Indian Institute of Technology Kharagpur, West Bengal, India

Ubonrat Siripatrawan
Chualalongkorn University, Department of Food Technology, Faculty of Science, Bangkok, Thailand

Paula Speranza
University of Campinas (UNICAMP), Department of Food and Nutrition, Food Engineering Faculty, Campinas, São Paulo, Brazil

Dongxiao Sun-Waterhouse
School of Chemical Sciences, University of Auckland, Auckland, New Zealand

Robert Thangjam
Mizoram University, Department of Biotechnology, School of Life Sciences, Mizoram, India

Sabu Thomas
Mahatma Gandhi University, International and Inter University Centre for Nanoscience and Nanotechnology; Mahatma Gandhi University, School of Chemical Sciences, Kottayam, Kerala, India

Shobha A. Udipi
SNDT Women's University, Department of Food Science and Nutrition, Mumbai, Maharashtra, India

Hisauyki Uneyama
Chemosensory Research Group, Frontier Research Labs., Institute for Innovation, Ajinomoto Co., Inc., Kawasaki, Japan

María Vargas
Universitat Politècnica de València, Research Institute of Food Engineering for Development, Valencia, Spain

Alejandra Vargas-Segura
Autonomous University of Coahuila, Nanobioscience Group, Chemistry Faculty, Col. Republic East, Saltillo, Coahuila, Mexico

Chi Chiu Wang
The Chinese University of Hong Kong, Department of Obstetrics and Gynaecology, Faculty of Medicine, Prince of Wales Hospital, Shatin, New Territories, Hong Kong

Geoffrey I.N. Waterhouse
School of Chemical Sciences, University of Auckland, Auckland, New Zealand

SERIES FOREWORD

The emergence of nanotechnology has reached impressive heights in recent years and the development of special nanodevices and nanomaterials has found intriguing applications in agriculture and food sector. Most of the investigated nanotechnological approaches initially aimed to solve evolving problems in the agri-food industry in order to impact on the economic potential. Soon after the implementation of new technologies and approaches that were using nanostructured materials, the worldwide concern was rapidly extended to numerous applications that could be developed by using the science of nanosized materials. Smart materials, biosensors, packaging materials, nutraceuticals, and nanodevices have been designed to address numerous agri-food related issues with direct impact in health, economy, ecology, and industry. As the engineering of nanostructures has constantly progressed and extended its applications, there is virtually unlimited potential in this sector. However, the widely differing opinions on the applicability and usefulness of nanotechnology between both specialists and the general public has hampered progress. The main concern manifested by people is related to the potential risk for health and the environmental impact of the recently developed nanoengineered materials and devices. Therefore, current approaches are strictly considering these concerns when designing nanotechnological solutions for agriculture and food sectors.

This multivolume series was developed by the constant need to discover current inquiries and approaches on the field of agri-food science and also to learn about the most recent progress, approaches, and applications that have emerged through nanotechnology.

As agriculture is the backbone of most developing countries, nanotechnology has the potential to revolutionize the agriculture and food sector by promoting productivity through genetic improvement of plant and animal foods. It can also ensure the delivery of drugs, genes, and pesticides to specific sites at cellular levels in targeted plants and animals, by limiting side effects. Nanotechnology can be used to evaluate gene expression under different stress condition for both plant and animal foods through the development of nanoarray-based gene-technologies. Additionally, this technology can detect fertilizers, pesticides with high precision by smart nanosenors for an adequate management of the natural resources. Moreover, numerous industrial-related applications with direct impact on economy have emerged. For example, nano- and micro-structured arrays can detect the early presence of pathogens, contaminants, and food spoilage factors. Other applications for this technology are smart integration systems for food processing and packaging, as well as nanoemulsion-based decontaminants for food equipment and storage compartments, and nanoparticles that facilitate the bioavailability and delivery of nutrients directly to cells.

The potential benefits of nanotechnology for agriculture, food, fisheries, and aquaculture were identified and supported by many countries, which invested a significant amount of money in the development of applications. Also, numerous campaigns are currently trying to increase awareness on the developing process and recent technologies in order to influence the acceptance of customers. Although nanoagri-food industrialized concept could help to find a sustainable solution for the current global food crisis, the offered advantages should balance the concerns regarding soil, water, environment, and health related issues that such approach could bring.

The series entitled *Nanotechnology in the Agri-Food Industry* brings comprehensive and recent knowledge regarding the impact of the science of nanometer-sized materials on the field of agriculture and food industry, but also discuss the current inquiries regarding risks of these applications in all relevant fields such as environment and health, aiming to increase awareness to a wider amount of readers.

Dr Alina Maria Holban
Department of Microbiology and Immunology, Faculty of Biology, University of Bucharest, Romania

Dr Monica Cartelle Gestal
Harvill Lab, University Of Georgia, Athens, GA, United States

SERIES PREFACE

About the Series (Volumes I–X)

In a permanently changing society, health and well being remain the key drivers for the food industry. Despite the technological progress made in the agri-food industry, a true food crisis emerges in several areas of the globe. This can be explained by insufficient food but mostly by inadequate food for a very distinct range of consumers. In this context, innovative technologies represent the core throughout the whole food chain from raw materials/ingredient sourcing, food processing, quality control of finished products, and packaging. Nanotechnology, coupled with novel interdisciplinary approaches and processing methods, has enabled some important advances recently flourishing in many of these areas. The science of nanosized materials can improve and even resolve the huge challenges faced by the food and bioprocessing industries for developing and implementing systems that can produce qualitative and quantitative foods that are safe, sustainable, environment friendly, and efficient. This emerging tool finds its applications in various fields and represents an endless approach for the development of innovative strategies in food development, processing, and packaging.

This multivolume set aims to bring together the most recent and innovative applications of nanotechnology in the agri-food industry, but also to present the future perspectives in the design of new or alternative foods.

The series contains 200 chapters organized in 10 volumes, prepared by outstanding research groups that made significant impacts on the field of nanotechnology and food-related research sectors. This comprehensive set represents an updated and highly structured material for undergraduate and postgraduate students in food science, biotechnological, engineering fields, but also a valuable resource of recent scientific progress, along with most known applications of nanomaterials on the food industry to be used by researchers, engineers, and academia. Moreover, novel opportunities and ideas for developing or improving technologies in the agri-food industry by innovative companies, biotechnological industries, and other economical structures are highlighted and their potential is widely dissected. This series may be also valuable for the wide audience interested in recent nanotechnological progress in the agri-food field worldwide.

These 10 volumes cover almost all aspects related to the applications of *Nanotechnology in the Agri-Food Industry* and are named as:

Volume I Novel Approaches
Volume II Encapsulations
Volume III Emulsions
Volume IV Nutraceuticals
Volume V Nutrient Delivery
Volume VI Food Preservation
Volume VII Food Packaging
Volume VIII NanoBioSensors
Volume IX Water Purification
Volume X New Pesticides and Soil Sensors

Each volume contains 20 chapters, which were carefully composed and illustrated to highlight the most innovative and intensively investigated applications of nanotechnology on particular wide interest domains of the agri-food industry field.

Alexandru Mihai Grumezescu
University Politehnica of Bucharest,
Department of Science and Engineering
of Oxide Materials and Nanomaterials,
Faculty of Applied Chemistry and
Materials Science, Bucharest, Romania
http://grumezescu.com/

VOLUME PREFACE

The first volume of the series, entitled *Approaches of Nanotechnology in Food*, represents a summary of the most recent advances made on the field of nanostructured materials that significantly impacted on agri-food industry. Since the current food market needs innovation, nanotechnology, coupled with novel interdisciplinary approaches and processing methods, has enabled important advances that have the potential to revolutionize the agri-food sector. Nanotechnology can serve to resolve challenges faced by the food and bioprocessing industries for developing and implementing systems that can produce qualitative and quantitative foods that are safe, sustainable, and ecofriendly. Volume I contains 20 chapters, prepared by outstanding international researchers from Brazil, China, Germany, Greece, India, Iran, Israel, Italy, Japan, Mexico, New Zealand, Pakistan, Spain, Taiwan, Thailand, and Turkey.

In Chapter 1, *An Update of Definitions and Regulations Regarding Nanomaterials in Foods and Other Consumer Products*, Claudia Baumung et al. present an up-to-date overview regarding the regulations for nanomaterials in the European Union and critically discuss if the rules are adequate for consumer protection. So far only France has implemented an obligatory registration for the manufacture and sale of synthetic nanomaterials.

Dongxiao Sun-Waterhouse and Geoffrey I.N. Waterhouse, in Chapter 2, entitled *Recent Advances in the Application of Nanomaterials and Nanotechnology in Food Research*, highlight recent research that has synergistically combined nanotechnology, physical chemistry, and food science to develop smart real-time sensors for adulterant/metabolite detection (nanosensors), intelligent delivery systems for functional foods and nutraceuticals (nanocarriers), and novel food-processing methods that yield "nanoenabled foods."

Chapter 3, prepared by Anna Ilyina et al., entitled *Magnetic Separation of Nanobiostructured Systems for Innovation of Biocatalytic Processes in Food Industry*, present an analysis of methods for enzyme immobilization onto magnetic nanoparticles. It is hoped that the use of magnetic nanosupports will lead to the development of more efficient, economical, and environmentally friendly technological processes applied to the food industry.

In Chapter 4, *Natural Materials as Additives in Food Emulsions*, Virangkumar Nanubhai Lad et al. provide the state-of-the-art review of some natural materials with their reported characteristics beneficial for food preservation, food processing, and end usage. Fundamentals of rheology of food emulsions and gels have also been discussed.

Anupam Roy et al., in Chapter 5, *Self-Assembled Carbohydrate Nanostructures: Synthesis Strategies to Functional Application in Food*, report the novel approaches on the different self-assembly strategies followed along with its functional applications in the food sector. A discussion of assembling strategies, behavior, and structure–function relationship associated with varying shapes and sizes is made to give a broader direction to assembly strategy, its enhancement, and control.

Chapter 6, *Nanotechnology and Wine*, prepared by Antonio Morata et al., highlights new approaches related to nanotechnology for winemaking in order to improve wine quality and safety, as well as to facilitate wine analysis. Among other implementations of nanotechnology in enology, the classification of wines using an electronic tongue, the nanoliquid chromatography, the use of silver nanoparticles as antimicrobial agents and the magnetic separation of yeasts in sparkling wines, are also discussed.

Gülşen Akın Evingür and Önder Pekcan, in Chapter 7, entitled *Effect of Content and Temperature on the Phase Transitions of Polymer Composites Doped by Kappa Carrageenan and Alginate*, present the current knowledge about polymers used in the food industry to produce milk gels and to stabilize products, sauces, syrups, and toppings for ice cream. The effect of contents and temperature on gelation, drying, swelling, and elasticity behaviors are also discussed.

Seyed Mohammad Ali Razavi and Asad Mohammad Amini, in Chapter 8, *Starch Nanomaterials: A State-of-the-Art Review and Future Trends*, summarize different preparation methods, characterizations, modifications, and applications of starch nanomaterials. Additionally, the research trend and prospects of these nanomaterials are proposed by specifically considering the potential applications in food science.

Robin Augustine et al., in Chapter 9, *Monitoring and Separation of Food-Borne Pathogens Using Magnetic Nanoparticles*, discuss exploitation of magnetic nanoparticles for the monitoring and separation of various pathogens in the processed food and raw food materials including milk, meat, fruits, and vegetables has been detailed. The state-of-the-art of sensor-based monitoring of microorganisms, strategies adopted to enhance the sensitivity of such devices, recent advancements in the functionalization of magnetic nanoparticles for the specific detection and separation of various pathogens are also discussed.

Muhammad Issa Khan et al., in Chapter 10, *Nanotechnology in Healthier Meat Processing*, present new concepts about the role of nanotechnology in providing functional ingredients and meat processing, nanostructured emulsions, biopolymeric nanoparticles, and regulations associated with the safety of nanostructured meat products.

Alberto Jiménez et al., in Chapter 11, *Antimicrobial Nanocomposites for Food Packaging Applications: Novel Approaches*, reveal the current knowledge about nanocomposites for antimicrobial food-packaging applications, taking into account both biodegradable (polylactides, starch, proteins, etc.) and nonbiodegradable polymers. The production techniques, the variety of utilized nanomaterials, and the last applications to food products are reviewed in a comprehensive way. In addition, regulatory aspects related to nanocomposites are discussed.

Bing Huei Chen and Baskaran Stephen Inbaraj, in Chapter 12, *Nanomaterial-Based Sensors for Mycotoxin Analysis in Food*, report an up-to-date overview about the developments on nanomaterial-based sensors developed for analysis of mycotoxins in food.

Monica Gallo et al., in Chapter 13, *Applications of Chitosan as a Functional Food*, reviews chitosan as long-term soluble dietary fiber to improve digestion and aid the development of intestinal microflora with significant benefits to the health of the organism.

Ubonrat Siripatrawan, in Chapter 14, *Active Food Packaging from Chitosan Incorporated with Plant Polyphenols*, report an up-to-date overview regarding the applications of chitosan films and coatings incorporated with plant polyphenols for preserving qualities and prolonging the shelf life of food and agricultural products.

Aditya S. Kulkarni et al., in Chapter 15, entitled *Applications of Nanotechnology in Nutrition: Potential and Safety Issues*, highlight the work carried out in the area of nanotechnological approaches for food science and nutrition.

Gene Chi Wai Man et al., in Chapter 16, *Nanoencapsulation of Green Tea Catechins and its Efficacy*, review the current approach on using nanotechnology to incorporate epigallocatechingallate as a drug delivery system and its efficacy to improve our health.

Konstantinos G. Gatos, in Chapter 17, *Potential of Nanomaterials in Food Packaging*, presents nanomaterials of different composition and shape, which have been proposed for plastic packaging in order to enhance specific properties and assist the food industry.

Danielle Branta Lopes et al., in Chapter 18, *A New Approach for Flavor and Aroma Encapsulation*, discuss the novelties and the most used technologies for flavor encapsulation and its applications, presenting the advantages and disadvantages and critically assessing its use, especially in food systems, highlighting potential future developments in this rapidly growing area.

Sophiya Devi Lamabam and Robert Thangjam, in Chapter 19, *Emerging Trends in the Application of Nanobiosensors in the Food Industry*, emphasize the developments in the use of nanomaterial-based biosensors and their applications in food industry.

Nobuyuki Sakai et al., in Chapter 20, *Psychological and Physiological Bases of Umami Taste Perception as Related to Nutrition*, introduce Maslow's Motivational Theory to explain human eating behavior, especially on starting and stopping of eating. Then physiological and psychological mechanisms underlying umami taste perception are introduced. In each section, applications and implementations of the findings from studies on regulations of hunger, satiety, and body weight are discussed. The authors recommend Washoku as a healthy diet for regulating hunger, satiety, and body weight.

Alexandru Mihai Grumezescu
University Politehnica of Bucharest,
Department of Science and Engineering
of Oxide Materials and Nanomaterials,
Faculty of Applied Chemistry and
Materials Science, Bucharest, Romania
http://grumezescu.com/

1

AN UPDATE OF DEFINITIONS AND REGULATIONS REGARDING NANOMATERIALS IN FOODS AND OTHER CONSUMER PRODUCTS

Claudia Baumung, Tabea Pflaum, Kerstin Schöberl, Evamaria Kratz, Dirk W. Lachenmeier

Chemisches und Veterinäruntersuchungsamt Karlsruhe, Karlsruhe, Baden-Württemberg, Germany

1 Introduction

Following the introduction of nanotechnology into consumer products (Cushen et al., 2012), there has been increasing tension within the literature about whether nanotechnology regulations are sufficient or not (see overview in Corley et al., 2013). Environmental groups and consumer organizations have strongly argued for stricter regulations for nanomaterials, while other scholars have opposed regulatory change because the emerging technologies and their risks were judged as not significantly different from conventional products (Corley et al., 2013). In a survey of US researchers the strongest gaps in nanomaterials regulations were seen in the areas of bioengineering, medicine, synthetic biology, and cosmetics (Corley et al., 2013). The interest of researchers about this topic is shown by a considerable rise of number in publications mentioning the keywords nanotechnology and regulations, beginning in 2002 (Fig. 1.1).

Nanotechnology regulation is certainly challenging because the term incorporates a broad, diverse range of materials, technologies, and products, with an even greater spectrum of potential risks and benefits (Marchant et al., 2009).

The European Union does not provide a horizontal legislation that would provide a common framework for nanomaterials

Novel Approaches of Nanotechnology in Food. http://dx.doi.org/10.1016/B978-0-12-804308-0.00001-7

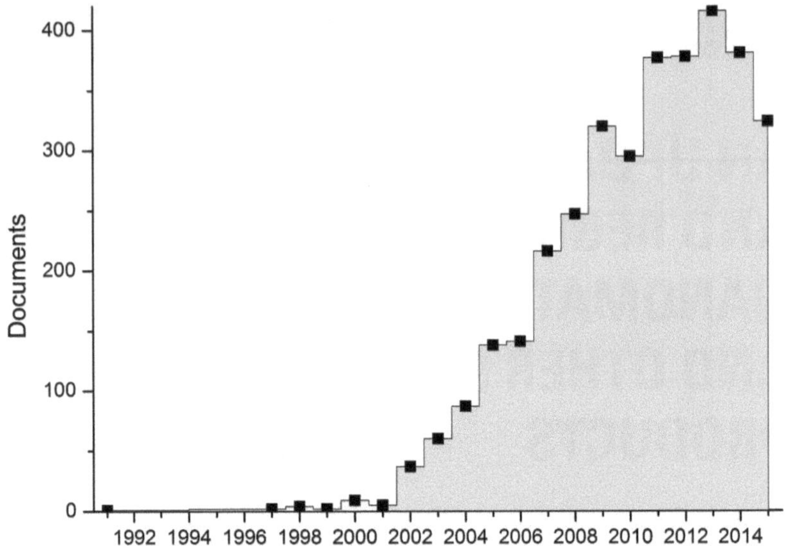

Figure 1.1. Number of scientific publications 1991–2015 mentioning nanotechnology and regulations. Scopus, Elsevier B.V.; search term: "(nanotechnology OR nanomaterial*) AND (policy OR regulation*)," search conducted Feb. 9, 2016.

in all kinds of consumer products. Only a recommendation No. 2011/696/EU on the definition of nanomaterial has been published by the European Commission (EU, 2011a). The recommendation provides the following definition:

> *'Nanomaterial' means a natural, incidental or manufactured material containing particles, in an unbound state or as an aggregate or as an agglomerate and where, for 50% or more of the particles in the number size distribution, one or more external dimensions is in the size range 1 nm–100 nm. In specific cases and where warranted by concerns for the environment, health, safety or competitiveness the number size distribution threshold of 50% may be replaced by a threshold between 1 and 50%.*

The definition of the recommendation should be used as a reference for determining whether a material should be considered as a "nanomaterial" for legislative and policy purposes in the European Union.

An in-depth review about the challenges and best practices for measuring nanomaterials according to the EU definition was provided by Brown et al. (2013). Especially the conversion of mass- or volume-based measurements to the number distribution according to the definition is difficult, due to a large potential error (Fig. 1.2). Particle size distributions from Nanoparticle Tracking Analysis (NTA) were suggested as possibility to assign the

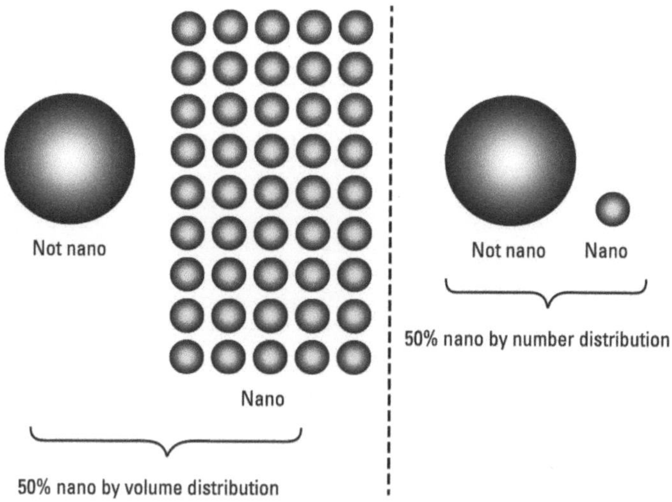

Figure 1.2. Schematic illustration of the difference between nanomaterial definitions based on volume and number distributions. Reproduced from Environmental Health Perspectives (Brown et al., 2013).

status of materials according to the EU's definition (see examples in Fig. 1.3) (Nanosight, 2011).

In the absence of a specific horizontal legislation, each specific product legislation may provide (or not provide) some requirements for nanomaterials. This chapter summarizes the definitions and regulations sorted according to product group. The research for regulations was conducted in May 2015.

There is limited previous literature available about the topic of definitions and regulations of nanotechnology. A comprehensive review comparing the legislations in the United States, Japan, Australia, and the United Kingdom was provided by Bowman and Hodge (2007), but due to the rapidly emerging technology and corresponding policy, the text is mostly outdated.

For an earlier review, which also comprises aspects of risk assessment and analytical techniques, the article of Bleeker et al. (2013) is highly recommended. An overview of regulations is provided in Table 1.1. The chronology of nanomaterials regulation in the European Union was summarized by Eisenberger et al. (2010) and by Kurath et al. (2014) with focus on the German-speaking countries.

2 Nanoregulations in Different Fields

2.1 Chemicals

The REACH (Registration, Evaluation, Authorisation and Restriction of Chemical substances) regulation (EC) No. 1907/2006

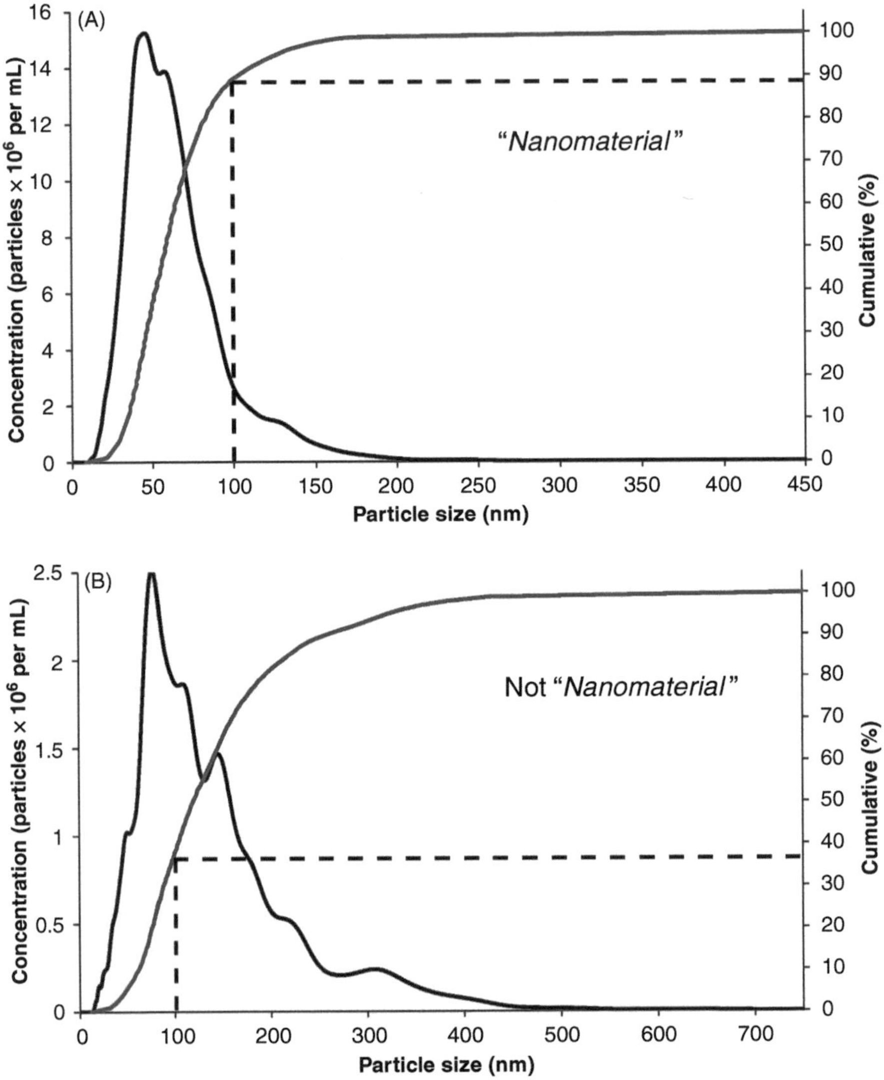

Figure 1.3. Examples for particle size distribution (PSD) profiles determined using nanoparticle tracking analysis (NTA). According to the EU's definition, sample (A) would be considered nanomaterial (89% of sample below 100 nm) and sample (B) would not be considered nanomaterial (38% of sample below 100 nm). Reproduced from Nanosight White Paper (Nanosight, 2011) with kind permission from Malvern Instruments GmbH, Herrenberg, Germany.

Table 1.1 State of the Art of Regulatory Frameworks That Deal or Potentially Have to Deal With Nanomaterials

Legislation	Definition[a]	Label[b]	Specific Provisions	Further Discussion/ Development Anticipated On[c]
Biocides[d]	Yes	Yes	Separate assessment	Guidance
Plant protection products[e]	No	No	None	Guidance
Cosmetics[f]	Yes	Yes	Separate assessment	Guidance
Food Information to consumers[g]	Yes	Yes	None	None
Food contact materials[h]	No[i]	No	Separate assessment	Guidance
Novel foods/feeds[j]	Yes	Yes	Separate assessment	Guidance
Food additives[k]	No	No	Separate assessment	Reevaluation of authorized food additives; guidance
Medicinal products	No[l]	No	None	Definition; guidance
Medical devices	No	No	None	Definition, risk classification, specific provisions, guidance[m]
Chemicals: REACH[n]	No	No	None	Additional or adaptation of legislation and guidance
Chemicals: classification, labeling and packaging[o]	No	No	None	Additional or adaptation of legislation and guidance
Occupational health and safety[p]	No	No	None	Guidance and OELs[q]

[a] In case the specific legislation includes a definition of "nanomaterial," this is indicated by "Yes."
[b] In case the specific legislation (will) require(s) that the use of nanomaterials is indicated on the label, this is indicated by "Yes."
[c] Issues to be considered in these discussions/developments are further elaborated on in Bleeker et al. (2013).
[d] This refers to the new biocides regulation (EU, 2012b).
[e] Plant protection products (Regulation (EC) No. 1107/2009 (EU, 2009b).
[f] Cosmetics regulation [EC No. 1223/2009 (EU, 2009c)].
[g] Regulation (EU) No. 1169/2011 (EU, 2011c).
[h] This refers to regulation (EC) No. 10/2011 (EU, 2011b).
[i] Nanoforms are mentioned but the term is not defined.
[j] This refers to the new draft regulation on novel foods.
[k] Regulation (EC) No. 1331/2008 (EU, 2008c) and related regulations.
[l] The legislation does not include a definition, but the European medicines agency does describe nanotechnology on its website as the use of tiny structures—less than 1000 nm across—that are designed to have specific properties.
[m] The EC new and emerging technologies working group recommended the addition of "all devices incorporating or consisting of particles, components or devices at the nanoscale" in the highest risk class (N&ET 2007); currently a revision of the regulatory framework for medical devices is being carried out.
[n] Registration, evaluation, authorization, and restriction of chemicals [Regulation (EU) No. 1907/2006; (EU, 2006)].
[o] Classification, labeling, and packaging [Regulation (EU) No. 1272/2008; (EU, 2008b)].
[p] Occupational health and safety (ie, EU, 1989, 1998, 2004a, 2006, 2008b).
[q] Occupational exposure limits.

Reprinted from Bleeker et al. (2013) with permission from Elsevier.

is the framework for the harmonized European regulations on chemicals (EU, 2006). Nanomaterials directly fall under the definition of substance in the REACH regulation. This means, however, that possible nanospecific properties of the substance may not be considered because the substance as such is regulated, but not its particle size. This means that carbon nanotubes or nanosized TiO_2 could be registered in theory under the same registration as carbon black or micrometer-sized TiO_2 (Linkov et al., 2009). However, the European Commission has stated that under REACH "manufacturers, importers and downstream users have to ensure that their nanomaterials do not adversely affect human health or the environment" (see REACH implementation report on substance identification of nanomaterials, JRC, 2011). Two detailed monographs about specific advice on fulfilling information requirements (Hankin et al., 2011) and on exposure assessment and hazard/risk characterization (Aitken et al., 2011) were published for nanomaterials under REACH.

In the United States, the US Environmental Protection Agency (EPA) is the primary regulatory agency with governing authority over nanotechnology (Wang et al., 2013). Similar to REACH, the EPA considers nanomaterials as "chemical substances" regulated by the Toxic Substances Control Act (TSCA) (US EPA, 2015). The inclusion into EPA's regulation of new chemical substances ensures that nanoscale materials receive appropriate regulatory review (US EPA, 2015).

2.2 Food

The General Food Law Regulation (EC) No. 178/2002 (EU, 2002) states that "food shall not be placed on the market if it is unsafe" (Article 14 (1) of Regulation (EC) No. 178/2002). In accordance with this Regulation, each food business operator is fully responsible for the safety of his products. Although the term "nanomaterial" is not specifically mentioned in this Regulation, this general principle of food law applies to food containing nanoscale ingredients as well as to food in general.

Nanoscale food ingredients, which are added as food additives for technological purposes, fall within the scope of Regulation (EC) No. 1333/2008 (EU, 2008d). According to this Regulation, food additives may only be placed on the market and used under specified conditions if they have been authorized after a safety assessment and if they are listed in the Union lists. The food additives that were permitted for use prior to the entry into force of Regulation (EC) No. 1333/2008 are at this time being reassessed by the European Food Safety Authority (EFSA) according to

Regulation (EU) 257/2010 (EU, 2010). This reevaluation "also covers any nano-related issues, which may be addressed in a revision of the conditions of use, where appropriate." For the food additives calcium carbonate (E170) and vegetable carbon (E153), EFSA recommended to define the particle size in the product specifications. The following additives, which might be in a nanoscale form, will be evaluated by:

1. 31 Dec. 2015: Titanium dioxide (E171), iron oxides and hydroxides (E172), silver (E174), gold (E175)
2. 31 Dec. 2016: Silicon dioxide (E551)
3. 31 Dec. 2018: Calcium silicate (E552), magnesium silicate (E553a), talc (E553b)

When a food additive is already approved for use and listed in a Community list and there is a notable change in the production process, which leads to a change in particle size (eg, through nanotechnology), the food additive generated by the new method shall be "considered as a different additive." When this occurs, "a new entry in the Community list or a change in the specifications is necessary before the additive can be placed on the market" (EU, 2008d). This will ensure that a safety assessment is carried out for each nanoscale food additive before it can be placed on the market.

Nanomaterials which are used for other than technological purposes may fall within the scope of Regulation (EC) No. 258/97 concerning novel food ingredients (EU, 1997). The term "nanomaterial" is not explicitly mentioned therein, however, this Regulation applies for foods and food ingredients that have not been used for human consumption to a significant degree within the Community before this Regulation entered into force, and when a production process not currently used has been applied to them, which gives rise to significant changes in their composition or structure affecting their nutritional value, their metabolism, or the level of undesirable substances. Nanomaterials falling within the scope of this Regulation have to undergo an authorization process including a safety assessment. The Novel Food Regulation is currently under revision. According to the revised version, food consisting of engineered nanomaterials shall always be considered as novel food. This would ensure that each nanoscale food or food ingredient would have to undergo an appropriate safety assessment before being authorized. Moreover, the revised version shall include a definition for "engineered nanomaterial."

In accordance with Regulation (EU) No. 1169/2011 (EU, 2011c) on "the provision of food information to consumers," "all ingredients present in the form of engineered nanomaterials shall be clearly indicated in the list of ingredients. The names of such ingredients shall be followed by the word 'nano' in brackets." In those cases in which Regulation (EU) No. 1169/2011 allows to omit the list of ingredients (eg, if the largest surface of the packaging has an area of less than 10 cm^2), no labeling of nanoscale ingredients is required. Furthermore, ingredients present in the form of engineered nanomaterials do not have to be labeled, if they fall under the ingredients for which an omission from the list of ingredients is allowed (eg, food additives that were contained in an ingredient of that food and serve no technological function in the finished product). In addition, it can be concluded from the aforementioned provision that only ingredients that are *present in the food* as engineered nanomaterials have to be labeled as "nano" ingredients. Thus, if an ingredient is added as a nanoscale material that changes to a non-nanoscale form in the food (eg, by solution), the name of the ingredient does not have to be followed by the word "nano" in brackets.

Article 2(2)(t) of this Regulation (EU, 2011c) provides the following definition for "engineered nanomaterial":

> *'Engineered nanomaterial' means any intentionally produced material that has one or more dimensions of the order of 100 nm or less or that is composed of discrete functional parts, either internally or at the surface, many of which have one or more dimensions of the order of 100 nm or less, including structures, agglomerates or aggregates, which may have a size above the order of 100 nm but retain properties that are characteristic of the nanoscale.*
>
> *Properties that are characteristic of the nanoscale include:*
> *(i) those related to the large specific surface area of the materials considered; and/or*
> *(ii) specific physico-chemical properties that are different from those of the non-nanoform of the same material.*

This definition explicitly refers to "intentionally produced" nanomaterials, expressing that natural or incidental nanomaterials shall be excluded from the labeling requirements. Moreover, this definition does not contain any threshold concerning the particle size distribution as is the case with the definition according to Commission Recommendation 2011/696/EU (EU, 2011a). In the last part of the definition, characteristic properties of the nanoscale materials are emphasized. These may include, for instance, optical properties, which can be illustrated by the following example: titanium dioxide is authorized as a white food color (E 171). However, when titanium dioxide is produced in a nanoscale form, the white color disappears and a transparent form of titanium dioxide is obtained, which can be used as a transparent UV filter.

According to Article 18(5) of Regulation (EU) No. 1169/2011, the definition of "engineered nanomaterial" referred to in Article 2(2)(t) of this Regulation can be adjusted and adapted to "technical and scientific progress or to definitions agreed at international level" (EU, 2011c). On this basis, the European Commission suggested the following revised and more specific definition for "engineered nanomaterial" in the Delegated Regulation (EU) No. 1363/2013 (EU, 2013):

> *'engineered nanomaterial' means any intentionally manufactured material, containing particles, in an unbound state or as an aggregate or as an agglomerate and where, for 50% or more of the particles in the number size distribution, one or more external dimensions is in the size range 1 nm to 100 nm.*
>
> *By way of derogation:*
> *(a) food additives covered by the definition set out in the first paragraph shall not be considered as engineered nanomaterials, if they have been included in the Union lists referred to in Article 4 of Regulation (EC) No. 1333/2008 (EU, 2008d);*
> *(b) fullerenes, graphene flakes, and single wall carbon nanotubes with one or more external dimensions below 1 nm shall be considered as engineered nanomaterials.*
>
> *For the purposes of the definition set out in the first paragraph:*
> *(i) 'particle' means a minute piece of matter with defined physical boundaries;*
> *(ii) 'agglomerate' means a collection of weakly bound particles or aggregates where the resulting external surface area is similar to the sum of the surface areas of the individual components;*
> *(iii) 'aggregate' means a particle comprising of strongly bound or fused particles;*
> *(iv) 'intentionally manufactured' means that the material is manufactured to perform/fulfil a specific function or purpose.*
>
> **EU (2013)**

On the one hand, this definition was adapted to the definition provided in Recommendation 2011/696/EU (EU, 2011a). On the other hand, the definition adopted by ISO was taken into account, according to which "engineered nanomaterial" is "nanomaterial designed for a specific purpose or function." Another important point in this definition is that nanoscale food additives that were permitted for use prior to the entry into force of Regulation (EC) No. 1333/2008 "should not be considered as engineered nanomaterials," which means that such additives would not have to be labeled as nano ingredients. However, this delegated Regulation, including the aforementioned definition, was rejected by the European Parliament and another definition shall be prepared by the European Commission.

2.3 Food Contact Materials

The use of nanoparticles in food contact materials offers many applications. For example, improved mechanical strength and improved barrier properties against moisture and gases such as oxygen and carbon dioxide can be obtained by adding certain nanoparticles to food packaging materials. Moreover, nanoparticles can be used to obtain active and intelligent food contact materials (Cushen et al., 2012).

According to Regulation (EC) No. 1935/2004 (EU, 2004b) on "materials and articles intended to come into contact with food," "materials and articles, including active and intelligent materials and articles, shall be manufactured in compliance with good manufacturing practice so that, under normal or foreseeable conditions of use, they do not transfer their constituents to food in quantities which could:

1. endanger human health; or
2. bring about an unacceptable change in the composition of the food; or
3. bring about a deterioration in the organoleptic characteristics thereof."

This provision applies to all food contact materials including those containing nanomaterials, even though nanomaterials are not explicitly mentioned in Regulation (EC) No. 1935/2004.

Regulation (EU) No. 10/2011 on "plastic materials and articles intended to come into contact with food" refers specifically to nanomaterials, stating that "substances in nanoform shall only be used if explicitly authorized and mentioned in the specifications in Annex I" (EU, 2011b). Therefore, an authorization that is based on the risk assessment of the conventional particle size of a substance does not cover the nanoparticles of this substance. The following substances are currently listed in Annex I of Regulation (EU) No. 10/2011 using the term "nano" and may be used under specific conditions:

1. titanium nitride (nanoparticles)
2. (butadiene, ethyl acrylate, methyl methacrylate, styrene) copolymer crosslinked with divinylbenzene (in nanoform)
3. (butadiene, ethyl acrylate, methyl methacrylate, styrene) copolymer not crosslinked (in nanoform)
4. (butadiene, ethyl acrylate, methyl methacrylate, styrene) copolymer crosslinked with 1.3-butanediol dimethacrylate (in nanoform)

Furthermore, the following nanoscale substances are listed without explicitly using the term "nano":

1. kaolin (particles "thinner than 100 nm")
2. carbon black (primary particles of 10–300 nm, which are aggregated to a size of 100–1200 nm which may form agglomerates within the size distribution of 300 nm to the mm size)
3. silicon dioxide (primary particles of 1–100 nm, which are aggregated to a size of 0.1–1 μm, which may form agglomerates within the size distribution of 0.3 μm to the mm size)

In Regulation (EC) No. 450/2009 (EU, 2009a) "on active and intelligent materials and articles intended to come into contact with food" it is stated that nanoparticles should always be assessed on a case-by-case basis as regards their risk, even if they are applied on the external surface of the package material and are not directly in contact with the food. According to this Regulation as well as to Regulation (EU) No. 10/2011, nanoparticles in food contact materials are excluded from the functional barrier concept. In accordance with this concept, it is allowed to use certain nonauthorized substances, provided that they are located behind a functional barrier and fulfill certain criteria, for example, their migration has to remain below a given value. However, the use of nanomaterials in food contact materials is not subject to any labeling requirements.

2.4 Feed

According to Regulation (EC) No. 178/2002 (EU, 2002), "feed shall not be placed on the market or fed to any food-producing animal if it is unsafe" (Article 15(1) of Regulation (EC) No. 178/2002). As described earlier for food, this Regulation states that each feed business operator is responsible for the safety of his products. In Regulation (EC) No. 767/2009 (EU, 2009) on the placing on the market and use of feed, these principles are extended to feed for nonfood producing animals. Although the term "nanomaterial" is not explicitly mentioned in these Regulations, these provisions apply to each animal feed regardless of whether it contains nanomaterials.

Feed additives may only be placed on the market and used if their safety has been assessed and if they have been authorized in accordance with Regulation (EC) No. 1831/2003 (EU, 2003). Regulation (EC) No. 429/2008 (EU, 2008a) specifies the data, which have to be submitted for the authorization of a feed additive. The term "nanomaterial" is not mentioned in these Regulations, however, it is stated that the application documents shall include a detailed description of the manufacturing process of the additive.

Hence, the presence of nanoparticles is taken into account in the safety assessment. Nevertheless, there exists no labeling requirement for nanomaterials in feed.

2.5 Pesticides and Biocides

Nanotechnology has a wide application range in agriculture. Nanoscale compounds may be used as carriers for controlled release of pesticides or herbicides, or the active ingredients may be on nanoscale (Ali, 2014). An overview about nanopesticides is available by Kah et al. (2013). Kookana et al. (2014) have provided an overview about the gaps in risk assessment of nanopesticides and suggested the use of the precautionary principle until the properties and behavior of nanopesticides are better understood.

The Biocidal Products Regulation (BPR, Regulation (EU) 528/2012) concerns the placing on the market and use of biocidal products, which are used to protect humans, animals, materials, or articles against harmful organisms such as pests or bacteria, by the action of the active substances contained in the biocidal product (EU, 2012b). This Regulation aims to improve the functioning of the biocidal products market in the European Union, while ensuring a high level of protection for humans and the environment (EU, 2012b). The BPR defines nanomaterials as follows:

> *'Nanomaterial' means a natural or manufactured active substance or non-active substance containing particles, in an unbound state or as an aggregate or as an agglomerate and where, for 50% or more of the particles in the number size distribution, one or more external dimensions is in the size range 1–100 nm. Fullerenes, graphene flakes and single-wall carbon nanotubes with one or more external dimensions below 1 nm shall be considered as nanomaterials. For the purposes of the definition of nanomaterial, 'particle', 'agglomerate' and 'aggregate' are defined as follows: 'particle' means a minute piece of matter with defined physical boundaries, 'agglomerate' means a collection of weakly bound particles or aggregates where the resulting external surface area is similar to the sum of the surface areas of the individual components, 'aggregate' means a particle comprising strongly bound or fused particles.*
>
> **EU (2012b)**

The BPR then has some specific requirement on nanomaterials. The approval of an active substance shall not cover nanomaterials except where explicitly mentioned. For authorization, the risk to human health, animal health, and the environment has to be assessed separately when nanomaterials are used in that product. A simplified authorization procedure may not be made when the biocidal product does contain any nanomaterials. The labeling must include the name of all nanomaterials contained in the biocidal products, followed by the word "nano" in brackets, and any specific related risks.

2.6 Cosmetics

In cosmetic products, nanotechnology has been applied for quite some time. The best-known application is the use of nanoscale UV filters such as titanium dioxide or zinc oxide in sunscreens. Due to its nano size, titanium dioxide is transparent—in contrast to microsized TiO_2, which shows a white coloring—and therefore the nanomaterial does not leave a white layer on the skin after application. Even in decorative cosmetics nanoparticles are used, for example, carbon black as pigment in mascara and eyeliner. Nanosized carbon black adheres as a result of its large surface very well to the eyelashes (Budde, 2013).

The EC cosmetics Regulation No. 1223/2009 (EU, 2009c) was the first regulation providing a definition on nanomaterials. Article 2 paragraph 1 k defines a nanomaterial as an "insoluble or biopersistent and intentionally manufactured material with one or more external dimensions, or an internal structure, on the scale from 1 to 100 nm." This clearly excludes soluble nanoscale systems such as nanoemulsions, liposomes, and nanosomes. This definition has to be perceived as a provisional definition, which will be adapted to the state of the art as soon as a uniform definition for nanomaterial is established (Whereas (29) and Article 2 paragraph 3 (Mildau and Huber, 2010)).

Every cosmetic product that is intended to be available on the EU market has to be notified to the EU Commission's online portal CPNP (cosmetic product notification portal) prior to placing the cosmetic product on the market. Within the scope of this notification the presence of a nanoscale ingredient has to be specified by notifying the kind of nanomaterial, for example, titanium dioxide. This general notification obligation for cosmetic products is regulated in Article 13 of the cosmetics regulation. Additionally (to the product's notification in Article 13), the responsible person has to notify cosmetic products containing nanomaterials to the Commission (Article 16). The information notified to the Commission shall include at least the following:

1. the identification of the nanomaterial including its chemical name (IUPAC);
2. the specification of the nanomaterial including size of particles, physical and chemical properties;

3. an estimate of the quantity of nanomaterial contained in cosmetic products intended to be placed on the market per year;
4. the toxicological profile of the nanomaterial;
5. the safety data of the nanomaterial relating to the category of cosmetic product, as used in such products;
6. the reasonably foreseeable exposure conditions.

In the case of concerns about a nanomaterial and its use in cosmetic products, the Commission can ask the Scientific Committee for Consumer Safety (SCCS) to assess the substance. This information shall be published by the Commission.

To verify the cosmetic product's safety, the responsible person has to set up a safety report of the cosmetic product in which the composite is assessed with reference to the toxicological profile of all ingredients. "Particular consideration shall be given to any possible impacts on the toxicological profile due to—amongst other things—particle sizes, including nanomaterials." (Article 10 in conjunction with Annex I).

A further aspect of the regulation of nanomaterials in the cosmetics legislation is the labeling of nanoscale ingredients. Nanomaterials have to be marked in the list of ingredients with the suffix "(Nano)," for instance, "Titanium Dioxide (Nano)" (Article 19 paragraph 1 g).

2.7 Medicinal Products and Devices

In the United States several nanomedicines are currently in approval; some of them are already placed on the market (Bawa, 2011, 2013). The regulatory situation of medicinal products and medical devices relating to nanomaterials is worldwide in its infancy (Bawa, 2011, 2013; Eisenberger et al., 2010). However, there is a proposal for a regulation of the European Parliament and the Council on medical devices existing since 2012 (EU, 2012a). This proposal includes a definition of nanomaterial corresponding to the recommendation No. 2011/696/EU of the European Commission. But up to now it is not in force. At the moment there are several directives valid in Europe that apply for manufacturing and approval of pharmaceuticals and medical devices and are implemented into national legislations. Therein, extensive procedures of evaluation and admission are considered. A special regulation of nanomaterials, including an approval, does not exist. Nanomedicines are regulated via what is already on the books (Bawa, 2011, 2013). That means all drugs have to be safe irrespective of the presence or absence of nanoscale ingredients. Size-related unique nano properties are so far insufficiently considered as are probable consequences or risks concerning the environment or the human being. Manufacturers are also not required to list nanoingredients on product labels. This lack of regulation in the medicines field is a worldwide phenomenon (Bawa, 2011, 2013). So it remains to be seen if or when the EU proposal may enter into force and which changes it will achieve.

2.8 Other Products

There are various products available on the market, which may contain nanomaterials, such as wall paint that remains in position during the application and stops running down the wall or water-repellent ceramics with so-called lotus effect. Another example is nanoscale pigments used in tattoo colors. At the moment there is no regulation about nanomaterials in such kinds of consumer products. These products have to be safe but no particular requirements exist. A declaration of nanomaterials is not required.

Contrarily, there have been cases reported where "nano" has been labeled as a marketing claim despite the absence of nanomaterials in the product (Wolinsky, 2006). Some consumers apparently may prefer nano in certain instances such as shoe polish or clothing. For this reason, clear regulations on nano labeling may be preferable for all kinds of consumer products to avoid misleading claims. If the labeling of nanomaterials according to the definition of the recommendation No. 2011/696/EU of the European Commission had to be registered in the product safety regulations, such compounds would have to be labeled as a nanomaterial, for example.

2.9 Environment

The environmental legislation for the regulatory control of nanomaterials has been reviewed in a report commissioned by DG Environment of the European Commission (Ganzleben et al., 2011). The report concluded that no specific requirements are implemented and nanomaterials are treated like any other compound, for example, in the waste, sewage sludge, or water framework directives. The authors specifically pointed out that the water legislation is limited in providing control of nanomaterials as pollutants in surface waters, groundwater, and drinking water. Given the limitations in scientific knowledge (eg, regarding the lack to detect and monitor nanomaterials in the environment, and a lack of reliable ecotoxicology data), recourse to the precautionary principle was advised for the regulation of nanomaterials in the environment.

3 Conclusions

There clearly has been a lag between the rapid nanotechnology deployment in industrial, commercial, and consumer settings and its regulations (Malloy, 2011). However, our review shows that the regulator has caught up in some areas, especially products that may directly reach the end-consumer such as foods or cosmetics. In some respects, the regulator may have even outpaced the science, for example, when detection methods to control the regulations are still unavailable (Brown et al., 2013). In summation of the EU regulations, it might have been preferable to include the definition of nanomaterials into a real horizontal directive, and not only a recommendation. The current situation leads to the fact that the definition of nanomaterials has to be included in each specific product directive. The lack of consistent implementation of definitions was pointed out by Jahnel (2015). For example, the BPR includes "natural, incidental and manufactured materials," while the food information and cosmetics regulations restrict the definition to "intentionally manufactured (or produced) materials" (Jahnel, 2015).

It is difficult to foresee how nanoproducts will be finally regulated. It is beyond all questions that "nanogovernance" is immediately needed to provide clarity and legal certainty to manufacturers, policy makers, health-care providers, enforcement authorities, and the consumer (Bawa, 2013).

References

Aitken, R.A., Bassan, A., Friedrichs, S., Hankin, S.M., Hansen, S.F., Holmqvist, J., Peters, S.A.K., Poland, C.A., Tran, C.L., 2011. Specific Advice on Exposure Assessment and Hazard/Risk Characterisation for Nanomaterials under REACH. European Commission, Brussels, Belgium.

Ali, M.A., Rehman, I., Iqbal, A., Din, S., Rao, A.Q., Latif, A., Samiullah, T.R., Azam, S., Husnain, T., 2014. Nanotechnology, a new frontier in agriculture. Adv. Life Sci. 1, 129-138.

Bawa, R., 2011. Regulating nanomedicine: can the FDA handle it? Curr. Drug Deliv. 8 (3), 227–234.

Bawa, R., 2013. FDA and nanotech: baby steps lead to regulatory uncertainty. In: Bagchi, D., Bagchi, M., Moriyama, H., Shahidi, F. (Eds.), Bio-Nanotechnology: a Revolution in Food, Biomedical and Health Sciences. Blackwell Publishing Ltd, Oxford, UK.

Bleeker, E.A., de Jong, W.H., Geertsma, R.E., Groenewold, M., Heugens, E.H., Koers-Jacquemijns, M., van de Meent, D., Popma, J.R., Rietveld, A.G., Wijnhoven, S.W., Cassee, F.R., Oomen, A.G., 2013. Considerations on the EU definition of a nanomaterial: science to support policy making. Regul. Toxicol. Pharmacol. 65 (1), 119–125.

Bowman, D.M., Hodge, G.A., 2007. A small matter of regulation: an International review of nanotechnology regulation. Columbia Sci. Technol. Law Rev. 8, 1.

Brown, S.C., Boyko, V., Meyers, G., Voetz, M., Wohlleben, W., 2013. Toward advancing nano-object count metrology: a best practice framework. Environ. Health Perspect. 121 (11–12), 1282–1291.
Budde, A., 2013. Nanotechnologie. Damit das make-up nicht verläuft. Available from: http://www.zeit.de/2013/45/nanotechnologie-kosmetik-kennzeichnung
Corley, E.A., Kim, Y., Scheufele, D.A., 2013. The current status and future direction of nanotechnology regulations: a view from nano-scientists. Rev. Policy Res. 30 (5), 488–511.
Cushen, M., Kerry, J., Morris, M., Cruz-Romero, M., Cummins, E., 2012. Nanotechnologies in the food industry—recent developments, risks and regulation. Trends Food Sci. Technol. 24 (1), 30–46.
Eisenberger, I., Nentwich, M., Fiedeler, U., Gazsó, A., Simkó, M., 2010. Nano regulation in the European Union. Nano Trust Dossiers 17, 1–6.
EU, 1989. Council Directive 89/391/EEC of 12 June 1989 on the introduction of measures to encourage improvements in the safety and health of workers at work. Off. J. Eur. Comm. L183, 1–8.
EU, 1997. Regulation (EC) No. 258/97 of the European Parliament and of the European Council of 27 January 1997 concerning novel food ingredients. Off. J. Eur. Comm. L43, 1–6.
EU, 1998. Council Directive 98/24/EC of 7 April 1998 on the protection of the health and safety of workers from the risks related to chemical agents at work (fourteenth individual Directive within the meaning of Article 16(1) of Directive 89/391/EEC). Off. J. Eur. Comm. L131, 11–23.
EU, 2002. Regulation (EC) No. 178/2002 of the European Parliament and of the council of 28 January 2002 laying down the general principles and requirements of food law, establishing the European Food Safety Authority and laying down procedures in matters of food safety. Off. J. Eur. Comm. L31, 1–24.
EU, 2003. Regulation (EC) No. 1831/2003 of the European Parliament and of the council of 22 September 2003 on additives for use in animal nutrition. Off. J. Eur. Union L268, 29–43.
EU, 2004a. Directive 2004/37/EC of the European Parliament and of the Council of 29 April 2004 on the protection of workers from the risks related to exposure to carcinogens or mutagens at work (Sixth individual Directive within the meaning of Article 16(1) of Council Directive 89/391/EEC). Off. J. Eur. Union L229, 23–34.
EU, 2004b. Regulation (EC) No. 1935/2004 of the European Parliament and the Council of 27 October 2004 on materials and articles intended to come into contact with food and repealing Directives 80/590/EEC and 89/109/EEC. Off. J. Eur. Union L338, 4–17.
EU, 2006. Regulation (EC) No. 1907/2006 of the European Parliament and of the Council of 18 December 2006 concerning the Registration, Evaluation, Authorisation and Restriction of Chemicals (REACH), establishing a European Chemicals Agency, amending Directive 1999/45/EC and repealing Council Regulation (EEC) No. 793/93 and Commission Regulation (EC) No. 1488/94 as well as Council Directive 76/769/EEC and Commission Directives 91/155/EEC, 93/67/EEC, 93/105/EC and 2000/21/EC. Off. J. Eur. Union L136, 3–280.
EU, 2008a. Commission Regulation (EC) No. 429/2008 of 25 April 2008 on detailed rules for the implementation of Regulation (EC) No. 1831/2003 of the European Parliament and of the Council as regards the preparation and the presentation of applications and the assessment and the authorisation of feed additives. Off. J. Eur. Union L133, 1–65.
EU, 2008b. Regulation (EC) No. 1272/2008 of the European Parliament and of the Council of 16 December 2008 on classification, labelling and packaging of

substances and mixtures, amending and repealing Directives 67/548/EEC and 1999/45/EC, and amending Regulation (EC) No. 1907/2006. Off. J. Eur. Union L353, 1–1355.

EU, 2008c. Regulation (EC) No. 1331/2008 of the European Parliament and of the Council of 16 December 2008 establishing a common authorisation procedure for food additives, food enzymes and food flavourings. Off. J. Eur. Union L354, 1–6.

EU, 2008d. Regulation (EC) No. 1333/2008 of the European Parliament and of the council of 16 December 2008 on food additives. Off. J. Eur. Union L354, 16–33.

EU, 2009a. Commission Regulation (EC) No. 450/2009 of 29 May 2009 on active and intelligent materials and articles intended to come into contact with food. Off. J. Eur. Union L135, 3–11.

EU, 2009b. Regulation (EC) No. 1107/2009 of the European Parliament and of the Council of 21 October 2009 concerning the placing of plant protection products on the market and repealing Council Directives 79/117/EEC and 91/414/EEC. Off. J. Eur. Union L309, 1–50.

EU, 2009c. Regulation (EC) No. 1223/2009 of the European Parliament and of the Council of 30 November 2009 on cosmetic products. Off. J. Eur. Union L342, 59–209.

EU, 2010. Commission Regulation (EU) No. 257/2010 of 25 March 2010 setting up a programme for the re-evaluation of approved food additives in accordance with Regulation (EC) No. 1333/2008 of the European Parliament and of the Council on food additives. Off. J. Eur. Union L80, 19–27.

EU, 2011a. Commission recommendation of 18 October 2011 on the definition of nanomaterial (2011/696/EU). Off. J. Eur. Union L275, 38–40.

EU, 2011b. Commission regulation (EU) No. 10/2011 of 14 January 2011 on plastic materials and articles intended to come into contact with food. Off. J. Eur. Union L12, 1–89.

EU, 2011c. Regulation (EU) No. 1169/2011 of the European Parliament and of the Council of 25 October 2011 on the provision of food information to consumers, amending Regulations (EC) No. 1924/2006 and (EC) No.1925/2006 of the European Parliament and of the Council, and repealing Commission Directive 87/250/EEC, Council Directive 90/496/EEC, Commission Directive 1999/10/EC, Directive 2000/13/EC of the European Parliament and of the Council, Commission Directives 2002/67/EC and 2008/5/EC and Commission Regulation (EC) No. 608/2004. Off. J. Eur. Union L304, 18–63.

EU, 2012a. Proposal for a Regulation of the European Parliament and of the Council on Medical Devices, and Amending Directive 2001/83/EC, Regulation (EC) No. 178/2002 and Regulation (EC) No. 1223/2009. COM(2012) 542 Final. European Commission, Brussels, Belgium.

EU, 2012b. Regulation (EU) No. 528/2012 of the European Parliament and of the Council of 22 May 2012 concerning the making available on the market and use of biocidal products. Off. J. Eur. Union L167, 1–123.

EU, 2013. Commission Delegated Regulation (EU) No. 1363/2013 of 12 December 2013 amending Regulation (EU) No. 1169/2011 of the European Parliament and of the Council on the provision of food information to consumers as regards the definition of "engineered nanomaterials". Off. J. Eur. Union L343, 26–28.

Ganzleben, C., Pelsy, F., Hansen, S.F., Corden, C., Grebot, B., Sobey, M., 2011. Review of Environmental Legislation for the Regulatory Control of Nanomaterials: Final Report. Technical University of Denmark, Lyngby, Denmark.

Hankin, S.M., Peters, S.A.K., Poland, C.A., Hansen, S.F., Holmqvist, J., Ross, B.L., Varet, J., Aitken, R.J., 2011. Specific Advice on Fulfilling Information Requirements for Nanomaterials under REACH. European Commission, Brussels, Belgium.

Jahnel, J., 2015. Addressing the challenges to the risk assessment of nanomaterials. In: Dolez, P.I. (Ed.), Nanoengineering. Elsevier, Amsterdam, pp. 485–521.
JRC, 2011. REACH Implementation Project. Substance Identification of Nanomaterials. European Commission, Joint Research Centre, Brussels, Belgium.
Kah, M., Beulke, S., Tiede, K., Hofmann, T., 2013. Nanopesticides: state of knowledge, environmental fate, and exposure modeling. Crit. Rev. Environ. Sci. Technol. 43 (16), 1823–1867.
Kookana, R.S., Boxall, A.B., Reeves, P.T., Ashauer, R., Beulke, S., Chaudhry, Q., Cornelis, G., Fernandes, T.F., Gan, J., Kah, M., Lynch, I., Ranville, J., Sinclair, C., Spurgeon, D., Tiede, K., Van den Brink, P.J., 2014. Nanopesticides: guiding principles for regulatory evaluation of environmental risks. J. Agric. Food Chem. 62 (19), 4227–4240.
Kurath, M., Nentwich, M., Fleischer, T., Eisenberger, I., 2014. Cultures and strategies in the regulation of nanotechnology in Germany, Austria, Switzerland and the European Union. Nanoethics 8 (2), 121–140.
Linkov, I., Satterstrom, F.K., Monica, J.C., Foss Hansen, S., Davis, T.A., 2009. Nano risk governance: current developments and future perspectives. Nanotech. Law Bus. 6 (2), 203–220.
Malloy, T.F., 2011. Nanotechnology regulation: a study in claims making. J. Scholarly Persp. 7 (01), 25–39.
Marchant, G.E., Sylvester, D.J., Abbott, K.W., 2009. What does the history of technology regulation teach us about nano oversight? J. Law Med. Ethics 37 (4), 724–731.
Mildau, G., Huber, B., 2010. The new EC cosmetics Regulation 1223/2009—contents and first explanations. SOFW J. 136 (3), 39–59.
N&ET, 2007. Report on nanotechnology to the medical devices expert group: findings and recommendations. Available from: http://ec.europa.eu/growth/sectors/medical-devices_old/scientific-technical-assessment/working-group/files/entr-2007-net-wg-report-nano-final_en.pdf
Nanosight, 2011. The EC Definition of a Nanomaterial: Potential Measurement Methodologies. White paper. NanoSight Ltd, Minton Park, UK.
US EPA, 2015. Control of nanoscale materials under the Toxic Substances Control Act. Available from: http://www.epa.gov/oppt/nano/
Wang, J., Gerlach, J.D., Savage, N., Cobb, G.P., 2013. Necessity and approach to integrated nanomaterial legislation and governance. Sci. Total Environ. 442, 56–62.
Wolinsky, H., 2006. Nanoregulation. EMBO Rep. 7 (9), 858–861.

2

RECENT ADVANCES IN THE APPLICATION OF NANOMATERIALS AND NANOTECHNOLOGY IN FOOD RESEARCH

Dongxiao Sun-Waterhouse, Geoffrey I.N. Waterhouse*
School of Chemical Sciences, University of Auckland, Auckland, New Zealand

1 Introduction

Global food challenges for humans in the 21st century are rapid population growth, population health and its maintenance, food supply and safety, and sustainable use of resources. Novel advances in science and technology are essential to realizing economically viable solutions to these challenges. Nanoscience and nanotechnology have great potential for adding value to the food chain through transforming the entire food industry in areas such as food production, processing, analysis and packaging, enabling future populations to achieve target nutritional outcomes or health benefits.

Since the term "nanotechnology" was first coined by Physicist Richard Feynman in 1959, various definitions for nanomaterials, nanoparticles, and nanotechnology have been proposed. However, there is no internationally agreed definition for regulatory uses of nanomaterials. The general term of nanotechnology is "the manipulation or self-assembly of very small particles (typically 1–100 nm) including individual atoms, molecules or molecular clusters, to create materials, systems, devices and processes with new or vastly different properties," although the upper limit of 100 nm is currently under review (EFSA, 2009). The nano terms were introduced to reflect the unique properties and often peculiar behaviors of materials with lengths of 10^{-7}–10^{-9} m. Many micron-sized objects or materials containing a nanoscale fraction

*Lead author

Novel Approaches of Nanotechnology in Food. http://dx.doi.org/10.1016/B978-0-12-804308-0.00002-9

such as nanofibers or nanowhiskers, also have at least one dimension on the 1–100 nm length scale and are generally classified as nanomaterials. For food systems, the anatomical and physiological properties of 1–100 nm need to be considered in product development. Even particles sized between several hundred nanometers up to 20 μm possess size-dependent reactivity and bioavailability (Garnett and Kallinteri, 2006) and health risks (Gatti and Rivasi, 2002). Accordingly, the US Food and Drug Administration (FDA) chose not to provide a specific size-based definition (FDA, 2007) for nanomaterials, and instead used a more general definition of "particles with dimensions less than one micrometer (ie, <1000 nm) that exhibit unique properties not recognized in micron or larger sized particles" (FDA, 2006). The Joint Research Council published a report in 2012 entitled "Requirements on Measurements for the Implementation of the EC Definition of the Term Nanomaterial," in which the specific surface area measured as BET surface area was suggested to "classify positively a material as a nanomaterial."

Nature (including the living organisms) is full of naturally occurring nanoscale objects or structures such as DNA (2–3 nm), proteins (~1–5 nm for those of M_w 5–500 kDa), viruses (~20–300 nm), spoilage bacteria (~1000 nm) and a red blood cell (~6000–8000 nm). Almost all consumer foods contain nanoscale sugars, amino acids, peptides, proteins, carbohydrates, and fats in varying proportions. Since the roadmap on nanotechnology application to the agricultural and food industries was introduced by the United States Department of Agriculture (USDA) in 2003, many countries including the United States, Japan, European Union, China, India, South Korea, Iran and Thailand have invested heavily in food nanotechnology research and development (R&D). Rapid advancements in nanosciences and nanotechnologies have been achieved in the forms of materials, ingredients, processing tools, foods and other consumer products.

At the nanoscale, the influence of gravity is remarkedly reduced while weak molecular interactive forces such as van der Waals forces and hydrogen-bonding dominate, and surface interactions govern the properties of materials. Accordingly, nanomaterials often display vastly different physiochemical properties (perceived as novel characteristics) compared with their respective bulk materials, including optical response, mechanical strength, electrical conductance, magnetism, solubility and chemical reactivity, which can be utilized in applications involving adsorption, filtration, separation, purification, immobilization and encapsulation. Surface area and porosity are the two most important physical properties of nanomaterials that influence bulk quality, utility, reactivity and stability characteristics. As particle size is reduced, the specific surface area (surface area per unit mass) increases in an inversely proportional manner. Manipulating the porosity of nanomaterials is an effective approach to increase surface area. High surface area systems can be attained by fabricating small particles/clusters where the surface-to-volume ratio of each particle is high, or by creating materials where the internal surface area (pores) is high compared to the amount of bulk support material. Nanoparticles have a high surface-to-volume ratio, which generally increases surface reactivity, dissolution rate and bioavailability but alters toxicity profile.

Nanomaterials are ubiquitous in industry and the consumer sector. Inorganic nanomaterials including transition metals and metal oxides/nitrides/sulfides/selenides are increasingly being used. Examples include titanium dioxide (TiO_2) as a whitening agent (Chen et al., 2015a), silver nanoparticles as antimicrobial agents or catalysts (Waterhouse et al., 2004), nickel oxide as a co-catalyst (Chen et al. 2015b), zinc oxide for UV sensing platforms (Allen et al., 2011), iron nanoparticles embedded in silicon dioxide for magnetic sensing (Leveneur et al., 2011), aluminum oxide for sorption and optical sensing elements (Waterhouse et al., 2015) or gold nanoparticles for textile coloration or H_2 production from biofuels (Murdoch et al., 2011). Examples of successful applications include smart sensing chips or nano(bio)sensors for food safety analysis, point-of-care diagnostics, nanobarcodes for food authenticity, antimicrobial agents for food packaging, handling tools or contact materials, nano TiO_2 coatings for photocatalytic sterilization of product surfaces or drinking water, and nanoporous sorbents/catalysts for water purification via removal of contaminants like pesticides, viruses and bacteria (Chaudhry et al., 2010; Sharma et al., 2013; Yabutani et al., 2014). Organic materials are also used as target-specific nanoparticles including biopolymers like proteins and carbohydrates (which are nontoxic and sustainable thus causing minimal negative impact on human health or the environment). For instance, protein or starch-based nanoparticles are used as encapsulating materials in drug or bioactive delivery systems for increased efficacy. Nanocomposites from traditionally harvested plant materials (eg, wheat straw or soy bean hulls) exhibit unique chemical, physical, electrochemical, thermal and mechanical properties, which are utilized for biosensors.

Nanotechnology is exerting significant impacts on food safety and security by revolutionizing computing, communication, and data processing, leading to electronic device miniaturization, improved tools or matrices for product labeling, packaging, authentification and traceability detection. Nanoparticles possess an intrinsic advantage in their ability to access different parts of the human body (eg, cells, tissues, and organs). Materials with size less than 300 nm can enter cells directly, while nanomaterials with size less than 70 nm can be taken up by cell nuclei (Geiser et al., 2005). This highlights both the favorable aspects (ie, improved body uptake) and potentially undesirable aspects (ie, potential toxicity to humans) of nanoparticles/nanocomposites. The size, chemical composition, shape, surface structure and charge, catalytic properties, extent of particle aggregation or disaggregation of nanomaterials, all contribute to their potential toxicity (Brunner et al., 2006). Hence, the decision to use nanomaterials in foods and food processing requires careful consideration of the possible deleterious consequences and fate of the nanomaterials under human physiological conditions and post-processing.

This chapter presents an overview of some important applications of nanotechnology in the food sector. Particular emphasis will be given to "nanosensors," "nanocarriers," and "nanoenabled foods" (with several case studies provided). Nanotechnologies applied to the farming practice such as nanosized agrochemicals and pesticides, although important to the food chain, are also out of the scope of this chapter. Information about the regulatory assessments, despite being a hot literature topic, is only covered briefly here.

2 Nanosensors

Biosensors are effective compact devices used for the noninvasive detection of a specific analyte or multiple analytes. Typically nanosensors use a biomimetic recognition element or biologically derived material (nucleic acid, enzyme, antibody, receptor, tissue, cell, ions, proteins, etc.) in combination with a transduction system to produce a measurable signal as a result of the changes in physical-chemical properties (eg, pH, electron, heat, potential, mass, or optical properties) caused by the interaction between a selective sensing element and the analyte of interest under mild temperature and pH conditions (Wilson and Gifford, 2005). These sensors are finding increasing application for food-quality evaluation, clinical analysis, and environmental control purposes.

The introduction of nanotechnology into biosensors, through incorporating nanomaterials such as magnetic nanoparticles, carbon nanotubes, nanorods, quantum dots, nanowires or nanochannels, has led to new devices ("nanosensors") for fast detection of biological, chemical, physical, electrical and microbiological changes in the environment, foods or humans (Velasco-García and Mottram, 2003). Intelligent nanosensors can readily achieve very low detection limits, for example, surface-enhanced Raman spectroscopy (SERS) can detect analytes at concentrations as low as 1 part per billion (ppb), with single molecule detection being possible under ideal conditions. Nanosensors function via electrochemical redox properties, electrocatalytic activity, conductivity or optical properties (eg, light absorption, reflectance, fluorescence/phosphorescence, bio/chemiluminescence, Raman scattering or refractive index modulation), to measure directly individual macro-/microcomponents including bacteria or contaminants (Merkoçi et al., 2010). Electrochemical biosensors may be preferrable to optical devices considering cost, size, ease of operation, user-friendly applicability, sensitivity, linear range of detection, disposability and incorporating options (Wang, 2006).

A key application of nanosensors to the food sector is their use for "real-time" and "online" monitoring of crop growth and harvest through detecting temperature variation, moisture level, soil nutrient contents, presence of pesticides, herbicides, insects, and diseases, and so on (Somerset et al., 2006). Immunoassay nanosensors in "dipstick format" are attracting interest for pesticide detection due to their advantages in cost, sensitivity and screening speed, for example, the detection range of dichlorodiphenyltrichloroethane (DDT) pesticide via a lateral flow immunoassay based on Au nanoparticles is 27–1000 ng mL^{-1} (Lisa et al., 2009). Furthermore, the increasing global complexity of the food chain prompts the development of more effective tracking systems for farming, processing, transportation, distribution and consumption to ensure food safety. For example, efficient identification by traceability systems enables the immediate removal of any tainted products, thereby avoiding food poisoning outbreaks and expensive product recalls. Nanosensors integrated with advanced IT systems can enable efficient daily monitoring of human metabolism and wellness via an early diagnosis of many diseases like arteriosclerosis, cancer, diabetes, Parkinson's and Alzheimer's, and ensure the safety and quality of food products and their packaging via almost instantaneous sensing of toxins, pollutants, adulterants, contaminants, chemicals, allergens, enzymes, pathogens or bacteria. Such nanosensors can simultaneously detect different substances in foods or human body fluids (eg, enzymes, biomarkers, precursors, metabolites, and associated electron or energy transfer phenomena). In fact, nanomaterials have already been used to develop food safety and quality indicators, for example, those for monitoring the integrity of packages via detecting leaks, the freeze-thaw-refreeze processes via detecting temperature variations, and food microbiological safety via detecting deterioration products. Nowadays, intelligent packaging is often equipped with nanosensors that can not only monitor the internal or external conditions of food products and containers throughout the supply chain, but also release a food preservative when necessary. The fusion of nanotechnology and nanotronics further enables the Radio Frequency Identification Tagging (RFIT) and Nucleic Acid Engineered Nanobio Barcode (NAENB) technology to track food products in the supply chain in a cheaper and more efficient manner. A visual quality label that can indicate the freshness of poultry products is a good example (ie, the label color changes due to the reaction between the silver nanolayer and the hydrogen sulfide generated when meat starts to deteriorate) (Smolander et al., 2004).

Nanobiocatalysis, in which enzymes are incorporated into functionalized nanoscale environments (ie, nanostructured supports like nanoporous media, nanofibers, carbon nanotubes, and nanoparticles), is developing rapidly. Catalytic biosensors can employ complete cells, isolated enzymes and subcellular organelles or tissue containing specific enzymatic systems (eg, double membrane organelles containing enzymatic systems like complete chloroplasts or mitochondria) (Velasco-García and Mottram, 2003). In practice, enzymatic biosensors are developed mostly using commercially available enzyme preparations that require no coenzymes or cofactors to improve the selectivity and overcome the slow response of microbial sensors (Laschi et al., 2000). Novel enzyme biosensors for sugar detection include those for analyzing glucose with glucose oxidase (GOx), lactose with β-galactosidase and GOx, lactulose with β-galactosidase and fructose dehydrogenase, and D-glucose with pyranose oxidase (PyOx) (Ozdemir et al., 2010; Serna-Cock et al., 2009; Yabutani et al., 2014). Such biosensors are also applied to analytes like choline, sorbitol, ethanol, polyphenols, vitamins, hypoxanthine, microorganisms, or heavy metals (Dimcheva et al., 2002; Kulys and Vidziunaite, 2003; Liu et al., 2006; Niculescu et al., 2003; Panfili et al., 2000; Saidman et al., 2000; Tsai et al., 2003).

Ensuring the stability and activity of enzymes over a long period of time remains challenging. Enzyme immobilization represents an efficient solution. Immobilized enzymes possess distinct advantages over those in free form in greater productivity per unit enzyme, feasibility of automated production due to ease of reusing the enzyme, and precise control over the reaction and product purity. Ideal supports for enzymes (eg, size ranging from several to tens of nanometers) are nanoporous media with high surface area including nanofibers or nanotubes. The nanoscale environment exerts a significant impact on enzyme stability and activity. For example, single-walled carbon nanotubes cause chymotrypsin to

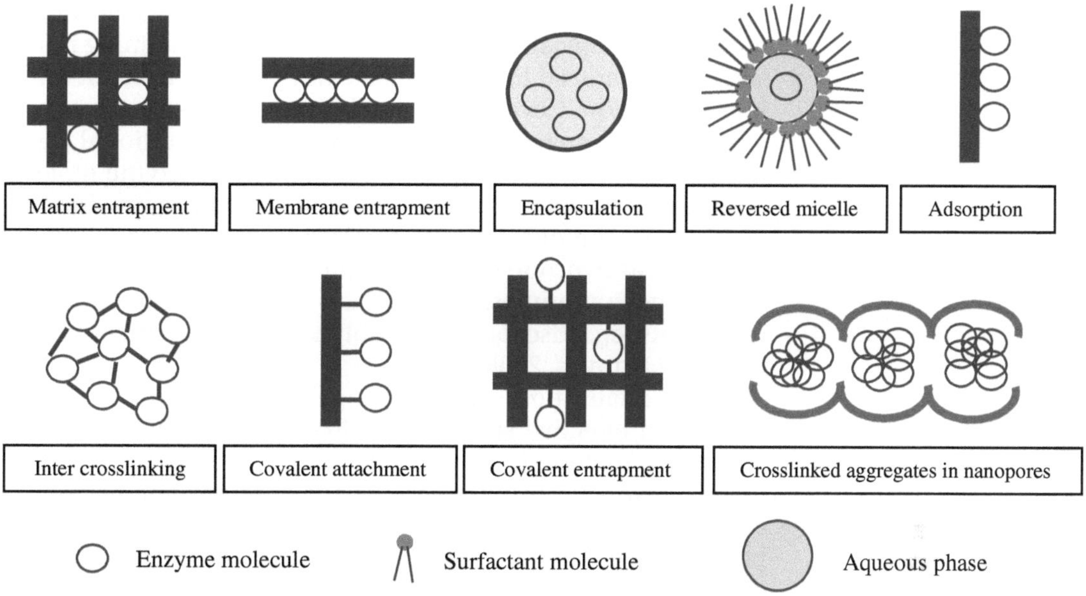

Figure 2.1. Schematic respresentation of the different approaches used for the nanoscale immobilization of an enzyme: Entrapment (matrix, membrane/micelle, encapsulation), adsorption, crosslinking.

unfold, yet retain well the 3-D conformation of soybean peroxidase (Karajanagi et al., 2006). Common methods for enzyme immobilization are physical adsorption onto a solid surface, entrapment in a matrix, crosslinking with bi-/multifunctional reagents, covalent binding to a surface, use of a membrane matrix, self-assembly of biomolecules and electrochemical polymerization (Fig. 2.1) (Campuzano et al., 2007). These approaches allow effective enzyme loading and high enzyme activity while reducing enzyme leaching to different extents. Adsorption of enzymes to a support via electrostatic, biospecific, hydrophobic interactions or van der Waals attractive forces is a common approach, with advantages of simplicity and enzyme reuse and a drawback of potential enzyme desorption and activity loss over time. Matrix-related methods for enzyme immobilization including matrix entrapment, encapsulation, or inclusion by biopolymers or inorganic materials (eg, silica, celite, agarose, Sepharose), often employ one or more enzyme-support interaction mechanism. Enzyme crosslinking or enzyme aggregate coatings can reduce the denaturation of enzyme under rigorous reaction conditions and during repetitive uses, thus maximizing the industrialization potential of enzyme (Kim et al., 2007). Porous sol–gel entrapment immobilization involves siloxane polymer chains attaching to the enzyme bound firmly to an inorganic oxide support. The entrapped enzyme can still freely interact with external chemical species or analytes. Self-organized films (eg, biological amphiphilic membranes containing multiple layers or organized lipid-like molecules) are also used for enzyme immobilization. Covalently immobilized enzymes confer specific advantages in enzyme catalytic activity and operational stability (Pavlidis et al., 2012). Covalent conjugation of enzyme molecules onto carboxylated nanotubes via amide bondings provides more durable attachment, because the bonds created are far stronger than typical hydrophobic/electrostatic interactions between enzymes and nanotubes. In this case, additional noncovalent attachments (eg, adsorption) possibly occur during immobilization (Gao and Kyratzis, 2008).

Functionalization of conducting carbon nanotubes with proteins is an effective strategy for biosensor development. Porous structures including nanocellular pores, membrane layers, or mesh networks can function as "nanosized enzyme reactors" for continuous enzymatic reactions, enabling excellent biosensing performance. The curvature of nanomaterials affects the immobilization efficiency, catalytic behavior, and enzyme secondary structure. The surface chemistry of the functionalized nanomaterials governs their dispersability and interactions with enzymes. Noble metals (gold, silver, and platinum) are usually used as self-assembled films or as discrete particles to facilitate good control over the orientation and distribution of immobilized enzymes, producing highly ordered immobilization structures and reproducible enzyme-containing electrodes. Au, Pt, and Pd nanoparticles have good biocompatibility and can enhance and stabilize the immobilized enzyme for high-throughput analysis through increasing the electroactive area of the electrode. Smart design of the nanoscale environment of each enzyme along with good choices of nanoparticles (ie, the type, size, and charge) are essential for creating an enzyme immobilization system with high performance and stability. Decreased enzyme stability is typically seen on larger nanoparticles with smaller surface curvature (eg, ribonuclease A on silica nanoparticle, Shang et al., 2007) or high negative charges (eg, chymotrypsin for anionic functionalized gold nanoparticles, Fischer et al., 2002). Enzyme nanoentrapment is commonly used to integrate magnetic nanoparticles and enzymes, in which an organic–inorganic hybrid polymer network with a thickness of a few nanometers or less is constructued over the surface of enzyme that largely retains its native structure. Direct attachments between the enzyme and polymer network are sometimes employed, in which the pore size should be tailored to confine the enzyme (thereby retarding or even preventing enzyme unfolding and denaturation), and the nanoscale interactions

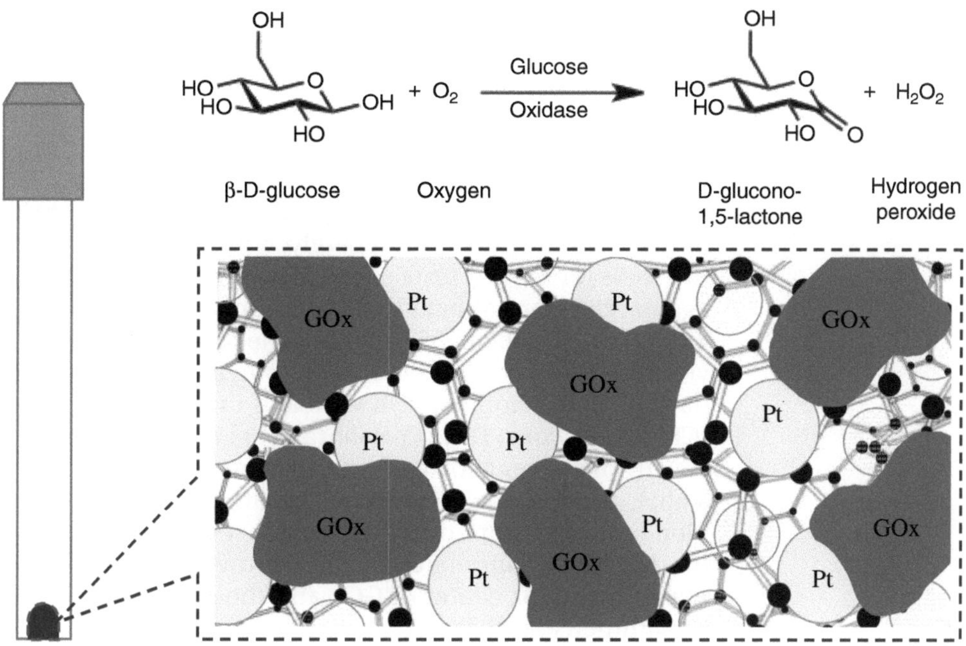

Figure 2.2. Schematic representation of glucose oxidase (GOx) immobilized on a platinum nanoparticle functionalized semi graphitic carbon support. The GOx/Pt/C electrocatalyst was packed into a glassy carbon electrode cavity.

between enzyme and its surrounding environment should be optimized (to realize synergic enhancements in analysis sensitivity and stability). The use of thin and porous polymer network nanostructures for enzyme immobilization is especially desirable, offering advantages such as large surface area (high reaction rates due to facile access and transport of analytes to and from active sites), uniform and tailorable particle size distributions, rapid electron transfer (to generate highly intense, specific and stable detection signals) and potential for magnetic separation of the enzyme from its dispersion medium (Wang, 2006).

GOx is a substrate specific enzyme that can be manufactured in large quantities. Fig. 2.2 shows a schematic representation of a biosensor fabricated by immobilization of GOx on a platinum nanoparticle-containing porous carbon (Pt/C) electrode. The Pt/C support was prepared by carbonization of a freeze-dried agarose gel containing K_2PtCl_4 at 800°C, before GOx was dispersed over the Pt/C support and the resulting GOx/Pt/C electrocatalyst was packed into the cavity of a glassy carbon working electrode. This represents a simple but effective approach for creating a sensitive amperometric glucose biosensor, which was developed jointly by Yabutani's group (Tokushima University, Japan) and the Waterhouse group (University of Auckland, New Zealand). The GOx/Pt/C electrode system detected glucose with good operational characteristics and high sensitivity (ie, response time <2 min, sensitivity of 1.9 mA M^{-1}, and a linear response to glucose concentration up to 10 mM) (Yabutani et al., 2014). It is well known that successful diagnosis and continuous monitoring of diabetes mellitus and hyperglycemia rely largely on easy-to-use tools for regular tests of blood glucose levels (ie, comparison against the normal healthy human range of 80–120 mg dL^{-1}, or 4.4–6.6 mM). Since 1962, various electrocatalytic detection devices based on consumed oxygen or hydrogen peroxide evolved, have been developed for in vitro and in vivo blood sugar testing (Clark and Lyons, 1962; O'Halloran et al., 2001). Despite the progress in this area, gaps and obstacles remain to achieve a highly stable, reliable, "painless," and "minimally invasive" continuous blood glucose monitoring device. Most current research focuses on the use of nanomaterials for improved electrical contact between the redox center of GOx and the electrocatalyst supports. Voltammetry is the core electro analytical method for detecting sucrose and other analytes. Among the different forms of voltammetry (including polarography, linear sweep, differential staircase, normal/reverse/differential pulse), cyclic voltammetry (CV) based on the redox potential of particular electrochemical reactions in analyte solutions is one of the most popular methods. Enzymes typically studied in the field of bioelectrochemistry are oxidoreductases such as GOx and dehydrogenases because of their good electron-transfer behaviors. The catalytic domain (ie, enzyme) in the nanobiosensors is typically connected by a flexible electrically conductive linker to their noncatalytic domain (usually the electrode). In order to obtain an efficient electron transfer coupling between the redox centers of enzyme and the electrode, good contact of the enzyme with the electrode surface is critical. Other factors, such as the distance between the redox site and the electrode surface, the reorganization energy associated with the oxidized/reduced forms of the redox site, and the thermodynamic driving force of electron transfer, also contribute to sensor's function and sensitivity. For the GOx immobilized Pt/C electrode shown in Fig. 2.2, agarose, an inexpensive commercially available thermosensitive sol-gel material, was used as the carbon source for making the Pt/C support. Freeze-dried agarose gels containing K_2PtCl_4 were first prepared via a novel one-pot process using an aqueous agarose solution (typical concentration 1 2 wt%) and K_2PtCl_4 aqueous solution (typical concentration 100 mM). Carbonization at 400 or 800°C under an

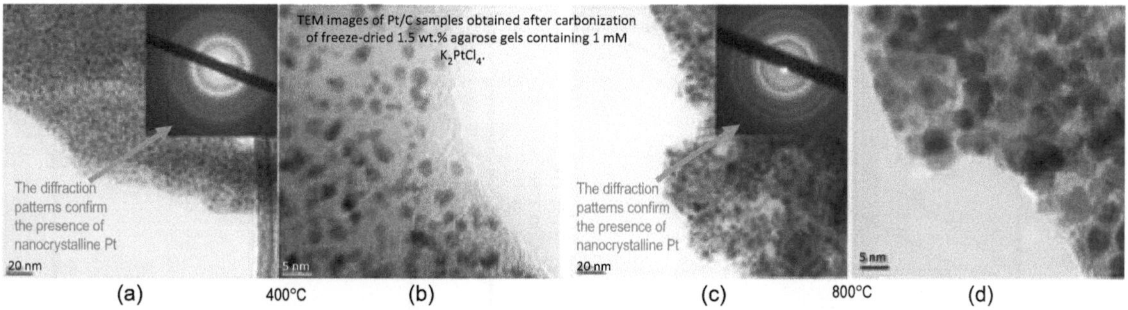

Figure 2.3. Transmission electron microscopy (TEM) for Pt/C electrode (4 wt% Pt/C) prepared at carbonization temperatures of (a), (b) 400 and (c), (d) 800°C (in both cases, the concentration of K_2PtCl_4 and agar concentration in the precursor gel was 1 mM and 2 wt%, respectively). The insets show the selected area diffraction patterns. Reproduced from Yabutani et al. (2014) with permission from Springer.

nitrogen atmosphere transformed the agarose to carbon and reduced Pt(II) to Pt nanoparticles. The direct one-pot synthesis of Pt/C materials provided a semi-graphitic carbon network. The Pt content in the final Pt/C products varied 0–20 wt%, and could be tuned through adjusting the ratio between agarose and K_2PtCl_4 in the freeze-dried gels. The porous nature of the obtained Pt/C materials allowed physical adsorption and entrapment of GOx, while also allowing good access to the analyte. The operation of the biosensor involved (1) oxidation of glucose in the presence of oxygen on GOx catalytic sites, and (2) transfer of resulting electrons to the Pt/C support and subsequently to the electrode surface producing a measurable current. The approach used for the synthesis of the Pt/C materials is likely widely applicable, and could similarly be used to construct other carbon supported noble metal catalysts and electrocatalysts, including Pd/C and Au/C.

Fig. 2.3 reveals the physicochemical characteristics Pt/C electrodes used in the GOx/Pt/C experiments. The Pt/C materials are composed of nanocrystalline graphite-like carbon and contain small Pt nanoparticles (mean size 2 ± 1 nm for samples carbonized at 400°C, and 6 ± 3 nm for samples carbonized at 800°C). The narrow Pt particle size distributions and uniform Pt dispersion within the carbon matrix are a feature of this synthesis method. The selected area diffraction patterns (inserts) of both samples display a well-defined ring pattern, confirming the presence of metallic Pt nanoparticles with a face-centred cubic structure. An increase in the size of Pt nanoparticles with increasing temperature was observed, which agrees with the previous findings (ie, due to aggregation of the smaller particles to minimize their surface free energy) (He et al., 2004). N_2 physisosorption analyses revealed the Pt/C materials had high surface areas (350–500 $m^2\ g^{-1}$) and

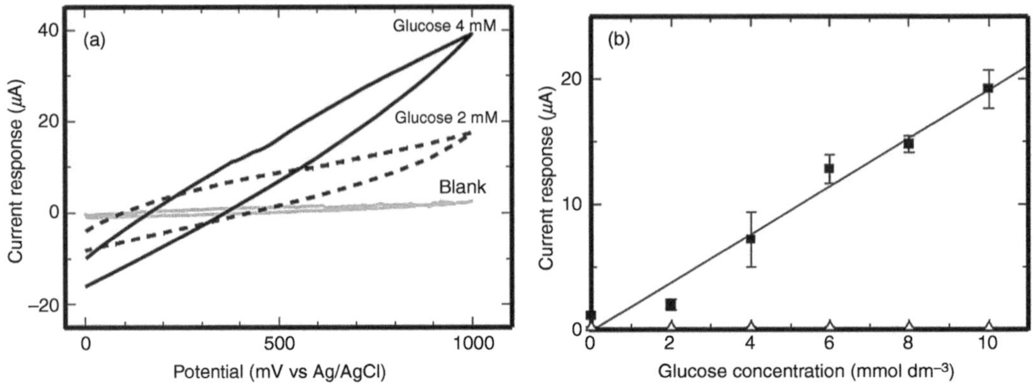

Figure 2.4. Cyclic voltammograms (a) and calibration curve (b) for glucose oxidase immobilized Pt/C electrode (4 wt% Pt/C) at different sucrose concentrations. The Pt/C support was prepared at a carbonization temperature of 800°C (the concentration of K_2PtCl_4 and agarose concentration in the precursor gel were 1 mM and 2 wt%, respectively). The symbols ■ and Δ indicate the current response observed at 400 mV versus Ag/AgCl for GOx/Pt/C and GOx/C electrodes, respectively. Reproduced from Yabutani et al. (2014) with permission from Springer.

inherent micro- and meso-porosity. Fig. 2.4a shows the response of a GOx/Pt/C electrode to glucose concentration. A linear current increase was observed with increasing glucose concentration (Fig. 2.4b), which was due to the hydrogen peroxide generated by the catalytic reaction between GOx and glucose in the presence of oxygen. The response curves GOx/Pt/C (4 wt% Pt/C) and GOx/C electrodes at 400 mV versus Ag/AgCl (Fig. 2.4b) confirm that the system sensitivity was enhanced by several orders of magnitude by the presence of Pt nanoparticles. The slope and correlation coefficient of the calibration curve for the GOx/Pt/C electrode in this case were 1.92 mA M^{-1} and 0.99, respectively, while the detection limit based on the signal to noise ratio of 3 was 1.4×10^{-6} M and the response time was <2 min. The linear response over the glucose concentration range (ie, 0–10 mM) for the GOx/Pt/C electrode covers the boundary blood glucose levels for healthy people (<7.8 mmol L^{-1}) and diabetes patients (>7.8 mmol L^{-1}) 2 h after consumption of a meal.

Increasing concerns about food safety and traceability worldwide motivate the development of advanced analytical techniques for selective detection and quantification of adulterants. Surface-enhanced fluorescence (SEF) (Le Ru et al., 2007), SERS (Le Ru and Etchegoin, 2009), and surface plasmon resonance spectroscopy (Haes and Van Duyne, 2005) have emerged as powerful newcomers to this field due to their high sensitivity for low concentration analytes. Plasmon resonances in metallic nanostructures and in particular unique optical properties of coinage metals like gold and silver at nanoscale particle sizes are utilized. Raman spectroscopy and fluorescence techniques utilize complicated optical processes involving at least two photons at different wavelengths (one from the incoming laser beam and the other(s) reemitted to give information about the analyte). SERS enhancement factors for optimized Au and Ag substrates could be as high as 10^{10}–10^{11}, which allows for single molecule detection in some cases. However, low reproducibility of SERS substrates is a current obstacle to the widespread deployment of SERS in routine quantitative analyses. Thus, the development of stable and sensitive SERS active substrates becomes a priority.

Stimulation of the localized surface plasmon resonance (LSPR) of noble metal nanoparticles (especially gold and silver) by visible light at the resonance frequency excites valence electrons in Au and Ag, causing a collective excitation of valence electrons in response to the oscillating electric field of the incident light. This results in large localized electric fields generated within a few nanometers of the metal nanoparticles. When a molecule experiences these strong electric fields, specific Raman active modes of the molecule are enhanced enormously (ie, enhancement factors $> 10^5$ over conventional Raman spectroscopy are realized), thus allowing for the detection of low concentration analytes. The Local Field Intensity Enhancement Factor (LFIEF) for the electromagnetic field at the surface is dependent on the size, shape, and loading density of the metal nanoparticles. Shape, size, dielectric environment of the isolated nanoparticles, in particular the separation between the metal nanoparticles in aggregated structures, allow the manipulation of optical response and sensitivity of SERS substrates. More than one resonance condition is realized for a given nanoparticle shape (metal nanorods have both a transverse and a longitudinal localized surface plasmon resonance). The magnitude of the plasmon-induced electric fields generated depends on particle size, with substrates containing Au or Ag nanoparticles of size ~20–30 nm generally giving the largest signal enhancement factors. Au nanoparticles have a comparable LFIEFs to Ag nanoparticles, but are more advantageous in relative ease of preparation, high long-term (chemical) stability, and better biocompatibility with other molecules.

The development of robust SERS substrates that can be used multiple times is challenging. Au and Ag nanoparticles are thermodynamically unstable because of their high surface-to-volume ratios, and in solution will aggregate irreversibly to reduce their surface free energy. This aggregation alters the size and optical plasmonic properties of the Au and Ag nanoparticles in a manner detrimental to SERS analyses (ie, shifts the plasmon

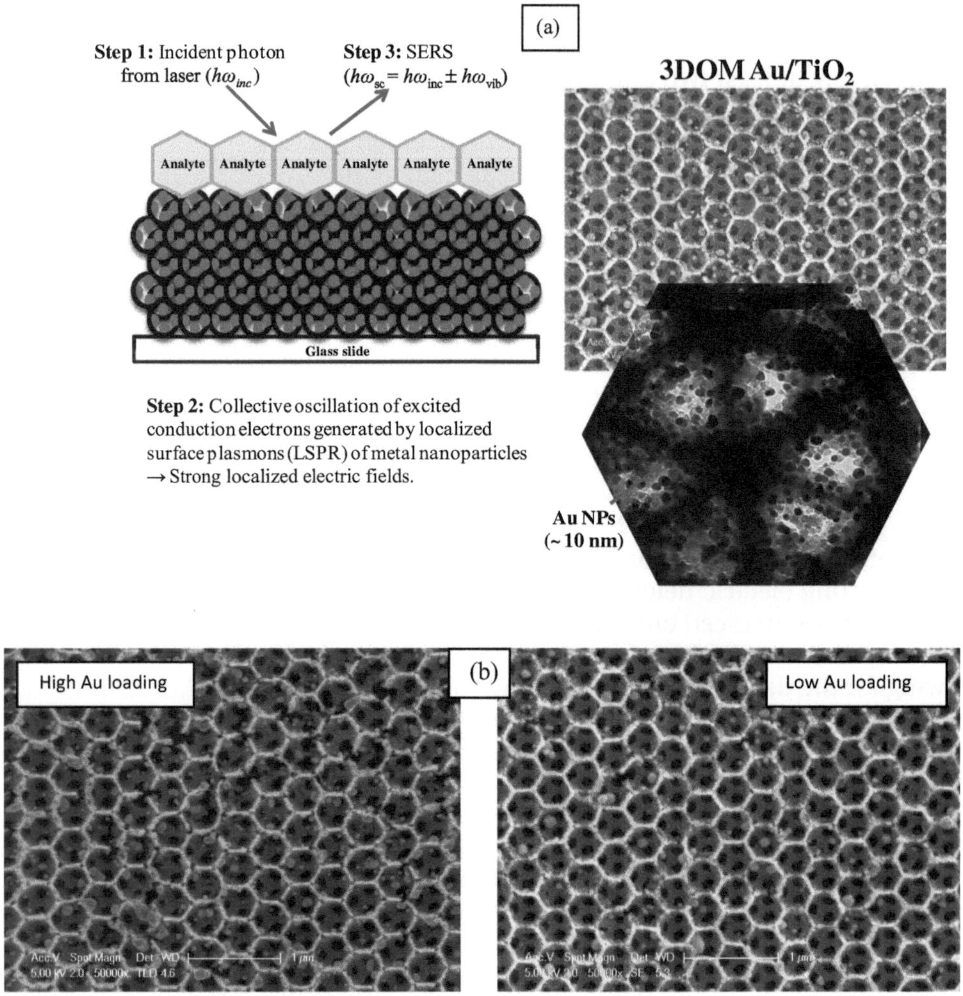

Figure 2.5. Novel three-dimensional SERS substrates with gold (Au) nanoparticles (diameter 1–20 nm) immobilized on the ordered macroporous (3DOM) TiO2 photonic-crystal supports, for detecting melamine in aqueous media including in milk (a). Different Au loadings can be used (b), and the pH-dependence (c) and solvent refractive index sensing (d) are possible for detection.

resonance, while large aggregated particles will sediment rapidly). Metal nanoparticles can be stabilized against aggregation in aqueous media by surface functionalization (steric or electrostatic stabilization), which can be used as a means to improve the detection selectivity of small molecules. An alternative approach for preventing aggregation is to immobilize the Au or Ag nanoparticles on a 2D or 3D support, thereby significantly reducing nanoparticle aggregation. Fig. 2.5a demonstrates a SERS biosensor developed for melamine detection in milk or aqueous

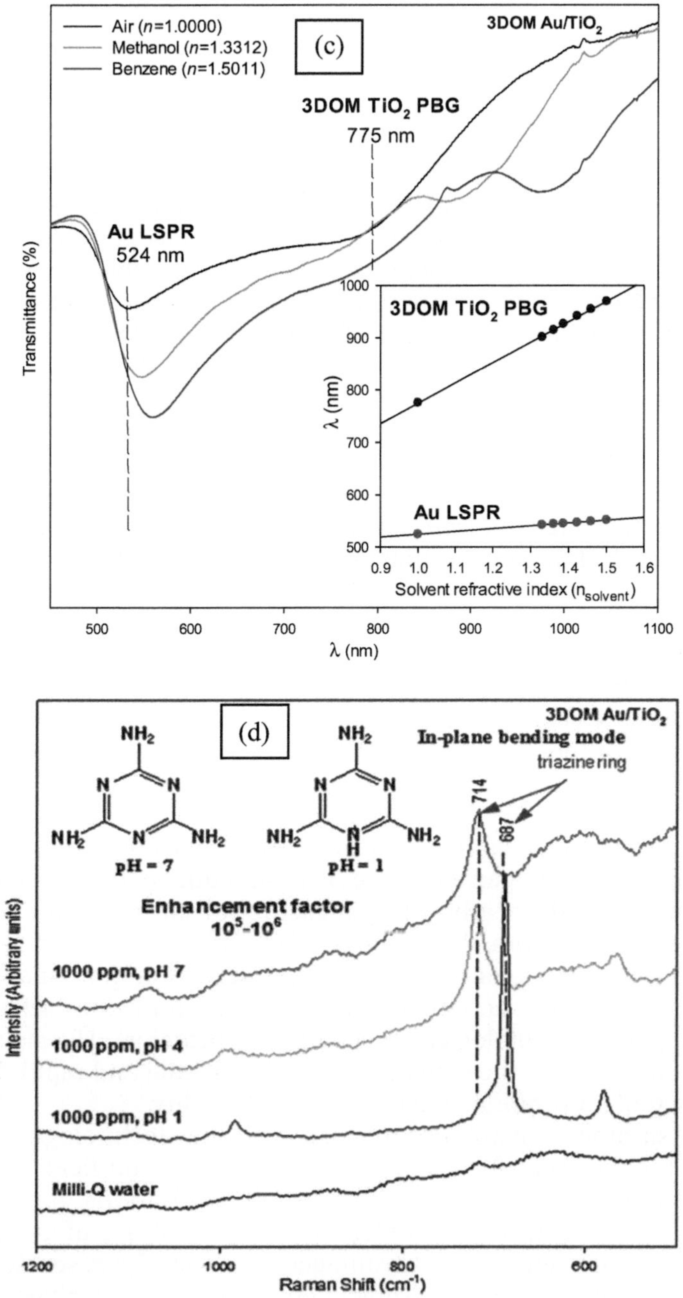

Figure 2.5 *(cont.)*

solutions by the Waterhouse research group at the University of Auckland (Hsieh et al., 2014a,b). The biosensor was based on novel three-dimensional (3D) SERS substrates comprising gold (Au) nanoparticles (diameter 1–20 nm) immobilized on 3D ordered macroporous (3DOM) TiO_2 photonic-crystal supports. A series of SERS substrates can be fabricated with gold nanoparticle loadings up to 80 wt% Au (Fig. 2.5b). For conventional 2D planar SERS substrates, the maximum number of SERS hot spots possible is limited to a single Cartesian plane. However, for the substrates shown in Fig. 2.5a, the hotspots are spatially distributed in 3D, at lengths approaching the active volume of the Raman microprobe. Accordingly, Au/3DOM TiO_2 substrates demonstrate excellent SERS activities with a high sensitivity and low detection limit (enhancement factors up to 10^7, allowing the detection of melamine in aqueous media at concentrations as low as 0.1 ppm). The large enhancement in Raman signal due to the presence of immobilized gold nanoparticles results from electromagnetic (a dominant factor) and chemical (a contributing factor) enhancement mechanisms (Kneipp et al., 1999). The SERS spectra (Fig. 2.5c) show the various pH-dependent forms of melamine, thus the SERS method allows quantification of spectroscopic fingerprinting of melamine. At pH 1, the endocyclic ring nitrogen of melamine is protonated, producing a very different SERS spectrum to those obtained at high pHs for the nonprotonated forms of melamine. The inherent structural and optical properties of the Au/3DOM TiO_2 photonic crystal substrate can also be used to determine the solvent refractive index via two different optical mechanisms. Both the Au LSPR and photonic bandgap along the [111] direction in the 3DOM support red-shifted linearly with increasing solvent refractive index (Fig. 2.5d). Therefore, the Au/3DOM TiO_2 substrates herein represent a multifunctional optical platform that could potentially be improved by Au nanoparticle functionalization to allow selective binding to target biological molecules (eg, proteins). The stability and robustness of the system suggests a versatile and rapid diagnostic tool for probing food and beverage quality.

In summary, nanobiosensors are versatile technologies that meet the demands for user friendliness and "in-the-field" devices for environmental, clinical, and food applications, providing rapid, sensitive and stable detection of a wide range of analytes. Enzymatic biosensors are advantageous in specificity, sensitivity, rapidity, stability, and cost-effectiveness, especially for targeting specific compounds in complex matrices. Advances in materials science, enzyme immobilization technology, molecular biology, genomics, microbial biodiversity, and analytical techniques, could further enhance the feasibility of nanobiosensors through decreasing the cost and improving the performance. However, concerns still exist on the toxicity of some nanomaterials included in the biosensors for in situ or real-time analysis of food systems (although this is not an issue for analysis of samples obtained from foods).

3 Nanocarriers

There now exists a huge market for food products enhanced with nutrients and bioactives, and there are increasing pressures to divert fruit and vegetable production toward the energy sector (eg, corn sugars for ethanol manufacture that can be blended with petrol). Hence, maximizing the utilization of food resources and optimizing the delivery of food bioactives to consumers are critical. Further, the eating habits of modern consumers are driven not only by necessity but also by pleasure. Tailored bioactive delivery systems can minimize adverse sensory effects including those nutritionally critical substances (eg, bitter polyphenol antioxidants) on consumer acceptance. Nanoencapsulation has been used to enable the production of "healthy" foods and sports drinks with improved bioavailability and optimal hydration, through delivering specific substances such as vitamins, minerals, flavors, antioxidants, omega-3 fatty acids, proteins, and nucleic acids in an easily absorbable nanoparticulate form. Nanocarriers refer to the nanoscale structures tailored to carry specific substance(s) (which could also be at a nanoscale). Nanoencapsulation offers advantages similar to microencapsulation, offering greater dispersion and bioabsorption, though at the same time potentially creating new health hazards to humans. The absorption, distribution, metabolism, and excretion of encapsulated substances can be quite different to that of conventional bulk equivalents. This is especially true for nanoencapsulated substances (which can enter the body and bloodstream via direct mechanisms rather than via absorption in the gastrointestinal tract and then into the blood circulatory system).

A wide range of nanodelivery systems have been explored for transporting/translocating target substances, including micelle-based systems (eg, surfactant micelles, bilayers, and reverse micelles) (Cabral et al., 2011), liposomes (Rashidinejad et al., 2014), (bio)polymeric nanoparticles (Sun-Waterhouse et al., 2013; Waterhouse et al., 2014), solid lipid nanoparticles (Frederiksen et al., 2003), nanoscale complexes (Davidov-Pardo et al., 2015; Sun-Waterhouse and Waterhouse, 2015), α-lactalbumin nanotubes (Graveland-Bikker et al., 2009), carbon nanotubes

(Kam and Dai, 2005) and chitosan-based particulate systems or lipid droplets (Klinkesorn and McClements, 2010). These systems protect the substances of interest, through encapsulating them in nanosized vessels and thus minimizing environmental stresses and processing impacts, while improving solubility and dispersability in aqueous based systems and allowing their controlled/targeted release. Further, immobilization or encapsulation of enzymes by nanoporous media enables highly stable enzyme systems for different applications and can protect enzymes against proteases. For example, lipase with entrapped magnetic nanoparticles possesses good resistance to proteolytic digestion, due to steric hindrances caused by crosslinked enzyme aggregates within nanoporous media (Kim et al., 2007). Besides protection, bioavailability improvement offered by nanocarriers is enhanced through their controlled release properties (which enables precise delivery of substances of interest to target sites in the human body). The release of the encapsulated substances from the core or matrix of nanocarriers can be triggered by changes in environmental conditions (eg, pH, enzymes, and/or salt concentration), interactions with specific cells or organs in the body, or even a controlled external stimuli (eg, electrical stimulation) (Sharma et al., 2013; Shinde et al., 2014; Sun-Waterhouse et al., 2014a). Different modes of action allow diverse controlled release mechanisms including immediate, sustained or triggered release (Chaudhary et al., 2012; Deshkar et al., 2013; Kojima et al., 2008), as well as parenteral delivery (Davis, 2008), ocular delivery (Di Colo et al., 2001), osmotic delivery (Waterman et al., 2011) or gastroretentive delivery (Jagdale et al., 2013). These different delivery mechanisms present new opportunities for nanocarrier systems in food and nutraceutical applications.

Fig. 2.6 demonstrates a nanocarrier system developed at the University of Auckland, in which the intrinsically conducting polymer polypyrrole (PPy) was fabricated in the form of a 3D ordered macroporous inverse opal thin film with high surface area for controlled drug release triggered by an external electrical stimulus. The macroporous PPy inverse opal thin films were fabricated by electrodeposition of PPy through the interstitial voids of a 3D colloidal crystal template composed of poly(methyl methacrylate) colloids of diameter ~400 nm. The PMMA template was then removed by wet chemical etching to leave the PPy inverse opal (which consists of an ordered array of macropores in a PPy matrix). Risperidone (an antipsycotic drug) was then loaded into the PPy inverse opal structure, after which a nonporous PPy overlayer was added by electrodeposition. PPy undergoes linear actuation under an applied potential, which stimulated the release of

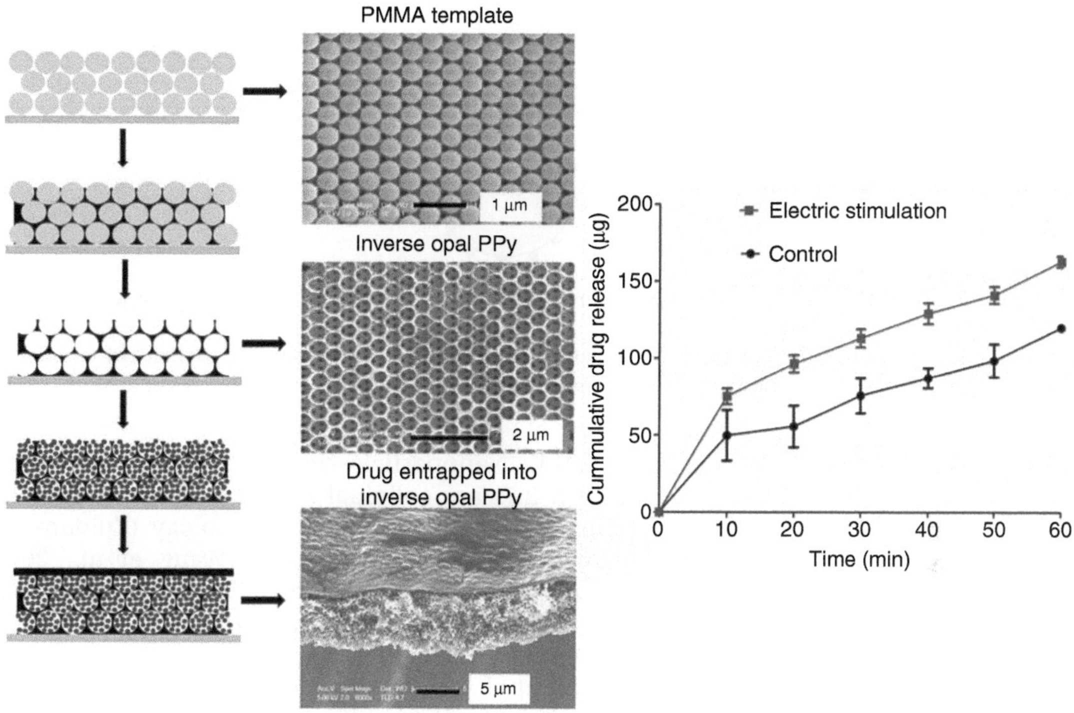

Figure 2.6. Schematic representation of inverse opal PPy thin film fabrication and drug loading (left), micrographs of PMMA colloidal crystal template, inverse opal PPy and entrapped drug (middle), and drug release as a function of time. Reproduced from Sharma et al. (2013) with permission from Elsevier.

the adsorbed/encapsulated Risperidine. Fig. 2.6 shows the passive release and stimulated release profiles of the PPy-based drug delivery system. Such electrically stimulated delivery implants allow high drug loading and easy-to-regulate dose profiles compared to conventional nonporous PPy films.

A physical barrier between the core substance(s) and the environment is specially created via encapsulation to protect the core against adverse environments, so as to minimize undesired impacts on the core substance caused by changes in oxygen concentration, moisture, heat, light, pH, metal ion catalysis, and so on. Encapsulated substances such as proteins, vitamins, omega-3 fatty acids, phytosterols, flavonoids, enzyme, probiotics, and prebiotics can be used as "ingredients" to produce processed foods. This approach improves the delivery, bioaccessibility, and bioavailability of bioactive substances in consumer foods for better nutrition and health outcomes (eg, blood pressure suppressing, energy boosting, and promotion of gut health, brain health and antiinflammatory effects) (Rashidinejad et al., 2014; Scrinis and Lyons, 2007;

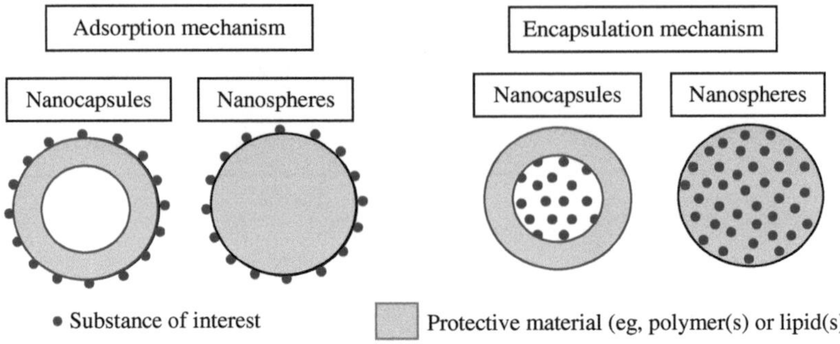

Figure 2.7. Schematic representation of different encapsulation approaches used to deliver substances of interest through nanostructures or nanoparticles.

Shinde et al., 2014). Nanoparticles including nanocapsules and nanospheres (Fig. 2.7) are colloidal particles sized 10–1000 nm and can be prepared using emulsion technology (Zuidam and Shimoni, 2010), coacervation technique (Wang et al., 2008), interfacial polymerization (Aboubakar et al., 1999), inclusion complexation (Lira et al., 2009), denaturation/desolvation of natural proteins or carbohydrates (Kreuter, 1994), nanoprecipitation technique (Anand et al., 2010), solvent displacement techniques (solvent evaporation or solvent diffusion) (Dandekar et al., 2010), supercritical fluid techniques (Reis et al., 2006), along with traditional drying techniques such as spray drying (Ferreira et al., 2007) and freeze drying (Bejrapha et al., 2010). The nature of the core substance(s) determines the selection of encapsulation and processing technologies. For solid lipid nanoparticles, hot high-pressure homogenization above the melting point of the lipid core ("melt emulsification") is often employed, whereas cold high pressure homogenization is often used for thermolabile bioactives (Liedtke et al., 2000).

Polymeric encapsulation systems have been developed for controlled release of active species for food and agrochemical applications. Polymers, including natural or synthetic polysaccharides, proteins, and other organic polymers such as lignins, chitosan, alginate, gelatin, collagen, dextran, cyclodextrin, polyethylene glycol, methyl cellulose, hydroxy propyl methyl cellulose, sodium carboxymethyl cellulose, polyethylene oxide, polyethylene, polypropylene or polyglycolic acid are used as solubilizers, stabilizers, or mechanical supports, allowing specifically tailored processing or health functionalities (Shah et al., 2014; Sun-Waterhouse and Waterhouse, 2015). For dietary products, toxicity of the polymer materials must be taken into consideration. Coating polymer formulations should be generally insoluble at salivary pH (~6.8), and either dissolve readily at gastric fluid pH (~1.2) or be insoluble in gastric fluid but soluble in intestine digesta. In this way, the release of bitter bioactives in the oral cavity is prevented after ingestion while still allowing their release in the stomach or small intestine (where they are expected to be metabolized). Quite a few polymers such as plant gums, modified celluloses, and gelatine also offer taste-masking effects especially when factors like particle size, flow characteristics, moisture sensitivity, swellability, bioadhesion (like mucoadhesivity), film/gel forming ability, long-term stability, and temperature of processing are well tailored (Sharma and Lewis, 2010). High molecular weight polymers can generate complex networks to delay the release of soluble and insoluble encapsulated substances from encapsulant matrix. The surface of nanocapsules can be modified using substances such as surfactants (eg, PEG) to reduce the undesirable interactions with reticuloendothelial system in the human body (Kreuter, 1991). New Zealand research groups have examined various biopolymeric delivery systems to carry phytochemical antioxidants, edible oils with high unsaturated lipids, and probiotic bacteria (Shinde et al., 2014; Waterhouse et al., 2014). Sun-Waterhouse et al. (2013) employed different gastroresistant biopolymers including dietary fiber oligo-/poly-saccharides (ie, sodium alginate, methyl β-cyclodextrin, inulin, or hydroxypropylmethyl cellulose) to encapsulate a phenolic antioxidant, quercetin, using spray drying technology. The polymeric matrices were effective vectors for carrying the antioxidant directly to the intestine thereby improving antioxidant bioavailability.

Another popular encapsulation strategy is the use of liposome nanocarriers to enhance delivery efficiencies. Liposomes are widely used to produce food, pharmaceutical, and cosmetic products, as liposomes can encapsulate both water- and lipid-soluble compounds and impart protective barrier function on their core substances during digestion, for example, resistance to enzymes, water, acid, digestive juices, bile salts, and/or microorganisms in the mouth, stomach, and intestine. Liposome encapsulation allows the core species to have different pH and ionic strength from that of the continuous phase, inside the liposome vesicles made up of polar lipids (eg, phospholipids in the forms of egg yolk/soybean lecithins). This technology is especially useful for delivering bitter bioactives like quinine, because the bitterness is also masked. Phospholipids are amphipathic and in aqueous media can form complex structures with different characteristics, for example, various degrees of rigidity or fluidity and permeability or stability, and different charge through using more unsaturated phospholipids (eg, phosphatidylcholine species from

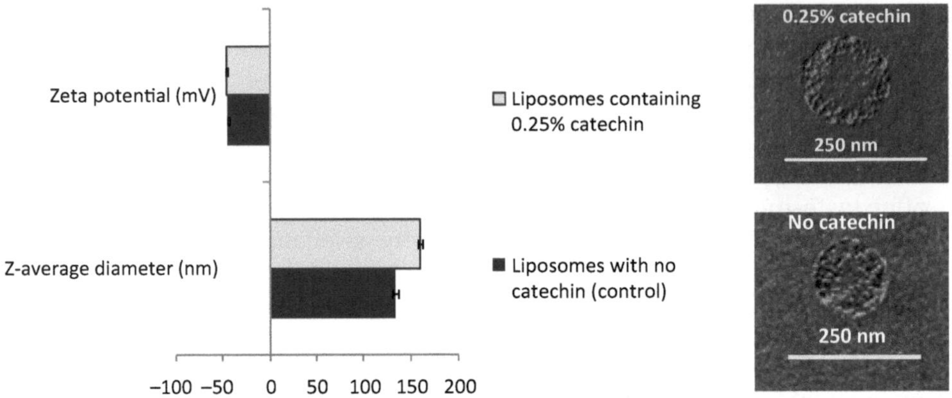

Figure 2.8. Liposomes containing no catechin or 0.25% catechin. After Rashidinejad et al. (2014).

egg or soybean) or more saturated phospholipids (eg, dipalmitoyl phosphatidylcholine with long acyl chains) (Gabizon et al., 1998). There exist a wide range of methods to produce liposomes with particular characteristics, including cross-flow filtration, detergent dialysis, dual asymmetric centrifugation, freeze drying, high-pressure extrusion, membrane contactor technology, high-speed or high-pressure homogenization, high shear stirring, microfluidization, reversed phase evaporation, solvent-injection, (ultra) sonication, and supercritical fluid technology (Huang et al., 2014). The obtained liposomes are generally in the form of spherical vesicles sized from 30 nm to several micrometers. Fig. 2.8 demonstrates a successful liposome nanocarrier developed for making functional cheese. Rashidinejad et al. (2013, 2014, 2015) found that addition of catechin to soy lecithin liposomes altered the size of liposome. Catechin encapsulated in the liposomes was retained completely in the cheese curd, rather than partitioning into the whey or adsorbing onto the surface of the nanocapsules. Conversely, adding catechin directly to the milk for cheese making caused a significant catechin loss in the whey upon separation from the curds during cheese manufacture. Adding catechin did not alter the surface charge of the liposome vesicles (remaining negatively charged), although significantly ($p < 0.05$) increased the Z-average diameter of the liposomes. The phase transition temperature of the liposomes with/without loaded catechin was higher than 50°C. The presence of catechin in liposome increased the phase transition temperature, suggesting that the catechin-containing liposomes are stable from oxidation at room temperature and up to 50°C. Fortification of a low-fat cheese with catechin encapsulated in such liposomes, led to a significant increase in total phenolic content and antioxidant activity measured within the digesta matrix (ie, cheese after 90 days ripening at 8°C followed by in vitro gastrointestinal digestion studies). Thus, such stable liposome nanocarriers represent a simple and effective approach to deliver antioxidants and their bioactivity in a cheese matrix.

Nanocarriers can be developed to deliver multiple substances that do not normally mix well (eg, water-soluble and fat-soluble ingredients), or only mix at specific sites and under particular conditions (eg, enzymes and substrates/cofactors). Solid hydrophobic nanosphere systems (in which a mixture of hydrophobic materials are encapsulated in moisture-/pH-sensitive bioadhesive shell) offer advantages such as ease of handling, dose control, retention of volatiles, taste masking, moisture/pH/heat controlled release, enhanced stability, protection against oxidation, bioavailability and efficiency (Shefer and Shefer, 2003). Oil-in-water nanoemulsions are important carrier systems in this regard as they are physiologically tolerable in the human body (Klang and Benita, 1998). However, rapid degradation of the ingested emulsion droplets by lipases takes place in human gastrointestinal tract and after entering the blood circulation system. Thus, further modification of conventional oil-in-water emulsions is required, for example, via introduction of multiple or multilayer emulsion delivery systems, in which the core substance can be encapsulated in different locations inside the W/O/W or O/W/O type emulsions, or nanostructured shell layers with tailored characteristics (eg, specific polyelectrolytes) (Balcão et al., 2014; Ray et al., 2015).

Nanocarrier systems can be designed to impart not only functional properties but also novel sensory attributes to foods and beverages, for example, offering distinct optical characteristics to clear beverages or slimming sensation and strong chocolate flavor to cocoa nanoclusters. The barrier properties of nanostructured delivery systems can further be manipulated to create protective films and packaging for food products. Nanocomposites, via incorporation into barrier materials, can reinforce the protection of the enclosed sensitive substance(s) through blocking the surface pores to limit oxygen permeability and hinder the penetration of other unwanted substances (in gas or liquid). For example, polymer composites (eg, plastic or starch-based films) incorporating nanometals, nanosilica, or metal oxides (eg, nanosilver or nanozinc oxide), act as UV absorbers, antimicrobial agents, surface biocides, abrasion resistance or mechanical strength enhancers for novel biodegradable and active packaging or edible films (Azeredo et al., 2012; Zhang et al., 2007).

Edible coatings and films are currently used on fresh and processed foods including fruits, vegetables, meats, confectionery, deep fried foods and bakery products, to help preserve food quality and reduce nonbiodegradable waste (typically generated by conventional food packaging systems like the synthetic plastics). Various edible coatings have been developed with different functions as preservative barriers (against moisture, gas, lipid and microbial attack/growth) and/or sensory enhancers (for better consumer appeal) (Rossman, 2009). In general, lipid-based coatings perform well as moisture barriers but are poor at preventing gas transfer or confering mechanical strength. However, specificially tailored edible films containing nanoscale internal structures and complementary materials could overcome most of these drawbacks. The use of nanostructured materials together with appropriate additives can greatly enhance the efficiency and magnitude of their target functionalities. In addition to antioxidants and antimicrobial agents (which are conventionally added to the coatings to reduce product deterioration and spoilage), other substances are also incorporated to offer packaged food products with improved nutritional value (eg, adding vitamins and minerals), desirable sensory properties (eg, adding specific colors), and reduced undesired attributes (eg, unpleasant tastes and odors). The characteristics (eg, composition, thickness, structure, and surface properties) of edible coating films depends on the nature of the film-forming materials (eg, polysaccharides, proteins, and lipids), processing methods and their end use. For example, alginate has been widely used in edible coatings due to its excellent film forming properties (Sun-Waterhouse et al., 2011a; Waterhouse et al., 2014) and compatibility with many other biopolymers (Sun-Waterhouse et al., 2011b, 2014a). Further, polysaccharide-based coatings often possess drawbacks such as poor moisture barrier properties (due to their hydrophilic nature associated with the abundant hydroxyl groups). Incorporation of nanoscale materials/particles such as nanosized cellulose whiskers and montmorillonite, can help overcome these drawbacks via chemical and/or physical effects (eg, reactions for antioxidant regeneration, electrostatic interactions, steric protection), mechanical protection (eg, reduced injury) and modification of water vapor barrier properties (eg, reduced desiccation) (Azeredo et al., 2012). Dipping and spraying processes are commonly employed for creating coating film(s) on an object (Fig. 2.9). While the characteristics (eg, composition, concentration, pH, ionic strength, dielectric constant, temperature, swellability, wettability, and viscoelasticity) of the coating solutions are critical to the ultimate functionality, the number of coatings also exerts impact. The interactive forces involved in coating formation include electrostatic attraction, hydrophobic interactions, hydrogen bonding. In particular, nanolaminates composed

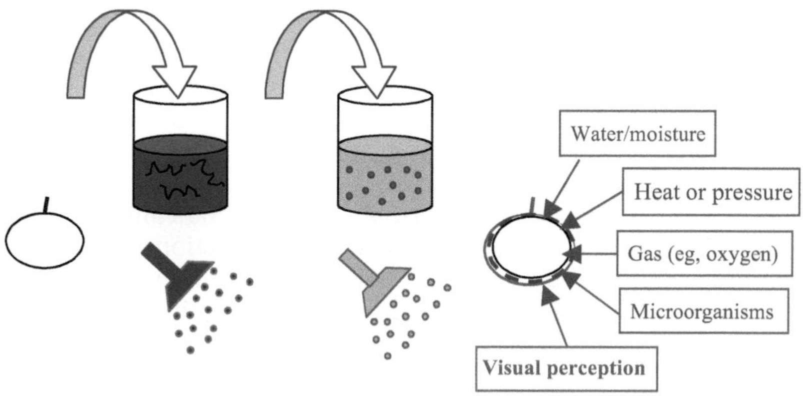

Figure 2.9. Schematic representation demonstrating the use of nanolaminates to coat an object (like a fruit) through a sequential dipping or spraying process.

of two or more layers of well-oriented nanosized thin laminae that are physically or chemically bonded to each other (eg, a "bricks and mortar" architecture), possess distinct and often desirable mechanical and transport properties (Schmidt, 2013). Layer-by-layer (LbL) assembly is a powerful method to produce nanolaminates, by which polyelectrolytes and other charged substances are deposited via electrostatic attraction onto the oppositely charged surfaces, forming interfacial films with tailored charateristics (eg, thickness ranging from 1 to 100 nm per layer).

Developing smart packaging to extend product shelf-life has been a long-standing R&D focus. Nanotechnology has revolutionized packaging R&D through enabling the selective modification of the gas permeation behavior, chemical, mechanical, thermal and microbial barrier properties of packaging materials (Lun et al., 2011; Waterhouse et al., 2003). A notable example was the "nanocomposite containing clay nanoparticles" developed by Voridan and Nanocor (which is used for beer bottles to minimize loss of carbon dioxide from the beer and to prevent the ingress of oxygen into the bottle). The use of nanobiosensors for signaling microbiological and biochemical changes and nanobarcodes for tagging and security are further examples of smart packaging systems. Nanofibrils of perylene-based fluorophores are used to detect gaseous amines released as fish spoils. An "electronic tongue" was developed by Kraft Foods (which is composed of an array of nanosensors for sensing gases released by food as an indicator of freshness), or the "hybrid system" developed by Bayer Polymers (which is essentially a light, strong heat resistant packaging film containing silicate nanoparticles to prevent contents from drying out or degradation by oxygen). Polymer nanocomposites incorporating silica, metal, and/or metal oxide nanoparticles have been used for packaging like plastics, polyethylene (PE) and polyvinylchloride (PVC). These packaging structures block UV radiation and eliminate gas, moisture or unwelcome scents, while offering antimicrobial and antioxidative protection, mechanical strength, temperature/abrasion stability, and enhanced mechanical, barrier, optical, thermal, flame-resistant and recycling properties (Momin et al., 2013). For instance, functionalized nanomaterials containing silver/MgO/ZnO as the antimicrobial agent, TiO_2 as UV absorber in transparent plastics, nanotitanium nitride for mechanical strength, nanosilica for hydrophobic surface coatings, nanoclay as gas barrier and Fe_3O_4/Fe_2O_3 for magnetic properties (Deshpande et al., 2012; Gupta et al., 2012; Waterhouse and Waterland, 2007). Nanocomposites within polymer matrices such as highly porous fibers of collagen, zein and cellulose can facilitate the filler matrix interactions due to the large surface area and can complicate the permeation path of foreign substances within nanoscale pores (Floros, 2010). Nowadays, nanoenabled functionalized food packaging is becoming a commercial reality (ie, occupying ~19% of market share in 2015). Many surface functionalized nanomaterials are being developed from agricultural and forestry resources, by-products or wastes and inorganic materials. Nanoscale lipid structures have been used for water-repellent surfaces, and nanofiltration systems have been created using porous silica or regenerated cellulose membranes for "self-cleaning" surfaces and "self-clarifying" functions (eg, for beers and wines). However, concerns about the safety of using inorganic nanomaterials for food packaging have also been raised and are under investigation from R&D and regulatory organizations (EFSA, 2009).

4 Nanoenabled Foods

Foods are a complex matrix composed of diverse components such as large polymers (eg, proteins and carbohydrates) and micronutrients (eg, vitamins and minerals) that differ in chemical composition, molecular structure and physical state. Naturally, water and air contain nanoparticles, and food contains various nanoscale substances such as sugars, amino acids, fats, enzymes and microorganisms from food ingredients. Such nanoscale constituents generally confer little or no harmful effect on humans after consumption. However, addition of additives to modify food stability, texture and processability could alter the safety status of foods. Thus, only limited number (<1000) of engineered nanoenabled foods are currently in the market, with the majority occurring as premarket model foods.

Food manufacturing and in-mouth/gastrointestinal processing could modify the chemical and physical properties of food, resulting in complex interactions between food components. Food structure can be tailored to optimize its breakdown during digestion and ultimately optimize nutritional profiles as a result of tailored nutrient release and bioaccessibility. The control of digestion and nutrient/bioactive release from a food matrix has been proven an alternative approach to combat diet-related problems/diseases. And the integration of food processing with nanotechnology enables such an approach. Over the past decade, R&D attempts to apply nanotechnology to food development have focused mainly on improving sensory qualities, nutrient uptake, and bioactive stability and availability. Questions have arisen concerning processing efficiency, handling stability, and consumption safety. Nanomaterials can be created by breaking up bulk materials using mechanical methods such as milling, grinding, or high shear homogenization (so-called top-down methods). Alternatively, they can be built or by self-assembly of small molecules through chemical and/or physical processes such as nanofabrication or crystal growth ("bottom up" methods). These two approaches have their pros and cons, for example, the former has the limitation for mass production in an industrial setting due to the constraints of speed and frequency, while the latter requires substantial knowledge of the intrinsic properties and compatibility of various materials used for synthesis (this approach is generally more cost effective and simpler). Moreover, both approaches share common technical challenges in producing nanomaterials or nanostructures in a uniform or reproducible form.

Food color, flavor, texture and functionality can be manipulated dramatically at the nanoscale. Customized nanoemulsions or nanostructures enable texture improvements as thickeners and emulsifying/foaming agents, for example, nanoemulsions can be used in mayonnaise, puddings, processed meat products and sauces to achieve desirable textures to consumers. Taste improvements are also possible aided by nanotechnology, for example, nanoscale or near nanoscale salt particles could allow less salt to be used in foods without compromising taste due to greater accessibility to tongue receptors. It is noted that the issue of food bitterness is becoming more challenging because of consumers' demand for healthier foods with reduced amounts of sugar, fat, and sodium (while these substances can help suppress bitterness). A recent example are functional foods that exert health-promoting properties through their high molecular weight phytochemicals (eg, bitter and astringent polyphenols) (Jaeger et al., 2009; Sun-Waterhouse and Wadhwa, 2013). Bitter compounds are commonly hydrophobic in nature. Molecules like lipoprotein, cyclodextrins, and cyclofructans can block the sites targeted by bitter substances on the taste receptor membrane without affecting responses to salts, acids, sugars or sweet amino acids (Mori et al., 2006; Sun-Waterhouse and Wadhwa, 2013). It is possible to minimize bitter taste by preventing contact between the bitter molecules and taste receptors through creating complex nanostructures (eg, the W/O/W or O/W/O type multiple emulsions or water-insoluble gel) that provide steric hindrance or complex the bitter compounds (to retard the release and diffusion of bitter compounds during in-mouth processing or reduce the speed of interactions with the bitter taste receptors) (Sun-Waterhouse and Wadhwa, 2013).

Ingredients in nanoparticulate form to facilitate protection or absorption are demanded for specialty products including nano formulated products with improved safety, efficacy and nutrition (the so-called next generation of "functional foods," "interactive foods," or "on demand foods") (Chaudhry et al., 2010; Sun-Waterhouse and Waterhouse, 2015). The major aims of R&D in this regard are (1) controlling the release of bioactives and nutrients in foods for improved nutritional/health properties, and (2) improving sensory and physical properties while minimizing potentially unhealthy food components (eg, salt, fat, and some food additives). This includes the use of nanoemulsions and polymeric networks (which can regulate food changes and changes of their components from food production to food digestion), or nanocapsules to accommodate substances of interest (eg, water-soluble/insoluble antioxidants and probiotic bacteria) before addition to foods or conventional food processing (Fig. 2.10). The target substances would remain and be released after consumption according to the changes in human digestive conditions (eg, temperature, pH, or ionic composition) (Wang et al., 2013; Waterhouse et al., 2014).

Nanoenabled food production is growing steadily, due to the successes brought about by intense R&D from global large food manufacturers such as Kraft, Nestle, Altria, Heinz, and Unilever. Good examples include Toddler Health's fortified chocolate and vanilla "nutritional drinks" containing SunActive® iron particles of size 300 nm (ie, marketed as "an all-natural balanced nutritional drink for children from 13 months to 5 years"), Tip-Top's Up bread fortified with the nanocapsules containing omega-3 fatty acids rich tuna fish oil in Western Australia (to deliver efficiently health-promoting omega-3 lipids while avoiding unpleasant sensation of fish oil taste in the mouth), Aquanova's nanosized micelles ("NovaSol") containing two active substances CoQ10 (for fat reduction), and alpha lipoic acid (for satiety) to enable weight management, Oilfresh 1000 catalytic antioxidant device marketed

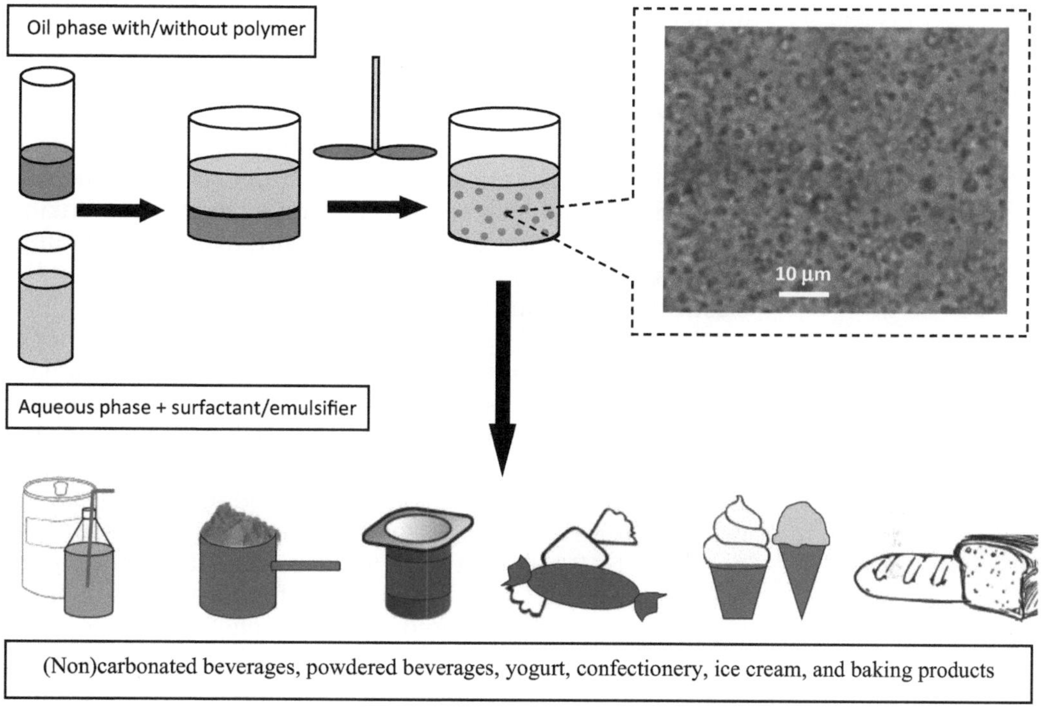

Figure 2.10. Scheme demonstrating the use of an emulsion containing a significant amount of water-insoluble phytochemical extract for producing beverage drinks, spray dried powders, ice creams, yogurts, confectioneries, and baking products.

in the United States, which keeps deep-frying oil fresh significantly longer, RBC Life Sciences' nanoceuticals™ Slim Shake Chocolate containing nanoClusters™ to enhance wetness and taste (for individuals on low sugar diet), and BASF's Solu™ E 200 BG clear sports beverages containing nano vitamin E or LycoVit 10% nanoscale lycopene food additive (size <200 nm, to provide bright red color and potent antioxidant activity). Furthermore, nanotechnology has the potential to change food manufacturing processes and improve food safety. Microfluidization, microparticulation, high hydrostatic pressure processing, high pressure valve homogenization, dry ball milling, dry jet milling, and ultrasound emulsification valve homogenization, are practical technological approaches to create nanostructures (Dissanayake and Vasiljevic, 2009). Food components experience different kinds of intermolecular and intramolecular interactions in a food matrix. Nanoparticles interact with other substances present in foods such as proteins, lipids, carbohydrates, enzymes, nucleic acids, antioxidants. Thus, tailored food formulation and processing, especially those that could influence the architecture of the nanostructure in foods, can modulate potential chemical and physical reactions and eventually impact food quality, digestibility, and bioactive delivery to humans. Sun-Waterhouse et al. (2014b) found that the matrix formed by carboxymethyl cellulose (CMC) and apple fiber could protect sensitive polyphenol antioxidants against degradation, and the stability, mobility and release were controlled by the crosslinking and "loop" conformation of CMC, along with hydrogen bonding, van der Waals forces, hydrophobic effects, steric hindrance and viscoelastic effects among CMC, apple fiber, and polyphenols.

Nanostructured hydrogels are highly hydrated polymeric networks containing interconnected porous nanostructures via physical or covalent linkages between the hydrogel polymer chains and/or between nanoparticles and the polymers (Gaharwar et al., 2014). Hydrogels are very porous (ie, typical pore size is 100–1000 nm, while the dimension of fibril fragments inside the hydrogels is 50–100 nm). Xerogels usually have a high porosity (15–50%, pore size 1–10 nm) and large surface area (150–900 $m^2\ g^{-1}$). It is feasible to create nanostructures or food structures containing nanoparticles with tailored physical, chemical, electrical and biological functionality (eg, gastric retention to improve bioavailability). The selection of suitable type and concentration of polymers and nanoparticles as well as fabrication conditions (eg, pH, temperature, homogenization speed) should be based on human gastric emptying and motility physiology (ie, considering one or more mechanisms such as flotation, mucoadhesion, sedimentation, expansion and size/shape modification) (Arora et al., 2005; Dimic-Misic et al., 2014). Spherical hollow polymeric structures are the most desirable for this purpose, which could be generated using swellable pH-dependent low-density polymers (such as methylcellulose, alginate, and chitosan) along with various effervescent compounds (such as sodium bicarbonate, calcium carbonate, tartaric acid, and citric acid). The New Zealand researchers (Shinde et al., 2014) created such a nanostructure using an alginate solution to "entrap" and "carry" probiotic bacteria *Lactobacillus acidophilus*, which was further protected via coextrusion encapsulation using an alginate solution of twice the concentration as that used for the core nanostructure (Fig. 2.11). Such an encapsulated nanostructure that prevented the loss of probiotic bacteria in a model milk drink during storage at 4°C, could be used for producing other milk drinks enhanced with different probiotic bacteria and polyphenol antoxidants. Waterhouse et al. (2014) and Wang et al. (2013) encapsulated the quercetin antioxidant and edible oils (in the form of nanoemulsion) inside a coextruded polymeric shell of fiber polysaccharides (Fig. 2.12). Such a

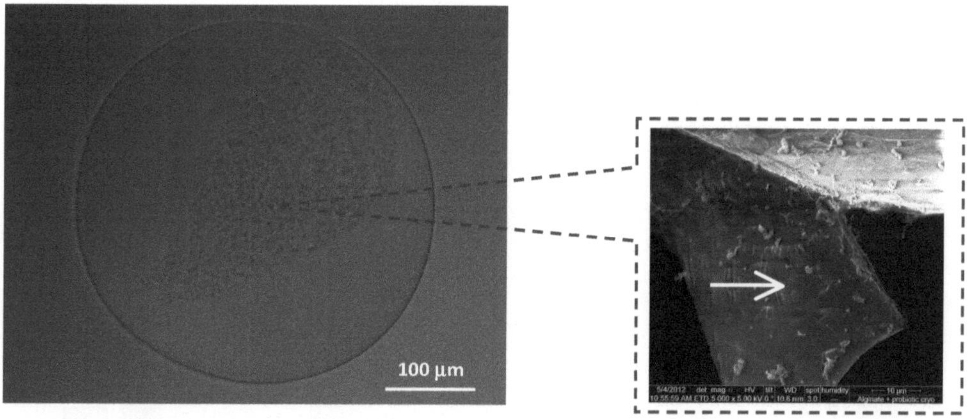

Figure 2.11. Coextrusion encapsulation of nanostructure carrying probiotic bacteria *Lactobacillus acidophilus* (which solid white arrow points to).

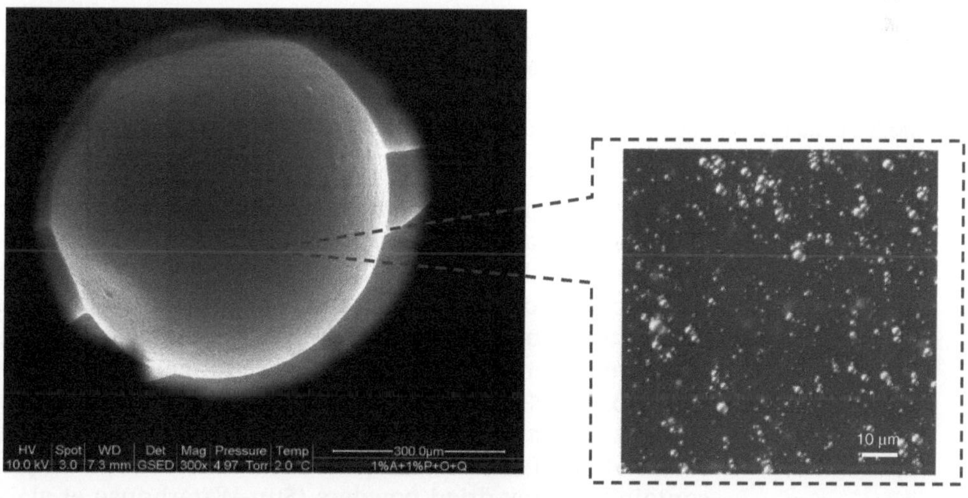

Figure 2.12. Nanoemulsion containing edible oil and quercetin antioxidant can be protected in a biopolymeric network prepared using alginate and high methoxyl pectin.

food system preserved both quercetin and edible oil while combining multinutritional advantages from fiber polysaccharides, flavonoid and unsaturated lipids, thereby presenting a favorable approach for functional food applications.

Spray drying is a cost-effective process and widely used in food manufacturing due to high availability of equipment and low-cost operation. Encapsulation by spray drying can provide accurate control of powder characteristics, for example, dissolution efficiency, encapsulation efficiency, storage stability, and nutritional

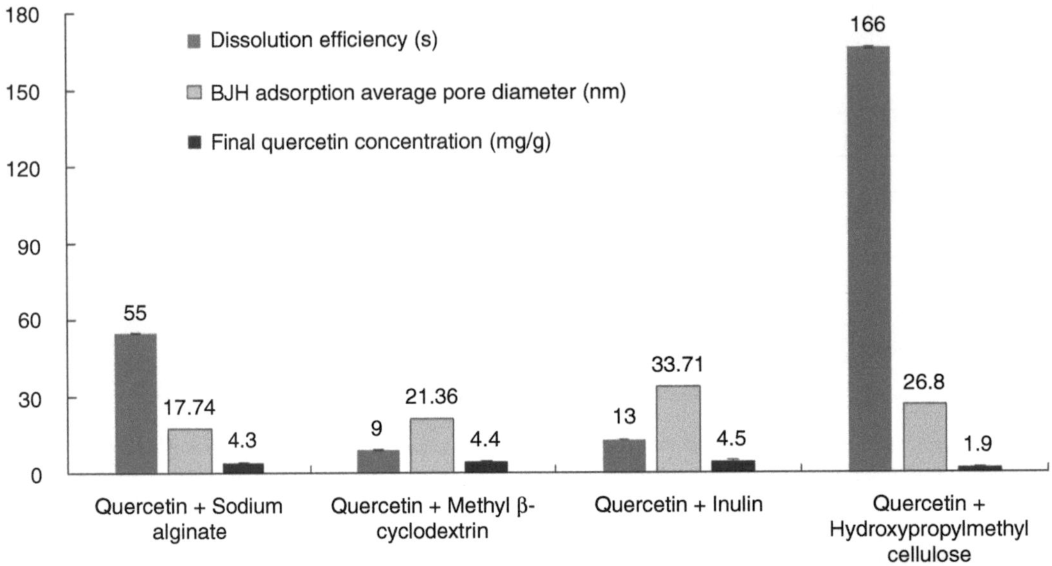

Figure 2.13. The dissolution efficiency, BJH adsorption average pore diameters, and quercetin concentration of the quercetin-containing spray-dried powders produced using sodium alginate, methyl *β*-cyclodextrin, inulin, or hydroxypropylmethyl cellulose. After Sun-Waterhouse et al. (2013).

value, through monitoring the feed (eg, solids content and degree of homogenization; to enable nanosized emulsion/suspension droplets typically in the size range from 200 to 800 nm) and spray drying conditions (eg, drying temperature and type of atomizing device; to enable dried powders that are sized >20 μm and contain core material still within nanosize range) (Jafari et al., 2008). Fig. 2.13 shows the dissolution efficiency, BJH adsorption average pore diameters (in nm, determined by N_2 physisorption measurements at 77 K) and loaded quercetin concentration of quercetin-containing spray-dried powders (Sun-Waterhouse et al., 2013b). Results show that such powder delivery systems are feasible for producing solid products as functional ingredients or supplements with dual health benefits of phenolic antioxidant and dietary fiber to consumers. The particle size of the polymeric encapsulation system influences the release profile from the hydrophilic matrix due to the differences in surface area-to-volume ratio. In practice, a number of potent antioxidants such as quercetin have low solubility in water and low stability thereby limiting their uses in food nutraceuticals and pharmaceutical industries. The combined use of spray-drying encapsulation and polymer encapsulant formula improves the dissolution and stability, while also introducing controlled release and nutritional advantages (Sansone et al. 2011;

Sun-Waterhouse and Waterhouse, 2015). This architecture (ie, "nanosized core with micron-sized outer network") could minimize the potential hazards of nanoparticles, for example, accidental inhalation of powder into the lungs during manufacturing, and mistransit or premature absorption of nanosized substances during transportation along the gastrointestinal tract before reaching the target absorption site. Therefore, in situ encapsulation by spray drying allows a mild "one-step" encapsulation process and has strong industry relevance.

Although nanotechnology may hold promise for food supply, quality control and innovation, it also brings challenges associated with safe use/consumption by humans. The use of nanotechnologies in food and agricultural applications, including incorporating nanomaterials into foods and beverages, food packaging, and edible food coatings, presents a whole new range issues associated with food safety, environment, ethics, and regulation. Thus, action should be taken to reduce such risks to food producer, processors and consumers. Mandatory labeling of nanomaterials was established in the cosmetics industry in early 2010, yet mandatory compliance and an associated regulatory framework in the food sector has still not been implemented. Obstacles preventing the rapid establishment or implementation of internationally agreed mandatory regulations in this regard largely relate to the scarcity of basic toxicological data on nanotechnology/nanomaterials and lack of critical knowledge of toxicokinetics of nanomaterials in the human body (ie, knowledge gaps related to the behavior, fate, and effects of nanosized foods including nanoparticles entering the body via oral and non oral routes). The extremely small size of nanomaterials/nanoparticles could alter digestive events following ingestion (including absorption, distribution, metabolism, and excretion/elimination). The differences of nanomaterials in size, structure, solubility, persistence, and so on, would further complicate the way in which nanomaterials behave and the risks upon consumer exposure to such insoluble and probably biopersistent nanoparticles (EFSA, 2008).

Understanding of the pharmacokinetics of nanoscale materials is the key to their safety assessment. Dissolution, which is a dynamic process changing with the local environment, is a critical factor for the biological fate of nanoscale particles during ingestion, transportation, residence in the stomach and intestine, and passing across the gastric mucosa to the apical surface of the epithelia for uptake (Borm et al., 2006). Particles differing in size and charge would diffuse through the intestinal mucus at different rates, for example, cationic nanoparticles could be entrapped within the negatively charged mucus while carboxylated (anionic) microparticles could diffuse through to the epithelial surface much more easily (Szentkuti, 1997). Concerns on the risk to consumer health or the environment could be less serious for food packaging incorporating nanomaterials in fixed, bound or embedded form, while issues could be severe when consumers are exposed to the engineered nanomaterials in a free form, for example, free nanoparticles in foods/beverages or free nanopesticides released deliberately into the environment. A well-known example is TiO_2 nanoparticles commonly used in tooth pastes and some foods, which can harm the liver, spleen, kidneys, lung, and brain via inhalation, intravenous injection, oral or dermal exposure (Shi et al., 2013). To assure long-term safe consumption, various initiatives have been avocated by regulatory authorities, clinicians, and consumer scientists from different countries in order to examine thoroughly the potential toxicological and microbiological hazards of nanoenabled foods. Examples include:

- The Food and Agricultural Organization of the United Nations and the World Health Organization (FAO/WHO) indicated the need to address the possible risks of substances containing nanosized materials to human health and environment, before being used in foods or food processing. FAO/WHO developed and finalized a technical paper entitled "State of the art on the initiatives and activities relevant to risk assessment and risk management of nanotechnologies in the food and agriculture sectors" (which was released in Jul. 2013).
- In 2009, EFSA advised the European Commission (EC) of the risks possibly brought by the application of nanoscience and nanotechnologies in the food and feed industries, and indicated that "case-by-case differences do exist, uncertainties for risk assessment of nanotechnologies for food and feed applications were due to presently limited information." EC has considered regulations suffice for nanotechnologies evident by a review performed in 2011 to "encourage the development of new types of foods and food production techniques such as nanotechnologies" (although this amendment ultimately collapsed thus nanofoods remain unregulated, for example, for European labeling). Approval for the use of nanotechnology to develop novel foods and processes would be required under the "Novel Foods Regulation" (Regulation (EC) No. 258/97) to ensure products are safe. The Food Standards Agency (FSA) is the body in the United Kingdom responsible for the assessment of novel foods through consulting an independent advisory committee ACNFP (Advisory Committee on Novel Foods and Processes). FSA published a report of consumers' views on the use of nanotechnology in food and food packaging. ACNFP

would carry out assessments following guidance on the risk assessment of engineered nanomaterials in food, food contact materials, animal feed, and pesticides issued by the EFSA in 2011. However, the Agency would only do so if it is asked.

- The US FDA regulates "products" not "technologies" employed. The Nanotechnology Task Force formed by FDA in 2006 indicated that nanoscale materials present regulatory challenges similar to those posed by products using other emerging technologies. In 2011, FDA indicated the need to consider the actual novel chemical or biological properties of nanoscale materials (rather than just one dimension strictly ranged in 1–100 nm) and proposed the inclusion of materials with a dimension up to 1 μm but possessing properties similar to those of 1–100 nm (ie, this size range is in line with the definition of nanomaterials proposed by International Organization for Standardization (ISO). In 2012, the FDA provided guidance on the use of nanotechnology in the food and cosmetics sectors, indicating that nanoenabled food items would not automatically be granted "generally recognized as safe" (GRAS) status. FDA regulates a wide range of products including foods, cosmetics, drugs, devices, veterinary, and tobacco products, some of which may utilize nanotechnology or contain nanomaterials. In 2014, FDA issued one draft and three final guidance documents related to the use of nanotechnology in regulated products including food ingredients intended for use in food for animals (FDA, 2014).
- In Canada, any new additive or food product claimed as a "novel food" needs premarket safety assessments prior to inclusion into the food supply. Health Canada has reviewed regulations associated with nanomaterial-containing or nanoenabled food products such as the Food Additive Regulations, Novel Food Regulations, and Food and Packaging Materials Regulations. An overall framework approach to the products and substances generated by the application of nanotechnology including food, has been proposed by Health Canada.
- In Australia and New Zealand, foods produced using materials, processes or handling tools involving nanotechnology or nanomaterials would be regulated in the same way as "novel foods," "food additives," "processing aids" or "nutritive substances" under the Australia New Zealand Food Standards Code. Any new foods manufactured using nanotechnologies that may present safety concerns will be subject to a comprehensive scientific safety assessment prior to legal supply. The Food Standards Australia New Zealand (FSANZ) understands that materials at very small scales (such as those of nanoscale) could perform differently from their larger counterparts, and

has adopted a range of strategies to review and manage continually potential risks associated with nanotechnologies in foods to ensure the public are not exposed to any health or safety risk. The lack of an agreed definition of nanotechnology or nanoscale materials does not alter the requirements proposed by FSANZ, and thus applicants must provide this information for all particulate material whether it be of nanoscale or microscale dimensions. This avoids the concept of a "one size fits all definition," that is, a size range of 1–100 nm may fail in capturing any human health and safety risks posed by particulate material in the submicron size range (Maynard, 2011).

As for any new technology or food products, consumer perception or acceptance is another important aspect that determines the potential market of nanoenabled foods. Uncertainties and lack of knowledge of potential impacts of new technologies, or lack of a clear communication of risks and benefits, can lead to rejection by the public. Thus, the success of nanotechnology ultimately relies on consumer acceptance and regulatory support. However, in reality, public knowledge about nanotechnology is very limited. Consumers are still hesitant to buy nanoenabled foods or food with nanoenabled packaging as they have concerns about long-term safety and the impact on the environment. Moreover, nanofood processing and nanofoods could challenge the traditional concepts of healthy eating. For example, micronutrients and bioactives such as vitamins and antioxidants that have traditionally been sourced through consumption of fresh fruits and vegetables, could be enriched in nanofortified processed foods like confectionery. Although these nanotechnology/nanomaterial-enabled approaches may enhance the delivery and uptake of desired nutrients and bioactives, consumers do need time to adapt their mind-set to accept the nanoenabled foods. Further, consumers have the right to make informed food choices and must have access to the relevant information. They desperately seek knowledge about the nature or impact of biotransformations on the safety of nanoenabled food products, and are uncomfortable about foods or contacting materials with "unlabeled nanotechnology/nanomaterials." But the currently available data do not provide any conclusive evidence that nanoenabled foods or contacting materials are more or less dangerous than their conventional counterparts. The only way to increase consumer confidence on nanotechnology and derived products is the openness and transparency by the food industry to declare the use of nanotechnology/nanomaterials under the appropriate regulatory frameworks and stringent safety assessment on minimal long term unforeseen adverse effects upon exposure to nanomaterials by legislation authorities.

In the future, "greener" or "more environmentally friendly" nanoparticles should be targeted to increase consumer acceptance. Production of nanoparticles under ecofriendly conditions is of significance to address growing concerns about the overall toxicity of nanoparticles for medical and biotechnological applications. For example, a miracle soybean Glycine max with a single soybean protein of 51 kDa molecular weight holds the promise for large-scale biological synthesis of functionalized and hybrid silver nanoparticles with improved stability using a soybean reservoir rich in isoflavonoids (Sasikala et al., 2012). Such silver nanoparticles coated with soybean antioxidative phytochemicals have great potential in nanoparticles-based pharmaceuticals, cosmetics, food, agriculture, and health applications requiring antimicrobial, antiinflammatory, antiviral, antiangiogenesis, and antiplatelet activity (Panacek et al., 2009; Ramirez et al., 2009).

Robust analysis and characterization techniques can accelerate the success of integrating nanotechnology into the food sector. Accurate characterization of nanomaterials in complex food matrices and packaging is essential for setting up validated analytical methods for regulatory purposes as well as for the creation of products with maximized benefits but minimal hazards. This remains highly challenging for products containing nanomaterials/nanoparticles, as the methodologies currently available are mainly those that have traditionally been used for substances greater than nanoscale and then adopted for the nanomaterials. To address the roles or track the fate of nanoscaled susbtances, in situ high-throughput characterization techniques for all substances simultaneously present are required (although in reality these techniques remain rare). Practically speaking, all these methodologies must be subject to optimization, calibration and validation in different matrices. In addition to biological assays [eg, polymerase chain reaction (PCR) and enzyme immunoassay (EIA)], coventional analytical techniques used for characterizing products involving nanomaterials include various microscopy techniques (eg, scanning electron microscopy, atomic force microscopy, confocal laser scanning microscopy), separation techniques (eg, field flow fractionation, ultra-centrifugation/filtration, size exclusion chromatography, high-performance liquid chromatography), and characterization techniques (eg, UV-visible spectrophotometry, differential scanning calorimetry, powder X-ray diffraction, fourier transform infrared (FTIR), circular dichroism, nuclear magnetic resonance, electron spin resonance, photon correlation spectroscopy, mass spectrometry, laser doppler velocimetry).

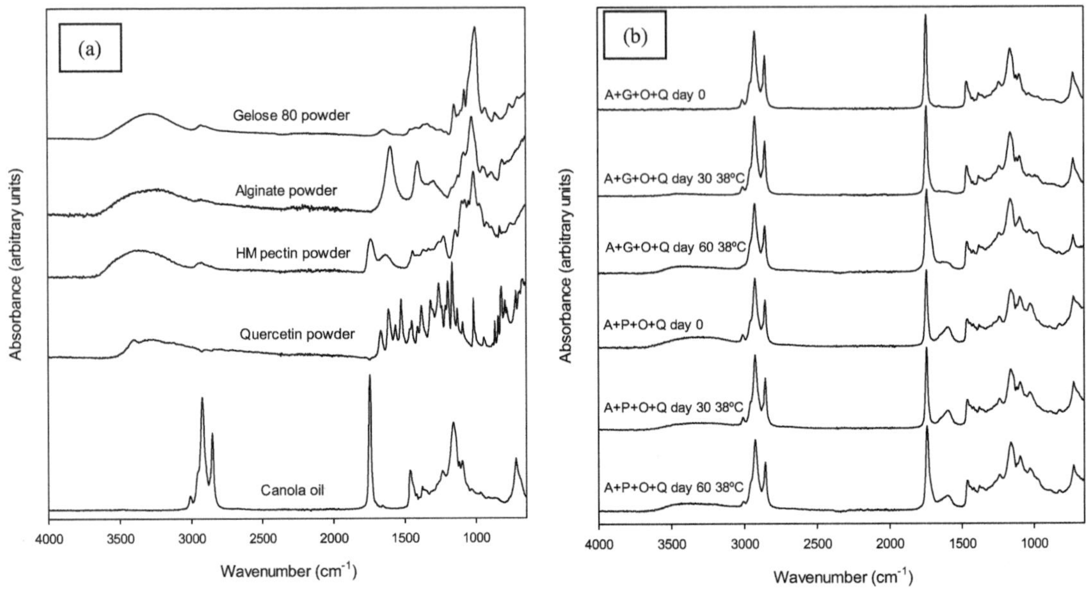

Figure 2.14. Normalized FT-IR absorbance spectra of (a) commercially available Gelose 80 starch, alginate, high methoxyl pectin, quercetin chemical, and canola oil; (b) changes over a 60-day storage at 38°C for quercetin-fortified canola oil beads encapsulated with alginate-pectin (A + P + O + Q) or alginate-Gelose 80 starch (A + G + O + Q).

Fig. 2.14 shows the uses of FTIR to demonstrate the chemical composition and functional groups of several ingredients, and changes caused by certain process or environmental conditions in the finished products made with these ingredients.

5 Conclusions

The food industry is under intense pressure to ensure food safety while satisfying consumer demand for novel health and well-being products. Global R&D and regulatory efforts have shown the great potential of nanotechnology throughout the whole supply chain, from on farm to table. One can design and create structures, devices and systems with unique functional properties through monitoring the size and shape of specific material down to nanoscale, to develop "nanoenabled foods" with target nutritional or health efficacy, and novel nanosensors and nanocarriers for food monitoring and security. A clear regulatory landscape calibrated to standards and open communication with consumers will become increasingly important. It is anticipated that a clear, fit-for-purpose definition of nanotechnology and clear labeling of nanoenabled food products will be implemented for regulatory purposes.

References

Aboubakar, M., Puisieux, F., Couvreur, P., Vauthier, C., 1999. Physico-chemical characterization of insulin-loaded poly(isobutylcyanoacrylate) nanocapsules obtained by interfacial polymerization. Int. J. Pharm. 183, 63–66.

Allen, M.W., Zemlyanov, D.Y., Waterhouse, G.I.N., Metson, J.B., Veal, T.D., Mcconville, C.F., Durbin, S.M., 2011. Polarity effects in the x-ray photoemission of ZnO and other wurtzite semiconductors. Appl. Phys. Lett. 98 (10), 101906.

Anand, P., Nair, H.B., Sung, B., Kunnumakkara, A.B., Yadav, V.R., Tekmal, R.R., Aggarwal, B.B., 2010. Design of curcuminloaded PLGA nanoparticles formulation with enhanced cellular uptake, and increased bioactivity in vitro and superior bioavailability in vivo. Biochem. Pharmacol. 79 (3), 330–338.

Arora, S., Javed, A., Ahuja, A., Khar, R.K., Baboota, S., 2005. Floating drug delivery systems: a review. AAPS Pharm. Sci. Tech. 6 (3), E372–E390.

Azeredo, H.M.C., Miranda, K.W.E., Ribeiro, H.L., Rosa, M.F., Nascimento, D.M., 2012. Nanoreinforced alginate–acerola puree coatings on acerola fruits. J. Food Eng. 113 (4), 505–510.

Balcão, V.M., Glasser, C.A., Chaud, M.V., del Fiol, F.S., Tubino, M., Vila, M.M., 2014. Biomimetic aqueous-core lipid nanoballoons integrating a multiple emulsion formulation: a suitable housing system for viable lytic bacteriophages. Colloids Surf. B. 123, 478–485.

Bejrapha, P., Min, S.G., Surassmo, S., Choi, M.J., 2010. Physicothermal properties of freeze-dried fish oil nanocapsules frozen under different conditions. Drying Tech. 28 (4), 481–489.

Borm, P., Klaessig, F.C., Landry, T.D., Moudgil, B., Pauluhn, J., Thomas, K., Trottier, R., Wood, S., 2006. Research strategies for safety evaluation of nanomaterials, part V: role of dissolution in biological fate and effects of nanoscale particles. Toxicol. Sci. 90 (1), 23–32.

Brunner, T., Piusmanser, P., Spohn, P., Grass, R., Limbach, L., Bruinink, A., Stark, W., 2006. In vitro cytotoxicity of oxide nanoparticles: comparison to asbestos, silica, and the effect of particle solubility. Environ. Sci. Tech. 40, 4374–4381.

Cabral, H., Matsumoto, Y., Mizuno, K., Chen, Q., Murakami, M., Kimura, M., Terada, Y., Kano, M.R., Miyazono, K., Uesaka, M., Nishiyama, N., Kataoka, K., 2011. Accumulation of sub-100 nm polymeric micelles in poorly permeable tumours depends on size. Nat. Nanotechnol. 6, 815–823.

Campuzano, S., Gamella, M., Serra, B., Reviejo, A.J., Pingarron, J.M., 2007. Integrated electrochemical gluconic acid biosensor based on self-assembled monolayermodified gold electrodes. Application to the analysis of gluconic acid in musts and wines. J. Agric. Food Chem. 55, 2109–2114.

Chaudhry, Q., Castle, L., Watkins, R. (Eds.), 2010. Nanotechnologies in Food. Royal Society of Chemistry Publishers, Cambridge, UK.

Chaudhary, S.A., Chaudhary, A.B., Mehta, T.A., 2012. Formulation development and optimization of Polyox based quick dissolving film of quetiapine. J. Pharm. Bioall. Sci. 4 (1), S19–S20.

Chen, W.-T., Chan, A., Jovic, V., Sun-Waterhouse, D., Murai, K., Idriss, H., Waterhouse, G.I.N., 2015a. Effect of the TiO_2 crystallite size TiO_2 polymorph and test conditions on the photo-oxidation rate of aqueous methylene blue. Top. Catal. 58 (2–3), 85–102.

Chen, W.-T., Chan, A., Sun-Waterhouse, D., Moriga, T., Idriss, H., Waterhouse, G.I.N., 2015b. Ni/TiO_2: a promising low cost photocatalytic system for solar H_2 production from ethanol-water mixtures. J. Catal. 326, 43–53.

Clark, L.C., Lyons, C., 1962. Electrode systems for continuous monitoring in cardiovascular surgery. Ann. N.Y. Acad. Sci. 102, 29–45.

Dandekar, P.P., Jain, R., Patil, S., Dhumal, R., Tiwari, D., Sharma, S., Vanage, G., Patravale, V., 2010. Curcumin-loaded hydrogel nanoparticles: application in anti-malarial therapy and toxicological evaluation. J. Pharm. Sci. 99 (12), 4992–5010.

Davidov-Pardo, G., Pérez-Ciordia, S., Marín-Arroyo, M.R., McClements, D.J., 2015. Improving resveratrol bioaccessibility using biopolymer nanoparticles and complexes: impact of protein–carbohydrate maillard conjugation. J. Agric. Food Chem. 63 (15), 3915–3923.

Davis, M.E., 2008. Nanoparticle therapeutics: an emerging treatment modality for cancer. Nat. Rev. Drug Discovery 7 (9), 771–782.

Deshkar, S., Pawar, M., Shirsat, A., Shirolkar, S., 2013. Development of sustained release tablet of mebeverine hydrochloride. J. Pharm. Educ. Res. 4 (1), 64–69.

Deshpande, R.J., Waterhouse, G.I.N., Jameson, G.B., Telfer, S.G., 2012. Photolabile protecting groups in metal–organic frameworks: preventing interpenetration and masking functional groups. Chem. Commun. 48 (10), 1574–1576.

Di Colo, G., Burgalassi, S., Chetoni, P., 2001. Gel-forming erodible inserts for ocular controlled delivery of ofloxacin. Int. J. Pharm. 215 (1–2), 101–111.

Dimic-Misic, K., Ridgway, C., Maloney, T., Paltakari, J., Gane, P., 2014. Influence on pore structure of micro/nanofibrillar cellulose in pigmented coating formulations. Transport Porous Med. 103, 155–179.

Dimcheva, N., Horozova, E., Jordanova, Z., 2002. An amperometric xanthine oxidase enzyme electrode based on hydrogen peroxide electro reduction. Z. Naturforsch C. 57 (9–10), 883–889.

Dissanayake, M., Vasiljevic, T., 2009. Functional properties of whey proteins affected by heat treatment and hydrodynamic high-pressure shearing. J. Dairy Sci. 92, 1387–1397.

EFSA. 2008. 21st list of substances for food contact materials, scientific opinion of the panel on food contact materials, enzymes, flavourings and processing aids (CEF) (question No EFSA-Q-2005-151, EFSA-Q-2006-324, EFSA-Q-2006-323), adopted on 27 November 2008. The EFSA J., 888–890, 1–14.

EFSA (European Food Safety Authority), 2009. Scientific opinion on "The potential pisks arising from nanoscience and nanotechnologies on food and feed safety," Scientific Opinion of the Scientific Committee, adopted on 10 February 2009, EFSA J., 958, 1–39.

FDA. 2006. Nanoscale materials nomination and review of toxicological literature. Dec 8, 2006. Prepared by the chemical selection working group, U.S. Food and Drug Administration. <https://ntp.niehs.nih.gov/ntp/htdocs/chem_background/exsumpdf/nanoscale_materials_508.pdf>

FDA, 2007. Nanotechnology: a report of the U.S. Food and Drug Administration, Nanotechnology Task Force. July 25, 2007. <http://www.fda.gov/downloads/ScienceResearch/SpecialTopics/Nanotechnology/ucm110856.pdf>

FDA, 2014. Guidance for industry use of nanomaterials in food for animals, June 2014. The U.S. Department of Health and Human Services Food and Drug Administration Center for Veterinary Medicine.

Ferreira, I., Rocha, S., Coelho, M., 2007. Encapsulation of antioxidants by spray-drying. Chem. Eng. Trans. 11 (9), 713–717.

Fischer, N.O., McIntosh, C.M., Simard, J.M., Rotello, V.M., 2002. Inhibition of chymotrypsin through surface binding using nanoparticle-based receptors. Proc. Natl. Acad. Sci. USA 99, 5018–5023.

Floros, J.D., 2010. New and emerging applications of nanotechnology in our food supply. IFT Int. Food Nanosci. Conf. 17, 25–28.

Frederiksen, H.K., Kristensen, H.G., Pedersen, M., 2003. Solid lipid microparticle formulations of the pyrethroid gamma-cyhalothrin-incompatibility of the lipid and the pyrethroid and biological properties of the formulations. J. Control Release 86 (2–3), 243–252.

Gabizon, A., Goren, D., Cohen, R., Barenholz, Y., 1998. Development of liposomal anthracyclines: from basics to clinical applications. J. Control Release 53, 275–279.
Gaharwar, A.K., Peppas, N.A., Khademhosseini, A., 2014. Nanocomposite hydrogels for biomedical applications. Biotech. Bioeng. 111 (3), 441–453.
Gao, Y., Kyratzis, I., 2008. Covalent immobilization of proteins on carbon nanotubes using the cross-linker 1-ethyl-3-(3-dimethylaminopropyl) carbodiimide—a critical assessment. Bioconjugate Chem. 19 (10), 1945–1950.
Garnett, M., Kallinteri, P., 2006. Nanomedicines and nanotoxicology: some physiological principles. Occup. Med. 56, 307–311.
Gatti, A., Rivasi, F., 2002. Biocompatibility of micro- and nanoparticles. Part I: in liver and kidney. Biomaterials 23, 2381–2387.
Geiser, M., Rothen-Rutlshauser, B., Knapp, N., Schurch, S., Kreyling, W., Schulz, H., Semmler, M., Im, H., Heyder, J., Gehr, P., 2005. Ultrafine particles cross cellular membranes by non-phagocytic mechanisms in lungs and in cultured cells. Environ. Health Perspect. 113 (11), 1555–1560.
Graveland-Bikker, J.F., Koning, R.I., Koerten, H.K., Geels, R.B.J., Heeren, R.M.A., de Kruif, C.G., 2009. Structural characterization of alpha-lactalbumin nanotubes. Soft Matter 5 (10), 2020–2026.
Gupta, A.S., Deshpande, R.K., Liu, L., Waterhouse, G.I.N., Telfer, S.G., 2012. Porosity in metal–organic frameworks following thermolytic postsynthetic deprotection: gas sorption, dye uptake and covalent derivatisation. Cryst. Eng. Comm. 14 (18), 5701–5704.
Haes, A.J., Van Duyne, R.P., 2005. A unified view of propagating and localized surface plasmon resonance biosensors. Anal. Bioanal. Chem. 379, 920–930.
He, J., Kunitake, T., Nakao, A., 2004. Facile fabrication of composites of platinum nanoparticles and amorphous carbon films by catalyzed carbonization of cellulose fibers. Chem. Comm. 4, 410–411.
Hsieh, P.-H., Chan, A., Sun-Waterhouse, D., Williams, D.E., Waterhouse, G.I.N., 2014. Novel SERS substrates for the identification of adulterants in milk. The Australian & NZ Institutes of Physics 38th Annual Condensed Matter and Materials Meeting at Waiheke Island Resort, 4–7 Feb, 2014, Auckland, New Zealand.
Hsieh, P.-H., Chan, A., Sun-Waterhouse, D., Williams, D.E., Waterhouse, G.I.N., 2014b. Novel SERS Substrates for the detection of adulterants in aqueous systems, in: the 2014 NZIFST conference, 1–3 July, Christchurch, New Zealand.
Huang, Z., Li, X., Zhang, T., Song, Y., She, Z., Li, J., Deng, Y., 2014. Progress involving new techniques for liposome preparation. Asian J. Pharm. Sci. 9 (4), 176–182.
Jaeger, S.R., Axten, L.G., Wohlers, M.W., Sun-Waterhouse, D., 2009. Polyphenol-rich beverages: insights from sensory and consumer science. J. Sci. Food Agric. 89 (14), 2356–2363.
Jafari, S.M., Assadpoor, E., Bhandari, B., He, Y., 2008. Nanoparticle encapsulation of fish oil by spray drying. Food Res. Int. 41 (2), 172–183.
Jagdale, S.C., Bari, N.A., Kuchekar, B.S., Chabukswar, A.R., 2013. Optimization studies on compression coated floating-pulsatile drug delivery of bisoprolol. BioMed Res. Int. 2013, 11, Article ID 801769.
Kam, N.W.S., Dai, H., 2005. Carbon nanotubes as intracellular protein transporters: generality and biological functionality. J. Am. Chem. Soc. 127, 6021–6026.
Karajanagi, S.S., Yang, H., Asuri, P., Sellitto, E., Dordick, J.S., Kane, R.S., 2006. Protein-assisted solubilization of singlewalled carbon nanotubes. Langmuir 22, 1392–1395.
Kim, M.I., Kim, J., Lee, J., Jia, H., Na, H.B., Youn, J.K., Kwak, J.H., Dohnalkova, A., Grate, J.W., Wang, P., Hyeon, T., Park, H.G., Chang, H.N., 2007. Crosslinked enzyme aggregates in hierarchically-ordered mesoporous

silica: a simple and effective method for enzyme stabilization. Biotechnol. Bioeng. 96, 210–218.

Klang, S., Benita, S., 1998. Design and evaluation of submicron emulsions as colloidal drug carriers for intravenous administration. Benita, S. (Ed.), Submicron Emulsions in Drug Targeting and Delivery, Drug Targeting and Delivery, vol. 9, Harwood Academic Publishers, Amsterdam, pp. 119–152.

Klinkesorn, U., McClements, D.J., 2010. Impact of lipase, bile salts, and polysaccharides on properties and digestibility of tuna oil multilayer emulsions stabilized by lecithin-chitosan. Food Biophys. 5 (2), 73–81.

Kneipp, K., Kneipp, H., Itzkan, I., Dasari, R.R., Feld, M.S., 1999. Ultrasensitive chemical analysis by Raman spectroscopy. Chem. Rev. 99, 2957–2976.

Kojima, H., Yoshihara, K., Sawada, T., 2008. Extended release of a large amount of highly water-soluble diltiazem hydrochloride by utilizing counter polymer in polyethylene oxides (PEO)/polyethylene glycol (PEG) matrix tablets. Eur. J. Pharm. Biopharm. 70, 556–562.

Kreuter, J., 1991. Nanoparticle-based dmg delivery systems. J. Controlled Release 16, 169–176.

Kreuter, J., 1994. Nanoparticles. In: Kreuter, J. (Ed.), Colloidal Drug Delivery Systems, vol. 66, Drugs and the Pharmaceutical Sciences. Marcel Dekker, New York, Basel, and Hong Kong, pp. 219–342.

Kulys, J., Vidziunaite, R., 2003. Amperometric biosensors based on recombinant laccases for phenols determination. Biosen. Bioelectron. 18, 319–325.

Laschi, S., Fránek, M., Mascini, M., 2000. Screenprinted electrochemical immunosensors for PCB detection. Electroanalysis 12, 1293–1298.

Le Ru, E.C., Etchegoin, P.G., Grand, J., Félidj, N., Aubard, J., Lévi, G., 2007. The mechanisms of spectral profile modification in surface enhanced fluorescence. J. Phys. Chem. C 44, 16076–16079.

Le Ru, E.C., Etchegoin, P.G., 2009. Principles of Surface-Enhanced Raman Spectroscopy and Related Plasmonic Effects. Elsevier, Amsterdam.

Leveneur, J., Waterhouse, G.I.N., Kennedy, J., Metson, J.B., Mitchell, D.R.G., 2011. Nucleation and growth of Fe nanoparticles in SiO_2: A TEM, XPS, and Fe L-edge XANES investigation. J. Phys. Chem. C 115 (43), 20978–20985.

Liedtke, S., Wissing, S., Müller, R.H., Mäder, K., 2000. Influence of high pressure homogenisation equipment on nanodispersions characteristics. Int. J. Pharm. 196, 183–185.

Lira, M.C.B., Ferraz, M.S., da Silva, D.G.V.C., Cortes, M.E., Teixeira, K.I., Caetano, N.P., Sinisterra, R.D., Santos-Magalhaes, N.S., 2009. Inclusion complex of usnic acid with β-cyclodextrin: characterization and nanoencapsulation into liposomes. J. Inclusion Phenom. Macro. Chem. 64 (3–4), 215–224.

Lisa, M., Chouhan, R.S., Vinayaka, A.C., Manonmani, H.K., Thakur, M.S., 2009. Gold nanoparticles based dipstick immunoassay for the rapid detection of dichlorodiphenyltrichloroethane: an organochlorine pesticide. Biosens. Bioelectron. 25, 224–227.

Liu, S., Chen, Y., Liu, Z., Hu, X., Wang, F., 2006. A highly sensitive resonance rayleigh scattering method for the determination of vitamin B1 with gold nanoparticles probe. Microchim. Acta 154, 87–93.

Lun, D.J., Waterhouse, G.I.N., Telfer, S.G., 2011. A general thermolabile protecting group strategy for organocatalytic metal-organic frameworks. J. Am. Chem. Soc. 133 (15), 5806–5809.

Maynard, A.D., 2011. Don't define nanomaterials. Nature 475, 31.

Merkoçi, A., Ambrosi, A., De la Escosura, A., Pérez, B., Guix, M., Maltez, M., Marín, S., 2010. Nanomaterials for electroanalysis. In: Meyers, R.A. (Ed.), Encyclopedia of Analytical Chemistry. Wiley, New York, pp. 1–26.

Momin, J.K., Jayakumar, C., Prajapati, J.B., 2013. Nutrition in food and science, potential of nanotechnology in functional foods. Emir. J. Food Agric. 25 (1), 10–19.

Mori, H., Nishioka, M., Nanjo, F., 2006. Cyclodextrin and cyclofructan for inhibition of bitterness and astringency of foods and beverages. Japan Patent 2006 067, 895.

Murdoch, M., Waterhouse, G.I.N., Nadeem, M.A., Metson, J.B., Keane, M.A., Howe, R.F., Llorca, J., Idriss, H., 2011. The effect of gold loading and particle size on photocatalytic hydrogen production from ethanol over Au/TiO_2 nanoparticles. Nat. Chem. 3 (6), 489–492.

Niculescu, M., Mieliauskiene, R., Laurinavicius, V., Csöregi, E., 2003. Simultaneous detection of ethanol, glucose and glycerol in wines using pyrroloquinoline quinone dependent dehydrogenases based biosensors. Food Chem. 82, 481–489.

O'Halloran, M.P., Pravda, M., Guilbault, G.G., 2001. Prussian Blue bulk modified screen-printed electrodes for H_2O_2 detection and for biosensors. Talanta 55, 605–611.

Ozdemir, C., Yeni, F., Odaci, D., Timur, S., 2010. Electrochemical glucose biosensing by pyranose oxidase immobilized in gold nanoparticle-polyaniline/AgCl/gelatin nanocomposite matrix. Food Chem. 119, 380–385.

Panfili, G., Manzi, P., Compagnone, D., Scarciglia, L., Palleschi, G., 2000. Rapid assay of choline in foods using a microwave hydrolysis and a choline biosensor. J. Agric. Food Chem. 48, 3403–3407.

Panacek, A., Kolar, M., Vecerova, R., Prucek, R., Soukupova, J., Krystof, V., Hamal, P., Zboril, R., Kvitek, L., 2009. Antifungal activity of silver nanoparticles against Candida spp. Biomaterials 30, 6333–6340.

Pavlidis, I.V., Vorhaben, T., Tsoufis, T., Rudolf, P., Bornscheuer, U.T., Gournis, D., Stamatis, H., 2012. Development of effective nanobiocatalytic systems through the immobilization of hydrolases on functionalized carbon-based nanomaterials. Bioresour. Technol. 115, 164–171.

Ramirez, I.M., Bashir, S., Luo, Z., Liu, J.L., 2009. Green synthesis and characterization of polymer-stabilized silver nanoparticles. Colloid. Surface B 73, 185–191.

Rashidinejad, A., Birch, E.J., Sun-Waterhouse, D., Everett, D.W., 2013. Effects of catechin on the phenolic content and antioxidant properties of low-fat cheese. Int. J. Food Sci. Technol. 48 (12), 2448–2455.

Rashidinejad, A., Birch, J.E., Sun-Waterhouse, D., Everett, D.W., 2014. Delivery of green tea catechin and epigallocatechin gallate in liposomes incorporated into low-fat hard cheese. Food Chem. 156, 176–183.

Rashidinejad, A., Birch, J.E., Sun-Waterhouse, D., Everett, D.W., 2015. Total content and antioxidant properties of hard low-fat cheese fortified with catechin as affected by in vitro gastrointestinal digestion. LWT—Food Sci. Technol. 62 (1), 393–399.

Ray, M., Gupta, R., Rousseau, D., 2015. Properties and applications of multilayer and nanoscale emulsions. In: Sabliov, C.M., Chen, H., Yada, R.Y. (Eds.), Nanotechnology and Functional Foods: Effective Delivery of Bioactive Ingredients. Wiley, Chichester, UK.

Reis, C.P., Neufeld, R.J., Ribeiro, A.J., Veiga, F., 2006. Nanoencapsulation I. Methods for preparation of drug-loaded polymeric nanoparticles. Nanomedicine: Nanotechnol. Biol. Med. 2, 8–21.

Rossman, J.M., 2009. Edible films and coatings for food applications. In: Embuscado, M.E., Huber, K.C. (Eds.), Edible Films and Coatings for Food Applications. Springer Science Business Media, US.

Saidman, S.B., Lobo-Castañón, M.J., Miranda-Ordieres, A.J., Tuñón-Blanco, P., 2000. Amperometric detection of D-sorbitol with NAD + -D-sorbitol dehydrogenasse modified carbon paste electrode. Anal. Chim. Acta 424, 45–50.

Sansone, F., Picerno, P., Mencherini, T., Villecco, F., D'Ursi, A.M., Aquino, R.P., Lauro, M.R., 2011. Flavonoid microparticles by spray-drying: influence of enhancers of the dissolution rate on properties and stability. J. Food Eng. 103, 188–196.

Sasikala, D., Govindaraju, K., Tamilselvan, S., Singaravelu, G., 2012. Soybean protein: a natural source for the production of green silver nanoparticles. Biotechnol. Bioprocess Eng. 17, 1176–1181.

Schmidt, D.F., 2013. Nanolaminates—bioinspired and beyond. Mater. Lett. 108, 328–335.

Scrinis, G., Lyons, K., 2007. The emerging nano-corperate paradigm: nanotechnology and the transformation of nature, food and agri-food systems. Int. J. Sociol. Food Agric. 15 (2.), 22–44.

Serna-Cock, L., Zetty-Arenas, A.M., Ayala-Aponte, A., 2009. Use of enzymatic biosensors as quality indices: a synopsis of present and future trends in the food industry. Chilean J. Agric. Res. 69 (2), 270–280.

Shah, K.R., Chaudhary, S.A., Mehta, T.A., 2014. Polyox (polyethylene oxide) multifunctional polymer in novel drug delivery system. Int. J. Pharm. Sci. Drug Res. 6 (2), 95–101.

Shang, W., Nuffer, J.H., Dordick, J.S., Siegel, R.W., 2007. Unfolding of ribonuclease A on silica nanoparticle surfaces. Nano Lett. 7, 1991–1995.

Sharma, S., Lewis, S., 2010. Taste masking technologies: a review. Int. J. Pharm. 2 (2), 6–13.

Sharma, M., Waterhouse, G.I.N., Loader, S.W.C., Garg, S., Svirskis, D., 2013. High surface area polypyrrole scaffolds for tunable drug delivery. Int. J. Pharm. 443, 163–168.

Shefer, A., Shefer, S., 2003. Multi component biodegradable bioadhesive controlled release system for food products. U.S. patent 6, 589, 562BI.

Shi, H., Magaye, R., Castranova, V., Zhao, J., 2013. Titanium dioxide nanoparticles: a review of current toxicological data. Part. Fibre Toxicol. 10, 15.

Shinde, T., Sun-Waterhouse, D., Brooks, J., 2014. Co-extrusion encapsulation of probiotic Lactobacillus acidophilus alone or together with apple skin polyphenols: an aqueous and value-added delivery system using alginate. Food Bioprocess Technol. 7 (6), 1581–1596.

Smolander, M., Hurme, E., Koivisto, M., Kivinen, S., 2004. PCT international patent application WO 2004/102185 A1, 2004.

Somerset, V.S., Klink, M.J., Sekota, M.M.C., Baker, P.G.L., Iwuoha, E.I., 2006. Polyaniline-Mercaptobenzothiazole biosensor for organophosphate and carbamate pesticides. Anal. Lett. 39, 1683–1698.

Sun-Waterhouse, D., Zhou, J., Miskelly, G.M., Wibisono, R., Wadhwa, S.S., 2011a. Stability of encapsulated olive oil in the presence of caffeic acid. Food Chem. 126 (3), 1049–1056.

Sun-Waterhouse, D., Penin-Peyta, L., Wadhwa, S.S., Waterhouse, G.I.N., 2011b. Storage stability of phenolic-fortified avocado oil encapsulated using different polymer formulations and co-extrusion technology. Food Bioprocess Technol. 5 (8), 3090–3102.

Sun-Waterhouse, D., Wadhwa, S.S., 2013. Industry relevant approaches for minimising the bitterness of bioactive compounds in functional foods: a review. Food Bioprocess Technol. 6 (3), 607–627.

Sun-Waterhouse, D., Wadhwa, S.S., Waterhouse, G.I.N., 2013. Spray-drying microencapsulation of polyphenol bioactives: a comparative study using different natural fibre polymers as encapsulants. Food Bioprocess Technol. 6, 2376–2388.

Sun-Waterhouse, D., Wang, W., Waterhouse, G.I.N., 2014a. Canola oil encapsulated by alginate and its combinations with starches of low and high amylose content: effect of quercetin on oil stability. Food Bioprocess Technol. 7 (8), 2159–2177.

Sun-Waterhouse, D., Bekkour, K., Wadhwa, S.S., Waterhouse, G.I.N., 2014b. Rheological and chemical characterization of smoothie beverages containing high concentrations of fibre and polyphenols from apple. Food Bioprocess Technol. 7 (2), 409–423.

Sun-Waterhouse, D., Waterhouse, G.I.N., 2015. Spray-drying of green or gold kiwifruit juice-milk mixtures; novel formulations and processes to retain natural fruit colour and antioxidants. Food Bioprocess Technol. 8 (1), 191–207.

Szentkuti, L., 1997. Light microscopical observations on luminally administered dyes, dextrans, nanospheres and microspheres in the pre-epithelial mucus gel layer of the rat distal colon. J. Controlled Release 46, 233–242.

Tsai, H.C., Doong, R.A., Chiang, H.C., Chen, K.T., 2003. Sol-gel derived urease-based optical biosensor for the rapid determination of heavy metals. Anal. Chim. Acta 481, 75–84.

Wang, P., 2006. Nanoscale biocatalyst systems. Curr. Opin. Biotechnol. 17, 574–579.

Wang, J.C., Chen, S.H., Xu, Z.C., 2008. Synthesis and properties research on the nanocapsulated capsaicin by simple coacervation method. J. Disper. Sci. Technol. 29 (5), 687–695.

Wang, W., Waterhouse, G.I.N., Sun-Waterhouse, D., 2013. Co-extrusion encapsulation of canola oil with alginate: effect of quercetin addition to oil core and pectin addition to alginate shell on oil stability. Food Res. Int. 54 (1), 837–851.

Waterhouse, G.I.N., Bowmaker, G.A., Metson, J.B., 2003. Oxygen chemisorption on an electrolytic silver catalyst: a combined TPD and Raman spectroscopic study. Appl. Surf. Sci. 214 (1–4), 36–51.

Waterhouse, G.I.N., Bowmaker, G.A., Metson, J.B., 2004. Influence of catalyst morphology on the performance of electrolytic silver catalysts for the partial oxidation of methanol to formaldehyde. Appl. Catal. A – Gen. 266 (2), 257–273.

Waterhouse, G.I.N., Waterland, M.R., 2007. Opal and inverse opal photonic crystals: fabrication and characterization. Polyhedron 26 (2), 356–368.

Waterhouse, G.I.N., Wang, W., Sun-Waterhouse, D., 2014. Stability of canola oil encapsulated by co-extrusion technology: effect of quercetin addition to alginate shell or oil core. Food Chem. 142, 27–38.

Waterhouse, G.I.N., Chen, W.-T., Chan, A., Jin, H., Sun-Waterhouse, D., Cowie, B., 2015. Structural, optical and catalytic support properties of γ-Al_2O_3 inverse opals. J. Phys. Chem. C 119 (12), 6647–6659.

Waterman, K.C., Goeken, G., Konagurthu, S., Likar, M.D., MacDonald, B.C., Mahajan, N., Swaminathan, V., 2011. Osmotic capsules: a universal oral, controlled-release drug delivery dosage form. J. Controlled Release 152, 264–269.

Wilson, G.S., Gifford, R., 2005. Review biosensors for real-time in vivo measurements. Biosens. Bioelectron. 20, 2388–2403.

Velasco-García, M.N., Mottram, T., 2003. Biosensor technology addressing agricultural problems. Review paper. Biosyst. Eng. 84, 1–12.

Yabutani, T., Waterhouse, G.I.N., Sun-Waterhouse, D., Metson, J.B., Iinuma, A., Thuy, L.T.X., Yamada, Y., Takayanagi, T., Motonaka, J., 2014. Facile synthesis of platinum nanoparticle-containing porous carbons, and their application to amperometric glucose biosensing. Microchim. Acta 181 (15–16), 1871–1878.

Zhang, L., Jiang, Y., Ding, Y., Povey, M., York, D., 2007. Investigation into the antibacterial behaviour of suspensions of ZnO nanoparticles (ZnO nanofluids). J. Nanopart. Res. 9 (3), 479–489.
Zuidam, N.J., Shimoni, E., 2010. Overview of microencapsulation use in food products or processes and methods to make them. In: Zuidam, N.J., Nedovic, V.A. (Eds.), Encapsulation Technique for Active Food Ingredients and Food Processing. Springer, New York, pp. 3–29.

3

MAGNETIC SEPARATION OF NANOBIOSTRUCTURED SYSTEMS FOR INNOVATION OF BIOCATALYTIC PROCESSES IN FOOD INDUSTRY

Anna Ilyina*, Rodolfo Ramos-González*, Alejandra Vargas-Segura*, Jaquelina Sánchez-Ramírez*, Sócrates A. Palacios-Ponce*, José L. Martínez-Hernández*, Elda P. Segura-Ceniceros*, Juan C. Contreras-Esquivel, Cristóbal N. Aguilar-González****

**Autonomous University of Coahuila, Nanobioscience Group, Chemistry Faculty, Col. Republic East, Saltillo, Coahuila, Mexico; **Autonomous University of Coahuila, Food Research Department, Chemistry Faculty, Col. Republic East, Saltillo, Coahuila, Mexico*

1 Introduction

Enzymes are fundamental biocatalysts in the proper functioning of any biological system. Due to their protein nature and complexity of their three-dimensional structures, which are able to selectively interact with substrates, these biomolecules act as catalysts of chemical reactions. Enzymes function is to decrease the activation energy of the reactions to increase thereby the reaction rate. These catalysts do not alter the energy balance of the reactions, nor alter the equilibrium of the reaction. Furthermore, enzymes are allowed to develop low temperature processes, do not generate toxic and polluting wastes, by-products and emissions in comparison with conventional chemical processes. So, they are widely used in food industry (Tomotani and Vitolo, 2007).

Despite these, enzymes' practical applications face many challenges, such as sensibility to pH and temperature changes, or other factors such as low stability and difficulties with their recovering and

Novel Approaches of Nanotechnology in Food. http://dx.doi.org/10.1016/B978-0-12-804308-0.00003-0

reuse in a reaction system that causes high operating costs. These characteristics sometimes prevent the efficient operation as well as technical and economic advantages of enzymes use, especially in a wide variety of biotransformation reactions (Cao et al., 2012; Demir et al., 2011). There are diverse techniques to improve the enzyme stability, such as immobilization, enzymatic modification, protein engineering, among which the immobilization is the most common method used to improve enzyme properties (Cao et al., 2012).

In recent years, nanoparticles have been used as supports for enzymes immobilization because they provide a large surface area, improving the catalytic efficiency of the immobilized enzymes. However, the recovery of nanoparticles with immobilized enzymes is usually limited and difficult (Ansari and Husain, 2012; Ashtari et al., 2012; Cao et al., 2012; Hong et al., 2009; Jiang et al., 2009). To simplify the recovery process, magnetic nanoparticles have been used because of their chemical and physical properties (Netto et al., 2013). Magnetic nanoparticles have been applied to selectively attach, manipulate, and transport targeted species to a desired location under the application of an external magnetic field. Because of their size, magnetic nanoparticles feature a superparamagnetic behavior, offering an immense potential in a diversity of applications in their bare form or through attachment on their surface of a molecule or functional group for a specific application. The magnetic nanoparticles possess attractive properties such as catalysts of chemical reactions, magnetic resonance imaging, biomedicine, sensors, magnetic recording, nanofluids, environment remediation, and so on. Most applications are based on the possibility of developing magnetic separation in several processes. Some biocatalysts (enzymes and microbial cells) applied in food industry can be immobilized on magnetic nanocarriers to be reused in repeated cycles of catalysis (Sánchez-Ramírez et al., 2014).

2 Magnetic Nanoparticles Behavior

Understanding of nanoparticles magnetic properties is a central topic for magnetic materials development. Magnetic nanoparticles themselves are used as the active component of a great variety of systems: ferrofluids, recording tapes, flexible disk recording media, as well as biomedical materials and catalysts. These diverse technological applications are the focus of many researches in areas of materials, chemistry, physics, geology, medicine, and biology (Kodama, 1999).

The magnetic behavior of nanoparticles is strongly related to their dimensions. The small size of magnetic particles, in a region less than a certain critical diameter, causes that the total magnetic energy becomes so small that there is no magnetic wall within the particle, and a single magnetic domain particle is considered. This critical diameter is approximately, $2A^{1/2}/M_S$, where A = exchange constant, M_S = moment per unit volume (Frei et al., 1957), otherwise, for typical material parameters it is 10–100 nm. The behavior of single domain particles can be described by assuming that all the atomic moments are rigidly aligned as a single spin. The aforementioned are the essence of the theory of superparamagnetism (Kodama, 1999). Superparamagnetism often arises on nanomagnetic substance. It appears when the substance is composed of nanomagnetic nanoparticles or magnetic fine particles existing in the metal. Superparamagnetic substances would not show magnetism without external magnetic fields because the magnetic moment easily changes the direction with thermal vibration. They behave as paramagnetic substances only with external magnetic fields (Hosokawa et al., 2007). This property can be applied to the development of processes involving magnetic separation of nanoparticles with immobilized enzymes.

3 Enzymes Immobilization

Enzymes are natural catalysts; they are biocompatible, biodegradable, and they are derived from renewable resources. Enzymes, such as carbohydrases, proteases, lipases, lysozymes, and oxidoreductases, are widely used in the food industry due to their excellent catalytic activity under relatively low temperatures (Cao et al., 2012). However, industrial application of enzymes is often hampered by a lack of long-term operational stability and difficulties in the recovery and reuse of enzymes (Sheldon and van Pelt, 2013) that sometimes may be improved by means of their immobilization. Enzyme immobilization techniques are strategies focused on retaining an enzyme on a surface or into a support. In theory, they are motivated by being able to reuse the enzyme and/or constrain the enzyme to a particular area of a reactor or device (Moehlenbrock and Minteer, 2011).

The enzyme immobilization combines the high and specific activity of the enzymes with the chemical and mechanical stability of the support materials. The aim of the immobilization is to maintain the enzyme bonded and/or trapped in a physical media that allows them to retain its catalytic activity (Wang and Caruso, 2005). Enzyme immobilization presents several advantages including a more convenient handling of the enzyme, as a solid rather than a liquid formulation that provides its facile separation from the product, thereby minimizing or eliminating protein contamination of the product (Sheldon and van Pelt, 2013). Moreover, enzymes immobilized on nanoparticles showed a broader working pH and temperature range and a higher thermal stability than the native enzymes (Ansari and Husain, 2012). On the other hand, from an industrial point of view, the essential properties are simplicity, profitability, and stability of immobilized enzymes (Cao, 2006). The most representative methods of enzymes immobilization are described in Table 3.1.

4 Magnetic Nanosupports

With the rapid development of nanotechnology, magnetic nanoparticles are currently being widely studied. It has been known that the physicochemical properties of magnetic nanoparticles can be vastly different from those of the corresponding bulk material (García-Cerda, 2004; Homaei et al., 2013; Issa et al., 2013). Recently, there has been important progress in the development of technologies in the field of magnetic supports (Ding et al., 2015; Jafari Khorshidi et al., 2015; Wang et al., 2015; Zhang et al., 2015).

The magnetic separation is an emerging technology that uses magnetism to the efficient separation of micro/nano- and ferro/paramagnetic particles from a diversity of chemical and biological samples. Magnetic nanosupports show the obvious feature of being ferro- and paramagnetic, which allows them to recover the immobilized enzyme preparation from a reactor by the application of a magnetic field to the slurry, without the need of decantation, centrifugation, or filtration (Zucca and Sanjust, 2014). An important property necessary for the magnetic nanoparticles used as carriers is the superparamagnetism, which means that the particles are attracted by a magnetic field, but retain no residual magnetism after the field is removed. This is important because suspended superparamagnetic particles can be removed from solutions using an external magnet, but they do not agglomerate after removal of the external magnetic field (Homaei et al., 2013; Karimi et al., 2011; Nuryono et al., 2014; Wahajuddin and Arora, 2012).

Magnetic nanoparticles normally consist of magnetic elements (eg, iron, nickel, cobalt, gadolinium, and dysprosium) and their chemical composites. In order to apply the magnetic nanoparticles in the food industry, the most important point is that they should be innocuous and not toxic. Therefore, the magnetic materials most utilized in food applications are iron oxides. Magnetic iron oxide nanoparticles, based on magnetite (Fe_3O_4) and maghemite (γ-Fe_2O_3), have been used in several fields due to their properties, which include large specific surface areas, no toxicity,

Table 3.1 Enzyme Immobilization Methods

Methods	Fundamentals	Advantages	Applications	References
Irreversible				
Covalent binding	Covalent attachment of enzymes to supports occur owing to interaction between their side chain amino acids and different active functional groups of carrier.	Strength of the binding, minimization of the catalyst leakage, and stabilization of the enzyme.	Immobilization of antibodies, proteases, and oxidases. Application in catalytic processes and biosensors.	Datta et al. (2013), Homaei et al. (2013), Zucca and Sanjust (2014)
Entrapment	Entrapment is caging of enzymes by covalent or noncovalent bonds within gels or fibers.	No chemical modification of the enzyme. The enzyme should retain catalytic activity under the conditions of polymerization of the support.	Xenobiotics degradation, biosensors, biofuel cells, dyes decolorization, and paper and textile industries.	Datta et al. (2013), Fernández-Fernández et al. (2013), Zucca and Sanjust (2014)
Crosslinking	Based on the occlusion of an enzyme within a polymeric network that allows the substrate and products to pass through retaining the enzyme.	Stabilization of the enzyme, no support needed, and minimization of catalyst leakage.	Synthetic chemistry, biosensors, and biomedical applications.	Guisan (2013), Jegan Roy and Emilia Abraham (2004), Zucca and Sanjust (2014)
Reversible				
Adsorption	Involves passive adsorption onto hydrophobic surfaces or electrostatic interactions with charged surfaces.	Easy and costless to be performed and no chemical modification of the enzyme needed.	Scaffold for mesoporous carbon materials and other porous carriers.	Homaei et al. (2013), Zucca and Sanjust (2014)

and excellent biocompatibility (Wei et al., 2012). They do not retain residual magnetism after the removal of external magnetic field, which allows simple separation under an external magnetic field (Cao et al., 2012). Table 3.2 shows some examples of iron oxides nanoparticles applications as magnetic nanocarriers for enzymes and other biochemical substances.

Table 3.2 Application of Iron Oxide Nanoparticles as Magnetic Carrier for Biomolecules

Magnetic Carrier	Attached Biomolecule	References
Magnetite (Fe_3O_4)	Cellulase	Garcia et al. (1989), Jordan et al. (2011)
Maghemite (γ-Fe_2O_3)	*Candida rugosa* lipase	Dyal et al. (2003)
γ-Fe_2O_3/Fe_3O_4	Glucose oxidase	Nomura et al. (2004), Rossi et al. (2004)
Magnetite	Starch	Dung et al. (2009)
Fe_3O_4	β-Galactosidase	Panesar et al. (2010)
Iron oxide-silica	Lysozyme	Valdés-Solís et al. (2009)
Fe_3O_4	*Bacillus alcalophilus* α-amylase	Mukherjee et al. (2010)
Iron oxide	Chitosan	Kievit et al. (2009)
Fe_3O_4	*Bacillus stearothermophilus* aminopeptidase II	Huang et al. (2010a)
Fe_3O_4	Alginate	Ma et al. (2007)
Fe_3O_4–SWCNTs	Amyloglucosidase	Goh et al. (2012)
Fe_2O_3	Dopamine	Xu et al. (2004)
Fe_3O_4	Plasma proteins	Liu et al. (2010)
Amylose-coated Fe_3O_4	Chitosanase	Kuroiwa et al. (2008)
Fe_3O_4	*Pycnoporus sanguineus* laccase	Wang et al. (2008)
Fe_3O_4	Horseradish peroxidase	Chalkias et al. (2008)
Fe_3O_4–CNTs	Tyrosinase	Cheng et al. (2009)
Fe_3O_4–APTES–PEG	Xanthine oxidase	Villalonga et al. (2011)
Fe_3O_4	*Pseudomonas putida* esterase	Shaw et al. (2006)
Silica-coated magnetic nanoclusters	Lactate dehydrogenase	Yusdy et al. (2009)

Besides iron oxides, other systems or compositions of magnetic nanoparticles can be synthesized and applied as supports for biocatalysts. Some examples include ferrites of cobalt, manganese, nickel, and magnesium, FePt, γ-Fe_2O_3, cobalt, iron, nickel, α-Fe, FePt, $CoPt_3$, and FeCo (Beveridge et al., 2011; Lu et al., 2007). Several methods are reported, including coprecipitation, micellar synthesis, thermal decomposition and/or reduction, hydrothermal synthesis, sol–gel, and laser pyrolysis techniques, to reach the synthesis of good-quality magnetic nanoparticles (García-Cerda, 2004; Lu et al., 2007; Osuna et al., 2012). Coprecipitation is the most representative and convenient way to obtain iron oxides from aqueous Fe^{2+}/Fe^{3+} cations solution by the posterior addition of an alkali (generally NaOH) and/or a reducing agent, performed at room or elevated temperatures (Osuna et al., 2012).

5 Immobilization of Enzymes Applied for Food Processing on Magnetic Nanoparticles

Immobilized enzymes have been applied in industry and medicine, and they are objects of various researches and developments. In any commercial application, the choice of a specific immobilized enzyme or immobilization method must be based on an analysis of all advantages and disadvantages between using free and immobilized enzyme (Homaei et al., 2013). Roughly speaking, all kinds of enzymes have an application within the food and industry areas. Some enzymes applied in the food industry are listed in Table 3.3.

5.1 Hydrolases

This class of enzymes performs the hydrolysis of specific substrates. For example, amylases and amyloglucosidase hydrolyze the starch present in plants, cellulase performs the hydrolysis of cellulose into β-glucose, chitosanase causes endohydrolysis of β-1,4-linkages present in chitosan, and lipase is an enzyme that hydrolyzes lipids, catalyzes the hydrolysis of carboxylic acid esters to carboxylic acids and alcohols (Netto et al., 2013).

5.1.1 Amylases

This class of enzymes is employed in the food industry to hydrolyze starch molecules to low molecular weight of diverse products such as dextrins, oligosaccharides, maltose, and glucose molecules (Lin et al., 2009; Mukherjee et al., 2010). Due to their poor stability, these enzymes cannot be used repeatedly. This can be avoided by enzyme immobilization on magnetic nanosupports. For instance, Lin et al. (2009) used superparamagnetic iron oxide (SPIO) nanoparticles in order to isolate or attach β-amylase obtained from soybean. They prepared the SPIO nanoparticles by coprecipitation method. Mukherjee et al. (2010) performed a study to optimize the *B. alcalophilus* α-amylase immobilization on iron oxide magnetic nanoparticles (MNPs), in order to decrease the cost of α-amylase industrial application.

In many cases, immobilization on nanoparticles with modified surface has been implemented in order to achieve chemical

Table 3.3 Enzymes Applied in Food Industry

Classes	Enzymes	Applications	References
Hydrolases	Amylases	Starch processing industry/protein purification	Lin et al. (2009), Mukherjee et al. (2010)
	Galactosidase	Production of lactose free milk products	Bayramoglu et al. (2007)
	Glucoamylase	Production of glucose from starch	Wang et al. (2013)
	Invertase	Sucrose conversion into glucose and fructose	Pospiskova et al. (2013)
	Lactase	Hydrolysis of lactose high concentrations	Dekker (1989)
	Pectinase	Used in the fruit-processing industry to increase yields, improve liquefaction, clarification and filterability of juices, maceration and extraction of plant tissues	Liu et al. (2013)
	Proteases	Used to tenderize meat and meat products, in brewing industry for clarifying juice and beer, in dairy industry for cheese, in backing industry, and in the extraction of flavor and color compounds from plants	Homaei et al. (2013)
Isomerases	Glucose isomerase	Glucose isomerization to fructose, production of high fructose syrup	Lee et al. (2013)
Lyases	Acetolactate decarboxylase	Precursor for leucine and valine synthesis that are essential for yeast growth, important enzyme for beer fermentation	Qian et al. (2015)
Oxidoreductases	Glucose oxidase	Dough strengthening	Kirk et al. (2002)
	Laccase	Clarification of juices, flavor enhancer (beer); wastewater bioremediation	Fernandes (2010), Minussi et al. (2002)
	Lipoxygenase	Protein hydrolysis, milk clotting, low-allergenic infant-food formulation, enhanced digestibility and utilization, flavor improvement in milk and cheese, meat tenderizer, prevention of chill haze formation in brewing	Fernandes (2010)
Transferases	Cyclodextringlycosyl-transferase	Cyclodextrin production	Biwer et al. (2002)
	Fructosyltransferase	Fructose oligomers synthesis; tequila production	Arrizon et al. (2012), Fernandes (2010)
	Transglutaminase	Modification of viscoelastic properties, dough processing, meat processing; to improve the quality of flour, the texture and volume of bread	Fernandes (2010), Kieliszek and Misiewicz (2014)

attachment of the enzyme. This could be performed by the use of a compatible molecule, which will serve as coupling agent between the magnetic nanosupports and enzymes. Iron oxide nanoparticles are surface modified with epichlorohydrin in order to get a major attachment of the amylase onto nanosupport (Lin et al., 2009; Mateo et al., 2000). Polymer-coated magnetic nanoparticles possess both potent magnetic susceptibility and a reactive polymer surface with large surface area, on which enzymes can be immobilized (Namdeo and Bajpai, 2009).

Several polymers have been used as surface modifiers of magnetic nanoparticles. Cellulose is one of the most abundant renewable resource and is a biodegradable, ecofriendly, and nontoxic polymer (Alam et al., 2014). Namdeo and Bajpai (2009) prepared cellulose-coated magnetite nanoparticles obtained by coagulation of aqueous solution of cellulose containing magnetite nanoparticles. After that, α-amylase was attached to the nanoparticles with cellulosic surface by a covalent bonding.

A useful property for the magnetic nanoparticles is the capability to recuperate the enzymatic catalyst by the application of an external magnetic field. This system can be used in starch degradation industries. Mizuki et al. (2010) immobilized α-amylase extracted from *Bacillus licheniformis* onto the surface of superparamagnetic nanoparticles covered with dextran, and they investigated the effect of a rotational magnetic field on the enzyme activity. It was found that the enzyme was still active after its immobilization in the absence of an external magnetic field, and that the enzyme activity increased in a rotational magnetic field to achieve a maximum value at frequency of 5 Hz.

Khan et al. (2012) used Fe_2O_3 magnetic particles as nanosupports to immobilize porcine pancreatic α-amylase. They applied this system to hydrolyze starch. The enzyme immobilization was performed by an adsorption mechanism. Immobilized α-amylase was assayed to study effects of various denaturing agents. The comparison of free and immobilized enzymes was carried out in the presence and absence of Na^+, K^+, Mg^{2+}, Cu^{2+}, Co^{2+}, Ba^{2+}, Zn^{2+}, Ca^{2+}, and Fe^{3+} ions, and urea. In the case of the metal ions, the Cu^{2+} caused the maximum enzymatic inhibition, while in the presence of Na^+, K^+, Ca^{2+}, and Co^{2+} increase of relative activity of immobilized α-amylase was detected. Moreover, the immobilized enzyme showed better stability than free α-amylase, even at high concentrations of urea (~4.0 M); meanwhile, free enzyme lost about 80% of activity, immobilized α-amylase retained ~40% of activity. These results confirm that the enzyme immobilization on nanoparticles allows its stabilizing, even to the conformational changes.

5.1.2 *Galactosidases*

This enzyme is commonly named lactase. It is able to hydrolyze the *β*-glycosidic bond formed between a galactose and its organic moiety. Its application in the food industry is to break down lactose from milk to make it suitable for persons who have lactose intolerance, and also is used in the manufacture of ice cream (Netto et al., 2013).

Lactase immobilization onto a magnetic support has been well known for years. Dekker (1989) immobilized lactase by covalent bonding to polyethyleneimine-glutaraldehyde-activated magnetite. In this case, the magnetic support was surface modified by a treatment with polyethyleneimine and crosslinked with glutaraldehyde. The lactase activity was decreased at 80%. An advantage was that the immobilized lactase was more heat stable and efficient than the free enzyme. Although 80% of the lactase activity was lost on immobilization, the magnetite-lactase system hydrolyzed high concentrations of lactose (up to 25% w/v).

Bayramoglu et al. (2007) prepared novel magnetic beads from glycidylmethacrylate and methylmethacrylate (GMA–MMA) via suspension polymerization in the presence of ethylenedimethylmethacrylate as crosslinker and $FeCl_2$ as iron oxide precursor. The magnetic poly(GMA–MMA) beads were used for the covalent immobilization of *β*-D-galactosidegalactohydrolase via glutaricdialdehyde activation. The immobilized enzyme retained much of their activity in wider ranges of temperature and pH than the ones with free form. Also was found that the *β*-galactosidase immobilized on the magnetic poly(GMA–MMA) beads presented an increase in enzyme thermostability. After a reactor probe for about 60 h at 35°C, the immobilized lactase lost only 12% of its activity. This procedure could be applied at industrial level in order to stabilize the enzymes and also to recover the catalyst by the application of an external magnetic field. Other enzymes, such as trypsin, have been immobilized onto the magnetic poly(GMA–MMA) system (Bayramoğlu et al., 2008).

Various researchers reported the use of magnetic chitosan derivatives in order to obtain a better attachment of enzymes on magnetic nanosupports (Gregorio-Jauregui et al., 2012; Ju et al., 2012; López et al., 2013; Osuna et al., 2012; Zang et al., 2014; Zhang et al., 2015). Frequently the processes for magnetic chitosan particles preparation are complicated, and not applicable for large-scale synthesis. Nevertheless, Gregorio-Jauregui et al. (2012) and Osuna et al. (2012) obtained a chitosan-coated magnetic magnetite–maghemite mixture nanoparticles prepared in one step.

Wide interest to chitosan as coating agent is related to its properties and the presence of amine- and alcohol as chemical groups. Chitosan has good biocompatibility and is biodegradable, is nontoxic, is hydrophobic, presents a mechanical stability and availability of reactive functional groups to realize chemical modifications. Moreover, it is an inexpensive material enabling low-cost carriers to prepare for large-scale applications (Pan et al., 2009; Pospiskova and Safarik, 2013).

Pospiskova and Safarik (2013) developed low-cost and easy-to-prepare procedures for the large-scale preparation of magnetic chitosan microparticles for the immobilization of lactase. In this case, the lactase maintained 86–94% of its initial activity after immobilization, and retained nearly 100% after various catalytic cycles.

Pan et al. (2009) developed a novel and efficient method for immobilization of β-D-galactosidase from *Aspergillusoryzae* using magnetic Fe_3O_4–chitosan nanoparticles as carrier. The immobilized β-D-galactosidase showed higher activity in wider ranges of temperature and pH than that of its free form. Immobilized enzyme retained 92% of its initial activity after 15 cycles of utilization. This method is recommended for the production of galactooligosaccharides from lactose.

5.1.3 Invertases

Invertase, also known as β-D-fructofuranosidefructohydrolase, β-fructofuranosidase, sucrase, saccharase. This enzyme catalyzes the hydrolysis of sucrose and related glycosides. It is commercially important due to its use for the hydrolysis of sucrose and is widely employed in food and beverage industries. Invertase is utilized to manufacture artificial honey, plasticizing agents, applied in cosmetics, pharmaceutical, and paper industries as well as in biosensor for sucrose detection (Kotwal and Shankar, 2009; Kulshrestha et al., 2013).

In order to improve the recovering of invertase and to evaluate its stability and activity, Uzun et al. (2010) synthesized polyamidoamine (PAMAM) dendrimer on the surface of magnetite nanoparticles to enhance an invertase immobilization. It is important to mention that the quantity of immobilized enzymes on the (PAMAM) dendrimer-magnetite nanoparticles was up to 2.5 times higher in comparison with the magnetite nanoparticles modified only with amino silane. The use of a surface-hyperbranched nanoparticles allows to anchor more enzyme molecules onto the particle surface. Although the invertase affinity to sucrose decreased after immobilization, the stability of the immobilized enzyme toward pH (enzyme activity from 3.0 to 9.0), temperature, and storage was enhanced.

5.1.4 Pectinases

Pectinases perform the breakdown of pectin, a complex polysaccharide found in the cell walls of plants. Its function is to hydrate and as cementing material for the cellulosic network (Netto et al., 2013; Thakur et al., 1997). These enzymes are widely utilized in the brewing industry to help the breakdown of plant materials and to help the extraction of flavors from the mash. Also, pectinases are used in the wine industry to clarify wines in order to prevent a haze or slight cloudiness (Netto et al., 2013).

Adsorption is a simple method which involves either electrostatic or ionic interactions between a charged protein and oppositely charged support. Bahrami and Hejazi (2013) reported an electrostatic adsorption of pectinase on negatively charged Fe_3O_4 nanoparticles by docusate sodium salt (AOT). They suggested that the use of an anionic surfactant, in this case AOT, stabilizes the magnetic nanoparticles and promotes the electrostatic immobilization of the positively charged enzyme onto the negative surface of the modified magnetic nanoparticles. An important result was the easy recovery of the immobilized enzyme by means of magnetic separation. The efficiency of the enzyme after 6 cycles was for about 90%. For this system, the activity of the enzymes remains similar to its free counterparts at different pHs and temperatures. Indeed, the electrostatic adsorption, not as a covalent immobilization method, allows the liberation of the enzyme at elevated pH conditions. This may be considered for development of drug delivery systems.

Seenuvasan et al. (2013) performed the covalent binding of polygalacturonase (pectinase) onto amino functionalized silica-coated magnetic Fe_3O_4 nanoparticles (CSMNPs). The maximum percentage of immobilized enzyme was 98%. Carrier saturation with enzymes led to the decrease the relative enzymatic activity. They attributed this to steric hindrance between enzyme molecules, which blocked active sites. Immobilization allowed to modify the activity as temperature and pH function. This system was applied to clarify apple juice. The results showed the effectiveness of the reuse of the enzymes in 12 cycles. Immobilized pectinase was suitable for juices treatment in industrial applications.

Lei et al. (2009) immobilized pectinase by covalent bond on Fe_3O_4/SiO_2-*g*-poly(PSStNa) nanocomposite microspheres. An improved storage stability as well as an enhanced activity and stability at higher temperatures (50–60°C) and wider pH range (3.0–4.7) were observed for the immobilized enzyme. The relative activity of the pectinase were improved. Pectinase immobilization led to stabilizing the enzyme due to the covalent bonding between the pectinase molecule and the magnetic nanosupport. The authors suggested that enzymes immobilization allowed to economically viable process with their repeated use following their easy separation with an application of an external magnetic field.

A multilayer assembly of polyelectrolyte/Fe_3O_4, in order to immobilize pectinase, was developed by Lei et al. (2007). Pectinase retained 80% of activity compared with the free enzyme. The enzyme relative activity remained at 73% after 25 batch reactions. This system permitted a magnetic handling and transport of the immobilized pectinase. It should be mentioned that the stability achieved by this method makes economically viable the use/reuse of expensive enzymes.

5.1.5 Proteases

Proteases hydrolyze peptide bonds (formed between amino acids). There are several sources of proteases, for example, papain, ficin, and bromelain, from plants and trypsin, pepsin, and rennin from animals. Proteases are industrial employed as meat tenderizing, as chillproofing of beer, for production of protein hydrolyzates. They are also used in the baking industry to modify wheat gluten. Some enzymes are utilized for the precipitation of the case in curd in cheese manufacturing (Sato, 1962).

Li et al. (2010) performed an exhaustive study focused on the general functionalization strategies for magnetic nanosupports. In this study, six different functional organosilanes with amine, aldehyde, carboxylic, epoxy, mercapto, and maleimide groups have been grafted onto maghemite nanoparticles surface. These groups have different abilities for biomolecules recognition. After functionalization of the magnetic γ-Fe_2O_3/Fe_3O_4 nanoparticles, a protease (trypsin) has been immobilized onto the surface modified of nanoparticles. The highest trypsin activity was achieved in the presence of –COOH and –SH groups. The best enzyme stability at 45°C was observed for enzyme immobilized by interaction with –SH groups.

5.1.6 Lipases

Lipases (triacylglycerol acylhydrolase, EC 3.1.1.3) hydrolyze triglycerides into diglycerides, monoglycerides, fatty acids, and glycerol (Dyal et al., 2003). They recognize different substrates including natural oils, synthetic triglycerides, as well as fatty acids esters. Moreover, they also can catalyze esterification, interesterification, acidolysis, alcoholysis, and aminolysis (Dyal et al., 2003). Reactions catalyzed by these enzymes are applied to food manufacturing in the fats and oil industry, dairy industry, pharmaceuticals, and bakery industry. Lipases are utilized for flavor modification in dairy products and processing of meat, milk products, fruit, baked foods, vegetables, and beer. These enzymes are applied in egg yolk treatment for baby food, mayonnaise, and other emulsifiers production, in lecithin modification, and for vegetable oils refining. These enzymes have activity in watery media and nonaqueous media. They can catalyze reverse reaction. Innovative processes for biodiesel production, biopolymer synthesis, treatment of fat-containing effluents, as well as enantiopure synthesis of nutraceutical and pharmaceuticals agents, also have been developed (Aravindan et al., 2007).

Osuna et al. (2015) reported *Aspergillus niger* lipase immobilization by covalent binding on chitosan-coated magnetic nanoparticles (ChCMNP) functionalized with glutaraldehyde and glycidol as coupling agents reacted with hydroxyl and amino groups, respectively (Osuna et al., 2015). Immobilized enzyme showed activities of 309.54 ± 2.01 and 266.20 ± 2.78 U g^{-1}-support for each studied system. Immobilization enhanced the enzyme stability against changes of pH and temperature, compared to free lipase. Finally, immobilized derivatives retained above 80% of its initial activity after 15 hydrolytic cycles and showed high storage stability for 50 days. Immobilized derivatives were characterized by K_m values 5 times lower than free lipase, which means that the affinity of the immobilized enzyme for its substrate is higher than the free enzyme (Osuna et al., 2015).

Silva et al. (2012) immobilized *Candida antarctica* lipase in chitosan supports using glycidol and glutaraldehyde as activators obtaining activity of 27.59 and 388.30 U g^{-1}, respectively. Instead, Rodrigues et al. (2008) reported 20 U g^{-1} of support and 300–400 U g^{-1} support for *C. antarctica* lipase immobilized in chitosan spheres using glycidol and glutaraldehyde.

5.2 Isomerases

These enzymes convert a molecule from one isomer to another, this means that the product has the same molecular formula although different spatial structure.

5.2.1 Glucose Isomerase

Glucose isomerase carried out the reversible enzymatic isomerization of D-glucose and D-xylose to D-fructose and D-xylulose, respectively. This enzyme is also recognized as D-glucose isomerase, D-xylose isomerase, and D-xylose ketolisomerase (Bhosale et al., 1996). Glucose isomerase is an industrially important enzyme due to its application in the production of high-fructose corn syrup, as well as in the fructose production by isomerization of glucose to fructose, which is applied as an alternative to cane sugar (Bhosale et al., 1996; Olsen, 2002).

Lee et al. (2013) developed a method to immobilize glucose isomerase onto Fe_3O_4-loaded mesoporous silica nanoparticles. The maximum yield of fructose was approximately near 51%. The enzyme activity was maintained for 5 cycles without a decrease in the product yield. The immobilized glucose isomerase activity was very similar to the free enzyme. The use of Fe_3O_4-loaded mesoporous silica nanoparticles as glucose isomerase hosts provides both excellent recyclability and stability to the enzyme.

Magnetic micro- and nanoparticles of iron oxides were synthesized and applied for the magnetic modification of a commercial *Streptomyces murinus* glucose isomerase immobilized onto solid carrier granules (Pospiskova and Safarik, 2015). In this study, initially magnetic iron oxides were synthesized, and then nonmagnetic granules were modified with the iron oxides. Immobilized enzyme activity was not affected by the subsequent magnetic modification, the magnetic form of the immobilized isomerase retained 100% of enzymatic activity. The magnetized immobilized glucose isomerase was stable during 8 reaction cycles, for which the enzyme lost only 3% of its activity. This method opens the possibility to magnetize systems with enzymes immobilized onto nonmagnetic supports to make them magnetically recoverable and reusable.

5.3 Lyases

Lyases are defined as enzymes cleaving C–C, C–O, C–N other bonds often forming a new double bond or a new ring structure (Mahdi and Kelly, 2001).

5.3.1 Acetolactate Decarboxylase

Acetolactate decarboxylase catalyzes the conversion of both enantiomers of acetolactate to the (*R*)-enantiomer of acetoin (Marlow et al., 2013). This enzyme is utilized in the brewing industry to accelerate the beer fermentation/maturation (Dulieu et al., 2000).

Qian et al. (2015) synthesized surface-functionalized paramagnetic nanoparticles to use them as nanosupports for α-acetolactate decarboxylase immobilization. Paramagnetic Fe_3O_4 nanoparticles have been synthesized by one-step hydrothermal reduction method. The diameter average of the particles was 100 nm. The paramagnetic nanoparticles were surface modified with tetraethyl orthosilicate (TEOS) in order to introduce highly reactive silanols on their surface. After that, $Fe_3O_4 \times SiO_2$ nanoparticles were treated with (3-aminopropyl) triethoxysilane (APTES) to introduce amino (–NH_2)groups. The terminal NH_2 groups allowed the covalent bonding of α-acetolactate decarboxylase to nanoparticles surface. The immobilized enzyme showed better tolerance to higher temperatures and pHs in comparison with the free enzyme. After 5 cycles, immobilized α-acetolactate decarboxylase maintained its activity. The immobilized enzyme exhibited satisfactory activity/reusability to be applied in the brewing industry.

5.4 Oxidoreductases

Oxidoreductases are enzymes that catalyze the reactions of oxidation/reduction, involving the transfer of hydrogen or electrons and oxygen atoms to the substrate (Netto et al., 2013; Whitaker, 2002). This class of catalysts is important for the food industry due to their influence in the taste, texture, appearance, shelf life, nutritional value, and process tolerance of food products (Laane et al., 2003).

5.4.1 Glucose Oxidase

This enzyme catalyzes the oxidation of β-D-glucose to D-glucono-δ-lactone and hydrogen peroxide (Netto et al., 2013). In food industry, glucose oxidase has been used as a component to transform glucose to gluconic acid in order to prevent the Maillard coloration in some foods. Moreover, this enzyme is used to remove oxygen and prevent oxidation of the food components. In bakery it is applied to improve the dough quality and the baking properties (Vroemen, 2002). It is widely used in wine production, tinned foods, textile bleaching and as a glucose biosensor. Enzyme immobilization technology allows the reuse of enzymes and improves their stability.

Glucose oxidase enzyme (EC 1.1.3.4) was immobilized onto aminated-$CoFe_2O_4/SiO_2$ nanoparticles via crosslinking with glutaraldehyde by Wang et al. (2011). First, the cobalt ferrite was synthesized by coprecipitation method. Then, the obtained $CoFe_2O_4$ nanoparticles were surface treated with tetraethyl orthosilicate (TEOS) and (3-aminopropyl) triethoxysilane (APTES) in order to add amino groups to the nanoparticles surface. The obtained $CoFe_2O_4$ nanoparticles showed a core/shell structure coated with amorphous silica and superparamagnetic response. Finally, the glucose oxidase was immobilized via covalent crosslinking with glutaraldehyde onto aminated-$CoFe_2O_4/SiO_2$ nanoparticles. The immobilized glucose oxidase exhibited improved thermal, storage, and operation stability. The immobilized enzyme maintained 57% of the activity after 7 cycles of reuse. These nanoparticles are a good alternative to immobilize glucose oxidase to be used/reused in industrial catalytic processes.

Huang et al. (2010b) prepared magnetic core-shell (Fe_3O_4/SiO_2) nanoparticles activated with glutaraldehyde. The nanoparticles were used as nanocarrier for glucose oxidase immobilization. The optimum pH of glucose oxidase activity was changed from pH 6 with free enzyme to wider pH range 5.5–7.5 with immobilized enzyme with optimum value at 7. The optimum temperature of glucose oxidase reaction was also improved from 30°C for free enzyme to 50°C for immobilized enzyme, which retained residual activity to 80°C. The immobilized glucose oxidase maintained ~75% of its activity after 1 month of storage at 4°C, while the free enzyme retained only ~60%. The immobilized glucose oxidase was reused in six consecutive reaction cycles maintaining 60% of its initial activity. These core-shell magnetic nanoparticles are an effective nanocarrier to improve operational glucose oxidase properties and may be recommended for application with other enzymatic systems used in the food industry.

5.4.2 Laccase

Laccase (benzenediol:oxygen oxidoreductase) is a type of copper-containing polyphenol oxidase produced by many plants and fungal strains. Laccase oxidizes methoxyphenols, aminophenols, substituted monophenols, diamines, and a considerable range of other compounds (Flurkey, 2002; Minussi et al., 2002; Netto et al., 2013). Laccase has been used as for bioremediation of food industry wastewater due to its ability to degrade phenolic compounds. Moreoever, this enzyme is used in the brewing to degrade the polyphenols in the wastewaters, and as wine stabilizer eliminating polyphenols (Minussi et al., 2002).

Kalkan et al. (2012) coated magnetite nanoparticles with chitosan and used them as nanosupports to immobilize *Trametes versicolor* laccase. The coating of Fe_3O_4 nanoparticles with chitosan was achieved by reversed phase suspension technique. Laccase was immobilized onto Fe_3O_4-chitosan nanoparticles either by adsorption or by covalent binding. The immobilized enzyme was more stable at different ranges of temperature and pH in comparison with the free laccase. After 30 cycles of repeated use, immobilized laccase maintained more than 71% of activity. This system is a useful method to be applied in the food industry in terms of magnetic enzymatic recuperation, catalytic activity and reuse capacity.

The development of magnetic mesoporous silica spheres (MSS) was achieved by Zhu et al. (2007). *T. versicolor* laccase was immobilized onto MSS with large pore size by adsorption and covalent bonding methods. The thermal and pH stabilities of immobilized laccase were improved in comparison with free enzyme. Laccase immobilized though covalent bonding retained about 70% of activity after 10 cycles of repeated use. An important data were the leaching observed in the system obtained by physical adsorption. This means that the immobilization is not stable enough to be used in systems of magnetic separation. Otherwise, the enzyme immobilized by covalent attachment method has no leaching, which makes it a useful system to enzyme magnetic recovering/reusability.

Wang et al. (2010) developed magnetic mesoporous silica nanoparticles and used them as nanocarriers for laccase immobilization. These magnetic nanoparticles were functionalized/chelated with Cu^{2+}. Due to this, the ion was known as an important part of laccase active site; immobilization was carried out through affinity adsorption on metal to be compared with adsorption on unfunctionalized carriers. Magnetic particles functionalized with copper ions exhibited higher capacity to enzyme adsorption [98.1 mg (particles g)$^{-1}$ and 92.5% for activity recovery] than particles without functionalization. Optimum temperature for laccase catalyzed reaction was shifted from 20°C (for the free enzyme) to 60°C (for immobilized derivative), that was explained by multipoint chelate interaction between nanoparticles and laccase molecules. Immobilization onto this magnetic nanostructured carrier allowed increasing laccase storage stability; free laccase lost its activity after 7 weeks, while the immobilized enzyme remained 92% of its activity. Moreover, after 10 cycles of reuse immobilized laccase maintained 87% of its initial activity. Magnetic mesoporous silica nanoparticles developed by Wang et al. is a promising material for the application as carrier for immobilization of different enzymatic systems, which may be utilized in the food industry.

5.5 Transferases

This class of enzymes removes groups from one substrate (donor) molecule and transfers the group to another substrate (acceptor) molecule (Netto et al., 2013; Whitaker, 2002). Transferases have been used to synthesize fructose oligomers, or fructose oligosaccharides, which serve as dietary fiber. Moreover, the transferases have been applied in the bakery industry to improve the texture of gluten-free breads for celiac disease patients (Simpson et al., 2012).

5.5.1 Cyclodextringlycosyltransferase

Cyclodextringlycosyltransferase catalyzes four different transformations: cyclization, coupling, disproportionation, and hydrolysis reactions. This enzyme produces nonreducing cyclic dextrins from starch, amylose, and other polysaccharides. In the presence of starch as substrate, main products are cyclodextrins and highly branched dextrins (Biwer et al., 2002; Dijkhuizen and van der Veen, 2002; Ibrahim et al., 2014). Cyclodextringlycosyltransferase is applied for cyclodextrins production, which are important for food and pharmaceutical industries. For example, they have a texture-improving effect on meat and pastry products (Dijkhuizen and van der Veen, 2002).

Ibrahim et al. (2014) covalent immobilized *Amphibacillus* sp. cyclodextringlucanotransferase onto amino-functionalized magnetic double mesoporous core-shell silica nanospheres. They parted from the synthesis of magnetite (Fe_3O_4) nanoparticles using the solvothermal method developed by Deng et al. (2005). Then, Fe_3O_4 nanoparticles were coated with tetraethyl orthosilicate (TEOS) to protect iron oxide core from any extreme conditions. The magnetic core-shell silica nanospheres were functionalized with 3-aminopropyltriethoxysilane (APMS) to introduce amino ($-NH_2$) groups over the magnetic double mesoporous core-shell silica nanospheres. Finally, the *Amphibacillus* sp. cyclodextringlucanotransferase was immobilized onto the amino-functionalized magnetic double mesoporous core-shell silica nanospheres by physical adsorption and covalent attachment. *Amphibacillus* sp. cyclodextringlucanotransferase immobilized by covalent attachment showed higher immobilization yield and loading efficiency compared to enzyme immobilized by physical adsorption. Thus, the covalent attachment was more effective for the enzyme immobilization. Immobilized cyclodextringlucanotransferase showed an improvement of thermal and pH of stability once immobilized. Moreover, the immobilized enzyme exhibited good stability, retaining 86 and 56% of the initial activity after reutilizations for 5 and 10 cycles, respectively. This nanosupport can be utilized as magnetic nanocarriers to recover biocatalysts from reaction mixture.

The immobilization of cyclodextringlycosyltransferase from *Bacillus lehensis* onto Fe_3O_4 nanoparticles has been reported by Blanco et al. (2013). Magnetite nanoparticles were synthesized by hydrothermal method. After that, the particles were treated with (3-aminopropyl)triethoxysilane (APTES) in order to add amino groups to the nanoparticles surface. Finally, *B. lehensis* cyclodextringlycosyltransferase was immobilized by covalent binding onto functionalized magnetic nanosupports. The thermal properties were improved. Optimal temperature of enzymatic reaction was increased from 55 to 70°C, compared with the free cyclodextringlycosyltransferase. The immobilized enzyme maintained 16% of activity after 4 cycles of reuse. This method allowed the magnetic separation of the immobilized *B. lehensis* cyclodextringlycosyltransferase.

6 One-Step Method to Obtain Chitosan-Coated Magnetic Nanoparticles and Their Application for Bioethanol Production From Agave Leaves

This section describes results obtained by our research group to apply chitosan-coated magnetic nanoparticles (ChCMNP) synthetized by one-step coprecipitation method, for utilization of agave leaves—by-product of tequila and sotol industry (Mexican alcoholic drink) in production of second-generation bioethanol.

ChCMNP were obtained at 50°C in a one-step method comprising coprecipitation in the presence of low chitosan content (Osuna et al., 2012). ChCMNP showed high magnetization and superparamagnetism. They were composed of a core of 9–10 nm of average diameter and a very thin chitosan layer (0.3 nm) in accordance with transmission electron microscopy measurements. The results from Fourier transform infrared spectrometry demonstrated that ChCMNP were obtained and those from thermogravimetric analysis allowed to determine that they were composed of 95 wt% of magnetic nanoparticles and 5 wt% of chitosan. However, only 60% of amino groups on ChCMNP surface were available for chemical interactions, probably due to a certain degree of nanoparticle agglomeration during the redispersion. The very low weight ratio chitosan to magnetic nanoparticles obtained in this study, 0.053, and the high yield of the precipitation reactions (≈97%) are noticeable.

Generally, nanomaterials have many novel properties compared with the bulk materials, and the magnetic nanoparticles are not the exception, they have many unique magnetic properties such as superparamagnetic comportment, high coercivity, low Curie temperature, and high magnetic susceptibility (Wu et al., 2008). Magnetic iron oxides, such as maghemite (γ-Fe_2O_3) and magnetite (Fe_3O_4) are widely known materials due to their properties like superparamagnetism, their low toxicity, and its great surface area (Gregorio-Jauregui et al., 2012). Magnetic nanoparticles have been used as magnetic fluids, data storage, for catalysis, and bioapplications. Currently, they are also utilized in important bioapplications, including magnetic bioseparation and detection of biological entities, clinic diagnosis and therapy, and magnetic fluid hyperthermia, targeted drug delivery and biological labels (Wu et al., 2008). However, it is crucial to improve the biocompatibility of the nanoparticles surface, besides protect them to avoid its oxidation or degradation. This could be remediated by coating the nanoparticles surface with biopolymers such as chitosan.

Chitosan is the most important derivative of chitin, which is obtained by partial deacetylation of chitin by alkaline treatment or by enzymatic hydrolysis (Ravi Kumar, 2000; Rinaudo, 2006). It has been identified as a versatile biopolymer, alkaline, nontoxic, hydrophilic, biocompatible, and biodegradable, which is used for different applications in a variety of industrial fields such as agriculture, food, and pharmaceutical (Gregorio-Jauregui et al., 2012; Laurent et al., 2008). Free amino and hydroxyl groups are responsible for the reactivity of this polymer (Gregorio-Jauregui et al., 2012). Nowadays, the development of magnetic nanoparticles encapsulated or surface modified in/with chitosan are of great interest.

These ChCMNP were applied for laccase immobilization used for lignin removal from agave leaves. Lignin removal is an important technical issue for paper manufacturing and a key challenge for the conversion of lignocellulosic feedstock into liquid transportation fuels such as ethanol. In our recent study, *T. versicolor* laccase was immobilized on ChCMNP by covalent attachment via glutaraldehyde reaction. Enzyme immobilization yield was 82%, and 89% of its initial activity was retained. Immobilized enzyme showed better thermal and storage stability than the free enzyme, increase of K_m value only at 1.25 times, and its activity behavior as pH and temperature function was maintained. Immobilized laccase retained 50% of its activity after 5 cycles of application in the reaction with catechol used as substrate.

The free and immobilized laccase were applied in order to degrade lignin in cactus *Agave atrovirens* leaves, which previously were pretreated by autoclaving for 30 min. Treatment with free laccase applied for 12 h at 3 U (solid vegetal material g^{-1}) led to raise the yield of enzymatic hydrolysis (with Celluclast at 15 U g^{-1}) at 11%. However, immobilized laccase application led to decreasing the yield of cellulose hydrolysis at 16% under the same conditions. Moreover, interaction between enzymatic magnetic preparation and lignocellulosic material was observed that hindered magnetic separation of immobilized enzyme. Thus, application of laccase immobilized on magnetic nanoparticles coated with chitosan does not lead to an advantage in the treatment of cactus *A. atrovirens* leaves previous to cellulose hydrolysis, unlike to free enzyme. However, immobilized enzyme may be applied for several biotechnological processes where magnetic separation may be useful.

The ethanol production from lignocellulosic biomass of agave cactus leaves is considered as a viable option because the agave plant has a high biomass yield, low requirements of cultivation, and efficient use of water. Moreover, it generates as residue of the

Mexican alcoholic drinks industry. However, cellulose saccharification by enzymatic hydrolysis remains as a process with high consumption of expensive enzymes. In our recent study *Trichoderma reesei* cellulase preparation (Celluclast) was covalently immobilized on ChCMNP by means of glutaraldehyde activation of carrier. An average diameter of magnetic nanoparticles before and after enzyme immobilization was 8.49 and 10.02 nm, respectively. Superparamagnetic properties were detected for both nanostructured preparations. The immobilized enzyme retained about 37% of the initial activity, showed better thermal and storage stability as free enzyme. Similar effect of pH and temperature on enzyme activity was observed. Immobilized cellulase retained about 80% of its activity for 15 cycles of hydrolysis of carboxymethylcellulose and was easily separated from reaction media by means of magnetic field. However, in the reaction with this substrate K_m value of immobilized enzyme was increased at 8 times. The immobilized enzyme was able to hydrolyze lignocellulosic material of *A. atrovirens* leaves with yield close to detected with free enzyme. However, less conversion was observed with microcrystalline cellulose. Immobilized preparation was reused in vegetal material conversion up to 4 cycles with 50% activity decrease. This provides an opportunity to reduce the enzyme consumption during lignocellulosic material saccharification for bioethanol production.

The interaction between biological and nanostructured systems can be applied to development of biotechnological processes for the enzyme and fermentation products production. The knowledge generated about using nanostructured systems such as ChCMNP for microorganism immobilization is limited. In our recent study effects of different conditions on the immobilization of *Trichoderma* sp. fungi spores on ChCMNP and adsorption isotherm properties were defined. For microorganism immobilization an exploratory Plackett–Burman design with seven variables at two levels (sample at 1E + 06− 1E + 07 spores mL^{-1}, ChCMNP at 10–100 μL; physiological solution as media at 5–10 mL, agitation at 100–200 rpm, temperature at 25–30°C, time for 0–1.5 h, pH at 5–7). To obtain fungal spores, the strain was propagated in flasks containing PDA at 28°C for 7 days. Adsorption isotherms were obtained under conditions of saturation with ChCMNP suspended in saline solution in the presence of different concentrations of inoculum at 30°C, 200 rpm for 1.5 h. No immobilized spores were quantified in a Neubauer chamber. Statistical analysis of data was performed using a statistical software STATISTICA 7 and Microsoft Excel. It was demonstrated that variables significantly influenced on spores' adsorption were: amount of inoculum, ChCMNP amount, and agitation rate. Adsorption kinetics detailed a system saturation phase at 1.5 h. The data could be linearized using the Langmuir and Freundlich models with correlation coefficient of 0.83 and 0.91, respectively. According to Langmuir model, maximum adsorption capacity (q_{max}) was estimated as 2.50E + 09 spores g^{-1} ChCMNP. Magnetic separation of immobilized fungus may be applied for biomass reuse in different biotechnological process.

7 Conclusions

The present literature analysis demonstrates that magnetic nanoparticles are widely applied to biocatalysts immobilization. Magnetic nanobiocatalytic systems are an evolving innovative field, which synergistically assimilates nanotechnological and biotechnological developments. Actually, real advances are achieved to improve activity, stability, and enzymes capability for engineering performances in food industry bioprocesses. Enzymes immobilization on functional magnetic nanomaterials allows its using as carriers, which facilitate the separation of the biocatalyst under an external magnetic field. Strategies for enzymes immobilization are varied from physical adsorption, covalent binding, crosslinking, to specific ligands application. Excellent results are revealed for catalytic activity and stability enhancement due to conformational changes allowed by magnetic nanoenvironments. Reutilization of magnetic nanoparticles into catalytic processes is an important factor to defray the high operational costs associated with enzyme production and nanocarrier synthesis. It is hoped that the use of magnetic nanosupports will lead to the development of more efficient, economical, and environmentally friendly technological processes applied to the food industry.

Acknowledgments

The authors would thank the Mexican Council of Science and Technology (CONACyT) for its financial support by Grant No. 213844 (PDCPN2013-01 CONACyT-Mexico). Also, the authors acknowledge CONACyT for the undergraduate and graduate scholarships, and for the financial support under the program "Cátedras-CONACyT".

References

Alam, M., Akram, D., Sharmin, E., Zafar, F., Ahmad, S., 2014. Vegetable oil based eco-friendly coating materials: a review article. Arab. J. Chem. 7, 469–479.

Ansari, S.A., Husain, Q., 2012. Potential applications of enzymes immobilized on/in nano materials: a review. Biotechnol. Adv. 30, 512–523.

Aravindan, R., Anbumathi, P., Viruthagiri, T., 2007. Lipase applications in food industry. Indian J. Biotechnol. 6, 141–158.

Arrizon, J., Morel, S., Gschaedler, A., Monsan, P., 2012. Fructanase and fructosyltransferase activity of non-Saccharomyces yeasts isolated from fermenting musts of Mezcal. Bioresour. Technol. 110, 560–565.
Ashtari, K., Khajeh, K., Fasihi, J., Ashtari, P., Ramazani, A., Vali, H., 2012. Silica-encapsulated magnetic nanoparticles: enzyme immobilization and cytotoxic study. Int. J. Biol. Macromol. 50, 1063–1069.
Bahrami, A., Hejazi, P., 2013. Electrostatic immobilization of pectinase on negatively charged AOT-Fe_3O_4 nanoparticles. J. Mol. Catal. B Enzym. 93, 1–7.
Bayramoglu, G., Tunali, Y., Arica, M.Y., 2007. Immobilization of β-galactosidase onto magnetic poly(GMA–MMA) beads for hydrolysis of lactose in bed reactor. Catal. Commun. 8, 1094–1101.
Bayramoğlu, G., Yılmaz, M., Şenel, A.Ü., Arıca, M.Y., 2008. Preparation of nanofibrous polymer grafted magnetic poly(GMA-MMA)-*g*-MAA beads for immobilization of trypsin via adsorption. Biochem. Eng. J. 40, 262–274.
Beveridge, J.S., Stephens, J.R., Williams, M.E., 2011. The use of magnetic nanoparticles in analytical chemistry. Annu. Rev. Anal. Chem. (Palo Alto Calif) 4, 251–273.
Bhosale, S.H., Rao, M.B., Deshpande, V.V., 1996. Molecular and industrial aspects of glucose isomerase. Microbiol. Rev. 60, 280–300.
Biwer, A., Antranikian, G., Heinzle, E., 2002. Enzymatic production of cyclodextrins. Appl. Microbiol. Biotechnol. 59, 609–617.
Blanco, K.C., dos Santos, F.J., Bernardi, N.S., Júnior, M.J., Monti, R., Contiero, J., 2013. Reuse of cyclodextrin glycosyltransferase through immobilization on magnetic carriers. Enzym. Eng. 2, 111.
Cao, L., 2006. Introduction: immobilized enzymes: past, present and prospects. Carrier-Bound Immobilized Enzymes. Wiley-VCH Verlag GmbH & Co. KGaAPL Weinheim, FRG, pp. 1–52.
Cao, M., Li, Z., Wang, J., Ge, W., Yue, T., Li, R., Colvin, V.L., Yu, W.W., 2012. Food related applications of magnetic iron oxide nanoparticles: enzyme immobilization, protein purification, and food analysis. Trends Food Sci. Technol. 27, 47–56.
Chalkias, N.G., Kahawong, P., Giannelis, E.P., 2008. Activity increase of horseradish peroxidase in the presence of magnetic particles. J. Am. Chem. Soc. 130, 2910–2911.
Cheng, Y., Liu, Y., Huang, J., Li, K., Xian, Y., Zhang, W., Jin, L., 2009. Amperometric tyrosinase biosensor based on Fe_3O_4 nanoparticles-coated carbon nanotubes nanocomposite for rapid detection of coliforms. Electrochim. Acta 54, 2588–2594.
Datta, S., Christena, L.R., Rajaram, Y.R.S., 2013. Enzyme immobilization: an overview on techniques and support materials. 3 Biotech 3, 1–9.
Dekker, R.F., 1989. Immobilization of a lactase onto a magnetic support by covalent attachment to polyethyleneimine-glutaraldehyde-activated magnetite. Appl. Biochem. Biotechnol. 22, 289–310.
Demir, A.S., Talpur, F.N., Betul Sopaci, S., Kohring, G.W., Celik, A., 2011. Selective oxidation and reduction reactions with cofactor regeneration mediated by galactitol-, lactate-, and formate dehydrogenases immobilized on magnetic nanoparticles. J. Biotechnol. 152, 176–183.
Deng, H., Li, X., Peng, Q., Wang, X., Chen, J., Li, Y., 2005. Monodisperse magnetic single-crystal ferrite microspheres. Angew. Chem. Int. Ed. 44, 2782–2785.
Dijkhuizen, L., van der Veen, B., 2002. Cyclodextrin glycosyltransferase. In: Whitaker, J.R., Voragen, A.G.J., Wong, D.W.S. (Eds.), Handbook of Food Enzymology. CRC Press, Boca Raton, FL.
Ding, S., Cargill, A.A., Medintz, I.L., Claussen, J.C., 2015. Increasing the activity of immobilized enzymes with nanoparticle conjugation. Curr. Opin. Biotechnol. 34, 242–250.

Dulieu, C., Moll, M., Boudrant, J., Poncelet, D., 2000. Improved performances and control of beer fermentation using encapsulated α-acetolactate decarboxylase and modeling. Biotechnol. Prog. 16, 958–965.
Dung, T.T., Danh, T.M., Hoa, L.T.M., Chien, D.M., Duc, N.H., 2009. Structural and magnetic properties of starch-coated magnetite nanoparticles. J. Exp. Nanosci. 4, 259–267.
Dyal, A., Loos, K., Noto, M., Chang, S.W., Spagnoli, C., Shafi, K.V.P.M., Ulman, A., Cowman, M., Gross, R.A., 2003. Activity of *Candida rugosa* lipase immobilized on γ-Fe_2O_3 magnetic nanoparticles. J. Am. Chem. Soc. 125, 1684–1685.
Fernandes, P., 2010. Enzymes in food processing: a condensed overview on strategies for better biocatalysts. Enzyme Res. 2010, 1–19.
Fernández-Fernández, M., Sanromán, M.Á., Moldes, D., 2013. Recent developments and applications of immobilized laccase. Biotechnol. Adv. 31, 1808–1825.
Flurkey, W., 2002. Laccase. In: Whitaker, J.R., Voragen, A.G.J., Wong, D.W.S. (Eds.), Handbook of Food Enzymology. CRC Press, Boca Raton, FL.
Frei, E.H., Shtrikman, S., Treves, D., 1957. Critical size and nucleation field of ideal ferromagnetic particles. Phys. Rev. 106, 446–455.
Garcia, A., Oh, S., Engler, C.R., 1989. Cellulase immobilization on Fe_3O_4 and characterization. Biotechnol. Bioeng. 33, 321–326.
García-Cerda, L.A., 2004. Magnetic nanocomposites: preparation and characterization of co-ferrite nanoparticles in a silica matrix. J. Alloy. Compd. 369, 148–151.
Goh, W.J., Makam, V.S., Hu, J., Kang, L., Zheng, M., Yoong, S.L., Udalagama, C.N.B., Pastorin, G., 2012. Iron oxide filled magnetic carbon nanotube-enzyme conjugates for recycling of amyloglucosidase: toward useful applications in biofuel production process. Langmuir 28, 16864–16873.
Gregorio-Jauregui, K.M., Pineda, M.G., Rivera-Salinas, J.E., Hurtado, G., Saade, H., Martinez, J.L., Ilyina, A., López, R.G., 2012. One-step method for preparation of magnetic nanoparticles coated with chitosan. J. Nanomater. 2012, 1–8.
Guisan, J.M. (Ed.), 2013. Immobilization of Enzymes and Cells. Humana Press, Totowa, NJ, pp. 1–375.
Homaei, A.A., Sariri, R., Vianello, F., Stevanato, R., 2013. Enzyme immobilization: an update. J. Chem. Biol. 6, 185–205.
Hong, R.Y., Li, J.H., Zhang, S.Z., Li, H.Z., Zheng, Y., Ding, J.M., Wei, D.G., 2009. Preparation and characterization of silica-coated Fe_3O_4 nanoparticles used as precursor of ferrofluids. Appl. Surf. Sci. 255, 3485–3492.
Hosokawa, M., Nogi, K., Naito, M., Yokoyama, T., 2007. Nanoparticle Technology Handbook, second ed. Elsevier, Amsterdam, The Netherlands.
Huang, C.-L., Cheng, W.-C., Yang, J.-C., Chi, M.-C., Chen, J.-H., Lin, H.-P., Lin, L.-L., 2010a. Preparation of carboxylated magnetic particles for the efficient immobilization of C-terminally lysine-tagged *Bacillus stearothermophilus* aminopeptidase II. J. Ind. Microbiol. Biotechnol. 37, 717–725.
Huang, J., Zhao, R., Wang, H., Zhao, W., Ding, L., 2010b. Immobilization of glucose oxidase on Fe_3O_4/SiO_2 magnetic nanoparticles. Biotechnol. Lett. 32, 817–821.
Ibrahim, A.S.S., Al-Salamah, A.A., El-Toni, A.M., El-Tayeb, M.A., Elbadawi, Y.B., 2014. Cyclodextrin glucanotransferase immobilization onto functionalized magnetic double mesoporous core–shell silica nanospheres. Electron. J. Biotechnol. 17, 55–64.
Issa, B., Obaidat, I.M., Albiss, B.A., Haik, Y., 2013. Magnetic nanoparticles: surface effects and properties related to biomedicine applications. Int. J. Mol. Sci. 14, 21266–21305.
Jafari Khorshidi, K., Lenjannezhadian, H., Jamalan, M., Zeinali, M., 2015. Preparation and characterization of nanomagnetic cross-linked cellulase aggregates for cellulose bioconversion. J. Chem. Technol. Biotechnol. 91, 539–546.

Jegan Roy, J., Emilia Abraham, T., 2004. Strategies in making cross-linked enzyme crystals. Chem. Rev. 104, 3705–3722.

Jiang, Y., Guo, C., Xia, H., Mahmood, I., Liu, C., Liu, H., 2009. Magnetic nanoparticles supported ionic liquids for lipase immobilization: enzyme activity in catalyzing esterification. J. Mol. Catal. B Enzym. 58, 103–109.

Jordan, J., Kumar, C.S.S.R., Theegala, C., 2011. Preparation and characterization of cellulase-bound magnetite nanoparticles. J. Mol. Catal. B Enzym. 68, 139–146.

Ju, H.-Y., Kuo, C.-H., Too, J.-R., Huang, H.-Y., Twu, Y.-K., Chang, C.-M.J., Liu, Y.-C., Shieh, C.-J., 2012. Optimal covalent immobilization of α-chymotrypsin on Fe_3O_4-chitosan nanoparticles. J. Mol. Catal. B Enzym. 78, 9–15.

Kalkan, N.A., Aksoy, S., Aksoy, E.A., Hasirci, N., 2012. Preparation of chitosan-coated magnetite nanoparticles and application for immobilization of laccase. J. Appl. Polym. Sci. 123, 707–716.

Karimi, M.A., Mohammadi, S.Z., Mohadesi, A., Hatefi-Mehrjardi, A., Mazloum-Ardakani, M., Sotudehnia Korani, L., Askarpour Kabir, A., 2011. Determination of silver(I) by flame atomic absorption spectrometry after separation/preconcentration using modified magnetite nanoparticles. Sci. Iran. 18, 790–796.

Khan, M.J., Husain, Q., Azam, A., 2012. Immobilization of porcine pancreatic α-amylase on magnetic Fe_2O_3 nanoparticles: applications to the hydrolysis of starch. Biotechnol. Bioprocess Eng. 17, 377–384.

Kieliszek, M., Misiewicz, A., 2014. Microbial transglutaminase and its application in the food industry. A review. Folia Microbiol. (Praha) 59, 241–250.

Kievit, F.M., Veiseh, O., Bhattarai, N., Fang, C., Gunn, J.W., Lee, D., Ellenbogen, R.G., Olson, J.M., Zhang, M., 2009. PEI-PEG-chitosan-copolymer-coated iron oxide nanoparticles for safe gene delivery: synthesis, complexation, and transfection. Adv. Funct. Mater. 19, 2244–2251.

Kirk, O., Borchert, T.V., Fuglsang, C.C., 2002. Industrial enzyme applications. Curr. Opin. Biotechnol. 13, 345–351.

Kodama, R.H., 1999. Magnetic nanoparticles. J. Magn. Magn. Mater. 200, 359–372.

Kotwal, S.M., Shankar, V., 2009. Immobilized invertase. Biotechnol. Adv. 27, 311–322.

Kulshrestha, S., Tyagi, P., Sindhi, V., Yadavilli, K.S., 2013. Invertase and its applications—a brief review. J. Pharm. Res. 7, 792–797.

Kuroiwa, T., Noguchi, Y., Nakajima, M., Sato, S., Mukataka, S., Ichikawa, S., 2008. Production of chitosan oligosaccharides using chitosanase immobilized on amylose-coated magnetic nanoparticles. Process Biochem. 43, 62–69.

Laane, C., Bruggeman, Y., Winkel, C., 2003. Applications of oxidoreductases in foods. In: Whitaker, J.R., Voragen, A.G.J., Wong, D.W.S. (Eds.), Handbook of Food Enzymology. CRC Press, Boca Raton, FL.

Laurent, S., Forge, D., Port, M., Roch, A., Robic, C., Vander Elst, L., Muller, R.N., 2008. Magnetic iron oxide nanoparticles: synthesis, stabilization, vectorization, physicochemical characterizations, and biological applications. Chem. Rev. 108, 2064–2110.

Lee, Y.-C., Chen, C.-T., Chiu, Y.-T., Wu, K.C.-W., 2013. An effective cellulose-to-glucose-to-fructose conversion sequence by using enzyme immobilized Fe_3O_4-loaded mesoporous silica nanoparticles as recyclable biocatalysts. ChemCatChem 5, 2153–2157.

Lei, Z., Bi, S., Hu, B., Yang, H., 2007. Combined magnetic and chemical covalent immobilization of pectinase on composites membranes improves stability and activity. Food Chem. 105, 889–896.

Lei, Z., Ren, N., Li, Y., Li, N., Mu, B., 2009. Fe_3O_4/SiO_2-*g*-PSStNa polymer nanocomposite microspheres (PNCMs) from a surface-initiated atom transfer radical polymerization (SI-ATRP) approach for pectinase immobilization. J. Agric. Food Chem. 57, 1544–1549.

Li, D., Teoh, W.Y., Gooding, J.J., Selomulya, C., Amal, R., 2010. Functionalization strategies for protease immobilization on magnetic nanoparticles. Adv. Funct. Mater. 20, 1767–1777.
Lin, J., Lin, Y.S., Kuo, S.T., Jiang, C.M., Wu, M.C., 2009. Purification of soybean amylase by superparamagnetic particles. Food Chem. 117, 94–98.
Liu, S., Han, Y., Qiao, R., Zeng, J., Jia, Q., Wang, Y., Gao, M., 2010. Investigations on the interactions between plasma proteins and magnetic iron oxide nanoparticles with different surface modifications. J. Phys. Chem. C 114, 21270–21276.
Liu, Y., Li, Q., Feng, Y.-Y., Ji, G.-S., Li, T.-C., Tu, J., Gu, X.-D., 2013. Immobilisation of acid pectinase on graphene oxide nanosheets. Chem. Pap. 68, 732–738.
López, R.G., Pineda, M.G., Hurtado, G., de León, R.D., Fernández, S., Saade, H., Bueno, D., 2013. Chitosan-coated magnetic nanoparticles prepared in one step by reverse microemulsion precipitation. Int. J. Mol. Sci. 14, 19636–19650.
Lu, A.-H., Salabas, E.L., Schüth, F., 2007. Magnetic nanoparticles: synthesis, protection, functionalization, and application. Angew. Chem. Int. Ed. 46, 1222–1244.
Ma, H., Qi, X., Maitani, Y., Nagai, T., 2007. Preparation and characterization of superparamagnetic iron oxide nanoparticles stabilized by alginate. Int. J. Pharm. 333, 177–186.
Mahdi, J.G., Kelly, D.R., 2001. Lyases. In: Rehm, H.-J., Reed, G. (Eds.), Biotechnology Set. Wiley-VCH Verlag GmbH, Weinheim, Germany, pp. 41–171.
Marlow, V.A., Rea, D., Najmudin, S., Wills, M., Fülöp, V., 2013. Structure and mechanism of acetolactate decarboxylase. ACS Chem. Biol. 8, 2339–2344.
Mateo, C., Abian, O., Fernandez-Lafuente, R., Guisan, J.M., 2000. Increase in conformational stability of enzymes immobilized on epoxy-activated supports by favoring additional multipoint covalent attachment. Enzyme Microb. Technol. 26, 509–515.
Minussi, R.C., Pastore, G.M., Durán, N., 2002. Potential applications of laccase in the food industry. Trends Food Sci. Technol. 13, 205–216.
Mizuki, T., Watanabe, N., Nagaoka, Y., Fukushima, T., Morimoto, H., Usami, R., Maekawa, T., 2010. Activity of an enzyme immobilized on superparamagnetic particles in a rotational magnetic field. Biochem. Biophys. Res. Commun. 393, 779–782.
Moehlenbrock, M.J., Minteer, S.D., 2011. Introduction to the field of enzyme immobilization and stabilization. In: Shelley, D., Minteer, S.D. (Eds.), Methods in Molecular Biology. Springer Science + Business Media, LLC, Clifton, NJ, pp. 1–7.
Mukherjee, A.K., Kumar, T.S., Rai, S.K., Roy, J.K., 2010. Statistical optimization of *Bacillus alcalophilus* α-amylase immobilization on iron-oxide magnetic nanoparticles. Biotechnol. Bioprocess Eng. 15, 984–992.
Namdeo, M., Bajpai, S.K., 2009. Immobilization of α-amylase onto cellulose-coated magnetite (CCM) nanoparticles and preliminary starch degradation study. J. Mol. Catal. B Enzym. 59, 134–139.
Netto, C.G.C.M., Toma, H.E., Andrade, L.H., 2013. Superparamagnetic nanoparticles as versatile carriers and supporting materials for enzymes. J. Mol. Catal. B Enzym. 85–86, 71–92.
Nomura, A., Shin, S., Mehdi, O.O., Kauffmann, J.M., 2004. Preparation, characterization, and application of an enzyme-immobilized magnetic microreactor for flow injection analysis. Anal. Chem. 76, 5498–5502.
Nuryono, N., Rosiati, N., Rusdiarso, B., Sakti, S.C., Tanaka, S., 2014. Coating of magnetite with mercapto modified rice hull ash silica in a one-pot process. Springerplus 3, 515.
Olsen, H.S., 2002. Enzymes in starch modification. In: Whitehurst, R.J., Law, B.A. (Eds.), Enzymes in Food Technology. CRC Press LLC, Boca Raton, FL, pp. 200–228.

Osuna, Y., Gregorio-Jauregui, K.M., Gaona-Lozano, J.G., De La Garza-Rodríguez, I.M., Ilyna, A., Barriga-Castro, E.D., Saade, H., López, R.G., 2012. Chitosan-coated magnetic nanoparticles with low chitosan content prepared in one-step. J. Nanomater. 2012, 1–7.

Osuna, Y., Sandoval, J., Saade, H., López, R.G., Martinez, J.L., Colunga, E.M., Segura, E.P., Arévalo, F.J., Zon, M.A., Fernández, H., Ilyina, A., 2015. Immobilization of *Aspergillus niger* lipase on chitosan-coated magnetic nanoparticles using two covalent-binding methods. Bioprocess Biosyst. Eng. 38, 1437–1445.

Pan, C., Hu, B., Li, W., Sun, Y., Ye, H., Zeng, X., 2009. Novel and efficient method for immobilization and stabilization of β-D-galactosidase by covalent attachment onto magnetic Fe_3O_4–chitosan nanoparticles. J. Mol. Catal. B Enzym. 61, 208–215.

Panesar, P.S., Kumari, S., Panesar, R., 2010. Potential applications of immobilized (galactosidase in food processing industries). Enzyme Res. 2010, 473137.

Pospiskova, K., Safarik, I., 2013. Low-cost, easy-to-prepare magnetic chitosan microparticles for enzymes immobilization. Carbohydr. Polym. 96, 545–548.

Pospiskova, K., Safarik, I., 2015. Low-temperature magnetic modification of sensitive biological materials. Mater. Lett. 142, 184–188.

Pospiskova, K., Prochazkova, G., Safarik, I., 2013. One-step magnetic modification of yeast cells by microwave-synthesized iron oxide microparticles. Lett. Appl. Microbiol. 56, 456–461.

Qian, S., Wang, C., Wang, H., Yu, F., Zhang, C., Yu, H., 2015. Synthesis and characterization of surface-functionalized paramagnetic nanoparticles and their application to immobilization of α-acetolactate decarboxylase. Process Biochem. 50, 1388–1393.

Ravi Kumar, M.N., 2000. A review of chitin and chitosan applications. React. Funct. Polym. 46, 1–27.

Rinaudo, M., 2006. Chitin and chitosan: properties and applications. Prog. Polym. Sci. 31, 603–632.

Rodrigues, D.S., Mendes, A.A., Adriano, W.S., Gonçalves, L.R.B., Giordano, R.L.C., 2008. Multipoint covalent immobilization of microbial lipase on chitosan and agarose activated by different methods. J. Mol. Catal. B Enzym. 51, 100–109.

Rossi, L.M., Quach, A.D., Rosenzweig, Z., 2004. Glucose oxidase-magnetite nanoparticle bioconjugate for glucose sensing. Anal. Bioanal. Chem. 380, 606–613.

Sánchez-Ramírez, J., Martínez-Hernández, J.L., Segura-Ceniceros, E.P., Contreras-Esquivel, J.C., Medina-Morales, M.A., Aguilar, C.N., Iliná, A., 2014. Artículo de revisión: inmovilización de enzimas lignocelulolíticas en nanoparticulas magnéticas. Quimica Nova 37, 504–512.

Sato, A., 1962. Proteolytic enzymes. Naika 9, 328–336.

Seenuvasan, M., Malar, C.G., Preethi, S., Balaji, N., Iyyappan, J., Kumar, M.A., Kumar, K.S., 2013. Fabrication, characterization and application of pectin degrading Fe_3O_4-SiO_2 nanobiocatalyst. Mater. Sci. Eng. C Mater. Biol. Appl. 33, 2273–2279.

Shaw, S.-Y., Chen, Y.-J., Ou, J.-J., Ho, L., 2006. Preparation and characterization of *Pseudomonas putida* esterase immobilized on magnetic nanoparticles. Enzyme Microb. Technol. 39, 1089–1095.

Sheldon, R.A., van Pelt, S., 2013. Enzyme immobilisation in biocatalysis: why, what and how. Chem. Soc. Rev. 42, 6223–6235.

Silva, J.A., Macedo, G.P., Rodrigues, D.S., Giordano, R.L.C., Gonçalves, L.R.B., 2012. Immobilization of *Candida antarctica* lipase B by covalent attachment on chitosan-based hydrogels using different support activation strategies. Biochem. Eng. J. 60, 16–24.

Simpson, B.K., Rui, X., XiuJie, J., 2012. Enzyme-assisted food processing. In: Boye, J.I., Arcand, Y. (Eds.), Green Technologies in Food Production and Processing. Springer Science + Business Media, LLC, New York, NY, pp. 327–361.
Thakur, B.R., Singh, R.K., Handa, A.K., Rao, M.A., 1997. Chemistry and uses of pectin—a review. Crit. Rev. Food Sci. Nutr. 37, 47–73.
Tomotani, E.J., Vitolo, M., 2007. Production of high-fructose syrup using immobilized invertase in a membrane reactor. J. Food Eng. 80, 662–667.
Uzun, K., Çevik, E., Şenel, M., Sözeri, H., Baykal, A., Abasıyanık, M.F., Toprak, M.S., 2010. Covalent immobilization of invertase on PAMAM-dendrimer modified superparamagnetic iron oxide nanoparticles. J. Nanoparticle Res. 12, 3057–3067.
Valdés-Solís, T., Rebolledo, A.F., Sevilla, M., Fuertes, A.B., Valle-Vigón, P., Bomatí-Miguel, O., Tartaj, P., 2009. Preparation, characterization, and enzyme immobilization capacities of superparamagnetic silica/iron oxide nanocomposites with mesostructured porosity. Chem. Mater. 21, 1806–1814.
Villalonga, R., Villalonga, M.L., Díez, P., Pingarrón, J.M., 2011. Decorating carbon nanotubes with polyethylene glycol-coated magnetic nanoparticles for implementing highly sensitive enzyme biosensors. J. Mater. Chem. 21, 12858.
Vroemen, A., 2002. Glucose oxidase. In: Whitaker, J.R., Voragen, A.G.J., Wong, D.W.S. (Eds.), Handbook of Food Enzymology. CRC Press, Boca Raton, FL.
Wahajuddin, Arora, S., 2012. Superparamagnetic iron oxide nanoparticles: magnetic nanoplatforms as drug carriers. Int. J. Nanomed. 7, 3445–3471.
Wang, Y., Caruso, F., 2005. Mesoporous silica spheres as supports for enzyme immobilization and encapsulation. Chem. Mater. 17, 953–961.
Wang, F., Guo, C., Liu, H.Z., Liu, C.Z., 2008. Immobilization of *Pycnoporus sanguineus* laccase by metal affinity adsorption on magnetic chelator particles. J. Chem. Technol. Biotechnol. 83, 97–104.
Wang, F., Guo, C., Yang, L., Liu, C.-Z., 2010. Magnetic mesoporous silica nanoparticles: fabrication and their laccase immobilization performance. Bioresour. Technol. 101, 8931–8935.
Wang, H., Huang, J., Wang, C., Li, D., Ding, L., Han, Y., 2011. Immobilization of glucose oxidase using $CoFe_2O_4/SiO_2$ nanoparticles as carrier. Appl. Surf. Sci. 257, 5739–5745.
Wang, J., Zhao, G., Li, Y., Liu, X., Hou, P., 2013. Reversible immobilization of glucoamylase onto magnetic chitosan nanocarriers. Appl. Microbiol. Biotechnol. 97, 681–692.
Wang, J., Zhao, G., Li, Y., Peng, X., Wang, X., 2015. Preparation of amine-functionalized mesoporous magnetic colloidal nanocrystal clusters for glucoamylase immobilization. Chem. Eng. J. 263, 471–478.
Wei, Y., Han, B., Hu, X., Lin, Y., Wang, X., Deng, X., 2012. Synthesis of Fe_3O_4 nanoparticles and their magnetic properties. Procedia Eng. 27, 632–637.
Whitaker, J.R., 2002. What enzymes do and why they are highly specific and efficient catalysts. In: Whitaker, J.R., Voragen, A.G.J., Wong, D.W.S. (Eds.), Handbook of Food Enzymology. CRC Press, Boca Raton, FL.
Wu, W., He, Q., Jiang, C., 2008. Magnetic iron oxide nanoparticles: synthesis and surface functionalization strategies. Nanoscale Res. Lett. 3, 397–415.
Xu, C., Xu, K., Gu, H., Zheng, R., Liu, H., Zhang, X., Guo, Z., Xu, B., 2004. Dopamine as a robust anchor to immobilize functional molecules on the iron oxide shell of magnetic nanoparticles. J. Am. Chem. Soc. 126, 9938–9939.
Yusdy, Patel, S.R., Yap, M.G.S., Wang, D.I.C., 2009. Immobilization of L-lactate dehydrogenase on magnetic nanoclusters for chiral synthesis of pharmaceutical compounds. Biochem. Eng. J. 48, 13–21.

Zang, L., Qiu, J., Wu, X., Zhang, W., Sakai, E., Wei, Y., 2014. Preparation of magnetic chitosan nanoparticles as support for cellulase immobilization. Ind. Eng. Chem. Res. 53, 3448–3454.

Zhang, W., Qiu, J., Zang, L., Sakai, E., Feng, H., 2015. Preparation of functionalized magnetic silica nanospheres for the cellulase immobilization. Nano 10, 1550013.

Zhu, Y., Kaskel, S., Shi, J., Wage, T., van Pée, K.-H., 2007. Immobilization of *Trametes versicolor* laccase on magnetically separable mesoporous silica spheres. Chem. Mater. 19, 6408–6413.

Zucca, P., Sanjust, E., 2014. Inorganic materials as supports for covalent enzyme immobilization: methods and mechanisms. Molecules 19, 14139–14194.

4

NATURAL MATERIALS AS ADDITIVES IN FOOD EMULSIONS

Virangkumar Nanubhai Lad, Zagabathuni Venkata Panchakshari Murthy

Sardar Vallabhbhai National Institute of Technology, Department of Chemical Engineering, Surat, Gujarat, India

1 Introduction

Emulsions contain at least two immiscible liquid phases and surface active agent(s) depending on the requirements. The stability of emulsions can be tuned by using appropriate surface active agents(s) and/or stabilizer(s), which are responsible for altering the interfacial tension between continuous phase and dispersed phase. Oil-in-water (O/W) emulsions are of very much importance in the food industry. They are also used for encapsulation of lipophilic functional components like antimicrobials, nutraceuticals, and vitamins in the food industry (McClements, 2005). Many food products like ice creams, pastries, yogurt, chocolates, and other confectionary items find their market acceptance based on the sensory perceptions during their end usage, which depends on the deformation of these materials under the condition of the use. Many of these products are based on the O/W or water-in-oil (W/O) emulsion formulations. Rheology is a useful tool to predict the flow behavior and the sensory effects observed during the use of such emulsions (Barnes, 1994; Tadros, 2004). Thanks to the recent awareness of the people for the use of natural materials in various products that lead the industry to explore the compatibility of suitable materials from natural origin that can possibly replace the synthetic chemicals used in such products in order to produce nonallergic, herbal, and more nutrient products. The major challenges in the production of these materials are the proper selection of suitable emulsifiers, stabilizers, preservatives, texturizers, thickeners, and so on, on the one hand, and controlling the microstructure, polydispersity, and rheology, on the other hand (Savary et al., 2013).

Novel Approaches of Nanotechnology in Food. http://dx.doi.org/10.1016/B978-0-12-804308-0.00004-2

Maintaining the stability, polydispersity, and rheology of O/W emulsions and emulsified soft materials are major challenging research problems for many decades, and researchers have proposed various emulsifiers and stabilizers in order to increase stability. Sometimes, the ingredients used in the preparation of colloidal dispersions for certain health care products also offer side effects and their use must be restricted from dermatological point of view. Interfacial interactions among the constituents of the emulsions may lead to coalescence of the droplets of the dispersed phase, flocculation, creaming, and Ostwald ripening, which ultimately results in destabilization of the emulsions. Besides the use of chemicals as emulsifiers, colloidal particles are also being used to enhance the stability to emulsions. It is really important to study the effects of interfacial interactions in emulsion systems in order to design the emulsions or soft food materials.

2 Importance of Use of Natural Materials for Food Emulsions

Often the use of many additives required for this purpose has to pass through the compatibility tests, clinical tests, environmental impact assessment, and so on. The usage of many synthetic chemicals in processing of various health-care products, including pharmaceuticals, food, and cosmetics is restricted due to dermatological view point, hygienic nature of final products, potential allergic symptoms, and so on, and must comply with the standards specified by the competent authorities, for example, Food and Drug Administration, irrespective of their emulsification or stabilizing power. Often the preparation of food emulsions with tailored rheology is one of the major engineering problems in food industries. Besides the synthesis of many emulsifiers, stabilizers, and rheology modifier-chemicals in past few decades, many materials from the natural origin have also been explored for their composition and useful characteristic properties of their ingredients by the researchers from various fields. Hence the study of evaluating the suitability of some of these naturally available materials is found appealing to provide the significant contribution to the scientific and technical community by identifying specific implications of some of these easily available, less costly, and biodegradable natural materials, which can also add their nutrient value to the final products. It is found that some naturally available materials are the potential candidates to replace the synthetic surface active agents, stabilizers, rheology modifiers, and so on, while offering emulsions with reasonable stability and consistency. Coconut milk, coconut milk protein, juice of coriander, aloe vera gel, turmeric, and split Bengal gram have been proved to be effective additives for preparation of O/W emulsions (Lad and Murthy, 2012, 2013a,b, 2014), which can be used for food products.

3 Basics of Emulsions

Emulsions are liquid–liquid dispersions consisting two or more liquid immiscible phases (Pal, 2011). Emulsions often contain one or more surface active agent(s) that serve the purpose of emulsification and/or stabilization. Emulsifiers are surface active agents facilitating easy formation of emulsions by lowering the interfacial tension between the liquid phases. Stabilizers prevent the coalescences among the droplets of dispersed phase thereby prevent the phase separation, and stabilize the emulsions. Depending on the character of the dispersed phase and continuous phase, the emulsions are classified as O/W or W/O emulsions. Droplets of organic oil phase are dispersed in the aqueous phase in O/W emulsions whereas aqueous droplets are dispersed in continuous organic phase in W/O emulsions. Emulsions are also categorized as simple emulsions and multiple emulsions. The W/O and O/W aforementioned defined emulsions are simple emulsions. The stability of O/W emulsions is governed by steric stabilization and electrostatic repulsion while in W/O emulsions-only steric forces are expected to play the vital role for stabilization due to low electrical conductivity of the continuous oil phase (Ushikubo and Cunha, 2014).

The emulsions of emulsions are known as double emulsions or multiple emulsions where the globule of such emulsion itself contains one or more smaller droplets inside. They are classified as oil-in-water-in-oil (O/W/O) and water-in-oil-in-water (W/O/W) emulsions. The former is prepared by dispersing O/W emulsion into another continuous oil phase while W/O emulsion is dispersed in another aqueous phase in order to prepare W/O/W emulsions. Comparatively low shear is used in the second emulsification step to prevent expulsion of internal droplets to the continuous phase in case of multiple emulsion preparation (Pal, 2011). Fig. 4.1 explains the schematic classification of emulsions.

3.1 Interactions Between Droplet Phase and Continuous Phase

Interfacial interactions in emulsions affect the structure, stability, and rheology that depend on the strength and nature of the interactions between dispersed phase and continuous phase. Due

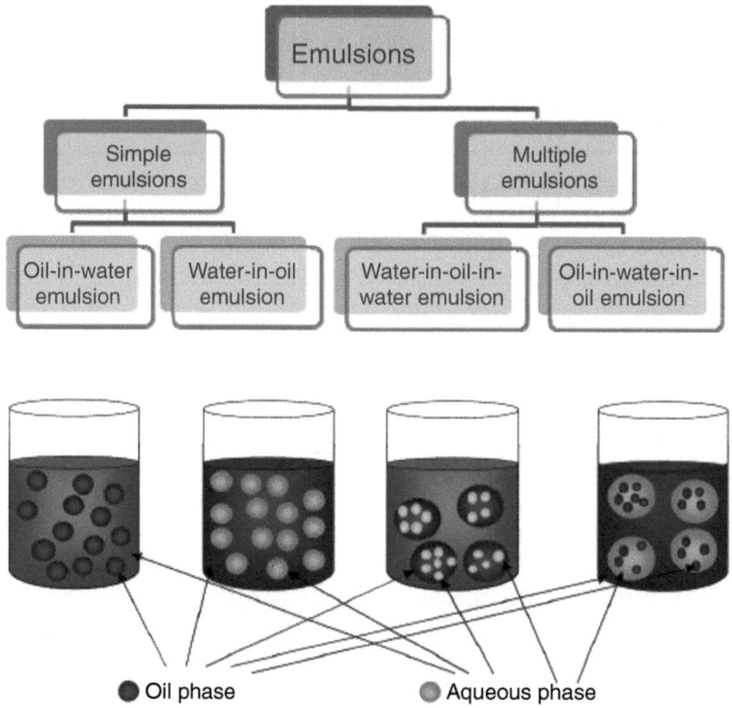

Figure 4.1. Classification of emulsions.

to their widespread occurrence and broad range of chemical engineering applications in the field of food industry, pharmaceutical industry, petroleum industry, agricultural formulations, cosmetics and personal care products, emulsions have long been of great research interest to chemical engineers. Many applications of emulsions require reasonable stability and concentration of dispersed phase. Coalescence, creaming, flocculation, and Ostwald ripening are the major phenomena observed in emulsions leading to destabilization of emulsions. Often, the concentrated emulsion is dispersed in another liquid and the system sometimes needs further homogenization.

Coalescence and creaming are the main cause of instability of emulsions, which ultimately result in separation of continuous and dispersed phases. State of aggregation of dispersed particles, rheology, and size distribution of dispersed droplets in the emulsions are used to understand the tendency of coalescence of droplets and predicting the stability of emulsions. Increase in the coalescence rate decreases the stability of emulsions. Coalescence rate increases with high ionic concentration, which decrease the thickness of the double layers around emulsion droplets. At high electrolyte concentrations, the particles tend to coagulate and form large aggregates so that the particles can no longer reside at the interface that leads to destabilization of the emulsions (Tambe and Sharma, 1994). Bridging flocculation, that is, adsorption of macromolecules to more than one droplet of the dispersed phase in the emulsion cause the aggregation of droplets, is one of the phenomena leading to destabilization of emulsions. Nonadsorbing polymers also destabilize emulsions by depletion flocculation (Dickinson, 1992).

3.2 Size Distribution of Dispersed Droplets in Emulsions

Properties of emulsions like viscosity, long-term stability, resistance to creaming, degradation rates, texture, optical appearance, physiological efficiency, and chemical reactivity depends on the droplet size distribution in the continuous phase (Mason et al., 1996; Fernandez et al., 2004; Jurado et al., 2007). Median size of droplets, Sauter diameter, and the diameter of the largest stable droplet give approximate idea for characterization of the size distribution of the emulsions but actually they do not give exact information of system. For O/W emulsions, Jurado et al. (2007) derived an equation for volume fraction of oil phase (ε_{oil}) as presented by Eq. (4.1):

$$\varepsilon_{\text{oil}} = n_{\text{B}} \frac{\pi}{6} d_{3,2}^{3} \frac{\Gamma[(n+4)/m]}{\Gamma[(n+1)/m]} k^{-3/m} \tag{4.1}$$

Interfacial area per unit volume of O/W emulsion (a_{oil}) can be found from following equation:

$$a_{\text{oil}} = \frac{6\varepsilon_{\text{oil}}}{d_{3,2}} \tag{4.2}$$

where, $d_{3,2}$ is Sauter diameter given by Eq. (4.3):

$$d_{3,2} = \frac{\sum n_i d_i^3}{\sum n_i d_i^2} \tag{4.3}$$

here, n_i is the number of droplets between two consecutive diameters, d_i is the corresponding droplet diameter. The normalized diameter is δ' is presented by the following Eq. (4.4):

$$\delta' = \frac{d_i}{d_{3,2}} \tag{4.4}$$

Number of droplets per unit volume is given by Eq. (4.5):

$$n_B = \sum n_i = K \int_0^\infty \delta'^n \exp(-k\delta'^m) \ d\delta' \tag{4.5}$$

Here, *K*, *k*, *m*, and *n* are empirical parameters. The variable *k* and *m* are defined as under:

$$y = k\delta'^m \tag{4.6}$$

$$\delta = \left(\frac{y}{k}\right)^{1/m} \tag{4.7}$$

$$K = n_B \frac{mk^{(n+1)/m}}{\Gamma[(n+1)/m]} \tag{4.8}$$

where, Γ is the gamma function defined by

$$\Gamma(x) = \int_0^\infty y^{x-1} \exp(-y) dy \tag{4.9}$$

Type of emulsifying agents, and the energy input during the emulsification decides the droplet size distribution of the dispersed phase in continuous phase in the emulsions. Under the action of acoustic radiation stresses, the emulsified droplets deform from their spherical shape when one-dimensional acoustic field is applied to the emulsion. The motion of emulsified droplets in response to applied force is sensitive to their shape which increases as the droplets are in close proximity to each other. Pangu and Feke (2007) theoretically predicted the relative trajectory of a drop pair in an acoustic field due to various body and interdroplet forces. They applied one-dimensional standing ultrasonic wave field to an isothermal emulsion of spherical droplets in immiscible fluid in absence of surfactants and found that the transport and coalescence of micron scale emulsions was induced by the application of the acoustic field.

3.3 Droplets Coalescence

The strength of the thin liquid film of continuous phase separating the dispersed droplets in the emulsions govern the coalescence of droplets in emulsions. Rupture of this thin film of continuous phase in the emulsion is enhanced by decrease of interfacial tension, increase of the film diameter, and the reduction in the strength of interdroplet repulsions. The O/W emulsions stabilized by proteins, impart not too low interfacial tension but due to their sufficient surface-activity, small droplets are produced with thick adsorbed layer resulting in strong repulsion between droplets by steric stabilization. Protein films are highly viscoelastic and tend to dampen down surface fluctuations that inhibit the film rupture (Dickinson, 1992). Coalescence occurs primarily after flocculation resulting in the build-up of three-dimensional aggregated structures. In stable emulsions, the rate of coalescence is such that if two isolated drops approach and pass one another within a short time, it is most unlikely that they will coalesce. Coalescence within the three-dimensional aggregated structures reduces the number of drops per unit volume and increases the drop diameter (Sherman, 1973). Gibbs–Marangoni effect enhances the stability of liquid lamellae between flocculated droplets. The net force F for two droplets approaching each other was given by Sherman (1973) as presented by the following expression:

$$F = F_H + F_{DL} - F_{VW} + F_D \pm F_G \tag{4.10}$$

Here, F_H is the hydrodynamic force causing the flow of continuous phase fluid from between the droplets and distortion of their surfaces. F_{DL}, F_{VW}, F_D, and F_G are electrostatic double layer repulsion force, van der Waals attraction force, the net deceleration produced by the continuous phase and the gravitational force and respectively. Probability of droplets coalescence is the highest at the points where the net force acting on them attains its maximum value. In the absence of interfacial viscoelasticity, the adsorbed emulsifier film ruptures more easily due to compressive force which results in a higher rate of droplet coalescence.

3.4 Competition of Droplets Breakup and Coalescence

Controlling the interfacial interactions, droplet size distribution and stability of emulsions are the most focused area of research globally among the researchers engaged in the field of Interfacial Science and Engineering. Deformation and breakup of liquid droplets have been studied since more than half a century (Karam and Bellinger, 1968). The competing phenomena of coalescence and breakup of droplets of the dispersed phase governs the stability of emulsions (Kumar et al., 1998; Hagesaether et al., 2002; Desnoyer et al., 2003). Levich (1962) proposed that under turbulent deformation, the droplet breakup occur if size of droplets is more or equal to the length scale of turbulence. However, this phenomenon depends on the kinematic viscosity and local rate of energy dissipation per unit mass of fluid. Later, Kumar et al. (1991, 1992) have experimentally shown that during turbulent breakup, as the holdup of dispersed phase increases, the value of maximum stable droplet size increases exhibiting a peak and again decreases. They proposed the equation to find maximum diameter (d_{max}) of stable droplet in a stirred agitated vessel as under:

$$d_{max} = \frac{Ca\gamma'}{113.2N^{1.2}\mu^{0.8}}\left(\frac{1}{L^2\rho}\right)^{0.2} \tag{4.11}$$

The capillary number *Ca* presented in Eq. (4.11), is given by:

$$Ca = \frac{\dot{\varepsilon} d_{max}\eta}{2\gamma'} \tag{4.12}$$

where, $\dot{\varepsilon}$ is the total strain rate, γ' is the interfacial tension, η is the apparent viscosity, ρ is the density, N is the speed of stirrer in revolutions per second and L is the length of blade. Kumar et al. (1993) have also proposed a mechanism to predict the maximum size of such stable droplet.

4 Coalescence of Droplets During Emulsification

The size of droplets of dispersed phase decreases with the emulsification time but during the emulsification, small droplets also tend to coalesce. Stable droplet size or steady state distribution of droplet size is obtained when these two processes attain the state of equilibrium. The breakup of droplets and coalescence depend on size of droplets and also on the physical properties of the dispersed and continuous phases. Mainly droplets breakage is done by turbulent deformation and viscous shear (Luo and Svendsen, 1996; Kumar et al., 1998). The electrostatic repulsion between charged surfaces in an electrolyte decreases the rate of coalescence. The steric repulsion due to physical barriers at the droplet surface also plays an important role to suppress the rate of droplet coalescence. A study on bitumen emulsion (Yeung et al., 2003) suggests that when two oil droplets come together on direct head-on approach, oil droplets do not prone to coalesce. On the other hand oblique encounter of the droplets causes shearing motion resulting in coalescence.

The study on coalescence frequency of droplets of lighter dispersed phase in emulsions under shear flow reveals that the prediction of the average coalescence probability requires full-size distribution of droplets, which can be interpreted as partially mobile deformable interfaces or fully immobile deformable interfaces (Korobko et al., 2005). Renardy (2007) has done experiments for droplet formation to study the effects of confinement and inertia, and found that the inertia can enhance elongation of the droplet leading to breakup. Norton et al. (2009) have conducted the study to explore the effects of various emulsifiers on droplet size of emulsion and found that with their myritol emulsion containing Tween 20 or Tween 80, the back reaction of nanoemulsion coalescence governed the final stable size of the droplets, which was independent of the initial energy input to the emulsification.

Empirical equations to estimate the minimum size of stable droplets, maximum size of stable droplets, coalescence frequencies, and size distributions in turbulent dispersions have been proposed by Ceylan et al. (2003), and Kelbaliyev and Ceylan (2005). Tcholakova et al. (2007) studied the droplet-size distribution of daughter droplets generated during emulsification in turbulent flow, and compared their results critically with that of other researchers' experimental results and with the theoretical model. It has been found that the daughter droplet size distribution depends remarkably on the viscosity of the dispersed phase. Much smaller droplets form in case of high viscous oils whereas in the case of oils of low viscosity, droplets with volume twice smaller than that of the breaking drop produce with high probability (Tcholakova et al., 2007).

4.1 Antifoaming Effects in Emulsions

Foaming is one of the major engineering problems in many practical applications. Typical examples of technological importance where foaming creates problems include fermentation in food and drug manufacturing, pulp and paper manufacturing, oil recovery, bottle filling of beverages, pharmaceuticals, and so on. Antifoams or defoamers are used to suppress the formation of foam or to destroy the already formed foam. Various nonpolar and polar oils like mineral and silicone oils, fatty acids, alcohols and esters, thioethers, tributylphosphates, nonionic surfactants above their cloud points, and so on are used as antifoam components (Garrett, 1993; Denkov, 1999, 2004).

4.2 Using Sonication for Emulsification

Li and Fogler (1978a) have explained the concept of acoustic emulsification illustrating that the ultrasonic waves cause an interfacial instability at oil–water interfaces. Mechanical vibration, acoustic streaming, and shock waves are generated when ultrasound is passed through liquid. Microbubbles generated in the liquid grow and ultimately collapse in microsecond time interval.

The transient cavitation bubbles produced during sonication collapse with the generation of high pressure shock waves. This phenomenon is almost adiabatic and generates very high local temperatures (3,000 K to more than 10,000 K) and pressures (more than 100 MPa) within bubbles for a short period of time (Mason and Lorimer, 1988; Young, 1994; Sonawane et al., 2010; Ashokkumar, 2011). Leighton (1994) presented following equation to calculate maximum theoretical temperature T_{max} within cavitation bubble assuming adiabatic compression:

$$T_{max} = T_0 \left\{ \frac{P_m[(c_p / c_v) - 1]}{P_v} \right\} \tag{4.13}$$

where, T_0 is the ambient solution temperature, P_m is the sum of hydrostatic and acoustic pressure in the liquid, P_v is the pressure in the bubble at its maximum size, which is generally taken as equal to the vapor pressure of liquid. c_p and c_v are specific heat at constant pressure and constant volume respectively for the gas/vapor mixture.

Cucheval and Chow (2008) have demonstrated the emulsion formation by power ultrasound incorporated by high speed cavitation for soybean oil-in-water system using Tween 80 as a surfactant. Lad and Murthy (2012) have also demonstrated the comparison of effects of sonication by various modes on the stability of O/W emulsions. Kentish et al. (2008) have successfully prepared flaxseed O/W emulsions using Tween-40 surfactant with a mean droplet size as low as 135 ± 5 nm using ultrasonic power at 20–24 kHz. Though the ultrasound is used for emulsification using a variety of surfactants since many decades (Li and Fogler, 1978b; Behrend and Schubert, 2000), the exploration of the intermolecular interaction of coconut milk protein with the continuous and droplet phases subjected to sonication is indeed challenging. Having high medicinal and nutrient value of coconut milk, it is advantageous to explore the possibility to replace conventional emulsifiers and stabilizers for emulsions used in cosmetics, pharmaceutical and food emulsions with reasonable stability.

4.3 Interfacial Interactions in Pickering Emulsions

The emulsions stabilized by insoluble solid particles having colloidal size range are known as Pickering emulsions. Presence of colloidal particles provides steric hindrance to the coalescence between droplets in the emulsion, and thus modifies the rheological properties of interfacial region that offer stability to the emulsion. Formation of the dense layer of particles at the oil–water interface governs the effectiveness of colloidal particles in stabilizing emulsions.

Concentration of adsorbed particles at the interface controls the properties of the interfacial region. Higher concentration of colloidal particles causes viscoelastic behavior of colloid-laden interface, which enhances the stability by increasing the magnitude of steric hindrance, and by retarding the liquid drainage between droplets in the emulsion (Tambe and Sharma, 1994). Accumulation of the particles at the oil–water interface in the form of a densed-packed layer results in a thick particle layer. This dense layer prevents the droplet flocculation and coalescence by steric mechanism, hence ultimately enhances the stability of emulsions.

Fine solid particles like silica, carbon black, and so on can be used to stabilize emulsions if they are capable of accumulating at the oil–water interface by forming a suitable film of enough strength at the interface to prevent droplets from coalescing (Pickering, 1907; Briggs, 1921; Yan et al., 2001; Frelichowska et al., 2009). Briggs (1921) found that even though O/W emulsions can be stabilized by silica, and W/O emulsions by carbon black, the mixture of silica and carbon black cannot stabilize any emulsion. The particulate material present in food emulsions accumulates at oil–water interfaces, and stabilizes the emulsions or foams. The particle size varies from a few nanometer to tens of micrometers depending on the systems (Dickinson, 2009). Orientation of the solid particles at the interface is important for predicting the interaction forces. The magnitude of the steric barrier depends on intricacy in removing the particles from the interface. If the particle layer lies strongly on the convex side of the dispersed droplets, which is the characteristic of the particles wetted by continuous phase, the steric barrier becomes more effective to impart stability to the emulsions (Dickinson, 2009). Particle bilayer arrangement occurs in various ways due to the approaching of particle-coated droplets. It is the stability of the thin film of continuous phase between the particle-coated droplets, which determines the effectiveness of a physical barrier, and prevents the coalescence (Lopetinsky et al., 2006). Tadros and Vincent (1983) correlated Gibbs free energy change for moving the spherical solid particle from interface to the oil or aqueous bulk phase at equilibrium, with the stability of solid stabilized emulsions. As pointed out by Tambe and Sharma (1994) in their very informative article, the minimum free energy required to move a spherical particle of radius r from the interface to the oil phase is given by:

$$\Delta G_{\mathrm{w}} = \pi r^2 \gamma'_{\mathrm{ow}} (1 - \cos\theta)^2 \qquad (4.14)$$

Similarly, the minimum free energy ΔG_o required to move a spherical particle of radius r from the interface to the oil phase is given by:

$$\Delta G_o = \pi r^2 \gamma'_{ow} (1+\cos\theta)^2 \tag{4.15}$$

where θ is the contact angle and γ'_{ow} is the oil–water interfacial tension. Additional barrier to coalescence is provided by the particles present at the interface, and having contact angle 90 degrees, which cause the coalescence of droplets more difficult. Schulman and Leja (1954) have investigated that the solid particles can be held at the interface if the contact angle at the oil–water–solid interface, measured across the aqueous phase is slightly less than 90 degrees, leading to stable O/W emulsions. Similarly, W/O emulsions can be stabilized by the particles having contact angles slightly greater than 90 degrees. Hydrophobic particles disperse in the oil phase and form an unstable and poor oil continuous dispersion while hydrophilic particles disperse in aqueous phase making unstable and poor water continuous emulsion (Tambe and Sharma, 1994). Adjustment of surfactant to particle ratio determines the liquid-like or solid-like behavior of the oil–water interfacial layer (Dickinson, 2009). Gautier et al. (2007) have found that the pH and ionic strength influence the condition whether the droplet surface is coated by a dense layer of particles or is only sparse covered.

Stepwise thickness transition and the contact angles of particle-containing liquid films can be predicted from the size and interactions of colloidal particles. Basheva et al. (2007) have determined the particle properties form the characteristics of the liquid films. (Nikolov et al. (1988, 1989); Nikolov and Wasan, 1989) have found that the layer-by-layer thinning of an ordered structure of spherical micelles within the liquid film causes the stepwise transition in the film thickness. This was also successfully shown by Denkov et al. (1996a,b) by freezing the foam films at various stages of stratification. Chen et al. (2011) have used oxygen plasma treated carbon nanotubes and prepared stable O/W pickering emulsions. Frost et al. (2012) have studied the effects of solid particles on emulsions containing ionic liquid as the continuous phase and found the presence of droplet bridging over a wide range of particle size in the emulsion samples.

Effectiveness of colloidal particles in stabilizing emulsions is more affected by the particle size and their ability to reside at the interface. It has been found that the stability of emulsion increased with decreasing particle size until a critical particle size below which the Brownian effects are significant enough to affect the partitioning of particles at interface. Below this critical size of particles, the stability of emulsion decreases with decrease in particle size. Further, the interaction between the particles is also a significant factor determining stability of emulsion as it depends on the strength of interfacial film on the droplets in emulsions (Tambe and Sharma, 1994). Dyab et al. (2009) have prepared anisotropic magnetic microparticles with diameters ranging from tens to hundreds of micrometers. They prepared a magnetic oil-in-water emulsion using a polymerizable oil and a polymerizable anionic surfactant. Such kind of anisotropic colloids can be used for control of suspension rheology and optical properties, stabilization of emulsions, engineering of biomaterials and complex colloidal composites (Dyab et al., 2009).

5 Rheology of Food Emulsions

Fluids like liquids, gases, and vapors deform continuously under the action of shear stress, while solids respond elastically to the deformation. Newtonian fluids dissipate energy through viscous flow. Solids store energy and provide a spring-like elastic response. There is a group of materials having complex viscoelastic nature, which exhibit solid-like and fluid-like responses depending on the time scale at which the sample is probed (Larson, 1999). Polymers, surfactant solutions, colloidal dispersions—suspensions, emulsions, foams, gels, and so on have mechanical properties between those of elastic solids and viscous liquids, and constitute another category of materials known as complex fluids. It's a challenge to fundamental and applied research to understand their behavior under deformation (Laurati et al., 2009). Many complex fluids like pastes and polymer solutions exhibit solid-like and liquid-like behaviors under the action of varying shear stress, and thus characterized by their viscoelastic properties (Coussot, 2005). Understanding the rheology of such complex fluids is very crucial to design the handling devices and processing equipments, and to predict the flow behavior and sensory effects during their usage. Many applications food emulsions, nanoparticle synthesis, and other colloidal dispersions are associated with an important physical phenomenon of sol–gel transition. Due to the gelling ability of some polysaccharides and proteins, used in food products, many rheological studies have been done on food hydrocolloids in past couple of decades, which helps to confer textural and mechanical properties to the final products (Gallegos and Franco, 1999).

Rheology of emulsions is a direct manifestation of the interaction forces present in the emulsion system. Measurements of shear stress as a function of shear rate, measurement of strain as a function of time at a constant applied stress, and oscillatory rheological measurements (Larson, 1999) are popular techniques. Viscosity and elasticity of bulk phases, and interfacial rheology of interfacial thin film provide the information of coalescence of droplets in emulsions. Rheological properties of emulsions strongly depend on the volume fraction of dispersed phase ϕ in the emulsions. The relative viscosity η_r for very dilute emulsions having volume fraction of dispersed phase $\phi < 0.001$, is given by the Einstein equation (Tadros, 1994) presented by Eq. (4.16):

$$\eta_r = 1 + 2.5\phi \quad (4.16)$$

The relative viscosity becomes a more complex function of the volume fraction of the dispersed phase in emulsion as its value increases. In such cases, the relative viscosity is given by:

$$\eta_r = 1 + k_1\phi + k_2\phi^2 + k_3\phi^3 + \ldots \quad (4.17)$$

In Eq. (4.17), the constant k_1 is the Einstein's constant, which is taken as equal to 2.5 for hard spherical particles. k_2 accounts for hydrodynamic interaction between droplets due to overlap of the associated flow patterns, and its value is taken as 6.2. The hydrodynamic interaction term describes the viscosity of the dispersions up to volume fraction equal to 0.2 above which higher order interaction terms are necessary. The rheology of emulsions is also affected by the viscosity of the dispersed droplets. Viscosity of dispersed phase is of significant importance when the viscosity of the droplets is comparable with, or lower than that of the dispersion medium. The Eq. (4.17) was modified for droplets in a liquid medium by Taylor (1932) assuming that the emulsifier film around the droplets does not prevent the transmission of tangential and normal stresses from the continuous phase to the disperse phase. In this condition, there is an absence of any slippage at the oil–water interface. These stresses reduce the flow patterns around the droplets by producing fluid circulation within the droplets. The equation, originally modified by Taylor, has been well discussed by Tadros (1994) in his review paper, which is presented by Eq. (4.18):

$$\eta_r = 1 + 2.5\left(\frac{\eta_d - 0.4\eta_c}{\eta_d + \eta_c}\right)\phi \quad (4.18)$$

where, η_d and η_c are viscosities of dispersed phase and continuous phase respectively. The viscosity of dispersed phase is much greater than that of the continuous phase in many O/W emulsions, hence the Eq. (4.18) reduces to the Einstein equation presented earlier. The viscosity of dispersed phase is much smaller than that of continuous phase in foams, and the coefficient in the Einstein equation becomes 1. As depicted by Tadros (1994), these equations are valid only for deformable droplets and cannot be used directly for surfactant or polymer film present at the interfaces where interfacial tension gradient or surface viscosity governs the behavior of emulsions as that of solid–liquid dispersions.

5.1 Behavior of Gels for Food Applications

The gel is a colloidal material containing a macromolecular network or fibers composed of variety of materials imparting structural coherence of a solid even though major portion of its composition is liquid (Saunders et al., 2009). The sol, which is the suspension of colloidal particles or molecules in the liquid, is mixed with another liquid or the gelation agent to form a continuous three-dimensional network, is known as a gel. Polymer gels are crosslinked polymer networks having ability to change their volume by absorption of solvent and responsive to the external stimuli. Addition of physical or chemical crosslinker or supramolecular association leads to entanglements in polymeric matrix, which results in gel. Generally macrogels are prepared at room temperature while microgels are synthesized at 70°C. The nature of the polymeric constituents and the final structure requirements are the main factors that govern the method of preparation of gels (Suarez et al., 2009). The gels exhibit the property of mechanical rigidity because they consist of small proportions of solids dispersed in relatively large proportions of liquid. The textural properties of gels can be obtained by rheological measurements and microstructural analysis. Sensory perception of gels used for food products and cosmetics applications are offered by the deformation and fracture characteristics of the gels. One of the important parameters of a solid-like gel is the storage modulus G', also known as elastic modulus having a plateau extending to times at least of the order of seconds. Another parameter of interest is the viscous modulus or loss modulus G'', which is much smaller than the storage modulus in the plateau region. Gels formed by denatured proteins are composed of a three-dimensional network caused by the molecular interaction forces of attraction and repulsion. These interaction forces are regulated by pH and ionic strength, which control the interaction with the polymeric solute molecules.

The water-holding and textural properties of the gel have direct applications to the food quality when they are used as food gels. Small deformation of the gel network is characterized by rheological properties. Nonhomogenities and defects in the network structure of the gels are characterized by the fracture properties. Protein concentration, solvent quality, heating time, and addition of other macromolecules or filler particles are the responsible parameters for the change in such characteristics of the protein gels (Dickinson, 2012). Chaterji et al. (2007) have classified smart polymeric gels based on the structural properties into superporous hydrogels, shape-memory gels, and artificial protein hydrogels. They also categorized the smart polymeric gels on the basis of stimuli response based on their sensitivity toward temperature, pH, glucose, glutathione, antigens, and so on. Different drying conditions result in the formation of the dried gel with unique final microstructure depending on the drying conditions. Carter and Norton (2007) discussed various characteristics of different types of dried gels. Aerogels are prepared in an autoclave by removing the liquid through hypercritical evacuation. They have the network consisting of very high porosity. The void fraction in the aerogels may be as high as 95%. Cryogels are freeze-dried gels and thus possess very finely divided powdered form. Sonogels are prepared by ultrasound treatment in the range of 20 kHz followed by autoclave treatment. The ultrasound incorporates the formation of multicomponent gels. Microgel is a polymer colloid that swells in the solvent. Electrostatic repulsion in the ionic microgel containing at least one comonomer leads to interparticle interactions in microgel dispersions depends on the volume fraction of polymer within the particle ϕ' and the swelling ratio q. Swelling ratio is defined as the cube of the ratio of diameter of swollen particle d' to that of collapsed particle d'_c (Saunders et al., 2009):

$$q = \frac{1}{\phi'} = \left(\frac{d'}{d'_c}\right)^3 \tag{4.19}$$

The swelling ratio can be changed by changing pH and temperature of the system. Swelling can be explained by the change in Gibbs free energy dG and osmotic pressure. The change of Gibbs free energy results from the mixing of an initially unstrained network immersed in a solvent, which is written mathematically presented by Fernandez-Barbero et al. (2009), which is based on the chemical potential difference between solvent inside the gel and that of the pure solvent:

$$dG = \Delta V dP - \Delta S dT + \Delta\mu_s dN_s \tag{4.20}$$

where ΔV and ΔS are the changes of volume and entropy after swelling respectively, P is the pressure inside the gel, T is the absolute temperature, N_s is the number of solvent molecules in the network and $\Delta\mu_s$ is the difference of the chemical potential of solvent inside the gel and that of pure solvent. Now, osmotic pressure is given by following Eq. (4.21):

$$\Pi = -\left(\frac{\partial G}{\partial V}\right)_T \quad or \quad \Pi = \frac{\phi}{V}\left(\frac{\partial G}{\partial \phi}\right)_T \tag{4.21}$$

For a positive value of osmotic pressure which means the higher pressure inside the gel than the outer pressure, the gel swells due to swelling of network structure. If the pressure inside the gel is less than that of outer pressure, the gel shows deswelling. The osmotic pressure becomes zero at the equilibrium condition. The solvent does not transfer across the gel–solvent interface at zero osmotic pressure. Addition of polymer in the microgel dispersion, results in the changing of equilibrium conditions due to inclusion of an external osmotic pressure. Microgel swelling also depends on the molecular weight of the added polymer (Fernandez-Barbero et al., 2009). The micromechanic response of the gels has been strongly correlated with the cluster-like and string-like nature of the gel structure (Lee and Furst, 2008). The elasticity of the gel depends on the interparticle bonds and the connectivity of the gel network (Laurati et al., 2009). Lefebvre et al. (1998) have found that the critical protein concentration in heat-treated bovine serum albumin systems affects the steady-state viscosity and steady state compliance. The rheology of protein gels has also found to be affected by the type of salt, ionic strength, and pH (Neiser et al., 1998; Verheul and Roefs, 1998; Puppo and Anon, 1998, 1999). Cheddadi et al. (2011) have analyzed the foam flow to correlate with their model and concluded that the yield strain is the major dimensionless parameter for characterizing the materials. Coussot et al. (2006) have studied the aging and solid–liquid behavior in pastes.

It is necessary to characterize the rheological response of such complex fluids by measuring the shear modulus as a function of frequency. Rheological measurements of gels provide valuable insight of the structural arrangement of the molecules or the colloidal particles in the gels. The time dependent stress σ is linearly proportional to the strain γ (Larson, 1999):

$$\gamma(t) = \gamma_0 \sin(\omega t) \tag{4.22}$$

$$\sigma(t) = \gamma_0[G'(\omega)\sin(\omega t) + G''(\omega)\cos(\omega t)] \tag{4.23}$$

Here, G' is the storage modulus also known as elastic modulus as it is the response in phase with the applied stress. Basically it is a measure of the storage of elastic energy by the material. G'' is the response out of phase with applied stress, in phase with the strain rate. It is known as the viscous modulus or loss modulus. Loss modulus is a measure of viscous dissipation of energy. ω is the frequency of oscillation. Storage modulus is the measure of deformation energy stored in the sample during shear, representing the elastic behavior of the material. On the other hand the loss modulus takes account of the deformation energy used up in the sample during the shear and lost to the material afterward representing the viscous behavior of the material. If the storage modulus is much greater than loss modulus, the material behaves more like a solid with the elastic deformation. The gels maintain their resistance to deform at lower oscillatory stress below the critical value. Beyond the critical value of oscillatory stress, gels undergo rapid transition from high complex modulus to the low modulus. The magnitude of the complex modulus can be found using Eq. (4.24):

$$|G^*| = \sqrt{(G')^2 + (G'')^2} \tag{4.24}$$

Complex viscosity is the ratio of complex modulus to the frequency of oscillation.

$$\eta^* = \frac{G^*}{\omega} \tag{4.25}$$

Another measure of the stress required to induce flow in the material filled in the container is the yield stress below which viscoelastic materials like creams, lotions and ointments do not flow significantly. If loss modulus is much greater than storage modulus, it indicates that the energy used to deform the material dissipates viscously, and material exhibits fluid-like behavior. The ratio of loss modulus to storage modulus is known as loss factor represented by tan δ. Zero value of loss tangent signify the elastic response whereas infinite value of loss tangent signify viscous response. For a phase angle having an intermediate value between 0 and 90 degrees, the material is said to be viscoelastic. The phase angle δ corresponds to the elastic structure within the material. High value of phase angle indicates small extent of elastic behavior. Rheological characteristics of gels are found by frequency sweep tests, temperature sweep tests, and time sweep tests as discussed by Tabilo-Munizaga and Barbosa-Canovas (2005). In the frequency sweep test the elastic modulus and viscous modulus are measured as a function of frequency at constant temperature.

The moduli are determined as a function of temperature and time at fixed frequency of oscillations during temperature sweep test and time sweep test respectively. The stress σ, which is the function of time t in the viscoelastic material, can be expressed by modified Maxwell model incorporating initial modulus of the material σ_0, equilibrium modulus σ_e and the relaxation time τ (Tabilo-Munizaga and Barbosa-Canovas, 2005):

$$\sigma(t) = \sigma_e + (\sigma_0 - \sigma_e)\exp\left(\frac{-t}{\tau}\right) \tag{4.26}$$

Microstructured gels with tunable rheological properties can be prepared by regulating the gelation processes. Microstructure of the gels can be modified by providing shear during gelation, which can result in distinct elastic properties of the gels formed (Sworn et al., 1995; Altmann et al., 2004). Caggioni et al. (2007) have done experiments on the gel formation by quiescently cooling the gel during gelation without applying any shear and by applying shear at constant rate during gelation. They have found that the gel formed without any applied shear during gelation yields sharply and shows slow recovery whereas smother transition from elastic-like to fluid-like behavior was exhibited by shear cooled gel in addition to faster recovery of elastic properties after yielding. They also found that uniformly embedded colloidal tracers into high density polymer regions results if the tracers are added before gelation whereas addition of tracer after gelation restricts the uniform distribution.

5.2 Microstructure and Rheology of Food Gels

Many applications of various types of gel require gels with certain physical properties, strength, responsive characteristics which can be predicted by the microstructure and rheology. Associative interactions between the constituents in the gel decide the thermodynamic compatibility or the stabilization of gel. Sometimes a gel contains more than 99% water. The gel structure determines whether the water could be held firmly or whether the gel behaves like a sponge with water squeezed out. The diffusion of water and other water soluble compound and texture of gel depends on the microstructure of the gel. Gelation properties of individual biopolymers and the interactions among them, in addition to the processing condition of the gel system containing two biopolymer components drives its microstructure. Kasapis (2008) presented a review on phase separation in such gel systems. Firoozmand et al. (2009) have prepared a model gel system containing gelatin, oxidized starch, sugar, and water to study the microstructure and rheology of phase separated gels and found that the microstructure and rheology of gel depend on the composition of the ingredients and thermal treatment of the material prior to its ultimately attainable structure by gelation. They found that the elastic modulus of gelatin-starch gel is strongly influenced by the extent of crosslinking. They did not find any straightforward relationship between rheological properties and microstructure of the model gel but observed bicontinuous phase separation due to the presence of liquid-like starch rich domains interspersed with elastic gelatin-rich domains. They found a decrease of the overall elastic modulus with increase of starch content in the gel influenced by the degree of phase separation.

Microstructure and rheology of egg yolk was studied by Cordobes et al. (2004), and found continuous denaturation of egg yolk protein above 60°C. They concluded that at 60°C temperature the protein aggregation and gel network formation were very rapid, thus moderate degree of protein denaturation lead to its aggregation. They found that increase in protein concentration altered the moduli independently of the temperature range, but crossover temperature reduced slightly with increasing the solid concentration. They observed the retardation in the structural development of the gel network and decrease in the gel strength upon dilution of the egg yolk, which reduces protein concentration in the gel. Dang et al. (2009) have studied the rheology and microstructure of suspension containing crosslinked waxy maize starch and whey protein. Starch granules loosen crystallinity and absorb a large amount of water and swell upon heating of starch in water dispersion resulting in enhancement of viscosity. Kaur et al. (2008) have studied the microstructure and rheology of starch—cassia gum and found the variation of storage modulus and loss modulus with structural transitions. During heating their control starch dispersions, they observed maxima in storage modulus and loss modulus when plotted against temperature. Exhibition of maxima in storage modulus and loss modulus is due to the formation of network of swollen starch granules. After attaining the maximum value, increase in temperature destructs the gel structure. In highly amylase corn, potato, and normal starch gels the increase in the storage modulus and loss modulus, and decreased in loss tangent are more clearly noticeable during early stages of storage. Amylose retrogradation during early hours of storage of gel is responsible for sharp increase of storage and loss modulus of the starch gels. Waxy and modified waxy corn starch gels exhibit less changes in their rheological parameters throughout storage period whereas high amylose corn, potato, and normal corn starch gels with added cassia gum shows less changes in their rheological parameters during initial hours of storage. Kaur et al. (2008) concluded that the addition of the cassia gum retards the amylose retrogradation and thus leads to further increase in storage and loss moduli of starch gels.

Colloidal particulate gels can form in the isoelectric regions far below the denaturation temperature of the proteins. Transparent fine gels need processing conditions outside the isoelectric region at a sufficiently high net charge. Conformation changes can be triggered by denaturation in molecular gels of globular proteins. Denaturation, conformational changes, special sequences of polypeptide chain, and role of special bonds like disulphide bonds have been well investigated for proteins including soy proteins, β-lactoglobulin, bovine, serum albumin, ovalbumin, and so on (Hoffmann and van Mil, 1997; Verheul et al., 1998). Dynamic rheological study of corn starch and sugar composites (Chang et al., 2004) revealed that the storage modulus decreased with the increase of sugar concentration. It has been found that for the swollen gel, the storage modulus is higher than the loss modulus, and the elastic behavior of the gel predominates over its viscous behavior. Ottone and Deiber (2005) have studied the rheology of gelatin gels theoretically for finite deformations assuming the gel as the elastic solid and found that gelatin gels exhibited a strong nonlinear strain energy density which were represented by two types of shear moduli, one related to physical crosslinks in the polymer network, and the other related to the growth of junction zones.

Ottone et al. (2005) also modeled gelatin gels under finite deformation considering viscoelastic solid model taking into account a degree of physical crosslinks at a given maturation time incorporating viscoelastic relaxation and average degree of physical crosslinks in partially generated permanent network acting in the short-term mechanical response. Their model can be used to evaluate the effect of shear rate on the gelatin gels for dynamic transforming of the polymeric links and the junction zones, via nuclei formation and growing, which is the cause of viscoelastic response of the elastic matrix. van Vliet et al. (2004) have discussed the rheology and structure of milk protein gels formed by aggregated casein, gels of pure whey proteins and gels containing both of these two. Casein micelles present in the aggregated form in milk are having size range from 20 to 300 nm. Casein gels can be formed by enzymatic action, by acidification or by combination of acidification and enzymatic action. Even though casein gels can be considered as particle gels, their behavior is not similar to that of gels formed from hard particles, which cause strand deformation. Excessive rearrangements of particles in the gel network before and during gelation are the governing factors identified for whey separation (Lucey, 2001). Destabilization of native tertiary folding of proteins induced by addition of chemicals, hydrostatic pressure, thermal treatment and partial enzymatic hydrolysis results in a stable network (van Vliet et al., 2004). Electrostatic and hydrophobic interactions along with the action of disulphide between protein molecules are responsible for whey protein gelation. Lee and Lucey (2003) have found that lower preheating temperature, higher incubation temperature during preparation of yogurt gels, lead to decreased storage modulus and result in more whey separation. Fat present in the dairy products plays an important role of the structure and properties of the gelled products. In low-fat products, the microparticles of the fat act as nonactive fillers in the gel structure, and do not interact with the protein network (Janhoj et al., 2006). Lee and Lucey (2006) studied the physical and sensory properties of stirred yogurts and concluded that increasing shear rate reduces apparent viscosity of stirred yogurt, exhibiting the shear thinning behavior. They observed a branched and crosslinked protein structure in set yogurts preheated at 85°C, while thinner strands and clusters in the protein network was observed for set yogurts preheated at 75°C. They found that the microstructure of the stirred yogurts exhibited more dense protein aggregates caused by stirring and destruction of the initial network and formation of new weak aggregates due to collision of particles. The structure of colloidal gels depends on the interactions between the particles. Bernal spiral structures of the gel are produced by the particles having short-ranged attraction and long-ranged repulsion (Campbell et al., 2005; Sciortino et al., 2005).

6 Natural Materials for Emulsification and Stabilization of Emulsions

Stability of O/W emulsions and polydispersity of emulsions are major research problems since many decades. Researchers have proposed various emulsifiers and stabilizers in order to increase stability (Avranas and Stalidis, 1991; Jumaa and Muller, 1998; Vleeschauwer and Meeren, 1999; Chen et al., 2000; Wiacek et al., 2002; Filley, 2003; Hsu and Nacu, 2003; Palanuwech and Coupland, 2003; Klinkesorn and Namatsila, 2009; Stamkulov et al., 2009; Kang et al., 2011). Many naturally derived materials, for example, milk protein (Granger et al., 2005; Singh, 2011), have been proved to be efficient in producing emulsions with enhanced stability.

Many plant extracts have also been found to affect the interfacial properties and stability of emulsions (Chakrabarty et al., 2010; Karunasawat and Anprung, 2010; Georgiadis et al., 2011; Kotsiou et al., 2011; Ramful et al., 2011; Lad and Murthy, 2014).

6.1 Use of Proteins in Emulsification and Stabilization of Food Emulsions

Proteins are very important class of molecules involved in the formation and stabilization of many food and cosmetic emulsions (Dickinson, 2010). Soy protein (Jang et al., 2005; Tang et al., 2011), whey protein (Damodaran and Anand, 1997; Viljanen et al., 2005; Ma et al., 2011), sodium caseinate (Dickinson et al., 1991; Jang et al., 2005; Kellerby et al., 2006; Hu et al., 2011), casein (Dickinson, 2010; van der Ven et al., 2001), β-lactoglobulin (Das and Kinsella, 1989; Pongsawatmanit et al., 2006; Tcholakova et al., 2006; Ma et al., 2011; Zhai et al., 2011), bovine serum albumin (Livney, 2010), and so on, are found to be suitable candidates as emulsifying agents. However, high sensitivity to temperature, pH, and ionic strength of protein stabilized emulsions has offered a constraint for their use in many food products (McClements, 2004). Kong et al. (2008) studied the stability of emulsions prepared by protein and sucrose ester and discussed the micellar interaction in emulsion containing protein. Dunlap and Côté (2005) found that increasing the size of polysaccharide conjugated with β-lactoglobulin resulted in more stable emulsions. Usages of protein hydrolysates (van der Ven et al., 2001; Dexter and Middelberg, 2008), mixture of protein and hydrocolloids (Hattori et al., 1997; Diftis and Kiosseoglou, 2003; Kim et al., 2003), as well as blending of other emulsifiers with proteins (Akoh, 1992; Riscardo et al., 2003) have been found to be beneficial for protein stabilized emulsions against environmental stresses.

6.2 Coconut Milk Protein for Food Emulsions

Fresh coconut milk mainly contains water (about 54%), 35–37% fat, 2–4% protein, 2–5% carbohydrate, and nonfat solid matters (Kwon et al., 1996; Peamprasart and Chiewchan, 2006; Tasakul and Chaisawang, 2006; Raghavendra and Raghavarao, 2010). Tangsuphoom and Coupland (2008) have reported improved stability of coconut milk emulsions homogenized with various surface active agents like sodium caseinate, whey protein isolate (WPI), sodium dodecyl sulfate (SDS) and polyoxyethylene sorbitan monolaurate. Consequently, they (Tangsuphoom and Coupland, 2008) have found that the homogenized coconut milk prepared without additives destabilized by freeze–thaw between −20 and −10°C but not by chilling up to 5°C. They also found that the emulsions prepared using WPI exhibited appreciable coalescence and phase separation after being heat treated at 90°C (Tangsuphoom and Coupland, 2009). Onsaard et al. (2006) have studied corn O/W emulsions prepared by coconut milk protein (CMP) using high pressure valve homogenization and found that CMP can be effectively used for stabilizing fairly viscous emulsions like sauces, desserts, and salad dressings.

6.3 An Important Natural Gel: Aloe Vera Gel

Aloe barbadensis Miller (aloe vera) is a member of the Aloaceae (formerly a member of Liliaceae) plant family (Ozsoy et al., 2009). Recently, many commercial food-product producers, and confectioners have initiated the use of aloe vera in their productions due to its high nutrient value. Further, aloe vera juice is also being marketed nowadays as a source of nutrient supplement and even as a medicament (Lad and Murthy, 2013a). As mentioned in Lad and Murthy (2013a), aloe vera gel contains highly nutritious bioactive constituents including anthraquinones, alkaloids, enzymes, glycoproteins, vitamins, and polysaccharides along with mucilaginous polysaccharides, which are the major responsible constituents required for effective treatment for skin care as well as other health-care and cosmetic applications (Choi & Chung, 2003; Duansak et al., 2003; Eshun & He, 2004; Ni et al., 2004; Talmadge et al., 2004; Moghaddasi & Verma, 2011). Hu et al. (2003) and Wu et al. (2006) have found that the polysaccharides present in aloe vera gel have anti oxidant effects. Chithra et al. (1998) demonstrated the positive effects of aloe vera on the healing of dermal wounds in diabetic rats. West and Zhu (2003) have found that the gradual delivery of aloe vera gel to the dry—wrinkled skin helps in improving skin integrity and reducing wrinkle. The composition of the typical gels has been presented in the literature (Pierce, 1983; Femenia et al., 1999; Vogler & Ernst, 1999; Hu et al., 2003).

6.4 Use of Coriander (*Coriandrum sativum* L.) in Food Emulsions

Coriander (*C. sativum* L.) belongs to *Apiaceae* genus and Umbelliferae family having many useful ingredients having medicinal value (Rajeshwari and Andallu, 2011). Melo et al. (2005) have reported the characterization of antioxidant compounds in aqueous coriander extract and found that phenolic acids were responsible for antioxidant activity of the same. The seeds of coriander possess a pleasant flavor due to the presence of up to 1% essential oil in it, which contains monoterpenoid linalool as the major ingredient. As explained in Lad and Murthy (2013b), the antioxidant activity of the extracts of coriander has also been reported by Wangensteen et al. (2004). The extracted essential oil from coriander not only used as a flavoring agent in the liquor, cocoa, and chocolate industries but it is also used in medicines for indigestion, rheumatism, joint pains (Wangensteen et al., 2004) and as a carminative agent. As discussed in Lad and Murthy (2013b), many researchers have reported the antidiabetic (Gallagher et al., 2003), antibacterial (Kubo et al., 2004; Matasyoh et al., 2009), anticancerus and antimutagenic (Chithra and Leelamma, 2000), antiaging, antiulcer, antitumor (Rajeshwari and Andallu, 2011) characters of its essential oil.

7 Conclusions

Starting with the fundamentals of emulsions we presented the importance of the use of natural materials for preparation of food emulsions. Current trends in the selection of the diet habits of the people of developed and developing countries reveal the high potential demand of organic foods in the coming days. Current awareness of the mankind for health and hygiene attracted food industry to produce foods with high nutrient values containing minimal synthetic additives. Coconut milk protein, coconut milk, and coriander have been reported to improve the emulsification and stability of emulsions. Aloe vera gel and juice, Bengal gram, and turmeric microparticles have been effectively used for improving the rheology of emulsions for food applications. Appropriate use of these naturally available high nutrient materials can effectively replace the use of artificial stabilizers, preservatives, thickeners, texturizers, and rheology modifiers. The food products containing these natural materials may serve the purpose of hygienic, herbal, and high nutrient supplementary diet imparting better mouth feeling and sensory perceptions resulting in improved customer acceptance.

To make the chapter self-contained, we discussed fundamental concepts of emulsions, emulsification, and interactions within emulsified systems and rheology of the same in various sections in sufficient detail. Further research in identification of other natural materials for their potential use as rheology modifiers, texturizers, antifoaming agents, emulsifiers, and stabilizers can open the way to improve physical properties, stability, rheology, sensory effects, and eventually marketability of final food products.

Notations

CMP	Coconut milk protein
c_p	Specific heat at constant pressure, J kg^{-1} K^{-1}
c_v	Specific heat at constant volume, J kg^{-1} K^{-1}
D	Translational diffusion coefficient, m^2 s^{-1}
D	Diameter of droplet, m
$d_{3,2}$	Sauter mean diameter, m
$d_{4,3}$	Volume weighted mean diameter, m
d_h	Hydrodynamic diameter of droplet, m
d_i	Diameter of droplet in *i*th fraction
d_{max}	Maximum droplet diameter, m
$d_{(v,0.5)}$	Median for size distribution of droplets
d'	Diameter of swollen particle, m
d'_c	Diameter of collapsed particle, m
F	Force, N
dG	Gibbs free energy change
G^*	Complex modulus, Pa
G'	Storage modulus, Pa
G''	Loss modulus, Pa
G	Gravitational acceleration, m s^{-2}
HLB	Hydrophilic–lipophilic balance
K	Empirical parameter in Eqs. (4.5) and (4.8)
k	Empirical parameter in Eqs. (4.1)–(4.8)
m	Empirical parameter in Eqs. (4.1)–(4.8)
N	Speed of agitator in revolutions per second
N_s	Number of solvent molecules
n	Empirical parameter in Eqs. (4.1)–(4.8)
n_i	Number of droplets two consecutive diameters
O/W	Oil-in-water
P	Pressure, Pa
P_m	Sum of hydrostatic and acoustic pressure in the liquid, Pa
P_v	Pressure in the bubble at its maximum size, Pa
r	Radius of particle, m
ΔS	Change of entropy, J K^{-1}
T	Absolute temperature, K

T_0	Solution temperature, K
T_{max}	Maximum theoretical temperature within cavitation bubble, K
t	Time, s
W/O	Water-in-oil
γ	Strain
$\dot{\gamma}$	Shear rate, s^{-1}
γ'	Interfacial tension, N m^{-1}
γ'_{ow}	Oil–water interfacial tension, N m^{-1}
δ	Phase angle
δ'	Normalized diameter, m
$\dot{\varepsilon}$	Total strain rate, s^{-1}
ε_{oil}	Volume fraction of oil phase in emulsion
η	Apparent viscosity, Pa.s
η^*	Complex viscosity, Pa.s
η_c	Viscosity of continuous phase, Pa.s
η_d	Viscosity of dispersed phase, Pa.s
η_r	Relative viscosity
θ	Contact angle
k	Boltzmann's constant
$\Delta\mu_s$	Difference of the chemical potential, J
Π	Osmotic pressure, Pa
ρ	Density, kg m^{-3}
σ	Shear stress, Pa
σ_0	Initial modulus of the material
σ_e	Equilibrium modulus of the material
τ	Relaxation time, s
ϕ	Volume fraction
ϕ'	Volume fraction of polymer
ω	Frequency of oscillation, s^{-1}

References

Akoh, C.C., 1992. Emulsification properties of polyesters and sucrose ester blends. 1. Carbohydrate fatty-acid polyesters. J. Am. Oil Chem. Soc. 69, 9–13.

Altmann, N., Cooper-White, J.J., Dunstan, D.E., Stokes, J.R., 2004. Strong through to weak 'sheared' gels. J. Non-Newtonian Fluid Mech. 124, 129–136.

Ashokkumar, M., 2011. The characterization of acoustic cavitation bubbles—an overview. Ultrason. Sonochem. 18, 864–872.

Avranas, A., Stalidis, G., 1991. Interfacial properties and stability of oil-in-water emulsions stabilized with binary mixtures of surfactants: II. Effect of a cationic surfactant on a nonionic surfactant stabilized emulsion. J. Colloid Interf. Sci. 143, 180–187.

Barnes, H.A., 1994. Rheology of emulsions—a review. Colloids Surf. A 91, 89–95.

Basheva, E.S., Kralchevsky, P.A., Danov, K.D., Ananthapadmanabhan, K.P., Lips, A., 2007. The colloid structural forces as a tool for particle characterization and control of dispersion stability. Phys. Chem. Chem. Phys. 9, 5183–5198.

Behrend, O., Schubert, H., 2000. Influence of continuous phase viscosity on emulsion by ultrasound. Ultrason. Sonochem. 7, 77–85.

Briggs, T.R., 1921. Emulsions with finely divided solids. J. Ind. Eng. Chem. 13, 1008–1010.

Caggioni, M., Spicer, P.T., Blair, D.I., Lindberg, S.E., Weitz, D.A., 2007. Rheology and microrheology of a microstructured fluid: the gellan gum case. J. Rheol. 51, 851–865.

Campbell, A.I., Anderson, V.J., van Duijneveldt, J.S., Bartlett, P., 2005. Dynamical arrest in attractive colloids: the effect of long-range repulsion. Phys. Rev. Lett. 94, 208301.

Carter, C.B., Norton, M.G., 2007. Ceramic Materials Science and Engineering. Springer, New York, NY.

Ceylan, S., Kelbaliyev, G., Ceylan, K., 2003. Estimation of the maximum stable drop sizes, coalescence frequencies and the size distributions in isotropic turbulent dispersions. Colloids Surf. A 212, 285–295.

Chakrabarty, K., Saha, P., Ghoshal, A.K., 2010. Separation of lignosulfonate from its aqueous solution using emulsion liquid membrane. J. Membr. Sci. 360, 34–39.

Chang, Y.H., Lim, S.T., Yoo, B., 2004. Dynamic rheology of corn starch-sugar composites. J. Food Eng. 64, 521–527.

Chaterji, S., Kwon, I.K., Park, K., 2007. Smart polymeric gels: redefining the limits of biomedical devices. Prog. Polym. Sci. 32, 1083–1122.

Cheddadi, I., Saramito, P., Dollet, B., Raufaste, C., Graner, F., 2011. Understanding and predicting viscous, elastic, plastic flows. Eur. Phys. J. E. 34, 1–15.

Chen, C.M., Lu, C.H., Chang, C.H., Yang, Y.M., Maa, J.R., 2000. Influence of pH on the stability of oil-in-water emulsions stabilized by a splittable surfactant. Colloids Surf. A 170, 173–179.

Chen, J., Vogel, R., Werner, S., Heinrich, G., Clausse, D., Dutschk, V., 2011. Influence of the particle type on the rheological behaviour of Pickering emulsions. Colloids Surf. A 382, 238–245.

Chithra, V., Leelamma, S., 2000. *Coriandrum sativum*—effect on lipid metabolism in 1,2-dimethyl hydrazine induced colon cancer. J. Ethnopharmacol. 71, 457–463.

Chithra, P., Sajithlal, G.B., Chandrakasan, G., 1998. Influence of aloe vera on the healing of dermal wounds in diabetic rats. J. Ethnopharmacol. 59, 195–201.

Choi, S., Chung, M.H., 2003. A review on the relationship between aloe vera components and their biologic effects. Semin. Integr. Med. 1, 53–62.

Cordobes, F., Partal, P., Guerrero, A., 2004. Rheology and microstructure of heat-induced egg yolk gels. Rheol. Acta 43, 184–195.

Coussot, P., 2005. Rheometry of Pastes, Suspensions and Granular Materials. John Wiley & Sons, Hoboken.

Coussot, P., Tabutear, H., Chateau, X., Tocquer, L., Ovarlez, G., 2006. Aging and solid or liquid behavior in pastes. J. Rheol. 50, 975–994.

Cucheval, A., Chow, R.C.Y., 2008. A study on the emulsification of oil by power ultrasound. Ultrason. Sonochem. 15, 916–920.

Damodaran, S., Anand, K., 1997. Sulfhydryl-disulfide interchange-induced interparticle protein polymerization in whey protein-stabilized emulsions and its relation to emulsion stability. J. Agric. Food Chem. 45, 3813–3820.
Dang, H.V., Loisel, C., Desrumaux, A., Doublier, J.L., 2009. Rheology and microstructure of cross-linked waxy maize starch/whey protein suspensions. Food Hydrocolloids 23, 1678–1686.
Das, K.P., Kinsella, J.E., 1989. pH dependent emulsifying properties of β-lactoglobulin. J. Dispersion Sci. Technol. 10, 77–102.
Denkov, N.D., 1999. Mechanisms of action of mixed solid-liquid antifoams. 2. Stability of oil bridges in foam films. Langmuir 15, 8530–8542.
Denkov, N.D., 2004. Mechanisms of foam destruction by oil-based antifoams. Langmuir 20, 9463–9505.
Denkov, N.D., Yoshimura, H., Nagayama, K., 1996a. Method for controlled formation of vitrified films for cryo-electron microscopy. Ultramicroscopy 65, 147–158.
Denkov, N.D., Yoshimura, H., Nagayama, K., Kouyama, T., 1996b. Phys. Rev. Lett. 76, 2354–2357.
Desnoyer, C., Masbernat, O., Gourdon, C., 2003. Experimental study of drop size distributions at high phase ratio in liquid–liquid dispersions. Chem. Eng. Sci. 58, 1353–1363.
Dexter, A.F., Middelberg, A.P.J., 2008. Peptides as functional surfactants. Ind. Eng. Chem. Res. 47, 6391–6398.
Dickinson, E., 1992. Interfacial interactions and the stability of oil-in-water emulsions. Pure Appl. Chem. 64, 1721–1724.
Dickinson, E., 2009. Hydrocolloids and emulsion stability. In: Phillips, G.O., Williams, P.A. (Eds.), Handbook of Hydrocolloids. second ed. Woodhead Publishing, Cambridge, pp. 23–49.
Dickinson, E., 2010. Flocculation of protein-stabilized oil-in-water emulsions. Colloids Surf. B 81, 130–140.
Dickinson, E., 2012. Emulsion gels: the structuring of soft solids with protein-stabilized oil droplets. Food Hydrocoll. 28, 224–241.
Dickinson, E., Evison, J., Owusu, R.K., 1991. Preparation of fine protein-stabilized water-in-oil-in-water emulsions. Food Hydrocoll. 5, 481–485.
Diftis, N., Kiosseoglou, V., 2003. Improvement of emulsifying properties of soybean protein isolate by conjugation with carboxymethyl cellulose. Food Chem. 81, 1–6.
Duansak, D., Somboonwong, J., Patumraj, S., 2003. Effects of aloe vera on leukocyte adhesion and TNF-alpha and IL-6 levels in burn wounded rats. Clin. Hemorheol. Microcirculation 29, 239–246.
Dunlap, C.A., Côté, G.L., 2005. β-Lactoglobulin – dextran conjugates: effect of polysaccharide size on emulsion stability. J. Agric. Food Chem. 53, 419–423.
Dyab, A.K.F., Ozmen, M., Ersoz, M., Paunov, V.N., 2009. Fabrication of novel anisotropic magnetic microparticles. J. Mater. Chem. 19, 3475–3481.
Eshun, K., He, Q., 2004. Aloe vera: a valuable ingredient for the food, pharmaceutical and cosmetic industries—a review. Crit. Rev. Food Sci. Nutr. 44, 91–96.
Femenia, A., Sanchez, E.S., Simal, S., Rossello, C., 1999. Compositional features of polysaccharides from aloe vera (*Aloe barbadensis* Miller) plant tissues. Carbohydr. Polym. 39, 109–117.
Fernandez, P., Andre, V., Reiger, J., Kuhnle, A., 2004. Nano-emulsion formation by emulsion phase inversion. Colloids Surf. A 251, 53–58.
Fernandez-Barbero, A., Suarez, I.J., Sierra-Martin, B., Fernandez-Nieves, A., de las Nieves, F.J., Marquez, M., Rubio-Retama, J., Lopez-Cabarcos, E., 2009. Gels and microgels for nanotechnological applications. Adv. Colloid Interf. Sci. 147–148, 88–108.

Filley, J., 2003. Stable water-oil emulsions from the new insoluble surfactant sodium 5-(1-dodecylaminocarbonyl) picolinate. J. Colloid Interf. Sci. 266, 438–441.
Firoozmand, H., Murray, B.S., Dickinson, E., 2009. Microstructure and rheology of phase-separated gels of gelatin + oxidized starch. Food Hydrocoll. 23, 1081–1088.
Frelichowska, J., Bolzinger, M.A., Chevalier, Y., 2009. Pickering emulsions with bare silica. Colloid Surf. A 343, 70–74.
Frost, D.S., Schoepf, J.J., Nofen, E.M., Dai, L.L., 2012. Understanding droplet bridging in ionic liquid–based Pickering emulsions. J. Colloid Interf. Sci. 383, 103–109.
Gallagher, A.M., Flatt, P.R., Duffy, G., Abdel-Wahab, Y.H.A., 2003. The effects of traditional antidiabetic plants on in vitro glucose diffusion. Nutr. Res. 23, 413–424.
Gallegos, G., Franco, J.M., 1999. Rheology of food, cosmetics and pharmaceuticals. Curr. Opin. Colloid Interf. Sci. 4, 288–293.
Garrett, P.R. (Ed.), 1993. Defoaming: Theory and Industrial Applications. Marcel Dekker, New York, NY.
Gautier, F., Destribats, M., Perrier-Cornet, R., Dechézelles, J.-F., Giermanska, J., Heroguez, V., Ravaine, S., Leal-Calderon, F., Schmitt, V., 2007. Pickering emulsions with stimulable particles, from highly—to weakly covered interfaces. Phys. Chem. Chem. Phys. 9, 6455–6462.
Georgiadis, N., Ritzoulis, C., Sioura, G., Kornezou, P., Vasiliadou, C., Tsioptsias, C., 2011. Contribution of okra extracts to the stability and rheology of oil-in-water emulsions. Food Hydrocoll. 25, 991–999.
Granger, C., Barey, P., Veschambre, P., Cansell, M., 2005. Physicochemical behavior of oil-in-water emulsions: influence of milk protein mixtures, glycerol ester mixtures and fat characteristics. Colloids Surf. B 42, 235–243.
Hagesaether, L., Jakobsen, H.A., Svendsen, H.F., 2002. A model for turbulent binary breakup of dispersed fluid particles. Chem. Eng. Sci. 57, 3251–3267.
Hattori, M., Ogino, A., Nakai, H., Takahashi, K., 1997. Functional improvement of β-lactoglobulin by conjugating with alginate lyase–lysate. J. Agric. Food Chem. 45, 703–708.
Hoffmann, M.A.M., van Mil, P.J.J.M., 1997. Heat-induced aggregation of β-lactoglobulin: role of free thiol group and disulfide bonds. J. Agric. Food Chem. 45, 2942–2948.
Hsu, J.P., Nacu, A., 2003. Behavior of soybean oil-in-water emulsion stabilized by nonionic surfactant. J. Colloid Interf. Sci. 259, 374–381.
Hu, Y., Xu, J., Hu, Q., 2003. Evaluation of antioxidant potential of aloe vera (*Aloe barbandensis* Miller) extracts. J. Agric. Food Chem. 51, 7788–7791.
Hu, M., Li, Y., Decker, E.A., Xiao, H., McClements, D.J., 2011. Impact of layer structure on physical stability and lipase digestibility of lipid droplets coated by biopolymer nanolaminated coatings. Food Biophys. 6, 37–48.
Jang, W., Nikolov, A., Wasan, D.T., Chen, K., Campbell, B., 2005. Effect of protein on the texture of food emulsions under steady flow. Ind. Eng. Chem. Res. 44, 4855–4862.
Janhoj, T., Petersen, C.B., Ipsen, R., Frost, M.B., 2006. Sensory and rheological characterization of low-fat stirred yoghurt. J. Texture Stud. 37, 276–299.
Jumaa, M., Muller, B.W., 1998. The stabilization of parenteral fat emulsion using non-ionic ABA copolymer surfactant. Int. J. Pharm. 174, 29–37.
Jurado, E., Bravo, V., Camacho, F., Vicaria, J.M., Fernandez-Arteaga, A., 2007. Estimation of the distribution of droplet size, interfacial area and volume in emulsions. Colloids Surf. A 295, 91–98.
Kang, W., Xu, B., Wang, Y., Li, Y., Shan, X., An, F., Liu, J., 2011. Stability mechanism of W/O crude oil emulsion stabilized by polymer and surfactant. Colloids Surf. A 384, 555–560.

Karam, H.J., Bellinger, J.C., 1968. Deformation and breakup of liquid droplets in a simple shear field. Ind. Eng. Chem. Fund 7, 576–581.
Karunasawat, K., Anprung, P., 2010. Effect of depolymerized mango pulp as a stabilizer in, oil-in-water emulsion containing sodium caseinate. Food Bioprod. Process 88, 202–208.
Kasapis, S., 2008. Phase separation in biopolymer gels: a low-to high-solid exploration of structural morphology and functionality. Crit. Rev. Food Sci. Nutr. 48, 341–359.
Kaur, L., Singh, J., Singh, H., McCarthy, O.J., 2008. Starch-cassia gum interactions: a microstructure—rheology study. Food Chemistry 111, 1–10.
Kelbaliyev, G., Ceylan, K., 2005. Estimation of the minimum stable drop sizes, break-up frequencies, and size distributions in turbulent dispersions. J. Dispersion Sci. Technol. 26, 487–494.
Kellerby, S.S., Gu, Y.S., McClements, D.J., Decker, E.A., 2006. Lipid oxidation in a menhaden oil-in-water emulsion stabilized by sodium caseinate cross-linked with transglutaminase. J. Agric. Food Chem. 54, 10222–10227.
Kentish, S., Wooster, T.J., Ashokkumar, M., Balachandran, S., Mawson, R., Simons, L., 2008. The use of ultrasonics for nanoemulsion preparation. Innov. Food Sci. Emerg. Technol. 9, 170–175.
Kim, H.J., Choi, S.J., Shin, W.S., Moon, T.W., 2003. Emulsifying properties of bovine serum albumin–galactomannan conjugates. J. Agric. Food Chem. 51, 1049–1056.
Klinkesorn, U., Namatsila, Y., 2009. Influence of chitosan and NaCl on physicochemical properties of low-acid tuna oil-in-water emulsions stabilized by non-ionic surfactant. Food Hydrocoll. 23, 1374–1380.
Kong, Y., Nikolov, A., Wasan, D., Ogawa, A., 2008. Emulsion texture and stability: role of surfactant micellar interactions in the presence of proteins. Ind. Eng. Chem. Res. 47, 9108–9114.
Korobko, A.V., van Den Ende, D., Agterof, W.G.M., Mellema, J., 2005. Coalescence in semiconcentrated emulsions in simple shear flow. J. Chem. Phys. 123, 1–11.
Kotsiou, K., Tasioula-Margari, M., Capuano, E., Fogliano, V., 2011. Effect of standard phenolic compounds and olive oil phenolic extracts on acrylamide formation in an emulsion system. Food Chem. 124, 242–247.
Kubo, I., Fujita, K., Kubo, A., Nihei, K., Ogura, T., 2004. Antibacterial activity of coriander volatile compounds against *Salmonella choleraesuis*. J. Agric. Food Chem. 52, 3329–3332.
Kumar, S., Kumar, R., Gandhi, K.S., 1991. Alternative mechanisms of drop breakage in stirred vessels. Chem. Eng. Sci. 46, 2483–2489.
Kumar, S., Kumar, R., Gandhi, K.S., 1992. A multistage model for drop breakage in stirred vessels. Chem. Eng. Sci. 47, 971–980.
Kumar, S., Kumar, R., Gandhi, K.S., 1993. A simplified procedure to predict d_{max} in stirred vessels. Chem. Eng. Sci. 48, 3092–3096.
Kumar, S., Ganvir, V., Satyanand, C., Kumar, R., Gandhi, K.S., 1998. Alternative mechanisms of drop breakup in stirred vessels. Chem. Eng. Sci. 53, 3269–3280.
Kwon, K.S., Bae, D., Park, K.H., Rhee, K.C., 1996. Aqueous extraction and membrane techniques improve coconut protein concentrate functionality. J. Food Sci. 61, 753–756.
Lad, V.N., Murthy, Z.V.P., 2012. Enhancing the stability of oil-in-water emulsions emulsified by coconut milk protein with the application of acoustic cavitation. Ind. Eng. Chem. Res. 51, 4222–4229.
Lad, V.N., Murthy, Z.V.P., 2013a. Rheology of *Aloe barbadensis* Miller: a naturally available material of high therapeutic and nutrient value for food applications. J. Food Eng. 115, 279–284.

Lad, V.N., Murthy, Z.V.P., 2013b. Alteration of interfacial stability of oil-in-water emulsions using bio-derived additives. J. Dispersion Sci. Technol. 34, 78–83.

Lad, V.N., Murthy, Z.V.P., 2014. Effects of interfacial interactions of solid microparticles on the emulsion containing ionic liquid. J. Ind. Eng. Chem. 20, 274–277.

Larson, R.G., 1999. The Structure and Rheology of Complex Fluids. Oxford University Press, New York, NY.

Laurati, M., Petekidis, G., Koumakis, N., Cardinaux, F., Schofield, A.B., Brader, J.M., Fuchs, M., Egelhaaf, S.U., 2009. Structure, dynamics, and rheology of colloid-polymer mixtures: from liquids to gels. J. Chem. Phys. 130, 134907.

Lee, M.H., Furst, E.M., 2008. Response of a colloidal gel to a microscopic oscillatory strain. Phys. Rev. E 77, 041408.

Lee, W.J., Lucey, J.A., 2003. Rheological properties, whey separation, and microstructure in set-style yogurt: effects of heating temperature and incubation temperature. J. Texture Stud. 34, 515–536.

Lee, W.J., Lucey, J.A., 2006. Impact of gelation conditions and structural breakdown on the physical and sensory properties of stirred yogurts. J. Dairy Sci. 89, 2374–2385.

Lefebvre, J., Renard, D., Sanchez-Gimeno, A.C., 1998. Structure and rheology of heat-set gels of globular proteins. I. Bovine serum albumin gels in isoelectric conditions. Rheol. Acta 37, 345–357.

Leighton, T.G., 1994. The Acoustic Bubble. Academic Press, London.

Levich, V.G., 1962. Physicochemical Hydrodynamics. Prentice-Hall, Englewood Cliffs, New Jersey.

Li, M.K., Fogler, H.S., 1978a. Acoustic emulsification—1. The instability of the oil–water interface to form the initial droplets. J. Fluid Mech. 88, 499–511.

Li, M.K., Fogler, H.S., 1978b. Acoustic emulsification. Part 2: breakup of the large primary oil droplets in a water medium. J. Fluid Mech. 88, 513–528.

Livney, Y.D., 2010. Milk proteins as vehicles for bioactives. Curr. Opin. Colloid Interf. Sci. 15, 73–83.

Lopetinsky, R.J.G., Masliyah, J.H., Xu, Z., 2006. Solids-stabilized emulsions: a review. In: Binks, B.P., Horozov, T.S. (Eds.), Colloidal Particles at Liquid Interfaces. Cambridge University Press, Cambridge, pp. 186–224.

Lucey, J.A., 2001. The relationship between rheological parameters and whey separation in milk gels. Food Hydrocoll. 15, 603–608.

Luo, H., Svendsen, H.F., 1996. Theoretical model for drop and bubble break-up in turbulent dispersions. AIChE J. 42, 1225–1233.

Ma, H., Forssell, P., Partanen, R., Buchert, J., Boer, H., 2011. Improving laccase catalyzed cross-linking of whey protein isolate and their application as emulsifiers. J. Agric. Food Chem. 59, 1406–1414.

Mason, T.J., Lorimer, P.J., 1988. Sonochemistry: Theory, Applications and Uses of Ultrasound in Chemistry. Ellis Horwood, Chichester.

Mason, T.G., Krall, A.H., Gang, H., Bibette, J., Weitz, D.A., 1996. Monodisperse emulsions: properties and uses. In: Becher, P. (Ed.), Encyclopaedia of Emulsion Technology, vol. 4, Marcel Dekker, New York, NY, pp. 299–336.

Matasyoh, J.C., Maiyo, Z.C., Ngure, R.M., Chepkorir, R., 2009. Chemical composition and antimicrobial activity of the essential oil of *Coriandrum sativum*. Food Chem. 113, 526–529.

McClements, D.J., 2004. Protein-stabilized emulsions. Curr. Opin. Colloid Interf. Sci. 9, 305–313.

McClements, D.J., 2005. Food Emulsions: Principles, Practices, and Techniques, second ed. CRC Press LLC, Boca Raton, FL.

Melo, E.A., Filho, J.M., Guerra, N.B., 2005. Characterization of antioxidant compounds in aqueous coriander extract (*Coriandrum sativum* L.). Lebensm. Wiss. u. Technol. 38, 15–19.
Moghaddasi, M.S., Verma, S.K., 2011. Aloe vera their chemical composition and applications: a review. Int. J. Biol. Med. Res. 2, 466–471.
Neiser, S., Draget, K.I., Smidsrod, O., 1998. Gel formation in heat treated bovine serum albumin–sodium alginate systems. Food Hydrocoll. 12, 127–132.
Ni, Y., Turner, D., Yates, K.M., Tizard, I., 2004. Isolation and characterization of structural components of aloe vera L. leaf pulp. Int. Immunopharmacol. 4, 1745–1755.
Nikolov, A.D., Wasan, D.T., 1989. Ordered micelle structuring in thin films formed from anionic surfactant solutions: I. Experimental. J. Colloid Interf. Sci. 133, 1–12.
Nikolov, A.D., Wasan, D.T., Kralchevsky, P.A., Ivanov, I.B., 1988. Ordered structures in thinning micellar foam and latex films. In: Ise, N., Sogami, I. (Eds.), Ordering and Organisation in Ionic Solutions. World Scientific, Singapore, pp. 302–314.
Nikolov, A.D., Kralchevsky, P.A., Ivanov, I.B., Wasan, D.T., 1989. Ordered micelle structure in thin films formed from anionic surfactant solutions: II. Model development. J. Colloid Interf. Sci. 133, 13–22.
Norton, I.T., Spyropoulos, F., Cox, P.W., 2009. Effect of emulsifiers and fat crystals on shear induced droplet break-up, coalescence and phase inversion. Food Hydrocoll. 23, 1521–1526.
Onsaard, E., Vittayanont, M., Srigam, S., McClements, D.J., 2006. Comparison of properties of oil-in-water emulsions stabilized by coconut cream proteins with those stabilized by whey protein isolate. Food Res. Int. 39, 78–86.
Ottone, M.L., Deiber, J.A., 2005. Modeling the rheology of gelatin gels for finite deformations. Part 1. Elastic rheological model. Polymer 46, 4928–4937.
Ottone, M.L., Peirotti, M.B., Deiber, J.A., 2005. Modeling the rheology of gelatin gels for finite deformations. Part 2. Viscoelastic solid model. Polymer 46, 4938–4949.
Ozsoy, N., Candoken, E., Akev, N., 2009. Implications for degenerative disorders antioxidative activity, total phenols, flavonoids, ascorbic acid, β-carotene and α-tocopherol in aloe vera. Oxidative Med. Cell. Longev. 2, 99–106.
Pal, R., 2011. Rheology of simple and multiple emulsions. Curr. Opin. Colloid Interf. Sci. 16, 41–60.
Palanuwech, J., Coupland, J.N., 2003. Effect of surfactant type on the stability of oil-in-water emulsions to dispersed phase crystallization. Colloids Surf. A 223, 251–262.
Pangu, G.D., Feke, D.L., 2007. Droplet transport and coalescence kinetics in emulsions subjected to acoustic fields. Ultrasonics 46, 289–302.
Peamprasart, T., Chiewchan, N., 2006. Effect of fat content and preheat treatment on the apparent viscosity of coconut milk after homogenization. J. Food Eng. 77, 653–658.
Pickering, S.U., 1907. Emulsions. J. Chem. Soc. 91, 2001–2021.
Pierce, R.F., 1983. Comparison between the nutritional contents of the aloe gel from conventionally and hydropobically grown plants. Erde Int. 1, 37–38.
Pongsawatmanit, R., Harnsilawat, T., McClements, D.J., 2006. Influence of alginate, pH and ultrasound treatment on palm oil-in-water emulsions stabilized by β-lactoglobulin. Colloids Surf. A 287, 59–67.
Puppo, M.C., Anon, M.C., 1998. Effect of pH and protein concentration on rheological behavior of acidic soybean protein gels. J. Agric. Food. Chem. 46, 3039–3046.
Puppo, M.C., Anon, M.C., 1999. Rheological properties of acidic soybean protein gels: salt addition effect. Food Hydrocoll. 13, 167–176.

Raghavendra, S.N., Raghavarao, K.S.M.S., 2010. Effect of different treatments for the destabilization of coconut milk emulsion. J. Food Eng. 97, 341–347.
Rajeshwari, U., Andallu, B., 2011. Medicinal benefits of coriander (*Coriandrum sativum* L.). Spatula DD 1, 51–58.
Ramful, D., Aumjaud, B., Neergheen, V.S., Soobrattee, M.A., Googoolye, K., Aruoma, O.I., Bahorun, T., 2011. Polyphenolic content and antioxidant activity of *Eugenia pollicina* leaf extract in vitro and in model emulsion systems. Food Res. Int. 44, 1190–1196.
Renardy, Y., 2007. The effects of confinement and inertia on the production of droplets. Rheol. Acta 46, 521–529.
Riscardo, M.A., Franco, J.M., Gallegos, C., 2003. Influence of composition of emulsifier blends on the rheological properties of salad dressing-type emulsions. Food Sci. Technol. Int. 9, 53–63.
Saunders, B.R., Laajam, N., Daly, E., Teow, S., Hu, X., Stepto, R., 2009. Microgels: from responsive polymer colloids to biomaterials. Adv. Colloid Interf. Sci. 147–148, 251–262.
Savary, G., Grisel, M., Picard, C., 2013. Impact of emollients on the spreading properties of cosmetic products: a combined sensory and instrumental characterization. Colloids Surf. B 102, 371–378.
Schulman, J.H., Leja, J., 1954. Control of contact angles at the oil-water-solid interfaces. Emulsions stabilized by solid particles ($BaSO_4$). Trans. Faraday Soc. 50, 598–605.
Sciortino, F., Tartaglia, P., Zaccarelli, E., 2005. One-dimensional cluster growth and branching gels in colloidal systems with short-range depletion attraction and screened electrostatic repulsion. J. Phys. Chem. B 109, 21942–21953.
Sherman, P., 1973. Rheology of interfaces and emulsion stability. J. Colloid Interf. Sci. 45, 427–429.
Singh, H., 2011. Aspects of milk-protein-stabilised emulsions. Food Hydrocoll. 25, 1938–1944.
Sonawane, S.H., Neppolian, B., Teo, B., Grieser, F., Ashokkumar, M., 2010. Ultrasound assisted preparation of semiconductor/polymer photoanodes and their photoelectrochemical properties. J. Phys. Chem. C 114, 5148–5153.
Stamkulov, N.S., Mussabekov, K.B., Aidarova, S.B., Luckham, P.F., 2009. Stabilisation of emulsions by using a combination of an oil soluble ionic surfactant and water soluble polyelectrolytes. I: emulsion stabilisation and interfacial tension measurements. Colloid Surf. A 335, 103–106.
Suarez, I.J., Sierra-Martin, B., Fernandez-Barbero, A., 2009. Swelling of ionic and non-ionic minigels. Colloids Surf. A 343, 30–33.
Sworn, G., Sanderson, G.R., Gibson, W., 1995. Gellan gum fluid gels. Food Hydrocoll. 9, 265–271.
Tabilo-Munizaga, G., Barbosa-Canovas, G.V., 2005. Rheology for the food industry. J. Food Eng. 67, 147–156.
Tadros, T.F., 1994. Fundamental principles of emulsion rheology and their applications. Colloids Surf. A 91, 39–55.
Tadros, T.F., 2004. Application of rheology for assessment and prediction of the long-term physical stability of emulsions. Adv. Colloid Interf. Sci. 108, 227–258.
Tadros, T.F., Vincent, B., 1983. Emulsion stability. In: Becher, P. (Ed.), Encyclopaedia of Emulsion Technology, vol. 1, Marcel Dekker, New York, NY, pp. 129–285.
Talmadge, J., Chavez, J., Jacobs, L., Munger, C., Chinnah, T., Chow, J.T., Williamson, D., Yates, K., 2004. Fractionation of aloe vera L. inner gel, purification and molecular profiling of activity. Int. Immunopharmacol. 4, 1757–1773.
Tambe, D.E., Sharma, M.M., 1994. The effect of colloidal particles on fluid–fluid interfacial properties and emulsion stability. Adv. Colloid Interf. Sci. 52, 1–63.

Tang, C.H., Chen, L., Foegeding, E.A., 2011. Mechanical and water-holding properties and microstructures of soy protein isolate emulsion gels induced by $CaCl_2$, glucono-δ-lactone (GDL), and transglutaminase: influence of thermal treatments before and/or after emulsification. J. Agric. Food Chem. 59, 4071–4077.
Tangsuphoom, N., Coupland, J.N., 2008. Effect of surface-active stabilizers on the microstructure and stability of coconut milk emulsions. Food Hydrocoll. 22, 1233–1242.
Tangsuphoom, N., Coupland, J.N., 2009. Effect of thermal treatments on the properties of coconut milk emulsions prepared with surface-active stabilizers. Food Hydrocoll. 23, 1792–1800.
Tasakul, A., Chaisawang, P., 2006. Thermophysical properties of coconut milk. J. Food Eng. 73, 276–280.
Taylor, G.I., 1932. The viscosity of a fluid containing small drops of another fluid. Proc. R. Soc. London, Ser. A 138, 41–48.
Tcholakova, S., Denkov, N.D., Sidzhakova, D., Campbell, B., 2006. Effect of thermal treatment, ionic strength, and pH on the short-term and long-term coalescence stability of β-lactoglobulin emulsions. Langmuir 22, 6042–6052.
Tcholakova, S., Vankova, N., Denkov, N.D., Danner, T., 2007. Emulsification in turbulent flow: 3. Daughter drop-size distribution. J. Colloid Interf. Sci. 310, 570–589.
Ushikubo, F.Y., Cunha, R.L., 2014. Stability mechanisms of liquid water-in-oil emulsions. Food Hydrocoll. 34, 145–153.
van der Ven, C., Gruppen, H., de Bont, D.B.A., Voragen, A.G.J., 2001. Emulsion properties of casein and whey protein hydrolysates and the relation with other hydrolysate characteristics. J. Agric. Food Chem. 49, 5005–5012.
van Vliet, T., Lakemond, C.M.M., Visschers, R.W., 2004. Rheology and structure of milk protein gels. Curr. Opin. Colloid Interf. Sci. 9, 298–304.
Verheul, M., Roefs, S.P.F.M., 1998. Structure of whey protein gels, studied by permeability, scanning electron microscopy and rheology. Food Hydrocoll. 12, 17–24.
Verheul, M., Roefs, S.P.F.M., de Kruif, C.G., 1998. Kinetics of heat-induced aggregation of β-lactoglobulin. J. Agric. Food Chem. 46, 896–903.
Viljanen, K., Kylli, P., Hubbermann, E.M., Schwarz, K., Heinonen, M., 2005. Anthocyanin antioxidant activity and partition behavior in whey protein emulsion. J. Agric. Food Chem. 53, 2022–2027.
Vleeschauwer, D.D., Meeren, P.V., 1999. Colloid chemical stability and interfacial properties of mixed phospholipid-non-ionic surfactant stabilized oil-in-water emulsions. Colloids Surf. A 152, 59–66.
Vogler, B.K., Ernst, E., 1999. Aloe vera: a systematic review of its clinical effectiveness. Br. J. Gen. Pract. 49, 823–828.
Wangensteen, H., Samuelsen, A.B., Malterud, K.E., 2004. Antioxidant activity in extracts from coriander. Food Chem. 88, 293–297.
West, D.P., Zhu, Y.F., 2003. Evaluation of aloe vera gel gloves in the treatment of dry skin associated with occupational exposure. Am. J. Infect. Control 31, 40–42.
Wiacek, A.E., Chibowski, E., Wilk, K., 2002. Studies of oil-in-water emulsion stability in the presence of new dicephalic saccharide-derived surfactants. Colloids Surf. B 25, 243–256.
Wu, J.H., Xu, C., Shan, C.Y., Tan, R.X., 2006. Antioxidant properties and PC12 cell protective effects of APS-1, a polysaccharide from aloe vera var. chinensis. Life Sci. 78, 622–630.
Yan, N., Gray, M.R., Masliyah, J.H., 2001. On water-in-oil emulsions stabilized by fine solids. Colloids Surf. A 193, 97–107.

Yeung, A., Moran, K., Masliyah, J., Czarnecki, J., 2003. Shear-induced coalescence of emulsified oil drops. J. Colloid Interf. Sci. 265, 439–443.
Young, F.R., 1994. Cavitation. McGraw–Hill, London.
Zhai, J., Wooster, T.J., Hoffmann, S.V., Lee, T.H., Augustin, M.A., Aguilar, M.I., 2011. Structural rearrangement of β-lactoglobulin at different oil–water interfaces and its effect on emulsion stability. Langmuir 27, 9227–9236.

5

SELF-ASSEMBLED CARBOHYDRATE NANOSTRUCTURES: SYNTHESIS STRATEGIES TO FUNCTIONAL APPLICATION IN FOOD

Anupam Roy*, Shanker Lal Shrivastava*, Santi M. Mandal**

**Agricultural and Food Engineering Department, Indian Institute of Technology Kharagpur, West Bengal, India; **Vidyasagar University, Department of Microbiology, Midnapore, West Bengal, India*

1 Introduction

The increasing shift toward preventive therapies (Goetzel, 2009), disposable income, health-care spending (Thorpe, 2005), and self-medication (Raja and Arunachalam, 2011) necessitates the development of new nutritionally balanced healthy food formulations and nutraceutical food matrix (Dickinson and Leser, 2007). Better understanding of optimum diets, health claims of specific food and food ingredients are attracting food manufacturers to fortify existing foods or to design new functional foods offering health benefits (Wildman et al., 2006). Most of the bioactive molecules face the problem about poor water solubility, bioavailability, reactivity, and instability during handling, processing, and storage. Also, during digestion in the intestinal tract, functional-food ingredients most often degrade and reduce their bioavailability. A harsh condition in gastrointestinal tracts, that is, wide pH variation, enzymes most often reduce the activity of the bioactive compounds. Thus, the search for the proper protective carrier and effective delivery agents enhancing stability and bioavailability during harsh processing, handling, storage, and movement through the gastrointestinal tract is much more important.

In the transformation of the nutrient and bioactive molecule, its liberation during digestion and absorption into the cells strongly

Novel Approaches of Nanotechnology in Food. http://dx.doi.org/10.1016/B978-0-12-804308-0.00005-4

depends on the grafting carrier matrix. Thus, the choice of a matrix and its interaction with bioactive molecule or nutrient is a key issue. Additionally, this matrix may also serve added benefits preventing the loss of the part, or adverse interactions with inhibitors of absorption (Garti, 2008). The carrier matrix should also be food grade, biodegradable, and have an ability to withstand harsh processing, storage, and distribution (Fathi et al., 2014). It is quite interesting that fast release and absorption of the essential nutrient and bioactive compounds into the body does not necessarily equate to "improved" function and "better" health (Omenn et al., 1996). In some instances, the same loading but slow release and absorption, that is, slow release carbohydrate foods, offers profound effects on subsequent health outcomes, reducing overloading of transport systems, or causing undesirable excursions in plasma and tissue concentration (Faulks et al., 2008). Even more food research is searching for better delivery systems. Optimization of delivery agents with key food ingredients is important to enhance the consumer sensory experience by improving consumer appeal on flavor, color, texture, or consistency (Reineccius, 2013). Carbohydrates are important part of a diet, offering body fuel and maintaining physical activity and organ functions. They are mostly biocompatible in nature, finds greater extent of biodegradability, and possesses a high potential to be modified to achieve the required properties (Fathi et al., 2014).

Carbohydrates can be found as simpler sugar or in complex forms. Due to the complex structure of the carbohydrate, polysaccharides from whole grain cereals, whole meal bread/pasta, brown rice, fruits, and vegetables take longer digestion and absorption time.

Some complex carbohydrates are more often seen to be resistant against digestion by human digestive enzymes, but fermentable in the colon (Anderson et al., 2009). These complex carbohydrates are termed as dietary fibers and offer several health benefits. Solubility, slow digestion, absorption, specific release, and even targeted delivery (eg, in colon) make carbohydrate polysaccharide matrix a unique promising agent in solubilizing and protecting the nutrient and bioactive compound (Shibakami et al., 2015; Corona-Hernandez et al., 2013). Apart from biocompatibility and biodegradability, the ability to modify the structure of polysaccharides to a higher extent makes it unique (Fathi et al., 2014). In most cases, the unique physical properties gained from the carbohydrate improve the physical properties of the food formulations. Polysaccharides generally alter the rheological properties of the food matrix offering textural changes (Ishihara et al., 2011). It increases water retention and forms gel resulting in the thickening of food matrix. Other applications include stabilization of foam, emulsions, and suspended particulate materials, offering a good mouth feel, prevention or reduction of ice crystal formation in frozen foods, and interacting with other biomolecules.

In the process of molecular interaction of the bioactive molecule with carbohydrate, multicomponent compounds reassociate themselves at the molecular level by a variety of weaker (noncovalent) interactions to form a larger functional unit. The process is generally termed as self-assembly (Stoddart et al., 2000) and offers the basis behind the carbohydrate matrix and bioactive compound interactions. The weak physical forces between self-assembled entities may also disrupt under thermodynamic or mechanical stresses. Attention is also given in reinforcing nanoscale structures via covalent bonding.

In this chapter, we mainly focus on the different self-assembly strategies followed for the food applications. A discussion of assembling strategies, behavior, and structure–function relationship associated with varying shapes and sizes is drawn to give a deeper insight into the assembly strategy, its enhancement and control.

2 Self-Assembly, a Bottom-Up Approach

Fabrication of biomaterial can be done by stripping down a complex entity into component parts or assembling molecule to molecule/atom to atom to form complex supramolecular architectures (Zhang, 2003). The first strategic method is termed as top-down, whereas the next one is called bottom-up. In the bottom-up fabrication process, chemical complementary and structural compatibility binds or self-assembles building blocks together by weak or covalent interactions. Thus, self-assembly is a generic term that explains the formation of an organized structure or pattern from spontaneous reassociation of disordered systems of pre-existing components (eg, atoms, molecules, nanoparticles) via specific local interactions (Whitesides and Boncheva, 2002). Self-assembly as a "bottom-up" approach is a challenging task. It is dependent on the self-assembling properties of individual molecules or particles. The spontaneous assembling is strongly dependent on the thermodynamics, force balance, and interaction patterns. Generally a variety of weaker noncovalent interactions, including van der Waals, electrostatic, and hydrophobic interactions as well as hydrogen and coordination bonds, is formed between the assembled molecules, substrate surface, as well as between molecules in the adjacent layers. Thermodynamically, any material or system is stable only when it is in a state with the lowest Gibbs free energy. Therefore, there is a strong tendency for a solid or a liquid to minimize the total surface energy (Cao and Wang, 2011). From the thermodynamic point of view, self-assembly is an equilibrium process representing a balance between repulsive and attractive forces between the entities that relate minimization of Gibbs free energy (Whitesides and Boncheva, 2002). It makes the entities closer to a thermodynamic equilibrium state. So the interference of external force is not generally thought of as it might increase the Gibbs free energy (O'Mahony, 2011).

Self-assembled molecular building blocks are associated with each other by means of weaker bonding. This bonding has a strong dependency on balancing the forces of attraction and repulsion between the molecular building blocks that form supramolecular structures (Padua and Wang, 2012). Balancing the forces of attraction and repulsion between the molecular building blocks is strongly influenced by a number of parameters. External parameters like change in pH, concentration, ionic strength, temperature, mechanical forces (pressure, shear, extension, and ultrasound), or electric and magnetic field strength have a strong influence on the self-assembly process relating to the structure–function relationship. The final structure relating to function is strongly dependent on the chemical, physical, and processing conditions applied to influence the balance of intramolecular and intermolecular forces between the components in the system (Padua and Wang, 2012). Free uncontrolled random movement of the reacting building blocks facilitates the self-assembly (Chen et al., 2014). Thus, the liquid environment is always preferable, facilitating flexibility in the architecture allowing the components to readjust their position within the structure (Whitesides and Boncheva, 2002).

Nature offers self-assemblies in material or biological origins. This includes well-ordered clay, photonic crystal to composites of organic and inorganic shell, pearls, corals, bones, teeth wood, silk, horn collagen, muscle fiber, extracellular matrix of the plant, and even in multifunctional macromolecular assemblies in biology, that is, hemoglobin, polymerases, ATP synthesis, and membrane channels (Zhang, 2003).

Self-assembly in the food system is not a new concept. It was in use in the food systems but explored in recent times with the advancement of scientific research. Examples of molecular assembly in the food systems include self-assembly of casein in cheese and yogurt, casein micelle in milk as a natural carrier of calcium, self-assembly of polysaccharides or gelatine in sauces and puddings for thickening. Engineered self-assembled food structures can also be effectively achieved by tuning the environmental conditions (ie temperature, pH, ionic strength, the

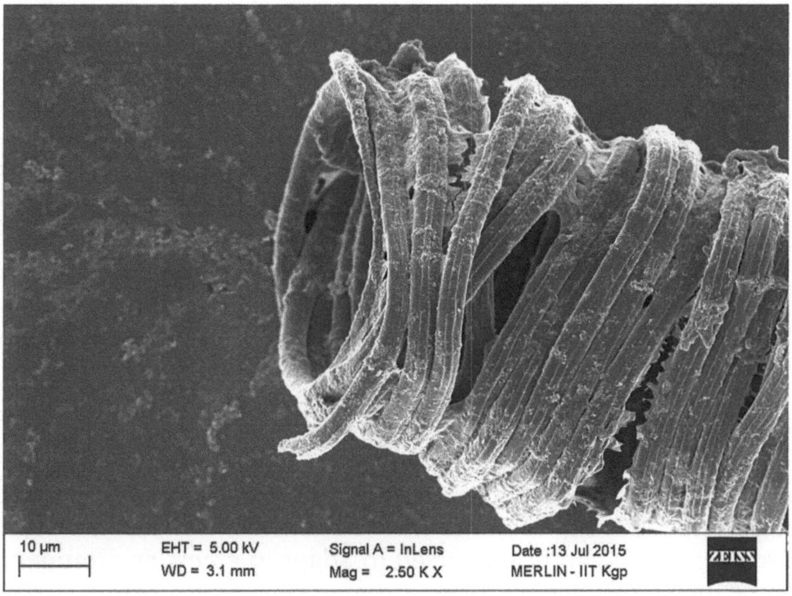

Figure 5.1. Scanning electron micrographs of natural polysaccharide. It is in the form of helical ribbon.

presence of specific substances or ions), processing condition (ie different applied forces), or the concentration of molecules/particles in the system. A strong understanding of the thermodynamic and molecular basis of food compound interactions is effective for self-assembly. To design novel systems with improved characteristics, food scientists and engineers should have a strong understanding of the thermodynamic and molecular basis of food compounds and their interactions. Assembled structures strongly relate to the physicochemical, nutritional and sensory quality of final products (Fig. 5.1).

3 Categorization of Self-Assembly on Thermodynamic and Molecular Basis

In self-assembly, interactions at molecular scales form nanoscale structure. The process of forming of such nanoscale structures does not require large amounts of energy. Instead the structures are formed via a series of optimized process utilizing the tendency of a system to minimize its overall free energy. As mentioned earlier, materials or systems are stable only when they are in a state with the lowest Gibbs free energy. Thus, the tendency of any system would be the reduction of Gibbs free energy or minimization of the total surface energies so that assembled systems are in a state closer to a thermodynamic equilibrium. Based on a thermodynamic equilibrium state, self-assembly is categorized as static or dynamic. When the system is in equilibrium or there is no relevant external force, building block forms a static self-assembled stable equilibrium state by minimizing the total surface energies (Ozin et al., 2009). The stability remains the same unless the environment is altered, leading to a different equilibrium state (Whitesides and Boncheva, 2002). Out of the equilibrium or in the presence of external forces and fields, dynamic self-assembly with the energy minimum is generally observed (Cademartiri and Ozin, 2009). The structure formed in dynamic self-assembly depends on the amounts of energy injected into the system. This situation with nonequilibrium structure or dynamic self-assembly retains its order as long as the system is dissipating energy (Huang, 2012). Dynamic self-assembly is an adaptive architecture. It can mutate itself to adapt to a new environment by occupying an energetic minimum that might be caused by the influx of energy in the system.

It is quite interesting that when the supply or flow of energy into the system stops, energetic minimum disappears. That leads the stable system to return into equilibrium (Ozin et al., 2009). Most common examples of dynamic self-assembly can be found in any living system. Here food, oxygen, and heat supply the source of energy keeping the body working. In the absence of energy, the system disassembles or ceases (Ozin et al., 2009).

Static self-assemblies are intended to get specific complex architectures relating to the specific functional activity like high surface areas for sensing, high strength for protective coating, low thermal conductivity for thermoelectrics (Cademartiri and Ozin, 2009). While static assembly is a cheaper way to fabricate complex structures that assign particular functional use. In sauces and pudding, polysaccharides or gelatine is used. Here, polysaccharide forms self-assembled structures with the other food ingredient during mixing and serves as a thickener.

Bartosz Grzybowsky's group at Northwestern University further subdivided static and dynamic self-assembly into coassembly, directed self-assembly and hierarchical self-assembly (Whitesides and Grzybowski, 2002). When two or more different components self-assemble within the same system contemporaneously and interdependently, a complex architecture is formed. It is then termed as coassembly. The components are usually segregated and not entirely mixed. The architecture formed from multicomponent assembly can never be generated by the isolated self-assembly of individual subunits. In directed self-assembly, external forces like order of mixing, pH, temperature, and ionic strength-time profiles is used to direct the self-assembly of the fundamental building blocks. In hierarchal self-assembly, fundamental building blocks self-assemble themselves with short-range forces. Such formed primary structures become the structural basis of these secondary structures followed by the highest level in the hierarchy by long-range forces (Whitesides and Grzybowski, 2002; Cademartiri and Ozin, 2009).

Based on the entity of building block or molecular unit and its arrangement, self-assembly can be simply divided into intermolecular and intramolecular self-assembly. Intermolecular self-assembly drives nanostructure formation, whereas intramolecular self-assembly results in folding of oligomers or polymers into nanostructures. A familiar example of intramolecular self-assembly is the folding of proteins, that is, spontaneous formation of globular structure by polypeptide chains whereas the spontaneous formation of the multiprotein complex like ribosome, cell membrane from various components are examples of intermolecular self-assembly (Sattler, 2010).

4 Self-Assembled Carbohydrate Matrix, A Unique Carrier for Functional Food Formulations

Protein, fat, and carbohydrates are the major molecular components of food. Carbohydrates and protein generally form natural nanostructures due to their structural features. Complex carbohydrates are high molecular weight chain–like structures with a diameter in nanorange. They are extensively used in the food sector as thickeners, stabilizers of dispersion (emulsion and foam), improving textural property, structuring (water and air), and partial or total replacement of fat, and so on. For most of the cases, the required functional properties are achieved due to their self-assembly nature when subjected to the food environment. They find local interactions with or without the external force and form self-assembly with other food compounds. The mixture state of multiple compounds tends to stabilize, thus tries to minimize the Gibbs free energy. To do this, an equilibrium process representing a balance between repulsive and attractive forces in between entities is formed, making a closer to a thermodynamic equilibrium state. The uniqueness of carbohydrates lies on its polar or charged group. This offers carbohydrates to bind with a variety of hydrophilic and hydrophobic bioactive ingredients. Thus, it can be used as a carrier to bind and entrap a wide range of bioactive material with targeted delivery. Polysaccharides can also be effectively used for surface medication like coatings improving the bioadhesion of bioactive compounds and targeted delivery.

Complex polysaccharides are generally soluble. Most of the bioactive compounds face problems regarding their solubility (eg, Curcumin from turmeric). Carbohydrates, specifically polysaccharides, form localized interactions with a wide range of bioactive compounds via their functional groups. This makes them versatile carriers to bind and entrap a variety of hydrophilic and hydrophobic bioactive food ingredients, improving solubility hence enhances bioavailability (Fathi et al., 2014).

Slow intestinal digestion rate of polysaccharide makes it unique and most favorable for targeted and slow release of bioactive compounds. Most cases of bioactive and nutrient components suffer the transport system by overloading and causing undesirable excursions in plasma and tissue concentration.

Release of compound of interest and nutrients can be delayed by a polysaccharide-bioactive compound localized interaction, as polysaccharide is digested slowly in the gastrointestinal tract. Extreme harsh acidic conditions and a lot pH variation followed by enzyme interaction greatly reduce the viability of probiotic bacteria. Protective carriers or encapsulation over probiotic bacteria extremely improves the probiotic viability. Here, also, polysaccharide layer-by-layer assembly finds significant results (Priya et al., 2011). More interestingly, as dietary fiber directly reaches to the colon, colon-specific delivery can also be done by polysaccharide assemblies. Polysaccharides are well known for their intrinsic functionalities that contribute to the texture of the food products. In addition, carbohydrates are temperature stable, hence under high temperature they offer a thermal protective shell to the bioactive ingredients (Fathi et al., 2014). Physicochemical properties of individual biopolymers are strongly related to self-assembly. This may be changed by critically controlling environmental factors, that is, ionic strength, pH, temperature, and presence of selected ions. The viscosity of a food has a strong relationship with the flavor perception. High viscous food delays the release of aroma and taste compounds. Thus, appropriate emulsion formulations are a key issue and can be effectively handled by the polysaccharide formulations. More food-flavor manufacturers are also searching for polysaccharide-based better delivery systems, improving and enhancing flavor quality and sensory experience.

In addition, there are several examples of complex carbohydrates offering bioactivity. Some of the benefits include increasing stool bulk and helping to regulate bowel movement, providing immunomodulatory activity, antithrombotic activity, free radicals eliminating activity, cholesterol-lowering effect, controlling blood glucose level and lipids, reducing hypertension, stimulating the immune system, and influencing probiotic growth (Liu et al., 2014). Thus, application of bioactive polysaccharide matrix may also serve functional benefits in addition to enhancing stability and bioavailability.

5 Carbohydrate-Based Different Self-Assembled Nanostructures

Chemical complementary and structural compatibility binds or self-assembles building blocks together by weak or covalent bonding interactions and form different hierarchical architectures with specific functional activity. Nanostructures can be subdivided based on their dimension and phase. It is also applicable to a self-assembled structure. Based on dimensions, nanoparticles are generally classified as zero-dimensional, one-dimensional, two-dimensional, and three-dimensional (Cao and Wang, 2011). Particles can also be classified based on phase as a single phase, multiphase, and composite. Organic nanoparticles pose rare zero-dimensional structure (metal nanoparticles form zero dimension). Self-assembled structure also falls under the same classification route. The vast majority of the food carbohydrates are one-dimensional. In a natural self-assembled polysaccharide matrix, one-dimensional growth is also seen. Dynamic self-assembly by electrospinning offers a self-assembled one-dimensional nanostructure offering exceptionally high surface areas that lead to high encapsulation efficiency (Rezaei et al., 2015). It applies a high voltage (usually 5–50 kV) to draw a fine electrically charged fiber or thread from a polymer solution (Schiffman and Schauer, 2008). Drawing a fine charged fiber can only be possible if the solution carries a charge. But most of the biopolymers, including carbohydrates, are seen to be neutral polymers in its electric pH (Schiffman and Schauer, 2008). Several efforts are made by chemical modifications or experimental conditions for electrospinning of uncharged biopolymers. Stijnman et al. (2011) have studied electrospinning using food-grade polysaccharides from plants and concluded that the concentration of the biopolymers should exceed a critical level. The solution should also exhibit a shear thinning behavior. They suggested that equipment settings and conductivity are of secondary importance and merely play a role in the optimization of the process (Stijnman et al., 2011). Electrospinning was effectively used as food grade nanofibrous thin films. Here the blend solution was prepared by incorporating cellulose acetate in 85% acetic acid and egg albumen in 50% formic acid by electrospinning. The mass percentage ratio of two different mixtures were further varied to get optimum results. Here conductivity and surface tension of the solutions were tuned using cellulose acetate and surfactant (Tween40®) (Wongsasulak et al., 2010). Cyclodextrin inclusion complex plays an important role in functional food delivery. Fatma Kayaci and Tamer Uyar in 2012 synthesized vanillin/cyclodextrin inclusion complex improving shelf life and temperature stability. They further electrospun this inclusion complex in polyvinyl alcohol (PVA) nanowebs for further stability and slow release of vanillin (Kayaci and Uyar, 2012).

In a separate approach, negatively charged electrospun cellulose acetate fibrous mats were prepared. This mat was further modified with layer-by-layer self-assembly techniques incorporating positively charged lysozyme–chitosan–organic rectorite composites and the negatively charged sodium alginate. Gained antimicrobial potential over electrospun cellulose acetate fiber suggests possible application as mats for pork preservation (Huang et al., 2012).

In another study, easy oxidation of epigallocatechin-3-gallate was retarded forming copper complexes by chelating copper ions, and effectively electrospun into PVA nanofibers into an antimicrobial nanofibrous membrane (Sun et al., 2011).

Preparation of electrospun starch acetate nanofibers, which were prepared using formic acid/water as a solvent, finds successful applications in drug delivery (Xu et al., 2009).

Recent time delivery of probiotic bacteria using a protective carrier is gaining more attention. In developing new probiotic food, protection of probiotic bacteria varying harsh environmental conditions are crucial as probiotic in general, and some strains in particular show lower resistances to different environmental conditions, such as oxygen, light, or temperature (Martín et al., 2014). Electrospinning probiotic delivery is also dependent on the nature of biopolymer conditions and modifications of the biopolymer. Agro waste dietary fiber was used with amending 8% poly(vinyl alcohol) for the preparation of electrospinning nanofiber encapsulated with *Lactobacillus acidophilus* (Fung et al., 2011). Bacteria-encapsulated electrospun nanofibers also provide good bacterial survivability (78.6–90%) under electrospinning conditions and even at refrigerated storage for 21 days. Thermal behavior of electrospun nanofibers offers possible thermal stability of probiotic bacteria in heat-processed foods. Amparo López-Rubio and coworkers have done a comparative study of

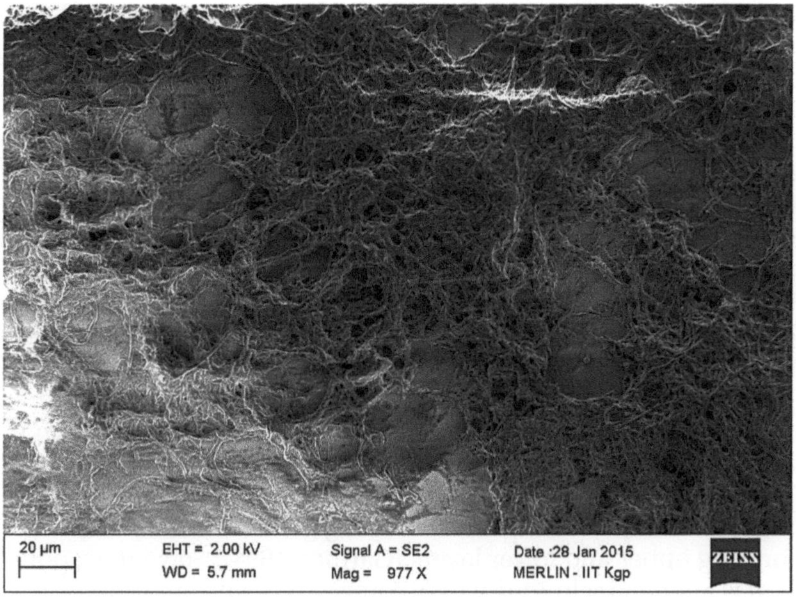

Figure 5.2. Distributed nanofiber in polysaccharide hydrogel.

electrospinning using both protein-based (whey protein concentrate) and a carbohydrate-based matrix (pullulan) to deliver *Bifidobacterium animalis* ssp. *lactis Bb12*. The viability of bacteria was seen to be higher in protein matrix as compared to carbohydrate matrix (López-Rubio et al., 2012). Most of the biopolymers form three-dimensional gel structures by via noncovalent bonds. Polysaccharide offers both weak (ie, xanthan) and strong gel (ie, agarose) (Ross-Murphy and Shatwell, 1993) structures. In some specific conditions like proper pH and temperature, the physical associations become strong enough to form strong gels (Shewan and Stokes, 2013). Incorporation of bioactive molecule in this three-dimensional network often solves the problem of bioactive delivery or even for specific targets. Functional food ingredients, that is, flavors (Heitfeld et al., 2008), polyphenols (Hu et al., 2008) are commonly used in hydrogel delivery (Fig. 5.2).

Phase separation or solvent evaporation, reverse micellar method is most often used to prepare the self-assembled structures. Upon mixing, materials of two different compositions may become completely miscible, or may separate into different phases. The relative strength of the materials and their intra-molecular interaction pattern plays the key role behind it. If two solution containing different molecules are mixed, and their interaction is greater than the interaction of individual liquids and molecules, then they will form a miscible liquid. But if the opposite becomes true, they would not be miscible and form phase separation (McClements, 2009). In mixed biopolymer systems, phase separation is very common due to strong thermodynamic instability among the different building blocks. For food systems, carbohydrate and proteins may undergo phase separation due to hydrophobicity or hydrophilicity arising from the molecular structure, conformation, and affinity to water (Moraru et al., 2002). Gelatin, a protein obtained from partial hydrolysis of collagen, is used to form phase-separated hydrogel with different carbohydrates (eg, maltodextrin, dextran, and sodium carboxymethyl cellulose) (Singh et al., 2014). Examples of carbohydrate phase separation is also seen. An example includes phase separation of amylose and amylopectin solution due to precipitation of amylopectin by capturing a lot of solvents (German et al., 1992). This forces amylase to aggregate. Dextran and locust bean phase separate in aqueous solutions forming gel rich and liquid phases forming upper and lower layers (Garnier, 1994). Solvent evaporation is another technique where compounds with appropriate solvent are mixed together. With the removal of a particular solvent, molecules form an assembled structure in the available matrix (Maruyama et al., 1994; Ichikawa et al., 2005). Foreign crosslinker such as gluteraldehyde is often used to do crosslinking to form the assembled structures (Sacchettini et al., 2001; Patenaude and Hoare, 2012). A similar approach called reserve micellar method is also used for forming self-assembled structures. Here generally surfactants are dissolved in organic solvents followed by the addition of aqueous solutions of biopolymer and encapsulent with constant stirring. Solvent evaporation, redispersion in water, salt precipitation followed by centrifugation leads to encapsulent-loaded nanoparticles (Vrignaud et al., 2012; Agnihotri et al., 2004). Dextran–doxorubicin conjugate using chitosan is an example of reserve micellar method (Mitra et al., 2001). Hydrogel is a method to form assembled structures. When the material stays in the gel environment, it poses three-dimensional structures. Crosslinking of xanthan chains by esterification at 165°C either in the presence of citric acid yield hydrogel with effective swelling (Bueno et al., 2013). Another example includes self-assembled nanofibers of carbohydrate-functionalized mono- and di(2,2′:6′,2″-terpyridinyl) arenes (Chan et al., 2009), multifunctional colloidal carbohydrate nanofiber electrolytes (Gökmen et al., 2015), and so on. Two-dimensional self-assembled nanostructures include thin film coating, laminates, and nanocomposites. Examples include biobased nanocomposites obtained from chitosan and cellulose nanocrystals via a covalent linkage (de Mesquita et al., 2012) (Fig. 5.3).

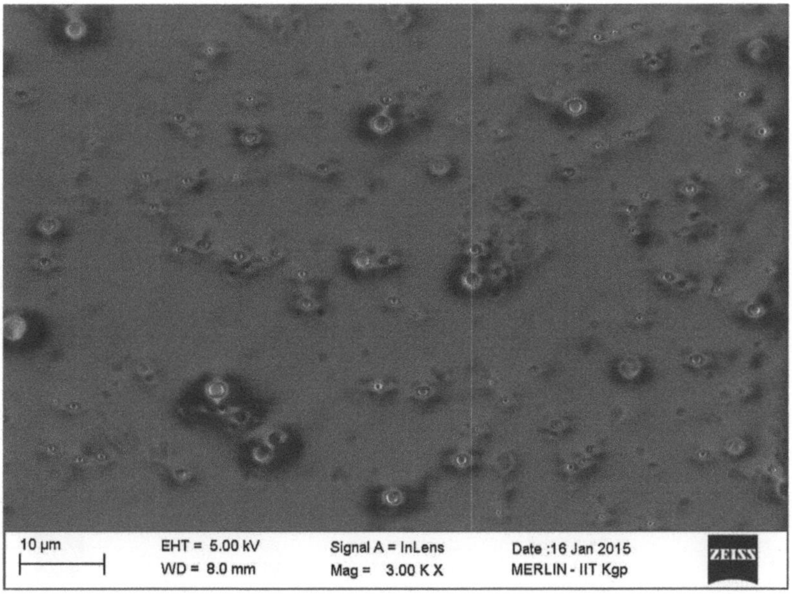

Figure 5.3. Polysaccharide micelle carrying bioactive molecules.

From localized self-assembled interactions carbohydrate-based three-dimensional micelle can be easily formed and targeted for several functional applications. Examples include amphiphilic chitosan-based polymeric self-assembly in aqueous media at low micromolar concentrations giving micellar clusters of 100–300 nm in size-enhancing hydrophobic drug bioavailability (Qu et al., 2006) and self-assembled carbohydrate-based micelles for lectin targeting (Dal Bó et al., 2011). Layer-by-layer assembly is an important technique for developing three-dimensional structures. Here, alternating multilayers of polyelectrolytes are built up through electrostatic deposition. The electrostatic attraction between negatively charged polyanions and positively charged polycations followed by complex formation is the key science behind the layer-by-layer assembly (Fig. 5.4).

In the food industry, layer-by-layer assembly is a newer concept and can be effectively used to prevent the texture degradation and encapsulate colors and flavors (McClements, 2009). Uta Klinkesorn et al. (2005a) effectively encapsulated emulsified tuna oil in two-layered interfacial membranes prepared by electrostatic layer-by-layer deposition. In another study, they showed increased oxidative stability of liquid and dried tuna oil-in-water emulsions with electrostatic layer-by-layer deposition (Klinkesorn et al., 2005b). Layer-by-layer assembly approach has recently been used to encapsulate probiotic *L. acidophilus* to enhance its survival

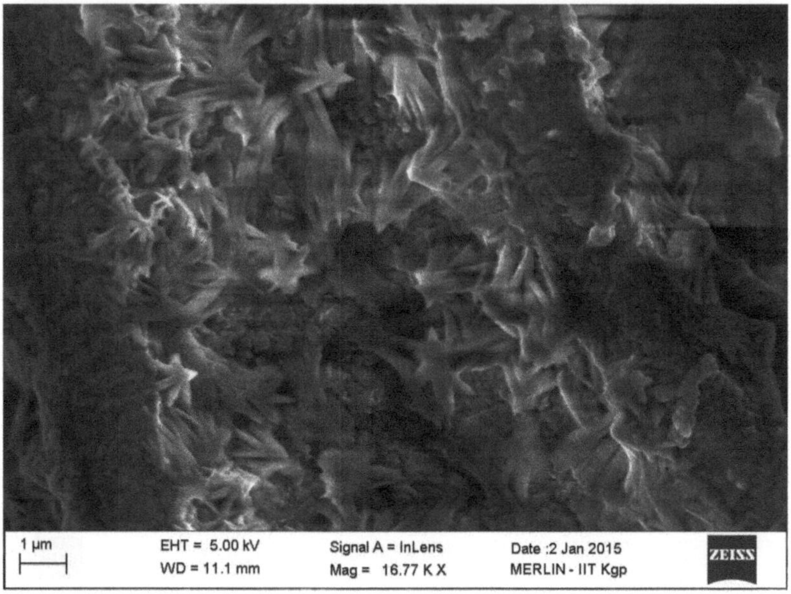

Figure 5.4. Polysaccharide-based nanocomposites.

in adverse conditions encountered in the gastrointestinal tract. With this approach, the survival of encapsulated cells in simulated gastric and intestinal fluids improved significantly (Priya et al., 2011).

6 Recent Advances in Carbohydrate-Based Self-Assembling Entities, its Preparation, and Applications

Foods are complex entities comprising a mixture of proteins, polysaccharides, lipids, surfactants, minerals, and nutraceuticals. For particular functional applications via self-assembly strategies, an experimental condition also varies significantly. Some complex polysaccharide carbohydrates have shown the potential application in self-assembly–based delivery. Here efforts have been given to arrange carbohydrates according to their origin that have been effectively used in self-assembly relating to their food applications.

7 Carbohydrates From Plant Origin

7.1 Starch

Starch is a low-cost carbohydrate abundantly found in plants serving as a major food reserve providing a bulk nutrient and energy source. It consists of a large number of glucose units joined by glycosidic bonds. Its morphologies and sizes in natural origin vary between 1 and 100 μm. It is a heteropolysaccharide having two distinct polysaccharides, that is, amylase and amylopectins in the ratio of 1:3. Amylases are straight chain polymer of D-glucose unit linked by (1–4)-glycosidic linkage. Amylopectins are branched chain polymer of D-glucose unit linked by (1–4)-glycosidic linkage at straight chain and (1–6)-glycosidic linkage in branched chain. Starch molecules often pose natural self-assembled structures.

Spherulite formation is an example of self-assembly, where the building blocks spontaneously arrange themselves into radially symmetric semicrystalline structures (Ma et al., 2011) offering additional benefits. Several approaches are made to form functionally active starch by spherulite formation. Starch spherulitic crystallization formation is induced by the formation of amylose inclusion complexes with fatty acids. This strategy follows heating of aqueous starch dispersion to 140°C in the presence of lipids followed by slow cooling (Fanta et al., 2002, 2005, 2006; Peterson et al., 2005; Shogren et al., 2006). In a simpler approach, spherulite is formed by the rapid heating of an aqueous starch dispersion above 170°C followed by rapid cooling (Nordmark and Ziegler, 2002; Nordmark, 2002; Ziegler et al., 2003, 2005). Starches do not offer controlled digestion and are digested either too rapidly or resistant to enzyme digestion, so that controlled and predetermined digestibility of the starch is very much opted. Researchers from Kansas State University offer a simpler novel approach combining enzymatic debranching, melting, and crystallization to produce spherulites from short linear α-1,4-linked glucans [short-chain amylose (SCA)]. Enzymatic debranched maize starch was obtained at 50°C and 25% solids in 0.01 M sodium acetate buffer. The mixture was then heated to 180°C followed by cooling and crystallization to form well-developed spherulites (Cai and Shi, 2013) offering controlled, predetermined digestibility.

Physiochemical properties make starch limited to different food applications. Several modifications are also done to overcome such drawbacks. Acid conversion, oxidization of starch, pyroconversion, or dextrinization are some common strategies used for modification of starch for better functionality. Quite similarly, modification of starch is also important for self-assembly.

Attachment of hydrophobic and/or hydrophilic groups to carbohydrate main structure alters the functional behavior in solutions leading to further opportunities to modulate self-assembly of polysaccharides. Examples include hydrophobic modification of hydroxyethyl starch (a water-soluble, semisynthetic polysaccharide that is used as a plasma volume expander and cryoprotectant) to produce a fully biodegradable amphiphilic polymer. Here, hydroxyethyl starch was esterified with lauric, palmitic, and stearic acids under mild reaction conditions using dicyclohexyl carbodiimide and dimethylaminopyridine. Aqueous self-assembly facilitates formation of micelles and polymeric vesicles. Hydrophobically modified hydroxyethyl starch may find possible applications as a potential drug carrier and encapsulated delivery (Besheer et al., 2007). In another study, hydrophobically modified starch was used to form micelles (diameter ~14 nm) to encapsulate curcumin. Upon encapsulation, curcumin showed increased solubility and anticancer activity (Yu and Huang, 2010).

Bimolecular assembly using starch matrix also finds promising application in controlled delivery and release of bioactive molecules. Efforts were made to produce starch piperinic esters by coupling a carboxyl group on the piperic acid and a hydroxyl group on the starch backbone. The synthesis was carried out using piperic acid from hydrolyzing piperine. It was further reacted with carbonyl diimidazole under N_2 at 70°C for 4 h to form an activated piperic acid. Activated piperic acid was reacted with water-soluble starch at 80°C under N_2 for 24 h to produce starch piperinic ester. This ester was found to produce self-assembled spherical micelle in aqueous solutions (Han et al., 2008). Starch, in particular amylose, is also capable of forming inclusion complexes with various molecules, for example, volatiles, fatty acids. Noncovalent binding between amylose and the guest molecule induces the formation of amylose helices. These amylose helices have a strong tendency to self-assemble via helix–helix associations into supramolecular structures (Conde-Petit et al., 2006; Lesmes et al., 2008; Yang et al., 2009).

7.2 Cyclodextrins

Cyclodextrins are a family of cyclic oligosaccharides consisting of 6 (α), 7 (β), 8 (γ), or 9 (δ) glucose monomers unit linked through α (1–4) glycosidic bonds with a hollow truncated conical structure (Del Valle, 2004). This is an enzymatic decomposition product of amylase, also commonly known as cycloamyloses, cyclomaltoses, and Schardinger dextrins (Eastburn and Tao, 1994). Cyclodextrin contains a relatively hydrophobic internal cavity that forms inclusion complexes with a variety of molecules including fats, flavors, and colors via molecular self-assembly. It is widely used in the food formulations for flavor protection or delivery and/or labile flavoring materials present in a multicomponent food system during the harsh processing storage and distribution (Astray et al., 2009). It also has a hydrophilic surface, containing primary and secondary hydroxyl groups. Inclusion complexation with cyclodextrins effectively facilitates conversion of volatile substances from gas or liquid state into solid powders. It is also used for protection against oxidative degradation, heat-induced changes, and light-induced decomposition of food flavor and bioactives. During thermal processes, that is, cooking, pasteurization β-cyclodextrins are found to be most effective at retaining the aroma compound (Jouquand et al., 2004; Bhandari et al., 2001). It removes polyphenol-oxidase from juices, protects phenolic compounds from enzymatic oxidation (Astray et al., 2009; Irwin et al., 1994).

Cyclodextrins inclusion complexes have also been used for taste modifications, elimination of bitter and disgusting tastes and odors of foods. The example includes deodorizing soybean milk, soy protein, peculiar fish odors (Astray et al., 2009), unpleasant off-flavor of rice (Astray et al., 2009). Even more, it is now being used for cholesterol sequestrant from animal products improving their nutritional characteristics. Silanization and β-cyclodextrins immobilization mediated immobilized β-cyclodextrins in glass beads resulted in 41% of cholesterol removal rate in milk with a recycling efficiency of almost 100% (Jung et al., 2008). In another case crosslinked β-cyclodextrin used for mayonnaise resulted in a decrease in oxidation, maintaining its physicochemical and sensory properties during storage (Jung et al., 2008). Five percent crosslinked β-cyclodextrins treated with butter and lard removes 91.2 and 93.0% cholesterol respectively (Jung et al., 2005). A similar reduction was also seen in cheese and eggs (Del Valle, 2004; Astray et al., 2009) (Fig. 5.5).

7.3 Inulin

Inulins are a group of the naturally occurring heterogeneous blend of fructose polymers consisting of β-(1-2)-D-fructofuranosyl links. It is mainly produced by the plant as storage carbohydrates. It offers several nutritional benefits including hypolipidemic effects, prebiotic properties, which influence gut microbiota, stimulate mineral (calcium, magnesium) absorption (Capriles and Arêas, 2013; Cherbut, 2002; Brennan et al., 2004). Inulin when hydrophobically modified offers self-assembly with functional benefit. The strategy followed for inulin is hydrophobic modifications. Kokubun et al. (2013) successfully modified inulin using a low-energy, environment friendly process by the derivatization of inulin with octenyl and dodecenyl succinic anhydrides in aqueous solution (Kokubun et al., 2013).

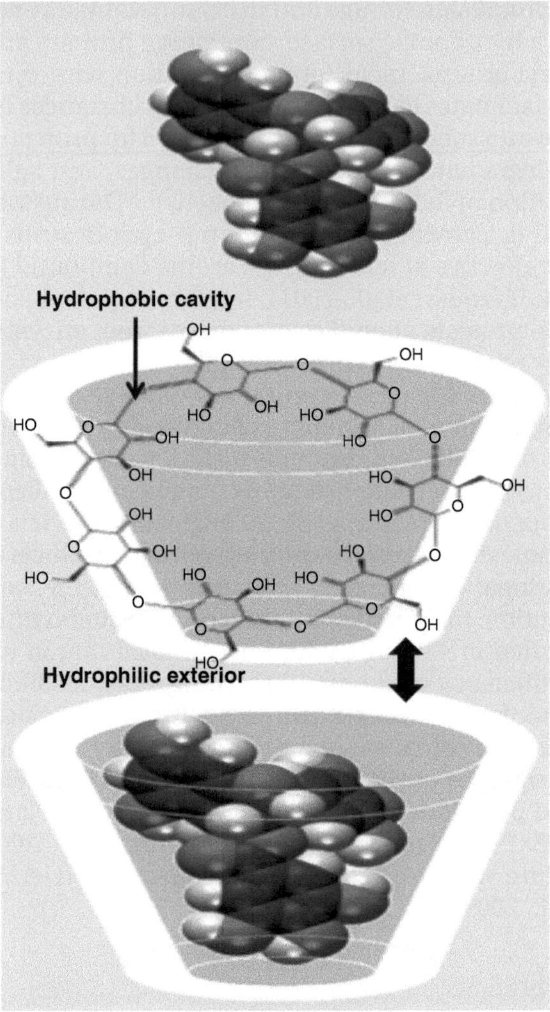

Figure 5.5. Schematic illustrations of the association of free cyclodextrin (CD) with bioactive molecule and formation of its inclusion complex.

In another work, hydrophobic modifications (alkenylated inulin) were carried out in aqueous solution using alkenyl succinic anhydrides with varying alkenyl chain lengths (C8–C18) (Han et al., 2015).

The derivatives form aggregates in solutions and stabilize emulsions and thus can be used as suitable candidates for stabilization of particulate dispersions and emulsions. They can also be effectively used in encapsulation and delivery of water-insoluble active compounds with potential application in food, cosmetic, personal care, and pharmaceutical formulations.

7.4 Cellulose

Cellulose is the world's most abundant naturally occurring polysaccharide consisting of a linear chain of several hundred to many thousands of β (1→4) linked D-glucose units. In spite of huge productivity of cellulose, indigestion pattern makes it less usable in human food and is consumed by animals only. Nevertheless, efforts are starting to make use of the cellulose to produce processed animal feed acceptable to both ruminants and nonruminants (Gentry et al., 1996; Kuhad et al., 2011). Besides, cellulose performs a critical function as dietary fiber. Self-assembly strategies are followed with the cellulose to improve its functional application. A major problem arises from the cellulose in its dissolution in common solvents, which is due to stiff cellulose molecular chains and close chain packing through numerous hydrogen bonds. Dynamic self-assembly induced rapid dissolution of cellulose at low temperatures is proposed where dissolution can be achieved in precooled (−12°C) 7 wt% NaOH–12 wt% urea aqueous solution within 2 min (Cai et al., 2008). Electrospinning technology can also be effectively used in the assembly of cellulose for different food applications (Rezaei et al., 2015).

7.5 Pectin

Pectin is an important food gelling agent, most commonly found in the cell walls of terrestrial plants. It is basically a heteropolysaccharide consisting of linearly connected α-D-(1→4) galacturonic acid residues. Its resistant nature against oral and stomach enzymes and specific degradation by gut microbiota (Salyers et al., 1977) can make it an effective carrier for acid-sensitive bioactives. But the problem arises from its high solubility that acts as a drawback in making protective shields over bioactives. Several strategies are followed to develop pectin derivatives, overcoming the drawbacks of solubility but keeping bacterial degradations the same (Sinha and Kumria, 2001). In another way, localized interactions with pectin were most often used in functional ingredient delivery. Heating the pectin and β-lactoglobulin solutions above the thermal denaturation temperature of the protein form nanoparticles of 150–500 nm range via electrostatic interaction (Jones and McClements, 2010). In a separate study, core–shell biopolymer particles were prepared by electrostatic deposition of anionic beet pectin onto amphoteric protein aggregates (Santipanichwong et al., 2008). Pectin-beta-lactoglobulin nanocomplexes are also found to be an efficient carrier for ω-3 polyunsaturated fatty acids (Zimet and Livney, 2009). In a recent study, pectin-based self-assembled nanostructures are formed using a mixture of aleuritic acid, tomato fruit cutin monomers, and 16-hydroxyhexadecanoic acid (7.5%, w/w) with pectin in aqueous solution (Guzman-Puyol et al., 2015).

8 Carbohydrates From Animal Origin

8.1 Chitosan

Chitosan is a natural polysaccharide mainly obtained from shells of shrimp and other sea crustaceans. It consists of varying amounts of (1–4)-glycosidic bonds linking *N*-acetyl-2-amino-2-deoxy-D-glucopyranose (glucosamine) and 2-amino-2-deoxy-D-glucopyranose (*N*-acetyl-glucosamine) (Ravi Kumar, 2000; Agrawal et al., 2010). Chitosan is widely used in the food industry due to its functional benefits. It shows antimicrobial action; acts as controlled releases medium of antioxidants, nutrients, flavors, and drugs. It is also used as an additive for clarification and deacidification of fruits and beverages, controls texture, and many more.

Cationic and hydrophobic sites of chitosan make it an intrinsic polyelectrolyte and amphiphile. Chitosan becomes highly positively charged in an acidic medium due to protonation of its amino groups. Besides, the intra- and intermolecular hydrogen bonds (due to the presence of the single bond —OH and single bond —NH_2 groups along the chitosan backbone) also contribute to the self-assembly. Cationic nature of the chitosan makes the favorable binding platform to anionic surfaces via electrostatic and hydrogen bonding. Binding of chitosan with oppositely charged biopolymers like alginate, pectin, xanthan, carrageenan, gum arabic, anionic lipids could provide interesting nanostructured delivery systems for various ingredients (Huang, 2012). An example includes formation of self-assembled biocompatible nanoparticles ranging from 123 to 350 nm using chitosan and modified lecithin for successful delivery of hydrophilic compounds (Chuah et al., 2009). Here, NH_3 groups of chitosan form electrostatic assemblies with the phosphate groups of lecithin. The nanoparticles exhibited excellent stability over an extended pH (3–6) and ionic strength range, and could be readily converted into a freeze-dried powder.

Also, chitosan chains can be trapped by the electrostatic attractions. This strategy was used to create stable and pH-sensitive nanogels by self-assembly of chitosan and ovalbumin. Upon heating, electrostatic attractions between chitosan and ovalbumin partially trap the chitosan chains in the nanogel core, while the rest of the chitosan chains form the shell of the nanogels. The nanogels did not change the size distribution after long-time storage and did not dissociate in the pH range of 2–10.5 (Yu et al., 2006). In a separate approach, a self-assembled chitosan-β-lactoglobulin core–shell nanoparticle as a biocompatible carrier for the oral administration of nutraceuticals was successfully prepared. It was found to be stable in simulated gastric fluid. Here uniform nanoparticles were prepared by ionic gelation with sodium tripolyphosphate. Electrostatic, hydrophobic interactions and hydrogen bonding between β-lactoglobulin and chitosan form the self-assembly (Chen and Subirade, 2005). Self-assembled complexes of chitosan with α-lactalbumin or with β-lactoglobulin have also been studied as potential delivery vehicles (Lee and Hong, 2009).

Ichikawa and coresearchers developed a novel preparation method for biocompatible nanoparticles of about 200 nm in mean size in high concentration by self-assembly of chitosan and carboxymethyl cellulose hydrolysates. They initially hydrolyzed chitosan and carboxymethyl cellulose making fragments of lower molecular weights. Finally nanoparticles were spontaneously formed only by mixing the two hydrolysate solutions that were stable, suspended over 1 week even under low pH (pH 3.0), high ionic strength (NaCl 1 M), or low temperature (4°C) conditions (Ichikawa et al., 2005). In a recent study, conjugated polymer/carboxymethyl chitosan grafted poly(*p*-dioxanone) nanomicelles are developed and used in functionalized indicator paper for fast and visual detection of a banned food dye (Chen et al., 2014). Since chitosan backbone has many functional groups, chemical modification can also be carried out for better self-assembly (Yang et al., 2014). Chitosan was modified by coupling with linolenic acid through the 1-ethyl-3-(3-dimethylaminopropyyl) carbodiimide-mediated reaction offering self-assembled nanoscale carrier for protein.

Self-aggregated nanoparticles exhibited an increased loading capacity of protein (Liu et al., 2005). Besides, many examples are also available on self-assembled chitosan nanomaterials for drug delivery applications (Yang et al., 2014).

9 Carbohydrates From Algal Origin

9.1 Sulfated Polysaccharides

Sulfated polysaccharides are most interesting group of polysaccharides found in marine algae. Interest in the sulfated polysaccharides, that is, fucoidans and laminarans in brown algae (Phaeophyceae), carrageenans in red algae (Rhodophyceae), and ulvans in green algae (Chlorophyceae) (Wrigstedt et al., 2010) in self-assembly applications is increasing due to their unique functional applications like binding with proteins. Their binding with proteins is due to self-assembly and finds applications in dairy and meat products.

Models studying the interactions of fucoidan with bovine serum albumin prove self-assembled interactions at varying pH and ionic strength (Kim and Shin, 2015). Carrageenans are another widely studied and used polysaccharide in the food industry for their gelling, thickening, and stabilizing properties. Carrageenans also strongly binds with food protein. In an in vitro model, they bind with protamine and form carrageenan/protamine polyelectrolyte nanoplexes (Dul et al., 2015).

9.2 Alginate

Alginate is another polysaccharide that forms self-assembled diarchy and finds applications in drug delivery. It is an anionic polysaccharide distributed widely in the cell walls of brown algae. It binds with water (200–300 times its own weight) and forms a viscous gum. Alginate does not provide any nutritional value, but its ability to form gel and stabilize aqueous mixtures, dispersions, and emulsion make its use as an additive. In the self-assembly process alginate serves as a base material. Several examples are showing self-assembled alginate matrix in delivering bioactive and drug molecules. Examples include capsaicin-loaded nanoemulsions stabilized with alginate and chitosan (Choi et al., 2011), *N*-trimethylchitosan/alginate layer-by-layer self-assembled fungal repellent coatings preventing biofilm (Jiang et al., 2015), layer-by-layer assembled minocycline-loaded chitosan/alginate multilayer on titanium substrates inhibiting biofilm formation (Lv et al., 2014).

In a recent study, alginate-based amphiphilic graft copolymers were effectively synthesized by a single electron transfer living radical polymerization forming stable micelles during polymerization induced self-assembly (Kapishon et al., 2015).

9.3 Agar

Agar refers to a class of galactan polysaccharides obtained from the cell walls of some species of red algae. It binds with water, forms thermoreversible gels, and is widely used in microbiological media preparations. In food, it finds application as a gelling and encapsulating agent for microorganisms in biochemical reactors, and foodstuffs like baked goods, icings, glazes, jelly confections, canned products, sherbets, ice creams, and cheeses. Agar matrix is also used for novel self-assembly studies.

Polymorphic transformation and self-assembly method offered roughness-controlled mannitol/LB Agar microparticles (Zhang et al., 2015) for effective mannitol delivery.

9.4 Microbial Polysaccharides

Production of polysaccharide is very common to bacteria, yeasts, and fungi. Although dextran is a bacterial polysaccharide, found first in commercial applications, but the addition of external microbial polysaccharide in food is still limited. Bacterial polysaccharides are generally produced during microbial growth in fermented food preparations (especially lactic acid bacteria mediated). Recent attention has shifted toward using microbial polysaccharides on developing advanced carbohydrate-based self-assembled structures for functional food ingredient and drug delivery. Marine gel biopolymers generally self-assemble and form the fibril forming network having a low to high degree of crosslinking (Pletikapić et al., 2014). Even more, formation of interlinked structure of bacterial biofilm is also an example of self-assembly in natural origin (Ploux et al., 2007; Bobeth et al., 2011; Bromley et al., 2015). Individually, bacterial polysaccharides are seen also to offer self-assembled nanostructures having functional applications.

Dextran, complex branched glucan, is composed of chains of varying length of glucose by mostly α-(1→6) and a small percentage of α-(1→3) linkages. It is also reported to have several medicinal benefits (Jordan, 1956). Hydroxy group in its structures forms covalent attachment to a variety of organic functional groups, especially hydrophobic compounds (Kaewprapan et al., 2012). Under the presence of crosslinker—*N,N'*-methylene bisacrylamide, self-assembly assisted graft copolymerization of a dextran fraction forms stabilized nanoparticles with pH-sensitivity and controllable dimensions (Tang et al., 2006). Amphiphilic block copolymer, composed of hydrophilic dextran sulfate as the targeting ligand and hydrophobic polycaprolactone as the hydrophobic segment, formed self-assembled nanoparticles via click chemistry. This development was targeted to rheumatoid arthritis treatment (Kim et al., 2013).

In another study, self-assembled anticancer multifunctional nanogel was formed having dextran–curcumin conjugates (Nagahama et al., 2015). In a recent study, stable soy β-conglycinin–dextran core–shell nanogels conjugates were prepared via self-assembly at the isoelectric point. Maillard dry-heating reaction was used to attach dextran to β-conglycinin. Formed conjugates were heated above the denaturation temperature at the isoelectric point (pH 4.8) so as to assemble nanogels with core–shell structures (Feng et al., 2015).

Xanthan, a food grade biopolymer obtained from plant pathogenic bacterium *Xanthomonas campestris*, is generally used as food additives for thickeners, stabilizers, rapid flavor releasers, and rheology modifiers. Attempts are made to modify the xanthan gum for functional applications. Palmitoylation of xanthan polysaccharide yields self-assembled microcapsules that effectively encapsulate cells in physiological conditions (Mendes et al., 2011). In another study, conjugation of phospholipid [1,2-dioleoyl-*sn*-glycero-phosphoetilamine (DOPE)] with the anionic xanthan gum spontaneously form self-assembled structures in mild aqueous conditions. Both strategies can be effectively used for cell-based therapies (Mendes et al., 2013). In another approach, xanthan gum was used to trigger the self-assembly of a positively charged multidomain peptide resulting in the formation of a nanofibrous matrix. Peptide/polysaccharide ratio was optimized for xanthan-peptide microcapsules preparations. Microfluidic device results in microcapsules with homogeneous size. Xanthan gum self-assembled materials with microfluidics are found to be effective in the fabrication of cytocompatible matrix for cell microencapsulation and delivery.

Another bacterial polysaccharide, curdlan, is also gaining attention in the self-assembly field due to its medicinal application. Structural modifications of curdlan lead to several functional applications. Incorporating hydrophobic moiety, sulfonylurea self-assembled hydrogel nanoparticles were effectively synthesized from curdlan focusing novel drug carriers for liver cancer (Na et al., 2000). Self-assembled cholesterol-conjugated carboxymethyl curdlan nanoparticles serve as a novel carrier of epirubicin (Li et al., 2010). In another study curdlan-*graft*-poly(ethylene glycol) was used to deliver the chemotherapeutic drug doxorubicin (Lehtovaara et al., 2012). Polysaccharide gellan gum from *Sphingomonas elodea* also shows effective potential for drug delivery. Self-assembly of gellan results in the formation of hybrid capsules (Ohkawa et al., 2004) and nanohydrogel (D'Arrigo et al., 2012).

10 Conclusions

The recent trend toward the development of modern nutraceutical and functional food makes demands on researchers for the integration of new techniques in food formulation. Self-assembly strategy is an effective technique for design and fabrication of delivery agents of bioactive molecules and can effectively serve to manipulate the interactions of food ingredients relating to the specific functional activity. Nutraceutical and functional food components significantly vary according to their structure, function, and physicochemical properties. Hence, the choice of a general delivery technique may not serve the associated surplus. Here, self-assembly plays a crucial role. As mentioned earlier, in self-assembly pre-existing components reassociate with a variety of weaker interactions and form larger structural units. Therefore, to manipulate or to control the structure, appropriate base matrix, environmental conditions, and their chemistry associated with physiological functions should be clearly understood. Carbohydrates, especially polysaccharides due to their structural and functional activity, serve as the appropriate base matrix. In the context of food, self-assembly has proven to be a fascinating technique and already has been used in nutraceutical and functional food delivery. In nutraceutical and functional food development, advancement in nanoscience and nanotechnology becomes a key tool to manipulate food polymers and their structures boosted with functional improvizations. The improvizations may provide a new way in proving efficient delivery, release, quality, texture, and shelf life for functional ingredients and products addressing safety. Disruption of weak physical forces between self-assembled entities is effectively reinforced via covalent bonding. Even more, specific research is going on in this particular problem. Although more research needs to be performed to address the key issues like assembly, its stability, functional improvization, and even nanoscale organization in mixed system. Parallel to laboratory experimental methods, computer simulation and theoretical modeling may help better in the key issues of assembly. In this chapter efforts are made to provide an outline of the recent advancement of carbohydrate-based self-assembly strategies for food applications. Different assembling strategies, behavior, and structure–function relationships associated with varying shapes and sizes may provide a basis for further research on healthier bioactive delivery via carbohydrate carrier with functionality.

References

Agnihotri, S.A., Mallikarjuna, N.N., Aminabhavi, T.M., 2004. Recent advances on chitosan-based micro- and nanoparticles in drug delivery. J. Control. Release 100 (1), 5–28.

Agrawal, P., Strijkers, G.J., Nicolay, K., 2010. Chitosan-based systems for molecular imaging. Adv. Drug Deliv. Rev. 62 (1), 42–58.

Anderson, J.W., et al., 2009. Health benefits of dietary fiber. Nutr. Rev. 67 (4), 188–205.

Astray, G., et al., 2009. A review on the use of cyclodextrins in foods. Food Hydrocoll. 23 (7), 1631–1640.

Besheer, A., et al., 2007. Hydrophobically modified hydroxyethyl starch: synthesis, characterization, and aqueous self-assembly into nano-sized polymeric micelles and vesicles. Biomacromolecules 8 (2), 359–367.

Bhandari, B., D'Arcy, B., Young, G., 2001. Flavour retention during high temperature short time extrusion cooking process: a review. Int. J. Food Sci. Technol. 36 (5), 453–461.
Bobeth, M., et al., 2011. Formation of tubes during self-assembly of bacterial surface layers. Langmuir 27 (24), 15102–15111.
Brennan, C.S., Kuri, V., Tudorica, C.M., 2004. Inulin-enriched pasta: effects on textural properties and starch degradation. Food Chem. 86 (2), 189–193.
Bromley, K.M., et al., 2015. Interfacial self-assembly of a bacterial hydrophobin. Proc. Natl. Acad. Sci. USA 112 (17), 5419–5424.
Bueno, V.B., et al., 2013. Synthesis and swelling behavior of xanthan-based hydrogels. Carbohydr. Polym. 92 (2), 1091–1099.
Cademartiri, L., Ozin, G.A., 2009. Concepts of Nanochemistry. John Wiley & Sons, Weinheim.
Cai, L., Shi, Y.-C., 2013. Self-assembly of short linear chains to A- and B-type starch spherulites and their enzymatic digestibility. J. Agric. Food Chem. 61 (45), 10787–10797.
Cai, J., et al., 2008. Dynamic self-assembly induced rapid dissolution of cellulose at low temperatures. Macromolecules 41 (23), 9345–9351.
Cao, G., Wang, Y., 2011. Nanostructures and Nanomaterials: Synthesis, Properties, and Applications. World Scientific, Singapore.
Capriles, V.D., Arêas, J.A.G., 2013. Effects of prebiotic inulin-type fructans on structure, quality, sensory acceptance and glycemic response of gluten-free breads. Food Funct. 4 (1), 104–110.
Chan, Y.-T., Moorefield, C.N., Newkome, G.R., 2009. Synthesis, characterization, and self-assembled nanofibers of carbohydrate-functionalized mono- and di(2,2':6',2"-terpyridinyl)arenes. Chem. Commun. (Camb.) 45, 6928–6930.
Chen, L., Subirade, M., 2005. Chitosan/beta-lactoglobulin core-shell nanoparticles as nutraceutical carriers. Biomaterials 26 (30), 6041–6053.
Chen, H., et al., 2014. Self-assembled conjugated polymer/carboxymethyl chitosan grafted poly(*p*-dioxanone) nanomicelles and their use in functionalized indicator paper for fast and visual detection of a banned food dye. Polym. Chem. 5 (14), 4251.
Chen, H.-L., et al., 2014. Fast algorithmic self-assembly of simple shapes using random agitation. DNA 2014: Proceedings of the 20th International Meeting on DNA Computing and Molecular Programming, (Kyoto, Japan, September 22–26, 2014), pp. 20–36.
Cherbut, C., 2002. Inulin and oligofructose in the dietary fibre concept. Br. J. Nutr. 87 (Suppl. 2), S159–S162.
Choi, A.-J., et al., 2011. Characterization of capsaicin-loaded nanoemulsions stabilized with alginate and chitosan by self-assembly. Food Bioprocess Technol. 4 (6), 1119–1126.
Chuah, A.M., et al., 2009. Formation of biocompatible nanoparticles via the self-assembly of chitosan and modified lecithin. J. Food Sci. 74 (1), N1–N8.
Conde-Petit, B., Escher, F., Nuessli, J., 2006. Structural features of starch-flavor complexation in food model systems. Trends Food Sci. Technol. 17 (5), 227–235.
Corona-Hernandez, R.I., et al., 2013. Structural stability and viability of microencapsulated probiotic bacteria: a review. Compr. Rev. Food Sci. Food Saf. 12 (6), 614–628.
D'Arrigo, G., et al., 2012. Self-assembled gellan-based nanohydrogels as a tool for prednisolone delivery. Soft Matter 8 (45), 11557.
Dal Bó, A.G., et al., 2011. Self-assembled carbohydrate-based micelles for lectin targeting. Soft Matter 7 (7), 3453.

de Mesquita, J.P., et al., 2012. Bio-based nanocomposites obtained through covalent linkage between chitosan and cellulose nanocrystals. Carbohydr. Polym. 90 (1), 210–217.
Del Valle, E.M.M., 2004. Cyclodextrins and their uses: a review. Process Biochem. 39 (9), 1033–1046.
Dickinson, E., Leser, M.E., 2007. Food Colloids: Self-Assembly and Material Science. Royal Society of Chemistry, Cambridge, UK.
Dul, M., et al., 2015. Self-assembled carrageenan/protamine polyelectrolyte nanoplexes-investigation of critical parameters governing their formation and characteristics. Carbohydr. Polym. 123, 339–349.
Eastburn, S.D., Tao, B.Y., 1994. Applications of modified cyclodextrins. Biotechnol. Adv. 12 (2), 325–339.
Fanta, G.F., Felker, F.C., Shogren, R.L., 2002. Formation of crystalline aggregates in slowly-cooled starch solutions prepared by steam jet cooking. Carbohydr. Polym. 48 (2), 161–170.
Fanta, G.F., et al., 2005. Crystalline particles formed in slowly-cooled cornstarch dispersions prepared by steam jet cooking. The effect of starch concentration, added oil and rate of cooling. Carbohydr. Polym. 61 (2), 222–230.
Fanta, G., et al., 2006. Effect of fatty acid structure on the morphology of spherulites formed from jet cooked mixtures of fatty acids and defatted cornstarch. Carbohydr. Polym. 66 (1), 60–70.
Fathi, M., Martín, Á., McClements, D.J., 2014. Nanoencapsulation of food ingredients using carbohydrate based delivery systems. Trends Food Sci. Technol. 39 (1), 18–39.
Faulks, R.M., Southon, S., Ottaway, P.B., 2008. Assessing the Bioavailability of Nutraceuticals. Food Fortification and Supplementation: Technological, Safety and Regulatory Aspects. Woodhead Publishing, Cambridge, UK, pp. 195–218.
Feng, J.-L., et al., 2015. Fabrication and characterization of stable soy β-conglycinin-dextran core-shell nanogels prepared via a self-assembly approach at the isoelectric point. J. Agric. Food Chem. 63 (26), 6075–6083.
Fung, W.-Y., Yuen, K.-H., Liong, M.-T., 2011. Agrowaste-based nanofibers as a probiotic encapsulant: fabrication and characterization. J. Agric. Food Chem. 59 (15), 8140–8147.
Garnier, C., 1994. Phase separation in dextran/locust bean gum mixtures. Carbohydr. Polym. 25 (3), 221–222.
Garti, N., 2008. Delivery and Controlled Release of Bioactives in Foods and Nutraceuticals. CRC Press, New York, NY.
Gentry, J.L., et al., 1996. Spent cellulose casings as potential feed ingredients for ruminants. J. Anim. Sci. 74 (3), 663–671.
German, M.L., et al., 1992. Structure formation in systems containing amylose, amylopectin, and their mixtures. Carbohydr. Polym. 18, 27–34.
Goetzel, R.Z., 2009. Do prevention or treatment services save money? The wrong debate. Health Aff. 28 (1), 37–41.
Gökmen, F.Ö., et al., 2015. Novel multifunctional colloidal carbohydrate nanofiber electrolytes with excellent conductivity and responses to bone cancer cells. Carbohydr. Polym. 133, 624–636.
Guzman-Puyol, S., et al., 2015. Pectin-lipid self-assembly: influence on the formation of polyhydroxy fatty acids nanoparticles. PloS One 10 (4), pe0124639.
Han, J., et al., 2008. Synthesis and characterization of starch piperinic ester and its self-assembly of nanospheres. J. Appl. Polym. Sci. 108 (1), 523–528.
Han, L., Ratcliffe, I., Williams, P.A., 2015. Self-assembly and emulsification properties of hydrophobically modified inulin. J. Agric. Food Chem. 63 (14), 3709–3715.

Heitfeld, K.A., et al., 2008. Temperature responsive hydroxypropyl cellulose for encapsulation. Mater. Sci. Eng. 28 (3), 374–379.
Hu, B., et al., 2008. Optimization of fabrication parameters to produce chitosan-tripolyphosphate nanoparticles for delivery of tea catechins. J. Agric. Food Chem. 56 (16), 7451–7458.
Huang, Q., 2012. Nanotechnology in the Food, Beverage and Nutraceutical Industries. Woodhead Publishing, Cambridge, UK.
Huang, W., et al., 2012. Layer-by-layer immobilization of lysozyme–chitosan–organic rectorite composites on electrospun nanofibrous mats for pork preservation. Food Res. Int. 48 (2), 784–791.
Ichikawa, S., Iwamoto, S., Watanabe, J., 2005. Formation of biocompatible nanoparticles by self-assembly of enzymatic hydrolysates of chitosan and carboxymethyl cellulose. Biosci. Biotechnol. Biochem. 69 (9), 1637–1642.
Irwin, P.L., et al., 1994. Binding geometry, stoichiometry, and thermodynamics of cyclomalto-oligosaccharide (cyclodextrin) inclusion complex formation with chlorogenic acid, the major substrate of apple polyphenol oxidase. Carbohydr. Res. 256 (1), 13–27.
Ishihara, S., et al., 2011. Swallowing profiles of food polysaccharide gels in relation to bolus rheology. Food Hydrocoll. 25 (5), 1016–1024.
Jiang, F., et al., 2015. *N*-trimethylchitosan/alginate layer-by-layer self assembly coatings act as “fungal repellents” to prevent biofilm formation on healthcare materials. Adv. Healthc. Mater. 4 (3), 469–475.
Jones, O.G., McClements, D.J., 2010. Biopolymer nanoparticles from heat-treated electrostatic protein-polysaccharide complexes: factors affecting particle characteristics. J. Food Sci. 75 (2), N36–N43.
Jordan, A., 1956. Dextran: its properties and use in medicine. J. Clin. Pathol. 9 (1), 82.
Jouquand, C., Ducruet, V., Giampaoli, P., 2004. Partition coefficients of aroma compounds in polysaccharide solutions by the phase ratio variation method. Food Chem. 85 (3), 467–474.
Jung, T.-H., et al., 2005. Properties of cholesterol-reduced butter and effect of gamma linolenic acid added butter on blood cholesterol. Asian Australas. J. Anim. Sci. 18 (11), 1646–1654.
Jung, T.-H., et al., 2008. Development of cholesterol-reduced mayonnaise with crosslinked β-cyclodextrin and added phytosterol. Korean J. Food Sci. Anim. Resour. 28 (2), 211–217.
Kaewprapan, K., et al., 2012. Macromolecular surfactants synthesized by lipase-catalyzed transesterification of dextran with vinyl decanoate. Carbohydr. Polym. 88 (1), 313–320.
Kapishon, V., et al., 2015. Polymerization induced self-assembly of alginate based amphiphilic graft copolymers synthesized by single electron transfer living radical polymerization. Biomacromolecules 16 (7), 2040–2048.
Kayaci, F., Uyar, T., 2012. Encapsulation of vanillin/cyclodextrin inclusion complex in electrospun polyvinyl alcohol (PVA) nanowebs: prolonged shelf-life and high temperature stability of vanillin. Food Chem. 133 (3), 641–649.
Kim, D.-Y., Shin, W.-S., 2015. Unique characteristics of self-assembly of bovine serum albumin and fucoidan, an anionic sulfated polysaccharide, under various aqueous environments. Food Hydrocoll. 44, 471–477.
Kim, S.-H., et al., 2013. Self-assembled dextran sulphate nanoparticles for targeting rheumatoid arthritis. Chem. Commun. (Camb.) 49 (88), 10349–10351.
Klinkesorn, U., Sophanodora, P., Chinachoti, P., Decker, E.A., et al., 2005a. Encapsulation of emulsified tuna oil in two-layered interfacial membranes prepared using electrostatic layer-by-layer deposition. Food Hydrocoll. 19 (6), 1044–1053.

Klinkesorn, U., Sophanodora, P., Chinachoti, P., McClements, D.J., et al., 2005b. Increasing the oxidative stability of liquid and dried tuna oil-in-water emulsions with electrostatic layer-by-layer deposition technology. J. Agric. Food Chem. 53 (11), 4561–4566.
Kokubun, S., Ratcliffe, I., Williams, P.A., 2013. Synthesis, characterization and self-assembly of biosurfactants based on hydrophobically modified inulins. Biomacromolecules 14 (8), 2830–2836.
Kuhad, R.C., Gupta, R., Singh, A., 2011. Microbial cellulases and their industrial applications. Enzyme Res. 2011, 280696.
Lee, A.-C., Hong, Y.-H., 2009. Coacervate formation of α-lactalbumin–chitosan and β-lactoglobulin–chitosan complexes. Food Res. Int. 42 (5–6), 733–738.
Lehtovaara, B.C., Verma, M.S., Gu, F.X., 2012. Synthesis of curdlan-graft-poly(ethylene glycol) and formulation of doxorubicin-loaded core-shell nanoparticles. J. Bioact. Compat. Polym. 27 (1), 3–17.
Lesmes, U., Barchechath, J., Shimoni, E., 2008. Continuous dual feed homogenization for the production of starch inclusion complexes for controlled release of nutrients. Innov. Food Sci. Emerg. Technol. 9 (4), 507–515.
Li, L., et al., 2010. Self-assembled nanoparticles of cholesterol-conjugated carboxymethyl curdlan as a novel carrier of epirubicin. Nanotechnology 21 (26), 265601.
Liu, C.-G., et al., 2005. Linolenic acid-modified chitosan for formation of self-assembled nanoparticles. J. Agric. Food Chem. 53 (2), 437–441.
Liu, J., Willför, S., Xu, C., 2014. A review of bioactive plant polysaccharides: biological activities, functionalizatison, and biomedical applications. Bioact. Carbohydr. Dietary Fibre 5 (1), 31–61.
López-Rubio, A., et al., 2012. Electrospinning as a useful technique for the encapsulation of living bifidobacteria in food hydrocolloids. Food Hydrocoll. 28 (1), 159–167.
Lv, H., et al., 2014. Layer-by-layer self-assembly of minocycline-loaded chitosan/alginate multilayer on titanium substrates to inhibit biofilm formation. J. Dent. 42 (11), 1464–1472.
Ma, U.V.L., Floros, J.D., Ziegler, G.R., 2011. Effect of starch fractions on spherulite formation and microstructure. Carbohydr. Polym. 83 (4), 1757–1765.
Martín, M.J., et al., 2014. Microencapsulation of bacteria: a review of different technologies and their impact on the probiotic effects. Innov. Food Sci. Emerg. Technol. 27, 15–25.
Maruyama, A., et al., 1994. Preparation of nanoparticles bearing high density carbohydrate chains using carbohydrate-carrying polymers as emulsifier. Biomaterials 15 (13), 1035–1042.
McClements, D.J., 2009. Structural design principles for improved food performance: nanolaminated biopolymer structures in foods. ACS Symp. Ser. 1007, 3–34.
Mendes, A.C., et al., 2011. Palmitoylation of xanthan polysaccharide for self-assembly microcapsule formation and encapsulation of cells in physiological conditions. Soft Matter 7 (20), 9647.
Mendes, A.C., et al., 2013. Fabrication of phospholipid-xanthan microcapsules by combining microfluidics with self-assembly. Acta Biomater. 9 (5), 6675–6685.
Mitra, S., et al., 2001. Tumour targeted delivery of encapsulated dextran–doxorubicin conjugate using chitosan nanoparticles as carrier. J. Control. Release 74 (1–3), 317–323.
Moraru, C.I., et al., 2002. Phase behavior of a meat-starch extrudate illustrated on a state diagram. J. Food Sci. 67 (8), 3026–3032.

Na, K., et al., 2000. Self-assembled hydrogel nanoparticles from curdlan derivatives: characterization, anti-cancer drug release and interaction with a hepatoma cell line (HepG2). J. Control. Release 69 (2), 225–236.
Nagahama, K., Sano, Y., Kumano, T., 2015. Anticancer drug-based multifunctional nanogels through self-assembly of dextran-curcumin conjugates toward cancer theranostics. Bioorg. Med. Chem. Lett. 25 (12), 2519–2522.
Nordmark, T., 2002. Spherulitic crystallization of gelatinized maize starch and its fractions. Carbohydr. Polym. 49 (4), 439–448.
Nordmark, T.S., Ziegler, G.R., 2002. Structural features of non-granular spherulitic maize starch. Carbohydr. Res. 337 (16), 1467–1475.
O'Mahony, C.T., Farrell, R.A., Goshal, T., Holmes, J.D., Morris, M.A., 2011. In: Moreno-Piraján, J.C. (Ed.), Thermodynamics—Systems in Equilibrium and Non-Equilibrium. InTech, Rijeka, Croatia.
Ohkawa, K., Kitagawa, T., Yamamoto, H., 2004. Preparation and characterization of chitosan-gellan hybrid capsules formed by self-assembly at an aqueous solution interface. Macromol. Mater. Eng. 289 (1), 33–40.
Omenn, G.S., et al., 1996. Effects of a combination of beta carotene and vitamin A on lung cancer and cardiovascular disease. N. Engl. J. Med. 334 (18), 1150–1155.
Ozin, G.A., et al., 2009. Nanofabrication by self-assembly. Mater. Today 12 (5), 12–23.
Padua, G.W., Wang, Q. (Eds.), 2012. Nanotechnology Research Methods for Foods and Bioproducts. Wiley-Blackwell, Oxford, UK.
Patenaude, M., Hoare, T., 2012. Injectable, mixed natural-synthetic polymer hydrogels with modular properties. Biomacromolecules 13 (2), 369–378.
Peterson, S.C., et al., 2005. Identification of complexed native lipids in crystalline aggregates formed from jet cooked cornstarch. Carbohydr. Polym. 61 (2), 162–167.
Pletikapić, G., et al., 2014. Self-assembly of polysaccharides gives rise to distinct mechanical signatures in marine gels. Biophys. J. 107 (2), 355–364.
Ploux, L., et al., 2007. Quantitative and morphological analysis of biofilm formation on self-assembled monolayers. Colloids Surf. B 57 (2), 174–181.
Priya, A.J., Vijayalakshmi, S.P., Raichur, A.M., 2011. Enhanced survival of probiotic *Lactobacillus acidophilus* by encapsulation with nanostructured polyelectrolyte layers through layer-by-layer approach. J. Agric. Food Chem. 59 (21), 11838–11845.
Qu, X., et al., 2006. Carbohydrate-based micelle clusters which enhance hydrophobic drug bioavailability by up to 1 order of magnitude. Biomacromolecules 7 (12), 3452–3459.
Raja, B.R., Arunachalam, K.D., 2011. Market potential for probiotic nutritional supplements in India. J. Bus. Manage. 5 (14), 5418–5423.
Ravi Kumar, M.N., 2000. A review of chitin and chitosan applications. React. Funct. Polym. 46 (1), 1–27.
Reineccius, G.A., 2013. Carbohydrates as flavor delivery systems. Functionalizing Carbohydrates for Food Applications. DEStech Publications, Lancaster.
Rezaei, A., Nasirpour, A., Fathi, M., 2015. Application of cellulosic nanofibers in food science using electrospinning and its potential risk. Compr. Rev. Food Sci. Food Saf. 14 (3), 269–284.
Ross-Murphy, S.B., Shatwell, K.P., 1993. Polysaccharide strong and weak gels. Biorheology 30 (3–4), 217–227.
Sacchettini, J.C., Baum, L.G., Brewer, C.F., 2001. Multivalent protein–carbohydrate interactions. A new paradigm for supermolecular assembly and signal transduction. Biochemistry 40 (10), 3009–3015.
Salyers, A.A., et al., 1977. Fermentation of mucin and plant polysaccharides by strains of bacteroides from the human colon. Appl. Environ. Microbiol. 33 (2), 319–322.

Santipanichwong, R., et al., 2008. Core-shell biopolymer nanoparticles produced by electrostatic deposition of beet pectin onto heat-denatured beta-lactoglobulin aggregates. J. Food Sci. 73 (6), N23–N30.
Sattler, K.D., 2010. Handbook of Nanophysics: Nanomedicine and Nanorobotics. CRC Press, Boca Raton.
Schiffman, J.D., Schauer, C.L., 2008. A review: electrospinning of biopolymer nanofibers and their applications. Polym. Rev. 48 (2), 317–352.
Shewan, H.M., Stokes, J.R., 2013. Review of techniques to manufacture micro-hydrogel particles for the food industry and their applications. J. Food Eng. 119 (4), 781–792.
Shibakami, M., et al., 2015. Preparation of transparent self-standing thin films made from acetylated euglenoid β-1,3-glucans. Carbohydr. Polym. 133, 421–428.
Shogren, R.L., Fanta, G.F., Felker, F.C., 2006. X-ray diffraction study of crystal transformations in spherulitic amylose/lipid complexes from jet-cooked starch. Carbohydr. Polym. 64 (3), 444–451.
Singh, V.K., et al., 2014. Gelatin-carbohydrate phase-separated hydrogels as bioactive carriers in vaginal delivery: preparation and physical characterizations. J. Appl. Polym. Sci. 131 (13), 1–11.
Sinha, V.R., Kumria, R., 2001. Polysaccharides in colon-specific drug delivery. Int. J. Pharm. 224 (1–2), 19–38.
Stijnman, A.C., Bodnar, I., Hans Tromp, R., 2011. Electrospinning of food-grade polysaccharides. Food Hydrocoll. 25 (5), 1393–1398.
Stoddart, J.F., Lindoy, L.F., Atkinson, I.M., 2000. Self Assembly in Supramolecular Systems. Royal Society of Chemistry, Cambridge, UK.
Sun, L., Zhang, C., Li, P., 2011. Characterization, antimicrobial activity, and mechanism of a high-performance (-)-epigallocatechin-3-gallate (EGCG)-CuII/polyvinyl alcohol (PVA) nanofibrous membrane. J. Agric. Food Chem. 59 (9), 5087–5092.
Tang, M., Dou, H., Sun, K., 2006. One-step synthesis of dextran-based stable nanoparticles assisted by self-assembly. Polymer 47 (2), 728–734.
Thorpe, K.E., 2005. The rise in health care spending and what to do about it. Health Aff. 24 (6), 1436–1445.
Vrignaud, S., et al., 2012. Design, optimization and in vitro evaluation of reverse micelle-loaded lipid nanocarriers containing erlotinib hydrochloride. Int. J. Pharm. 436 (1–2), 194–200.
Whitesides, G.M., Boncheva, M., 2002. Beyond molecules: self-assembly of mesoscopic and macroscopic components. Proc. Natl. Acad. Sci. USA 99 (8), 4769–4774.
Whitesides, G.M., Grzybowski, B., 2002. Self-assembly at all scales. Science 295 (5564), 2418–2421.
Wildman, R.E.C., Wildman, R., Wallace, T.C., 2006. Handbook of Nutraceuticals and Functional Foods, second ed. CRC Press, Boca Raton.
Wongsasulak, S., et al., 2010. Electrospinning of food-grade nanofibers from cellulose acetate and egg albumen blends. J. Food Eng. 98 (3), 370–376.
Wrigstedt, P., et al., 2010. Synthesis and antioxidant activity of hydroxycinnamic acid xylan esters. J. Agric. Food Chem. 58 (11), 6937–6943.
Xu, W., Yang, W., Yang, Y., 2009. Electrospun starch acetate nanofibers: development, properties, and potential application in drug delivery. Biotechnol. Prog. 25 (6), 1788–1795.
Yang, Y., Gu, Z., Zhang, G., 2009. Delivery of bioactive conjugated linoleic acid with self-assembled amylose-CLA complex. J. Agric. Food Chem. 57 (15), 7125–7130.
Yang, Y., et al., 2014. Advances in self-assembled chitosan nanomaterials for drug delivery. Biotechnol. Adv. 32 (7), 1301–1316.

Yu, H., Huang, Q., 2010. Enhanced in vitro anti-cancer activity of curcumin encapsulated in hydrophobically modified starch. Food Chem. 119 (2), 669–674.

Yu, S., et al., 2006. Stable and pH-sensitive nanogels prepared by self-assembly of chitosan and ovalbumin. Langmuir 22 (6), 2754–2759.

Zhang, S., 2003. Fabrication of novel biomaterials through molecular self-assembly. Nat. Biotechnol. 21 (10), 1171–1178.

Zhang, F., et al., 2015. Roughness-controlled self-assembly of mannitol/LB agar microparticles by polymorphic transformation for pulmonary drug delivery. Mol. Pharm. 12 (1), 223–231.

Ziegler, G.R., Nordmark, T.S., Woodling, S.E., 2003. Spherulitic crystallization of starch: influence of botanical origin and extent of thermal treatment. Food Hydrocoll. 17 (4), 487–494.

Ziegler, G.R., Creek, J.A., Runt, J., 2005. Spherulitic crystallization in starch as a model for starch granule initiation. Biomacromolecules 6 (3), 1547–1554.

Zimet, P., Livney, Y.D., 2009. Beta-lactoglobulin and its nanocomplexes with pectin as vehicles for ω-3 polyunsaturated fatty acids. Food Hydrocoll. 23 (4), 1120–1126.

6

NANOTECHNOLOGY AND WINE

Morata Antonio, Loira Iris, Suárez-Lepe José Antonio
enotecUPM, Polytechnic University of Madrid, Chemistry and Food Technology Department, Higher Technical College of Agronomic Engineers, Avenida Complutense S/N, Madrid, Spain

1 Introduction

Nanoscience is one of the most important researches in modern science. Nanoparticles have unique optical, electrical, catalytic, thermal, and magnetic characteristics due to their small size (1–100 nm). Also, regarding their cost of production, nanostructured materials seem a promising and cost-effective alternative to expensive analysis equipment (Di Fusco et al., 2010; Barthelmebs et al., 2011; Monošík et al., 2012a), often providing the opportunity to recycle.

Despite being a relatively modern science, nanotechnology is already applied (or at least is being widely studied) in the fields of biomedicine, environment, construction, food industry, and cosmetics. Along with the recent advances in information and communication technology (ICT), nanotechnology represents the industrial revolution of the 21st century.

In the food sector, silver nanomaterials are currently being used for water purification (Pradeep, 2009) and also in the design of new packaging materials with antimicrobial properties (Tankhiwale and Bajpai, 2010; Emamifar et al., 2011).

In this review some implementations of nanotechnology in the field of enology are collected and their convenience for winemaking is evaluated in order to improve wine quality and safety, as well as to facilitate wine analysis.

2 Nanotechnologies Applied to Winemaking

One of the most widespread applications of nanotechnology in the field of enology concerns wine chemical analysis. The development of analytical equipment incorporating nanoparticles or nanomaterials may improve the performance, and, in most cases, simplify the analysis methodology. Table 6.1 summarizes some

Novel Approaches of Nanotechnology in Food. http://dx.doi.org/10.1016/B978-0-12-804308-0.00006-6

Table 6.1 Some Applications of Nanomaterials in Enology and Their Outstanding Achievements

Nanomaterial	Device	Application	Remarks	References
Pure and composite nanostructured films of conducting polymers: 16-mer polyaniline, polypyrrole, and stearic acid	Electronic tongue	Sensorial analysis of wines	• Detect trace amounts of tastants and inorganic contaminants in liquid systems • Detect the suppression of sourness due to the addition of sucrose • Fast and simple detection of wine adulteration	Riul et al. (2003)
Langmuir–Blodgett films of conducting polymers and lipids and layer-by-layer films from chitosan	Electronic tongue	Red wine classification	• Distinguish tastes below the human threshold • Distinguish a large number of wine samples • High classification accuracy	Riul et al. (2004)
Gold nanoparticles	Tyrosinase biosensor	Bioelectrochemical polyphenols index determination in red and white wines	• Small volume of sample (270 μL) • Rapid estimation of TPI • No need of cleaning pretreatment	Sanz et al. (2005)
Multiwalled and single-walled carbon nanotubes	Laccase biosensor	Determination of polyphenol index in wine	• Fast response • Unaffected by interferences • Simple and rapid preparation • Low cost	Di Fusco et al. (2010)
Magnetic nanoparticles amino functionalized	Electrochemical method: Osteryoung square wave voltammetry (OSWV)	Determination of ochratoxin A in red wine grapes	• Rapid and sensitive determination • Low detection limit	Fernández-Baldo et al. (2010)

Multiwalled carbon nanotubes	Peroxidases biosensor	Total polyphenolic content estimation in wine and tea	• Good analytical performance • Short detection time • Small sample volume • High accuracy • High simplicity	Granero et al. (2010)
Multiwalled carbon nanotubes	Tyrosinase biosensor	Sensing phenolics in red wines		Kim et al. (2009, 2010)
Superparamagnetic nanoparticles	Electrochemical aptasensor	Sensitive detection of ochratoxin A	• Rapid and real-time monitoring • Long-term stability • Simple equipment • Highly sensitive • Cost-effective	Barthelmebs et al. (2011)
Multiwalled carbon nanotubes	Amperometric detector	Detection and quantification of polyphenols in white wines	• Wine samples can be analyzed directly • Highly stable analysis of samples with low detection potential	Moreno et al. (2011)
Multiwalled carbon nanotubes	Microbial biosensor	Sensing phenolic compounds in red wines	• Wide linear range • High sensitivity • Good stability	Kim et al. (2011)
Multiwalled carbon nanotubes	Electrochemical microbial biosensor	Analysis of phenolic compounds in red wine	• Good electron transfer material • High selectivity and sensitivity	Shin et al. (2011)
Carbon nanotubes	Modified carbon paste electrode	Determination of the antioxidant capacity for red and white wines through direct determination of gallic acid	• No interference of glucose and ascorbic acid	Souza et al. (2011)
Iron oxide carboxyl-modified magnetic nanoparticles	Electrochemical impedance spectroscopy immunosensor	Sensitive determination of ochratoxin A	• Label-free • Highly sensitive • Cost-effective • Fast • High specificity	Zamfir et al. (2011)
Gold nanoparticles	Laccase + long wavelength fluorimetry	Determination of polyphenolic content in red grape juice	• Use of long wavelength fluorimetry • Very low limit of detection • High selectivity	Andreu-Navarro et al. (2012)

(Continued)

Table 6.1 Some Applications of Nanomaterials in Enology and Their Outstanding Achievements (*cont.*)

Nanomaterial	Device	Application	Remarks	References
Silver nanoparticles/ zinc oxide nanoparticles	Laccase biosensor	Determination of total phenolic content in wine	• Low working potential • Short response time (8 s) • High sensitivity • High storage stability	Chawla et al. (2012)
Gold-chitosan nanocomposite	Electrochemical sensor	Determination of polyphenol index in wines	• Sensitive and selective	Curulli et al. (2012)
Multiwalled carbon nanotubes	Tyrosinase biosensor	Detection of phenolic compounds in red wine	• Use of bifunctionalized MWNT supports (good electron transfer material) • Low detection limit • Good stability	Lee et al. (2012)
Multiwalled carbon nanotubes	Amperometric biosensor	Determination of L-lactic acid in wines	• Able to determine low concentrations of lactic acid • Long-term storage stability • Low fabrication expenses • Good analytical performance • Monitoring malolactic fermentation	Monošík et al. (2012a)
Multiwalled carbon nanotubes	Amperometric biosensor	Determination of glycerol in wine	• Fast, sensitive, reusable biosensor • Monitoring glycerol during the winemaking process • Determination of triglycerides after their hydrolysis by lipase	Monošík et al. (2012b)
Magnetic nanoparticles (Fe_3O_4)	Electrochemical biosensor	Measurement of sulfite level in red and white wines	• Excellent electrocatalysis toward sulfite • Low detection limit • Low response time (2 s) • High storage stability • No interference by ascorbate, cysteine, fructose, and ethanol	Rawal et al. (2012)

Gold nanoparticles	Electrochemical biosensor	Detection of phenolic compounds in musts and wines	• Rapid and almost automated estimation of the "total phenolics" and "high antioxidant" fraction of must and commercial wines	Sánchez-Obrero et al. (2012)
Multiwalled carbon nanotubes	Amperometric detector	Continuous analysis of total polyphenols in white and red wines	• Low overpotentials, high sensitivity, and very high signal stability under a dynamic regime • Selective estimation of the easily oxidizable polyphenol fraction as well as the total polyphenol content • Fast, simple, and direct measurement	Arribas et al. (2013)
Silver nanoparticles	Electroanalytical sensor	Determination of caffeic acid in red wine		Karabiberoğlu et al. (2013)
Molecularly imprinted nanoparticles	Potentiometric sensor	Histamine quantification in wine and fish	• Short response time • Label-free detection/quantification of histamine • Minimal sample pretreatment	Basozabal et al. (2014)
Fullerene and gold nanoparticles	Enzymatic biosensor	Determination of polyphenol content in wines	• Fast amperometric response • Enhancement of the electrochemical properties • Good stability and reproducibility	Lanzellotto et al. (2014)
Gold nanoparticles	Electrochemical sensor	Determination of preservative sulfur dioxide in white wines	• Low cost • Ultra fast determination • No interference with polyphenols • High sensitivity • Small sample volumes	Schneider et al. (2014)
Fluorescent carbon nanodots	Fluorescent nanosensor	Sensitive and selective detection of tannic acid in red and white wines	• Rapid detection • High sensitivity and selectivity • Wide linear response range • Low cost	Ahmed et al. (2015)

(Continued)

Table 6.1 Some Applications of Nanomaterials in Enology and Their Outstanding Achievements (*cont.*)

Nanomaterial	Device	Application	Remarks	References
Terbium oxide nanoparticles	Fluorometric detector	Determination of polyphenols in wine	• Automatic determination • Short analyzing times • Low enzyme consumption • Low cost • Great overall sample throughput (35 samples h^{-1})	Godoy-Navajas et al. (2015)
Platinum nanoparticles supported on graphitized carbon nanofibers	Lactate biosensor	Detection of lactate in wines	• Simplicity and low-cost fabrication • Stable and sensitive • Repetitive use of the biosensor	Loaiza et al. (2015)
Gold nanoparticles	Voltammetric sensor	Analysis of organic acids and phenols in musts and wines	• High surface to volume ratios • Low limits of detection • Simultaneous detection of organic acids and phenols • Complete absence of interferences	Medina-Plaza et al. (2015)

applications of nanomaterials in enology and brings together the most outstanding achievements.

2.1 Sensors for Wine Compounds Analysis

Modern sensing technologies can use sensors or biosensors for the selective identification of several compounds in the wine matrix. One of the main features provided by nanomaterials resides in increasing the specific surface area of the electrode so that its sensitivity and selectivity are improved (Barroso et al., 2011). Sensitivity is a key requirement in sensor applications and, on the whole, the higher the film thickness the lower the device sensitivity (Riul et al., 2003). High sensitivity is achieved when interdigitated electrodes are coated with nanostructured thin films (~2 nm thick per deposited layer), which are able to detect very small changes in conductivity and dielectric properties of the materials comprising individual sensing units in contact with a liquid medium (Riul et al., 2004).

These nanomaterials can be incorporated into the electrode by different techniques suitable for each compound, among which are adsorption (eg, carbon nanotubes) (Barroso et al., 2011) and electrodeposition (eg, gold nanoparticles) (Sanz et al., 2005).

Chemically modified electrodes (CME) can be defined as electrodes made of a conducting or semiconducting material that are coated with a selected monomolecular, multimolecular, ionic, or polymeric film of a chemical modifier and that by means of faradaic (charge-transfer) reactions or interfacial potential differences (no net charge transfer) exhibit chemical, electrochemical, and/or optical properties of the film (Durst, 1997; Fig. 6.1). Thanks to the ability of new nanomaterials (eg, carbon nanotubes, graphene, metallic nanoparticles) incorporated into sensors to improve the electron-transfer process, in the past years stability and selectivity of electrochemical measurements has been significantly improved (Arribas et al., 2012).

Over the past few years, with the development of nanotechnology and biotechnology, different nanostructured materials have been tested according to their particular properties in order to facilitate and improve wine chemical analysis. Among them, carbon nanotubes (CNTs) are the most widely used nanomaterial to improve the performance of the sensing units, that is, the electrodes (Kim et al., 2009, 2010; Di Fusco et al., 2010; Shin et al., 2011; Lee et al., 2012; Monošík et al., 2012a,b). This 20-year-old material is an excellent promoter of the electron transfer and electrocatalytic activity, so it is not rare that it has been tested for the detection of uncountable molecules including antioxidants (Barroso et al., 2011).

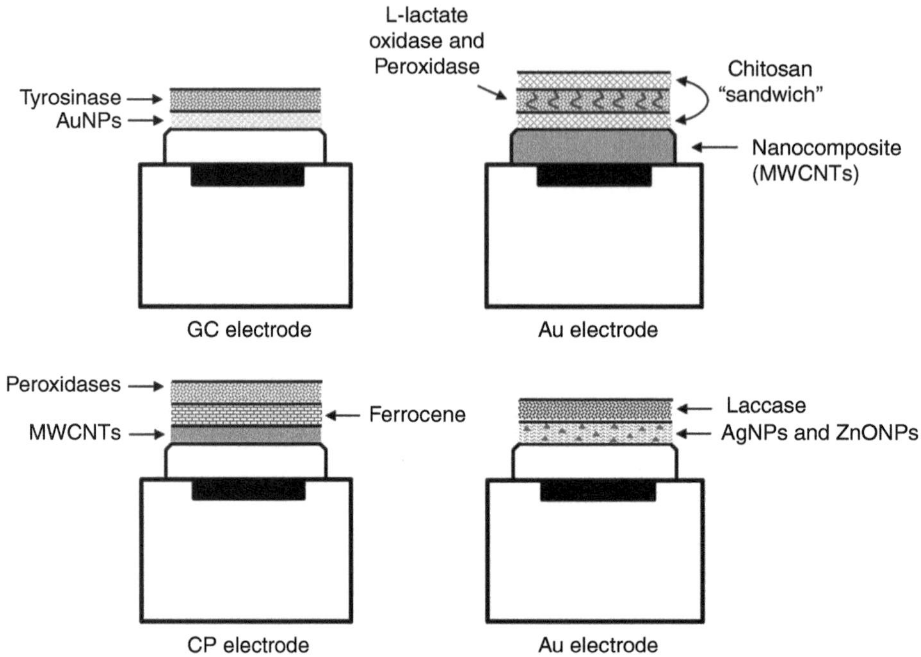

Figure 6.1. Examples of chemically modified electrodes with different nanomaterials.

CNTs can be considered as the result of folding graphene layers into carbon cylinders (Shin et al., 2011).

Colloidal gold nanoparticles have been widely used because of their ease of formation and great ability to modify the surface (Feng et al., 2007), their good compatibility with biological molecules (Qu et al., 2006), as well as their capability to act as nanoscale conduction centers, which facilitate the transfer of electrons between molecules (Jia et al., 2002). Moreover, use of gold nanoparticles (AuNPs) is playing an increasingly important role for the preparation of biosensors as they provide a stable surface for enzyme immobilization, and allow the electrochemical sensing to be performed without the need of external electron-transfer mediators (Sanz et al., 2005).

Furthermore, silver nanomaterials are antimicrobial toward a broad spectrum of Gram-negative and Gram-positive bacteria and also exert some antifungal and antiviral activities (Marambio-Jones and Hoek, 2010). These authors indicate that silver nanoparticles (AgNPs) may exert a similar level of control over the microbial populations in the must as sulfur dioxide, but showing no antioxidant activity. Moreover, recent studies indicate that silver nanoparticles exhibit a catalytic activity for the reduction of H_2O_2 (Welch et al., 2005).

Zinc oxide nanoparticles (ZnONPs) have been exploited as a potential material for biosensing because of their unusual properties, namely high surface areas for strong adsorption (high isoelectric point of 9.5), good biocompatibility, chemical stability, nontoxicity, and high electron communication. In addition, nanoporous ZnONPs also enhance the surface area for strong adsorption of biomolecules (Chawla et al., 2012).

Similarly, magnetic nanoparticles (MNPs) can be used as immobilization support in biosensing technology (Zamfir et al., 2011). A wide variety of surface functional groups can be attached to MNPs and these materials also show advantages in the process of antibodies and enzyme immobilization by increasing the surface area, the binding surface stability, and improving orientation of the immobilized antibody/enzyme. Another great advantage of MNPs is that they respond to an applied magnetic field and redisperse upon removal of the magnet. As nanostructured metal oxides, known to have a unique ability to promote faster electron transfer kinetics between electrode and the active site of the desired enzyme (Chawla et al., 2012), use of MNPs can improve the performance of immobilized biomolecules in analytical assay by providing a relatively large numbers of binding sites for biochemical reactions, which offer higher efficiency of interactions between the samples and reagents, resulting in faster assay kinetics (Barthelmebs et al., 2011; Rawal et al., 2012).

A biosensor is commonly defined as a device incorporating a bioreceptor connected to a transducer, which converts an observed response into a measurable signal proportional to analyte concentration, which then is conveyed to a detector (Choi et al., 2011). Fig. 6.2 represents the principle of the biosensors. By binding to a biocomponent (eg, enzyme, antigen, or antibody, plant, or animal tissue), which acts as the biological recognition element or the biochemical receptor, the electrode surface is modified and its sensitivity greatly enhanced (Durst, 1997).

Immobilization of enzymes such as tyrosinase (Sanz et al., 2005; Kim et al., 2009, 2010; Lee et al., 2012; Sánchez-Obrero et al., 2012), laccase (Di Fusco et al., 2010; Andreu-Navarro et al., 2012; Chawla et al., 2012), L-lactate oxidase (Monošík et al., 2012a; Loaiza et al., 2015) and/or peroxidase (Monošík et al., 2012b), onto electrodes modified with gold (Sanz et al., 2005; Andreu-Navarro et al., 2012; Sánchez-Obrero et al., 2012; Lanzellotto et al., 2014; Schneider et al., 2014; Medina-Plaza et al., 2015), silver (Chawla et al., 2012; Karabiberoğlu et al., 2013), or zinc oxide (Chawla et al., 2012) nanoparticles, or carbon nanotubes (Kim et al., 2009, 2010, and 2011; Di Fusco et al., 2010; Granero et al., 2010; Moreno et al., 2011;

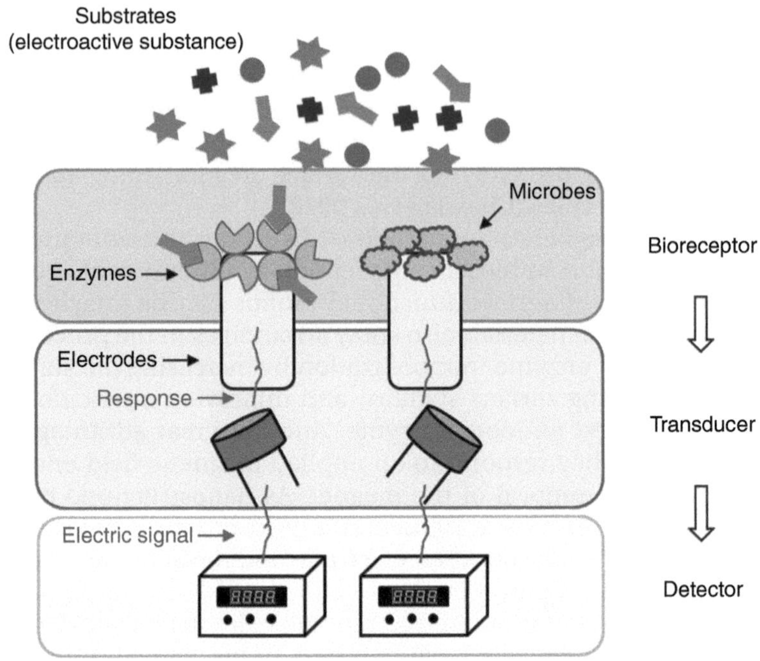

Figure 6.2. Schematic diagram for the principle of an enzyme-based and a microbe-immobilized biosensors.

Shin et al., 2011; Souza et al., 2011; Lee et al., 2012; Monošík et al., 2012a,b; Arribas et al., 2013), permits the construction of biosensors with a good analytical performance (concerning repeatability of the measurements, sometimes with no need of cleaning or sample pretreatment, stability with time, reproducibility in the construction of different biosensors, and sensitivity) for the electroanalytical detection (potentiometry, voltammetry, amperometry, impedance spectroscopy) of chemical compounds such as phenolics (eg, phenol, catechol, caffeic acid, gallic acid, *t*-resveratrol, ferulic acid, and *p*-coumaric acid), lactic acid, glycerol, both in must and in wine samples.

Tyrosinase (Tyr) is a monophenol mono-oxygenase, which catalyzes the oxidation of the phenol group to *o*-quinone at the expense of reducing oxygen to water, thus allowing a variety of phenolic compounds (eg, phenol, catechol, caffeic acid, chlorogenic acid, gallic acid, and protocatechualdehyde) to be used as substrates of this enzyme (Sanz et al., 2005; Kim et al., 2009).

Laccases (Lacc) are copper containing oxidoreductases detected in many plants and secreted by numerous fungi. They are able to oxidize many different substrates, for example, phenols and anilines, with the concomitant reduction of oxygen to water (Di Fusco et al., 2010). This makes these biocatalysts very useful for their application in several biotechnological processes. Recently, the utility of laccases has also been applied to nanobiotechnology, more specifically in the development of biosensors (Rodríguez Couto and Toca Herrera, 2006). Unlike other redox enzymes, laccase does not require hydrogen peroxide (H_2O_2) as a cosubstrate or any other cofactors for its catalytic action, therefore the construction of laccase based biosensors is much easier (Chawla et al., 2012).

Regardless the enzyme used, the immobilized enzymes may retain a high bioactivity when attached to the electrodes, giving rise to fast, stable, and sensitive responses (Sanz et al., 2005). Biosensor performance highly depends on enzyme adopted and their selectivity toward the analyte (Di Fusco et al., 2010).

Some authors have also tested a chitosan "sandwich" structure as immobilization method thus ensuring general improvements in the analytical performance and increments in the useful lifetime of the sensor (Curulli et al., 2012; Monošík et al., 2012a). Moreover, chitosan is a relatively cheap material that has good biocompatibility for proteins or enzymes (Kang et al., 2009).

During the past two decades, many authors worked on designing sensors and biosensors that incorporate nanomaterials to estimate the total content of polyphenols in wine (Sanz et al., 2005; Di Fusco et al., 2010; Granero et al., 2010; Kim et al., 2009, 2010, 2011; Moreno et al., 2011; Shin et al., 2011; Souza et al., 2011; Andreu-Navarro et al., 2012; Chawla et al., 2012; Curulli et al., 2012; Lee et al., 2012; Sánchez-Obrero et al., 2012; Arribas et al., 2013; Lanzellotto et al., 2014; Godoy-Navajas et al., 2015; Medina-Plaza et al., 2015). Other wine metabolites such as lactic acid (Monošík et al., 2012a; Loaiza et al., 2015), glycerol (Monošík et al., 2012b), and ochratoxin A (Fernández-Baldo et al., 2010; Barthelmebs et al., 2011; Zamfir et al., 2011) have been also analyzed to a lesser extent. A summary of some analytical characteristics of nanomaterial-based sensors and biosensors proposed for wine analysis is shown in Table 6.2.

Taking a look at some of the experiments made by different authors it is possible to have an idea of the magnitude and the advantages provided by the use of nanomaterials in the wine measuring equipment.

Sanz et al. (2005) developed a tyrosinase biosensor-based on the immobilization of the enzyme (Tyr) onto a glassy carbon electrode (GCE) modified with electrodeposited gold nanoparticles (Tyr-AuNPs-GCE) for bioelectrochemical polyphenols index determination in red and white wines. This biosensor was demonstrated to exhibit high catalytic activity toward some electrooxidation

Table 6.2 Some Analytical Characteristics of Nanomaterial-Based Sensors and Biosensors for Wine Analysis

Nanomaterial	Biosensor	Dp (mV)	Km App	Analytes	Linear Range	r^2	LOD	Repeatability RSD (%)	Sensitivity	Useful Lifetime	References
Gold nanoparticles	Tyr-AuNPs-GCE	−100	0.14 M [a] 0.12 M [b] 0.22 M [c] 0.44 M [d]	Phenol Catechol Caffeic acid Gallic acid	$0.01–0.4 \times 10^{-4}$ M [a] $0.005–0.5 \times 10^{-4}$ M [b] $0.02–2.0 \times 10^{-4}$ M [c] $0.25–9 \times 10^{-4}$ M [d]	0.9993 [a] 0.9998 [b] 0.9980 [c] 0.9980 [d]	2.1×10^{-7} M [a] 1.5×10^{-7} M [b] 6.6×10^{-7} M [c] 70×10^{-7} M [d]	3.6 ($n = 6$) [b]	112.14 μA mM^{-1} [a] 235.83 μA mM^{-1} [b] 26.82 μA mM^{-1} [c] 0.68 μA mM^{-1} [d]	18 days	Sanz et al. (2005)
Multiwalled carbon nanotubes	Tyr-MWCNTs-ITO	−1000 and 750	0.041 mmol L^{-1}	Phenol	0.01–0.08 mM	0.9768			2.925 A (mol $L^{-1})^{-1}$		Kim et al. (2009)
Multiwalled and single-walled carbon nanotubes	Lacc-MWCNTs-SPE	−100	133.3 ± 17.0 mg L^{-1}	Gallic acid	0.1–18.0 mg L^{-1}	0.9990	0.1 mg L^{-1}	3–4 ($n = 6$)	0.006 μA mg^{-1} L	10 days	Di Fusco et al. (2010)
Magnetic nanoparticles amino functionalized	Anti-OTA-$MNPs^2$-GCE	700–1,400		Ochratoxin A	0.5–50 μg kg^{-1}	0.997	0.02 μg kg^{-1}				Fernández-Baldo et al. (2010)
Multiwalled carbon nanotubes	PBHR–Fc–MW-CNTs-CPE	−50	2.44 ± 0.07 mmol L^{-1} [c] 1.24 ± 0.04 mmol L^{-1} [e]	Caffeic acid T-resveratrol	0.06–69 mg L^{-1} [c] 0.05–52 mg L^{-1} [e]	0.9999 [c] 0.9997 [d]	0.020 mg L^{-1} [c] 0.023 mg L^{-1} [d]	5.2 ($n = 6$) [c] 4.1 ($n = 6$) [e]	9.08 nA mg^{-1} L	5 days	Granero et al. (2010)
Multiwalled carbon nanotubes	Tyr-MWCNTs-ITO			Phenol	0.2–0.9 mM 0.1–0.5 mM						Kim et al. (2010)
Multiwalled carbon nanotubes	*Alka*-QD-MWNTs-GCE	500 and 700		Phenol	0.5–5.0 mM [q] 0.7–10 mM [r]	0.91756 [q] 0.95882 [r]				10 days [l]	Kim et al. (2011)
Superparamagnetic nanoparticles	DNA-apt-SpMNPs-SPE	250		Ochratoxin A	0.11–0.15 ng mL^{-1}		0.11 ng mL^{-1}	5.1 ($n = 5$)		4 weeks	Barthelmebs et al. (2011)
Multiwalled carbon nanotubes	MWCNTs-GCE	200		Caffeic acid Ferulic acid Gallic acids (+)-catechin	$10^{-5}–10^{-4}$ M		3.1 μM				Moreno et al. (2011)
Multiwalled carbon nanotubes	*Alka*-MWCNTs-CGE	300–400		Phenol	0.005–7.0 mM	0.99606					Shin et al. (2011)
Carbon nanotubes	CNTs-CPE	350		Gallic acid	$5.0 \times 10^{-7}–1.5 \times 10^{-5}$ mol L^{-1}		3.0×10^{7} mol L^{-1}				Souza et al., 2011

Iron oxide carboxyl-modified magnetic nanoparticles	Anti-OTA-MNPs[l]-AuE	−400		Ochratoxin A	0.01–5 ng mL^{-1}	0.9676	0.01 ng mL^{-1}	4.7 ($n=5$)			Zamfir et al. (2011)
Gold nanoparticles	Lacc-AuNPs			Gallic acid Catechol	0.13–5 μmol L^{-1} [d] 0.08–5 μmol L^{-1} [b]	0.999 [d] 0.998 [b]	0.04 μmol L^{-1} [d] 0.01 μmol L^{-1} [b]	3.9–5.4 ($n=11$) [d] 4.4–5.5 ($n=11$) [b]			Andreu-Navarro et al. (2012)
Silver nanoparticles/zinc oxide nanoparticles	Lacc-AgNPs/ZnONPs-AuE	220		Guaiacol	0.1–500 μM		0.05 μM		0.71 μA $μM^{-1}$ cm^{-2}	5 months [m]	Chawla et al. (2012)
Multiwalled carbon nanotubes	Tyr-GMA/VF-MWCNTs-GCE	−100		Phenol	0.1–20 mM		25 nM		0.187 A M^{-1} cm^{-2}	30 days [n]	Lee et al. (2012)
Multiwalled carbon nanotubes	LLO-PO-CH-MWCNTs-AuE	−50		Lactic acid	5–340 μM	0.993	1.66 μM	3.7 ($n=8$)	1.86 nA $μM^{-1}$	15 months [n]	Monošík et al. (2012a)
Multiwalled carbon nanotubes	MenzC-CH-MWCNTs-AuE	−50		Glycerol	5–566 μM	0.9986	2.24 μM	7.08 ($n=8$)	0.81 nA $μM^{-1}$	15 months [n]	Monošík et al. (2012b)
Magnetic nanoparticles (Fe_3O_4)	SOx-MNPs-AuE	200		Sulfite	0.5–1000 μM	0.997	0.15 μM			4 months	Rawal et al. (2012)
Gold nanoparticles	Tyr-AuNPs-PVC/TTF-TCNQ-GCE	0	114.0 μM 84.4 μM	Phenol Catechol	0.6–10.0 μM	0.9958 [a] 0.9990 [b]	7.9×10^{-7} M [a] 6.43×10^{-7} M [b]	2.2 ($n=5$)	62 ± 1.4 mA M^{-1} [a] 54.5 ± 0.6 mA M^{-1} [b]	2 weeks	Sánchez-Obrero et al. (2012)
Multiwalled carbon nanotubes	MWCNTs-GCE	700		Caffeic acid Gallic acid Ferulic acid *p*-Coumaric acid	0.1–100 μM [c] 0.1–100 μM [d] 0.1–100 μM [f] 0.2–100 μM [g]	0.9998 [c] 0.9995 [d] 0.998 [f] 0.997 [g]	0.07 μM [c] 0.05 μM [d] 0.08 μM [f] 0.09 μM [g]	2.8 ($n=10$) [c] 2.5 ($n=10$) [d] 5.3 ($n=10$) [f] 6.1 ($n=10$) [g]		4 days[o]	Arribas et al. (2013)
Silver nanoparticles	AgNPs-PTh-GCE			Caffeic acid	1.00×10^{-8} 4.83×10^{-6} M		5.3×10^{-9} M				Karabiberoğlu et al. (2013)
Molecularly imprinted nanoparticles	MIN-PVC-ISE		9.90×10^{-11} M	Histamine	10^{-6}–10^{-2} mol L^{-1}		1.12×10^{-6} mol L^{-1}	4.30–5.75		10 days	Basozabal et al. (2014)
Fullerene and gold nanoparticles	Lacc-AuNPs/FuNPs-SPE	−100	0.66 mM	Gallic acid	0.03–0.30 mmol L^{-1}	0.9998	0.006 mmol L^{-1}	2.2	9.09 μA mM^{-1}	120 days [p]	Lanzellotto et al. (2014)
Gold nanoparticles	AuNPs-IPE	470		Sulfur dioxide					0.477×10^{-3} μA L mol^{-1}	Single-use	Schneider et al. (2014)
Fluorescent carbon nanodots	FCN			Tannic acid	0.1–10 mg L^{-1}	0.994	0.018 mg L^{-1}				Ahmed et al. (2015)
Gold nanoparticles	AuNPs-LBE	−850 [hijk] 750 [c]		Lactic acid Malic acid Tartaric acid Citric acid Caffeic acid Gallic acid	10^{-3}–10^{-5} mol L^{-1} [c]	0.980 [h] 0.999 [i] 0.990 [j] 0.999 [k] 0.9968 [d]	17.63×10^{-6} mol L^{-1} [h] 5.91×10^{-6} mol L^{-1} [i] 4.73×10^{-6} mol L^{-1} [j] 5.62×10^{-6} mol L^{-1} [k] 0.67×10^{-6} mol L^{-1} [c] 0.51×10^{-6} mol L^{-1} [d]	<2 ($n=5$)			Medina-Plaza et al. (2015)

(*Continued*)

Table 6.2 Some Analytical Characteristics of Nanomaterial-Based Sensors and Biosensors for Wine Analysis (*cont.*)

Nanomaterial	Biosensor	Dp (mV)	Km App	Analytes	Linear Range	r^2	LOD	Repeatability RSD (%)	Sensitivity	Useful Lifetime	References
Terbium oxide nanoparticles	Lacc-Tb_4O_7NPs-HPTS			Gallic acid	0.5–12 μM	0.996	0.14 μM				Godoy-Navajas et al. (2015)
Platinum nanoparticles supported on graphitized carbon nanofibers	LLO-PtNPs/GCNFs-SPE	300		Lactic acid	0.01–2.0 mM	0.999	6.9 μM	4.9 (n = 10)	41,302 ± 546 μA M^{-1} cm^{-2}	3 months (room temperature)[n] 18 months (−20°C)[l]	Loaiza et al. (2015)

Dp Detection potential (mV); Km app, apparent Michaelis–Menten constant (M); LOD, Limit of detection; RSD, Relative standard deviation

Lacc, laccase; Tyr, tyrosinase; PBHR, peroxidases from *Brassica napus* hairy roots; Fc, Ferrocene; *Alka, Alkaligenes* spp.; GMA, glycidyl methacrylate; VF, vinyl ferrocene; LLO, L-lactate oxidase; PO, peroxidase; SOx, sulfite oxidase; MenzC, multienzyme complex consisting of kinase/creatine kinase/creatinase/sarcosine oxidase/peroxidase; CH, chitosan; Anti-OTA, Monoclonal antibody for OTA; DNA-apt, DNA aptamer; PVC/TTF-TCNQ, poly(vinyl chloride) and tetrathiafulvalene-tetracyanoquinodimethane salt; PTh, poly(thiophene); HPTS, 8-hydroxypyrene-3-sulphonate trisodium; AuNPs, gold nanoparticles; AgNPs, silver nanoparticles; ZnONPs, zinc oxide nanoparticles; CNTs, carbon nanotubes; MWCNTs, multiwalled carbon nanotubes; MNPs[1], magnetic nanoparticles made of iron oxide and carboxyl groups; MNPs[2], magnetic nanoparticles amino functionalized; SpMNPs, magnetic beads coated with streptavidin and carboxylic groups; QD, quantum dot; MIN, molecularly imprinted nanoparticles; FuNPs, fullerene nanoparticles; FCN, fluorescent carbon nanodots; Tb_4O_7NPs, terbium oxide nanoparticles; PtNPs, platinum nanoparticles; GCNFs, graphitized carbon nanofibers; GCE, glassy carbon electrodes; SPE, screen-printed electrode; AuE, gold electrode; CPE, carbon paste electrodes; ITO, indium tin oxide glass electrode; ISE, ion-selective electrode; LBE, Langmuir–Blodgett films electrode; IPE, inkjet-printed electrode.

[a] phenol
[b] catechol
[c] caffeic acid
[d] gallic acid
[e] *t*-resveratrol.
[f] ferulic acid
[g] *p*-coumaric acid
[h] lactic acid
[i] malic acid
[j] tartaric acid
[k] citric acid
[l] After this time period, the biosensor lost less than 5% of its initial activity.
[m] After this time period, the biosensor lost 25% of its initial activity (200 uses).
[n] After this time period, the biosensor lost 10% of its initial activity.
[o] After this time period, the biosensor lost 30% of its initial activity.
[p] After this time period, the biosensor lost 13% of its initial activity.
[q] CdS-MWNTs
[r] Cu_2S-MWNTs

reactions and exhibited a rapid response to the changes in the substrate concentration for all the phenolic compounds tested: phenol, catechol, caffeic acid, chlorogenic acid, gallic acid, and protocatechualdehyde. When compared with performance results reported in the literature for tyrosinase colloidal-gold-modified carbon paste electrodes, the linear range for phenol is wider, the sensitivity is higher, and the useful lifetime longer with the Tyr-AuNPs-GCE (See Table 6.2 for specific data). For TPI determination in wine samples, a 270 μL wine aliquot was directly added to 10 ml of a 0.1 mol L^{-1} phosphate buffer solution of pH 7.4, which was continuously stirred at a constant rate in the electrochemical cell. Then, amperometric measurements at −100 mV were carried out, and the estimation of the phenolic compounds content was performed by applying the standard additions method, which implied the addition of successive 20 μL aliquots of a 2.5×10^{-3} mol L^{-1} phenol stock solution. The content of phenolic compounds was then obtained as a phenol concentration, which was further expressed in milligrams per liter of caffeic acid (Sanz et al., 2005). A good correlation of the results ($r = 0.990$) was found when they were plotted versus those obtained by using the spectrophotometric method involving the Folin–Ciocalteau reagent. Briefly, this reagent by reacting with the phenol −OH group, produces a blue-color complex whose absorbance is read at 730 nm.

Di Fusco et al. (2010) developed and characterized the use of laccases from *Trametes versicolor* (TvL) and *Trametes hirsuta* (ThL) as biocatalytic components of electrochemical biosensors for the determination of polyphenol index in wines. The operating bases of such laccase-based detectors are shown in Fig. 6.3. The resulting method is characterized by a fast analytical response, is unaffected by interferences within the wine's matrix, and requires a low investment. By including multi- and single-walled carbon nanotubes into the device based on screen-printed electrodes (SPE), the linearity range was 0.1–18.0 mg L^{-1} with a limit of detection of 0.1 mg L^{-1} of gallic acid.

Granero et al. (2010) designed an amperometric biosensor based on peroxidases from *Brassica napus* hairy roots (PBHR) used to determine the total polyphenolic content in wine and tea samples. This biosensor comprising PBHR, ferrocene, and multiwalled carbon nanotubes embedded in a mineral oil deposited onto carbon paste electrodes, have some advantages over the commonly used Folin–Ciocalteu method for total polyphenol index measurement, such as a shorter detection time, a smaller sample volume, higher accuracy, and a high simplicity. In addition, colored samples can be directly used for the measurement without any pretreatment, and there is no interference from sulfur

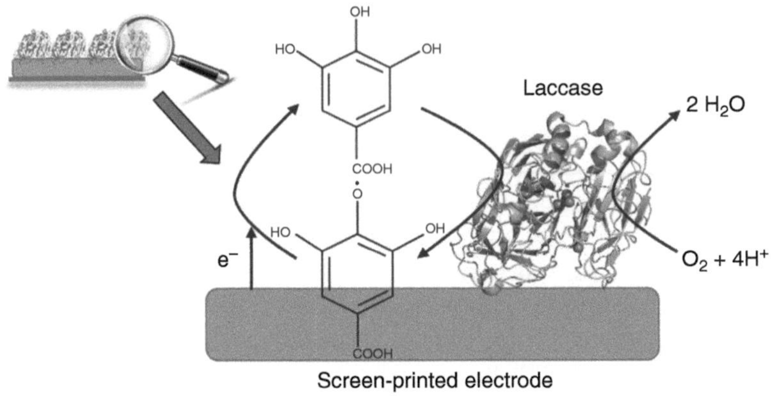

Figure 6.3. Electron transfer mechanism of gallic acid oxidation on a laccase biosensor.

dioxide, ascorbic acid, and glucose content. The limit of detection of this device is 0.02 mg L^{-1} both for caffeic acid and *t*-resveratrol, and the linear range (0.06–69 mg L^{-1} and 0.05–52 mg L^{-1} for caffeic acid and *t*-resveratrol, respectively) is greater in comparison with other laccase biosensors reported in literature. Similarly, Montereali et al. (2010) designed an amperometric biosensor for the detection of polyphenols in must and wine by immobilizing two enzymes, namely tyrosinase and laccase, on graphite screen printed electrodes modified with ferrocene. In both biosensors, ferrocene is acting as a redox mediator.

A microbial biosensor was fabricated by Kim et al. (2011) for sensing phenolic compounds in red wines. The sensor comprises a composite thin film of a quantum dot-modified multiwall carbon nanotube (QD-MWCNTs) covering the surface of glassy carbon electrodes and cells of *Alkaligenes* spp. immobilized outward. QD-MWCNTs film acts both as electron transfer mediator and as microbe immobilization support. *Alkaligenes* spp. is known as a potent phenol removing bacteria, able to catalyze the conversion of phenolic substrates to catechol and then o-quinone, allowing the detection by cyclic voltammetry. The sensing ranges of this microbial biosensor based on CdS-MWCNT and Cu_2S-MWCNT supports were 0.5–5.0 mM and 0.7–10 mM for phenol in phosphate buffer solution, respectively. In a similar study, Shin et al. (2011) developed an electrochemical microbial biosensor by immobilization of the same microorganism (*Alkaligenes* spp.), but this time including amine groups in the formulation of the composite, in order to incorporate a functional group onto MWCNTs surface that provides ionic properties and unpaired electron, and thereby increases its solubility and enhances its binding affinity to various biomolecules, such as enzymes, microbes, and proteins. The sensing efficiency of the biosensor was checked by measuring phenol content in phosphate buffer solution and a linear range of 0.005–7.0 mM was obtained. Using a calibration curve for phenol (selected as phenolic standard) constructed under the optimal experimental conditions, both microbial sensors (Kim et al., 2011; Shin et al., 2011) were able to determine the total content of phenolic compounds in commercial red wines.

Later on, Andreu-Navarro et al. (2012) proposed an enzymatic fluorimetric method for the determination of polyphenolic content in different foodstuff samples, among them, red grape juice. Unlike amperometric biosensors, in which the determination of total polyphenol content involves the detection of the reduction currents of such compounds when they are enzymatically oxidized to quinones or radicals, this novel method is based on the temporal inhibition caused by polyphenols on the oxidation of the long wavelength fluorophor indocyanine green (acting as enzyme substrate), in the presence of the enzyme laccase and positively charged gold nanoparticles (acting as enzyme inhibitor). Using a stopped-flow mixing device, all the reactants are placed in a system of syringes connected to a spectrofluorimeter in which the variation of the fluorescence signal of indocyanine green with time was monitored at $\lambda_{ex} = 764$ and $\lambda_{em} = 806$ nm. Due to the particular design of the equipment, the enzyme consumption is slightly higher than in the amperometric methods involving its immobilization on a suitable electrode, but this enzymatic fluorimetric method shows better selectivity and lower limits of detection. LODs ranged between 0.01 and 0.04 μmol L^{-1} for the five polyphenols studied (catechol, hydroquinone, hydroxyhydroquinone, gallic acid, and pyrogallol) after optimizing the method variables. In a similar way, Godoy-Navajas et al. (2015) studied the combination of terbium oxide nanoparticles (Tb_4O_7NPs) with the enzyme laccase and 8-hydroxypyrene-3-sulfonate trisodium (HPTS; fluorescent enzyme substrate; λ_{ex} 450 nm and λ_{em} 535 nm) for the fluorometric determination of polyphenol compounds in different types of wine (red, white, and sweet wines). Thanks to Tb_4O_7NPs, responsible for increasing the reaction rate of the system, it is possible to shorten the analysis time. Furthermore, the automation of the measurements using a microplate reader allows a great overall sample throughput (35 samples h^{-1}), instead of the 15 min per sample needed in the method using indocyanine green. Another improvement achieved over the previous method (Andreu-Navarro et al., 2012) is the use of wells with a total volume of 180 μL that results in cost-effectiveness, because the reagent consumption is lower (among them the enzyme and the fluorophor). The limit of detection was 0.14 μM of gallic acid (used as polyphenol standard), while the value obtained in the absence of Tb_4O_7NPs was 0.4 μM of gallic acid, which shows the positive effect of the NPs on the method. It was found that the proposed fluorometric method has better selectivity toward polyphenols compared to the reference Folin–Ciocalteu method.

Chawla et al. (2012) first investigated the analysis of total phenolic compounds in wine using an amperometric biosensor constructed by covalently immobilization of laccase onto a nanocomposite consisting of silver nanoparticles (AgNPs) and zinc oxide nanoparticles (ZnONPs) electrochemically deposited onto a gold electrode. On the one hand, ZnONPs provide interesting qualities for biosensors, such as high surface area for strong adsorption of biomolecules, good biocompatibility (nontoxicity), and high electron communication. On the other hand, AgNPs exhibit a catalytic activity for the reduction of H_2O_2. The combination of both types of nanoparticles in a single composite film led to improvements in the analytical performance of the biosensor in terms of low working potential (0.22 V), short response time (8 s), sensitivity (0.71 μA μM^{-1} cm^{-2}), and high storage stability (5 months).

Arribas et al. (2013) developed a highly stable amperometric detector based on MWCNTs-modified glassy carbon electrodes for the continuous determination of total polyphenols in white and red wines using a flow injection analysis (FIA) system. The MWCNT's layer coating the working electrode provides improvements in the electrocatalytic properties (decrease in the overpotentials) plus an enhanced signal stability and increased resistance to passivation. The limits of detection obtained for the four phenolic compounds studied (gallic, caffeic, ferulic, and p-coumaric acids) were below 0.1 μM. A selective detection of polyphenols with good sensitivity, wide linear range (0.1–100 μM) and excellent stability (4 days, with an average of 40 wine injections per day, before a decrease of 30% from the initial signal) was achieved. Furthermore, the developed detector is able to estimate the easily oxidizable polyphenol fraction of the wine by fixing the potential detection at 300 mV, instead of the 700 mV used for total polyphenol content analysis. Finally, to verify the electroanalytical performance of the detector, a total of 4 red wines and 5 white wines were measured, and although the polyphenol content estimated by the Folin–Ciocalteu method (reference method of the OIV) was notably higher than that obtained electrochemically, the correlation of spectrophotometric and electroanalytical signals was good (r^2 = 0.98059). Correlation with the results obtained using the I280 method (absorbance spectrophotometric measurement at 280 nm) was slightly higher (r^2 = 0.98773).

Lanzellotto et al. (2014) investigated the coupling of two nanostructured materials (fullerene and gold nanoparticles) for the construction of a laccase-based biosensor in order to analyze the polyphenol content of the wine. Besides the positive effect on the analytical performance provided by the gold nanoparticles, fullerenes are electroactive compounds that may serve as mediators in amperometric biosensors by increasing its sensitivity and allowing it to operate at lower potentials thus reducing the interferences from other wine compounds. The biosensor was able to detect polyphenols in the range from 0.03 to 0.3 mmol L^{-1} of gallic acid (selected as standard for polyphenols) with a good repeatability of the measurement (RSD = 2.2%) and a detection limit of 1.1 mg L^{-1}. When the doubly nanostructured biosensor was tested in wine samples, electrochemical results were reliable and in good agreement with those obtained with the Folin–Ciocalteau reference method.

A fluorimetric nanosensor was proposed by Ahmed et al. (2015) for sensitive and selective detection of tannic acid in red and white wines using fluorescent carbon nanodots (FCN; λ_{ex} 362 nm and λ_{em} 440 nm) without sample pretreatment and no interference of other wine compounds. This was the first time a fluorescence method for tannic acid analysis in wines using carbon nanodots was proposed. Carbon nanodots have unique optical properties, such as broad excitation spectra, tunable emission wavelength, and stable photoluminescence. In this work, FCN were passivated using polyethyleneglycol bis(3-aminopropyl) (PEGA) as amino precursor. Calibration curve was linear within the range 0.1–10 mg L^{-1} and with a detection limit of 0.018 mg L^{-1} tannic acid. The reliability of the method was tested by measurement of wine samples spiked with different tannic acid concentrations (0, 1, and 3 mg L^{-1}). Satisfactory recoveries were achieved in the range of 90–112.5%. In addition, it showed high sensitivity and low detection limit when compared with some electrochemical and optical methods.

The use of cell materials can be a mean to produce nanosensors in wines. Using fibroblast cells membrane-engineered in an alginate matrix it is possible to measure variations in the membrane potential as a response to the analyte concentration (Varelas et al., 2011; Fig. 6.4). Membranes were engineered by osmotic insertion of specific trichloroanisole (TCA) antibodies at a concentration of 0.5 μg L^{-1}. The sensor was very sensible detecting in few minutes (3–5 min) very low concentrations of TCAs (10^{-1} ppt). Also it shows high selectivity against other haloanisols and halophenols occurring in wine products. This nanosensor was able to detect TCAs at 1.02–12 ng L^{-1} in less than 5 min in real white wine samples.

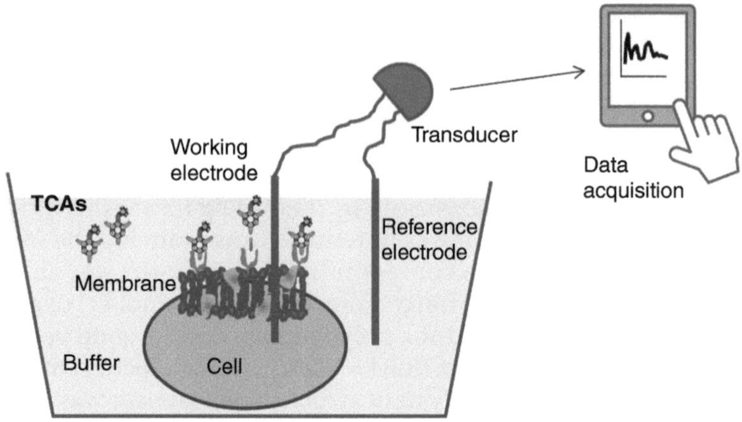

Figure 6.4. Device using engineered-membrane cells with antibody nanosensors to measure TCAs in wines.

The ochratoxin A (OTA) immunosensor designed by Zamfir et al. (2011) is based on the covalent immobilization of anti-OTA monoclonal antibody on the surface of magnetic nanoparticles (MNPs) and immobilization of these MNPs on the surface of gold working electrode. OTA maximum level allowed in wines by the European Union is 2 ppb (Commission, 2006). When analyzing a spiked white wine with known and increasing contents of OTA, they found a good correlation between the immunosensor response and ELISA test, without any interference from the real sample matrix. However, for the accurate determination of OTA in red wines with an immunosensor device, a simple sample pretreatment by pH adjustment and polyvinylpyrrolidone (PVP) addition is necessary. Otherwise, polyphenols may affect severely the antibody and immunosensor performances (Vidal et al., 2012). Previously, Fernández-Baldo et al. (2010) reported an electrochemical method based on modified magnetic nanoparticles and Osteryoung square wave voltammetry (OSWV), which had shown a rapid and sensitive determination of OTA in *Vitis vinifera* red grapes tissues with a detection limit of 0.02 μg kg^{-1}.

Similarly, Barthelmebs et al. (2011) tested a novel electrochemical DNA aptasensor, based on disposable screen-printed electrodes and using superparamagnetic nanoparticles, for the sensitive detection of OTA in wine samples. The detection limit achieved with the direct determination in which a fixed concentration of OTA and alkaline phosphatase-avidin (ALP-avidin) is involved, was 0.11 ng mL^{-1} of OTA, below the maximum limit set by the European legislation. Magnetic beads may improve the performance of immobilized biomolecules by providing a relatively large number of binding sites, thus favoring biochemical reactions and speeding up assay kinetics. As already mentioned, a simple pretreatment of the wine sample with PVP (dose: 10 g L^{-1}) is required to avoid interference of polyphenols in the measurement. The main advantages provided by this aptasensor over other analytical techniques, such as HPLC or immunosensors, are shorter analysis time, higher sensitivity, stability, and cost effectiveness. Moreover, the electrochemical devices may be miniaturized, making them portable.

An amperometric sulfite biosensor was fabricated by Rawal et al. (2012) through immobilization of a sulfite oxidase (SO_x) (EC 1.8.3.1) purified from *Syzygium cumini* leaves, onto carboxylated gold coated magnetic nanoparticles (Fe_3O_4) electrodeposited onto the surface of a gold electrode. This biosensor showed a 96.46% recovery of added sulfite in red wine and a good correlation ($r = 0.99$) with a standard DTNB (5.5′-dithio-bis-(2-nitrobenzoic acid)) method. Some advantages provided by this biosensor compared to others are its excellent electrocatalysis toward sulfite, lower detection limit (0.15 μM sulfite), higher storage stability (120 days), and no interference by ascorbate, cysteine, fructose, and ethanol.

The use of gold nanoparticles to increase the sensitivity of inkjet-printed silver electrodes in the determination of wine preservative sulfur dioxide was studied by Schneider et al. (2014). In the construction of this sensor, electrodes were fully inkjet-printed on a PET foil to develop a portable and single-use device with the advantage of small sample volumes, as a single drop is needed to record the electrochemical signals. Furthermore, the low-cost production of the sensor elements and the ultrafast determination of sulfur dioxide by cyclic voltammetry made it a reliable technique for simple and selective determination of SO_2 during transportation and storage of wine, as well as monitoring during winemaking. The AuNPs-based sensor showed excellent suitability for SO_2 analysis in white wines and no sensitivity toward polyphenols, compounds that typically generate interference problems in cyclic voltammetry measurements due to their close oxidation potentials.

Recently, Basozabal et al. (2014) created a potentiometric sensor that can selectively quantify histamine in the presence of other biogenic amines in real wine and fish matrices, with minimal sample pretreatment. This sensor consists of molecularly imprinted nanoparticles (MIN) with high affinity and specificity for histamine (including antihistamine antibodies), incorporated in PVC membranes and integrated in a potentiometric transducer. Its limit of detection is 1.12×10^{-6} mol L^{-1}, with a linear range which comprises frequent histamine contents in the food industry, between 10^{-6} and 10^{-2} mol L^{-1}.

Also, a sensor to imitate the astringency sensations produced in the mouth has been developed (Siem, 2014). The sensor uses salivary proteins to measure the interaction between them and certain molecules like phenols. Traditional assessment of astringency and dryness depends on the evaluation by a sensory panel and it is sensible to human inaccuracy. The sensor is a small plate coated with gold nanoparticles. These nanoparticles (AuNPs) work as nanooptic zooms focusing a beam of light, measuring nanoscale interactions (20 nm) among phenols and proteins.

In order to determine the concentration of L-lactic acid in food and wine samples, Monošík et al. (2012a) constructed an amperometric biosensor using nanocomposite electrodes consisting of multiwalled carbon nanotubes and by immobilization of two enzymes, L-lactate oxidase and peroxidase, within chitosan layers on the surface of gold electrodes. For the current measurement, a working potential of −50 mV was selected in order to achieve satisfactory sensitivity and avoid undesired interferences, particularly of polyphenols. After optimization of the assay conditions, the analytical features of this biosensor were: linear range of 5–340 mM, limit of detection of 1.66 μM, a response time of 65 s, good operational stability (no loss of sensitivity after 60 consecutive measurements), and an excellent storage stability (15 months of storage at room temperature). Unlike the assessed reference analytical methods for the quantification of lactic acid in food samples (enzymatic-spectrophotometric and standard HPLC analysis), the MWCNTs-based biosensor was able to determine very low concentrations of L-lactic acid (0.02–0.12 g L^{-1}). Furthermore, a good correlation was found between the results provided by the three analysis methods, enabling the use of the biosensor for monitoring malolactic fermentation. Following the same principle, Monošík et al. (2012b) designed a MWCNTs-based biosensor that incorporates a multienzyme cascade consisting of glycerol kinase, creatine kinase, creatinase, sarcosine oxidase, and peroxidase for the amperometric determination of glycerol in wine. The biosensor showed linearity over the range of 5–566 μM with a detection limit of 2.24 μM and a sensitivity of 0.81 nA μM^{-1}. The suitability of the biosensor to quantify glycerol was verified by measuring different red and white wine samples and subsequently comparing the results with reference enzymatic–spectrophotometric method. Only a small discrepancy in results was found in one of the 11 samples analyzed. In addition, this multienzymatic biosensor is very versatile because it can also be used for determination of triglycerides (after their hydrolysis by lipase) or other kinase substrates if glycerol kinase is replaced by the corresponding kinase, for example, determination of glucose and fructose using hexokinase.

Loaiza et al. (2015) proposed a hybrid nanomaterial consisting of graphitized carbon nanofibers (GCNFs) and platinum nanoparticles (PtNPs) for the amperometric detection of lactate in various types of wines (positively tested in red, white, and rosé). The resulting lactate biosensor, constructed using covalent immobilization of lactate oxidase, show good reproducibility (RSD 4.9%, $n = 10$) and high sensitivity (41,302 ± 546 $\mu A\ M^{-1}\ cm^{-2}$), with a good limit of detection (6.9 μM). Furthermore, it is reusable for several measurements with a long useful lifetime, retaining 95% of its initial activity after 18 months at −20°C (storage conditions) or 90% after 3 months at room temperature. The main advantage of using platinum nanoparticles instead of bulk platinum is their greater surface area, which results in several electroanalytical improvements, significantly increasing the sensitivity of the biosensor when mixed with GCNFs (responsible for promoting the electron-transfer reactions) in order to create a chemically modified electrode.

An electroanalytical sensor for determination of caffeic acid in red wine was fabricated by modifying glassy carbon electrode with silver nanoparticles and poly(thiophene) (Karabiberoğlu et al., 2013). Linearity ranged from 1.00×10^{-8} to 4.83×10^{-6} M with a detection limit of 5.3×10^{-9} M caffeic acid.

In 2015, for the first time dodecanethiol-functionalized gold nanoparticles immobilized in Langmuir–Blodgett (LB) films have been used as voltammetric sensors for detection of organic and phenolic acids both in must and wine samples (Medina-Plaza et al., 2015). Thanks to the electrocatalytic effect of gold nanoparticles (increased peak currents) and to the large specific surface of the LB films, it was possible to improve the analytical performance of the sensor. Moreover, the presence of gold nanoparticles increased drastically the sensitivity toward the organic acids (tartaric, malic, lactic, and citric) and the phenolic compounds (caffeic and gallic acids) tested. A very interesting fact is that the chemically modified electrode was able of simultaneously determine acidity and caffeic acid in the range of concentrations usually found in wines with no interferences.

2.2 Electronic Tongue for Wine Classification

The development of artificial taste sensors (also called "electronic tongues") able to differentiate tastes, mimicking the human tongue, is a feasible endeavor. This device allows a quantitative treatment of "tastes" and does not suffer from lack of sensitivity due to long term exposure to a certain substance, unlike the case of humans (Riul et al., 2004).

Through the combination of supramolecular thin films of conducting polymers with lipid-like material and using impedance spectroscopy technology, it is possible to identify taste flavors at levels below the human threshold, being even feasible the sensing of nonelectrolytic substances such as sucrose. Moreover, conducting polymers might increase the sensitivity and selectivity of devices due to their unique properties, serving, as well, as a matrix to immobilize certain molecules or suppress interfering reactions (Riul et al., 2003).

Each composite nanostructured film composed of different materials (~12-nm-thick each) works as individual sensing units and is characterized by the production of singular electric signals. Therefore, based on this feature, the electronic device is able to create a "fingerprint" for the wine samples analyzed. Furthermore, it is possible to obtain a specific response of the sensor by modifying the different layers covering the working electrodes (Riul et al., 2003).

Riul et al. (2003) checked the performance of the device by analyzing different red wines with regard to vintage and adulteration criteria. Two Cabernet Sauvignon wines, from 1999 and 2000 vintages, and a table wine with added sugar and conservatives of the same winemaker were analyzed. All three red wines were readily characterized by the sensor. Since the materials used for the construction of the sensing units do not interact specifically with the analyzed solutions, the electrical pattern observed in the sensor may be associated with different taste sensations, which allows the identification of taste.

Later on, Riul et al. (2004) could identify distinct clusters in principal component analysis (PCA) plots for six types of red wine, using impedance spectroscopy as the principle of detection (ac measurement). The ac impedance was recorded with a Solartron 1260A. The responses were analyzed at 1 kHz, because this frequency region is the most sensitive to changes in the film properties. These wines were classified according to their vintage, vineyard, brand, and storage conditions. Moreover, when data was treated with artificial neural networks (ANNs), this artificial tongue could identify wine samples stored under different conditions.

2.3 Nanoliquid Chromatography for Proteins and Biogenic Amines Analysis

Nanoliquid chromatography (nano-LC), first introduced by Karlsson and Novotny (1988), is a new miniaturized separation method offering the possibility to achieve good resolutions and high efficiency in a short analysis time making use of minute volumes of both samples and reagents.

So far, nano-LC has been applied in different areas: pharmaceuticals, biomedicine, environment and recently food and agrochemicals (Fanali et al., 2013). Separation is carried out in capillaries of narrow internal diameter (10–100 μm) that operate at nanoliter flow rates using injection volumes of few nanoliters (Chervet et al., 1996).

The main advantage of using smaller i.d. columns in HPLC is the increased detection sensitivity that can be obtained as a result of reduced sample dilution (Rieux et al., 2011). The smaller the column radius, the lower the dilution factor, with a resulting increase in analyte detectability. In addition, the low flow rate used in nano-LC facilitates easy coupling to mass spectrometry for specific detection (Hernández-Borges et al., 2007).

As for its application in wine analysis, nano-LC has been used in the determination of proteins (Kwon 2004) and biogenic amines (Hernández-Borges et al., 2007).

In combination with mass spectrometry, Kwon (2004) could identify 20 proteins from a Sauvignon Blanc wine including five proteins derived from the grape, 12 from yeast, 2 from bacteria, and 1 from fungi.

Hernández-Borges et al. (2007) proposed a method for the simultaneous analysis of 10 biogenic amines (ethanolamine, methylamine, tryptamine, 2-phenylethylamine, putrescine, cadaverine, histamine, tyramine, spermidine, and spermine) in wines by nanoliquid chromatography (nano-LC) using UV detection and a capillary bidentate C18 column of 100 μm i.d. The mobile phase composed of acetonitrile, water, acetic acid, and triethylamine (TEA) mixture was pumped at a low flow rate (634 nL min^{-1}) and using a trapping column the injection volume could be set at 1 μL. Limits of detection (LODs) achieved ranged between 18.3 and 48.3 ng mL^{-1}; while calibration curves showed good linearity ($r^2 > 0.9924$).

The proposed method combined a sample pretreatment using polyvinylpyrrolidone (PVP; dose: 50 g L^{-1}), a postderivatization cleanup by means of an offline solid-phase extraction (SPE) procedure using C18 cartridges, as well as an online cleanup and preconcentration step using a C18 trapping column.

Derivatization with dansyl-chloride (known as dansilation) is required for biogenic amines detection because they generally exhibit little, if any, ultraviolet (UV) or visible light absorption, fluorescence, or electrochemical activity. Moreover, a solid-phase extraction procedure was carried to extract the analytes from wine samples. As well, C18 cartridges were used to increase the selectivity of the method and to provide a preconcentration of the analytes, avoiding the presence of interfering compounds of the matrix. Moreover, a PVP treatment of the samples should be carried out in order to remove the substances that interfere in the derivatization and subsequent quantification of the derivatives (Hernández-Borges et al., 2007).

Repeatability assays showed that the accuracy of the nano-LC-UV method is good both for intraday and day-to-day analysis. Besides, it was proved that after more than 500 injections of real wine samples, the column is still appropriate for the determination of biogenic amines.

2.4 Silver Nanoparticles as Antimicrobial Agent

Nanoparticles (NPs) can be of natural origin or produced by industry and they are ubiquitous in the environment (Smita et al., 2012). Silver and gold NPs with antimicrobial activity can be synthesized using cell-free extracts of *Candida albicans* (Ahmad et al., 2013; Fig. 6.5a). The effectivity of this colloidal biometallic NPs have been verified against both *Staphylococcus aureus* and *Escherichia coli*. NPs behave as effective growth inhibitors. The results have shown higher effectivity for silver nanoparticles and especially in *E. coli* (gram negative bacteria). Both species showed alterations in their growth curves with an extension of the lag phases. The mechanism of the inhibitory action remains misunderstood, it looks like the differences between the negatively charged microbial cell and the positively charged nanoparticles is essential for the antimicrobial activity (Sondi and Salopek-Sondi, 2004). Accumulation of AgNPs in bacterial cell walls increases the permeability affecting strongly the viability (Sondi and Salopek-Sondi, 2004). Moreover, interactions between AgNPs and sulfhydryl groups of proteins produce R-S-S-R bonds affecting the cell wall electronic carriers and

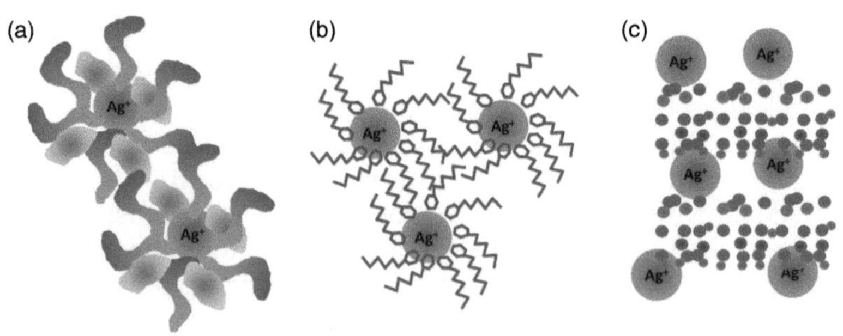

Figure 6.5. Ag-NPs formed with biologic fragments from microorganisms or plants extracts (a), organic molecules (b), silicates such as kaolin (c). These kinds of nanostructures can be also formed with other metals such as Au, Cu, or Ti.

the respiration mechanisms (Siva Kumar et al., 2004). The ability of AgNPs to penetrate inside the cell facilitates DNA damage (Morones et al., 2005). The effectiveness of AgNPs produced with polyethylene glycol or glutathione has also been studied in wine lactic acid bacteria and acetic bacteria (García-Ruiz et al., 2015; Fig. 6.5b). AgNPs forming colloidal structures with silicates have been also proposed as an alternative tool to sulfur dioxide to control microbial developments in wines (Izquierdo-Cañas et al., 2012; Garde Cerdán et al., 2014; Fig. 6.5c). Depending on the organic fraction, silver nanoparticles show a variable effect on different bacteria strains or species. AgNPs have also been used in the control of *Brettanomyces* (Brett) yeast in wines showing some effectivity (Communication, 2015). Kaolin-silver complex looks selective against Brett showing low toxicity against *S. cerevisiae.*

Copper NPs are also highly toxic against many bacteria and fungus by several mechanisms such us membrane disruption, blocking of metabolic pathways, DNA damage, and complexation with proteins (Cioffi et al., 2005; Ruparelia et al., 2008; Ren et al., 2009). The main problem with CuNPs is their low stability under oxidative conditions evolving to copper oxides. Use of reductive NPs or configurations that isolate copper from oxygen can be some successful ways to protect them and minimize these alterations.

2.5 Magnetic Separation of Yeasts in Sparkling Wines

Recently, Berovic et al. (2014) proposed a novel method for the magnetic separation of exhausted yeast biomass from sparkling wine by making the cells responsive to a magnetic field through the surface adsorption of superparamagnetic iron oxide maghemite (γ-Fe_2O_3) nanoparticles. These nanoparticles are coated with a thin layer of silica and grafted with (aminoethylamino)propylmethyldimethoxysilane (APMS) that provides a positive charge able to create a strong and lasting interaction with the negatively charged cell surface.

Moreover, electrostatic interactions established between the nanoparticles and the cell walls promote cell clustering, which strongly increases the efficiency of the magnetic separation (Fig. 6.6).

Regarding the effects on the wine yeast metabolism during primary fermentation, the presence of magnetic nanoparticles stimulates the rate of cell multiplication. With magnetized yeast cells a maximum biomass of 4.22 g L^{-1} was obtained after 88 h, while in the control comparable amount of 4.14 g L^{-1} was obtained after 120 h (Berovic et al., 2014). Likewise, a faster and more intensive fermentation process was observed with the magnetized

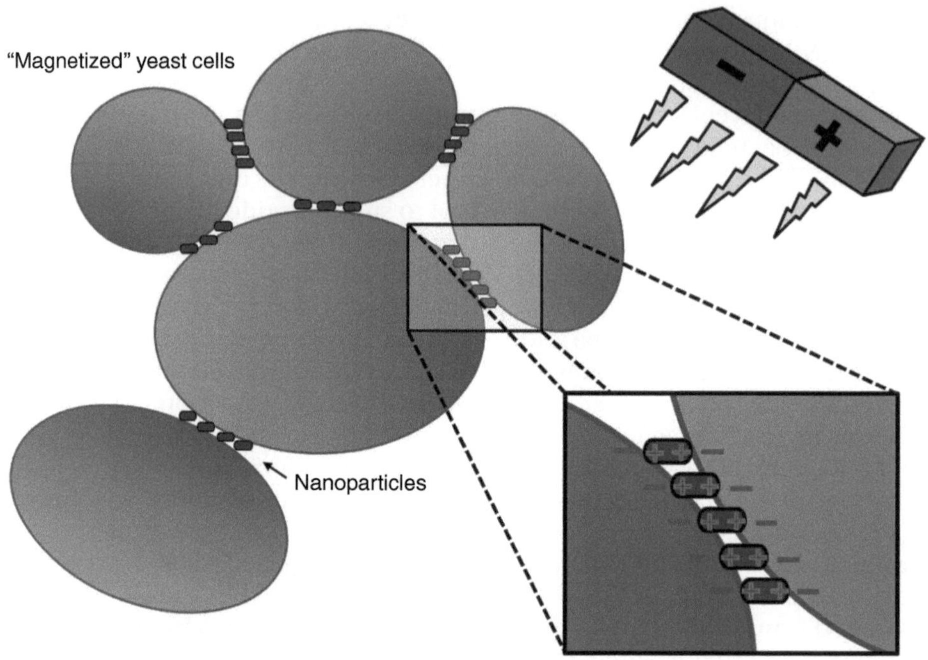

Figure 6.6. Electrostatic interactions established between nanoparticles and yeast cells.

biomass (nanoparticles-to-yeast ratio of 1:10), probably due to higher amounts of free iron Fe^{+3} in the fermentation substrate, but no changes in the final wine composition were found. Moreover, there were no remarkable differences in the carbon dioxide pressure produced during the secondary fermentation.

It worth mentioning that iron-oxide nanoparticles are considered to be nontoxic and were even approved for in vivo medical applications by the American Food and Drug Administration (FDA) (Thanh, 2012). Currently, according to EU regulations, the maximum permissible iron concentration in white wine is 10 mg L^{-1}.

This new separation method is inexpensive, simple, and very effective regarding the time and energy consumption when compared to the classic method. It was awarded at the fourth International Technology Transfer Conference 2011 in Ljubljana and patent protected (Berovic et al., 2010).

2.6 Others

2.6.1 Nanoencapsulation of Wine Flavors

As experimental proposal, the company NextNature has developed a programmable wine, referred to as "nano wine," by means of nanoencapsulation technology. Nano wine is a basic Merlot grape product that is mixed with millions of flavor-containing nanocapsules. When exposed to microwaves these capsules release their flavor in a different manner depending on the wattage and duration of exposure. Thus, it is possible to create a different type of wine with vanilla, truffle, wood, or pepper notes. Inactivated nanocapsules move unnoticed through the body, while the opened capsules alter the taste, smell, and color of the wine (NanoWine, 2015).

2.6.2 Package Nanotechnologies

Silver nanoparticles-storage containers can reduce bacterial growth by 99.9% (Sozer and Kokini, 2012a).

2.6.3 Nanosensors and RFID Systems

Connecting specific nanosensors able to detect microbial contaminants in wines (ie, *Brettanomyces* cells in barrels) also toxic compounds, off-flavor or heavy metals (ie, biogenic amines, ethylphenols, lead) to radiofrequency identification systems (RFID) could be a fast method to trace the evolution of alterations or contaminations in wines during barrel aging or bottle storage. These nanosensors-RFID connected can send a Wi-Fi signal to the mobile phone of the enologist when the amount detected is above the acceptable microbial count or the legal limit or sensory threshold in toxic molecules or off-flavors. These techniques are cheap, easy to implement, fast, and allow a good traceability during aging, storage, and transportation (Wang et al., 2006; Sozer and Kokini, 2012b).

3 Food Safety

The long-term effect on human health of colloidal nanoparticles must be better studied. Although penetration of nanoparticles into the human body appears to be low, long-term exposure can produce accumulation because of the limited excretion rate of nanoparticles, and so accumulative effects rather than acute toxicity is probable (Smolkova et al., 2015). Nowadays, low information is reported about biological effects of long-term exposure to nanoparticles. Silver NPs frequently generate reactive oxygen species that are able to induce pro-inflammatory processes and oxidative stress (Gaillet and Rouanet, 2015).

A recommendation has been published on the definition of a nanomaterial (2011/696/EU, European Commission, 2011) and more recently a draft law on novel food has been produced

(COM/2013/0435, European Commission, 2013). Moreover, the EFSA Scientific Committee has published a document assessing the risks and repercussions of the application of nanotechnologies in the food chain (Scientific and Committee, 2011).

4 Conclusions and Future Prospects

In the past decade, nanotechnology has gained much attention in enology and was used in several analytical applications, such as sensors and biosensors construction, in the production process as antimicrobial agent being a promising alternative to the use of SO_2, or in the selective extraction of yeast cells after secondary fermentation in sparkling wines, among other applications. There are multiple ways to harness the advantages offered by the different nanostructured materials in the field of winemaking, which have already been discussed. However, due to the great versatility of nanomaterials, there is still much potential to explore and discover, when it comes to develop new applications to improve and facilitate the production process as well as grape, must, and wine chemical analysis.

In the interaction between nanotechnology and wine, the next step is to further investigate the possibility of sensing multiple analytes of different nature by means of integration of several nanomaterials onto single chips, thus enabling the simultaneous and accurate detection of various wine chemical compounds.

Moreover, as demonstrated, working at nanoscale level with materials that offer interesting features such as magnetism, and with trapping techniques such as encapsulation, leads to the emergence of original ideas and fosters the development in this field of research. For instance, an interesting idea would be to develop a feeding system for the yeasts based on nutrient nanoencapsulation that somehow may be released according to the existing demand at each time.

References

Ahmad, T., Wani, I.A., Manzoor, N., Ahmed, J., Asiri, A.M., 2013. Biosynthesis, structural characterization and antimicrobial activity of gold and silver nanoparticles. Colloids Surf. B Biointerfaces 107, 227–234.

Ahmed, G.H.G., Laíño, R.B., Calzón, J.A.G., García, M.E.D., 2015. Fluorescent carbon nanodots for sensitive and selective detection of tannic acid in wines. Talanta 132, 252–257.

Andreu-Navarro, A., Fernández-Romero, J., Gómez-Hens, A., 2012. Determination of polyphenolic content in beverages using laccase, gold nanoparticles and long wavelength fluorimetry. Anal. Chim. Acta 713, 1–6.

Arribas, A.S., Martínez-Fernández, M., Chicharro, M., 2012. The role of electroanalytical techniques in analysis of polyphenols in wine. TrAc Trend. Anal. Chem. 34, 78–96.
Arribas, A.S., Martínez-Fernández, M., Moreno, M., Bermejo, E., Zapardiel, A., Chicharro, M., 2013. Analysis of total polyphenols in wines by FIA with highly stable amperometric detection using carbon nanotube-modified electrodes. Food Chem. 136 (3), 1183–1192.
Barroso, M.F., De-Los-Santos-Alvarez, N., Delerue-Matos, C., Oliveira, Maria Beatriz Prior Pinto, 2011. Towards a reliable technology for antioxidant capacity and oxidative damage evaluation: electrochemical (bio) sensors. Biosens. Bioelectron. 30 (1), 1–12.
Barthelmebs, L., Hayat, A., Limiadi, A.W., Marty, J., Noguer, T., 2011. Electrochemical DNA aptamer-based biosensor for OTA detection, using superparamagnetic nanoparticles. Sensor. Actuat. B Chem. 156 (2), 932–937.
Basozabal, I., Guerreiro, A., Gomez-Caballero, A., Goicolea, M.A., Barrio, R.J., 2014. Direct potentiometric quantification of histamine using solid-phase imprinted nanoparticles as recognition elements. Biosens. Bioelectron. 58, 138–144.
Berovic, M., Makovec, D., Boskovic, S., 2010. Procedure for the magnetic separation of yeast biomass from sparkling wine. WO 2012082079 A1.
Berovic, M., Berlot, M., Kralj, S., Makovec, D., 2014. A new method for the rapid separation of magnetized yeast in sparkling wine. Biochem. Eng. J. 88, 77–84.
Chawla, S., Rawal, R., Kumar, D., Pundir, C.S., 2012. Amperometric determination of total phenolic content in wine by laccase immobilized onto silver nanoparticles/zinc oxide nanoparticles modified gold electrode. Anal. Biochem. 430 (1), 16–23.
Chervet, J., Ursem, M., Salzmann, J., 1996. Instrumental requirements for nanoscale liquid chromatography. Anal. Chem. 68 (9), 1507–1512.
Choi, S., Chung, D., Kwen, H., 2011. Fabrication of biosensors using vinyl polymer-grafted carbon nanotubes. In: Serra, P.A. (Ed.), New Perspectives in Biosensors Technology and Applications. InTech, Vienna, Austria. pp. 245–268.
Cioffi, N., Torsi, L., Ditaranto, N., Tantillo, G., Ghibelli, L., Sabbatini, L., Bleve-Zacheo, T., D'Alessio, M., Zambonin, P.G., Traversa, E., 2005. Copper nanoparticle/polymer composites with antifungal and bacteriostatic properties. Chem. Mater. 17, 5255–5262.
Commission Regulation (EC) No. 1881/2006, 2006. Official Journal of the European Union L 364/5. http://eur-lex.europa.eu/LexUriServ/LexUriServ.do?uri=OJ:L:2006:364:0005:0024:EN:PDF
OIV Communication, 2015. Group of experts in wine microbiology. International Organization of Vine and Wine. Must and wine treatment with kaolin silver complex. OENO-TECHNO 13-535 and 536. Paris.
Curulli, A., Di Carlo, G., Ingo, G.M., Riccucci, C., Zane, D., Bianchini, C., 2012. Chitosan stabilized gold nanoparticle-modified au electrodes for the determination of polyphenol index in wines: a preliminary study. Electroanalysis 24 (4), 897–904.
Di Fusco, M., Tortolini, C., Deriu, D., Mazzei, F., 2010. Laccase-based biosensor for the determination of polyphenol index in wine. Talanta 81 (1–2), 235–240.
Durst, R., 1997. Chemically modified electrodes: recommended terminology and definitions (IUPAC recommendations 1997). Pure Appl. Chem. 69 (6), 1317–1324.
Emamifar, A., Kadivar, M., Shahedi, M., Soleimanian-Zad, S., 2011. Effect of nanocomposite packaging containing Ag and ZnO on inactivation of *Lactobacillus plantarum* in orange juice. Food Control 22 (3), 408–413.

European Commission, 2011. Commission recommendation of 18 October 2011 on the definition of nanomaterial. Official Journal of European Union 2011/696/EU, pp. 38–40.
European Commission, 2013. Proposal for a regulation of the European Parliament and of the council on novel foods, COM(2013), pp. 894.
Fanali, C., Dugo, L., Dugo, P., Mondello, L., 2013. Capillary-liquid chromatography (CLC) and nano-LC in food analysis. TrAC Trends Anal. Chem. 52, 226–238.
Feng, H., Wang, H., Zhang, Y., Yan, B., Shen, G., Yu, R., 2007. A direct electrochemical biosensing platform constructed by incorporating carbon nanotubes and gold nanoparticles onto redox poly (thionine) film. Anal. Sci. 23 (2), 235–239.
Fernández-Baldo, M.A., Bertolino, F.A., Messina, G.A., Sanz, M.I., Raba, J., 2010. Modified magnetic nanoparticles in an electrochemical method for the ochratoxin A determination in *Vitis vinifera* red grapes tissues. Talanta 83 (2), 651–657.
Gaillet, S., Rouanet, J.M., 2015. Silver nanoparticles: their potential toxic effects after oral exposure and underlying mechanisms—a review. Food Chem. Toxicol. 77, 58–63.
García-Ruiz, A., Crespo, J., López-de-Luzuriaga, J.M., Olmos, M.E., Monge, M., Rodríguez-Álfaro, M.P., Martín-Álvarez, P.J., Bartolome, B., Moreno-Arribas, M.V., 2015. Novel biocompatible silver nanoparticles for controlling the growth of lactic acid bacteria and acetic acid bacteria in wines. Food Control 50, 613–619.
Garde Cerdán, T., López, R., Garijo Jiménez, M.P., González Arenzana, L., Gutiérrez Viguera, A.R., López Alfaro, I., Santamaría Aquilué, M.P., 2014. Application of colloidal silver versus sulfur dioxide during vinification and storage of Tempranillo red wines. Austr. J. Grape Wine Res. 20, 51–61.
Godoy-Navajas, J., Aguilar-Caballos, M.P., Gómez-Hens, A., 2015. Automatic determination of polyphenols in wines using laccase and terbium oxide nanoparticles. Food Chem. 166, 29–34.
Granero, A.M., Fernández, H., Agostini, E., Zón, M.A., 2010. An amperometric biosensor based on peroxidases from *Brassica napus* for the determination of the total polyphenolic content in wine and tea samples. Talanta 83 (1), 249–255.
Hernández-Borges, J., D'Orazio, G., Aturki, Z., Fanali, S., 2007. Nano-liquid chromatography analysis of dansylated biogenic amines in wines. J. Chromatogr. A 1147 (2), 192–199.
Izquierdo-Cañas, P.M., García-Romero, E., Huertas-Nebreda, B., Gómez-Alonso, S., 2012. Colloidal silver complex as an alternative to sulphur dioxide in winemaking. Food Control 23, 73–81.
Jia, J., Wang, B., Wu, A., Cheng, G., Li, Z., Dong, S., 2002. A method to construct a third-generation horseradish peroxidase biosensor: self-assembling gold nanoparticles to three-dimensional sol-gel network. Anal. Chem. 74 (9), 2217–2223.
Kang, X., Wang, J., Wu, H., Aksay, I.A., Liu, J., Lin, Y., 2009. Glucose oxidase–graphene–chitosan modified electrode for direct electrochemistry and glucose sensing. Biosens. Bioelectron. 25 (4), 901–905.
Karabiberoğlu, Ş.U., Ayan, E.M., Dursun, Z., 2013. Electroanalysis of caffeic acid in red wine and investigation of thermodynamic parameters using an Ag nanoparticles modified poly (thiophene) film glassy carbon electrode. Electroanalysis 25 (8), 1933–1945.
Karlsson, K.E., Novotny, M., 1988. Separation efficiency of slurry-packed liquid chromatography microcolumns with very small inner diameters. Anal. Chem. 60 (17), 1662–1665.

Kim, K., Kang, H., Lee, J., Choi, S., 2009. Fabrication of a multi-walled nanotube (MWNT) ionic liquid electrode and its application for sensing phenolics in red wines. Sensors 9 (9), 6701–6714.
Kim, K., Lee, J., Robards, K., Choi, S., 2010. Immobilization of tyrosinase in carboxylic and carbonyl group-modified MWNT electrode and its application for sensing phenolics in red wines. J. Nanosci. Nanotechnol. 10 (6), 3790–3798.
Kim, S., Kwen, H., Choi, S., 2011. Fabrication of a microbial biosensor based on QD-MWNT supports by a one-step radiation reaction and detection of phenolic compounds in red wines. Sensors 11 (2), 2001–2012.
Kwon, S.W., 2004. Profiling of soluble proteins in wine by nano-high-performance liquid chromatography/tandem mass spectrometry. J. Agric. Food Chem. 52 (24), 7258–7263.
Lanzellotto, C., Favero, G., Antonelli, M., Tortolini, C., Cannistraro, S., Coppari, E., Mazzei, F., 2014. Nanostructured enzymatic biosensor based on fullerene and gold nanoparticles: preparation, characterization and analytical applications. Biosens. Bioelectron. 55, 430–437.
Lee, Y., Chung, D., Oh, S., Choi, S., 2012. Introduction of bifunctional group onto MWNT by radiation-induced graft polymerization and its use as biosensor-supporting materials. J. Nanomater. 2012, 7.
Loaiza, O.A., Lamas-Ardisana, P.J., Añorga, L., Jubete, E., Ruiz, V., Borghei, M., Cabañero, G., Grande, H.J., 2015. Graphitized carbon nanofiber–Pt nanoparticle hybrids as sensitive tool for preparation of screen printing biosensors. Detection of lactate in wines and ciders. Bioelectrochemistry 101, 58–65.
Marambio-Jones, C., Hoek, E.M., 2010. A review of the antibacterial effects of silver nanomaterials and potential implications for human health and the environment. J. Nanopart. Res. 12 (5), 1531–1551.
Medina-Plaza, C., García-Cabezón, C., García-Hernández, C., Bramorski, C., Blanco-Val, Y., Martín-Pedrosa, F., Kawai, T., de Saja, J.A., Rodríguez-Méndez, M.L., 2015. Analysis of organic acids and phenols of interest in the wine industry using Langmuir–Blodgett films based on functionalized nanoparticles. Anal. Chim. Acta 853, 572–578.
Monošík, R., Stred'anský, M., Greif, G., Šturdík, E., 2012a. A rapid method for determination of l-lactic acid in real samples by amperometric biosensor utilizing nanocomposite. Food Control 23 (1), 238–244.
Monošík, R., Ukropcová, D., Stred'anský, M., Šturdík, E., 2012b. Multienzymatic amperometric biosensor based on gold and nanocomposite planar electrodes for glycerol determination in wine. Anal. Biochem. 421 (1), 256–261.
Montereali, M., Della Seta, L., Vastarella, W., Pilloton, R., 2010. A disposable Laccase–Tyrosinase based biosensor for amperometric detection of phenolic compounds in must and wine. J. Mol. Catal. B Enzym. 64 (3), 189–194.
Moreno, M., Arribas, A.S., Bermejo, E., Zapardiel, A., Chicharro, M., 2011. Analysis of polyphenols in white wine by CZE with amperometric detection using carbon nanotube-modified electrodes. Electrophoresis 32 (8), 877–883.
Morones, J.R., Elechiguerra, J.L., Camacho, A., Holt, K., Kouri, J.B., Ramirez, J.T., Yacaman, M.J., 2005. The bactericidal effect of silver nanoparticles. Nanotechnology 16, 2346–2353.
NanoWine, 2015. Nano wine. The entire wine store in one bottle. http://www.nanosupermarket.org/nanowine
Pradeep, T., 2009. Noble metal nanoparticles for water purification: a critical review. Thin Solid Films 517 (24), 6441–6478.
Qu, L., Dai, L., Osawa, E., 2006. Shape/size-controlled syntheses of metal nanoparticles for site-selective modification of carbon nanotubes. J. Am. Chem. Soc. 128 (16), 5523–5532.

Rawal, R., Chawla, S., Pundir, C.S., 2012. An electrochemical sulfite biosensor based on gold coated magnetic nanoparticles modified gold electrode. Biosens. Bioelectron. 31 (1), 144–150.

Ren, G., Hu, D., Cheng, E.W.C., Vargas-Reus, M.A., Reip, P., Allaker, R.P., 2009. Characterisation of copper oxide nanoparticles for antimicrobial applications. Int. J. Antimicrob. Agents 33, 587–590.

Rieux, L., Sneekes, E., Swart, R., 2011. Nano LC: principles, evolution, and state-of-the-art of the technique. LCGC North America 29 (10), 926–934.

Riul, A., Malmegrim, R., Fonseca, F., Mattoso, L., 2003. An artificial taste sensor based on conducting polymers. Biosens. Bioelectron. 18 (11), 1365–1369.

Riul, A., de Sousa, H.C., Malmegrim, R.R., dos Santos, D.S., Carvalho, A.C., Fonseca, F.J., Oliveira, O.N., Mattoso, L.H., 2004. Wine classification by taste sensors made from ultra-thin films and using neural networks. Sensor. Actuat. B Chem. 98 (1), 77–82.

Rodríguez Couto, S., Toca Herrera, J.L., 2006. Industrial and biotechnological applications of laccases: a review. Biotechnol. Adv. 24 (5), 500–513.

Ruparelia, J., Chatterjee, A., Duttagupta, S., Mukherji, S., 2008. Strain specificity in antimicrobial activity of silver and copper nanoparticles. Acta Biomater. 4, 707–716.

Sánchez-Obrero, G., Mayén, M., Rodríguez-Mellado, J.M., Rodríguez-Amaro, R., 2012. New biosensor for phenols compounds based on gold nanoparticle-modified PVC/TTF-TCNQ composite electrode. Int. J. Electrochem. Sci. 7, 10952–10964.

Sanz, V.C., Mena, M.L., González-Cortés, A., Yanez-Sedeno, P., Pingarrón, J., 2005. Development of a tyrosinase biosensor based on gold nanoparticles-modified glassy carbon electrodes: application to the measurement of a bioelectrochemical polyphenols index in wines. Anal. Chim. Acta 528 (1), 1–8.

Schneider, M., Türke, A., Fischer, W., Kilmartin, P.A., 2014. Determination of the wine preservative sulphur dioxide with cyclic voltammetry using inkjet printed electrodes. Food Chem. 159, 428–432.

EFSA Scientific Committee, 2011. Guidance on the risk assessment of the application of nanoscience and nanotechnologies in the food and feed chain. Scientific Opinion. EFSA J. 9 (5), 2140.

Shin, S., Kwen, H., Choi, S., 2011. Fabrication of electrochemical microbial biosensor based on MWNT supports prepared by radiation-induced graft polymerization. Polym. Korea 35 (3), 216–222.

Siem A.-M., 2014. Nanotechnology makes your wine better, http://scitech.au.dk/roemer/sep14/nanoscience-makes-your-wine-better/

Siva Kumar, V., Nagaraja, B.M., Shashikala, V., Padmasri, A.H., Shakuntala Madhavendra, S., David Raju, B., Rama Rao, K.S., 2004. Highly efficient Ag/C catalyst prepared by electro-chemical deposition method in controlling microorganisms in water. J. Mol. Catal. A Chem. 223, 313–319.

Smita, S., Gupta, S.K., Bartonova, A., Dusinska, M., Gutleb, A.C., Rahman, Q., 2012. Nanoparticles in the environment: assessment using the causal diagram approach. Environ. Health 11, S13.

Smolkova, B., El Yamani, N., Collins, A.R., Gutleb, A.C., Dusinska, M., 2015. Nanoparticles in food. Epigenetic changes induced by nanomaterials and possible impact on health. Food Chem. Toxicol. 77, 64–73.

Sondi, I., Salopek-Sondi, B., 2004. Silver nanoparticles as antimicrobial agent: a case study on *E. coli* as a model for Gram-negative bacteria. J. Colloid Interface Sci. 275, 177–182.

Souza, L.P., Calegari, F., Zarbin, A.J., Marcolino-Júnior, L.H., Bergamini, M.F., 2011. Voltammetric determination of the antioxidant capacity in wine samples using a carbon nanotube modified electrode. J. Agric. Food Chem. 59 (14), 7620–7625.

Sozer, N., Kokini, J.L., 2012a. The applications of nanotechnology. In: Picó, Y. (Ed.), Chemical Analysis of Food: Techniques and Applications: Techniques and Applications. Academic Press: San Diego, California, USA. p. 146.

Sozer, N., Kokini, J.L., 2012b. The applications of nanotechnology. In: Picó, Y. (Ed.), Chemical Analysis of Food: Techniques and Applications: Techniques and Applications. Academic Press: San Diego, California, USA. p. 168.

Tankhiwale, R., Bajpai, S., 2010. Silver-nanoparticle-loaded chitosan lactate films with fair antibacterial properties. J. Appl. Polym. Sci. 115 (3), 1894–1900.

Thanh, N.T., 2012. Magnetic Nanoparticles: From Fabrication to Clinical Applications. CRC Press: Boca Raton, Florida, USA.

Varelas, V., Sanvicens, N., Marco, M.P., Kintzios, S., 2011. Development of a cellular biosensor for the detection of 2,4,6-trichloroanisole (TCA). Talanta 84, 936–940.

Vidal, J.C., Bonel, L., Ezquerra, A., Duato, P., Castillo, J.R., 2012. An electrochemical immunosensor for ochratoxin A determination in wines based on a monoclonal antibody and paramagnetic microbeads. Anal. Bioanal. Chem. 403 (6), 1585–1593.

Wang, N., Zhang, N., Wang, M., 2006. Wireless sensors in agriculture and food industry—recent development and future perspective. Comput. Electron. Agr. 50, 1–14.

Welch, C., Banks, C., Simm, A., Compton, R., 2005. Silver nanoparticle assemblies supported on glassy-carbon electrodes for the electro-analytical detection of hydrogen peroxide. Anal. Bioanal. Chem. 382 (1), 12–21.

Zamfir, L., Geana, I., Bourigua, S., Rotariu, L., Bala, C., Errachid, A., Jaffrezic-Renault, N., 2011. Highly sensitive label-free immunosensor for ochratoxin A based on functionalized magnetic nanoparticles and EIS/SPR detection. Sensor. Actuat. B Chem. 159 (1), 178–184.

7

EFFECT OF CONTENT AND TEMPERATURE ON THE PHASE TRANSITIONS OF POLYMER COMPOSITES DOPED BY KAPPA CARRAGEENAN AND ALGINATE

Gülşen Akın Evingür*, Önder Pekcan**

**Piri Reis University, Faculty of Engineering, Department of Electrical and Electronics Engineering, Tuzla- Istanbul, Turkey; **Kadir Has University, Faculty of Engineering and Natural Sciences, Department of Bioinformatics and Genetics, Cibali- Istanbul, Turkey*

1 Introduction

Three-dimensional networks can be hydrophilic and/or hygroscopic, where gelation, drying, swelling, and elasticity of these materials enclosed many fields in food technology. Water molecules affect the internal structures of the gel (Harmathy, 1969; Mikhailov, 1975).

A composite is a mixture of two or more different materials that produce a product that has different properties than its constituents. There are many polymer composite systems, including biopolymer-clay and hydrogel composites. Polyacrylamide (PAAm) hydrogels are produced by free radical crosslinking polymerization of acrylamide with *N, N′*-methylenebis (acrylamide) (BIS), where the polymerization occurs in aqueous solutions of the monomers. Acrylamide is a chemical that naturally forms in certain foods, particularly plant-based foods that are rich in carbohydrates and low in protein, during processing or cooking at high temperatures. Foods likely to contain acrylamide are found in wide-ranging contents in potato chips, french fries, cookies, breakfast cereals, bread, as well as other foods that are also processed at high temperatures such as coffee, roasted almonds, and grain-based coffee substitutes (Healthy Canada, www.hc-sc.gc.ca).

Polysaccharides are made up repeating galactose units and 3.6 anhydrogalactose both sulfated and nonsulfated. There are three

Novel Approaches of Nanotechnology in Food. http://dx.doi.org/10.1016/B978-0-12-804308-0.00007-8

main commercial classes of carrageenan κ, ι, and λ. The carrageenan family is known to form thermally reversible gels as a function of temperature and gel inducing agents (Webb et al., 1986; Abd El-Mohdy and Abd El-Reim, 2009). Carrageenans are sulfated D-galactans extracted from red algae. The κ-carrageenan (κC) is characterized by an alternating disaccharide of α-1, 3-linked D-galactopyranose and β-1-linked 3.6-anhydro-D-galactopyranose (Piculell, 1995). The effects of casein, κ-carrageenan and sugar contents on the mechanical properties of gels acidified by glucona-lactone (GDL) were investigated (Ribeiro et al., 2004). Gels containing sugars showed the lowest values of stress at rupture, indicating that this cosolute, even in small proportions, decreased the polysaccharide contribution to the formation of a gel network. Production of carbohydrates that dissolve in water is a growing industry. The three major groups of polycocolloids are alginates, carrageenan, and agar, and these are used for applications such as thickening aqueous solutions, forming gels and water soluble films, and stabilizing products such as ice cream, tooth paste, and mayonnaise (Holdt and Kraan, 2011). Sulfated polysaccharides (kappa and iota carrageenan) and sodium alginate added to blue whiting mince were subjected to their different pressure/heat treatments in order to determine the functionality of each one in mince gel (Perez-Mateos et al., 2002). The results indicate that the textural properties of gels can be modified not only by the addition of ingredients, but by different gelling treatments, which charge the behavior of both mince and ingredient.

Alginic acid is a biopolymer consisting of homopolymeric blocks of (1–4)-linked β-D-mannuronate and α-L-guluronate. The anionic form of alginic acid, alginate is widely used in various areas, including controlled release systems (Kikuchi et al., 1997; Işıklan et al., 2011), food applications (Bierhalz et al., 2012), and many other industrial applications. Alginic acid forms stable gels as metals interact ionically with blocks of uronic acid residues, resulting in the formation of a 3-dimensional network that is usually described as an "egg-box" structure (Papageorgiou et al., 2010; Grant et al., 1973). A calcium ion release from alginate gels is known and has been reported in the literature as it pertains to antibiotics (Anal and Stevens, 2005), proteins (Nochos et al., 2008), and vitamins (Abd El-Ghaffar et al., 2012).

The phase transitions of carrageenan in KCl solution and in pure water were investigated by the photon transmission technique (Kara and Pekcan, 2000a,b). The photon transmission technique was studied for the drying and swelling of polyacrylamide gels (Kara and Pekcan, 2001; Pekcan and Kara, 2001). Hysteresis (Kara et al., 2003a) and percolative behavior (Tari and Pekcan, 2004) of κ-carrageenan were studied by fluorescence method. The swelling kinetics of κC were also characterized by the same technique (Kara et al., 2003b). They mentioned that gels with low carrageenan content swell slower than gels with high carrageenan content, which may contain more double helices and more lattice dislocations. The drying of polyacrylamide (Aktaş et al., 2007a), κC at various temperatures (Tari and Pekcan, 2008), and polyacrylamide hydrogels of various crosslinker contents (Evingür et al., 2009) were investigated by the steady state fluorescence (SSF) technique. Polyacrylamide (PAAm) with crosslinker contents was performed by the same technique for the swelling kinetics at various temperatures and it was observed that the time constants decreased when the swelling temperature were increased (Aktaş et al., 2007b, 2009). On the other hand, when the crosslinker content was increased, cooperative diffusion coefficients decreased. The SSF technique was performed to study the swelling of κC at various temperatures at a molecular level (Tari and Pekcan, 2007). And also, the desorption coefficient of polyacrylamide composite gels prepared by various κC contents decreased (Evingür and Pekcan, 2011a) when κC contents in PAAm increased. Swelling activation energies, ΔE of PAAm-κC composite were exothermic between 30 and 40°C, and also endothermic between 40 and 60°C, respectively (Evingür and Pekcan, 2012a). As a result, PAAm-κC composite gels play an important role in biotechnology, drug delivery system, and so on.

The mechanical properties and the swelling behavior of aqueous poly(acrylamide) and poly(acrylamide-Na-acrylate) were investigated (Oppermann et al., 1985). The elastic modulus of such ionic gels did not show the predicted decrease with rising swelling ratio, but on the contrary increased up to a fivefold of its value. This behavior was explained by non-Gaussian behavior in the network chains. A review of modeling techniques for predicting the mechanical behavior of polymer nanocomposites was presented (Valavala and Odegard, 2005). A detailed discussion of computational chemistry and computational mechanics modeling techniques was given. Viscoelastic behavior of carrageenan/milk mixtures was studied at different type of carrageenan types (Langendorff et al., 2000). Iota and κC only have attractive interactions with casein micelles at temperatures below the coil-helix transition temperatures, because only the helix form has a high enough charge density. At higher temperatures the iota and κC/milk systems undergo depletion induced segregative phase separation. Sulfated polysaccharides(kappa- and iota- carrageenan) and sodium alginate added to blue whiting mince were subjected to three different pressure/heat treatments in order to determine the functionality of each one in mince gel

(Perez-Mateos et al., 2002). Higher pressure conditions (375 MPa, 37°C, 20 min) induced gels with the lowest hardness. The carrageenans (iota or kappa) appeared to form a reticular structure in the heat induced gels. The extraction parameters (pH, temperature, and extraction duration) of seaweeds had been systematically varied in order to produce κ/ι- hybrid carrageenans exhibiting a wide range of chemical properties (Hilliou et al., 2006).

This chapter is concerned with fluorescence, gravimetrical, and volumetric techniques of gelation, drying, swelling, and mechanical measurements of PAAm-κC, and alginate, contained with different metal ions, and surfactants, which are specifically designed for content and temperature variations that control and modify products for the food industry.

2 Theoretical Considerations

2.1 Gelation

Hydrogels have received considerable attention for the sol-gel phase transition process. Bethe lattice is a special lattice to define the sol-gel phase transition on the closed loop (Flory, 1941a,b,c; Stockmayer, 1943, 1944). On the other hand, alternative theory is the lattice percolation theory. Monomers are occupied the sites of a periodic lattice (Stauffer et al., 1982; Stauffer and Aharony, 1994). A bond between these lattice sites is formed randomly with probability p. The infinite cluster starts to form in the thermodynamic limit. p_c is called the percolation cluster in polymer language and also defined as the percolation threshold for a certain bond content. The critical exponents in these two theories are different from each other because of their universality. If the exponents γ and β are for the weight average degree of polymerization, DP_w, and the gel fraction G (average cluster size, S_{av}, and the strength of the infinite network P_∞ in percolation language) near the percolation threshold. They are defined as,

$$DP_w \propto (p_c - p)^{-\gamma} \quad p \to p_c^- \tag{7.1}$$

$$G \propto (p - p_c)^{\beta} \quad p \to P_c^+ \tag{7.2}$$

where the Flory-Stockmayer theory gives $\beta = \gamma = 1$, independent of the dimensionality while the percolation studies based on computer simulations give γ and β around 1.7 and 0.43 in 3-dimension (Stauffer et al., 1982; Stauffer and Aharony, 1994).

In our case, near the percolation threshold, the critical point $|p - p_c|$ is linearly proportional to the $|t - t_c|$ and the fluorescence intensity from the bonded pyranines monitors the DP_w and G for below and above the gel point, respectively (Yılmaz et al., 1998, 2002) where t_c is defined as the gel point. When $t < t_c$, the maximum fluorescence intensity, I_{max}, measures weight average degree of polymers. On the other hand, if $t > t_c$, the corrected intensity $I_{max} - I_{ct}$ measures only the gel fraction G. As a result, Eq. (7.1) and (7.2) can be summarized.

$$I_{max} \propto DP_w = C_+ (t_c - t)^{-\gamma} \quad t \to t_c^- \tag{7.3}$$

$$I_{ct} \propto DP_w = C_- (t_c - t)^{-\gamma'} \quad t \to t_c^+ \tag{7.4}$$

$$I' = I_{max} - I_{ct} \propto G = B(t - t_c)^{\beta} \quad t \to t_c^+ \tag{7.5}$$

where C_+, C_-, and B are the critical amplitudes. I_{ct} is the intensity from finite clusters distributed through the infinite network. γ, γ', and β are the critical exponents. (Stauffer et al., 1982, 1994; Sahimi, 1994; Gennes, 1979). The ratio of the critical amplitudes was discussed by (Aharony, 1980; Stauffer et al., 1982; Stauffer and Aharony, 1994).

2.2 Stern-Volmer Model

The kinetics of photophysical intermolecular deactivation process is defined by quenching mechanisms. In this model, the fluorescence intensity is written as (Birks, 1971),

$$\frac{I_0}{I} = 1 + k_q \tau_0 [Q] \tag{7.6}$$

where, τ_0 is the lifetime of the fluorescence probe with no quenching has taken place, k_q is the quenching rate constant, $[Q]$ is the quencher content and I_0 is the fluorescence intensity for the zero quencher content.

In our case, the amount of water diffusion, W is produced over the differential volume when $\tau_0 k_q [Q] \ll 1$

$$W = \left(1 - \frac{I}{I_0}\right) \frac{\upsilon}{k_q \tau_0} \tag{7.7}$$

where υ is the volume at the equilibrium swelling state, k_q is the infinity equilibrium value of water diffusion, W, and τ_0 is 5 ns for pyranine molecules. In our system, it is assumed that water molecules are the only quencher for the excited pyranine molecules.

2.3 Moving Boundary Model

In all cases considered the two regions are separated by a plane surface and diffusion takes place only in the direction perpendicular to this plane (Crank, 1975). The total amount, M_t of diffusing substance desorbed from the unit area of a plane sheet of thickness a at time t is given by the following relation

$$\frac{M_t}{M_f} = 2\left[\frac{D_d}{\pi a^2}\right]^{1/2} t^{1/2} \tag{7.8}$$

where D_d is a diffusion coefficient, and M_f is proportional to the amount of water molecules released, W at time, t.

2.4 Li–Tanaka Model

Volume phase transitions in gels may undergo from dry to swollen states either continuously, or discontinuously. The assumption in this theory is that the shear modulus, G is negligible compared to the osmotic bulk modulus. Li and Tanaka have developed the theory of the kinetics of the swelling (Tanaka, 1978). On the other hand, the shear modulus plays a significant role in the assumption by (Zrinyi et al., 1993).

The kinetics of the swelling of a gel are described the following relation

$$\frac{\partial \vec{u}}{\partial u} = D_0 \vec{\nabla}^2 \vec{u} \tag{7.9}$$

where $\vec{u}$ is the displacement vector, t is the time, and K is the bulk modulus, respectively (Li and Tanaka, 1990).

The swelling of a disc gel in axial and radial directions can be solved for small diffusion process (Li and Tanaka, 1990). Eq. (7.9) can be written in terms of water uptakes W and W_f at time t and equilibrium at large t.

$$\frac{W}{W_f} = 1 - B_1 \exp(-t/\tau_1) \tag{7.10}$$

where B_1 is related to the ratio, R (Li and Tanaka, 1990), which are determined by shear modulus and longitudinal osmotic modulus.

The cooperative diffusion coefficient, D_0 at the surface of a gel disc is given by the following relation

$$D_0 = \frac{3a_f^2}{\tau_I \alpha_1^2} \tag{7.11}$$

where α_1 is a function of R only (Li and Tanaka, 1990), a_f stands for the half thickness of the gel in the final equilibrium state, and $\tau_1 \approx (3/4 - R)^{-1}$ is the swelling time constant.

2.5 Fickian Model

The mathematical theory of diffusion in isotropic substances is based on the hypothesis that the rate of transfer of diffusing substances through the unit area of a section is proportional to the content gradient measured normal to the section, that is,

$$F = -D\frac{\partial C}{\partial z} \tag{7.12}$$

where F is the rate of transfer per unit area of the section, C is the content of diffusing substance, z is the space coordinate measured normal to the section, and D is called the diffusion coefficient. The negative sign in Eq. (7.12) presents the direction of diffusion. The differential equation of diffusion in one dimension can be derived with simple geometric considerations.

$$\frac{\partial C}{\partial t} = D\frac{\partial^2 C}{\partial^2 z} \tag{7.13}$$

Eqs. (7.12) and (7.13) are usually referred to as Fick's first and second laws. Diffusion occurs just in the z direction. When the boundary is kept at a constant content, C_0, the initial content being zero throughout the medium. Eq. (7.13) satisfies the boundary conditions which is for $t = 0$ at $z = 0$, $C = 0$ and for $t \geq 0$, at $z = a$, $C = C_0$.

Eq. (7.13) can be solved as

$$\frac{M_t}{M_\infty} = \frac{4}{\pi^{1/2}}\left(\frac{Dt}{a^2}\right)^{1/2} - \frac{Dt}{a^2} - \frac{1}{3\pi^{1/2}}\left(\frac{Dt}{a^2}\right)^{3/2} + \ldots \tag{7.14}$$

where M_t and M_∞ are the amount of material diffused in time t and at equilibrium, respectively, and a represents the thickness of the disc shape structure (Crank, 1975).

2.6 Elasticity

Food technologists involved in product development or quality assurance are interested in the mechanical properties of the gel in relation to shaping, stand up, handling, cutting/slicing, or eating characteristics. For most foods, the ratio stress and strain is not constant at large deformations that the stress–strain relation becomes nonlinear. The deviation from linearity is a product-dependent characteristic. In this nonlinear region, products can still be characterized by a modulus (Holdt and Kraan, 2011).

Comparison of theory with experiments and relationships between stress, strain, and molecular constitution of polymer networks was developed by (Erman and Flory, 1982). The compression measurements were analyzed through plots of the applied stress, τ, ($\tau = f/A$ with f being the acting force and A the cross section of the undeformed swollen specimen) versus $\lambda - \lambda^{-2}$, where $\lambda = L/L_0$ is the relative deformation caused by compression of the length of the swollen sample along the direction of the stress (L), with respect to the initial length (L_0) of the swollen but undistorted sample. They show the typical curvature at the smallest range of deformation that is generally assigned to imperfect geometry of the surface of the gel specimen (Valencia et al., 2009). The elastic modulus was determined by linear regression, as the slope of such plots as given in Eq. (7.15).

$$S \approx \frac{(f/A)}{\alpha - \alpha^{-2}} \tag{7.15}$$

3 Experimental

3.1 Preparation of PAAm-κC Composites

Composite gels were prepared from AAm (Acrylamide, Merck), MBA (*N*, *N'*-methylenebisacrylamide, Merck), APS (ammonium persulfate, Merck) and TEMED (tetramethylethylenediamine, Merck), which were dissolved in 5 ml distilled water at 80°C. κC contents changed between 0.5 and 3 (w/v)%. The initiator and pyranine (8-hydroxpyrene- 1.3.6 trisulfonic acid, trisodium salt, HPTS) contents were kept constant at 7×10^{-3} M and 4×10^{-4} M, respectively. After bubbling nitrogen for 10 min, each precomposite gel solution of 5 mL was poured into a quartz tube for gelation and injector for drying and swelling experiment, respectively (Aktaş et al., 2006).

3.2 Preparation of Alginate Beads

Alginate beads were prepared from alginic acid sodium salt (Viscosity of ~250 cps, Sigma-Aldrich Co., US) and calcium chloride dehydrate (J.T. Baker Co., Holland). Fluorescence measurements were performed with the use of a Shimadzu RF-5301PC spectrofluorometer (Japan) at room temperature to decide encapsulation efficiency. Sodium alginate was dissolved in 5 mL deionized water at a content of 2% (w/v), and the Py content was kept constant at 10^{-4} M in the solution. The solution was mixed and left for several hours to eliminate air bubbles. It was then added dropwise through a 0.8 cm inner diameter syringe to a calcium chloride solution of various contents (0.5–20% (w/v)). The addition of the alginate/pyranine solution was conducted 20 cm over the calcium chloride solution vessel in order to produce perfect spherical beads. The formed beads were filtered, washed three times with deionized water to make sure that the Py was completely removed from the surface of the beads and then placed on a Petri dish. The beads had an average diameter of 2.0 ± 0.1 mm.

In order to understand the effect of cation content on the beads containing anionic surfactant (SDS), alginate and SDS contents were kept constant (2% and 400 mmol/L, respectively) whereas content of both Ca^{2+} and Ba^{2+} was varied from 1% to 5% (w/v). The beads were produced and stored in the same way as earlier (Kaygusuz et al., 2014).

3.3 Fluorescence Measurement

Gelation (Aktaş et al., 2006), drying (Evingür and Pekcan, 2011a,b), swelling (Evingür and Pekcan, 2012a), and release process (Kaygusuz et al., 2014) were performed by fluorescence spectroscopy. During the measurement, the position of the composite was 90°C and slit widths were 5 nm. During in situ gelation experiments, excitation wavelength was 340 nm. On the other hand, emission intensity of the pyranine was monitored as a function of polymerization time. The dimensions of composites were prepared as 10 mm in diameter and 4 mm in thickness from the injector for drying and swelling experiments. Disc-shaped gel samples were placed on the wall of a square quartz cell filled with air and water for each experiment, respectively. After drying completed, a swelling experiment was started. And also, volumetric and gravimetric measurements were performed from each composite gel.

3.4 Mechanical Characteristics Measurement

After gelation, the composites prepared with various molar monomer percentages were cut into discs with 10 mm in diameter and 4 mm thickness. Before the compression measurements, the composites were maintained in water at different temperatures. After a final wash with distilled water of all samples was carried out for 1 week, at a desired temperature, they achieved swelling equilibrium, and unreacted monomers were removed. The mechanical experiments of the PAAm-κC composites were performed at temperatures of 25, 30, 40, 50, and 60°C, respectively. A Hounsfield H5K-S model tensile testing machine settled a crosshead speed of 1.0 cm min^{-1} and load cell of 5 N sensibility were used to perform uniaxial compression experiments on the samples of each type of composite (Evingür and Pekcan, 2015). Loss of water and changes in temperature during the measurements were not detected because the compression period was less than 1 min. During the compressive deformation stage, swelling was not realized.

4 PAAm-κC Composites

Food applications for carrageenans are many, including canned foods, dessert mousses, salad dressings, bakery fillings, ice cream, instant desserts, and canned pet foods. They are in the brewing industry for clarifying beer, wines, and honeys, although less commonly than previously (Holdt and Kraan, 2011). As a result, knowing the gelation, drying, swelling, and elasticity of PAAm-κC composites are important.

4.1 Gelation

The sol-gel transition was performed by *steady state* fluorescence (SSF) for PAAm-κC composites with various carrageenan contents. The sol-gel transition can be monitored by fluorescence spectroscopy and the universality of the transition can be tested as a function of gel fraction and time. Fig. 7.1 shows the fluorescence spectrum

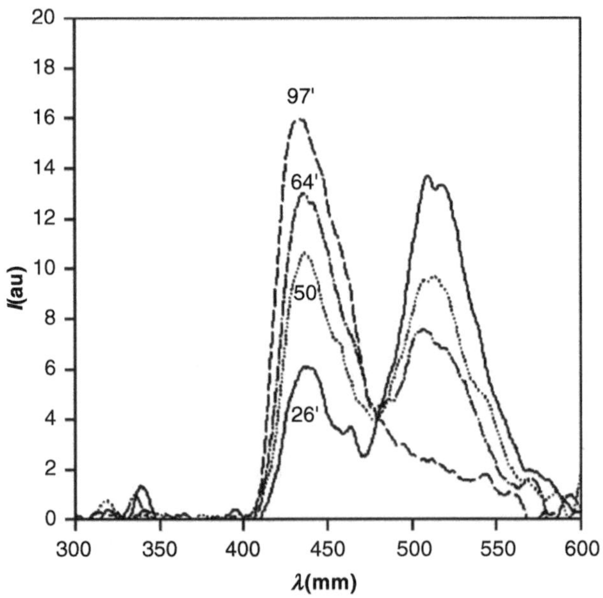

Figure 7.1. The fluorescence spectrum of PAAm-κC composite polymerization at different times (Aktaş et al., 2006).

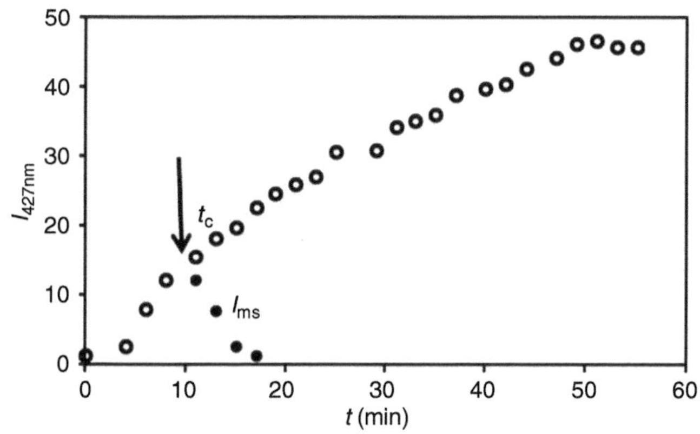

Figure 7.2. The fluorescence intensity versus time for PAAm-2.5(w/v)% of κC composite gels.

of polymerization of PAAm-κC composite at different times. In the spectrum, 512 nm-peak presents the behavior of free pyranine. During the course of AAm-κC polymerization, while the intensity of 427 nm peak increased, the intensity 512 nm peak decreased. At the end of the reaction the long wavelength peak disappears.

The gelation experiments of PAAm-κC polymerization were performed with various κC contents. During the experiment, 512 nm-peak does not shift but the intensity of 512 nm-peak decreases with reaction time. There are three functional groups in pyranine bonded to the polymeric network during the polymerization. The spectra from 400 to 427 nm is shifted because of the neutralization SO_3^- groups of pyranine. The cluster size increases before the gel point. On the other hand, it decreases above the gel point (Aktaş et al., 2006).

Fig. 7.2. exhibits to the fluorescence intensity versus time for 2.5 (w/v)% of AAm-κC composite gels. The values of t_c and β as a function of κC content were calculated from the log-log plots of the intensity summarized in Table 7.1.

4.2 Drying

The drying process of PAAm-κC composite gels was performed by steady state fluorescence techniques at various temperature and κC contents. The scattered light, I_{sc}, and fluorescence intensities, I, from these gels were monitored during drying at different temperatures. The emission spectra of pyranine from PAAm-κC composite gel with various (w/v)% of κC content, during drying in air at different temperatures, respectively. Because the drying temperature is increased, fluorescence intensity, I_{em} has increased

Table 7.1 Experimentally Measured Parameters for Various (w/v)% of κC Contents (Aktaş et al., 2006)

AAm [M]	(w/v) % κC	t_c ±2 (min)	C^-/C^+	β(±0.02)
2.0	0.5	6	1.0	0.75
			0.37	0.50
			0.28	0.50
			0.23	0.50
			0.1	0.66
2.0	1.0	8	1.0	1.00
			0.37	0.40
			0.28	0.42
			0.23	0.42
			0.1	0.40
2.0	1.5	12	1.0	0.70
			0.37	0.64
			0.28	0.40
			0.23	0.47
			0.1	0.75
2.0	2.0	16	1.0	1.30
			0.37	0.53
			0.28	0.56
			0.23	0.56
			0.1	0.57
2.0	2.5	22	1.0	1.00
			0.37	0.90
			0.28	0.80
			0.23	0.80
			0.1	0.80
2.0	3.0	19	1.0	0.72
			0.37	0.80
			0.28	0.80
			0.23	0.80
			0.1	0.00

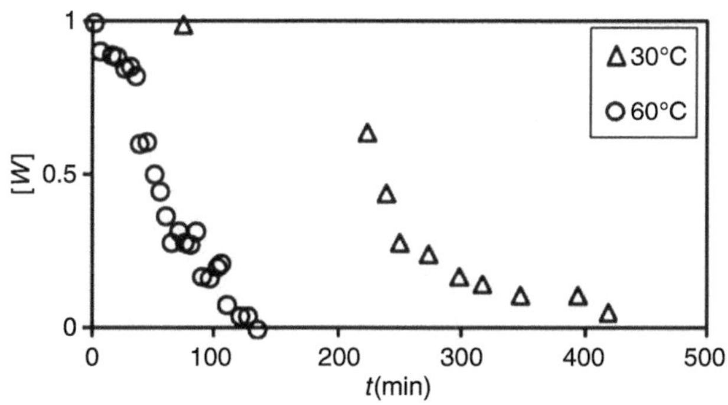

Figure 7.3. The plots of water release versus drying time, *t*, for PAAm-κC composite gel dried in air measured by fluorescence technique for 0.5 (w/v)% of κC content samples at 30 and 60°C temperatures, respectively.

with respect to the scattered light intensity, I_{sc}. When drying gels, the scattered light intensity declined in turbidity. To eliminate the turbidity effect, the corrected fluorescence intensity was presented as I_{em}/I_{sc} modeled by quenching mechanism because of excluding the structural waves (Evingür and Pekcan, 2011a,b; Kara and Pekcan, 2001). For the dry composite gel, τ_0 (=300 ns.) is known (Birks, 1971), then W can be calculated from fluorescence intensity. Fig. 7.3 shows the plots of W versus time, t for 0.5 (w/v)% of κC content.

The desorption coefficients, D_I were calculated from Eq. (7.8) and listed in Table 7.2. Then we assume that the diffusion coefficient of polymer segments in the gel is negligible compared to desorption coefficient, D_{dI} of water.

It is seen that values decreased while temperature is increased for all samples in Table 7.2. This result is presented when increasing temperature naturally causes increasing the Brownian motion of water molecules in the gel, which then helps some of them to break away from the gel surface resulting faster dry. Thus, D_{dI} values decline due to increasing κC content at a given temperature because of moisture absorbing capacity of κC.

It is clear that less κC content gel dries faster resulting large D_{dI} values compare to high κC content gel. Gravimetric and volumetric desorption coefficients, D_{dW} and D_{dV} are listed in Table 7.2.

As seen in Table 7.2 that D_d values measured by using volumetric and gravimetric techniques are 10–30 times smaller than the values measured by fluorescence technique, which measures the behavior of the pyranine molecules in the microstructure of the gel. On the other hand, volumetric and gravimetric techniques anticipate the information of the macroscopic (ie, bulk) behavior of the drying composite gel.

Table 7.2 Experimentally Measured Parameters of PAAm-κC Composite Gels for Various Temperatures and (w/v)% of κC Content during Drying Process (Evingür and Pekcan, 2011b)

T(°C)	(w/v) % κC	D_{dl}*10^{-9} (m^2/s)	D_{dW}*10^{-9} (m^2/s)	D_{dV}*10^{-9} (m^2/s)
30	0.5	30.16	3.54	3.11
	1	14.17	1.93	1.57
	1.5	8.21	1.27	1.19
	2	2.9	1.13	1.02
	2.5	0.82	0.91	0.91
	3	0.81	0.51	0.22
40	0.5	41.90	4.9	4.9
	1	40.5	4.65	3.73
	1.5	13.67	3.57	2.37
	2	8.16	3.45	2.32
	2.5	0.91	3.05	2.05
	3	0.85	2.82	1.25
50	0.5	120.87	6.1	5.2
	1	116.4	5.61	4.7
	1.5	37.30	5.23	4.53
	2	22.5	4.13	3.65
	2.5	6.97	3.9	3.11
	3	6.71	3.57	2.59
60	0.5	171	6.14	6.14
	1	131	5.8	5.53
	1.5	79.8	5.27	4.65
	2	28.4	4.43	4.25
	2.5	7.4	4.06	3.29
	3	6.9	3.89	2.97

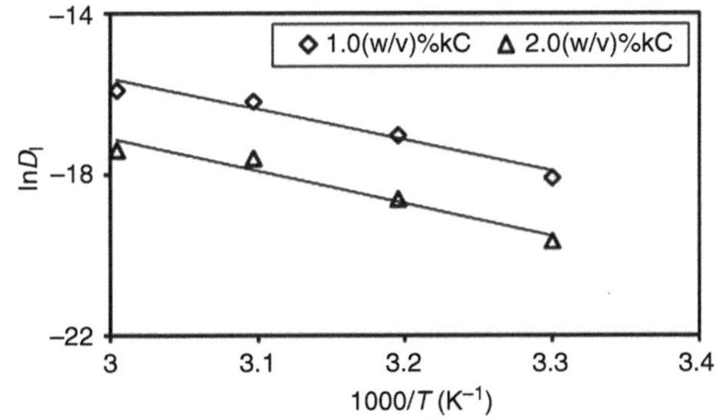

Figure 7.4. Linear regressions of desorption diffusion coefficients versus temperature measured by fluorescence techniques for 1.0 and 2.0 (w/v)% of κC content gel samples, respectively.

The relation of the energy, temperature, and desorption coefficients (D_{dI}, D_{dW}, and D_{dV}) can be given the following Arrhenius law:

$$\Delta E = -kT \ln\left(\frac{D}{D_0}\right) \tag{7.16}$$

where ΔE is the energy of drying, k is Boltzmann's constant, and D_0 is the desorption coefficients at $T = \infty$ for each technique. Fig. 7.4 shows the logarithmic form of Arrhenius's law for 1.0 and 2.0 (w/v)% of κC contents obtained by fluorescence techniques respectively, from which ΔE values are produced and listed in Table 7.3.

Table 7.3 Energy for Various (w/v)% of κC Content during Drying Process by Fluorescence, Gravimetrically, and Volumetric Techniques (Evingür and Pekcan, 2011b)

(w/v) % κC	0.5	1	1.5	2	2.5	3
ΔE_I(kJ/mol)	50.62	62.94	63.13	63.93	69.81	68.65
ΔE_W(kJ/mol)	15.28	28.63	38.12	35.12	38.85	52.01
ΔE_V(kJ/mol)	17.12	32.76	38.62	38.61	34.92	69.83

It is seen that the drying energies increase by increasing κC content, that is, when the high κC content gels are dried they need more energy. Although the gels with low water absorbing capacity possess large D_d values and needs smaller energy for drying, D_d values for high κC content gels are smaller and needs more energy for drying.

This study has shown a novel method for the study of drying kinetics of PAAm-κC composite gel in various κC contents. The desorption coefficients for the drying process were measured by a moving boundary model combined with Stern-Volmer kinetics. It is well understood that energy need for macroscopic drying is smaller than it is in microscopic drying.

4.3 Swelling

The real time monitoring of the pyranine (*Py*) fluorescence intensity, I decreased during the swelling experiments of PAAm-κC composites in water at various temperatures. The behavior of pyranine intensity was modeled by Stern-Volmer equation for low quenching efficiencies. On the other hand, the swelling time constants, τ_1, and cooperative diffusion coefficients, $D_{0,}$ were calculated from the Li-Tanaka equation. When the fluorescence intensity, I_{em}, decreased and the scattered light intensity, I_{sc}, increased during the swelling experiment in pure water.

When the gel swells, two junctions are located on neighboring lattice sites, connected and formed clusters of the first topological neighbors, which are called frozen blob clusters (Bastide and Leibler, 1988). During fluorescence measurement, the scattering light intensity comes from the contrast between frozen blob clusters and holes created by the dilution. As a result of separation of frozen blob clusters, the scattering intensity increases and the transmitted light intensity declines. Clusters lead to a strong increase of the scattering intensity, $I_{sc,}$ or decline in the transmitted light intensity, I_{tr}. The empirical formula was mentioned to define the fluorescence quenching mechanisms. As far as the correction of fluorescence emission is concerned, the empirical formula was mentioned (Aktaş et al., 2007a,b, 2009), to define the fluorescence quenching mechanisms. The aging of acrylamide gels and the swelling of PAAm gels with various crosslinker contents were performed by in situ photon transmission technique, respectively (Kara and Pekcan, 2001). The transmitted light intensity, I_{tr}, decreases. Therefore, the corrected fluorescence intensity was calculated to eliminate the effect of the turbidity of the gel on the fluorescence intensity and elaborate the Stern-Volmer model.

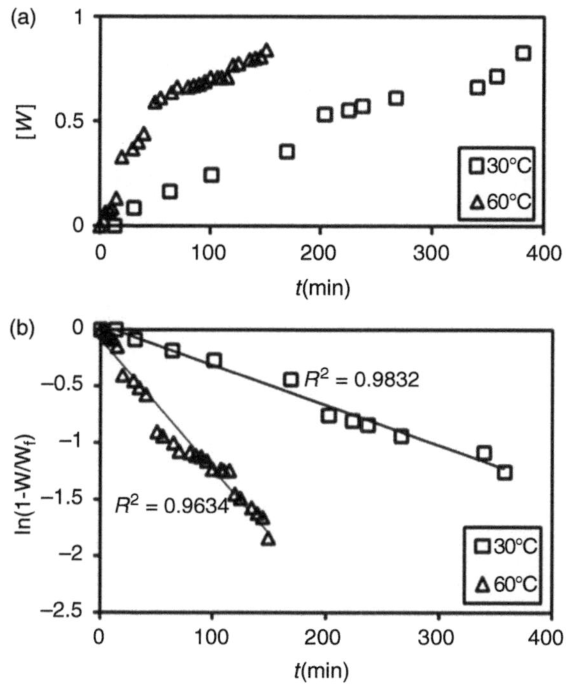

Figure 7.5. (a) The plots of fluorescence data by using Eq. (7.7) (b), linear regression of Fig. 7.5a by using Eq. (7.10), versus swelling time, *t*, for PAAm-κC composites swelling in water measured by fluorescence technique at 30 and 60°C for 0.5 (w/v)% of κC concentration samples, respectively.

Plots of water uptake (swelling), *W*, versus swelling time are shown in Fig. 7.5a. The logarithmic form of the data as presented in Fig. 7.5b was fitted to the relation produced from Eq. (7.10). Then, using Eq. (7.11), cooperative diffusion coefficients D_0 measured by fluorescence, gravimetric, and volumetric techniques were calculated for these disc-shaped hydrogels and found to be around 10^{-9} m^2 s^{-1}. Experimentally obtained τ_1 and D_0 values are summarized in Table 7.4. It should be noted that D_0 values increased when the temperature was increased to 40°C and then decreased to 60°C.

All the swelling time constants, τ_1 with respect to temperature measured by fluorescence, gravimetric, and volumetric techniques presented similar behaviors, that is, gel swells slower below 40°C during swelling process. On the other hand, maximum cooperative diffusion coefficients has been found at 40°C. This is because these gels were recognized to be influenced by the alkaline hydrolysis and are hydrolyzed to introduce some carboxylic groups into the chain.

Table 7.4 Experimentally Measured Parameters of PAAm-κC Composites for Various Temperature and (w/v)% of κC Content during Swelling Process (Evingür and Pekcan, 2012a)

T(°C)	(w/v) % κC	τ_{1l} (min)	D_{0l}*10^{-9} (m^2/s)	τ_{1w} (min)	D_{0w}*10^{-9} (m^2/s)	τ_{1v} (min)	D_{0v}*10^{-9} (m^2/s)
30	0.5	333.33	0.53	111.11	0.97	166.66	0.72
	1	1000.00	0.12	142.85	0.86	200.00	0.47
	1.5	500.00	0.25	142.85	0.90	166.66	0.99
	2	333.33	0.26	111.11	3.84	142.85	1.85
	2.5	200.00	0.33	90.90	4.16	125.00	3.69
	3	166.67	0.76	55.55	8.95	58.82	6.22
40	0.5	71.42	5.81	58.82	3.06	58.82	6.22
	1	32.57	6.29	43.47	3.44	55.00	6.23
	1.5	26.73	9.22	38.46	3.51	52.63	7.20
	2	25.00	10.05	34.48	5.60	37.03	7.38
	2.5	23.52	14.05	33.00	8.15	33.33	8.07
	3	22.72	22.03	20.00	39.98	27.77	22.14
50	0.5	91.74	4.73	62.50	1.47	62.50	6.17
	1	62.89	4.93	54.05	2.82	58.82	6.21
	1.5	57.14	8.76	41.66	3.25	54.05	6.67
	2	56.81	9.18	37.04	3.41	40.00	6.83
	2.5	48.30	12.72	35.71	4.38	37.00	7.52
	3	34.01	18.10	27.77	24.39	35.71	7.79
60	0.5	166.66	4.54	63.29	1.32	83.33	1.24
	1	142.85	4.69	62.50	2.11	62.50	1.94
	1.5	125.00	6.75	47.61	2.34	55.55	2.00
	2	111.11	8.41	47.16	2.82	41.66	4.06
	2.5	100.00	9.35	42.30	3.53	40.90	4.49
	3	37.03	13.95	28.54	5.24	40.00	4.60

Table 7.5 The Measured Energy of Endothermic (1) and Exothermic (2) Process during Swelling for Various (w/v)% of κC Content Gels by Fluorescence, Gravimetrically, and Volumetric Techniques, Respectively (Evingür and Pekcan, 2012a)

		(w/v) % κC					
	Process	**0.5**	**1**	**1.5**	**2**	**2.5**	**3**
ΔE_I (kJ/mol)	1	10.3	12.3	12.9	7.4	16.9	19.0
	2	−181.9	−300.9	−274.2	−277.8	−285.1	−255.9
ΔE_W (kJ/mol)	1	35.3	20.3	16.7	28.7	35.1	84.4
	2	−87.30	−105.3	−103.4	−28.6	−51.1	−113.7
ΔE_V (kJ/mol)	1	66.6	48.2	53.0	24.7	24.2	65.8
	2	−163.9	−196.5	−150.8	−105.1	−59.4	−96.5

Because of the existence of the charged units and mutual repulsion of the negative charges, water solubility and hydrodynamic volume of the chain increase (Bardajee et al., 2008; Dafader et al., 2009). As a result, low κC content swelled much slower than high κC content.

The dependence of temperature on the swelling of PAAm-κC composites predicts that D_0–$1000/T$ relation may obey the Arrhenius law as given in Eq. (7.16). The relation for the data by each techniques from which ΔE values are produced and listed in Table 7.5. ΔE_1 and ΔE_2 are endothermic and exothermic swelling activation energies, respectively.

The composites are defined exothermic in between 30 and 40°C. On the other hand, the composites behave endothermic in between 40 and 60°C. The composite swells slowly and releases some heat below 40°C; however, they absorb some heat above 40°C.

Here it is seen in Table 7.5 that exothermic and endothermic swelling activation energy values measured by using volumetric and gravimetric techniques are at least two times smaller and/or larger than the values measured by fluorescence technique because it can be possible to monitor the swelling at a molecular level. However, the information of bulk behavior is provided by the gravimetric and volumetric measurements.

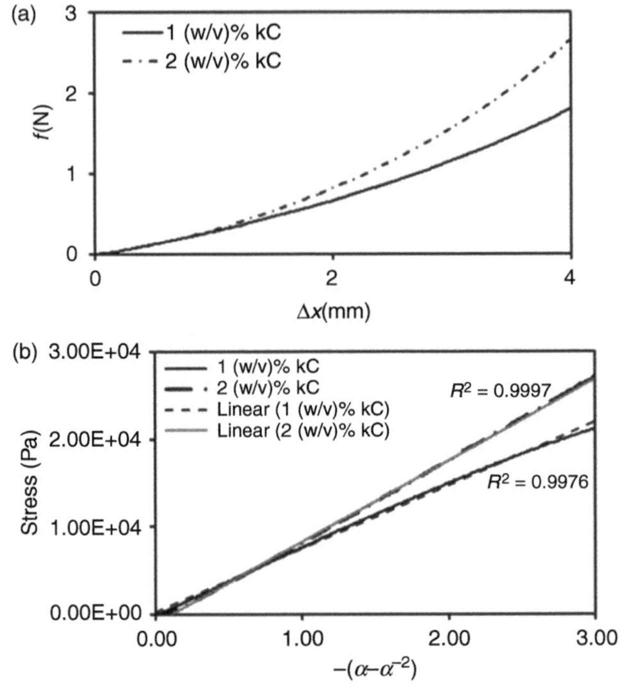

Figure 7.6. (a) The force f(N) and compression (mm) curves, and (b) stress and -(α-α^{-2}) curves for 1, and 2 (w/v)% of κC contents at 50°C, respectively.

4.4 Elasticity

Elasticity properties such as stress, strain, and compressive elastic modulus, S, of these composite gels were studied in various κC content and at several temperatures. The content and temperature dependence of the compressive elastic modulus, S, of the swollen PAAm-κC composite gels due to volume phase transition were produced by using compressive testing technique.

Forces (f) or loads corresponding to compression (mm) were obtained from the original curves of uniaxial compression experiments. The force, f (N) versus compression (mm) curves for 1, and 2 (w/v)% of κC contents at 50°C are shown in Fig. 7.6a, where it is seen that the repulsive force between monomers increases rapidly when the bond length is shorten with respect to the equilibrium position. It is also seen that when κC content is increased to higher values, then repulsive force increased as presented in Fig. 7.6a.

Stress (Pa)–(α-α^{-2}) plots were drawn by using the data obtained from the plots of f (N) versus compression curves for 1, and 2 (w/v)% κC contents at 50°C in Fig. 7.6b.

Table 7.6 Compressive Elastic Modulus at Different Temperatures and (w/v)% of κC Contents

Sample		Temperature				
PAAm	(w/v) (%) κC	25°C S(kPa)	30°C S(kPa)	40°C S(kPa)	50°C S(kPa)	60°C S(kPa)
	0.5	12.46	12.85	13.03	10.93	10.89
	1	7.60	7.31	6.93	7.34	8.53
	1.5	7.80	8.79	9.49	9.22	9.12
	2	8.10	9.49	9.65	9.29	9.30
	2.5	8.20	9.66	9.86	9.51	9.55
	3	9.54	10.10	10.80	10.90	10.90

It is also seen in Fig. 7.6b that 1 (w/v)% of κC content composite presents smaller initial slope than 2 (w/v)% of κC content composites and the stress of 2 (w/v)% of κC in the composite increases dramatically where the κC monomers are in charge of this behavior. In this case, it appears that an increase in the viscosity of the medium hinders the movement of free radicals and monomers molecules and enhanced the chance of chain transfer between PAAm and κC molecules (Pourjavadi and Mahdavinia, 2006), which play an important role for getting the different onset behavior.

Table 7.6 shows the plot of compressive elastic modulus, *S* versus (w/v)% κC content in the composite, where after compressive elastic modulus decreases up to1 (w/v)% of κC content, the modulus increases with increasing (w/v)% of κC content, which indicates that there is a sudden change in the material elasticity at 1 (w/v)% of κC content composite. The change in *S* shows that the copolymers have commanded a superelastic percolation network above 1 (w/v)% of κC content composite (Evingür and Pekcan, 2013). The reason behind the changes of compressive elastic modulus, *S* as shown in Table 7.6 can be explained by increasing κC content that enhances the hydrophilicity network below 1 (w/v)% of κC content.

In addition, the hydrophobicity of network importantly develops the stress relaxation of the composite above 1 (w/v)% of κC content due to its small size in the hydrated state when the double helices can cluster together; this promotes the aggregation, which improves the elasticity of the composite structure. This must suggest that the viscose medium has a network structure with flexible polymer chains just as swollen medium (Meena et al., 2006). Last, as predicted that the elastic properties of PAAm-κC composite are highly dependent on the (w/v)% of κC content, which directly influence monomer interactions between PAAm and κC in the composite. Such monomer interactions will play an important role in load transfer and interfacial bonding that regulates elastic properties of the composites. The variations in the κC content in the resultant composite could be the major reason for this phenomenon. In addition, lower κC (below 1 (w/v)% of κC) content enhances the hydrophilicity of the hydrogel in PAAm-κC composite. A further enhancement in κC content (above 1 (w/v)% of κC content), however, results in decreased absorbency (Meena et al., 2006). This is probably due to an increase in the viscosity of the medium, which hinders the movement of free radicals and monomers molecules and the enhanced chance of chain transfer between PAAm and κC monomer molecules (Pourjavadi and Mahdavinia, 2006). Thus, κC can show a significant effect in the PAAm-κC composite. This is due to its small part in the hydrated state when the double helices can cluster together; which results higher elasticity of the composite system (Kampf and Nussinovitch, 1997).

The force, $f(N)$ versus compression (mm) curves at 30 and 40°C for 2.5 (w/v)% of κC are presented in Fig. 7.7a. It is seen in Fig. 7.7a that when temperature increases for high κC content composites, then repulsive force, presenting larger numbers. This is because the repulsive interaction between monomers of composites of the swollen state is much stronger than that of the collapsed state. In other words, the composites are easily affected by deformation (Suzuki and Ishii, 1999).

Stress (Pa)–(α-α^{-2}) plots for 2.5 (w/v)% of κC content composites were drawn in Fig. 7.7b by using the data observed in the plots of $f(N)$ versus compression curves in Fig. 7.7a at 30, and 40°C, respectively. The stress for the 2.5 (w/v)% of κC composites at 40°C increases dramatically, From earlier observations it can be predicted that the stress as well as the composite content can affect the stability of stress-(α-α^{-2}) curve and thus enhances the transverse interchain interaction (Evingür and Pekcan, 2015).

The compressive elastic modulus, S of various (w/v)% of κC contents composites at several temperatures were produced from the linear regions, observed from the plots of stress-(α-α^{-2}) curves, which are shown in Table 7.6. The modulus for all composites increases up to 40°C, and then decreases between 40 and 60°C, κC presenting a maxima at 40°C. However, compressive elastic modulus presented a minimum at 40°C, for 1 (w/v)% of κC content sample.

The change in compressive elastic modulus in the initial stage may originate from the greater availability of monomer molecules

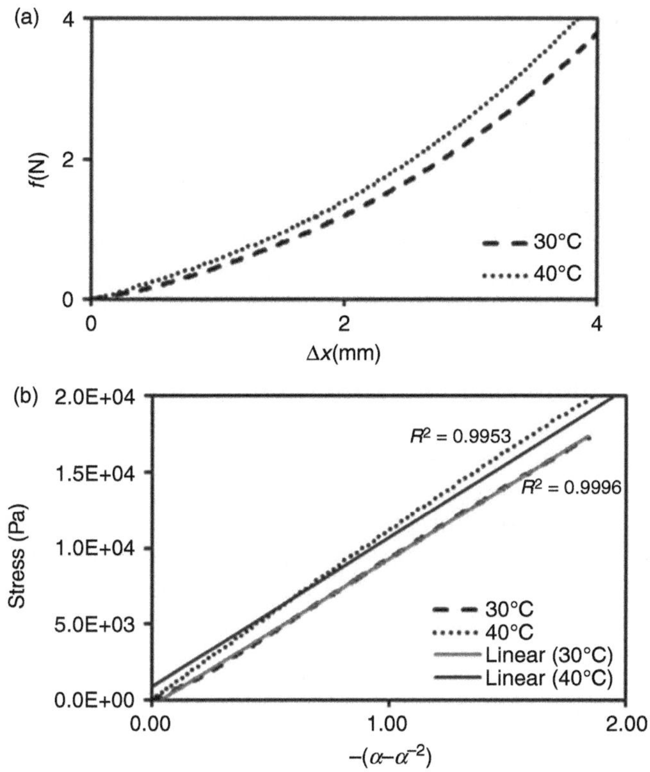

Figure 7.7. (a) The force f(N) and compression (mm) curves and (b) stress versus $-(\alpha-\alpha^{-2})$ curves for 2.5 (w/v)% of κC contents at 30 and 40°C, respectively.

in the vicinity of the chain propagating sites of κC in PAAm-κC composite. Because these gels were noticed to be inspired by the alkaline hydrolysis, which can be reasoned by the low osmotic pressure that enhances up to 40°C, then declines again from 40°C to 60°C. On the other hand, the existence of κC in the PAAm gel at 40°C generates larger water absorbing volumes, which then results in maximum compressive elasticity for below and above of 1 (w/v)% of κC content composites (Evingür and Pekcan, 2012a,b, 2013). In a consequence, the behavior of S for the 1 (w/v)% of κC content composite can be understood by the superelastic network property of the gel system (Evingür and Pekcan, 2013).

5 Alginates

Algins/Alginates are extracted from brown seaweed. Alginic acid leads to a decrease in the concentration of cholesterol, exerts an antihypertension effect, can prevent absorption of toxic chemical substances, and plays a major role as dietary fiber for the maintenance of animal and human health (Holdt and Kraan, 2011). The thickening property of alginate is useful in sauces and in syrups and toppings for ice cream. Edible dessert jellies can be formed from alginate-calcium mixtures. Alginate gels are used in restructured or reformed food products, and calcium alginate films and coatings have been used to help preserve frozen fish (McHugh, 2003). Because of them, it is significant to know the physical parameters such as encapsulation efficiency, desorption coefficients, and incubation time of calcium alginate beads.

5.1 Calcium Alginate Beads

Spherical alginate beads were prepared from sodium alginate and calcium ions by ionotropic gelation. Pyranine (Py) as a fluorescence probe was dissolved into the alginate solution.

The steady state fluorescence technique was performed for desorption of Py in water from the alginate beads crosslinked with calcium ions. The influence of $CaCl_2$ content on encapsulation efficiency (EE), desorption coefficients, and incubation time was observed.

5.1.1 Encapsulation Efficiency

The amount of Py that escaped into the calcium chloride solution and washing water during preparation was determined using the absorption at 290 nm via UV-VIS spectroscopy (Ashokkumar and Grieser, 1998). Py calibration plot procedures were conducted as follows: Various amounts of standard Py calibration solutions were prepared and their absorbance at 445 nm was measured using the Shimadzu UV-1800 spectrometer. After that, the absorbance values of the samples were also measured. Encapsulation efficiency of Py in the alginate beads was calculated by using Eq. (7.17)

$$EE\,(\%) = \frac{n_0 - (n_m + n_w)}{n_0} \times 100 \tag{7.17}$$

where n_0 is the initial amount of Py, and n_m and n_w are the amount of Py in the calcium chloride solution and washing water, following preparation of the beads, respectively. The units are in moles. The encapsulation efficiency of Py in the calcium alginate beads decreased with increasing $CaCl_2$, presented in Table 7.7. The EEs were obtained in the range of 16.92–63.52%; the beads crosslinked with 0.5% (w/v) $CaCl_2$ solution gave the lowest EE% of 16.92% and the beads crosslinked with 20% (w/v) $CaCl_2$ solution resulted in the highest EE% of 63.52%.

Table 7.7 Experimentally Measured Parameters for the Beads Crosslinked with Various % (w/v) $CaCl_2$ Solution (Kaygusuz et al., 2014)

	% (w/v) $CaCl_2$ Solution					
	0.5	1	3	5	10	20
Incubation time, t_0 (s)	300	1000	2100	2800	3600	4300
Encapsulation efficiency of pyranine (%)	16.92	25.15	42.48	52.18	59.84	63.52
Desorption coefficient, $D^* 10^{-12}$ (m^2/s)	62.17	59.41	52.78	47.75	42.98	41.83

5.1.2 Desorption Coefficients

To create a single layer structure, suitable amounts of wet beads were placed at the bottom of a quartz cell. The position of the single layer alginate beads in the cuvette with and without 10^{-4} M Py before and after desorption are given in Fig. 7.8a, b, respectively. Excitation wavelength and fluorescence emission wavelength were set to 340 nm and 512 nm, which were used to monitor the Py desorption into the desorption media (water) using the time-drive mode of the spectrofluorometer.

The fluorescence intensities at 512 nm, which were measured versus time are shown in Fig. 7.9a, b for the alginate beads crosslinked with 3, and 10% (w/v) $CaCl_2$ solutions, respectively. The Py intensity, I, during desorption, were normalized according to the highest intensity, I_∞, at the longest time. It was seen that the fluorescence intensity increased, indicating that Py molecules were released gradually from the beads crosslinked with various calcium ion contents as the desorption time was increased.

The curves in Fig. 7.9 show typical desorption behavior, and the desorption process can be treated using the Fickian diffusion model. It is interesting to note that the desorption process starts after a certain time, called the incubation time, t_0, at which desorption just starts to increase.

5.1.3 Incubation Time

It is the time limit that the bead absorbs enough amount of water and mobilize pyranines in the bead before Py desorption starts. Table 7.7 shows the variation of t_0 and calcium ion content in the

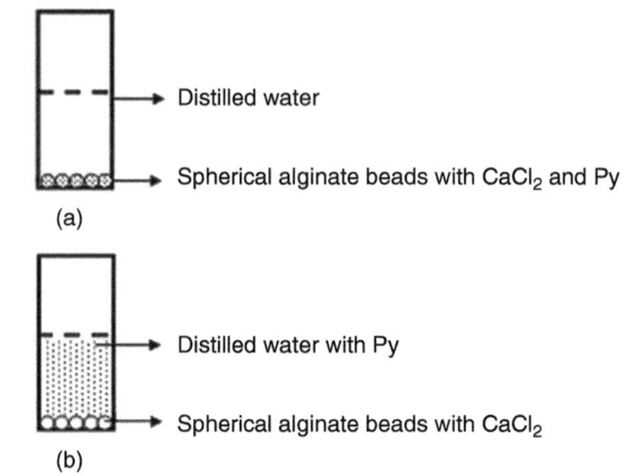

Figure 7.8. Position of alginate beads with Py on the bottom of the fluorescence cuvette (a) before and (b) after desorption process.

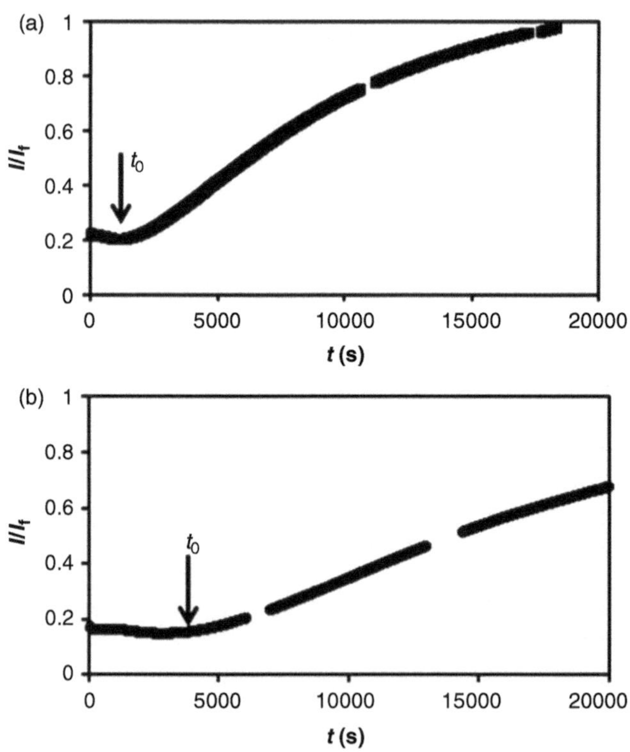

Figure 7.9. Normalized fluorescence intensity at 512 nm, I/I_∞, versus time for the beads crosslinked with (a) 3, and (b) 10% (w/v) $CaCl_2$ solution, respectively.

solutions. As can be seen in Table 7.7, when the calcium ion content was increased, t_0 increased as expected due to the degree of crosslinking in the alginate beads, that is, more ions create stronger gels, which kept the Py molecules in the alginate beads, resulting in a delay in the desorption process because of the egg-box junction of Ca^{2+} ions (Papageorgiou et al., 2010; Grant et al., 1973). That indicates a higher amount of crosslinking is related to amount of Ca^{2+} ions.

5.2 Alginate Beads with Different Metal Ions

Spherical alginate beads were prepared with P_y. The pyranine release from the alginate beads prepared with calcium, barium, and aluminum was performed by steady state fluorescence technique. The extent of crosslinking and the size of the cations are the causes of the behavior of D_{sl}, the initial amount of pyranine, and EE as shown in Fig. 7.10. Since Ba^{2+} and Ca^{2+} ions are divalent, their bonding to the alginate is expected to occur in a planar 2-dimensional manner inside the beads (Bajpai and Sharma, 2004).

5.3 Alginate with Nonionic and Ionic Surfactants

Alginate beads were prepared by nonionic Brij 35 and anionic sodium dodecyl sulfate (SDS) surfactants with two different types of cations (Ca^{2+} and Ba^{2+}). Compression measurements were performed to investigate the effect of the surfactant and cation types and their contents on Young's modulus of alginate beads as shown in Fig. 7.11.

Young's modulus was measured by using the Hertz theory, Eq. (7.18)

$$F = \frac{4R^{0.5}}{3}\frac{E}{1-\upsilon^2}\left(\frac{H}{2}\right)^{3/2} \tag{7.18}$$

Fig. 7.12a, b show the effect of surfactant types and contents on Young's modulus of the beads with varying contents of alginate solutions by crosslinking with 3% Ca^{2+} ions. As seen from Fig. 7.12a, Young's modulus of alginate beads decreases with Brij 35 content. This decrease is more apparent when alginate content increases. On the other hand, the value of Young's modulus of alginate beads increases with SDS contents for beads containing different alginate contents. A slight increase in Young's modulus for hydrophobically modified alginates have been reported by increasing SDS contents (Bu et al., 2007). It is important to note that Young's moduli for calcium crosslinked beads are in the range of 75–400 kPa, which are in accordance with the results demonstrated by Kaklamani et al. (2014), where they manufactured disc-shaped alginate hydrogels.

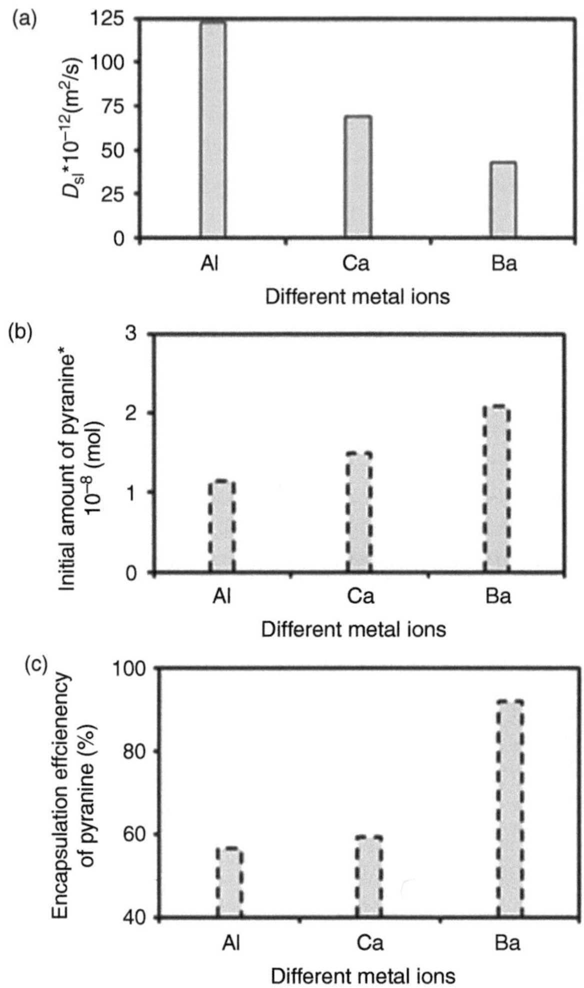

Figure 7.10. The graphical presentation of (a) slow release coefficient, (b) the initial amount of pyranine, and (c) encapsulation efficiency (EE) of the beads crosslinked with different metal (Al^{3+}, Ca^{2+}, and Ba^{2+}) ions, respectively.

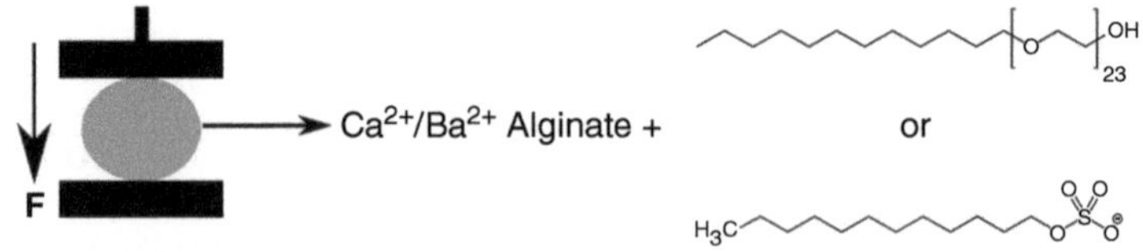

Figure 7.11. The position of alginate beads during the compression measurement.

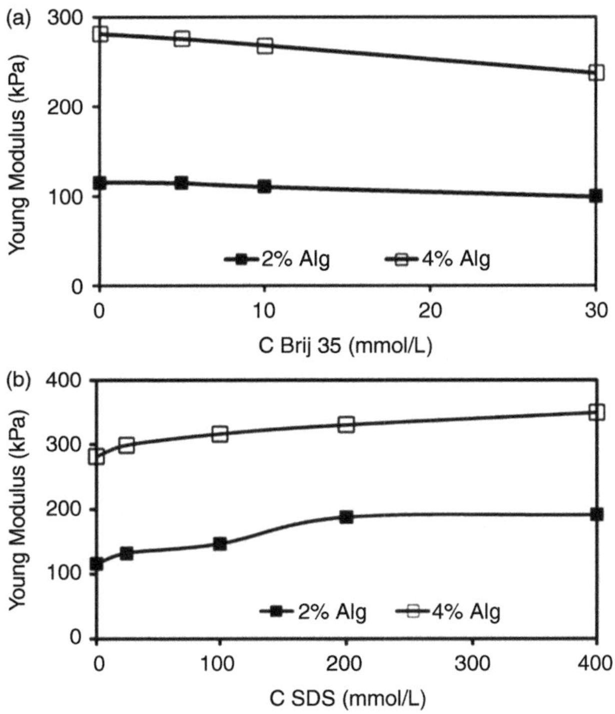

Figure 7.12. Effect of (a) Brij 35 and (b) SDS concentrations on Young's modulus of the 2 and 4% alginate beads with various alginate concentrations. Crosslinking ion: 3% (w/v) Ca^{+2}.

The effect of Brij 35 and SDS contents on Young's modulus of alginate beads crosslinked with Ba^{2+} ions (3%) is shown in Fig. 7.13a, b, respectively. As Young's modulus of the beads slightly decrease with Brij content, a considerable increase in the Young modulus is observed with increasing SDS content. Young's modulus of alginate beads containing SDS was approximately 2 times higher when compared to the modulus of beads containing Brij 35, reaching up to 500 kPa. Comparing Fig. 7.12a and Fig. 7.13b, SDS effect on the elastic modules of alginate beads are stronger for Ba-alginate beads compared to Ca-alginate beads. It should be noted that there was no barium gel formation for the formulations containing SDS and 1% (w/v) Ba^{2+}.

It was seen that the nonionic surfactant Brij 35 decreases Young's modulus of the beads. In contrast, Young's modulus of the beads containing anionic surfactant SDS showed the reverse behavior. On the other hand Young's modulus of the alginate beads showed higher values for larger alginate contents in both cases. For the case of Ba^{2+} Young's modulus of the beads with surfactant

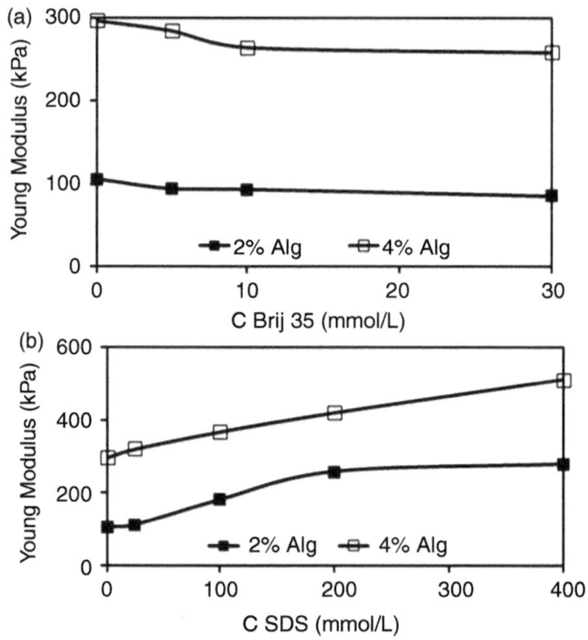

Figure 7.13. Effect of (a) Brij 35 and (b) SDS concentrations on the Young's modulus of the 2% and 4% alginate beads with various alginate concentrations. Crosslinking ion: 3% (w/v) Ba^{+2}.

SDS was found to be approximately two times higher than when compared with the modulus of beads with surfactant, Brij 35 (Kaygusuz et al., 2016).

6 Conclusions

We mentioned the effect of contents and temperature on gelation, drying, swelling, and elasticity behaviors of κC doped polyacrylamide gels in the first part of the chapter. The gelation percolation threshold occurs at 2.0% (w/v) κC with the critical exponents, which are closed to the theoretical predictions. The gel fraction exponent, β obeyed the percolation theory for below 2.0% (w/v) κC content. On the other hand, β agrees with the classical results were produced above 2.0% (w/v) κC content. In the drying process, desorption coefficient, D_d decreased as κC contents were increased for a given temperature. However, physical quantities of swelling process were calculated from Li-Tanaka model. Cooperative diffusion coefficient, D_0 showed exactly the reversed behavior in the swelling process. In the second part of the chapter, steady state fluorescence technique was used to study pyranine release from the alginate beads crosslinked with calcium, barium, and aluminum ions will be reported respectively.

It was found that the encapsulation efficiency and initial amount of pyranine of different cationic beads follow the order $Al^{3+} < Ca^{2+} < Ba^{2+}$. The slow release coefficients show the reverse order with respect to them. The results obtained can be explained in terms of the extent of crosslinking and the size of the cations involved in the beads. On the other hand, the highest % (w/v) $CaCl_2$ content produced the lowest desorption coefficient for the beads, which can be explained in terms of the increasing of crosslinking in the beads and the egg-box junction structure of Ca^{2+} ions. Last, the young modulus of alginate beads changed with the type of used surfactant and metal ions. Brij addition decreased the Young modulus of both calcium and alginate beads, however, SDS addition increased the Young modulus of both calcium and barium alginate beads. This increase is considerably higher for barium alginate beads. Moreover, the increase of barium ion content as gelation agent has a significant effect on the increase of Young modulus values.

Thus, we try to give the experimental evidences based on the mentioned theoretical models, Here, because of its sensitivity to environment and monitoring the gelation, drying, swelling, and desorption process at a molecular level, the fluorescence technique is greatly emphasized. However, the compressive measurements were introduced for deciding the modulus.

The knowledge about the relationship between microstructure of PAAm-κC composite gels and alginate beads crosslinked with calcium, barium, and aluminum ions, and their physical quantities of them should lead to optimized food production processes as well as to the development of new products with desirable properties.

References

Abd El-Ghaffar, M.A., Hashem, M.S., El-Awady, M.K., Rabie, A.M., 2012. pH-sensitive sodium alginate hydrogels for riboflavin controlled release. Carbohydr. Polym. 89, 667–675.

Abd El-Mohdy, H.L., Abd El-Reim, H.A., 2009. Radiation synthesis of kappa-carrageenan/acrylamide graft copolymers as superabsorbents and their possible applications. J. Polym. Res. 16, 63–72.

Aharony, A., 1980. Universal critical amplitude ratios for percolation. Phys. Rev. B. 22, 400–414.

Aktaş, D.K., Evingür, G.A., Pekcan, Ö., 2006. Universal behavior of gel formation from acrylamide-carrageenan mixture around the gel point: fluorescence study. J. Biomol. Struct. Dyn. 24 (1), 83–90.

Aktaş, D.K., Evingür, G.A., Pekcan, Ö., 2007a. Drying of PAAm hydrogels at various temperatures: a fluorescence study. J. Macromol. Sci. Part B: Phys. 46, 581–590.

Aktaş, D.K., Evingür, G.A., Pekcan, Ö., 2007b. Study on swelling of hydrogels(PAAm) at various temperatures by using fluorescence technique. J. Mat. Sci. 42, 8481–8488.

Aktaş, D.K., Evingür, G.A., Pekcan, Ö., 2009. A fluorescence study on swelling of hydrogels(PAAm) at various crosslinker contents. Adv. Polym. Tech. 28 (4), 215–223.

Anal, A.K., Stevens, W.F., 2005. Chitosan-alginate multilayer beads for controlled release of ampicillin. Int. J. Pharm. 290, 45–54.

Ashokkumar, M., Grieser, F., 1998. Sonophotoluminescence: pyranine emission induced by ultrasound. J. Chem. Soc. Chem. Comm., 561–562.

Bardajee, G.R., Prased, A., Sheikh, N., Amini, Fazl, M.S., 2008. Grafting of acrylamide onto kappa-carrageenan via γ-irradiation: optimization and swelling behavior. Radia. Phys. Chem. 77, 131–137.

Bastide, J., Leibler, L., 1988. Large-scale heterogeneities in randomly cross-linked networks. Macromolecules 21, 2647–2649.

Bajpai, S.K., Sharma, S., 2004. Investigation of swelling/degradation behavior of alginate beads crosslinked with Ca^{2+} and Ba^{2+} ions. React. Funct. Polym. 59, 129–140.

Birks, J.B., 1971. Photopyhsics of Aromatic Molecules. Wiley, New York.

Bierhalz, A.C.K., Silva, M.A., Kieckbusch, T.G., 2012. Natamycin release from alginate/pectin films for food packaging applications. J. Food Eng. 110, 18–25.

Bu, H., Kjoniksen, A.-L., Knudsen, K.D., Nyström, B., 2007. Characterization of interactions in aqueous mixtures of hydrophobically modified alginate and different types of surfactant. Colloids Surface A. 29 (1–3), 105–113.

Crank, J., 1975. The Mathematics of Diffusion. Clarendon Press, Oxford.

Dafader, N.C., Ganguli, S., Sattar, M.A., Haque, M.E., Akhtar, F., 2009. Synthesis of superabsorbent acrylamide/kappa-carrageenan blend hydrogel by gamma radiation. Malaysian Polym. J. 4 (2), 37–45.

Erman, B., Flory, P.J., 1982. Relationships between stress, strain, and molecular constitution of polymer networks. Comparison of theory with experiments. Macromolecules 15, 806–811.

Evingür, G.A., Aktaş, D.K., Pekcan, Ö., 2009. In situ steady state fluorescence (SSF) technique to study drying of PAAm hydrogels made of various cross-linker contents. Chem. Eng. Process. 48, 600–605.

Evingür, G.A., Pekcan, Ö., 2011a. Drying of polyacrylamide composite gels formed with various kappa-carrageenan content. J. Fluoresc. 21, 1531–1537.

Evingür, G.A., Pekcan, Ö., 2011b. Energy needs for drying of PAAm-κC composites prepared with various κC contents. E-Polymer 054, 1–11.

Evingür, G.A., Pekcan, Ö., 2012a. Temperature effect on the swelling of PAAm-κ-carrageenan composites. J. Appl. Polym. Sci. 123, 1746–1754.

Evingür, G.A., Pekcan, Ö., 2012b. PAAm-kappa carrageenan (κ-car) composites: drying and swelling with various κ-car contents. Acta Polonica A: Phys. 121 (1), 169–171.

Evingür, G.A., Pekcan, Ö., 2013. Superelastic percolation network of polyacrylamide (PAAm)-kappa carrageenan(κC) composite. Cellulose 20, 1145–1151.

Evingür, G.A., Pekcan, Ö., 2015. Elasticity study of polyacrylamide- κappa carrageenan composite prepared in various κC content and measured at several temperatures. Acta Physica Polonica A: Phys. 128 (3), 331–335.

Flory, P.J., 1941a. Molecular size distribution in three dimensional polymers. I. Gelation J. Am. Chem. Soc. 63, 3083–3090.

Flory, P.J., 1941b. Molecular size distribution in three dimensional polymers. II. Trifunctional branching units. J. Am. Chem. Soc. 63, 3091–3096.

Flory, P.J., 1941c. Molecular size distribution in three dimensional polymers. III. Tetrafunctional branching units. J. Am. Chem. Soc. 63, 3096–3100.

Gennes, P.G.d., 1979. Scaling Concepts in Polymer Physics. Cornell University Press, Ithaca, NY.
Grant, G.T., Morris, E.R., Rees, D.A., Smith, P.J.C., Thom, D., 1973. Biological interactions between polysaccharides and divalent cations: the egg-box model. FEBS Lett. 32, 195–198.
Harmathy, T.Z., 1969. Simultaneous moisture and heat transfer in porous systems with particular reference to drying. Ind. Eng. Chem. Fund. 8, 92–103.
Healthy Canada, www.hc-sc.gc.ca.
Hilliou, L., Larontonda, F.D.S., Sereno, A.M., Gonçalves, M.P., 2006. Thermal and viscoelastic properties of k/ι-hybrid carrageenan gels obtained from the Portuguese seaweed *Mastocarpus stellatus*. J. Agric. Food Chem. 54, 7870–7878.
Holdt, S.L., Kraan, S., 2011. Bioactive compounds in seaweed: functional food applications and legislation. J. App. Phys. Coll. 23, 543–597.
Işıklan, N., İnal, M., Kurşun, F., Ercan, G., 2011. pH responsive itaconic acid grafted alginate microspheres for the controlled release of nifedipine. Carbohydr. Polym. 84, 933–943.
Kaklamani, G., Chender, D., Grover, L.M., Adams, M.J., Bowen, J., 2014. Mechanical properties of alginate hydrogels manufactured using external gelation. J. Mech. Behav. Biomed. Mater. 36, 135–142.
Kampf, N., Nussinovitch, A., 1997. Rheological characterization of κ-carrageenan soy-milk gels. Food Hydrocolloids 11 (3), 261–269.
Kara, S., Pekcan, Ö., 2000a. Photon transmission technique for monitoring free radical crosslinking copolymerization in various crosslinker contents. Polymer 41, 3093–3097.
Kara, S., Pekcan, Ö., 2000b. In situ photon transmission technique for monitoring formation of hydrogels in real time at various water contents. Polymer 41, 6335–6339.
Kara, S., Pekcan, Ö., 2001. Photon transmission technique for monitoring drying process in acrylamide gels formed with various crosslinker contents. J. Appl. Polym. Sci. 80, 1898–1906.
Kara, S., Tamerler, C., Bermek, H., Pekcan, Ö., 2003a. Hysteresis during sol-gel and gel-sol phase transitions of κ-carrageenan: a photon transmission study. J. Bioact. Compat. Polym. 18, 33–44.
Kara, S., Tamerler, C., Arda, E., Pekcan, Ö., 2003b. Photon transmission study on swelling of κ-carrageenan gels prepared in various concentrations. Int. J. Bio. Macromol. 33, 235–243.
Kaygusuz, H., Erim Berker, F.B., Pekcan, Ö., Akın Evingür, G., von Klitzing, R., 2016. Surfactant and metal ion effects on the mechanical properties of alginate Int. J. Biol. Macromol.
Kaygusuz, H., Erim Berker, F.B., Pekcan, Ö., Akın Evingür, G., 2014. Cation effect on slow release from alginate beads: a fluorescence study. J. of Fluoresc. 24 (1), 161–167.
Kikuchi, A., Kawabuchi, M., Sugihara, M., Sakurai, Y., Okano, T., 1997. Pulsed dextran release from calcium-alginate gel beads. J. Control. Rel. 47, 21–29.
Langendorff, V., Cuvelier, G., Michon, C., Launay, B., Parker, A., De kruif, C.G., 2000. Effects of carrageenan type on the behavior of carrageenan/milk mixtures. Food Hydrocolloids 14, 273–280.
Li, Y., Tanaka, T., 1990. Kinetics of swelling and shrinking of gels. J. Chem. Phys. 92, 1365–1371.
McHugh, J., 2003. A guide to the seaweed industry, FAO Fichers technical paper, in alginate. Rome 441, 39–49.
Meena, R., Prasad, K., Mehta, G., Siddhanta, A.K., 2006. Synthesis of the copolymer hydrogel κ-carrageenan-graft-PAAm: evaluation of its absorbent and adhesive properties. J. Appl. Polym. Sci. 102, 5144–5152.

Mikhailov, M.D., 1975. Exact solution of temperature and moisture distributions in a porous half-space with moving evaporation front. Int. J. Heat Mass Transfer 18, 797–804.

Nochos, A., Douroumis, D., Bouropoulos, N., 2008. In vitro release of bovine serum albumin from alginate/HPMC hydrogel beads. Carbohydr. Polym. 74, 451–457.

Oppermann, W., Rose, S., Rehage, G., 1985. The elastic behaviour hydrogels. Br. Polym. J. 17 (2), 175–180.

Papageorgiou, S.K., Kouvelos, E.P., Favvas, E.P., Sapalidis, A.A., Romanos, G.E., Katsaros, F.K., 2010. Metal–carboxylate interactions in metal–alginate complexes studied with FTIR spectroscopy. Carbohydr. Res. 345, 469–473.

Pekcan, Ö., Kara, S., 2001. Photon transmission technique for monitoring swelling of acrylamide gels formed with various crosslinker contents. Polymer 42, 10045–10053.

Perez-Mateos, M., Solas, T., Montero, P., 2002. Carrageenans and alginate effects on properties of combined pressure and temperature in fish mince gels. Food Hydrocolloids 16, 225–233.

Piculell, L., 1995. In: Stephen, A.M. (Ed.), Food Polysacharides and Their Applications. Gelling Carrageenans. Marcel Dekker, New York.

Pourjavadi, A., Mahdavinia, G.R., 2006. Superabsorbency, pH-sensitivity and swelling kinetics of partially hydrolyzed chitosan-g-poly(acrylamide) hydrogels. Turk J. Chem. TUBITAK 30, 595–608.

Ribeiro, K.O., Rodrigues, M.I., Sabadini, E., Cunha, R.L., 2004. Mechanical properties of acid sodium caseinate-κ-carrageenan gels: effect of co-solute addition. Food Hydrocolloids 18, 71–79.

Sahimi, M., 1994. Application of Percolation Theory. Taylor and Francis, London.

Stauffer, D., Coniglio, A., Adam, M., 1982. Gelation and critical phenomena. Adv. Polym. Sci. 44, 103–158.

Stauffer, D., Aharony, A., 1994. Introduction to Percolation Theory, second ed. Taylor and Francis, London.

Stockmayer, W., 1943. Theory of molecular size distribution and gel formation in branched-chain polymers. J. Chem. Phys. 11, 45–54.

Stockmayer, W., 1944. Theory of molecular size distribution and gel formation in branched polymers II. General cross linking. J. Chem. Phys. 12, 125–129.

Suzuki, A., Ishii, T., 1999. Phase coexistence of neutral polymer gels under mechanical constraint. J. Chem. Phys. 110 (4), 2289–2296.

Tanaka, T., 1978. Collapse of gels and critical endpoint. Phys. Rev. Lett. 40, 820–823.

Tari, Ö., Pekcan, Ö., 2007. Swelling activation energy of κ-carrageenan in its gel state: a flourescence study. J. Appl. Polym. Sci. 106, 4164–4168.

Tari, Ö., Pekcan, Ö., 2004. A percolation approach for investigating the sol-gel phase transition of κ-carrageenan: a steady state fluorescence study. J. Bioact. Compat. Polym. 19 (6), 491–509.

Tari, Ö., Pekcan, Ö., 2008. Study of drying of κ-carrageenan gel at various temperatures using a fluorescence technique. Drying Tech. 26, 101–107.

Valavala, P.K., Odegard, G.M., 2005. Modeling techniques for determination of mechanical properties of polymer nanocomposites. Rev. Adv. Mater. Sci. 9, 34–44.

Valencia, J., Baselga, J., Pierola, I.F., 2009. Compression elastic modulus of neutral, ionic, and amphoteric hydrogels based on N-vinylimidazole. J. Polym. Sci. B Polym. Phys. 47, 1078–1087.

Webb, C., Black, G.M., Atkinson, B., 1986. Process engineering aspects of immobilized cell systems, first ed. Institution of Chemical Engineers, London.

Yılmaz, Y., Erzan, A., Pekcan, Ö., 1998. Critical exponents and fractal dimension at the sol-gel phase transition via in situ fluorescence experiments. Phys. Rev. E. 58, 7487–7491.

Yılmaz, Y., Erzan, A., Pekcan, Ö., 2002. Slow release percolate near glass transition. Eur. Phys. J.E. 9, 135–141.

Zrinyi, M., Rosta, J., Horkay, F., 1993. Studies on the swelling and shrinking kinetics of chemically crosslinked disk-shaped poly(vinyl acetate) gels. Macromolecules 26, 3097–3102.

8

STARCH NANOMATERIALS: A STATE-OF-THE-ART REVIEW AND FUTURE TRENDS

Seyed Mohammad Ali Razavi, Asad Mohammad Amini

Ferdowsi University of Mashhad (FUM), Food Hydrocolloids Research Centre, Department of Food Science and Technology, Mashhad, Iran

1 Introduction

Nowadays, the use of understandings gained through nanoscience techniques from both characterization and synthesis points of view provided the opportunity to identify or to develop novel materials with desired properties. Through different techniques developed in the field of nanotechnology, for example, top-down or bottom-up methods, not only the functional properties of traditional methods have been improved but also novel properties and characteristics have been emerged. The fabrication or isolation of nanomaterials from sustainable and naturally occurring sources extends the promising potentials of nanotechnology, especially in the field of food science. In this regard, various types of natural biopolymers have been investigated as base materials to fabricate different nanomaterials with applications in food packaging, delivery systems, emulsion stabilizing, fat mimic, and so on.

Starch is a natural polymer produced by many plants as the major energy reserve which is widely used in food, pharmaceutical, biomedical, and other applications because of its biocompatibility, biodegradability, nontoxicity, low price, and abundant sources. Starch is composed of amylose and amylopectin molecules organized as concentric semicrystalline multiscale structures, namely granules, of different sizes and shapes depending on the botanic origin (Le Corre et al., 2010; Morris, 2010; Lin et al., 2011a). Due to the unique characteristics of starch, a broad range of applications have been developed not only in the field of food science, but also in other nonfood systems. One of these special properties of starch is the possibility of modification to exert new functionalities and or to eliminate some applied limitations of native starch.

Novel Approaches of Nanotechnology in Food. http://dx.doi.org/10.1016/B978-0-12-804308-0.00008-X

The modification of native starch has been practiced through different chemical, physical, and enzymatic methods or their combinations.

Since starch granules are naturally consisted of numerous nanosize semicrystalline blocklets, it has gained special interests in the field of nanotechnology due to the low price, availability, biodegradability, being edible, and ease of modification. Different types of nanomeric materials, called nanomaterials hereinafter, have been fabricated from starches with various botanical origins and chemical compositions. Starch nanomaterials (SNMs) could be categorized to different types based on their morphology, physical properties, and so on, but unfortunately no distinct or concise classification has been made to date and there are overlaps which may cause confusions. Until present, different terms have been used for SNMs such as nanoparticles, nanocrystals, nanospheres, and nanogels, which do not provide a clear distinction. Therefore, we divided SNMs to nanocrystals (SNCs), nanoparticles (SNPs), and nanofibers (SNFs) based on the shape or crystalline nature in order to distinguish different types of particles as distinct as possible. This chapter reviews the preparation methods, morphology, and physicochemical properties of different starch nanomaterials, and their potential applications in food area.

2 Starch Granule Architecture

The composition of starch has been generally elucidated by studying the residues after hydrolysis, whether acidic or enzymatic. It is mainly composed of two homopolymers namely amylose and amylopectin. Amylose, with linear chain, is composed of α-(1→4) linked glucan with a degree of polymerization (DP) up to 600, which exists in the range of 18–28% in native granules. In contrast, amylopectin is a branched α-(1→4) linked glucan with α-(1→6) branch points (c. 5%), which is the major component of the granule (72–82%) with a molecular weight in the range of $50–500 \times 10^6$. Based on the sequential analyses, it has been revealed that the molecular chains can be classified as short chains ($12 < DP < 20$), long chains ($30 < DP < 45$), and very long chains having an average DP higher than 60 (Perez et al., 2009; Le Corre et al., 2010; Kim et al., 2015).

Based on microscopic observations (SEM, TEM, and polarized light microscopy), it has been observed that interior architecture of starch granule is arranged as concentric rings of stacks of alternating crystalline and amorphous lamellae with 100–400 nm thickness observed as an onion-like structure. Recent studies revealed that the crystalline lamellae are composed of densely packed

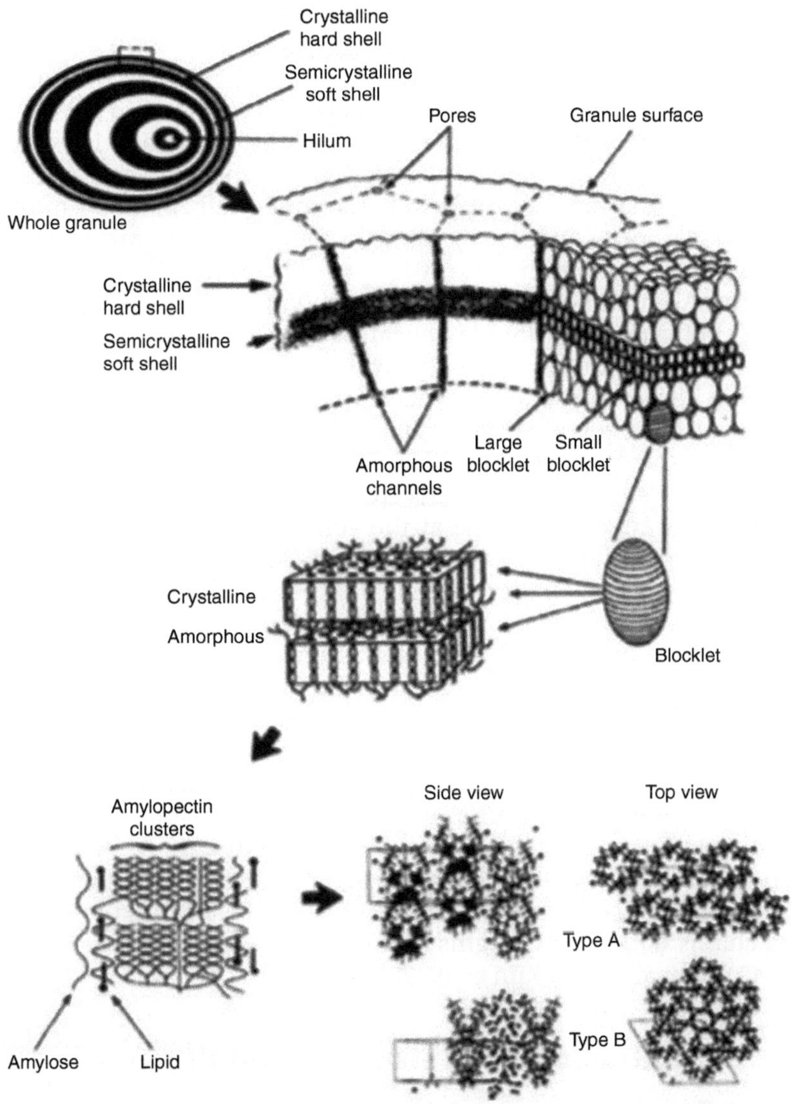

Figure 8.1. Starch granule structure overview with representation of A- and B-type crystalline patterns. Adapted from Gallant et al. (1997) with permission.

ellipsoid blocklets with diameters in the range of 20–500 nm depending on the botanical origin of starch (Kaur et al., 2007; Perez et al., 2009; Kim et al., 2015). A schematic representation of starch granule structure is presented in Fig. 8.1.

The widely accepted model for amylopectin structure is the cluster model, which represents that the branch points in amylopectin chain are in a clustered fashion rather than randomly distributed.

The crystalline lamellae with 5–7 nm thickness are believed to be mainly formed by short linear chains with $12 < DP < 70$ followed by the branch points with less crystallinity and 3–4 nm thickness (Perez et al., 2009). Based on wide-angle X-ray diffraction (WAXD), it has been revealed that the packing within crystalline lamellae of granule are arranged as two distinct allomorphs: monoclinic and hexagonal packing of parallel-stranded left-handed double helices namely A and B crystalline patterns (Fig. 8.1), respectively (Perez et al., 2009; Kim et al., 2015).

The A-type allomorphs are closely packed with water molecules between each double helical structure, whereas the B-type structures are more open, and water molecules formed by six double helices are located in the central cavity. In both A and B allomorphs, there is a pairing of double helices that corresponds to 1.1 times the distance between the axis of the two double helices. The C-type structures are believed to contain both kinds of polymorphs arranged as B-type at the center of granule and A-type in the surrounding area. Generally, most cereal starches exhibit A-type crystalline pattern, while B-type pattern has been observed mainly in tuber starches and high-amylose cereal starches. The C-type pattern is mainly characterized in legume starches, whereas V-type allomorph can only be found in amylose helical complex starches after starch gelatinization and complexing with lipid or related compounds.

It has been reported that the chain length of amylopectin defines the crystalline form of starches; B-type crystals formed by chains of $DP > 12$ while chains with $10 < DP < 12$ form A-type, and chains with $DP < 10$ do not form crystals. It has been related to the differences in loss of entropy on crystallization experienced by chains of different lengths (Kaur et al., 2007; Perez et al., 2009; Le Corre et al., 2010; Kim et al., 2015).

3 Preparation of Starch Nanomaterials

3.1 Starch Nanocrystals

3.1.1 Acid Hydrolysis

The main idea underlying the production of starch nanocrystals (SNCs) is to release the crystalline blocklets and/or the platelets inside them using different methods of hydrolysis, by analogy with cellulose nanocrystals. As a commonly used method of starch modification, acid hydrolysis has been utilized for a long time to prepare acid-thinned starch or to investigate the internal structure of starch granules from different botanical origins. It has been widely accepted that the hydrolysis of starch has mainly a two-stage profile, a fast initial stage presumably due to the hydrolysis of granule amorphous regions, and a second slower step related to the hydrolysis of granule crystalline lamellas (Hoover, 2000; Jayakody and Hoover, 2002; Le Corre et al., 2010). The slower rate of hydrolysis in second stage is believed to be due to the dense packing of amylopectin molecules in the crystalline domains that does not allow the readily penetration of acid. Also, a change in molecular conformation from chair to half-chair is required for breaking the glycosidic bonds, which is sterically impossible because of the rigidity of crystallites (French, 1984; Hoover, 2000; Li et al., 2007; Le Corre et al., 2010).

The first attempts to prepare crystalline particles from potato starch using acid hydrolysis date back to the works of Nägeli (1874) and Lintner (1886), which reported the production of water insoluble acid resistant residues. The term microcrystalline starch was first proposed by Dufresne et al. (1996) for crystalline residues obtained after hydrolysis of 5% (w/w) potato starch using 2.2 N hydrochloric acid at 35°C for 15 days. They reported that microparticles are composed of crystalline particles with a few tens of nanometer in diameter. The same procedure, except hydrolysis temperature of 30°C, was followed by Dufresne and Cavaille (1998) to prepare potato starch microcrystals. Based on TEM observations, Putaux et al. (2003) introduced the term nanocrystal for crystalline residues obtained from hydrolysis of waxy maize starch using the same protocol utilized by Dufresne et al. (1996) for 6 weeks. They reported that the individual particles were parallelepipedal platelets with a marked 60–65-degrees acute angles and a length of 20–40 nm and 6–8 nm thickness.

Angellier et al. (2004) investigated the influence of hydrolysis temperature and time, acid and starch concentration, and stirring speed on preparing SNCs from waxy maize starch using sulfuric acid instead of HCl. The authors reported that the optimal conditions of preparing SNCs consisted of 3.16 M H_2SO_4 concentration, temperature of 40°C, and 14.7% starch concentration. The yield of SNCs obtained by sulfuric acid hydrolysis were lower than HCl hydrolysis but with reasonably shorter hydrolysis time, c. 5 days instead of 6 weeks. Also, it has been claimed that the existence of sulfate groups on the surface of SNCs made them more stable in aqueous suspensions than those prepared by hydrochloric acid.

The optimized method proposed by Angellier et al. (2004) was a great success in the field of preparing SNCs with increased yield and significantly shorter production time. Many researchers followed their method to produce nanocrystals from starches of different botanical origins having different amylose contents and crystallinity.

The ultrastructure of Rhizoma Dioscorea starch with C-type crystalline pattern was investigated by different electron microscopy techniques through hydrolysis with 2.2 M hydrochloric acid at 35°C for up to 40 days. The authors reported that the SNCs were of face-centered cubic morphology with marked 60–65-degrees acute angles and constituting parallelepipedal blocks with a length of 20–40 nm and a width of 15–30 nm (Wang et al., 2008a).

In addition, morphological properties of pea starch SNCs prepared by hydrolysis with 2.2 M hydrochloric acid at 35°C for up to 25 days has been investigated by Wang et al. (2008b). The authors reported that the amorphous regions were located mostly in the center of C-type granule, which consisted of B-type polymorph, while A-type polymorph was surrounding them. The dimensions of crystalline lamellae observed to be about 20 nm.

The influence of botanical origin (waxy, normal, and high amylose maize, potato, and wheat) and amylose content (0, 25, and 70%) of starch on morphology and properties of SNCs was assessed by Le Corre et al. (2011b). They reported that the size and crystallinity of SNCs depend on amylopectin content, but the botanic origin had limited influence. They suggested that high-amylose starches are not recommended to prepare nanocrystals.

Kim et al. (2012) investigated the influence of various starches (waxy, normal, and high-amylose maize, mungbean, and potato) with different amylose content on the properties of SNCs prepared during 7 days by the method of Angellier et al. (2004). They concluded that the polymorph of starch affect the hydrolysis rate, which in turn is inversely related to the size of isolated SNCs.

As an attempt to shorten the preparation time of SNCs, Le Corre et al. (2012a) investigated the influence of higher levels of parameters that were optimized previously by Angellier et al. (2004). The authors reported that by increasing acid and starch concentration to 4.5 M and 40%, respectively, the hydrolysis time to prepare SNCs could be reduced to about 15 h with 7.6% yield and 64 nm particle size.

The preparation of SNCs from lima bean and jack bean starches by sulfuric acid hydrolysis was conducted by Oladebeye et al. (2013) and it was found that the nanocrystals were of V-type crystalline pattern with 10.8 and 13.5% yield, respectively. The SNCs were almost insoluble in organic solvents, so that lima bean nanocrystals have been more soluble than jack bean nanocrystals.

Xu et al. (2014) prepared SNCs from different starches (corn, barley, potato, tapioca, chickpea, and mungbean) and investigated their morphology and characteristics. The authors concluded that the shape and size of SNCs were similar but their yield varied over the range of 8.8–35.7% depending on the botanical origin; the lowest yield and crystallinity and the largest particle size were observed for potato starch nanocrystals.

It has been observed that by increasing the pH from 2.07 to 11.96, SNCs obtained by sulfuric acid hydrolysis of waxy maize starch were of smaller size and wider size distributions (Wei et al., 2014a). Also, it seems that the morphology of nanocrystals is dependent to the pH of dispersions, so that it has been observed that aggregated parallelepiped nanoplatelets (1.5 μm) were changed to monodispersed sphere-like particles with 50 nm size as the dispersion pH increased. It has been related to the increased repulsion forces among particles and partial dissolution of larger SNCs to smaller ones (Wei et al., 2014a).

The hydrolysis of potato starches with varying content of phosphate using 2.2 M hydrochloric acid at 25 and 45°C to about 80% solubilization to prepare SNCs was reported by Wikman et al. (2014). The authors reported that the molecular composition depends on hydrolysis temperature rather than the phosphate content, so that more branched dextrins will be present in SNCs when the higher temperature is used.

3.1.2 Combined Approaches

From previous section, it was evident that the most commonly used protocols in preparing nanocrystals from starch are based on acid hydrolysis because of its simplicity and controllability, but there are issues associated with the method; acid hydrolysis is laborious and time consuming, and the yield or recovery of SNCs is low (Le Corre et al., 2011a, 2012a; Kim et al., 2015; Romdhane et al., 2015a). These issues hinder the industrial production and applicability of SNCs, therefore many attempts have been conducted to overcome the mentioned issues, which are including the use of processes and/or methods other than acid hydrolysis or their combination with acid hydrolysis. The following are some of the methods studied and proposed to increase the yield or to shorten the preparation time.

Through α-amylase hydrolysis of waxy rice starch at 37°C and further sonication (20 kHz, 300 W) of hydrolysis residues in ethanol for 420 s, polydisperse SNCs were prepared in less than 24 h by Kim et al. (2008). The authors reported that hydrolysis for up to 9 h has not changed the crystalline structure, while excessive hydrolysis (24 h) or sonication increased the particle size and decreased the X-ray diffraction intensity.

Kim et al. (2009) prepared nanocrystals with V-type crystalline pattern from dextrins with different molecular sizes obtained after acid-alcohol hydrolysis of high-amylose maize starch through complexing with *n*-butanol. The dextrin–butanol complexes were further treated by α-amylase to selectively remove the amorphous matrices contained in the complexes to yield nanocrystals with 23–72 nm particle size.

Membrane filtration has been investigated as an approach to increase the recovery of SNCs and to scale-up the production process to meet the industrial demands. For the first time, increasing the recovery of wheat starch hydrolysates resulting from 1 day sulfuric acid hydrolysis using cross-flow membrane filtration was assessed by Le Corre et al. (2011a). The results showed that filtration using ceramic microfiltration membranes increased the yield and homogeneity of SNCs without destructing the crystalline structure, regardless of the pore size (0.1–0.8 μm) of membrane.

The potentials of enzymatic pretreatment (α- and β-amylase, glucoamylase) to reduce the preparation time of SNCs from waxy maize starch was investigated by Le Corre et al. (2012c). Among the investigated enzymes, glucoamylase has been the most efficient enzyme for producing SNCs without destructing the crystalline structure of starch while providing the same yield after 45 h rather than 5 days.

The influence of ultrasound treatment on the preparation of waxy maize starch SNCs by sulfuric acid hydrolysis was assessed by Kim et al. (2013a). The authors applied ultrasound (20 kHz, 750 W, 20 and 40% amplitude) in two different strategies: (1) sonication of starch suspension for 30 and 60 min once a day during hydrolysis at 40°C for up to 7 days and (2) treatment of starch hydrolysates obtained after 2 days with ultrasound (40–90% amplitude) for 2–3 min at 25°C. The first strategy increased the yield from 20 to 70% without significant destruction of crystalline structure of SNCs, but the second strategy transformed the microparticles to nanosize particles whereas reduced the crystallinity.

Kim et al. (2013b) combined the acid hydrolysis protocol of Angellier et al. (2004) with ultrasound treatment to prepare SNCs from waxy maize starch. They applied ultrasound at 20 kHz frequency with 750 W power for 3 min (60% amplitude in pulse mode of 30/10 s) at 4°C, 40°C, or a temperature cycle at 4 and 40°C (1 day at each temperature) on hydrolysates obtained from 6 days of acid hydrolysis. The authors reported that performing hydrolysis at 4°C followed by sonication produced 50–90 nm particles with 78% yield.

Recently, a cross-flow ceramic membrane microfiltration system with pore size of 1.4 and 0.8 μm has been used to continuously isolate the SNCs from hydrolysates obtained by acid hydrolysis of waxy maize starch (Romdhane et al., 2015a). In this regard, dilute dispersions of SNCs (0.01–0.04% w/w) have been filtered at a transmembrane pressure of 50–150 kPa and room temperature.

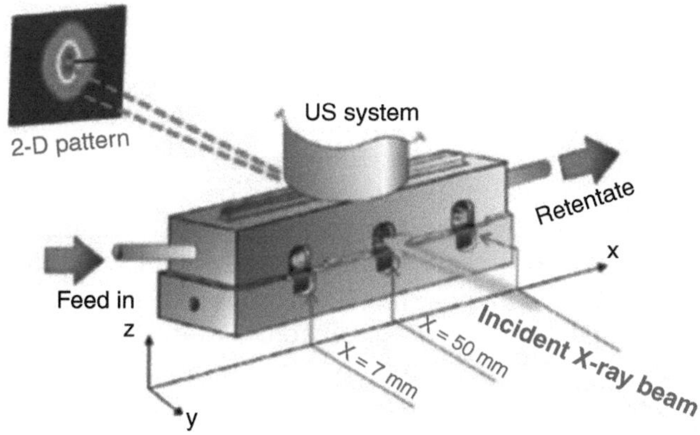

Figure 8.2. SAXS cross-flow ultrasound-assisted filtration cell. *US*, ultrasound. Adapted from Jin et al. (2015) with permission.

The authors reported that the 0.8 μm membrane has been less sensitive to the fouling and had higher permeate flux so that by microfiltration, the mean particle size of SNCs has been reduced to less than 300 nm with transmission yields of 6–37%.

Recently, Jin et al. (2015) investigated the influence of cross-flow ultrafiltration and real-time sonication (20 kHz, 2 W cm^{-2}) on concentration profile and structural organization of waxy maize starch SNCs using an in situ SAXS cross-flow ultrasound-assisted filtration cell system (Fig. 8.2). The authors concluded that ultrasound increased the performance of filtration system by increasing permeate flux through fouling inhibition and the structure of SNCs have been preserved at 10–60 nm length scale range.

3.1.3 Modification of Nanocrystals

As many other systems, the dimensions and polydispersity of nanocrystals are determinant on the stability of dispersions so that the particles should be as small as possible and of higher homogeneity to give their suspensions greater stability and reduce the possibility of forming aggregates. Unfortunately, obtaining well-dispersing and individual crystallites of SNCs are difficult due to the inherent tendency of nanocrystals to aggregate, which rises from the high number of free hydroxyl groups present on the surface of SNCs (Liu et al., 2009; Kim et al., 2015). Although the existence of sulfate esters on the surface of SNCs gives them stability in aqueous suspensions to some extent, but their presence may decrease thermal stability of SNCs, similar to cellulose nanocrystals, this is a property of the material that is important for some industrial applications. In this regard, many attempts have been made to produce more stable nanocrystals through different chemical modifications.

The surface modification of waxy maize starch SNCs has been performed through reacting with alkenyl succinic anhydride and phenyl isocyanate to prepare hydrophobic SNCs. The modification has led to a change in morphology, but the crystalline structure remained unchanged (Angellier et al., 2005a).

The waxy maize starch SNCs prepared by sulfuric acid hydrolysis have been modified through grafting stearic acid chloride and polyethylene glycol methyl ether respectively at 60°C for 4 days and 70°C for 7 days (Thielemans et al., 2006). Analysis of hydrophobic modified nanocrystals indicated that the crystalline structure has not been altered while the polar forces on the surface of nanocrystals have been reduced.

The nanocrystals prepared by 2.87 M sulfuric acid from normal corn starch at 45°C for a week have been modified through graft copolymerization with styrene in an aqueous emulsion system to prepare amphiphilic SNCs (Song et al., 2008). Although modification decreased the crystallinity of SNCs slightly, the starch-*g*-polystyrene nanocrystals (80–100 nm) had the possibility to be well dispersed in both polar and nonpolar solvents.

The polycaprolactone grafted nanocrystals were fabricated through ring-opening polymerization of potato starch SNCs prepared by HCl hydrolysis at 35°C for 15 days (Namazi and Dadkhah, 2008). The modification procedure was observed to retain the crystalline structure and morphology of nanocrystals.

The SNCs prepared from waxy maize starch by sulfuric acid hydrolysis has been modified as thermo-responsive hydrophobic nanocrystals of about 50 nm through crosslinking with 1,4-hexamethylene diisocyanate (HMDI) at ambient temperatures (Valodkar and Thakore, 2010).

By reacting the SNCs prepared by sulfuric acid hydrolysis from normal corn starch with acetic acid under sonication for a short period (c. 5 min) at temperature range of 40–65°C, acetylated nanocrystals with different degree of substitution (DS) values have been prepared (Xu et al., 2010). After esterification, spherical shape particles with increased size and a change in crystalline pattern from A- to V-type were achieved, whereas they exhibited enhanced hydrophobicity.

The modification of potato and waxy maize starches by adding fatty acid chlorides (C_8, C_{12}, and C_{16}) and dialysis method were conducted by Namazi et al. (2011) in order to produce hydrophobic SNPs with particle size of 360–500 nm. They reported that the crystalline structure of SNCs remained unchanged after modification and the modified nanocrystals had higher affinity toward organic solvents rather than aqueous medium.

As a method to improve the water dispersibility of waxy maize SNCs produced by acid hydrolysis, an attempt made based on crosslinking with sodium hexametaphosphate (pH 10.0) at 40°C (Ren et al., 2012). It was claimed that the modification produced stable (up to 72 h) and uniform aqueous suspension of SNCs through enhancing negative charges on the surface of particles while the crystalline structure has been preserved.

The crosslinking method introduced by Ren et al. (2012) was used by Jalali Jivan et al. (2013) to modify the SNCs (48 nm mean size) prepared from potato starch by hydrolyzing with HCl (2.2 or 3.7 M) at 35°C for 50 days. The authors reported that crosslinking resulted in forming a stronger lattice than their unmodified counterparts, resisting against the phase transitions induced by heating.

A process has been adapted to prepare oxidized SNCs from waxy maize starch through modifying previously prepared nanocrystals with varying concentrations (2–4% w/w) of sodium hypochlorite at pH 9.5 and 35°C (Wei et al., 2015). It has been reported that the content of carbonyl and carboxyl groups in oxidized SNCs increased with increasing oxidation level, but sulfate ester content decreased. The oxidation decreased the initial degradation temperature of SNCs in which the authors claimed that they can be used in dry process below 253°C to avoid degradation.

Recently, Wei et al. (2016) prepared 20 nm SNCs with improved dispersibility in aqueous medium using sodium hypochlorite oxidation. They reported that the oxidation increased carboxyl content on the surface of nanocrystals, characterized by zeta potential, resulting in a greater stability (20 days) of aqueous dispersions while the crystallinity of SNCs slightly decreased when the applied chlorine exceeded 2%.

3.2 Starch Nanoparticles

This category is the broadest class of SNMs with diverse fabrication methods and functionalities. The preparation methods make starch nanoparticles almost completely amorphous or with low degree of crystallinity, while some of nanoparticles are prepared from previously gelatinized starch. Different terms have been used to address them in the literature including nanogels, nanospheres, nanoparticles, and nanosponges, which we categorized as the general term starch nanoparticles (SNPs). Different chemical and physical methods have been developed to prepare SNPs as presented in the following.

Through precipitation with ethanol, SNPs were prepared from gelatinized normal corn starch and modified through crosslinking with citric acid (Ma et al., 2008). The authors reported that the modified SNPs with 50–100 nm particle size can resist gelatinization at temperatures up to 95°C.

By applying high-pressure (207 MPa) homogenization by passing high-amylose maize starch through a microfluidizer at a rate of 133 mL min^{-1}, SNPs with 10–20 nm particle size were prepared (Liu et al., 2009). The authors reported that by increasing the number of homogenization passes, the particle size has been reduced while the crystallinity decreased so that by 10 or more passes, the particles lost their crystalline structure. Also, the homogeneity, absolute zeta potential, and viscosity of particles were increased as the number of passes increased, indicating the formation of a suspension or stable gel composed of nanoparticles.

Based on the precipitation method, monodisperse nanospheres have been prepared from waxy maize starch acetate through drop wise addition of water to acetone dispersion of starch, in which the size of particles can be controlled by regulating the starch concentration, proportion of solvent and antisolvent, and the molecular weight and degree of substitution of starch acetate esters (Tan et al., 2009). The critical overlapping concentration of starch has been reported to be within the range of 8–10 mg mL^{-1} so that the interfacial effect between the two nonequilibrated liquid phases (water and acetone) controlled the formation and size of nanospheres based on their polarity. The higher the numbers of hydroxyl groups and molecular weight, the lower the size of nanospheres. Based on fluorescence spectroscopy characterizations, it has been suggested that these nanospheres could be used for the encapsulation of hydrophobic substances due to the existence of less polar microdomains within them (Tan et al., 2009).

The preparation of nanogel with particle size lower than 50 nm from 1% (w/v) carboxymethyl starch solution through a radiation crosslinking process using electron beam linear accelerator (5 MeV, 150 kW) with a 5 μs pulse width and 20 μs pulse duration at 5°C was introduced by Binh et al. (2012). They reported that 7–10 mg mL^{-1} solution of starch with absorbed irradiation dose of 10–16 kGy is suitable for increasing weight-average molecular weight while decreasing particle size and intrinsic viscosity.

Rutkaite et al. (2012) produced cationic SNPs by reacting hydroxyethyl starch with glycidyltrimethylammonium chloride (GTAC) in presence of sodium hydroxide at 45°C for 24 h to reach the DS of 0.33–0.64. They obtained spherical shape particles with average diameter of 75–302 nm and zeta potential in the range of +4 to +34 mV, which will be able to form stable aqueous dispersions.

Through dispersing the normal corn starch in an aqueous mixture of sodium hydroxide and urea and freezing (−15°C for 24 h), subsequent thawing (0–4°C), and stirring (30 min, 1000 rpm) at room temperature, nanosize particles (92–362 nm) have been prepared in short durations and with high yield (94%), which their particle size could be controlled by adjusting the ratio of alkali:urea and temperature (Zhang et al., 2013).

Bel Haaj et al. (2013) prepared SNPs from normal and waxy maize starch with 1.5% concentration using ultrasound treatment for up to 90 min at 8–10°C and reported that both starch types turned into nanoparticles (30–250 nm) after 75 min with high yield, while the crystalline structure has been disrupted by sonication.

As an approach to prepare amorphous SNPs with particle size of 41 and 31 nm, gamma irradiation with 20 kGy dose and 14 kGy h^{-1} radiation rate at room temperature was applied on previously gelatinized waxy maize and cassava starches, respectively (Lamanna et al., 2013).

A method to prepare SNPs at high yield (50.47–74.23%) with particle size in the range of 91–255 nm from normal and waxy maize starches (20% w/v) using polyoxometalate acid at 40°C for 0.5–12 h has been reported by Chen et al. (2014).

In a study, SNPs with particle size in the range of 135–155 nm were prepared based on the solvent displacement method from gelatinized native corn starch with varying concentrations (2.5–10% w/v) using mechanical stirring in sodium hydroxide at 25°C for 120 min, and subsequent precipitation with ethanol (Hebeish et al., 2014). Also, to investigate the influence of surfactant on the properties of SNPs, the same process has been used to prepare SNPs with addition of varying concentrations of Tween 80 (10–40% based on the weight of SNPs) prior to precipitation. The authors reported that in presence of 20% Tween 80 as surface-active agent, the particle size of SNPs decreased and the dispersibility has been improved.

By analogy to the ultrasound-assisted oxidation method used by Qin et al. (2011) and Qian et al. (2012) for cellulose nanocrystals, SNPs have been prepared from waxy maize starch (35% w/w) with low crystallinity (11.35–1.72%) and size range of 20–60 nm. The method was consisted of oxidation using sodium hypochlorite at pH 9.5 and 35°C followed by sonication (40 kHz, 500 W) at 5°C for 180 min. The process has been performed once or twice and the reported yield was 40 and 70%, respectively. Also, the authors performed the oxidation step using sequential TEMPO (2,2,6,6-tetramethyl-piperidine-1-oxyl) and hypochlorite addition and sonication by the same conditions. The achieved yield was reported to be 50% (Sun et al., 2014a).

As a combination of enzymatic and recrystallization methods, different concentrations (10–25% w/v) of waxy maize starch has been gelatinized in buffer solution with pH of 5.0 by heating, then adding the pullulanase (30 ASPU g^{-1} of dry starch) at 58°C and incubation for 8 h and inactivation by heating at 100°C for 30 min, storing at 4°C for 8 h, and neutralizing by washing. By air drying at 60°C for 48 h, microparticles have been produced, whereas freeze drying has led to nanoparticle preparation. By treating 15% starch with the aforementioned method, SNPs with 60–120 nm particle size, B+V crystalline type with 55.4% crystallinity, and 85% yield has been fabricated (Sun et al., 2014b).

The preparation of amylose nanoparticles using precipitation of gelatinized high-amylose maize starch by ethanol has been investigated by Don et al. (2015) and it has been reported that by elevating the gelatinization temperature to 90°C, increasing the amount of ethanol and the ratio of DMSO/water to 60:40, decreasing the concentration of amylose solution, and applying ultrasound during precipitation, the size of nanoparticles has been reduced to about 160 nm.

Hydrophobic nanospheres with particle sizes less than 200 nm have been prepared by partially hydrolyzing waxy maize starch (hydrochloric acid–ethanol treatment, 65°C), hydrophobic modification using octenylsuccinic anhydride (OSA) at 85°C, and fabricating SNPs using initial water dialysis method (Gu et al., 2015). The authors reported that the shape and size of SNPs depend on the DS value and concentration of OSA-starch.

3.3 Starch Nanofibers

During the past decade, the preparation of fibers from different polymers through electrospinning technique has been the subject of many studies. Electrospinning, being a century-old technique, is a straightforward and easy-to-scale process, for which its setup is generally composed of a solution reservoir with spinneret at its tip, a grounded collector, and a high voltage power supply. Usually, the polymer solution or melt is passed through the spinneret, a high voltage (100–500 kV m^{-1}) is applied between the solution and collector, and the fibers will be collected after evaporation of solvent. By increasing the voltage of electrostatic force, the droplets of polymer solution or melt deform and a Taylor cone forms. To form a jet from Taylor cone, the electrostatic field should be increased up to a critical value in which the surface tension of polymer solution ought to be overcome by repulsive electrostatic force. In this situation, the jet travels toward the collector or coagulation bath and solvent evaporates. The sufficient entanglement of polymer chains prevents the jet from breaking into droplets to form fibers (Huang et al., 2003; Kong and Ziegler, 2012, 2014; Sutjarittangtham et al., 2014). In the first phase of electrospinning process, the structural parameters of the solution at the tip of the spinneret such as concentration, homogeneity of the solution, gelling behavior, and surface tension determines the possibility to form a jet. The second phase is likely to be dominated by the elongation of the solution, where rheological aspects such as extensional viscosity and polymer entanglement concentrations as well as Rayleigh instability, governed by the surface tension, are important (Stijnman et al., 2011).

The most widely studied polymers to fabricate fibers by electrospinning are synthetic rather than biopolymers. Many attempts have been made to prepare starch fibers by electrospinning, but the difficulties arise from their electrospinnability limited the successful applications. The key factor determining the electrospinnability of most biopolymers is to find an appropriate solvent capable of dissolving the polymer and promoting sufficient chain entanglements. The attempts to fabricate fiber from starch solubilized in water as the solvent were unsuccessful, which may be due to the lack of sufficient long-range chain entanglements in aqueous solution (Huang et al., 2003; Stijnman et al., 2011; Sutjarittangtham et al., 2014).

The influence of potato starch on preparing bicomponent nanofiber from polyvinyl alcohol has been investigated by Sukyte et al. (2010). It was observed that the maximum possible amount of starch to form nanofibers was 3%, but in presence of ethanol, the concentration of starch could be raised to 5%. Based on SEM images, it has been observed that by increasing the amount of starch added to the spinning solution, it was possible to form nonwoven material without spots but at the same time with a lower density of the number of nanofibers.

In a comprehensive study, Kong and Ziegler (2012) investigated the correlation between rheological properties of starches with different amylose content (0–80%) and botanical origin and the electrospinnability via extrapolation of the critical entanglement concentration. The authors used DMSO as solvent and reported that the crucial parameters in determining electrospinnability are starch concentration, molecular conformation, and shear viscosity. They observed that electrospinnability of starch depends on amylose content, so that as the amylose content decreased, the required starch concentration for effective electrospinning increased (Fig. 8.3). They concluded that starches with low amylose content do not form fibers through electrospinning.

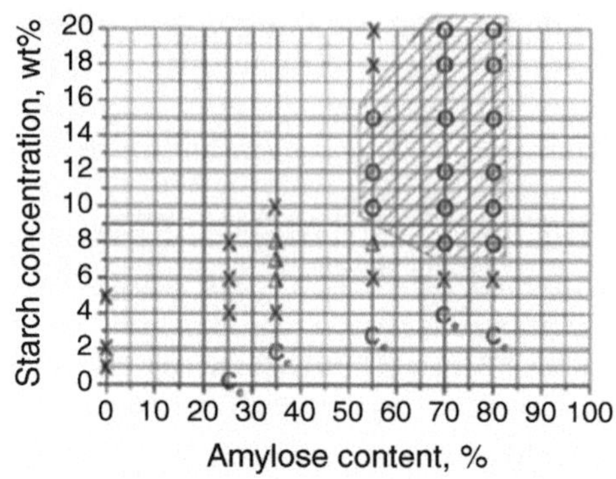

Figure 8.3. The relationship between electrospinnability (*shaded area*) and amylose content and concentration of starch in 95% DMSO. *Circles, triangles,* and *multiplication signs* represent good, poor, and no fiber formation, respectively. Adapted from Kong and Ziegler (2012) with permission.

As an attempt to improve the electrospinnability of tapioca starch, the dehydration of formed fibers was performed at −20°C in ethanol. The fabricated fibers from 4.5% starch were in micrometer range and reported to be of high water swelling ratio (Sutjarittangtham et al., 2014).

He and Wang (2014) reported the fabrication of starch nanofibers with about 12 nm diameter through electrospinning at 20 kV voltage, 2 mL h^{-1} flow rate, and 15 cm distance between the spinneret and the metal receiver. Although their prepared SNFs were somehow in nanometer range but they were inhomogeneous and many defects in the shape of knots existed.

Kong and Ziegler (2014) fabricated high-amylose maize starch fibers by a modified electrospinning technique, they called electro-wet-spinning. The process included solubilizing starch in DMSO/water mixture to form continuous jet. The authors reported that smooth and uniform starch fibers were fabricated. Also, they conducted crosslinking using vapor-phase glutaraldehyde to improve wet stability and crystallinity, thus mechanical strength, of fibers.

The fibrous electrospun mats were prepared by electrospinning at room temperature from starch solubilized in DMSO and gelatinized by addition of hot water. The diameter of fabricated fibers were observed to be in the nanometer range and it was used to impregnate the poorly soluble drug carbamazepine. The results showed that the nanofibers successfully improved wettability and reduced the crystallinity of loaded drug (Meer et al., 2015).

4 Characteristics of SNMs

4.1 Morphology

The morphology of starch nanomaterials is believed to depend mainly on the chemical structure and botanical origin of starch and preparation method of SNMs (Le Corre et al., 2011b; Kim et al., 2012, 2015). Different techniques are available to characterize the nanomaterials, including: scanning electron microscopy (SEM), transmission electron microscopy (TEM), and atomic force microscopy (AFM). The choice of equipment and techniques rely on the type of nanomaterials to be characterized and their particle size.

The SNMs could be categorized to three distinct morphologies, platelet-like particles, spheroids, and fibrous materials; other types will be covered by one of these categories. Generally, the platelet-like morphology has been reported for SNCs, but there may be exceptions based on the starch botanical origin and/or fabrication method. The first evidence on nanocrystals obtained from disruption of waxy corn starch granules by acid hydrolysis for 6 weeks was reported by Putaux et al. (2003) based on TEM observations. They observed lamellar fragments consisting of a stack of elongated elements with a width of 5–7 nm on a longitudinal view, whereas the planar view showed a marked 60–65-degrees acute angles for parallelepipedal blocks with a length of 20–40 nm and a width of 15–30 nm for individualized platelets (Fig. 8.4).

In contrast to the platelet-like parallelepipedal morphology observed in case of waxy maize SNCs (Putaux et al., 2003; Angellier et al., 2004, 2005a; Thielemans et al., 2006), most researchers reported round edge, polygon, or irregular shape platelets

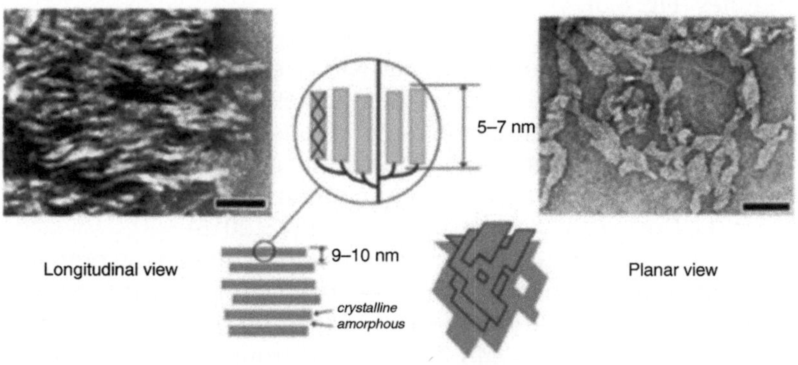

Figure 8.4. The morphology of starch nanocrystals in planar and longitudinal directions observed by TEM. Scale bars: 50 nm. Adapted from Putaux et al. (2003) with permission.

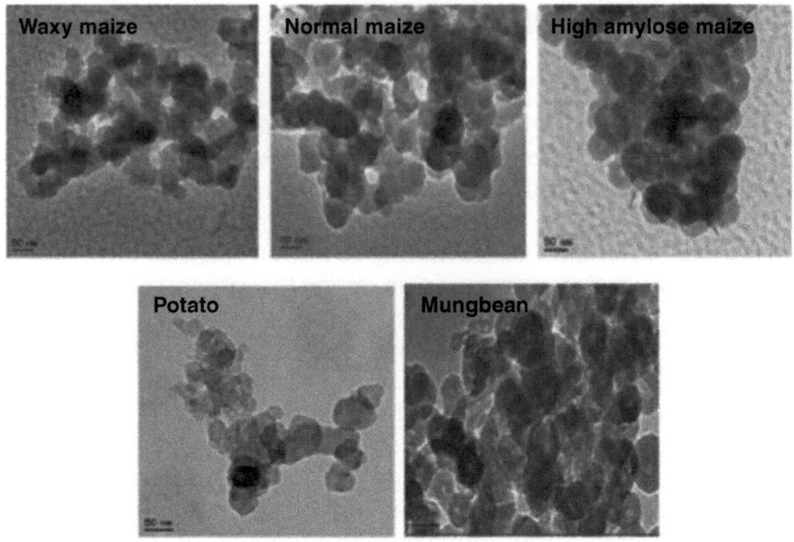

Figure 8.5. TEM morphology of starch nanocrystals prepared by acid hydrolysis from starches with different botanical origins and/or amylose content. Scale bars: 50 nm. Adapted from Kim et al. (2012) with permission.

(Le Corre et al., 2010, 2011b; Kim et al., 2012, 2015). TEM images of SNCs prepared from different starches with varying amylose content and different botanical origins are presented in Fig. 8.5.

It has been concluded that the crystalline structure of starch determines the morphology of SNCs (Le Corre et al., 2011b; Kim et al., 2015), so that platelets with acute edges are observed for SNCs prepared from starches with A-type crystalline pattern, while the round shape particles are mostly related to the B-type starches (Fig. 8.6).

Also, it has been observed that the crystalline pattern of the parent starch determines the particle size of SNCs, so that the nanocrystals obtained from A-type starches are generally of smaller particle size compared with nanocrystals prepared from B- and C-type starches. It has been related to the greater resistance of A-type starches to acid hydrolysis in comparison with other starch types (Le Corre et al., 2011b; Kim et al., 2012, 2015).

The morphology of SNPs are mainly dependent on the preparation and/or modification methods, so that a broad range of shapes could be found in the literature. As it can be seen from Fig. 8.7, different morphologies can be observed from platelet (Fig. 8.7a, Angellier et al., 2004; Fig. 8.7k, Kim et al., 2009) to polygon and irregular spheroids (Fig. 8.7b and h, Kim et al., 2008; Fig. 8.7d, Bel Haaj et al., 2013; Fig. 8.7j, Tan et al., 2009) and spherical shape

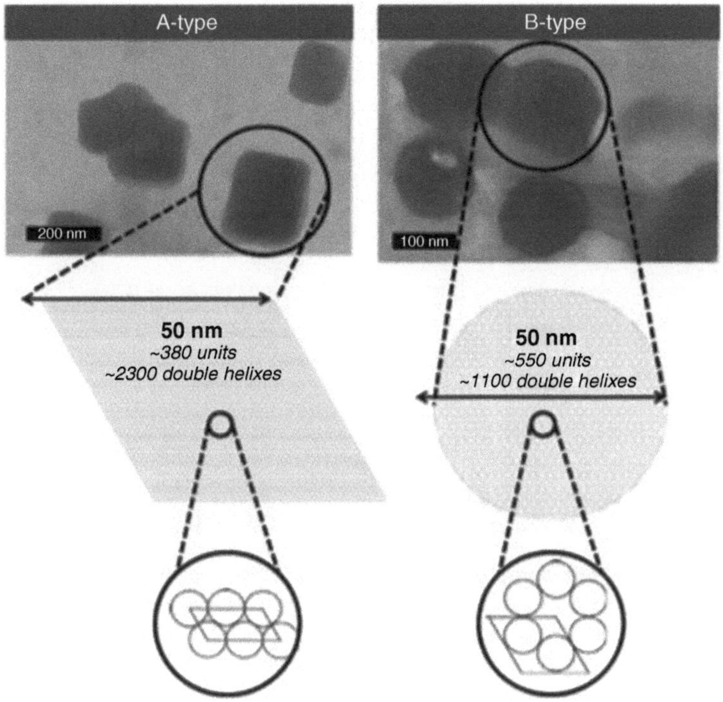

Figure 8.6. Morphological differences between nanocrystals prepared by acid hydrolysis from maize starches with A- and B-type crystalline pattern along with their amylopectin double helix packing configuration. Adapted from Le Corre et al. (2011b) with permission.

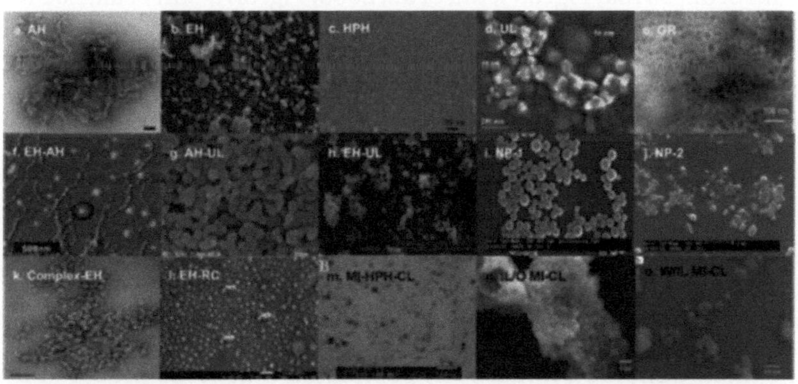

Figure 8.7. Morphology of starch nanoparticles prepared and/or modified by different methods. (a) AH, (b) EH, (c) HPH, (d) UL, (e) GR, (f) EH-AH, (g) AH-UL, (h) EH-UL, (i) NP-1, (j) NP-2, (k) Complex-EH, (l) EH-RC, (m) MI-HPH-CL, (n) IL/O MI-CL, and (o) W/IL MI-CL. AH, EH, HPH, UL, GR, NP, Complex, RC, and MI stand for acid hydrolysis, enzyme hydrolysis, high-pressure homogenization, ultrasound, gamma radiation, nanoprecipitation, complex formation, recrystallization, and microemulsion, respectively. Adapted from Kim et al. (2015) with permission.

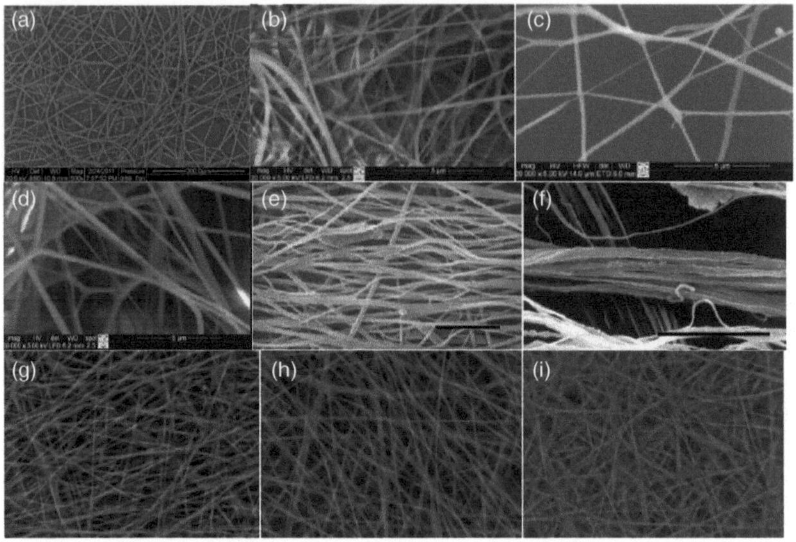

Figure 8.8. SEM micrographs of starch fibers. (a) Fiber from high-amylose maize starch (scale bar: 200 μm); (b–d) fibers from poly(vinyl alcohol) solutions containing 1, 3, and 5% potato starch, respectively (scale bar: 5 μm); (e) fibers of sodium starch glycolate (scale bar: 5 μm); (f) sodium starch glycolate fibers loaded with carbamazine (scale bar: 5 μm); (g–i) fibers of high-amylose maize starch prepared in 95, 90, and 85% DMSO, respectively. Adapted from Sukyte et al. (2010), Kong and Ziegler (2012, 2014), and Meer et al. (2015) with permission.

(Fig. 8.7c, Liu et al., 2009; Fig. 8.7f, Le Corre et al., 2012c; Fig. 8.7g, Kim et al., 2013a; Fig. 8.7i, Ma et al., 2008; Fig. 8.7l, Sun et al., 2014b; Fig. 8.7m–o, Shi et al., 2011; Zhou et al., 2014a,b).

There are scarce studies regarding the preparation of fibers from starch and the attempts generally led to preparation of microfibers rather than nanofibers. The SEM micrographs of some fibers prepared from different starch sources by electrospinning technique are presented in Fig. 8.8.

4.2 Surface Properties

It is widely accepted that the surface properties of SNMs are dependent on the method utilized for fabrication and/or modification. Generally, it has been postulated that SNCs prepared by sulfuric acid hydrolysis contain sulfate esters along with carboxyl groups, while sulfate groups are absent in SNCs prepared by hydrochloric acid (Wei et al., 2014a,b). It has been observed that the zeta potential of waxy maize SNCs prepared by sulfuric acid hydrolysis decrease from −6.7 to −34.5 mV as the dispersion pH increased from 2.07 to 11.96 so that by increasing pH, the particles were of smaller size and wider size distributions (Wei et al., 2014a). Also, it seems that the morphology of nanocrystals is dependent to the pH of dispersions, so that it has been observed that aggregated parallelepiped nanoplatelets (1.5 μm) were changed to monodispersed spherelike particles with 50 nm size as the dispersion pH increased. It has been related to the increased repulsion forces among particles and partial dissolution of larger SNCs to smaller ones. Therefore, the dispersibility and stability of aqueous suspensions of SNCs seems to be controlled by adjusting the dispersion pH so that the more stable particles could be achieved in pH range of 7.44–9.45 (Wei et al., 2014a).

Based on conductometric titration, it has been revealed that both weak acidic groups (carboxylic acids) in significant amounts (32–38 mmol kg^{-1}) and lower amounts of strong acidic groups as sulfate esters (13.2 mmol kg^{-1}) are present on the surface of waxy maize SNCs prepared by sulfuric acid hydrolysis (Romdhane et al., 2015b). It seems that the amount of sulfate groups present on the surface of nanocrystals is directly related to the hydrolysis time, while the content of weak acid groups are independent. Also, based on zeta potential measurements, it has been stated that the surface charge of nanocrystals decrease upon reducing the pH values below 4, remain constant at the range of 4–8, and increase at elevated values so that the absolute zeta potential values vary in the range of −6 to −29 mV, representing the high potential to form aggregates in aqueous medium. The stability of nanocrystals seems to be higher in the alkaline pH range due to the smaller size of particles. Similar behavior for the dependence of surface charge to pH has been observed in case of increasing ionic strength (Romdhane et al., 2015b).

4.3 Crystalline Properties

The crystallinity of nanomaterials is of great importance when it comes to the reinforcing characteristics required by some applications like fabrication of packaging materials. Generally, the most crystalline nanomaterials are SNCs, in which preserving the crystalline structure of native starch has to be guaranteed. On the other side, some modification methods of SNCs and almost majority of preparation techniques of SNPs resulted in a change in the crystalline type and/or loss of crystallinity.

Generally, the crystallinity and crystalline pattern of SNMs are characterized by X-ray diffraction (XRD) method, but some researchers postulated the changes in crystalline properties based on other techniques such as differential scanning calorimetry (DSC) and Fourier transform infrared spectroscopy (FTIR).

During the preparation of SNCs by acid hydrolysis, the amorphous regions of starch granule will be eliminated and the highly crystalline blocklets remain more or less intact, therefore, the crystallinity of SNCs increases in comparison with the source starch. This has been evidenced by many studies (Kim et al., 2008, 2012, 2013a,b, 2015; Le Corre et al., 2010, 2011b, 2012a,b,c; Le Corre and Angellier-Coussy, 2014). It should be noted that the extensive hydrolysis during preparing SNCs may lead to a loss of crystallinity, because after the amorphous regions of the granule, the crystalline domain will be destroyed by the act of acid. It has been revealed that the botanic origin of starch does not affect the crystallinity of SNCs prepared by acid hydrolysis, whereas the amylose content of starch has profound influence on the crystallinity so that from starches with lower amylose content, more crystalline SNCs can be prepared. In this regard, it has been suggested that high-amylose starches not to be used to prepare SNCs (Le Corre et al., 2011b; Kim et al., 2015). Generally, the value of crystallinity of SNCs vary in the range of about 30% to higher than 60% based on the preparation method and its duration, and/or the chemical structure of starch granule.

In some cases, the crystallinity or crystalline structure of nanomaterials have been diminished or changed by modification process utilized, which is completely dependent on the method type. Generally, it could be concluded that the loss of crystallinity mainly occurs through physical modification processes rather than chemical methods. Ultrasound treatment is the most widely physical method utilized for size reduction of SNCs, whether applied on the prepared nanocrystals or during the preparation. There is some evidence that sonication disrupts the crystalline structure of SNCs (Bel Haaj et al., 2013; Kim et al., 2008, 2013a,b).

On the contrary, based on the fabrication technique, SNPs are almost completely amorphous or with low crystallinity values. In some cases the SNPs are prepared from previously gelatinized starch, while some of fabrication processes, mostly physical, degrade the crystalline structure of granule. Different physical methods that have been reported to produce SNPs with amorphous structures include gamma irradiation (Lamanna et al., 2013), high pressure homogenization (Liu et al., 2009; Shi et al., 2011), extrusion (Song et al., 2011), and sonication (Bel Haaj et al., 2013). This could be attributed to the higher energy levels delivered by the physical treatments, which could exceed the bond energy of crystalline structures resulting in their disruption.

There are some instances that preparation processes, mostly those starting with gelatinized starch, could result in fabricating crystalline nanoparticles but with different crystalline patterns.

The examples of these kind of methods are precipitation (Ma et al., 2008), recrystallization (Sun et al., 2014b), and complex formation (Kim et al., 2009).

Therefore, the choice of method and parameter settings of the process are crucial to prepare SNMs with desired crystalline properties, and by precise tuning the parameters, nanomaterials with desired characteristics and functionalities can be tailored.

4.4 Thermal Properties

One of the most important material characteristics of nanomaterials, from industrial point of view, are thermal properties. Based on the crystalline and chemical structure of SNMs, a broad range of thermal behavior could be observed. The most common techniques to characterize thermal properties of nanomaterials are differential scanning calorimetry (DSC) and thermogravimetric analysis (TGA).

Thermal properties of starch granules are the most important characteristics in industrial applications, so that the gelatinization behavior determines the processing method and conditions, or final product characteristics. When it comes to the nanomaterials, the thermal transition temperature and thermal stability are determinant characteristics to select the field of application.

It has been revealed that the botanical origin of starch and its amylose content do not affect the thermal properties of SNCs prepared by acid hydrolysis. Based on DSC measurements, two thermal transition temperatures have been observed for SNCs, while starch granules generally show one transition temperature. The first peak has been attributed to the unpacking of the double helices resulting in melting of crystallites, while the second one is probably related to the unwinding of helices observed as the second stage of crystallite melting (Le Corre et al., 2012b). The TGA measurements revealed that the second peak correspond to the perfect crystallites melting which occurred before depolymerization.

Based on the value of first thermal transition temperature in excess water, it has been revealed that the B-type SNCs are more thermally stable than the A-type crystallites, which has been attributed to the longer amylopectin helices of B-type crystallites exhibiting the rigidity effect (Waigh et al., 2000; Le Corre et al., 2012b), whereas the second endothermic peak does not differ between the two crystalline types. In contrast, the thermal transition temperatures did not correlate with crystalline type or amylose content in dry state. On the contrary, there are reports that no endothermic peaks were found for SNCs prepared from potato and high-amylose maize (B-type) or mungbean (C-type) starches (Kim et al., 2012).

Based on the TGA characterizations, it has been revealed that the degradation temperature of SNC is affected by the size and surface charge (sulfate groups) of nanocrystals (Le Corre et al., 2012b). The thermal stability of SNCs are observed to be highly dependent on the presence of water due to its plasticizing effect, so that it has been suggested that the SNCs can be used in wet processes below 100°C, while in dry state they could be processed at temperatures up to 150–200°C (Le Corre et al., 2012b). Therefore, by changing the physical parameters and starch sources to prepare SNMs, thermal transition properties and their thermal resistance might be controlled (Kim et al., 2015).

4.5 Rheological Properties

Although different physicochemical properties of SNMs are studied extensively during the past decade, there is scarce information regarding the rheological properties of starch nanomaterials. Determining the rheological properties of SNMs would be of great value when dealing with the processing requirements for industrial applications. For example, when the SNCs ought to be added to composites, emulsions, reduced fat products, or as carriers of functional substances, it is highly required to characterize their flow behavior or thermal-rheological properties.

In contrast to native starch granules' pasting behavior, it was observed by Ma et al. (2008) that the SNPs prepared by extrusion do not show the characteristic gelatinization curve, also known as pasting curve, so that SNPs started to absorb water and dissolve at low temperatures with extremely low viscosities without any profound change in viscosity compared with starch granule gelatinization, but by increasing the temperature to 95°C the viscosity raised and experienced a plateau, while its viscosity continued to increase during the cooling step to 50°C (Fig. 8.9a). This may be related to the gelatinization of starch during preparation of SNPs because of high temperatures and the presence of strong plasticizing agent (glycerol). Also, very high shear loads induced during the extrusion inhibited the possible network formation through entanglement of amylose and amylopectin or swollen granules, which are the main reason for increasing viscosity in native starch. The steady shear flow behavior of SNPs showed that, compared to the native starch, the nanoparticles exhibit very weak shear-thinning behavior, in which at some concentrations the flow behavior approached to Newtonian (Fig. 8.9b).

Recently, the flow behavior properties of SNCs in relation to concentration, temperature, and the effect of membrane filtration and ultrasound has been investigated (Jin et al., 2015). The

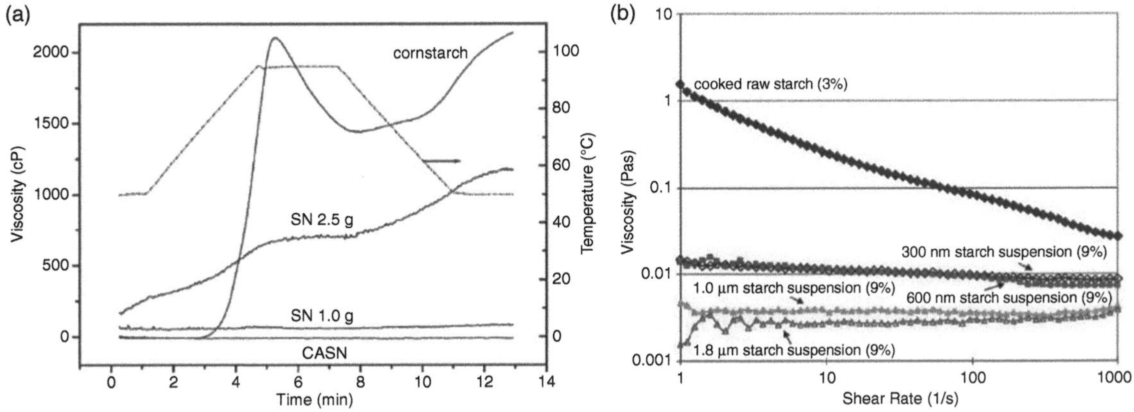

Figure 8.9. (a) Pasting behavior and (b) steady shear properties of corn starch nanoparticles. *CA* and *SN* stand for citric acid modified and starch nanoparticles, respectively. Adapted from Kim et al. (2015) with permission.

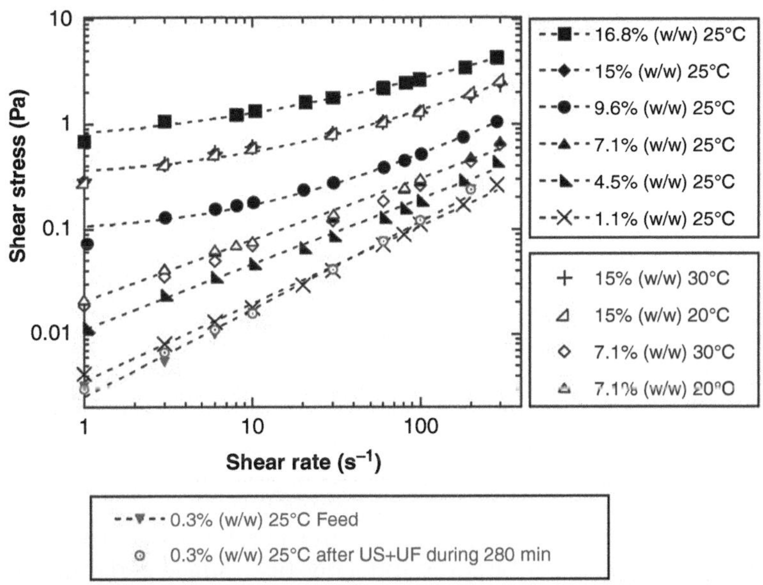

Figure 8.10. Flow behavior curve of SNCs suspensions prepared by a combination of acid hydrolysis, ultrafiltration, and ultrasound. Adapted from Jin et al. (2015) with permission.

flow curve of SNCs is presented in Fig. 8.10. It has been reported that SNCs exhibited shear-thinning behavior, while the feed suspension of filtration system showed Newtonian-like flow behavior. As expected, an increase in concentration increased the flow behavior index and consistency coefficient of Power-law model, showing that the pseudoplasticity of suspensions has been increased. It has been postulated that the clusters of SNCs at above a given volume fraction could connect to form a continuous network through hydrogen bonding, therefore, the more pronounced interaction between SNCs cause formation of a rigid network with increased viscosity and exhibiting yield stress (Jin et al., 2015).

5 Applications of SNMs

5.1 Packaging

The most widely studied application of SNMs is their incorporation in packaging films as reinforcing or filler components of film matrix, especially SNCs. The main advantage of SNCs in nanocomposite packaging materials, which are composed of at least two components of matrix and nanofillers, is in enhancement of mechanical, barrier, and thermal properties at rather low addition levels (≤5% weight basis) because of their partially crystalline nature. Also being prepared from natural sources, their biodegradability, renewability, and nontoxicity made SNCs as superior components of nanocomposite packaging fabrication (Angellier et al., 2005b; Sorrentino et al., 2007; Chen et al., 2008; Le Corre et al., 2010; Mary et al., 2013; Dufresne, 2014; Le Corre and Angellier-Coussy, 2014; Kim et al., 2015). Another characteristic of nanocomposites required for food packaging applications is the barrier properties, which are of great importance. In this regard, plate-shape nanoparticles (eg, SNCs) are preferred not only due to providing a parameter known as tortuosity that increases molecule migration pathways and consequently limits the permeability, but also retains the flexibility and transparency, which are essential for food packaging applications.

The incorporation of SNCs in nanocomposites with different synthetic and/or natural polymeric matrixes has been studied extensively by many researchers and it has been suggested that SNCs are good candidates as reinforcing nanofillers so that an increase in tensile strength, elastic modulus, glass transition temperature, and a decrease in elongation at break and water vapor permeability of composite packaging materials have been reported. The subject has been recently reviewed extensively and the reader is referred to the literature for more detailed information (Le Corre et al., 2010; Lin et al., 2011a; Mary et al., 2013; Dufresne, 2014; Le Corre and Angellier-Coussy, 2014; Kim et al., 2015).

Similar to the SNCs, the starch fibers could play an important role in reinforcing packaging materials due to their high area:volume ratio and mechanical strength (Huang et al., 2003; Kong and Ziegler, 2014). Also, the possibility of forming SNFs as woven and nonwoven mats makes them as promising candidates in fabrication of composite packaging, along with many other important applications.

5.2 Delivery Systems

Due to the diversity of SNMs and their characteristics, there are great potential to incorporate them in systems suitable for delivery of pharmaceutical, nutritional, and functional components. Of most promising nanomaterials for this purpose are SNPs, in which through different modification methods one can prepare tailor-made nanoparticles with desired properties to carry active components and deliver them to different locations in the human body. The increased surface area resulting from size reduction to nanometer range makes SNMs a valuable carrier, especially when smaller than 230 nm, because it has been reported that larger particles may congregate in body organs due to their capillary size (Kreuter, 1991; Kim et al., 2015).

Unfortunately, most of the works in the literature regard the use of nanoparticles fabricated for drug delivery and there is a lack of studies that focus on food-related delivery systems, so there is a great demand toward developing SNMs as carriers for delivering food ingredients, enzymes, and other active compounds related to the food industry.

In a study, the SNCs from pea starch were introduced to ionic crosslinked alginate-based microspheres for drug delivery purposes (Lin et al., 2011b). It has been reported that the incorporation of SNCs caused more consistent swelling patterns, higher encapsulation efficiency, and promising sustained release profiles of the drug through increasing mechanical strength and free diffusing routes of the drug molecules (Lin et al., 2011b, 2012; Dufresne, 2014). Also, SNPs of propyl-modified maize starch and dialdehyde starch have been used as drug carriers for controlled release (Yu et al., 2007; Santander-Ortega et al., 2010).

Similarly, the SNFs can be considered useful for encapsulation and delivery potentials through mixing unstable high-value components with biopolymers to be electrospun and/or load them into the prepared fibers. These fibers are foreseen to have very specific release properties, due to the specific geometrical dimensions (Stijnman et al., 2011; Meer et al., 2015). Also, these fibers can be considered as promising potentials for enzyme stabilization purposes due to their woven networks.

5.3 Emulsifying Agent

The small particle size, existence of sulfate and hydroxyl groups on the surface of SNCs along with the possibility and ease of performing modification on the surface, provides the potentials for SNMs to be used as promising emulsifying agents. Although there are some studies regarding the modification of SNMs as hydrophobic or amphiphilic nanomaterials, there are no publications regarding the emulsifying properties of starch nanomaterisls of any kind.

The only published work available is performed by Li et al. (2012) that investigated the emulsification of 50% paraffin O/W emulsion using SNCs (40–100 nm) prepared from waxy maize starch by sulfuric acid hydrolysis. The authors reported that the level of SNCs for emulsification has been about 0.02% by weight, and the particle size of dispersed phase droplets decreased by increasing the proportion of nanocrystals. The emulsions have been stable to coalescence over months. Also, it has been observed that the surface tension of SNC dispersions decreased profoundly by an increase in the concentration of nanocrystals (Kim et al., 2015). However, additional studies are needed for the utilization of SNPs in this respect.

5.4 Fat Mimics

As a promising potential application, through different preparation and/or modification methods, SNMs with fat-like properties could be tailored to be utilized as fat mimics and fat replacers. Through surface modification of SNMs, different hydrophilic and hydrophobic groups can be added to exert hydrophobic fat like properties. The fat replacer characteristic can be also exhibited by the emulsion stabilization properties of SNMs.

To date, no studies have been published to investigate the fat replacing characteristics of SNMs. Therefore, there is a great demand to carry out investigations regarding the matter.

6 Concluding Remarks and Future Trends

The nanometric size materials fabricated from different types of starches, that is, SNMs, through different methods, have opened new promising horizons in the field of food science and technology. Although few studies have been conducted in the field of food-related applications, the outstanding characteristics of SNMs could be utilized to fulfill the essential demands in this applied field. Different types of nanomaterials with diverse physicochemical and functional properties have been developed and introduced, in which these functionalities could be considered to a vast variety of applications, such as food packaging materials, delivery systems for active food compounds, functional and sensitive additives, emulsion stabilization, purification of unwanted substances, and fat replacing. It is highly demanded to develop and utilize SNMs in food technologies and systems, as their advances in other fields have been performed.

References

Angellier, H., Choisnard, L., Molina-Boisseau, S., Ozil, P., Dufresne, A., 2004. Optimization of the preparation of aqueous suspensions of waxy maize starch nanocrystals using a response surface methodology. Biomacromolecules 5 (4), 1545–1551.

Angellier, H., Molina-Boisseau, S., Belgacem, M.N., Dufresne, A., 2005a. Surface chemical modification of waxy maize starch nanocrystals. Langmuir 21, 2425–2433.

Angellier, H., Molina-Boisseau, S., Dufresne, A., 2005b. Mechanical properties of waxy maize starch nanocrystal reinforced natural rubber. Macromolecules 38 (22), 9161–9170.

Bel Haaj, S., Magnin, A., Petrier, C., Boufi, S., 2013. Starch nanoparticles formation via high power ultrasonication. Carbohydr. Polym. 92, 1625–1632.

Binh, D., Hong, P.T.T., Duy, N.N., Duoc, N.T., Dieu, N.N., 2012. A study on size effect of carboxymethyl starch nanogel crosslinked by electron beam radiation. Radiat. Phys. Chem. 81, 906–912.

Chen, G., Wei, M., Chen, J., Huang, J., Dufresne, A., Chang, P.R., 2008. Simultaneous reinforcing and toughening: new nanocomposites of waterborne polyurethane filled with low loading level of starch nanocrystals. Polymer 49 (7), 1860–1870.

Chen, L., Zhang, Z., Zhao, Z., Wang, X., Chen, X., 2014. Polyoxometalates acid treatment for preparing starch nanoparticles. Carbohydr. Polym. 112, 520–524.

Don, Y., Chang, Y., Wang, Q., Tong, J., Zhou, J., 2015. Effect of operating conditions on size and morphology of amylose nanoparticles prepared by precipitation. Starch 67, 365–372.

Dufresne, A., 2014. Crystalline starch based nanoparticles. Curr. Opin. Colloid Interface Sci. 19, 397–408.

Dufresne, A., Cavaille, J.-Y., 1998. Clustering and percolation effects in microcrystalline starch-reinforced thermoplastic. J. Polym. Sci. B 36, 2211–2224.

Dufresne, A., Cavaille, J.-Y., Helbert, W., 1996. New nanocomposite materials: microcrystalline starch reinforced thermoplastic. Macromolecules 29 (23), 7624–7626.

French, D., 1984. Organization of starch granules. In: Whistler, R.L., BeMiller, J.N., Paschall, E.F. (Eds.), Starch: Chemistry and Technology. Academic Press, New York, pp. 183–247.

Gallant, D.J., Bouchet, B., Baldwin, P.M., 1997. Microscopy of starch: evidence of a new level of granule organization. Carbohydr. Polym. 32, 177–191.

Gu, F., Li, B.-Z., Xia, H., Adhikari, B., Gao, Q., 2015. Preparation of starch nanospheres through hydrophobic modification followed by initial water dialysis. Carbohydr. Polym. 115, 605–612.

He, C.-H., Wang, X.-C., 2014. Edible starch for fabrication of nanoparticles by electrospinning. Adv. Mater. Res. 843, 49–52.

Hebeish, A., El-Rafie, M.H., El-Sheikh, M.A., El-Naggar, M.E., 2014. Ultra-fine characteristics of starch nanoparticles prepared using native starch with and without surfactant. J. Inorg. Organomet. Polym. Mater. 24 (3), 515–524.
Hoover, R., 2000. Acid-treated starches. Food Rev. Int. 16 (3), 369–392.
Huang, Z.-M., Zhang, Y.-Z., Kotaki, M., Ramakrishna, S., 2003. A review on polymer nanofibers by electrospinning and their applications in nanocomposites. Compos. Sci. Technol. 63, 2223–2253.
Jalali Jivan, M., Madadlou, A., Yarmand, M., 2013. An attempt to cast light into starch nanocrystals preparation and cross-linking. Food Chem. 141, 1661–1666.
Jayakody, L., Hoover, R., 2002. The effect of lintnerization on cereal starch granules. Food Res. Int. 35, 665–680.
Jin, Y., Hengl, N., Baup, S., Pignon, F., Gondrexon, N., Sztucki, M., Romdhane, A., Guillet, A., Aurousseau, M., 2015. Ultrasonic assisted cross-flow ultrafiltration of starch and cellulose nanocrystals suspensions: characterization at multi-scales. Carbohydr. Polym. 124, 66–76.
Kaur, L., Singh, J., Liu, Q., 2007. Starch: a potential biomaterial for biomedical applications. In: Mozafari, M.R. (Ed.), Nanomaterials and Nanosystems for Biomedical Applications. Springer, Dordrecht, The Netherlands, pp. 83–98.
Kim, J.-Y., Park, D.-J., Lim, S.-T., 2008. Fragmentation of waxy rice starch granules by enzymatic hydrolysis. Cereal Chem. 85 (2), 182–187.
Kim, J.-Y., Yoon, J.-W., Lim, S.-T., 2009. Formation and isolation of nanocrystal complexes between dextrins and *n*-butanol. Carbohydr. Polym. 78, 626–632.
Kim, H.-Y., Lee, J.H., Kim, J.-Y., Lim, W.-J., Lim, S.-T., 2012. Characterization of nanoparticles prepared by acid hydrolysis of various starches. Starch 64, 367–373.
Kim, H.-Y., Han, J.-A., Kweon, D.-K., Park, J.-D., Lim, S.-T., 2013a. Effect of ultrasonic treatments on nanoparticle preparation of acid-hydrolyzed waxy maize starch. Carbohydr. Polym. 93 (2), 582–588.
Kim, H.-Y., Park, D.J., Kim, J.-Y., Lim, S.-T., 2013b. Preparation of crystalline starch nanoparticles using cold acid hydrolysis and ultrasonication. Carbohydr. Polym. 98 (1), 295–301.
Kim, H.-Y., Park, S.S., Lim, S.-T., 2015. Preparation, characterization and utilization of starch nanoparticles. Colloids Surf. B 126, 607–620.
Kong, L., Ziegler, G.R., 2012. Role of molecular entanglements in starch fiber formation by electrospinning. Biomacromolecules 13, 2247–2253.
Kong, L., Ziegler, G.R., 2014. Fabrication of pure starch fibers by electrospinning. Food Hydrocoll. 36, 20–25.
Kreuter, J., 1991. Nanoparticle-based drug delivery systems. J. Control. Release 16 (1–2), 169–176.
Lamanna, M., Morales, N.J., Lis Garcia, N., Goyanes, S., 2013. Development and characterization of starch nanoparticles by gamma radiation: potential application as starch matrix filler. Carbohydr. Polym. 97, 90–97.
Le Corre, D., Angellier-Coussy, H., 2014. Preparation and application of starch nanoparticles for nanocomposites: a review. React. Funct. Polym. 85, 97–120.
Le Corre, D., Bras, J., Dufresne, A., 2010. Starch nanoparticles: a review. Biomacromolecules 11, 1139–1153.
Le Corre, D., Bras, J., Dufresne, A., 2011a. Ceramic membrane filtration for isolating starch nanocrystals. Carbohydr. Polym. 86, 1565–1572.
Le Corre, D., Bras, J., Dufresne, A., 2011b. Influence of botanic origin and amylose content on the morphology of starch nanocrystals. J. Nanopart. Res. 13, 7193–7208.
Le Corre, D., Bras, J., Choisnard, L., Dufresne, A., 2012a. Optimization of the batch preparation of starch nanocrystals to reach daily time-scale. Starch 64, 489–496.
Le Corre, D., Bras, J., Dufresne, A., 2012b. Influence of native starch's properties on starch nanocrystals thermal properties. Carbohydr. Polym. 87, 658–666.

Le Corre, D., Vahanian, E., Dufresne, A., Bras, J., 2012c. Enzymatic pretreatment for preparing starch nanocrystals. Biomacromolecules 13 (1), 132–137.
Li, W., Corke, H., Beta, T., 2007. Kinetics of hydrolysis and changes in amylose content during preparation of microcrystalline starch from high-amylose maize starches. Carbohydr. Polym. 69 (2), 398–405.
Li, C., Sun, P., Yang, C., 2012. Emulsion stabilized by starch nanocrystals. Starch 64, 497–502.
Lin, N., Huang, J., Chang, P.R., Anderson, D.P., Yu, J., 2011a. Preparation, modification, and application of starch nanocrystals in nanomaterials: a review. J. Nanomater. 2011, 1–14.
Lin, N., Huang, J., Chang, P.R., Feng, L., Yu, J., 2011b. Effect of polysaccharide nanocrystals on structure, properties, and drug release kinetics of alginate-based microspheres. Colloids Surf. B 85, 270–279.
Lin, N., Huang, J., Dufresne, A., 2012. Preparation, properties and applications of polysaccharide nanocrystals in advanced functional nanomaterials: a review. Nanoscale 4, 3274–3294.
Lintner, C.J., 1886. Studien uber diastase. J. Prakt. Chem. 34, 378–386.
Liu, D., Wu, Q., Chen, H., Chang, P.R., 2009. Transitional properties of starch colloid with particle size reduction from micro- to nanometer. J. Colloid Interface Sci. 339 (1), 117–124.
Ma, X., Jian, R., Chang, P.R., Yu, J., 2008. Fabrication and characterization of citric acid-modified starch nanoparticles/plasticized-starch composites. Biomacromolecules 9, 3314–3320.
Mary, S.K., Pothan, L.A., Thomas, S., 2013. Applications of starch nanoparticles and starch-based bionanocomposites. In: Dufresne, A., Thomas, S., Pothen, L.A. (Eds.), Biopolymer Nanocomposites: Processing, Properties, and Applications. John Wiley & Sons Inc., Hoboken, NJ, pp. 293–307.
Meer, T.A., Baig, M.S., Amin, P.D., 2015. Preparation and evaluation of carbamazepine loaded fibrous electrospun mats of starch. J. Pharm. Invest. 45, 301–309.
Morris, V.J., 2010. Natural food nanostructures. In: Chaudhry, Q., Castle, L., Watkins, R. (Eds.), Nanotechnologies in Food. RSC Publishing, Cambridge, pp. 50–68.
Nägeli, C.W., 1874. Beiträge zur näheren kenntnis der stärkegruppe. Ann. Chem. 173, 218–227.
Namazi, H., Dadkhah, A., 2008. Surface modification of starch nanocrystals through ring-opening polymerization of ε-caprolactone and investigation of their microstructures. J. Appl. Polym. Sci. 110, 2405–2412.
Namazi, H., Fathi, F., Dadkhah, A., 2011. Hydrophobically modified starch using long-chain fatty acids for preparation of nanosized starch particles. Sci. Iran. 18 (3), 439–445.
Oladebeye, A.O., Oshodi, A.A., Amoo, I.A., Abd Karim, A., 2013. Morphology: X-ray diffraction and solubility of underutilized legume starch nanocrystals. Int. J. Sci. Res. 2 (3), 497–503.
Perez, S., Baldwin, P.M., Gallant, D.J., 2009. Structural features of starch granules I. In: BeMiller, J., Whistler, R. (Eds.), Starch: Chemistry and Technology. Academic Press, UK, pp. 149–192.
Putaux, J.-L., Molina-Boisseau, S., Momaur, T., Dufresne, A., 2003. Platelet nanocrystals resulting from the disruption of waxy maize starch granules by acid hydrolysis. Biomacromolecules 4 (5), 1198–1202.
Qian, Y., Qin, Z., Vu, N.-M., Tong, G., Chin, Y.C.F., 2012. Comparison of nanocrystals from tempo oxidation of bamboo, softwood, and cotton linter fibers with ultrasonic-assisted process. Bioresources 7 (4), 4952–4964.
Qin, Z.-Y., Tong, G.-L., Chin, Y.C.F., Zhou, J.-C., 2011. Preparation of ultrasonic-assisted high carboxylate content cellulose nanocrystals by TEMPO oxidation. Bioresources 6 (2), 1136–1146.

Ren, L., Jiang, M., Wang, L., Zhou, J., Tong, J., 2012. A method for improving dispersion of starch nanocrystals in water through crosslinking modification with sodium hexametaphosphate. Carbohydr. Polym. 87, 1874–1876.
Romdhane, A., Aurousseau, M., Guillet, A., Mauret, E., 2015a. Cross flow microfiltration of starch nanocrystals suspensions. Can. J. Chem. Eng. 93, 412–418.
Romdhane, A., Aurousseau, M., Guillet, A., Mauret, E., 2015b. Effect of pH and ionic strength on the electrical charge and particle size distribution of starch nanocrystal suspensions. Starch 67, 319–327.
Rutkaite, R., Bendoraitiene, J., Klimaviciute, R., Zemaitaitis, A., 2012. Cationic starch nanoparticles based on polyelectrolyte complexes. Int. J. Biol. Macromol. 50, 687–693.
Santander-Ortega, M.J., Stauner, T., Loretz, B., Ortega-Vinuesa, J.L., Bastos-Gonzalez, D., Wenz, G., Schaefer, U.F., Lehr, C.M., 2010. Nanoparticles made from novel starch derivatives for transdermal drug delivery. J. Control. Release 141, 85–92.
Shi, A.-M., Li, D., Wang, L.-J., Li, B.-Z., Adhikari, B., 2011. Preparation of starch-based nanoparticles through high-pressure homogenization and miniemulsion cross-linking: influence of various process parameters on particle size and stability. Carbohydr. Polym. 83 (4), 1604–1610.
Song, S., Wang, C., Pan, Z., Wang, X., 2008. Preparation and characterization of amphiphilic starch nanocrystals. J. Appl. Polym. Sci. 107, 418–422.
Song, D., Thio, Y.S., Deng, Y., 2011. Starch nanoparticle formation via reactive extrusion and related mechanism study. Carbohydr. Polym. 85, 208–214.
Sorrentino, A., Gorrasi, G., Vittoria, V., 2007. Potential perspectives of bio-nanocomposites for food packaging applications. Trends Food Sci. Technol. 18 (2), 84–95.
Stijnman, A.C., Bodnar, I., Tromp, R.H., 2011. Electrospinning of food-grade polysaccharides. Food Hydrocoll. 25, 1393–1398.
Sukyte, J., Adomaviciute, E., Milasius, R., 2010. Investigation of the possibility of forming nanofibres with potato starch. Fibers Text. East. Eur. 18 (5), 24–27.
Sun, Q., Fan, H., Xiong, L., 2014a. Preparation and characterization of starch nanoparticles through ultrasonic-assisted oxidation methods. Carbohydr. Polym. 106, 359–364.
Sun, Q., Li, G., Dai, L., Ji, N., Xiong, L., 2014b. Green preparation and characterisation of waxy maize starch nanoparticles through enzymolysis and recrystallisation. Food Chem. 162, 223–228.
Sutjarittangtham, K., Jaiturong, P., Intatha, U., Pengpat, K., Eitssayeam, S., Sirithunyalug, J., 2014. Fabrication of natural tapioca starch fibers by a modified electrospinning technique. Chiang Mai J. Sci. 41 (1), 213–223.
Tan, Y., Xu, K., Li, L., Liu, C., Song, C., Wang, P., 2009. Fabrication of size-controlled starch-based nanospheres by nanoprecipitation. Appl. Mater. Interfaces 1 (4), 956–959.
Thielemans, W., Belgacem, M.N., Dufresne, A., 2006. Starch nanocrystals with large chain surface modifications. Langmuir 22, 4804–4810.
Valodkar, M., Thakore, S., 2010. Isocyanate crosslinked reactive starch nanoparticles for thermo-responsive conducting applications. Carbohydr. Res. 345, 2354–2360.
Waigh, T.A., Gidley, M.J., Komanshek, B.U., Donald, A.M., 2000. The phase transformations in starch during gelatinisation: a liquid crystalline approach. Carbohydr. Res. 328, 165–176.
Wang, S., Yu, J., Jin, F., Yu, J., 2008a. The new insight on ultrastructure of C-type starch granules revealed by acid hydrolysis. Int. J. Biol. Macromol. 43 (2), 216–220.

Wang, S., Yu, J., Yu, J., 2008b. The semi-crystalline growth rings of C-type pea starch granule revealed by SEM and HR-TEM during acid hydrolysis. Carbohydr. Polym. 74, 731–739.

Wei, B., Hu, X., Li, H., Wu, C., Xu, X., Jin, Z., Tian, Y., 2014a. Effect of pHs on dispersity of maize starch nanocrystals in aqueous medium. Food Hydrocoll. 36, 369–373.

Wei, B., Xu, X., Jin, Z., Tian, Y., 2014b. Surface chemical compositions and dispersity of starch nanocrystals formed by sulfuric and hydrochloric acid hydrolysis. PLoS One 9 (2), e86024.

Wei, B., Li, H., Tian, Y., Xu, X., Jin, Z., 2015. Thermal degradation behavior of hypochlorite-oxidized starch nanocrystals under different oxidized levels. Carbohydr. Polym. 124, 124–130.

Wei, B., Zhang, B., Sun, B., Jin, Z., Xu, X., Tian, Y., 2016. Aqueous re-dispersibility of starch nanocrystal powder improved by sodium hypochlorite oxidation. Food Hydrocoll. 52, 29–37.

Wikman, J., Blennow, A., Buleon, A., Putaux, J.-L., Perez, S., Seetharaman, K., Bertoft, E., 2014. Influence of amylopectin structure and degree of phosphorylation on the molecular composition of potato starch lintners. Biopolymers 101 (3), 257–271.

Xu, Y., Ding, W., Liu, J., Li, Y., Kennedy, J.F., Gu, Q., Shao, S., 2010. Preparation and characterization of organic-soluble acetylated starch nanocrystals. Carbohydr. Polym. 80, 1078–1084.

Xu, Y., Sismour, E.N., Grizzard, C., Thomas, M., Pestov, D., Huba, Z., Wang, T., Bhardwaj, H.L., 2014. Morphological, structural, and thermal properties of starch nanocrystals affected by different botanic origins. Cereal Chem. 91 (4), 383–388.

Yu, D.M., Xiao, S.Y., Tong, C.Y., Chen, L., Liu, X.M., 2007. Dialdehyde starch nanoparticles: preparation and application in drug carrier. Chin. Sci. Bull. 52 (21), 2913–2918.

Zhang, Z., Shan, H., Sun, J., Weng, Y., Wang, X., Xiong, J., Chen, L., Chen, X., 2013. Facile preparation of corn starch nanoparticles by alkali-freezing treatment. RSC Adv. 3, 13406–13411.

Zhou, G., Luo, Z., Fu, X., 2014a. Preparation and characterization of starch nanoparticles in ionic liquid-in-oil microemulsions system. Ind. Crop. Prod. 52, 105–110.

Zhou, G., Luo, Z., Fu, X., 2014b. Preparation of starch nanoparticles in a water-in-ionic liquid microemulsion system and their drug loading and releasing properties. J. Agric. Food Chem. 62 (32), 8214–8220.

9

MONITORING AND SEPARATION OF FOOD-BORNE PATHOGENS USING MAGNETIC NANOPARTICLES

Robin Augustine[*,], Ann Rose Abraham[**,‡], Nandakumar Kalarikkal[**,‡], Sabu Thomas[**,†]**

[*]Department of Materials Science and Engineering, Technion Israel Institute of Technology, Technion City, Haifa, Israel; [**]Mahatma Gandhi University, International and Inter University Centre for Nanoscience and Nanotechnology, Kottayam, Kerala, India; [†]Mahatma Gandhi University, School of Chemical Sciences, Kottayam, Kerala, India; [‡]Mahatma Gandhi University, School of Pure and Applied Physics, Kottayam, Kerala, India

1 Introduction

Infectious pathogenic bacteria cause a variety of ailments, including food poisoning, which leads to a significant public health risk. Since bacteria and other microbes exponentially multiply over time, the rapid detection of bacteria during the early stages of contamination is most important for preventing food-borne diseases. There has been an upsurge in craving for maintaining good health all over the world and this has called for research on advanced techniques for monitoring and continuous inspection of food distributed among the public. With ever-increasing threat of bioterrorism assaults of late times, microbial contamination of water and food is more of a security issue (Solodoukhina, 2011). Even small numbers of bacterial pathogens in raw food products and contaminated water impose a great risk to human health, and also show chances of extensive smash up. *Food security* is defined in the 1996 World Food Summit as a state that exists "when all people at all times have access to sufficient, safe, nutritious food to maintain a healthy and active life" (World Food Summit, 1996). Safe and harmless food is the vital privilege of any individual. Continuous monitoring of foodstuff is essential to keep a control over the quality of food distributed among the people.

Novel Approaches of Nanotechnology in Food. http://dx.doi.org/10.1016/B978-0-12-804308-0.00009-1

Food poisoning resulting from the consumption of food contaminated with pathogenic bacteria, viruses, or parasites, has become a risk to public health, with the *chicken shawarmas* and other fast-food items controlling diets of people and occupying a major position among the list in the menu card of restaurants. Nearly 2.2 million people, mostly children, die annually afflicted by food-borne and waterborne diarrheal diseases (Water Sanitation Health, 2015). The five major food-borne pathogens that affect human health are *E. coli* O157:H7, non-O157 STEC (shiga toxin-producing *E. coli*), *Salmonella* (nontyphoidal serotypes), *Listeria monocytogenes*, and *Campylobacter* (Yang and Bashir, 2008). The department of Food Safety and Zoonoses (FOS) of WHO provides guidance for the public and organizations on matters pertaining to the food safety. Its mission is to improve and ensure the health security and sustainable development by reducing the issues related to food-borne disease. The International Food Safety Authorities Network (INFOSAN), a combined program of the WHO and Food and Agriculture Organization of the United Nations (FAO) since 2004, assists countries to reinforce their foundations to handle food safety risks to achieve the target of preventing and controlling the international spread of infected food and food-borne diseases. WHO works intimately with the FAO to tackle food safety issues using innovative methods of hazard investigation, which provide efficient, technology-based tools to ensure food safety, for the benefit of both public health and economic development (FAO/WHO, 2010).

In food sectors, the critical examination of contaminants in the foodstuff and food components is of fundamental importance for ensuring food safety and quality because it poses a severe risk to public health. The question of food-borne pathogens has also captured the focus and interest of the scientific community, academia, and the food industry. The society has become increasingly aware and vigilant about the health risks posed by these food-borne pathogens. Incessant outbreaks of food-borne diseases have caused alarm in the government and pressures on regulatory and inspection agencies to find novel, advanced, scientific, technological, and facile methods for the monitoring of food-borne pathogens. Continuous monitoring of pathogens is the essential and foremost step to have a control over the presence of pathogens in the foodstuff and subsequent ailments. Efficient scrutiny and detection techniques are necessary to effectively control pathogens in food products. Despite the fact that the frequently employed techniques for food analysis are supportive, they are painstaking and time consuming. It is still a hurdle to achieve necessary detectability and desired specificity toward many specific pathogens from a heterogeneous group of pathogenic and nonpathogenic microorganisms. Thus, highly selective, rapid, and sensitive detection of food-borne pathogens still represents a great challenge to achieve a complete protection for final consumers. Thus, there is a great need in contemporary technological advancement for finding effective methods for microbial decontamination and rapid detection of the pathogens without time-consuming cell culture-based assays. Another criterion is that these technical methods developed from advancements in science and technology must be economical, but also precise, dependable, well-built, and well-fitting to the technological and financial background of the developing world.

Quick and sensitive nanoparticle mediated *monitoring* of food-borne pathogens is an effective method to deal with food contamination (Wang et al., 2007). Nanoparticles, especially antibody conjugated magnetic nanoparticles-based approaches, are key steps in the rapid detection and separation of food related diseases.

2 Food-Borne Diseases: Present Scenario

Bacteria are the major causative organisms of food-borne diseases. The Food Borne Diseases Active Surveillance Network (FoodNet) is an active population-based infections surveillance system of the Centers for Disease Control and Prevention (CDC) emerging infections program, collects data from various US states on diseases caused by enteric pathogens transmitted through food. Food-borne diseases caused by the bacteria *Campylobacter*, *E. coli* O157:H7, *Listeria*, *Salmonella*, *Vibrio*, *Shigella*, and *Yersinia*, and the parasites *Cyclospora* and *Cryptosporidium* are monitored through FoodNet. Every year, FoodNet studies and reports the number of people in the United States affected with food-borne infections from these pathogenic agents that have been invariably confirmed by laboratory tests. The 2013 data showed that food-borne infections continue to be an important public health problem and highlight the need to move forward with preventive measures (FoodNet MMWR Report, 2014). FoodNet identified 19,056 cases of culture-confirmed bacterial and laboratory-confirmed parasitic infection, 4200 annual hospitalizations, and also 80 deaths among 48 million residents of 10 US states (15% of the US population). The most common infection was caused by *Salmonella* accounting for 38% of reported infections, and the second by *Campylobacter* (35%). *Vibrio* accounted for 1.3% of reported infections. FoodNet data indicate that children younger than 5 years old have higher incidence rates of food-borne infection than any other age group. CDC estimates that *Salmonella* infection causes more hospitalizations and deaths than any other type of germ found in food.

Food-borne diseases usually arise from preparation, improper handling, or food storage. Adequate hygiene practices before, during, and after food preparation can reduce the chances of food-borne illness. Codex Alimentarius Commission, established in 1961 by the FAO, and coupled with WHO in 1962, disputes relating to food safety and consumer safeguard. The Codex Alimentarius international food standards has set goals to provide the healthiness and safe food products to consumers and ensure fair practices in the international food trade (Codex Alimentarius: How It All Began, FAO website). If whoever is involved in the food production, supply, and every key step in the food industry could follow these regulations, consumers could be confident about the safety and superiority of the food products that they obtain and importers can hope that the food product that they have ordered will be in agreement with their conditions. HACCP (Hazard Analysis and Critical Control Point) is a system of the US Food and Drug Administration (FDA) that checks food in each and every key critical point of food production and supply in order to maintain food safety (Hazard Analysis & Critical Control Points, FDA website). The HACCP system adopted by the Codex Alimentarius commission, recognizes thoroughly the hazards and measures for their control to ensure the safety of food. The functioning of HACCP is well suited with the accomplishment of quality management systems, such as the ISO 9000 series. The HACCP assists the food business operators to see how they handle food and establish strategies to confirm that the food produced/supplied is safe enough for consumption. HACCP addresses food safety through the analysis and monitoring of chemical, biological, and physical hazards during raw material harvesting, development, management handling, distribution, and utilization of the finished product.

2.1 Probable Risk Factors of Food Contamination

Raw food materials like cereals, legumes are all sources of food-borne pathogens. Water is the most important raw material in the food industry. Water activity (A_w) of foods is an important factor that determines their microbial susceptibility (Troller and Christian, 2012). Water activity should not be considered as the total moisture content but determines the lower limit of available water for microbial growth. In order to support the growth microorganisms like bacteria, yeast, and molds, require a certain amount of available water. Designing a product below a critical low level of water activity provides an effective means to control microbial growth. Microbes require water activity of 0.94 for their growth. Some species can even flourish in a media of water activity of 0.6 (Tapia et al., 2008).

Water activity depicts the energy status of water in a particular system. It is equal to the relative humidity of the air in equilibrium with a particular sample in an isolated sealed container. It is the ratio between vapor pressure of water in a sample and the vapor pressure of pure water at the sample temperature.

Water activity (A_W) is defined as,

$$A_W = \frac{P}{P_0}$$

P is partial vapor pressure above food sample, P_0 is vapor pressure of pure water.

Table 9.1 shows the water activity, microorganisms grow at a particular A_W and above, and the usual food products with this A_W. Meat, milk, eggs, vegetables, and fruits have water activity above 0.94. Many foodstuffs, like ready-to-eat foods that are consumable without cooking are available in the market. They are eaten up without heating to a temperature that kills the microbes. Raw ready-to-eat food products also suffer this danger. Fish and other seafood like cold-smoked fish, raw oysters, clams and mussels, and raw finfish have the possibility of being consumed by users in the same manner. These may enclose pathogens, due to many possibilities like harvest water contamination, poor aquaculture activities, improper handling, or poor sanitary practices during processing. For example, oysters, those gathered during the warm weather months, might contain *Vibrio vulnificus* or *Vibrio parahaemolyticus*. Raw finfish might contain *V. parahaemolyticus* and *Salmonella* spp. Pathogenic bacteria might also be introduced during processing, even after cooking. Microbes like *Vibrio vulnificus*, *V. parahaemolyticus*, and *Vibrio cholerae* non-O1 and non-O139 are associated with marine and estuarine species of fish and a major reason for food-borne diseases (Thompson et al., 2005).

The adequate nutrients necessary for growth of pathogenic bacteria are supplied by fish and fishery products. However, chemical and physical properties of the food product itself and that of packaging material could limit or enhance the growth of pathogenic bacteria and subsequent toxin formation (Dutta et al., 2009). Mostly temperature-abused cooked fish act as good sources of pathogenic bacteria, unless their development is not

Table 9.1 Water Activity Limits for the Common Bacteria and Fungi in Food Products Adapted from Beuchat (1981)

Water Activity (a_w)	Microorganisms Grow at This A_w and Above	Examples of Food Products
0.95	Pseudomonas, *Escherichia, Proteus, Shigella, Klebsiella, Bacillus, Clostridium perfringens*, some yeasts	Highly perishable foods (fresh and canned fruits, vegetables, meat, fish), milk, cooked sausages, breads, foods with up to 4 oz sucrose or 7% NaCl
0.91	*Salmonella, Vibrio parabaemolyticus, C. botulinum, Lactobacillus*, some molds	Some cheese (cheddar, Swiss, provolone), cured meat, fruit juice concentrates with 55% sucrose or 12% NaCl
0.87	Many yeasts, Candida, Torulopsis, Hansenula micrococcus	Fermented sausage, sponge cakes, dry cheese, margarine, foods with 65% sucrose or 15% NaCl
0.80	Most molds, most *Saccharomyces* spp., *Debaryomyces, S. aureus*	Most fruit juice concentrates, condensed milk, syrup, flour, high-sugar cakes, pulses containing 15–17% moisture
0.75	Most halophilic bacteria, *Mycotoxigenic aspergilli*	Jam, marmalade, glace fruits, marzipan, marshmallows
0.60	Osmophilic yeasts, few molds	Dried fruits with 15–20% moisture, caramel, toffee, honey
0.50		Noodles with 12% moisture, spices with 10% moisture
0.40		Whole egg powder with 5% moisture
0.03	-	Whole milk powder with 2–3% moisture, dehydrated soups

prohibited by preservation means like drying, salting, or acidification. Microbes like *V. vulnificus, V. cholera, V. parahaemolyticus*, and *L. monocytogenes* are those that flourish well in temperature-abused raw fish like for instance, raw molluscan shellfish (Swanson, 2011). Pathogenic bacteria may be introduced during processing, even after cooking. Well-made and hygienic sanitation programs diminish the introduction of bacteria into foods. However, these sanitation schedules would not be able to completely avoid the invasion of pathogenic bacteria. This necessitates strict and efficient monitoring techniques for control of bacteria.

2.2 Customary Microbial Detection Techniques

There are quite a lot of long-established techniques that facilitate the detection of microbes in food. Technologies like conventional microbiology assays, electrochemical biosensors, PCR, and enzyme- linked immunosorbent assay (ELISA) assist in the identification of pathogens in food products (Velusamy et al., 2010). The conventional culture and colony counting method that involves microbiological culturing and isolation of the pathogen, followed by confirmation by biochemical and serological tests, has been practically employed for detection and identification of pathogens in food products. It demands serious efforts and moreover, it is time consuming, though reliable results can be expected. The standard biological methods that are conventionally employed for the purpose of detection and identification of *E. coli* relying on the microbial culture takes broad assay times ranging from 1 to 7 days. It possesses reduced sensitivity and specificity due to lack of efficient selective enrichment medium for this particular strain of *E. coli*. The incapability of detecting toxins and the ability to sense only live cells briefs its shortcomings. Newly developed rapid assays for detecting *E. coli* with shortened detection time such as those based on diverse mechanisms, like PCR, immunoassay, optical methods, still takes prolonged time and shows reduced sensitivity. The waiting for 7–8 days to yield intended critical results by these customary procedures has aroused the necessity to seek out new rapid methods based on the use of novel technologies (Loaiza et al., 2008). In recent years, new surveillance techniques with better and more advanced quality, quantity, and speed of detection have emerged. Due to the unique properties exhibited by nanoparticles, the incorporation of magnetic nanoparticle into conventional detection techniques can make them simple, rapid, highly selective, and sensitive (Yang et al., 2008).

Magnetic nanomaterials, owing to their distinctive characteristic features, are being used for constructing impedimetric biosensors. Huang et al. (2010) used Fe_3O_4 nanoparticles to immobilize monoclonal antibodies in the fabrication of electrochemical impedimetric immunosensors for the fast identification of *Campylobacter jejuni* in the stool of diarrhea patients. Such systems can be modified and adopted for food-borne pathogen monitoring. The Fe_3O_4 nanoparticles-based immunosensor showed good performance owing to the simplicity of use, wide linear range, rapid response, acceptable reproducibility, and long stability. Sensing and detection of pathogenic *E. coli* O157:H7 bacteria by electrochemical impedance spectroscopy (EIS) was already reported (dos Santos et al., 2013).

Electrical/electrochemical impedance studies are being used for rapid detection of food borne pathogenic bacteria (Yang and Bashir, 2008). The impedance technique has been applied in the field of microbiology as a robust approach to monitor food-borne pathogenic bacteria. The integration of impedance measurement technology with biological recognition methods has made remarkable advancement in the development of impedance biosensors that are finding widespread use in microbial detection. The impedance technique, a kind of the electrochemical biosensors, has proved to be a hopeful method for food-borne pathogenic bacteria detection. It is handy, swift, and sensitive, and more notably it could be used for instant detection of pathogens (Wang et al., 2012). Based on the applicability of specific biorecognition elements, the impedance detection techniques are classified into two types. In the first technique, the impedance change caused by binding of targets to bio receptors (antibodies and nucleic acids) immobilized onto the electrode surface is measured, while the detection principle of the second type is based on the presence of metabolites produced by bacterial cells.

Another conventional immunomagnetic separation (IMS) of pathogens from food products is the use of Dynabeads®. Dynabeads are superparamagnetic spherical particles with a uniform size and a consistent, defined surface for the adsorption or conjugation of many different biological molecules like antibodies or living cells themselves. The technology behind Dynabeads was licensed to Dyno Industrier (1980) and this has been since used for the isolation and manipulation of biomolecules, including cells and pathogenic microorganisms (Neurauter et al., 2007). The performance of Dynabeads® Anti-Listeria (Dynal, Oslo, Norway) for the selective separation of *L. monocytogenes* and their applicability for direct detection of low number of *L. monocytogenes* from cheese was quiet successful. However, this technology could not separate and recover *L. monocytogenes* from the food particles in relatively concentrated bacterial suspensions (Uyttendaele et al., 2000). Bilir Ormanci et al. (2008) compared conventional cultivation and IMS methods for the specificity of isolation and sensitivity of *L. monocytogenes* from turkey meat. Day and Basavanna (2015) utilized a macrophage-based cell culture system for the isolation and enrichment of *L. monocytogenes* and *Listeria ivanovii* from infant formula and vegetables for further identification using the Luminex xMAP technique. Here, cultures of macrophage monolayers were exposed to food products and intentionally contaminated with *L. monocytogenes* or *L. ivanovii*. Then, magnetic microspheres, which were conjugated to listeria-specific antibody were used to specifically capture listeria bacteria from macrophages and subsequently analyzed using the Bio-Plex 200 analyzer. As few as 10 cfu mL^{-1} of *L. monocytogenes* was detected in all tested food products. The detection limit for *L. ivanovii* was 10 cfu mL^{-1} in infant formula and 100 cfu g^{-1} in leafy greens. An automated IMS and enzyme immunoassay (EIA) was applied for the detection of *Salmonella enterica* from poultry environmental samples. The analytical sensitivity and specificity of this technique for both the bacteria was above 90%. When postenrichment *S. enterica* cultures were used, the limit of detection of the assay was 10^4–10^6 cfu mL^{-1}. Application of IMS–EIA on naturally contaminated poultry environmental samples achieved excellent selectivity to these bacteria. However, this method took 48 h, including confirmatory identifications procedures (Leon-Velarde et al., 2009).

3 Emergence of Nanotechnology in Food Pathogen Monitoring

Nanotechnology is a fast evolving and hopeful innovative field, by means of which particles with at least one dimension less than 100 nm are developed and utilized in diverse applications, including but not limited to science, technology, engineering, and medicine. Nanotechnology has matured over time as an influential means to tailor the properties of materials. Nanodimensions of the nanomaterials provide specific physicochemical properties, such as strength, magnetism, chemical reactivity, electrical conductance, thermal, and optical properties that mimic neither the atom nor their bulk counterparts and that give rise to innovative applications. Reducing the particle size of materials is an efficient tool for improving their biological activity (Augustine et al., 2014c). The rapid development of nanostructured materials and nanotechnology has brought about a revolt in recent years, and revealed its vivid and far-reaching potential applications in day-to-day life. Nanosized materials have gained global research interest and brought into play applications in all areas of life. This technology exploiting nanomaterials has revealed extensive platforms for revolution in chemistry, biochemistry, biology, and medicine and hence coupled into them. It has proved its wide applications in biology and medicine, including drug delivery, protein purification, and medical imaging, bioseparation, clinical immunoassay, wound treatment, biomaterials, and tissue regeneration (Augustine and Rajarathinam, 2012; Augustine et al., 2014a,b,d,e, 2015).

Advances in nanotechnology and biotechnology recommend pioneering promise for the rapid and sensitive identification of harmful pathogens (Wang et al., 2011). Recently, nanobiosensor-based strategies have been adopted to detect food-borne pathogens (Vikesland and Wigginton, 2010). The fast advancement of bioelectronics to the molecular biology has promoted its manipulation for food safety and health-related applications. Today's lab-on-a-chip is all capable of recognition and medical diagnoses of pathogens causing diseases. Multiplex detection technologies like planar-array technologies for food analysis, with bioluminescent, fluorescent or chemiluminescent (CL) labels for detection, and direct (label-free) detection were extensively studied. Microarrays or similar multichannel platforms with short measurement times, reduced sample volumes, automation and high sensitivity are highly preferred for multianalyte screenings (Raz and Haasnoot, 2011).

To detect pathogenic microorganisms with high sensitivity, insitu gold nanoparticle (AuNP)–based technology has been developed. In this system, immuno-AuNP network-based ELISA biosensor integrated with a sample concentration step which relies on immunomagnetic separation and subsequent concentration of microbes. This biosensor was able to detect pathogens at extremely low numbers that is about 3 cells mL^{-1} of *E. coli* O157:H7 and *Salmonella typhimurium* in buffer solution and 15 cfu mL^{-1} of *S. typhimurium* and 3 cfu mL^{-1} of *E. coli* O157:H7 and in real sample conditions (Cho and Irudayaraj, 2013). By using combinations of multiple technologies like gold nanoparticle labeling, antibody affinity binding, and inductively coupled plasma mass spectrometry (ICPMS) detection technologies, it is possible to detect bacteria within a few minutes. Taking advantage of the signal amplification property of gold nanoparticles and owing to the high sensitivity of ICPMS detection, this combined assay was able to identify as few as 500 *E. coli* O157:H7 cells in 1 mL of sample (500 cfu mL^{-1}). Tests with nonpathogenic *E. coli* (DH5α, ATCC35218, and ATCC25922) showed high specificity of the assay for pathogenic *E. coli* O157:H7 (Li et al., 2010).

The field of nanobioelectronics, resulting from the merging of biology, nanotechnology, and electronics, is highly promising and has a major role in diverse fields, including but not limited to healthcare and food industries. The strategic developments in the field of nanoelectronics and biotechnology have opened up extensive opportunities in food pathogen monitoring. To address the need of the developing world, a paper-based bioassay using gold nanoparticle (AuNP) colorimetric probes have been developed. In the two individual prototype DNase I and adenosine-sensing assays, dark-color DNA-crosslinked AuNP aggregates were spotted on paper substrates. When target DNase I (or adenosine) solution was added to the spots, gold aggregates dissociated into dispersed AuNPs. This leads to the generation of an intense red color on paper within 1 minute. The combination of paper substrates and AuNP colorimetric probes makes this device inexpensive, portable, disposable, and easy to use (Zhao et al., 2008).

4 Application of Magnetic Nanoparticles in Food-Borne Pathogen Monitoring

The past decade witnessed increased developments and advancements in the field of nanosized magnetic particles. A large variety of appropriate methods have been utilized for the synthesis of magnetic nanoparticles of many different compositions, successful use of these magnetic nanoparticles in various fields, including magnetic fluids, catalysis, biotechnology/biomedicine, data storage, magnetic resonance imaging, and environmental remediation is highly dependent on the stability of the particles under a certain range of many different conditions (Lu et al., 2007). Magnetic nanoparticles form an emblematic group of nanoparticles that allow themselves to be manipulated by external magnetic field gradients. It is extremely beneficial to exploit the potential of magnetic nanoparticles (MNPs) for applications in the food industry. The size distribution, the shape, the surface reactivity, the ability to be easily manipulated are all key characteristics that multiply the glory of magnetic nanomaterials. Magnetic nanoparticles allow simple, quick, and low-cost separation of microbes from a relatively complex mixture by using an external magnetic field. The fact that magnetic separation allows speedy and cost-effective separation of magnetic carriers from the bacterial contaminated sample without any filtration or centrifugation procedures makes the exploitation of magnetic materials more preferred, advantageous, and practically effortless. Application of these technological principles to biological science call for high values of magnetization and size in the nanorange with narrow particle size distribution. This assures uniform material and elemental properties for the particles. Magnetic nanoparticles have made significant impact on detection of pathogens in the foodstuff. MNPs have been utilized to separate and concentrate specific target analytes from complex media composed of many biological and chemical entities. The application of MNPs is considered as a powerful technique for sample purification without complex instrumentation. Due to their robust superparamagnetic property, iron oxide nanoparticles have been utilized as a promising tracer for pathogen detection. Nuclear magnetic resonance (NMR) spectroscopy that measures the spin–spin relaxation time ($T2$) of water photons on MNP-tagged bacteria samples, which can be further enhanced by synthesizing iron-based MNP with high transverse relaxivity, is an influential measurement means for detection using magnetic particles.

There are many exceptional properties that make the nanoparticles promising in the food industry and monitoring of food-borne pathogens. Nanoparticles often possess astonishing properties as they are small enough to confine their electrons and produce quantum effects. The special properties that make nanoparticles relevant in the monitoring of food-borne pathogens and many other applications are high surface to volume ratio, high values of magnetization, superparamagnetic behavior, capability to be surface modified and safety or less toxicity.

4.1 High Surface/Volume Ratio

Nanoparticles are well established in the industry due to the unique physical and chemical properties they enjoy that are noticeably distinct from their bulk counterparts (Cao et al., 2012). The shortcoming of bulky particles is its small specific surface area available for the attachment of functional groups or immobilization of biomolecules. This is the motivation for relying on nanometer-sized particles, for the reason that they guarantee a satisfactorily high specific surface area available for the immobilization of reactive groups, enzymes, and other biologically active compounds and also boost the catalytic activity in phase transfer reactions. However, too miniature size of a particle may cause the reduction of magnetic property (Horak et al., 2007). High value of saturation magnetization and narrow particle size distribution are the requirements for bioseparation and detection applications. The magnetic susceptibility of magnetic microspheres needs to be as high as possible. In practice, a negotiation between active surface and sufficient magnetic properties must be achieved. The surface area to volume ratio increases as the particle size decreases. A clearer picture can be obtained from Fig. 9.1 where the surface area doubles when a cube divided into eight cubes having equal dimensions. The enhanced surface/volume ratio of nanoparticles offers more contact surface area for attaching biomolecules like antibodies and hence for achieving high sensitivity to pathogens. Atoms on the surface of a material are often more reactive than those in the center. Hence, a better surface area paves the way for more reactive material. Hence, the large surface-to-volume ratio of MNPs offers plentiful chemically active sites for biomolecule conjugation. It allows subtle design and surface engineering of the magnetic nanoparticles for various anticipated functions (Fang and Zhang, 2009).

The extremely high surface-to-volume ratio of nanoparticles offers more contact surface area that promises great adsorption capacity and efficiency for capturing pathogens and make them attractive in food-borne pathogen monitoring. The spatial dimension is much smaller than that of a bacterium, enables the attachment of numerous nanoparticles on a bacterial cell, and

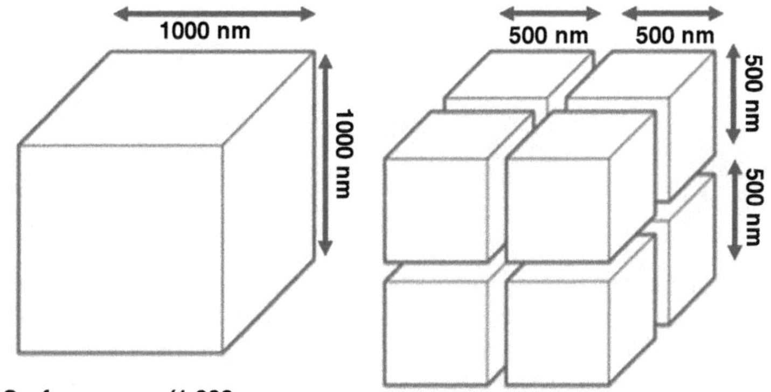

Figure 9.1. Scheme shows the increase in surface area with respect to the decrease in particle size.

assists in easy magnet-mediated partition of pathogens. Rapid kinetics of tiny nanoparticles in solutions when compared with their micrometer-sized counterparts, helps in quicker detection of pathogens (El-Boubbou et al., 2007).

Shan et al. (2014) showed that the difference in diameter of magnetic beads resulted in a significantly different coverage on the *L. monocytogenes* bacterial cells. For large magnetic beads, bacterial cells were bound to few magnetic beads. Magnetic beads with 180 nm size that showed high capture efficiency (CE) may be attributed to the large surface area of the nanoparticles available for immunoreaction. Fig. 9.2a clearly indicates that many magnetic beads with 180 nm size bound to the surface of a single cell. For biological binding reaction, small beads with high surface/volume ratio and high mobility have many chances to interact with bacterial cells in a solution, resulting in high CE.

4.2 High Magnetization Values

Nanomagnetism is an essential aspect of modern bacterial monitoring techniques. High values of magnetic susceptibility are the vital requirement for magnetic targeting and magnetic separation. Magnetic nanoparticles are very supportive in isolation of bacteria from food matrices and also help to concentrate isolated cells to a tiny volume. Iron oxide nanoparticles (IONPs) are highly magnetic materials with particle sizes within 100 nm. Magnetite,

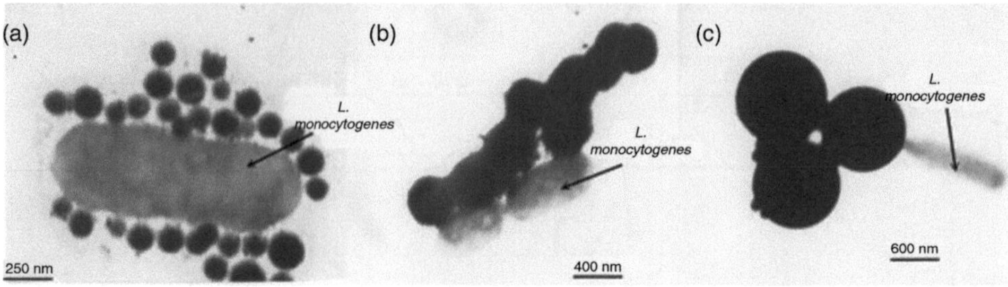

Figure 9.2. Transmission electron microscope observations of *Listeria monocytogenes* interacting with different sizes of magnetic beads. (a) 180 nm (30,000×), (b) 350 nm (20,000×), and (b) 1150 nm (10,000×). Reproduced with the permission from Shan, S., Zhong, Z., Lai, W., Xiong, Y., Cui, X., and Liu, D., 2014. Immunomagnetic nanobeads based on a streptavidin-biotin system for the highly efficient and specific separation of *Listeria monocytogenes*. Food Control, 45, 138–142. With permission from Elsevier.

Fe_3O_4, and maghemite, γ-Fe_2O_3 (oxidized form), the two main forms of iron oxide nanoparticles, have attracted great interest due to their super paramagnetic properties and their potential applications in many fields.

Magnetism originates from the spin of electron. Uncompensated electron spins are the reason why individual atoms possess local magnetic moments. A ferromagnetic crystal exhibits a stable and switchable magnetization that arises from the quantum mechanical phenomenon of exchange (Martin and Ramesh, 2012).

Ferrites are mixed metal oxides with iron oxides as their foremost constituent. Ferrites have been a focus of great interest in many technical and engineering fields. There are three different crystal types in which ferrites crystallize namely spinel, garnet and magnetoplumbite. Magnetite, Fe_3O_4 (Chockalingam et al., 2010), is a well-known magnetic iron oxide that has a cubic inverse spinel structure. The inverse spinel is then one in which oxygen ions form an fcc closed packed structure, the divalent (A^{2+}) metal ions occupy half of the number of octahedral sites and the trivalent (B^{3+}) metal ions are equally divided between the tetrahedral and the remaining octahedral sites. The hopping of electrons between Fe^{2+} and Fe^{3+} ions in the octahedral sites, cause magnetite to be a significant class of half-metallic materials. These magnetic nanoparticles are dispersed into suitable solvents, after proper surface coating, to form ferrofluids. Ferrofluids interact with an external magnetic field and promote separation of microbes. Even if cobalt and nickel are also highly magnetic as iron, they are noxious and easily oxidized.

The magnetic properties of nanomaterials have a great role to play in determining its monitoring and diagnostic efficiency. The magnetic properties of nanomaterials can be studied by techniques like vibration sample magnetometry (VSM) and superconducting quantum interference device (SQUID), which can determine feeble magnetic signals, even in the human body (Wu et al., 2015). The behavior of the material to the magnetic field is a very crucial factor. The ferromagnetic, paramagnetic, antiferromagnetic, and ferrimagnetic properties and magnetization, can be estimated from the hysteresis loops (M–H) and zero-field cooled/ field cooled (ZFC/FC, M–T) curves. The saturation magnetization (M_s), remnant magnetization (M_r), and coercivity (H_C) can be obtained from the hysteresis loops. The superparamagnetic nature of the magnetic nanoparticles can be identified from the Mossbauer spectra analysis.

4.3 Superparamagnetism

Superparamagnetism is a kind of magnetism that appears in minute ferromagnetic or ferrimagnetic nanoparticles. If the size of nanoparticles is small enough, magnetization can randomly flip direction under the influence of temperature. SPIONS are superparamagnetic iron oxide nanoparticles below a critical particle size of 15 nm that shows this phenomenon. These are single magnetic domain materials, that is, they hold a condition of uniform magnetization at any field and large values of saturation magnetization. The high values of magnetization under applied magnetic field, cause large difference in susceptibility between the particles and surrounding medium, and consequently microscopic magnetic field gradients. SPIONs are of meticulous inquisitiveness because they do not keep or hold any magnetism on elimination of magnetic field. In case of the magnetization curve (M–H curve) of superparamagnetic nanoparticles, the forward and backward magnetization curves overlap completely as they exhibit no (ignorable) hysteresis.

Biocompatible magnetic nanoparticles with superparamagnetic properties enjoy wide demand for pathogen monitoring and seclusion applications due to its appropriate surface designs, smaller size, and capability of attachment to targeting ligands, enzymes, or antibodies specific to pathogens (Ling et al., 2011). Due to their small size, these particles can be easily eliminated in the course of outpouring or renal clearance.

The ordering of magnetic moments in ferromagnetic, antiferromagnetic, and ferrimagnetic materials decreases with increasing temperature. Ferromagnetic and ferrimagnetic materials become disordered and lose their magnetization beyond the Curie temperature and antiferromagnetic materials lose their magnetization beyond the Néel temperature. Magnetite is ferrimagnetic at room temperature and has a Curie temperature of 850 K.

4.4 Surface Modifications

The ultimate efficacy of the solid particles in microbial detection is determined principally by the success of the controlled, optimal, and nontoxic surface coating or attachment of biomolecules, which enables pathogen identification and targeting. Though quantum dots (QDs), owing to their exceptional optical characteristics like wide excitation spectra, precise emissions, and quench-resistance, are treated as a substitute for conventional fluorophores in bioassays. However, QDs are smaller, more costly, and their surface modification is not trivial. The surface engineering of the nanoparticles is promising under defined, optimized conditions. Magnetic nanoparticles can bind with biorecognition molecules like, for instance, enzymes, antibody, aptamers, lipids, proteins, enzymes, and carbohydrates, according to the specific design for explicit interactions. Superparamagnetic nanoparticles can effectively be encapsulated or functionalized with antibodies that are specific to intended microbes or pathogens, in spite of the fact that the grain size range of nanomagnetic particles is characteristically one to two orders of magnitude slightly lesser as compared to that of the pathogens. Through appropriate surface alteration, nanomagnetic particles can be functionalized by special functional groups like —OH, —COOH, $—NH_2$, which facilitate further attachment of various bioactive molecules for special sensing or detection applications. Surface-activated nanoparticles are successfully used in techniques like magnetic separation, and biosensor-based multidetection systems. Magnetic nanoparticles experience lesser diffusion limit in solutions. The surface modification influences the particle mobility interrelated to particle size and solution viscosity by Stokes-Einstein equation (Miller, 1924), as given by

$$D = \frac{KT}{6\pi\eta r}$$

Where, D is the diffusion constant, r is the radius of a diffusing nanoparticle, T is the absolute temperature and η is the viscosity of the diffusion medium. These surface functionalized nanoparticles can thus be directed to an exact intended location like pathogens in a viscous food product or a microbial cell component using an external magnetic field.

4.5 Safety

Safety or toxicity of magnetic nanoparticles is an important factor of concern, for their food-related applications. Iron-based

MNPs are most utilized because iron has an inoffensive toxicity sketch. In addition, iron is a fundamental element in human bodies and can be added to the body's iron store after particle degradation. Despite the fact that Manganese is also another element that is important for survival in human bodies, its endurable value is much poorer than that of iron. Other highly magnetic elements like chromium, cobalt, nickel, and gadolinium are extremely toxic, and hence intolerable for the food industry and the environment.

Oxides of iron, for instance say ferrites, are highly magnetic, less toxic, biocompatible, and will not exhibit residual magnetism when the external magnetic field is removed, and these features make them graceful and strong candidates for the detection of food-borne pathogens. The unique physical and chemical properties of iron oxides, along with low toxicity level, particularly superparamagnetic magnetite (Fe_3O_4) nanoparticles make them the most the extensively exploited ones for applications in food analysis (Sundaram et al., 2012). The low systemic toxicity or cytotoxicity shown by superparamagnetic iron oxide nanoparticles at higher doses (>100 μg mL^{-1}) in the human body, make them superb candidates for pathogen detection and separation from food products.

5 Multifunctional Magnetic Nanoparticles

Great efforts were put into the progress of multifunctional nanoparticles that offer multiple functionalities that are not attainable in individual components. Multifunctional magnetic nanoparticles (Fang and Zhang, 2009), materials that concurrently hold more than one functional component form an imperative category of nanomaterials (Song et al., 2013). It is a lively research area with the potential to impact a wide range of technological applications. Functionalization of magnetic nanoparticles by affixing organic functional groups to them is the most probable strategy to attain multifunctional nanomaterials that proves useful in isolation/detection of hazardous pathogens (bacteria/viruses) (Gu et al., 2006).

Novel nanomaterials that merge multiple functions, for example, an optical signature with other relevant attributes like magnetism that allow optical addressability for tracking or monitoring of particles in addition to other properties have radically expanded the applicability of MNPs. A very useful strategy for imparting optical properties at the nanoscale is the integration of noble metals and their associated localized surface plasmons into the particle or structure. Magnetic-plasmonic nanoparticles (Brullot et al., 2012), combining magnetic and plasmonic (quantum of plasma oscillation) components, are promising candidates for wide range of applications including pathogen monitoring. Development of a thin layer of gold (Au) shell around a functional nanoparticle core provides a practical and highly general approach for providing optical specificity to almost any type of nanoparticle. Au shell layers attribute a strong plasmon resonant optical response to the nanoparticle, with a resonant frequency that is determined by geometry, the dielectric properties of the nanoparticle core, and the surrounding medium of the nanoparticle. In addition, the Au shell layer gives almost chemically inert surface layer that can be further functionalized to enhance solubility in various media, promote biocompatibility and preserve the properties of the core material.

Nanoparticle plasmonics is a fast growing and highly promising research area that deals with the synthesis and optical characterization of metal nanoparticles with tunable plasmon resonances over visible-near IR region of the spectrum. Nanoparticles with multifunctionalities like magnetization and absorption in the near-infrared (NIR) region of electromagnetic spectrum have warranted interest. These crossbreed nanomaterials have promising applications in targeted drug delivery, diagnosis, bioseparation, MRI, and NIR photo-thermal ablation. In addition, they have reasonable significance in monitoring of food-borne pathogens. Their tunable magnetic and plasmonic properties make them superior candidates for multiplex detection and separation of microbes.

6 Principle of Magnetic Nanoparticle-Based Separation of Food-Borne Pathogens

Magnetic nanoparticles conjugated with various biomolecules like antibodies, carbohydrates, antibiotics, and other binding proteins specific for pathogens, to enable the capture of a particular target pathogen. Magnetic nanoparticles and the ligand-receptor interactions are exploited for quick detection of microbes. While adding magnetic nanoparticles with appropriate bacterial ligands or functional groups, they will bind to the bacteria and form bacterium nanoconjugates. The ligands covalently bind to the magnetic nanoparticles and they are linked by the high affinity between magnetic nanoparticles and bacteria. The magnetic dipole interactions bring together the magnetic nanoparticles under an external magnetic field. Thus, the bacteria bound to magnetic nanoparticles (bacterium nanoconjugates) will also move

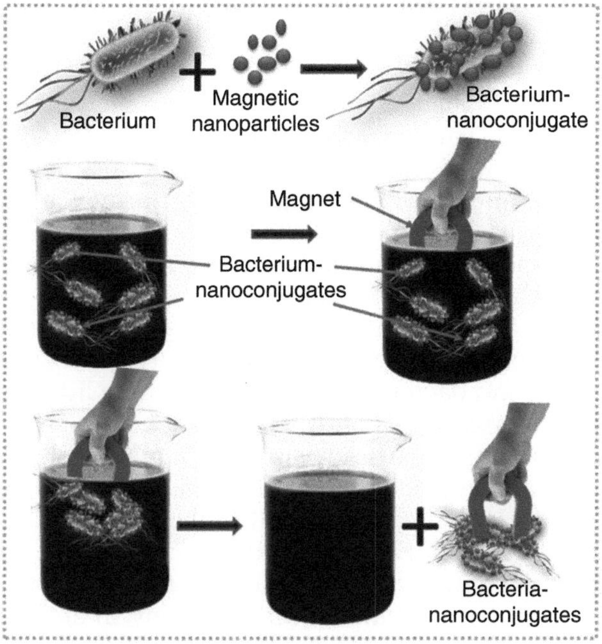

Figure 9.3. Scheme showing the magnetic nanoparticle based separation of bacteria from a contaminated solution.

toward the magnetic field (Fig. 9.3). The bacterium nanoconjugates attached to the magnet can be removed and decontaminated by appropriate sterilization techniques.

7 Magnetic Nanoparticle-Mediated Microbial Detection Techniques

Though there are techniques like PCR for sensitive and rapid detection of pathogens at small concentrations, without lingering procedures like incubation or amplification, there are no practical techniques to assess and detect bacteria at concentrations less than 10^2 cfu mL^{-1} (Chockalingam et al., 2010). In this scenario, BMNPs have a great role to play in monitoring pathogens at low concentrations. The success of isolation of pathogens in food using magnetic nanoparticles depends on smaller size nanoparticles, magnetic properties, and also the nature of attached biomolecules. The following sections will briefly discuss the recent advancements in magnetic nanoparticle-based approaches and technologies to detect and separate pathogenic microorganisms from food products.

7.1 Detection of Vancomycin-Resistant Enterococci (VRE) and other Gram-Positive Bacteria

Biofunctional magnetic nanoparticles can be used to confine and concentrate vancomycin-resistant enterococci (VRE) and other Gram-positive bacteria. Magnetic nanoparticles smaller than 10 nm assure excellent application performance because of their large surface/volume ratios and easy entry into cells, compared to magnetic beads in micrometer range used in biological separations (Gu et al., 2003). The magnetic dipole-dipole interactions that amass the magnetic nanoparticles under external magnetic field, is exploited here. The great affinity between magnetic nanoparticles and bacteria, mediated by definite ligand-receptor interactions that occur when the ligands are covalently bonded to the magnetic nanoparticles, also play a major role and assist in the nanoparticle mediated fast separation of pathogens. Even bacteria at very low concentrations (less than 10^1 cfu mL^{-1}) can be detected within an hour (Gu et al., 2003).

Fig. 9.4 a, b shows both the structure of the iron oxide-based FePt – Van nanoparticles linked with vancomycin and that of the control (FePt – NH_2 nanoparticles). From the figure, it is clear that vancomycin linked nanoparticles could attach to the bacteria and the nanoparticle-bacteria conjugates move toward the applied magnetic field. NH_2 functionalized control nanoparticles could not attach with the bacteria. Fig. 9.4c shows the binding of van to the terminal peptide, D-Ala-D-Ala, on the cell wall of a Gram-positive bacterium via hydrogen bonds.

A US patent has been filed for biofunctional magnetic nanoparticles for pathogen detection (Xu et al., 2010). This invention demonstrates a technique for detecting pathogens employing biofunctional magnetic nanoparticles. The principle employed or the steps involved in the process are: (1) adequate quantity of biofunctional magnetic nanoparticles are brought into contact with an appropriate amount of sample for a suitable period of time to allow the development of complexes between the pathogens in the sample and the nanoparticles; (2) An external magnetic field is used to amass the bacterial-nanoparticle complexes and the detection of the complexes.

7.2 Immunomagnetic Methods for Pathogen Detection and Separation

IMS has become a very well-admired approach for pathogen detection. It is appropriate to easily accomplish the task of fast isolation of the target organisms, by this physical selective

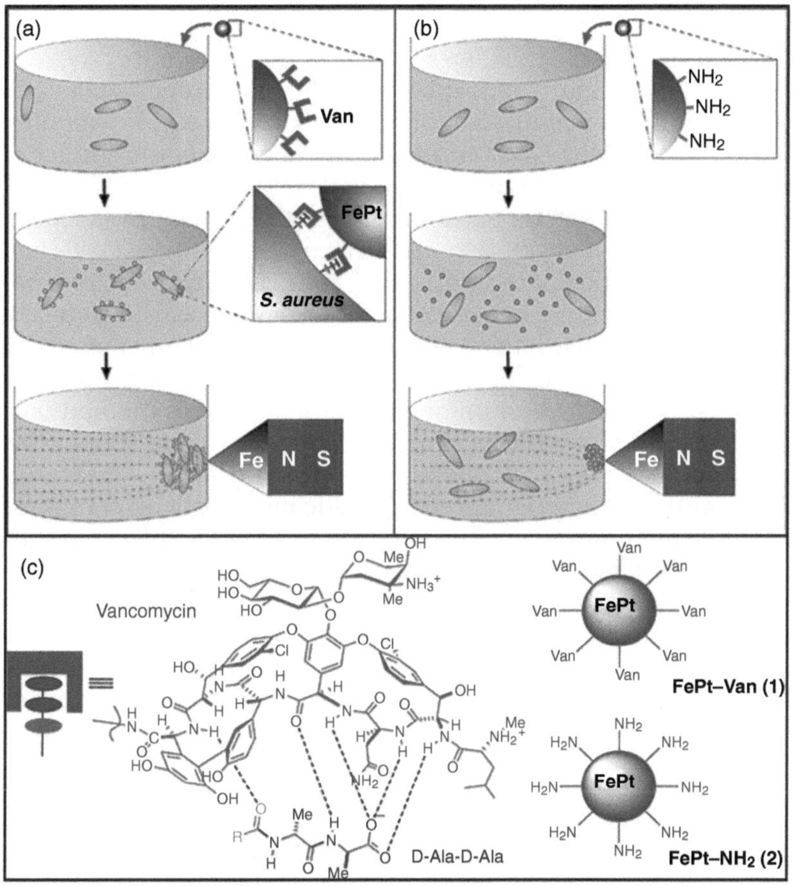

Figure 9.4. Schematic representation of the capture of bacteria by vancomycin-conjugated magnetic nanoparticles. (a) via a plausible multivalent interaction and the corresponding control experiment (b). (c) shows the interaction between Van and the terminal peptide, D-Ala-D-Ala, on the cell wall of a Gram-positive bacterium via hydrogen bonds. Reproduced with the permission from Gu, H., Ho, P.L., Tsang, K.W., Wang, L., and Xu, B., 2003. Using biofunctional magnetic nanoparticles to capture vancomycin-resistant enterococci and other Gram-positive bacteria at ultralow concentration. J. Am. Chem. Soc., 125(51), 15702–15703. Copyright (2015) American Chemical Society.

enrichment procedure when coupled with nanotechnology. Here, immunomagnetic nanoparticles are used to isolate bacteria from food products based on the antigen-antibody interactions taking place between the immobilized antibodies on the magnetic nanoparticles and the bacteria. Immunomagnetic nanoparticles are prepared by coupling magnetic particles with antibody specific to the microbes to be isolated from food products. Magnetic nanoparticles has opened new scope and pace in IMS of food-borne pathogens (Tamer et al., 2013). Assay sensitivity, rapidness, and accuracy are the advantages of using magnetic nanoparticles. Super paramagnetic nanoparticles (less than 10 nm in diameter) with a specific shape and size were prepared and coupled to various proteins. It directly traps target biomolecules from liquid media with the help of selective probe, such as antibodies or aptamers, immobilized onto magnetic nanoparticles.

7.2.1 Biofunctional Immunomagnetic Nanoparticles for the Detection of E. coli

The adenosine triphosphate (ATP) bioluminescence can estimate the concentration of viable bacteria within minutes. Biotin-labeled *anti-E.coli antibodies* are anchored onto surface of nano*magnetic* particles to fabricate BMNPs and then used to capture the target bacteria *E. coli* from samples (Cheng et al., 2009). The BMNPs strongly bind to the surface of entire *E. coli* cells and expose high capacity to imprison *E. coli*. The target bacteria become magnetic and hence could be separated easily from the sample solution with the assistance of an external magnetic field. The concentration of *E. coli* captured by the BMNPs was then detected by an ATP bioluminescence method. This method possesses elevated specificity, short detection limit, and small assay time, due to the remarkable features of the BMNPs. These anti-*E. coli* antibody functionalized magnetic nanoparticles combined with ATP bioluminescence was applied to detect *E. coli* in pasteurized milk and apple juice (Cheng et al., 2009). Transmission electron microscope (TEM) images of the *E. coli* captured by BMNPs (Fig. 9.5) showed that the BMNPs were firmly bound to the surface of entire *E. coli* cells and further confirmed by Fourier transform infrared (FTIR) spectrum and the ATP bioluminescence intensity reading. It is evident from the bioluminescence intensity readings

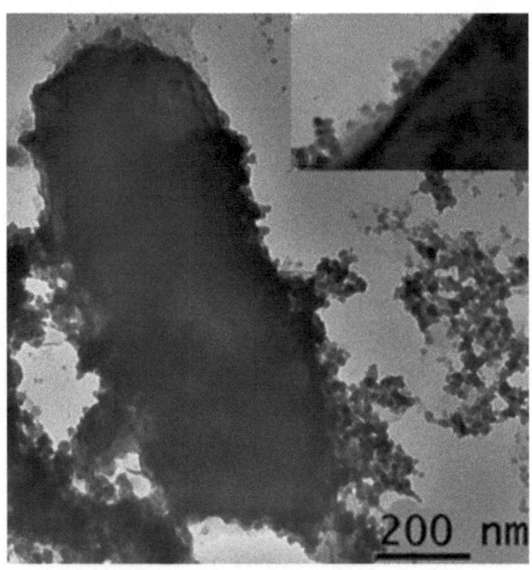

Figure 9.5. Electron microscopic image of *E. coli* binding with BMNPs. Inset is the magnified TEM image of *E. coli* binding with BMNPs. Reproduced with the permission from Cheng, Y., Liu, Y., Huang, J., Li, K., Zhang, W., Xian, Y., and Jin, L., 2009. Combining biofunctional magnetic nanoparticles and ATP bioluminescence for rapid detection of *Escherichia coli*. Talanta, 77(4), 1332–1336. With permission from Elsevier.

Table 9.2 Capture Efficiency of BMNPs to *E. coli*.

E. coli Sample (cfu mL^{-1})	ATP Bioluminescence of Total *E. coli* (RLU)	ATP Bioluminescence of *E. coli* Captured by BMNPs (RLU)	Capture Efficiency (%)
2×10^4	1,650	1,573	95.3
2×10^5	7,930	7,210	90.9
2×10^6	31,203	24,971	80.0

Capture efficiency was described as the ratio of *E. coli* captured by BMNPs to original *E. coli* sample in the form of ATP bioluminescence. ATP bioluminescence intensity was given as relative light units (RLUs).

Source: Reproduced with the permission from Cheng, Y., Liu, Y., Huang, J., Li, K., Zhang, W., Xian, Y., and Jin, L., 2009. Combining biofunctional magnetic nanoparticles and ATP bioluminescence for rapid detection of Escherichia coli. *Talanta, 77(4), 1332–1336. With permission from Elsevier.*

of the total *E. coli* in the sample and the *E. coli* captured by BMNDs that most of the bacteria were captured by BMNPs (Table 9.2). This method has thus proved its potential to be employed for testing of any other kind of food or drink sample.

7.2.2 Detection of E. coli *O157:H7 Using Antibody Conjugated Magnetic Iron-Dextran Nanoparticles*

Magnetic iron-dextran nanoparticles were used for the isolation of *E. coli* O157:H7 from food products (Duan et al., 2005). By integrating the magnetic particles with antibody specific to *E. coli* O157:H7, it is feasible to achieve the goal of rapid isolation of the target organisms. This method detects *E. coli* with elevated sensitivity and specificity. Magnetic iron-dextran nanoparticles were prepared and tagged with antiserum against *E. coli* O157: H7. *E. coli* O157:H7 can be isolated from food samples under optimum conditions, in about 15 min with high sensitivity of 10^1 cfu mL^{-1}. The technique is so simple in the sense that it can be accomplished in laboratory and does not require highly skilled professionals. Ferromagnetic iron-dextran nanoparticles can be synthesized using ferric ($FeCl_3 \cdot 6H_2O$) and ferrous ($FeCl_2 \cdot 4H_2O$) ions and conjugated with dextran polymers. Anti-O157:H7-specific antibody can be prepared and then a required amount of antibody can be added to the particle suspension and mixed thoroughly. Definite quantity of magnetic nanoparticles were dispersed into the sample of *E. coli* O157:H7 suspension, incubated, and rotated to bring the samples and the particles into closer contact. The particles were then magnetically separated by introducing a magnetic plate to the side wall of the sample for some time to concentrate the particles into a pellet. Numerous bacteria were attached to one nanomagnetic particle simultaneously and appear as one colony, forming a particle-bacteria complex. Magnetic fields of about 0.23 T could draw almost all the particle-bacteria complex to the side wall of the tube in about 2 min. The number of the bacteria in samples is then counted by standard procedure and confirmed by biochemical or molecular biological method.

The technique of IMS possesses the advantage in the sense that it gives fewer numbers of false negative results, compared to other isolation methods. Magnetic microspheres (MMS), which is a core of metal ions enclosed within macromolecule polymer. Magnetic iron-dextran nanoparticles are advancement over MMS that they are awfully stable and do not aggregate during bacteria isolating procedures. This is largely due to their small size and the substantial dextran coating, which surrounds the iron oxide core. Also, nonspecific binding of other bacteria to the particles was not observed.

7.2.3 Immunomagnetic Detection of Listeria *from Food Products*

L. monocytogenes is the bacterium that causes the infection called listeriosis. It is an acute fatal disease characterized by gastrointestinal problems to disorders of like meningitis, abortion of pregnancy, septicemia, and even death. Milk and other milk products like cheeses, ice cream, meat, and other ready-to-be-served food products, serve as medium for transfer and spread of listeriosis. *L. monocytogenes* that are found in most animals can grow at adverse conditions and also at refrigeration temperatures. This bacteria are responsible for an estimated 1600 illnesses and 260 deaths in the United States annually. Listeriosis is the third leading cause of death among food-borne bacterial pathogens, with fatality rates exceeding even *Salmonella* and *Clostridium botulinum*. The development of molecular methods and immuno-based techniques for detection of pathogenic *Listeria* in food products has been challenging due to the presence of assay inhibiting food components.

An immunomagnetic immunofluorescent method was developed for the rapid detection of *L. monocytogenes* by Duffy et al. (1997). This technique involved the enrichment of suspect sample at 30°C overnight. *L. monocytogenes* cells were isolated from the enriched sample using IMS and were subsequently visualized using an immunofluorescent microscopy technique. This approach was employed for the detection of listeria cells from pure culture, inoculated beef mince samples and naturally contaminated beef samples. Here, a detection level of approximately 1×10^3 cfu mL^{-1} was achieved. When compared with conventional detection techniques no false negatives or positives were recorded. Isolation of the listeria cells by surface adhesion involved the simple dipping of a membrane attached to a microscope slide into the enriched sample for 10 min.

Conventional multiplex polymerase chain reaction (mPCR) has been widely used for the simultaneous detection of various target bacteria in food products like vegetables. Enrichment periods required to improve the sensitivity of this method make it less convenient. Mao et al. (2016) reported, large-volume (10 mL) immunomagnetic separation (IMS) combined with mPCR for the fast detection of *L. monocytogenes* and *L. ivanovii* in lettuce without further enrichment process. Under optimum conditions, the capture efficiency of large-volume IMS for *L. monocytogenes* and *L. ivanovii* was greater than 90% when the content of target bacteria was less than 10^6 cfu mL^{-1} in pure culture, and was more than 80% when the amount was below 10^5 cfu mL^{-1} in the lettuce samples. Effective limits of detection of IMS combined with mPCR assay reached as low as 1.0 cfu mL^{-1} in pure culture and 10 cfu g^{-1} in lettuce. Fig. 9.6 gives a schematic illustration of IMS combined with mPCR assay for the simultaneous and fast detection of *L. monocytogenes* and *L. ivanovii*. Magnetic nanobeads (MNBs) were functionalized with streptavidin (SA-MNBs) and conjugated with biotinylated anti-listeria monoclonal antibodies (biotin- anti-listeria-mAbs). Resultant antibody-linked nanobeads were incubated with bacterial suspension, separated the bacteria and the bacterial DNA was amplified by PCR. Moreover, this IMS–mPCR method also showed excellent discrimination of other common nontarget pathogens, and to complete the assay

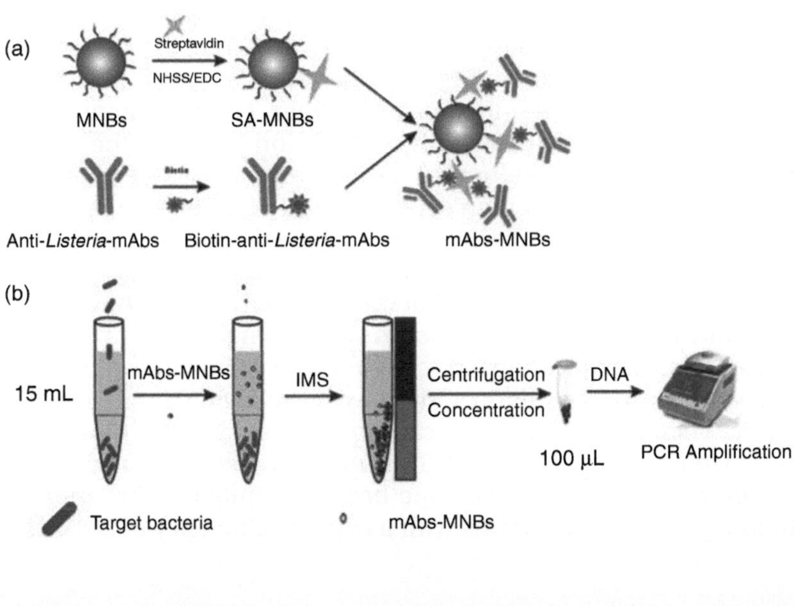

Figure 9.6. Schematic illustration of large-volume IMS combined with mPCR assay for the simultaneous detection of *L. monocytogenes* and *L. ivanovii*. Reproduced with permission from Mao, Y., Huang, X., Xiong, S., Xu, H., Aguilar, Z.P., and Xiong, Y., 2016. Large-volume immunomagnetic separation combined with multiplex PCR assay for simultaneous detection of *Listeria monocytogenes* and *Listeria ivanovii* in lettuce. Food Control, 59, 601–608. With permission from Elsevier.

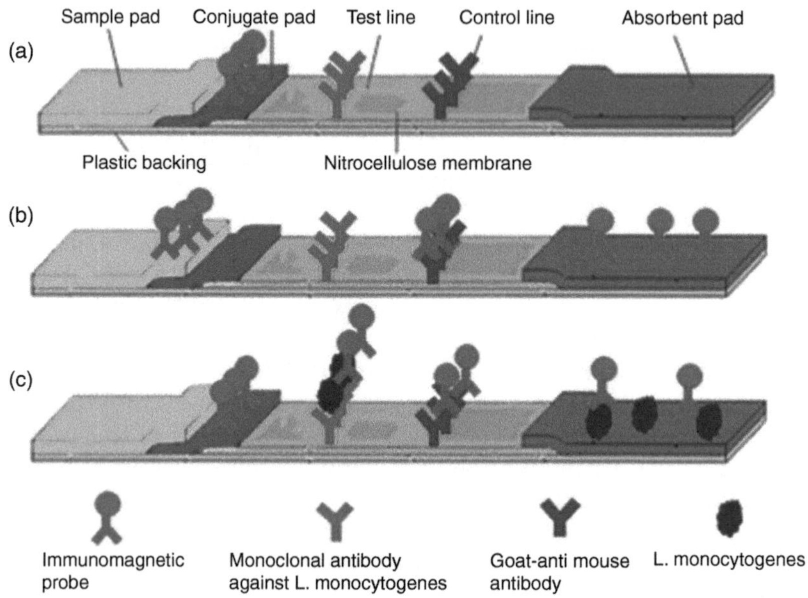

Figure 9.7. Schematic representation of the superparamagnetic LFIA system. Cartoon schematic showing the structure of the LFIA strip (a), negative result schematic when there was no *L. monocytogenes* in the sample (b), positive result schematic when there was *L. monocytogenes* in the sample (c). Reproduced with the permission from Shi, L., Wu, F., Wen, Y., Zhao, F., Xiang, J., and Ma, L., 2015. A novel method to detect *Listeria monocytogenes* via superparamagnetic lateral flow immunoassay. Anal. Bioanal. Chem., 407(2), 529–535. With permission from Springer.

only took a total time of less than 7 h, including sample preparation, large-volume IMS, and mPCR assay.

A strip test system combining immunomagnetic separation with lateral flow immunoassay (LFIA) was developed for the detection of *L. monocytogenes* (Shi et al., 2015). Here, a pair of matched monoclonal antibodies was utilized to fabricate a sandwich immunoassay, where superparamagnetic particles were coupled with one of the antibodies as a labeled antibody to bind with the target bacteria, while the other antibody was immobilized on the detection zone (Fig. 9.7). After an incubation for 20 min, the strips were analyzed by an instrument that could detect the magnetic signal of the immunocomplex in a magnetic field. The sensitivity determination showed that the limit of detection (LOD) of the superparamagnetic LFIA system for *L. monocytogenes* was 10^4 cfu mL^{-1}, which was much lower than conventional LFIA. This system could distinguish *L. monocytogenes* when *Salmonella, E. coli* O157:H7, or three types of harmless listeria strains were tested.

Shan et al. (2014) showed that the time required for the incubation and separation of *L. monocytogenes* using 0.1 mg of 180 nm immunomagnetic nanobeads coupled with monoclonal antibody using a streptavidin-biotin system was just 45 and 3 min, respectively. They could achieve a capture efficiency of 94.12% when the magnetic density was 1.5 T, using 1 mL of PBS as separation medium. The immunomagnetic beads exhibited highly specific binding with listeria strains with a capture efficiency of 90%. The system showed very low binding with nontarget bacteria with a capture efficiency of less than 5%. The capture efficiency of this method against *L. monocytogenes* in ground beef and milk samples were 25.1% and 91.2%, respectively.

7.2.4 Immunomagnetic Separation (IMS) Coupled with Polymerase Chain Reaction (PCR) to Detect Bacteria

Taking into advantage of nanoparticle-based IMS and real-time PCR has proved as efficient for the fast and quantitative detection of *L. monocytogenes*. Carboxyl modified magnetic nanoparticles were covalently attached to rabbit anti-*L. monocytogenes* through the amine groups and the nanoparticle-based IMS in combination with real-time PCR was used for detecting *L. monocytogenes* from intentionally contaminated milk. Calculated numbers of cells from the means of threshold cycles (C_T) of PCR amplification curves were further compared to those from conventional plate counts to determine the consistency of quantitative data. Cell numbers calculated from C_T values were 1.5–7 times higher than those derived from plate counts. Immunomagnetic nanoparticles in combination with real-time PCR has improved the sensitivity of *L. monocytogenes* detection from both nutrient broth and milk samples (Yang et al., 2007). Due to the demonstrated higher capture efficiencies and ability of removing PCR inhibitors by using IMNPs, the application of IMNPs in a continuous flow IMS format plus real-time PCR could achieve the long-term goal of rapid single bacterial cell detection in food products (Yang et al., 2007).

An IMS of bacteria directly from ham followed by extraction of DNA and subsequent detection using a novel multiplex PCR was reported by Hudson et al. (2001). The PCR approach utilized one primer pair targeted at the listeriolysin O gene of *L. monocytogenes* and the other pair for a region of the 23S rRNA genes of listeria, giving products of 706 and 239 bp, respectively. The combination of IMS and PCR was demonstrated to be capable of detecting as few as 1·1 *L. monocytogenes* cells g^{-1} in a 25-g ham sample within 24 h.

7.2.5 Immunomagnetic Detection of Salmonella *Bacteria*

A method for quick and sensitive detection of *Salmonella* bacteria making use of 3-D immunomagnetic flow assay was developed by Lee et al. (2014). In this method, binding between the antibody-functionalized magnetic nanoparticle clusters (AbMNCs) and *Salmonella* bacteria was facilitated by magnetically immobilizing the AbMNCs on the surface of a 3-dimensional microchannel in a hollow cylinder. The strength of the magnetic field was maximized by stacking permanent magnets with spacers inside the cylinder,

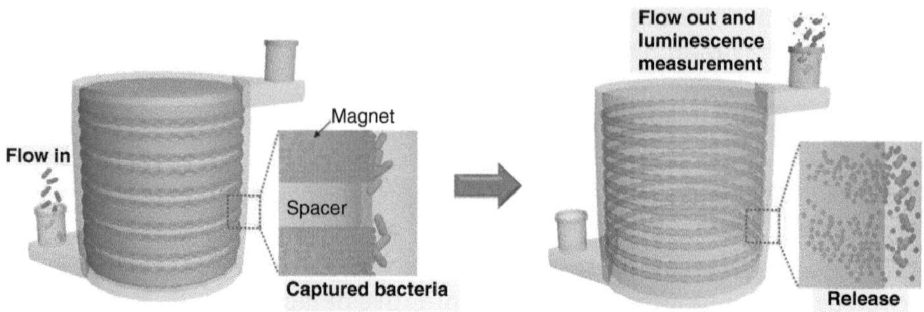

Figure 9.8. Schematic illustrations of 3D immunomagnetic flow assay. Reproduced with the permission from Lee, W., Kwon, D., Chung, B., Jung, G. Y., Au, A., Folch, A., and Jeon, S., 2014. Ultrarapid detection of pathogenic bacteria using a 3D immunomagnetic flow assay. Anal. Chem., 86(13), 6683–6688. Copyright (2015) American Chemical Society.

which allowed the capture of AbMNC-bacteria complexes. *Salmonella* bacteria were detected by this method based on a 3-D-printed microfluidic device and a luminescent ATP detection kit. A schematic representation of the microchannel device is shown in Fig. 9.8. The antibody-functionalized magnetic nanoparticle clusters (AbMNCs) were magnetically immobilized on the surface of a 3- dimensional microchannel in a hollow cylinder and thus binding between the AbMNCs and *Salmonella* bacteria was made possible (Lee et al., 2014). Presence of permanent magnets in a pile with spacers inside the cylinder maximizes the strength of the magnetic field and allowed for the capture of the AbMNC-bacteria complexes. The highly superior detection limit better than 10 cfu mL^{-1} that was achieved for the *Salmonella* bacteria in spiked lettuce solution, makes this technique highly advantageous. The most outstanding and notable feature is that the inspection of a 10 mL food sample was possible in just 3 min, including 24 s for the conjugation, 1 min for the rinsing, and 1 min for the detection steps. The assay time required to scrutinize a bigger volume of food sample solution was only 4 min, which includes 2 min for the conjugation step.

7.2.6 Biofunctionalized Magnetic Nanoparticles for the Immunomagnetic Detection of Multiple Bacteria

Anti-*E. coli* O157:H7 or anti-*S. typhimurium* antibodies functionalized magnetic nanoparticles that can selectively bind to their target microorganisms were used to isolate *E. coli* O157:H7 and *S. typhimurium* separately from a cocktail of bacteria and from food products. The pathogens were then detected using label-free IR spectrum. This method took a detection time less than 30 min in foodstuffs with a detection limit of 10^4–10^5 cfu mL^{-1}.

This combination of such functionalized magnetic nanoparticles and IR spectroscopy provides specificity through spectroscopic fingerprinting and high selectivity through bacteria specific antibodies with an in-built sample extraction step and could be applied for on-site food-borne pathogen monitoring (Ravindranath et al., 2009). Target differentiation achieved through this protocol will help to facilitate adaptation of various portable spectrometers integrated to a biosensing step to improve rapid pathogen detection and better protection for the food supply chain from future outbreaks.

7.3 Core-Shell Structured Iron Oxide-Gold (Fe_3O_4–Au) Magnetic Nanoparticles Modified with Self-Assembled Monolayers (SAMs) for *E. coli* Detection

Tamer et al. (2013) reported the preparation and characterization of spherical core-shell structured iron oxide-gold (Fe_3O_4–Au) magnetic nanoparticles, which were modified with two component self-assembled monolayers (SAMs) consisting of 3-mercaptophenylboronic acid (3-MBA) and 1-decanethiol (1-DT). Surface modification with SAMs against glucoside moieties on the surface of bacteria made efficient magnetic separation. Bacteria capturing efficiency by different molecular recognition agents 3-MBA, 1-DT and their combinations was studied. The best capturing efficiency of *E. coli* was achieved with the mixed monolayer of 3-MBA and 1-DT-modified nanoparticles. Moreover, molecular specificity and selectivity were also confirmed by comparing the surface-enhanced Raman scattering (SERS) spectrum of *E. coli*-nanoparticle conjugates with bacterial growth in culture media (Tamer et al., 2013).

7.4 Amine Functionalized Magnetic Nanoparticles to Capture Bacteria

Magnetic nanoparticles functionalized with amine groups (AF-MNPs) are highly efficient in capturing of bacteria. The surface layers of bacterial cells are negatively charged. Thus, positively charged amine-functionalized magnetic nanoparticles (AF-MNPs) will bind with bacterial cells (Huang et al., 2010). A schematic representation of the synthesis of AF-MNPs for pathogen separation is shown in Fig. 9.9. AF-MNPs could attach to at least eight diverse species of Gram-positive and Gram-negative bacteria. The pendant amine groups on the surface of the magnetic nanoparticle

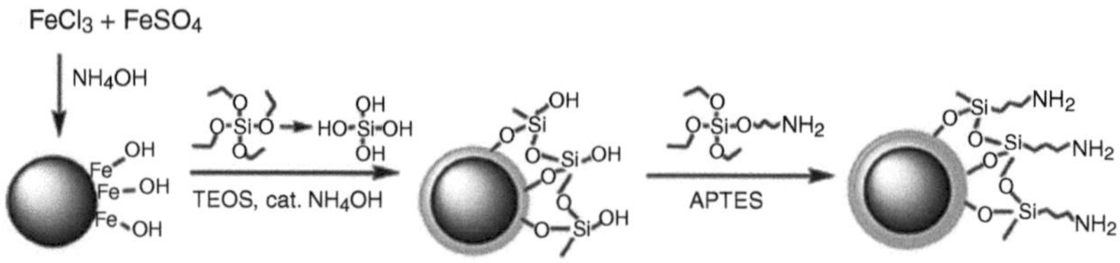

Figure 9.9. Schematic diagram for the synthesis of AF-MNPs for pathogen separation. Reproduced with the permission from Huang, Y.F., Wang, Y.F., and Yan, X.P., 2010. Amine-functionalized magnetic nanoparticles for rapid capture and removal of bacterial pathogens. Environ. Sci. Tech., 44(20), 7908–7913. Copyright (2015) American Chemical Society.

also provides hydrophobic interaction between bacteria and guarantee a high binding affinity to bacteria. The AF-MNPs have enormous potential for rapid removal of bacteria from water and foodstuffs with very high efficiency of 88.5–99.1%, respectively. Amine-functionalized magnetic nanoparticles (AF-MNPs) are thus enables the fast and binding and removal of bacterial pathogens from food products.

7.5 Immunomagnetic Detection Coupled with Surface-Enhanced Raman Scattering (SERS)

Methods combining IMS and surface-enhanced Raman scattering (SERS) was developed to enumerate *E. coli*. Magnetic surface-enhanced Raman scattering is yet another advanced rapid and sensitive detection method. Magnetic SERS nanoprobes has contributed greatly for the separation and detection of multiple pathogens in food products. In this method, silica-coated magnetic nanoparticles (MNPs) are used to capture pathogens and then pathogen specific SERS probes were coupled with corresponding antibodies. Core-shell structured nanomagnetic probes coated with silica (MNPs@SiO_2) of 100 nm in diameter were first prepared by the reverse microemulsion method and functionalized with specific pathogen antibodies to first capture the bacteria directly from a food matrix followed by detection using an optical approach enabled by SERS (Wang et al., 2011).

Gold nanoparticles incorporated with a Raman reporter paves the way for the detection of multiple pathogens in selected food matrices. Varying the kinds of Raman reporters on SERS probes enables the detection of multiple pathogens in selected food matrices. Superparamagnetic Fe_3O_4 nanoparticles (MNPs) are integrated into gold nanostructures to form multifunctional plasmonic magnetic nanoparticles. This helps to achieve both plasmonic and magnetic features in a single nanoparticle.

Controlling the size of gold nanoparticles helps to tune the plasmonic properties. Alteration of inter-particle distance by an external magnetic field helps to further modulate the plasmonic properties. Simultaneous fast concentration of bacterial cells is possible by multifunctional magnetic–plasmonic Fe_3O_4–Au core–shell nanoparticles (Au-MNPs) by applying an external point magnetic field. This allows for perceptive detection and subsequent identification of pathogenic bacteria using surface-enhanced Raman spectroscopy (SERS) (Bao et al., 2009). A magnetic– plasmonic Fe_3O_4–Au core–shell nanoparticle offers a platform for the fast and convenient concentration, sensitive detection, and specific separation of bacteria. The multifunctional Au- MNPs possess a strong magnetic property due to the superparamagnetic Fe_3O_4 core, and can be condensed to a small point by applying an external point magnetic field. Therefore, this platform holds a great importance for in-field applications in food safety, environmental monitoring, and chemical and biological threat detections.

Guven et al. (2011) have fabricated gold-coated magnetic nanoparticles by immobilizing biotin-labeled anti-*E. coli* antibodies on avidin functionalized magnetic nanoparticles and used in the separation and concentration of the *E. coli* cells. Raman labels have been developed using rod-shaped gold nanoparticles coated with 5.5-dithiobis-(2-nitrobenzoic acid) (DTNB) and further with a molecular recognizer. Then, these DTNB-labeled gold nanorods were interacted with gold-coated magnetic spherical nanoparticle-antibody-*E. coli* complex. A linear correlation between the concentration of bacteria and SERS signal was obtained. The selectivity of the developed immunoassay was examined with other bacteria like *Enterobacter dissolvens, Enterobacter aerogenes,* and *Salmonella enteriditis,* which did not produce any significant response. The ability of the immunoassay to specifically detect *E. coli* in real water samples was also done and the results were further compared with the obtained experimental results from plate-counting methods and found similar.

7.6 Magnetic Nanoparticles Coupled Matrix-Assisted Laser Desorption/Ionization Mass Spectrometry (MALDI-MS) for Pathogen Detection

Magnetic nanoparticles are so flexible and adaptable that they can couple with other analytical means for food-borne pathogen detection. Recently, magnetic nanoparticles are combined with matrix-assisted laser desorption/ionization mass spectrometry (MALDI-MS) to detect pathogenic bacteria in food products. The technique of affinity mass spectrometry was developed in an attempt to specifically concentrate minute quantities of bacteria from foodstuffs. MALDI-MS provides a straightforward means to identify microorganisms based on mass spectral fingerprinting. Immunomagnetic separation can also be coupled with MALDI-MS to successfully differentiate target bacteria from biological sample solutions (Lin et al., 2005).

Gram-positive bacteria, like *Staphylococcus aureus*, *Staphylococcus saprophyticus*, and *Enterococcus faecalis* present in food materials can be effectively detected using this technique. The vancomycin-immobilized magnetic nanoparticles perform the job of affinity probes for pathogens and bind to the cell surfaces of target bacteria through hydrogen bonds. Thus, the target bacteria become magnetized and can be isolated readily from the sample solution by employing an external magnetic field. MALDI-MS can be used as an analytical method to differentiate microorganism species based on their fingerprinting mass spectra, Vancomycin-immobilized magnetic nanoparticles are anchored on the surface of a Gram-positive bacterial cell (Fig. 9.10). Vancomycin has high specificity and hence binds to cell surface (the terminal D-Ala-D-Ala peptide unit of the cell wall) of the Gram-positive bacterium by

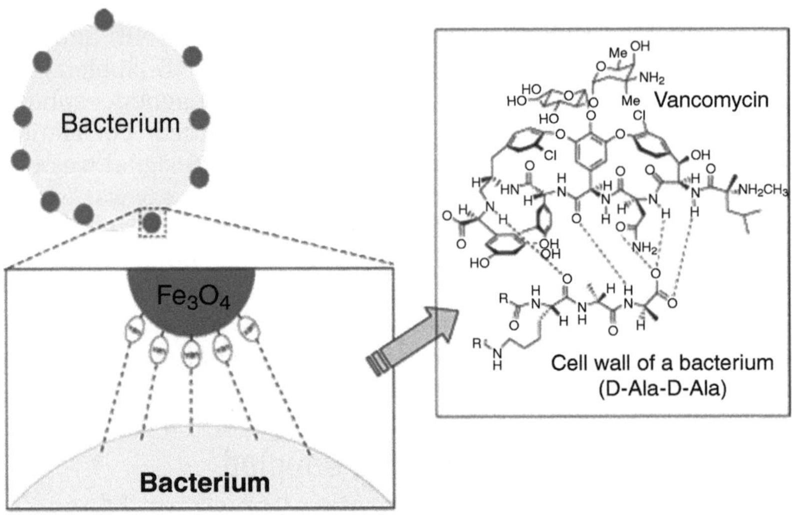

Figure 9.10. Cartoon illustrations of the proposed method for anchoring vancomycin-immobilized magnetic nanoparticles on the surface of a Gram-positive bacterial cell and the binding of vancomycin to the terminal of D-Ala-D-Ala units of the peptides on the cell wall of a Gram-positive bacterium. Reproduced with the permission from Lin, Y.S., Tsai, P.J., Weng, M.F., and Chen, Y.C., 2005. Affinity capture using vancomycin-bound magnetic nanoparticles for the MALDI-MS analysis of bacteria. Anal. Chem., 77(6), 1753–1760. Copyright (2015) American Chemical Society.

hydrogen bond formation. The immobilized magnetic nanoparticles as nanoscale probes to selectively trap traces of pathogens from biological fluids. These bacteria can be isolated from sample solutions by applying a magnetic field. The captured cells are then characterized by MALDI-MS (Lin et al., 2005).

Magnetite nanoparticles were functionalized by attaching human immunoglobulin (IgG) on their surfaces. The IgG-modified Fe_3O_4 nanoparticles, which bind specifically to IgG-binding regions of pathogens, serve as affinity probes to capture targeted bacteria from sample solutions (Ho et al., 2004). A readout of the captured bacteria can be generated using MALDI-MS.

7.7 Magnetic Nanoparticle-Based Biosensors Food Pathogen Monitoring

Developing innovative sensor-based technologies that enables fast, sensitive, and selective detection and quantification of pathogens is a high priority for scientists in academia and industries. The biosensors have a wide range of applications for fast, sensitive, specific, inexpensive, in-field, online and/or real-time detection of pathogens, toxins, proteins, and microbes in food and water (Singh, 2011). Thus, biosensors are excellent analytical tools for food-borne pathogen monitoring. Nanobiosensor specificity is achieved through the use of recognition elements that recognize antigens or other epitopes on the exterior of a pathogen. Sensitivity is achieved by signal transduction modalities that explicitly detect the interaction between the target pathogenic microorganism or its metabolites and the sensing element. In a nanomaterial-enabled biosensor, the recognition element is usually bound to the surface of the nanostructure, and the interaction of this conjugate with a pathogen is monitored by signal transduction technologies (Vikesland and Wigginton, 2010).

An impedance biosensor with magnetic nanoparticle-antibody conjugates (MNAC) integrated into a microfluidic flow cell with embedded gold interdigitated array microelectrode (IDAM) was developed to detect pathogenic bacteria in ground beef samples (Varshney et al., 2007). The flow cell consisting of a microchamber for detection and an inlet and an outlet microchannel was fabricated by bonding an indium-tin oxide interdigitated array microelectrode (IDAM) chip to a poly(dimethylsiloxane) (PDMS) microchannel. MNAC were prepared by conjugating streptavidin-coated magnetic nanoparticles with biotin-labeled polyclonal goat anti-*E. coli* antibodies and were used in the separation and concentration of target bacteria. The cells of *E. coli* O157:H7 present in a food product were first captured by the MNAC, separated, and concentrated by applying a magnetic field. After washing and suspending in mannitol solution, the bacterial suspension was injected through the microfluidic flow cell for impedance measurement. This biosensor was able to detect as low as 1.6×10^2 and 1.2×10^3 cells of *E. coli* O157:H7 cells present in pure culture of this bacteria and ground beef sample, respectively. Sample preenrichment, redox probes, secondary antibodies on the microelectrode surface were not required in this impedance biosensor. Only 35 min was required to complete the detection process from sampling to measurement.

Chen and co-workers developed an impedance biosensor system that combines immunomagnetic separation with urease catalysis for the sensitive detection of food-borne bacteria. Here *L. monocytogenes* was used as model bacterium and an immobilization-free microelectrode was used as the detector. A schematic representation of the sensor assembly is shown in Fig. 9.11. The monoclonal antibodies (MAbs) were tagged on magnetic nanoparticles (MNPs) with an average diameter of 180 nm by biotin–streptavidin system for providing specificity and for efficiently separating listeria cells. The urease and polyclonal antibodies (PAbs) were attached on the surface of comparatively smaller (20 nm) gold nanoparticles (AuNPs). Modified AuNPs will react with listera in the sample to form the MNP–MAb–*Listeria*–PAb–AuNP–urease sandwich complexes. After successive washing steps to remove the unreacted PAb and urease modified AuNPs and the excess conductive ions, the complexes were resuspended in the urea prepared in deionized water. The urease in the complexes could catalyze the hydrolysis of the urea into ammonium carbonate and subsequently this led to a considerable increase in the ionic strength of the media, which could be detected by the microelectrode and analyzed by using electrochemical impedance spectroscopy (EIS). The magnetic separation efficiencies for *L. monocytogenes* at the concentrations ranging from 3.0×10^1 to 3.0×10^4 cfu mL^{-1} were more than 95% for the pure cultures and more than 85% for the spiked lettuce samples. The lower detection limit for *L. monocytogenes* was 300 cfu mL^{-1} in both cases. This urease-based magnetic biosensor demonstrated for its potential to develop a cost-effective, simple, and sensitive method for quick detection of food-borne pathogens (Chen et al., 2015).

Studies worldwide have shown that *E. coli* are often present in fresh meat (Todd, 1996). For the rapid and specific detection of *E. coli* O157:H7 in beef samples, an impedance biosensor based on interdigitated array microelectrode (IDAM), which is coupled with magnetic nanoparticle–antibody conjugates (MNAC) was

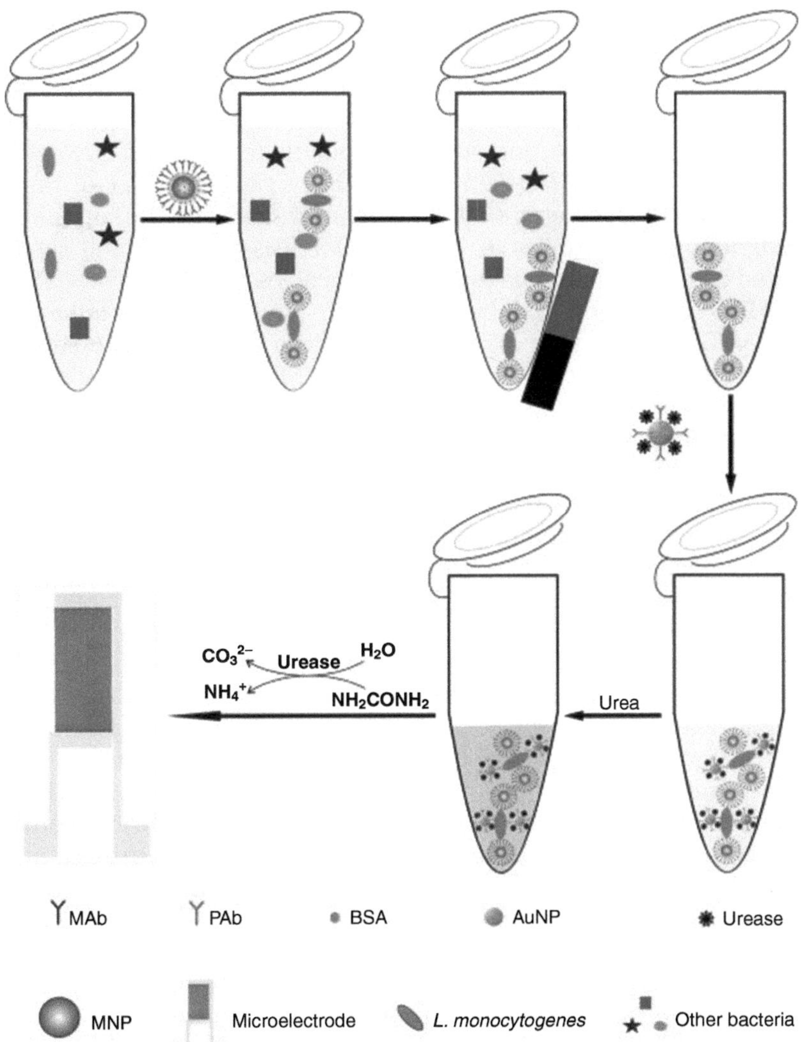

Figure 9.11. The principle of the impedance biosensor based on immunomagnetic separation and urease catalysis. Reproduced with the permission from Chen, Q., Lin, J., Gan, C., Wang, Y., Wang, D., Xiong, Y., ... and Wang, M., 2015. A sensitive impedance biosensor based on immunomagnetic separation and urease catalysis for rapid detection of *Listeria monocytogenes* using an immobilization-free interdigitated array microelectrode. Biosensors and Bioelectronics, 74, 504–511. With permission from Elsevier.

developed (Varshney et al., 2007). Here, biotin-labeled polyclonal goat anti-*E. coli* antibodies were immobilized on streptavidin-coated magnetic nanoparticles, which were effectively utilized to separate and concentrate *E. coli* O157:H7 from ground beef samples. In order to detect the presence of bacteria, magnitude of impedance and phase angle were measured. This biosensor was able to detect as low as 7.4×10^4 and 8.0×10^5 cfu mL^{-1} *E. coli* O157:H7 in pure culture and ground beef samples respectively. The task of concentrating microbial cells bound to MNAC in the active layer of IDAM above the surface of electrodes with the help of a magnetic field could improve the sensitivity of the impedance biosensor assembly. The entire detection process from sampling to measurement was completed within 35 min.

A quartz crystal microbalance (QCM) immunosensor based on beacon immunomagnetic nanoparticles (BIMPs), streptavidin–gold, and growth solution was developed to detect *E. coli* O157:H7. Polyclonal anti-*E. coli* O157:H7 antibody (target antibody, T-Ab) and biotin-IgG (beacon antibody, B-Ab) were attached on the surface of magnetic nanoparticles. *E. coli* O157:H7 was captured and separated by the antibody magnetic nanoparticles system (*E. coli* O157-BIMPs) in a sample, and the streptavidin–gold was further linked to *E. coli* O157-BIMPs by using a biotin–avidin system. After enlargement of gold particles on *E. coli* O157-BIMPs, they were collected using a magnet. The QCM immunosensor was fabricated with protein A from *S. aureus* and monoclonal anti-*E. coli* O157:H7 antibody. Presence bacteria-gold-immunomagnetic compounds decreased the immunosensor's resonant frequency. The frequency shift was correlated to the bacterial concentration. This biosensor was able to detect bacteria as low as 23 cfu mL^{-1} in phosphate-buffered saline and 53 cfu mL^{-1} in milk. The time required to complete the procedure for the detection of *E. coli* O157:H7 was 4 h. (Shen et al., 2011).

Bacillus anthracis is a spore-forming bacterium and a dangerous pathogen for causing the disease anthrax, is often contaminated with food products. *B. anthracis* bacteria can exist in the form of rod-shaped organisms and spores. Rod-shaped organisms grow and divide in a nutrient rich environment. At low nutritional state, the organisms turn into spores and survive for decades (Zhang et al., 2005). To detect *B. anthracis* endospores in contaminated food samples, an electrically active polyaniline coated magnetic (EAPM) nanoparticle-based biosensor has been proposed (Pal and Alocilja, 2009). Biologically modified EAPM nanoparticles could act as an immunomagnetic concentrator of *B. anthracis* spores from the samples of ground beef, lettuce, and whole milk. These are then directly applied to a direct-charge transfer biosensor for the detection. Biosensor could detect *B. anthracis* spores as low as 4.2×10^2 spores mL^{-1} from the above samples. The entire detection process was completed within 16 min.

8 Challenges in Magnetic Nanoparticle-Based Food-Borne Pathogen Detection

Monitoring of food-borne pathogens is not an easy task because of the need to recover microorganisms from food matrices (Cho et al., 2014). Specificity to distinguish many different pathogens in the real food samples is still a challenging task. On the other hand, like any other advanced technology the adverse effects of nanotechnology are also arguable. It is suspicious that minute traces of magnetic nanoparticles that assisted in the isolation of microbes may remain in the foodstuff if they are directly employed in the separation process. Nano Risk Framework, a Framework shaped in cooperation with organizations, including specialists in multidisciplinary fields, creates a data profiles of characteristics of nanomaterials, its inherent hazards, and exposure potential. FDA Nanotechnology Task Force (NTF) was formed in August 2006 to determine regulatory approaches. Unfailing and trustworthy tests to determine the presence of nanoparticles in food can be an effective solution to combat this issue. Extensive research must be carried out to evaluate the impact of magnetic nanoparticles for the assessment and monitoring of hazardous pathogens. It also requires increased involvement of industry, governmental and consumer organizations.

9 Conclusions

Magnetic nanoparticles have offered an exclusive opportunity to have command over the detection of food-borne pathogens by making the greatest possible advantage of magnetic force. Improved safety, worthiness, health profits, and increased efficiency offered by magnetic nanoparticles has stimulated the growth and promotion of such nanotechnology-based approaches. Increased surface area to volume ratio of nanomaterials offer added surface area for the attachment of bacteria and other pathogens on the nanoparticle surface. The magnetic nature of magnetic nanoparticles enables magnetic separation of the bacteria-nanoparticle conjugate, in presence of an external magnetic field, which is expedient, saves time, and improves efficiency. Magnetic nanoparticles have thus proved to be a boon to the food industry, too, favoring competent and fast capture of pathogens. Various governmental and nongovernmental organizations and agencies have come up, realizing its potential in food monitoring. Integration of multifunctional and nanosized magnetic materials to electronics will accelerate the encroachment of efficient miniaturization techniques and multiplex detection systems for the instantaneous and competent sensing of hazardous pathogens that cause severe danger to human health. The development of technologies and devices that enable further miniaturization, fast and efficient detection of pathogens will definitely have a great impact on food safety. Magnetic nanoparticles have proved themselves to be indispensable for efficient and rapid monitoring and separation of pathogens in food products and it calls for more intensive research to overcome some key issues that still exist.

References

Augustine, R., Rajarathinam, K., 2012. Synthesis and characterization of silver nanoparticles and its immobilization on alginate coated sutures for the prevention of surgical wound infections and the in vitro release studies. Int. J. Nano Dimen. 2 (3), 205–212.

Augustine, R., Dominic, E.A., Reju, I., Kaimal, B., Kalarikkal, N., Thomas, S., 2014a. Investigation of angiogenesis and its mechanism using zinc oxide nanoparticle-loaded electrospun tissue engineering scaffolds. RSC Adv. 4 (93), 51528–51536.

Augustine, R., Dominic, E.A., Reju, I., Kaimal, B., Kalarikkal, N., Thomas, S., 2014b. Electrospun polycaprolactone membranes incorporated with ZnO nanoparticles as skin substitutes with enhanced fibroblast proliferation and wound healing. RSC Adv. 4 (47), 24777–24785.

Augustine, R., Kalarikkal, N., Thomas, S., 2014c. A facile and rapid method for the black pepper leaf mediated green synthesis of silver nanoparticles and the antimicrobial study. Appl. Nanosci. 4 (7), 809–818.

Augustine, R., Kalarikkal, N., Thomas, S., 2014d. Advancement of wound care from grafts to bioengineered smart skin substitutes. Prog. Biomater. 3 (2–4), 103–113.

Augustine, R., Kalarikkal, N., Thomas, S., 2015. Electrospun PCL membranes incorporated with biosynthesized silver nanoparticles as antibacterial wound dressings. Appl. Nanosci., 1–8.

Augustine, R., Malik, H.N., Singhal, D.K., Mukherjee, A., Malakar, D., Kalarikkal, N., Thomas, S., 2014e. Electrospun polycaprolactone/ZnO nanocomposite membranes as biomaterials with antibacterial and cell adhesion properties. J. Polym. Res. 21 (3), 1–17.

Bao, F., Yao, J.L., Gu, R.A., 2009. Synthesis of magnetic Fe2O3/Au core/shell nanoparticles for bioseparation and immunoassay based on surface-enhanced Raman spectroscopy. Langmuir 25 (18), 10782–10787.

Beuchat, L.R., 1981. Microbial stability as affected by water activity. Cereal Foods World 26 (7), 345–349.

Bilir Ormanci, F.S., Erol, I., Ayaz, N.D., Iseri, O., Sariguzel, D., 2008. Immunomagnetic separation and PCR detection of *Listeria monocytogenes* in turkey meat and antibiotic resistance of the isolates. Br. Poult. Sci. 49 (5), 560–565.

Brullot, W., Valev, V.K., Verbiest, T., 2012. Magnetic-plasmonic nanoparticles for the life sciences: calculated optical properties of hybrid structures. Nanomed. Nanotechnol. Bio. Med. 8 (5), 559–568.

Cao, M., Li, Z., Wang, J., Ge, W., Yue, T., Li, R., William, W.Y., 2012. Food related applications of magnetic iron oxide nanoparticles: enzyme immobilization, protein purification, and food analysis. Trend. Food Sci. Technol. 27 (1), 47–56.

Chen, Q., Lin, J., Gan, C., Wang, Y., Wang, D., Xiong, Y., Wang, M., 2015. A sensitive impedance biosensor based on immunomagnetic separation and urease catalysis for rapid detection of *Listeria monocytogenes* using an immobilization-free interdigitated array microelectrode. Biosens. Bioelectron. 74, 504–511.

Cheng, Y., Liu, Y., Huang, J., Li, K., Zhang, W., Xian, Y., Jin, L., 2009. Combining biofunctional magnetic nanoparticles and ATP bioluminescence for rapid detection of *Escherichia coli*. Talanta 77 (4), 1332–1336.

Cho, I.H., Irudayaraj, J., 2013. In-situ immuno-gold nanoparticle network ELISA biosensors for pathogen detection. International Journal of Food Microbiology 164 (1), 70–75.

Cho, I.H., Radadia, A.D., Farrokhzad, K., Ximenes, E., Bae, E., Singh, A.K., Irudayaraj, J., 2014. Nano/micro and spectroscopic approaches to food pathogen detection. Ann. Rev. Anal. Chem. 7, 65–88.

Chockalingam, A.M., Babu, H.K.R.R., Chittor, R., Tiwari, J.P., 2010. Gum arabic modified Fe_3O_4 nanoparticles cross linked with collagen for isolation of bacteria. J. Nanobiotechnol. 8 (1), 30.

Codex Alimentarius: how it all began, Food and Agriculture Organization of the United Nations, http://www.fao.org/docrep/v7700t/v7700t09.htm.

Day, J.B., Basavanna, U., 2015. Magnetic bead based immuno-detection of *Listeria monocytogenes* and *Listeria ivanovii* from infant formula and leafy green vegetables using the Bio-Plex suspension array system. Food Microbiol. 46, 564–572.

dos Santos, M.B., Agusil, J.P., Prieto-Simón, B., Sporer, C., Teixeira, V., Samitier, J., 2013. Highly sensitive detection of pathogen *Escherichia coli* O157: H7 by electrochemical impedance spectroscopy. Biosens. Bioelectron. 45, 174–180.

Duan, H.L., Shen, Z.Q., Wang, X.W., Chao, F.H., Li, J.W., 2005. Preparation of immunomagnetic iron-dextran nanoparticles and application in rapid isolation of *E. coli* O157: H7 from foods. World J. Gastroenterol. 11 (24), 3660–3664.

Duffy, G., Sheridan, J.J., Hofstra, H., McDowell, D.A., Blair, I.S., 1997. A comparison of immunomagnetic and surface adhesion immunofluorescent techniques for the rapid detection of *Listeria monocytogenes* and *Listeria innocua* in meat. Lett. Appl. Microbiol. 24 (6), 445–450.

Dutta, P.K., Tripathi, S., Mehrotra, G.K., Dutta, J., 2009. Perspectives for chitosan based antimicrobial films in food applications. Food Chem. 114 (4), 1173–1182.

El-Boubbou, K., Gruden, C., Huang, X., 2007. Magnetic glyco-nanoparticles: a unique tool for rapid pathogen detection, decontamination, and strain differentiation. J. Am. Chem. Soc. 129 (44), 13392–13393.

Fang, C., Zhang, M., 2009. Multifunctional magnetic nanoparticles for medical imaging applications. J. Mater. Chem. 19 (35), 6258–6266.

FAO/WHO., 2010. FAO/WHO framework for developing national food safety emergency response plans, Food and Agriculture Organization of the United Nations and World Health Organization/Rome. http://www.fao.org/docrep/013/i1686e/i1686e00.pdf.

FoodNet MMWR Report, 2014. Foodborne Diseases Active Surveillance Network (FoodNet), Centers for Disease Control and Prevention 2014. http://www.cdc.gov/foodnet/index.html.

Gu, H., Ho, P.L., Tsang, K.W., Wang, L., Xu, B., 2003. Using biofunctional magnetic nanoparticles to capture vancomycin-resistant enterococci and other Gram-positive bacteria at ultralow concentration. J. Am. Chem. Soc. 125 (51), 15702–15703.

Gu, H., Xu, K., Xu, C., Xu, B., 2006. Biofunctional magnetic nanoparticles for protein separation and pathogen detection. Chem. Commun. (9), 941–949.

Guven, B., Basaran-Akgul, N., Temur, E., Tamer, U., Boyacı, I.H., 2011. SERS-based sandwich immunoassay using antibody coated magnetic nanoparticles for *Escherichia coli* enumeration. Analyst 136 (4), 740–748.

Hazard Analysis & Critical Control Points (HACCP), U.S. Food and Drug Administration. http://www.fda.gov/Food/GuidanceRegulation/HACCP.

Ho, K.C., Tsai, P.J., Lin, Y.S., Chen, Y.C., 2004. Using biofunctionalized nanoparticles to probe pathogenic bacteria. Anal. Chem. 76 (24), 7162–7168.

Horak, D., Babič, M., Mackova, H., Beneš, M.J., 2007. Preparation and properties of magnetic nano-and microsized particles for biological and environmental separations. J. Sep. Sci. 30 (11), 1751–1772.

Huang, J., Yang, G., Meng, W., Wu, L., Zhu, A., Jiao, X.A., 2010. An electrochemical impedimetric immunosensor for label-free detection of *Campylobacter jejuni* in diarrhea patients' stool based on O-carboxymethylchitosan surface modified Fe 3 O 4 nanoparticles. Biosens. Bioelectron. 25 (5), 1204–1211.

Huang, Y.F., Wang, Y.F., Yan, X.P., 2010. Amine-functionalized magnetic nanoparticles for rapid capture and removal of bacterial pathogens. Environ. Sci. Technol. 44 (20), 7908–7913.

Hudson, J.A., Lake, R.J., Savill, M.G., Scholes, P., McCormick, R.E., 2001. Rapid detection of *Listeria monocytogenes* in ham samples using immunomagnetic separation followed by polymerase chain reaction. J. Appl. Microbiol. 90 (4), 614–621.

Lee, W., Kwon, D., Chung, B., Jung, G.Y., Au, A., Folch, A., Jeon, S., 2014. Ultrarapid detection of pathogenic bacteria using a 3D immunomagnetic flow assay. Anal. Chem. 86 (13), 6683–6688.

Leon-Velarde, C.G., Zosherafatein, L., Odumeru, J.A., 2009. Application of an automated immunomagnetic separation–enzyme immunoassay for the detection of *Salmonella enterica* subspecies enterica from poultry environmental swabs. J. Microbiol. Methods 79 (1), 13–17.

Li, F., Zhao, Q., Wang, C., Lu, X., Li, X.F., Le, X.C., 2010. Detection of *Escherichia coli* O157:H7 using gold nanoparticle labeling and inductively coupled plasma mass spectrometry. Anal. Chem. 82 (8), 3399–3403.

Lin, Y.S., Tsai, P.J., Weng, M.F., Chen, Y.C., 2005. Affinity capture using vancomycin-bound magnetic nanoparticles for the MALDI-MS analysis of bacteria. Anal. Chem. 77 (6), 1753–1760.

Ling, B., Wen, Y., Yu, Z., Yu, Y., Yang, H., 2011. Multifunctional magnetic nanocomposites: separation, photodecomposition and Raman detection. J. Mater. Chem. 21 (12), 4623–4628.

Loaiza, O.A., Campuzano, S., Pedrero, M., Pingaron, J.M., 2008. Designs of *Enterobacteriaceae* Lac Z gene DNA gold screen printed biosensors. Electroanalysis 20 (13), 1397–1405.

Lu, A.H., Salabas, E.E., Schüth, F., 2007. Magnetic nanoparticles: synthesis, protection, functionalization, and application. Angewandte Chem. Int. Ed. 46 (8), 1222–1244.

Mao, Y., Huang, X., Xiong, S., Xu, H., Aguilar, Z.P., Xiong, Y., 2016. Large-volume immunomagnetic separation combined with multiplex PCR assay for simultaneous detection of *Listeria monocytogenes* and *Listeria ivanovii* in lettuce. Food Control 59, 601–608.

Martin, L.W., Ramesh, R., 2012. Multiferroic and magnetoelectric heterostructures. Acta Materialia 60, 2449–2470.

Miller, C.C., 1924. The Stokes-Einstein law for diffusion in solution. Proceedings of the Royal Society A: mathematical. Phys. Eng. Sci. 106 (740), 724–749.

Neurauter, A.A., Bonyhadi, M., Lien, E., Nøkleby, L., Ruud, E., Camacho, S., Aarvak, T., 2007. Cell isolation and expansion using Dynabeads®. Cell Separation. Springer, Berlin Heidelberg, pp. 41–73.

Pal, S., Alocilja, E.C., 2009. Electrically active polyaniline coated magnetic (EAPM) nanoparticle as novel transducer in biosensor for detection of *Bacillus anthracis* spores in food samples. Biosens. Bioelectron. 24 (5), 1437–1444.

Ravindranath, S.P., Mauer, L.J., Deb-Roy, C., Irudayaraj, J., 2009. Biofunctionalized magnetic nanoparticle integrated mid-infrared pathogen sensor for food matrixes. Anal. Chem. 81 (8), 2840–2846.

Raz, S.R., Haasnoot, W., 2011. Multiplex bioanalytical methods for food and environmental monitoring. TrAC Trend. Anal. Chem. 30 (9), 1526–1537.

Shan, S., Zhong, Z., Lai, W., Xiong, Y., Cui, X., Liu, D., 2014. Immunomagnetic nanobeads based on a streptavidin-biotin system for the highly efficient and specific separation of *Listeria monocytogenes*. Food Control 45, 138–142.

Shen, Z.Q., Wang, J.F., Qiu, Z.G., Jin, M., Wang, X.W., Chen, Z.L., Cao, F.H., 2011. QCM immunosensor detection of *Escherichia coli* O157: H7 based on beacon immunomagnetic nanoparticles and catalytic growth of colloidal gold. Biosens. Bioelectron. 26 (7), 3376–3381.

Shi, L., Wu, F., Wen, Y., Zhao, F., Xiang, J., Ma, L., 2015. A novel method to detect *Listeria monocytogenes* via superparamagnetic lateral flow immunoassay. Anal. Bioanal. Chem. 407 (2), 529–535.

Singh, R.P., 2011. Prospects of nanobiomaterials for biosensing. Int. J. Electrochem. 1–30. doi:10.4061/2011/125487.

Solodoukhina, D., 2011. Food safety and bioterrorism from public health perspective. Advances in Food Protection. Springer, Netherlands, pp. 17–25.

Song, G., Wang, Q., Wang, Y., Lv, G., Li, C., Zou, R., Hu, J., 2013. A low-toxic multifunctional nanoplatform based on Cu9S5@ mSiO2 core-shell nanocomposites: combining photothermal-and chemotherapies with infrared thermal imaging for cancer treatment. Adv. Func. Mater. 23 (35), 4281–4292.

Sundaram, P.A., Augustine, R., Kannan, M., 2012. Extracellular biosynthesis of iron oxide nanoparticles by *Bacillus subtilis* strains isolated from rhizosphere soil. Biotechnol. Bioprocess Eng. 17 (4), 835–840.

Swanson, K.M., 2011. Fish and seafood products. Microorganisms in Foods 8. Springer, US, pp. 107–133.

Tamer, U., Cetin, D., Suludere, Z., Boyaci, I.H., Temiz, H.T., Yegenoglu, H., Elerman, Y., 2013. Gold-coated iron composite nanospheres targeted the detection of *Escherichia coli*. Int. J. Mol. Sci. 14 (3), 6223–6240.

Tapia, M.S., Alzamora, S.M., Chirife, J., 2008. 10 effects of water activity (a w) on microbial stability: as a hurdle in food preservation. Water Activity in Foods: Fundamentals and Applications, Wiley, Hoboken, NJ, p. 239.

Thompson, J.R., Marcelino, L.A., Polz, M.F., 2005. Diversity, sources, and detection of human bacterial pathogens in the marine environment. Oceans and Health: Pathogens in the Marine Environment. Springer, US, pp. 29–68.

Todd, E.C., 1996. Epidemiology of foodborne diseases: a worldwide review. World health statistics quarterly. Rapport Trimestriel de Statistiques Sanitaires Mondiales 50 (1–2), 30–50.

Troller, J.A., Christian, J.H.B., 2012. Water Activity and Food. Academic Press, New York.

Understanding the Codex Alimentarius Preface, 2006. Third ed. World Health Organization and the Food and Agriculture Organization of the United Nations. http://www.fao.org/docrep/010/a0850e/a0850e00.htm.

Uyttendaele, M., Van Hoorde, I., Debevere, J., 2000. The use of immuno-magnetic separation (IMS) as a tool in a sample preparation method for direct detection of *L. monocytogenes* in cheese. Int. J. Food Microbiol. 54 (3), 205–212.

Varshney, M., Li, Y., Srinivasan, B., Tung, S., 2007. A label-free, microfluidics and interdigitated array microelectrode-based impedance biosensor in combination with nanoparticles immunoseparation for detection of *Escherichia coli* O157: H7 in food samples. Sensors Actuators B Chem. 128 (1), 99–107.

Velusamy, V., Arshak, K., Korostynska, O., Oliwa, K., Adley, C., 2010. An overview of foodborne pathogen detection: in the perspective of biosensors. Biotechnol. Adv. 28 (2), 232–254.

Vikesland, P.J., Wigginton, K.R., 2010. Nanomaterial enabled biosensors for pathogen monitoring: a review. Environ. Sci. Technol. 44 (10), 3656–3669.

Wang, L., Zhao, W., O'Donoghu, M.B., Tan, W., 2007. Fluorescent nanoparticles for multiplexed bacteria monitoring. Bioconjug. Chem. 18 (2), 297–301.

Wang, Y., Ravindranath, S., Irudayaraj, J., 2011. Separation and detection of multiple pathogens in a food matrix by magnetic SERS nanoprobes. Anal. Bioanal. Chem. 399 (3), 1271–1278.

Wang, Y., Ye, Z., Ying, Y., 2012. New trends in impedimetric biosensors for the detection of foodborne pathogenic bacteria. Sensors 12 (3), 3449–3471.

Water Sanitation Health, 2015. World Health Organization (WHO). http://www.who.int/water_sanitation_health/diseases/diarrhoea/en/.

World Food Summit, 1996. Rome, Declaration on World Food Security.

Wu, W., Wu, Z., Yu, T., Jiang, C., Kim, W.S., 2015. Recent progress on magnetic iron oxide nanoparticles: synthesis, surface functional strategies and biomedical applications. Sci. Technol. Adv. Mater. 16 (2), 023501.

Xu, B., Ho, P.L., Gu, H., 2010. US Patent No. 7,754,444. Washington, DC: US Patent and Trademark Office.

Yang, H., Li, H., Jiang, X., 2008. Detection of foodborne pathogens using bioconjugated nanomaterials. Microfluid. Nanofluid. 5 (5), 571–583.

Yang, H., Qu, L., Wimbrow, A.N., Jiang, X., Sun, Y., 2007. Rapid detection of *Listeria monocytogenes* by nanoparticle-based immunomagnetic separation and real-time PCR. Int. J. Food Microbiol. 118 (2), 132–138.

Yang, L., Bashir, R., 2008. Electrical/electrochemical impedance for rapid detection of foodborne pathogenic bacteria. Biotechnol. Adv. 26 (2), 135–150.

Zhang, X., Young, M.A., Lyandres, O., Van Duyne, R.P., 2005. Rapid detection of an anthrax biomarker by surface-enhanced Raman spectroscopy. J. Am. Chem. Soc. 127 (12), 4484–4489.

Zhao, W., Ali, M.M., Aguirre, S.D., Brook, M.A., Li, Y., 2008. Paper-based bioassays using gold nanoparticle colorimetric probes. Anal. Chem. 80 (22), 8431–8437.

10

NANOTECHNOLOGY IN HEALTHIER MEAT PROCESSING

Muhammad Issa Khan*, Amna Sahar, Ubaid ur Rahman***
**University of Agriculture, National Institute of Food Science and Technology, Faculty of Food, Nutrition and Home Sciences, Faisalabad, Pakistan;*
***University of Agriculture, Department of Food Engineering, Faculty of Agricultural Engineering and Technology, Faisalabad, Pakistan*

1 Introduction

Nanotechnology involves the manufacturing and use of nanosized materials (<100 nm) having unique and novel functional properties (Lövenstam et al., 2010; Gruère, 2012). It mainly deals with the production, categorization, handling and employment of nanoparticles (Weiss et al., 2006). These particles vary from the bulk materials in terms of their physicochemical, optical, catalytic, and mechanical properties. Nanomaterials also differ from other bulk ingredients in their strength, weight, absorption, stabilization, and functionality (Cockburn et al., 2012). These distinctive properties of nanoparticles make this technique very promising, which ensures the innovative developments in the domain of food processing, preservation, and packaging (Sozer and Kokini, 2009; Rhim et al., 2013). However, application of this generic technology in food manufacturing also imposes a significant impact on the properties of foods due to unique and modified properties of added nanoparticles (Cockburn et al., 2012). Even so, improved bioavailability and functionality of serviceable compounds and decreased level of bulk functional ingredients can be achieved by using nanomaterials (Weiss et al., 2006). For intense levels of sugar, salt, and preservatives can be minimized and color, flavor, texture, and sensory properties of food can be enhanced by using nanomaterials. Furthermore, absorption and delivery of nutrients and active ingredients can be expressively improved (Chaudhry and Castle, 2011). Nanotechnology also provides several other benefits like targeted delivery of active ingredients, improved stability of the bioactive compounds, and better antimicrobial

Novel Approaches of Nanotechnology in Food. http://dx.doi.org/10.1016/B978-0-12-804308-0.00010-8

properties against a wide range of pathogenic microorganisms that are usually resistant to traditional chemical antimicrobial agents (Duncan, 2011; Cockburn et al., 2012).

Nanotechnology encompasses the production and utilization of materials and structures at the nanoscale. There are a number of opportunities for use of nanotechnology in food systems. Utilization of nanotechnology is considered as a characteristic part of conventional foods and food processing, as the quality attributes of numerous foods depend on its nano components, for example, foams and nanoemulsions. Recent developments in science and technology lead the approach to produce and incorporate nanoparticles in food. These nanoparticles could be completely novel chemical structures or finely divided forms of present components. As many definitions are available in literature regarding nanotechnology, the particular National Nanotechnology Initiative calls the idea "nanotechnology" simply, when the idea consists of all of the following:

- Exploration and innovation at atomic and macromolecular levels (1–100)
- Production and utilization of structures, equipment, and systems that have novel functions and characteristics because of their smaller size
- Ability to manipulate or control matter at atomic scale

We are just at the start of nanotechnology development. Applications of nanotechnology in food are increasing overall and are expected to develop quickly in future. The materials created at nanoscale can be very thin coatings (one dimension), nanowires (two dimensions), or nanoparticles (three dimensions in the form of fine powder). Nanotechnology is not a new field and researchers have been continuously working for decades to make polymers in view of nanoscale subunits. Because of new discovered properties of the nanoparticles (NPs), numerous new items containing these NPs have been introduced in the food sector. Applications associated with NPs in electronics, drugs, fabrics, food, agriculture, defense, cosmetic products, along with other parts are now possible and applications are starting to influence the food sectors (Chen et al., 2006).

The impending advantages for customers and manufacturers of products containing nanomaterials are broadly emphasized. In agriculture and food sectors nanotechnology covers several features, including food safety measures, handling materials, ailment treatment, bioavailability, delivery systems, new instruments for cellular and molecular biology and materials to detect a number of pathogens (Maynard et al., 2006; Jasińska et al., 2010). Nanoparticles have colossal capability to be used in agro-nourishment generation, that is, handling of pesticides, composts, sustenance added substances, beauty care products, sustain (eg, vitamins), bundling, and materials. Nanoparticles have remarkable potential for usage in agro-food production such as fertilizers, pesticides, feed, vitamins, food additives, cosmetics, packaging systems, and textiles.

Nanotechnologies are important innovations of the 21st century and enormous research innovations have been made in this industry. Because applications with basic elements at nanoscale levels have biological, physical, and chemical properties which are considerably not the same as their macroscopic complements, nanotechnology can also be useful on different levels. Several studies in the field of biology, physics, engineering, and biochemistry drive the improvement and development of the nanotechnology field. Therefore, several industries for example, nutraceuticals and microelectronics, have been working for the development of a wide range of products with nanometer size. Although food processing industries have also started to investigate its applications in the development of native functional food products, nanotechnology demonstrates incredible potential.

Food experiences considerable changes after harvesting and processing that influence its natural and biochemical properties, so nanotechnology advancements in the domains of science and organic chemistry could impact the food industry in a positive way. Systems with basic elements in the nanometer size could influence many aspects such as molecular synthesis and food safety. Nanotechnology mainly focuses on manufacturing, description and management of materials, with sizes smaller than 100 nm. Structures of this size have been indicated to possess distinctive and novel functional characteristics. The use of the expression "nano" does permit to highpoint the way that procedures or material structures are composed and streamlined to utilize particular properties and behaviors with size of 10–7 to 10–9 m (Fig. 10.1).

2 Nanotechnology: Science and Technology

Nanotechnology can possibly affect numerous parts of food and agriculture system. Illustrations of nanotechnology as an apparatus for accomplishing further progressions in the nourishment business are as per the following:

- Augmented security of processing, production and delivery of food products by using nanosensors for the detection of pathogenic microorganisms and contaminants
- Tools to keep records of a particular product and shipments

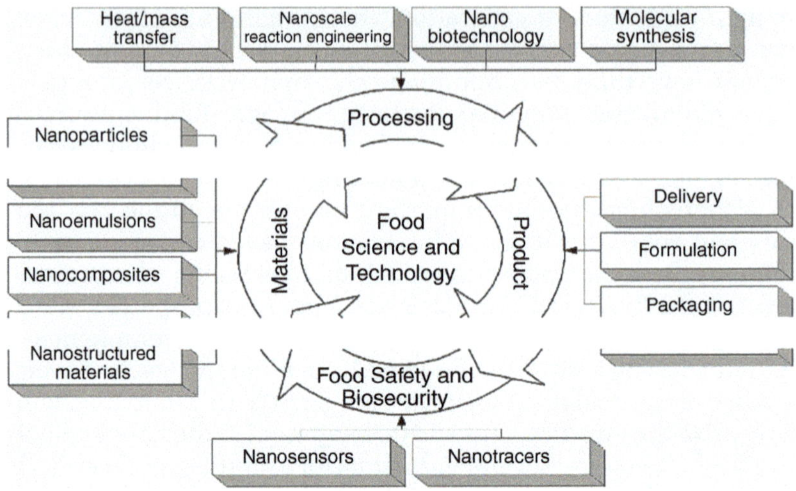

Figure 10.1. Application of nanotechnology in food science and technology.

- Programs that provide integration regarding sensing, reporting, and control of food products understanding that can increase usefulness and stability involving foods processing and traveling
- Encapsulation and shipping programs which ensure safety and supply functional food elements on their particular site of action

Nanotechnologies and nanoparticles will probably have an impact on the food chain from primary manufacturing to customer. Strategies to implement nanoscience toward the foods market may not be the same as these are more conventional applications involving nanotechnology. Food processing can be a multitechnological manufacturing market affecting numerous raw materials, high biosafety requirements, and directed innovative procedures. Four major areas within foods manufacturing might make use of nanotechnology:

1. Improvement of new functional products
2. Product development
3. Processing at nanoscale and microscale
4. Techniques and instrumentation pattern pertaining to enhanced foods basic safety and biosecurity

It is reported that there are numerous possible uses of nanotechnology within the foods systems. Application of nanotechnology in food packaging systems is considered a great innovation that can enhance the quality and safety of food products. For instance, intelligent packaging is usually responsive to the surroundings while active packaging can be used to interact with a food product.

Table 10.1 Applications of Nanotechnology in Food Industries

Technology	Advantages	Description
Nanostructures	Improved flavor, taste, and texture; salt and sugar reduction; enhanced bioavailability	Nanosized food ingredients and additives
Nanoencapsulation	Taste improvement; protection from degradation	Provide delivery systems for food supplements
Nanocomposites	Improve flexibility and barrier properties	Food packaging
Nutrient delivery	Targeted delivery of nutrients with improved bioavailability	Enzyme-assisted delivery; structural modifications
Membrane filtration	Higher quality food products	Effective separation of particles
Surface disinfectants	Engineered nanoparticles	Protection from pathogens
Nanolayers and inks	Nanolithography depositions	Prevention from adulteration, authenticity, traceability

It can also be seen that the global marketplace pertaining to nanofoods has been rapidly developing and is approximated from $20 billion in 2010 and $1 trillion by 2015 for the whole nanotechnology manufacturing. Possible uses of nanotechnology in food industries and associated benefits are depicted in Table 10.1.

3 Nanotechnology to Deliver Functional Ingredients

The particular nanosized self-assembled organized liquids (NSSL) technology enables encapsulation of cosmeceuticals, nutraceuticals, essential oils along with medication and cosmetics. Yet another edge to the NSSL technology is that it enables addition of insoluble chemical substances in to food and cosmetics products, important example is canola oil.

3.1 Technologies for the Delivery of Active Ingredients in Foods

The prime objective of using nanotechnology to deliver bioactive food components is to increase the performance of those elements in the food matrix. New well-designed elements tend to be significantly built into the food matrix development techniques (Haruyama, 2003). In addition to that, bioavailability and also the chance to distribute these kinds of ingredients in the biological systems are generally more than they will in their traditionally produced counterparts. Encapsulation is the separation of the bioactive components within the food product or service using food-grade resources. Formula features the manufacturing of the active component, typically at molecular and nanoscale levels through food-grade elements interacting with the active component. Control conditions may affect component activity. Novel microencapsulation techniques have also been introduced and utilized simply by professional research workers.

4 Nanotechnology in Meat Systems

Continuously increasing competition, awareness of health concerns and incessant demand for sustainable production have gotten the attention of meat processing industries to a gateway of innovation. Globally, the main concern of meat industries is to focus on developing new processes and innovative meat products to meet the growing demand for meat consumption. Nanotechnology has the potential to develop novel meat products with improved functionality, better preservation effects and reliable and smart packaging. Nanotechnology can also play a significant role to combat meat adulteration by ensuring meat authenticity. Potential benefits of using nanoparticles in meat systems include improvement in bioavailability of nutrients, provision of antimicrobial properties, improved sensory attributes, and targeted delivery of functional components. However, a number of bottlenecks are found in the practical applications of nanoparticles due to insufficient knowledge about the production strategies of several nanoingredients, their stability in meat products, and associated health risks after consuming products containing nanomaterials. For the fruitful utilization of nanotechnology in meat processing industries, these challenges need to be addressed properly. So far, intelligent and smart meat packaging is the most promising zone of nanotechnology application in meat processing but further investigation is required to study the health effects of migrated nanoparticles from packaging material to product. Moreover, governments, nanoparticle manufacturing industries, and legislative authorities can also play a role in deciding the future applications of nanotechnology.

Increasingly, consumers' demand for healthier and supreme quality meat products at reasonable prices has led the market of meat products toward change in the processing systems and

Table 10.2 Examples of Commonly Used Additives in Meat Products

Additives	Examples
Antioxidants	Butylated hydroxytoluene (BHT), butylated hydroxyanisole (BHA), tocopherols
Binders	Carrageenan, sodium caseinate
Thickeners	Gelatin
Humectants	Sodium salt, glycerine
Curing agents	Sodium erythorbate, sodium nitrite, sodium nitrate
Flavor enhancers	Monosodium glutamate
Tenderizing enzymes	Bromelin, ficin, papain
Sweeteners	Corn syrup

Source: USDA (2008).

choice of ingredients (Weiss et al., 2010). In addition, stresses on sustainable production of meat products for the improvement of human health have also got the attention of scientists for the development and upgradation of meat processing industries (Young et al., 2013). These factors have raised the expectations about the utilization of several ingredients with improved bioavailability and functionality to augment the quality of foods of animal origin (Olmedilla-Alonsoa et al., 2013). Examples of some extensively used food additives in poultry and meat products are depicted in Table 10.2.

Although these ingredients have been used extensively for the manufacturing of a wide range of meat products, however, due to growing health apprehensions focus has been shifted to the development of innovative meat products with potential health benefits like reduced saturated fats, salts, nitrites, and cholesterol. Studies have also proposed that these novel functional meat products containing modified ingredients of nanometer size should hold comparative sensory and gustatory properties as traditional meat products for getting maximum benefits from these products (Weiss et al., 2010).

Generally, functional bioactive components have been incorporated to the food products for the provision of additional health benefits and to cure or prevent several diet-related metabolic disorders (IFIC, 2006). Nonetheless, a wide range of hitches have been associated with the development, transport, and storage of foods enriched with bioactive compounds. As far as the wide range of conventional meat products is concerned, the impediments are likely far bigger in the meat industry. Meat products contain significant levels of proteins, fats, and mineral components so one of the major challenges for meat industries producing functional meat products is the low bioavailability of incorporated bioactive compounds. Accordingly, several alterations have been endeavored for the preparations of functional meat products, but these attempts have often led to undesirable impacts on organoleptic properties, moisture retention, and resistance to microbial spoilage (Weiss et al., 2010). Consequently, the main agenda of meat processors is to address and overcome these challenges to improve the quality of value added meat products and to win the confidence of consumers (Troy and Kerry, 2010). Hence, nanotechnology is considered as one of the process-based innovations that could have a momentous influence on the food industry particularly in the meat sector (Linton and Walsh, 2008).

Nanotechnology imposes the impact on the food processing industry by manufacturing nanosized particles with unique properties, by developing novel methods of food processing, by introducing novel functional products in the market, and by improving food safety and biosecurity (Moraru et al., 2003, 2009). Furthermore, varied approaches are required to incorporate nanomaterials in meat products in comparison with those strategies that are employed in conventional food processing techniques (Weiss et al., 2006). However, because of novel functional properties displayed by these nanocomponents, considerable beneficial modifications are estimated to be assisted in the manufacture, packaging, distribution, and storage of foods including meat products (Weiss et al., 2006; Duncan, 2011; Gruère, 2012). Contrarily, several lifestyle-related and environmental threats are also thought to be linked with the use of nanomaterials containing meat products due to the similar characteristics that offer potential remunerations (O'Brien and Cummins, 2010; Chaudhry and Castle, 2011). The indistinctness regarding the consumption of these functional ingredients possibly will enact a significant impact on the end user acceptance (Troy and Kerry, 2010).

4.1 Production of Nanomaterials for Use in Meat Products

Generally, two comprehensive approaches are involved in the production of nanomaterials around the globe, namely, top-down approach and bottom-up approach. In top-down techniques, inorganic materials are processed by using traditional approaches like grinding, milling, and sieving followed by several chemical changes (Cockburn et al., 2012). A common illustration of the top-down approach is homogenization that involves use of pressure to shorten the particle size of different constituents like fat globules. Milling process uses mechanical force for size reduction, which ultimately boosts up the functionality of ingredients (Cushen et al., 2012). On the other hand in bottom-up smaller molecules assemble together by self-organization, which results in the development of supramolecular structures possessing novel functional properties (Cockburn et al., 2012). Most common examples of bottom-up method are layer by layer deposition (LBLD) and solvent evaporation that are extensively used in food applications (Cockburn et al., 2012).

4.2 Potential Areas of Application in Meat Industry

According to a group of researchers, nanomaterials are mainly used in food systems for the production of food ingredients having nanostructure and play a vital role in delivery and supplementation of nutrients (Chaudhry et al., 2008). Dominant domains of application in meat processing industries comprise of reformulation by modifying and lowering the fat contents in meat products, minimizing the levels of salts like nitrates, sodium, and phosphate and incorporation of prebiotics, probiotics, and other functional ingredients for the development of healthier meat products. In addition to that, several other applications are enhancing the bioavailability of active ingredients and nutrients, development of healthier compounds during processing, and reduction of health threatening compounds while handling and storage (Olmedilla-Alonsoa et al., 2013).

A widespread range of substances has been found around the globe for potential use in the development of functional meat products. For example, citrus fiber, oat fiber, soy fiber, soy protein concentrate, carrageenan, and plasma proteins are used as fat replacers in meat products. Other potential domains of application in the meat industry include: modification lipid profile by using fish oil extract, linseed oil, and flaxseed; use of fruit pulp and edible seaweeds for salt reduction; utilizing spinach juice and celery for the reduction of nitrites; addition of rosemary extract and ascorbic acid as a potential source of antioxidants; and providing antimicrobial properties by incorporation of rosemary, nisin, and oregano oil (Weiss et al., 2010). Several benefits get through the aforementioned applications of nanotechnology, for instance, a variety of antimicrobial and antioxidant nanomaterials exist that can be used in the meat processing with the aim of extended shelf life and better functional attributes (Ozimek et al., 2010). Nonetheless, incorporation of these functional nanoingredients in the meat products entails strict deliberations (Augustin and Sanguansri, 2009).

4.3 Types of Nanomaterials

Functional properties of nanomaterials depend on their type and size (FSAI, 2008). Nanoparticles are usually manufactured into one-, two-, or three-dimensional structures. Most common types of nanomaterials are thin films, nanotubes, and nanoparticles. Classification of nanomaterials is not an easy task due to great diversity in their structural and functional properties. Additionally, structures naturally present in the food and produced as a result of traditional methods are not considered as nanoparticles. Only those structures can be classified in the category of nanoparticles which are intentionally manufactured at nanometer scale and hold unique properties (HOL, 2010).

4.3.1 Nanoscale Ingredients

Nanoscale ingredients can be extensively used to improve the taste profile of food and to modify the textural properties of altered food products. Moreover, ingredients produced by nanotechnology also contribute to the improvement of bioavailability of nutrients and other bioactive components (Chaudhry and Castle, 2011). The most important benefit of nanoscale materials is that they have the capacity to mask the adverse effects of unpleasant odors and undesirable flavors (Cushen et al., 2012). Usually, dry and wet milling processes are employed for the development of ultrafine, nanosized powder with a particle diameter in the range of 100 nm to 1 μm. These nanoparticles are obtained from organic compounds and are broadly used by the food industry for the development of functional foods in a cost effective manner (Borm et al., 2006; Zhu et al., 2010). A very common example of using nanoparticles in product development is nano green tea, which has shown significantly higher antioxidant potential in comparison with the traditional green tea due to its reduced particle size (Shibata, 2002). Another example is the development of ultrafine ground chitosan nanopowder that also has shown greater antimicrobial activity particularly against pathogenic bacteria and improved hypolipidemic potential (Zhu et al., 2010; Zhang et al., 2013). An additional beneficial approach is to reduce the size of ginger as microsized powder that increases penetrability and enhances tenderizing effect in meat products as compared to native ginger (Zhao et al., 2009). If these ingredients are further reduced to nanopowders by using nanotechnology, they may exhibit novel chemical and physical properties. Although nanoscale powders have enormous potential for use in meat and other food industries, there is a need to properly understand the nature of raw materials, modified behavior of nanoparticles, advancements and requirements for equipment for their successful application in food industries (Zhu et al., 2010). Nevertheless, manufacturing, marketing, and consumption of food products with added functional nanomaterials is still in the nascent stage, and is being expected a broad range of potential applications in the food systems by allying other related technological approaches (Cushen et al., 2012).

Nanostructures are components of one dimension with 1–100 nm. Nanostructures as well as nanomaterials show characteristics totally diverse from their bulk counterparts, for instance, mechanical properties, electrical conductivity (speed of moving electrons), thermal conductivity (flow of temperature), chemical reactivity (effectiveness of their response to other chemicals), magnetism (whether these are magnetic or not), and transparency. Developments in nanotechnology are generally starting to affect the food industry as well as associated sectors. This influences critical features of food safety practices of molecular structures of new ingredients and food products (Chen et al., 2006).

Nanotechnology permits to quantify and control at the nanoscale level to modify these properties and functionality in an advantageous manner. Advancements in nanotechnology are determined by basic and employed research in biology, chemistry, engineering, and physical science. Nanotechnology gives the apparatuses to measure and comprehend biosystems in science and innovation platform. The potential advantages of nanofoods are amazing. Promoters of the innovation guarantee improved food processing, improved flavor and nutrition, improved food safety and packaging, and increased production. In this present reality where a huge number of individuals starve every day, the increased production is sufficient to warrant overall support. For the past few years, the nourishment business has been putting many dollars in nanotechnology innovative work. In recent years, the food industry has been investing huge capital for the investigation and growth of nanotechnology in food products. The largest producers of world Nestle, Altria, H.J. Heinz, and Unilever, are pioneering the trail, while many small organizations take after their lead. Yet, regardless of the potential advantages, contrasted and other nanotechnology arenas, nanofoods do not get considerable marketing. Uses of nanotechnology in the food industry are starting to affect the vital aspects of food safety of ingredients and food products (Chen et al., 2006).

Nanofood growth is usually Kraft Ingredients, which had taken this industry's lead when it set up this Nanotek Consortium, in collaboration associated with 15 universities as well as national analysis labs, in 2000. Kraft's concentrate is usually in "interactive" foods. The products are going to be tailored to fit taste and consumer requirements. Beverages include products that will change color as well as tastes to be able to produce foods that may understand as well as alter any consumer's allergic reactions as well as dietary needs. Various other huge firms, for instance, Nestlé as well as Unilever, are generally discovering much better emulsifiers that may create food surface more uniform. In Australia, for example, nanocapsules containing omega-3 essential fatty acids are added to white bread. As indicated by the producer, addition of fish oil in the form of nanocapsules to Tip Top Bread ensures the availability of added supplements and prevention of developing fishy aroma in bread. Most of these American firms have an effect on the majority of the food industry's development and research. Summarizing, it is usually mentioned which the nanoscale measurement effect includes:

- Acknowledgment of scaled tools and structures while giving more usefulness
- Achievement of increased surface area to volume ration
- Advent of novel properties
- Incorporating beneficial changes in physical, chemical, functional, and structural properties

4.3.2 Nanoencapsulation

Functional ingredients are usually not added to the food products in the original form but several modifications have been performed to these ingredients prior to incorporation in the product to attain desirable functional properties (Weiss et al., 2006). For instance, bioactive components are mostly sensitive to high temperature and oxidation. Some active ingredients have poor solubility in the original form, therefore it is mandatory to modify them for their efficient absorption in the blood stream (Shimoni, 2009). Hence, a proper form and delivery system is required for the transportation of functional ingredients toward their target and to protect them from oxidative degradation. Besides, release and transportation of functional components can be synchronized by controlling the ionic strength and surrounding conditions such as temperature and pH. Care should also be taken into account that selected functional ingredients must be compatible with the quality aspects of food like taste, aroma, color, and texture. A variety of delivery systems exists for the transportation of active components in the body, however, only a few of them systems have a significant impact on food production. The most promising delivery systems include association colloids, biopolymeric nanoparticles, and nanoemulsions (Weiss et al., 2006).

4.3.2.1 Association Colloids

The colloid is a continuous setup of the substances made up of small particles which are uniformly dispersed. Association colloids contain particles having a very small particle size. Association colloids have been used for decades for the delivery of several polar and nonpolar functional components (Garti and Benichou, 2004; Golding and Sein, 2004; Garti et al., 2005; Bilska et al., 2009). Association colloids are transparent and typically range from 5 to 100 nm. Liquid crystals, surfactant micelles, reverse micelles, vesicles, and bilayers are all examples of association colloids. The major drawback of using association colloids to deliver functional ingredients is that they immediately dissociate after dilution. The significant point of interest of association colloid structures is that they appear spontaneously and tend to be thermodynamically good. The development of affiliation colloids is fixation driven. Solutions spontaneously dissociate when diluted. It is worthy to say that a huge amount of surfactants and to some extent cosurfactants for their formation which may cause difficulties with cost, lawfulness, and flavor.

Association colloids provide a stable system that allows uniform distribution of nanoparticles in a particular product. Good examples of association colloids are micelles and reverse micelles. This system involves the delivery of nanoparticles (5–100 nm) with novel properties by using polar, nonpolar, and amphiphilic materials, thereby enhancing the shelf stability of food products in addition to providing other health benefits (Flanagan and Singh, 2006). Studies elucidate that use of nanoparticle paprika oleoresin in chicken breast fillets can improve the marination and contribute to the improvement of sensory attributes of broiler meat (Yusop et al., 2012).

4.3.2.2 Biopolymeric Nanomaterials

Biopolymer nanomaterials also have a wide number of applications in the food systems particularly to extend the shelf life of foods. A very common example of biopolymeric compounds is polylactic acid (PLA), which is a synthetic food-grade biopolymer that has the potential to deliver functional ingredients in food systems (Weiss et al., 2006). Some other commonly used synthetic biopolymers include polyethylene glycol and polylactic-co-glycolic acid (PLGA) that also have wide applications in the regulation of proper delivery of added ingredients through biological systems (Gupta and Gupta, 2005). Antimicrobial potential of phenolic compounds is seen to be enhanced with the use of PLGA nanoparticles containing delivery systems in raw and processed chicken. Researchers revealed that phenolics like benzoic and vanillic acids were found to be very effective at lower concentrations to inhibit the growth of pathogenic microorganisms like *Salmonella typhimurium*, *Escherichia coli* O157:H7, and *Listeria monocytogenes* when embalmed in polylactic glycolic acid nanomaterials (Ravichandran et al., 2011). Contrarily, chitosan has been claimed as a best alternative for synthetic biopolymers for the encapsulation of functional ingredients (Weiss et al., 2006). Accordingly, in a work conducted by Abdou et al. (2012), chitosan nanoparticles having antimicrobial potential were deployed in fish coatings and a significant control in the growth of microorganisms was recorded in comparison with the uncoated fish fingers as well as those coated with commercial edible films, proposing a potential shelf life extension of fish fillets. Investigation of the rheological properties also showed that pseudoplastic coatings were used for all chitosan concentrations (Abdou et al., 2012).

Several protein- and carbohydrate-based biopolymers have also shown their potential to generate nanoparticles (Chang and Chen, 2005; Gupta and Gupta, 2005; Ritzoulis et al., 2005). Utilizing aggregative associations, a single biopolymer can be distributed into various small nanoparticles. Functional components are encapsulated by using nanoparticles and are released under particular environmental triggers.

Polylactic acid is the most common part of biodegradable biopolymeric nanoparticles. It is widely available and used as delivery systems for a number of functional ingredients such as proteins, laminations, vaccines, and pharmaceutics. It is immediately expelled from the circulatory system, staying segregated in the liver and kidneys. Because its function as a nanoparticle is to deliver functional ingredients to the body, polylactic acid needs a cooperative compound, for example, polyethylene glycol to be effective in such cases (Riley et al., 1999).

4.3.2.3 Nanoemulsion

Another important delivery system of functional compounds is nanoemulsion. In this system, active components can be introduced in the form of droplets either as continuous or interfacial phase. A major advantage of this delivery system is that it provides a pathway for more than one material, having antioxidant and antimicrobial utilities (Weiss et al., 2006). A common instance of nanoemulsions is nanostructured multilayer emulsion that permits the discharge of functional ingredients with the help of an external stimulus (McClements and Rao, 2011). However, limitations exist while using nanoemulsions in meat systems (Salminen et al., 2013). For intense, use of oil-in-water nanoemulsion in sausages can increase the chances of lipid oxidation during the storage of sausages. In a study, sunflower oil nanoemulsion was prepared and was incorporated in king mackerel steaks. The results demonstrated a significant reduction in the microbial load, improved organoleptic attributes, and an increase in the shelf life of fish meat during the storage for 48 h.

Emulsions of droplets having a diameter from 100 to 500 nm can be obtained by using high-pressure homogenizers or microfluidizers. In present literature emulsions are known as nanoemulsions. A lot of literature is available on application, production, and characterization of nanoemulsions (Jasińska et al., 2010). Functional ingredients are incorporated into the fine droplets, interfacial area, or continuous phase. Encapsulation of functional ingredients permits a slowdown process of compound degradation by modifying the characteristics of interfacial covering (McClements and Decker, 2000).

4.3.2.4 Nanostructured Multiple Emulsions

The utilization of various emulsions can make delivery systems novel with delivery and encapsulation. The most widely recognized examples are oil-in-water-in-oil (O/W/O) and water-in-oil-in-water (W/O/W) emulsions (Garti and Benichou, 2001, 2004). Functional ingredients are encapsulated inside the aqueous phase or oil phase, which enables it to develop a delivery system which contains a number of functional ingredients (Flanagan and Singh, 2006).

Association colloids and nanoemulsions are very cost-effective delivery systems. Another advantage of these systems is feasible production of these emulsions (Chaudhry et al., 2008). Apart from cost effectiveness and compatibility, a number of advantages and disadvantages are linked with each delivery system in terms of patterns of their discharge and encapsulation of bioactive ingredients (Weiss et al., 2006). Although several investigations have confirmed improved encapsulation with these delivery systems, however, it is quite critical to explore functional efficacy of these systems particularly nanoemulsions in complex food matrices, particularly meat and meat-based products (Salminen et al., 2013).

4.4 Potentiality of Nanomaterials to Deliver Functional Ingredients in Meat Products

4.4.1 Nanotubes and Nanofibers

Nanotubes and nanofibers are two important applications of nanotechnology. Nanofibers have only a few uses in the food industry because they are not produced from food grade components. Nanofibers have 10–1000 nm diameter. Developments are continuing in nanofiber production from food grade substances so their application in food industry will probably increase. Nanotubes are also predominately used for nonfood uses. Carbon nanotubes are used in catalytic reaction vessels and low resistance conductors. Under suitable conditions for some proteins, such as caseins and whey protein and meat proteins, can be constructed into nanotubes (Graveland-Bikker and de Kruif, 2006; Graveland-Bikker et al., 2006).

The awareness of utilizing nanotubes for the development of functional food products was originated from carbon nanotubes (CNTs). CNTs have a number of other nonfood applications (Weiss et al., 2006). CNTs exist in two forms namely, single-walled and multiple-walled that have potential applications in the domain of food safety to detect pathogenic microorganisms by immobilizing the antibodies, besides several other benefits. Sensitivity of immunosensors can be enhanced up to sixfold by increasing the surface area of nanotubes, which ultimately reduces the detection limit of *Staphylococcal enterotoxin* B (Yang et al., 2008). In addition, incorporation of CNTs along with allyl isothiocyanate improves the antimicrobial potential of cellulose-based food packaging. A group of researchers concluded that inclusion of CNTs in packaging materials inhibited salmonella growth in shredded cooked broiler meat during a storage time of 40 days (Dias et al., 2013).

Alfa-lactalbumin is present in milk and has the ability to assemble as a nanotube under certain conditions. This ability of α-lactalbumin makes it possible to use it as a nanotube for food processing by using the knowledge of nanotechnology. Encapsulation efficiency of active ingredients can be enhanced by increasing surface to volume ratio of the materials. These nanotubes have good mechanical and heat stress, as well as improved storage ability, viscosity, and gelation properties (Graveland-Bikker and de Kruif, 2006). Additionally, gelatin nanofibers are produced by electrospinning of gelatin, which has improved thickening and stabilizing properties as compared to bulk gelatin. These nanofibers can also be utilized as stabilizers in food emulsions (Okutan et al., 2014). Zein nanofibers are also used to improve the delivery of several antioxidants such as gallic acid (Neo et al., 2013).

4.4.2 Nanodispersions and Nanocapsules

The important food components including antimicrobials, antioxidants, food additives, and vitamins come in a variety of molecular and physical structures. They are seldom employed purely, but are generally used in delivery systems. A delivery system provides many functions such as transportation of the functional components to its functional site. These delivery systems must be compatible with food component characteristics such as shelf life, taste, and texture; in addition, they also protect food components from biological and chemical degradation and control release of functional ingredients under particular conditions. Because they can efficiently perform all these functions, nanodispersions and nanocapsules are perfect systems for functional ingredient's delivery. Most of these nanostructures include: colloids, nanoemulsions in addition to biopolymeric nanoparticles. A delivery system should conduct following unique functions:

- They can carry the functional ingredients to the site of activity
- They protect the ingredients from chemical or biological destruction throughout processing and storage processes and keep them in active state
- It controls the release of functional ingredients under specific conditions such as temperature, pH, and ionic strength
- It must be compatible with other ingredients in food system as well as with qualitative and physicochemical characteristics such as taste, texture, and shelf life

Properties of the delivery system are one of the most critical factors having an influence on efficiency of functional ingredients in most products. Numerous delivery systems continue to be designed to encapsulate functional food ingredients, as well as basic alternatives such as association colloids, emulsions, biopolymer matrices, and many others. Every delivery system provides a unique benefit and drawback for encapsulation, safety, delivery of components, along with cost, convenience, regulating status, biodegradability, in addition to biocompatibility.

4.5 Nanotechnology in Meat Packaging

Food packaging is very important for shelf-life extension and to preserve the organoleptic properties of food. Packaging materials must have proper physical, thermal, mechanical, and optical properties to attain maximum benefits. Packaging materials should also possess good antimicrobial and barrier properties against moisture, gases, and aroma. However, traditional packaging materials are not biodegradable, therefore serious environmental and health concerns are associated with their use in food systems. These packaging materials can be replaced with ecofriendly packaging materials produced by biopolymers but the major disadvantages of using this kind of packaging stuff are poor mechanical strength insufficient barrier properties (Rhim et al., 2013). Studies have also demonstrated that nanotechnology is very helpful for the growth and development of packaging industries (EFSA, 2009). Some key benefits of using nanotechnology for food packaging are improved mechanical, physical, heat-resistant, and barrier properties, along with enhanced biodegradability and flame retardancy while comparing with conventional packaging materials (Lee, 2010). Use of nanomaterials in food packaging also increases the antimicrobial efficiency of materials, and detects spoilage of food through nanosensors (de Azeredo, 2009).

4.5.1 Types of Nanopackaging

4.5.1.1 Improved Nanopackaging for Meat Products

Improved packaging includes the use of nanocomposites to provide good barrier properties. Nanocomposites are made by reinforcing polymer matrices in the nanoscale through fillers to attain optimum packaging efficiency (Silvestre et al., 2011). Some common examples of fillers include clay, silicates, cellulose microfibrils, and carbon nanotubes (de Azeredo, 2009). Silicates are the most broadly used fillers used in nanocomposites. Nanocomposites are classified on the basis of silicate arrangement as exfoliated, intercalated, or flocculated nanocomposites (Weiss et al., 2006). Polymers used for manufacturing nanocomposite-based packaging materials are polyamides, polyolefins, polystyrene, and nylon (Silvestre et al., 2011). However, increasing demand for recyclable packaging materials has directed the use of natural and synthetic biopolymers in packaging systems. Some commonly used natural biopolymers are cellulose, chitosan, and carrageenan, while examples of synthetic biopolymers are polyvinyl alcohol, polylactide (PLA), and polyglycolic acid (PGA) (Rhim et al., 2013). Among the most often used fillers used in the nanoscale production of packaging materials, montmorillonite is the most extensively exploited filler in nanocomposites due to several advantages like safety and cost effectiveness (Silvestre et al., 2011).

Barrier properties of packaging material can be enhanced by using nanocomposites, which enables oxygen to take a more convoluted path comparative to traditional food packaging (Silvestre et al., 2011). Studies also revealed that the use of nanoclays in vacuum-packaging of beef loins also improved barrier properties (Picouet et al., 2014). The nanosized clays are generally dispersed in a polymer matrix of polyamide 6 (PA6), which leads to the formation of a packaging film. This demonstrates the following advantages:

- Elevated oxygen barrier properties
- Improved ability to block UV rays
- Better stiffness

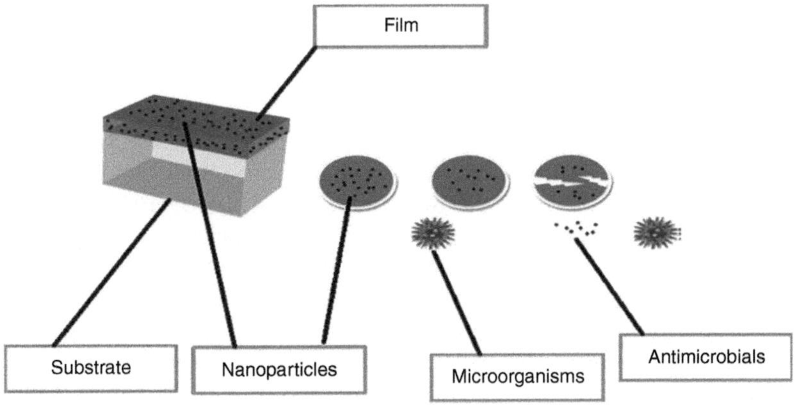

Figure 10.2. Animation nanopackaging coating film.

In a study, effectiveness of nanocomposite packaging was compared with commercial packaging. Results explicated similar effects on the beef quality with an added advantage of less thickness (Picouet et al., 2014). Likewise, Brody et al. (2008) stated that meat packed in a polyamide plastic film with added clay nanoparticles demonstrated longer shelf life as compared to the samples packed in conventional packaging due to improved oxygen, water and carbon dioxide barrier properties. Fig. 10.2 shows schematic diagram of typical antimicrobial nanopackaging coating film.

4.5.1.2 Active Nanopackaging

Active nanopackaging includes the use of metallic oxide nanoparticles in nanocomposites that are added in the packaging systems to improve preservation effects. In this technology, inorganic nanoparticles are incorporated into polymers to enhance antimicrobial properties of packaging materials. These nanoparticles can be used without difficulty and have improved functionality in comparison to molecular antimicrobials (Althues et al., 2007; Duncan, 2011). Nonetheless, their specific functions and aptness for packaging depends on their chemical composition. For intense, silver nanoparticles keep excellent antimicrobial potential while no antimicrobial properties can be observed with gold nanoparticles (Kim et al., 2007; Duncan, 2011). A number of proposed mechanisms are thought to be involved behind antimicrobial action of silver nanoparticles (Morones et al., 2005; Lok et al., 2007; Duncan, 2011). The main mechanistic approaches include:

- Cellular destruction by release of silver atoms from nanoparticles
- Cell toxicity due to discharge of silver ions from inside of the nanoparticles
- Structural demolition of cell membranes by certain nanoparticles

Silver nanoparticles have a wide spectrum of antimicrobial activity even at small concentrations (2–4 μg mL^{-1}). These nanoparticles are also quite effective against those pathogenic strains of microorganisms that are resistant to traditionally used chemical antimicrobials. Silver nanoparticles having size from 45 to 50 nm can significantly improve the shelf life of meat products. Likewise, TiO_2 nanoparticles also have the potential to tackle the microbial growth if incorporated in packaging materials. Another advantage of TiO_2 nanoparticles is that these particles protect packaged food items from damage caused by UV light (Duncan, 2011). In addition, several other features of nanoparticles such as shape, size, and surface charge are also very important to evaluate their antimicrobial efficiency (Duncan, 2011). One of the peers, Panea et al. (2013) showed that use of silver along with zinc oxide (ZnO) in the form of a nanocomposite have improved antimicrobial potential if incorporated in low-density polyethylene packaging systems. Consequences of aforementioned study revealed that pathogenic bacteria frequently present in meat and meat-based products like *E. coli, L. monocytogenes*, and *Pseudomonas aeruginosa* were significantly reduced by using this kind of packaging materials, based on a study of chicken breasts (Panea et al., 2013). In another experiment conducted by Fedotova et al. (2010), silver nanoparticles were added in sausage casings made of collagen and cellulose films. Results of the study showed that casings with added silver nanoparticles had strong antifungal and antibacterial effects.

4.5.1.3 Smart Nanopackaging

Smart nanopackaging involves the use of nanodevices combined with polymers for food packaging. These kinds of packaging materials are significantly used for the monitoring of packaged food and its surrounding environmental conditions during the transport and storage (Yam et al., 2005). In addition, smart packaging also guarantees the food authenticity by providing protection against counterfeiting. Likewise, smart packaging can also be used to improve the integrity of the packaging material. The inclusion of some nanodevices in the packaging systems is also helpful to trace the history of origin of raw material, processing place, time of packaging, suitable storage temperature, and expiration date of the product. Additionally, these nanodevices also have the ability to detect different microbes, chemicals, and toxins produced or already present in the food while being merged in the packaging materials (Silvestre et al., 2011).

4.6 Nanomaterials for Meat Packaging

4.6.1 Nanolaminates

A nanolaminate comprises two or more layers of material with nanometer measurements that are physically or artificially bonded to one another. Nanolaminates can provide food researchers a few points of interest for the production of edibles and have various beneficial applications in the food industry. Edible coatings are presently utilized on a wide range of foods products such as vegetables, fruits, meats, chocolate, confections, bakery items, and French fries (Morillon et al., 2002; Cagri et al., 2004).

These coatings act as lipid, moisture, and gas barriers. They can improve the appearance of food products and can also carry functional ingredients such as antimicrobials, antioxidants, colors, flavors, and nutrients to food systems. The fundamental properties of the edible coatings depend on the materials which are used for their formation. The structure, thickness, and characteristics of laminates can be altered in various ways such as adsorbing substances in the dipping solutions, the number of plunging steps used, the order used for the products is brought into the different plunging arrangements, the organization, and ecological conditions utilized such as ionic strength, temperature, and dielectric constant. The main thrust for surface adsorption of a substance would rely on the type of surface and the way of adsorption by substance and it could be: hydrogen bonding, electrostatic, hydrophobic interactions, thermodynamically incompatibility, and so on.

Nanolaminates are modified type of edible coatings that are used to cover food surfaces with the help of nanoscale materials (Weiss et al., 2006). Layer-by-layer (LbL) coating on food material having a certain electrostatic charge is known as one of the most prominent techniques that have been used in the modern times (Kotov, 2003). Controlling the thickness of coating (1–100 nm) is the major advantage of employing LbL deposition method (Weiss et al., 2006). These properties make LbL-deposited coatings more suitable for food as compared to traditional coatings (Kotov, 2003; Weiss et al., 2006). These types of coatings can also be used as a carrier of antimicrobial and antioxidant agents in addition to providing barrier properties against moisture and gases. However, properties of these edible coatings depend on the type and properties nanomaterials used for incorporation in the coating materials (Weiss et al., 2006). Different types of coating materials have been used for coating different food products. These coating materials might be proteins, lipid- or carbohydrate-based. Different functional properties can be observed in the edible coatings by incorporation various biopolymeric nanoparticles. For instance, films containing lipids provide excellent moisture barrier properties but have poor gas barrier functions and have deprived mechanical strength. Conversely, edible coatings based on proteins and polysaccharides offer good gas barrier properties but are not very effective against the transfer of moisture. Furthermore, use of nanolaminates in edible coatings for simultaneously getting the benefits of extended shelf life and provision of nutrition (Weiss et al., 2006).

4.6.2 Nanotracers in Meat Packaging

Because engineered nanomaterials are exposed to human beings, several hazards associated with nanoparticles need to be controlled for the reduction of risks allied to the consumption of nanomaterials. Nanotracers are very useful to ensure the safety and biosecurity of food by monitoring possible threats of exposure (Diallo and Brinker, 2011). Nanomonitors or nanotracers have diverse applications in quality control, environmental monitoring systems, and exposure assessment of nanoparticles (Philips Co. Inc., 2014). For example, aerasense, a nanotracer, has the ability to monitor and enumerate the size, concentration, and surface area of nanocomponents. Such devices can also be explored for exposure assessments of nanoparticles at personal level, identification of pollution and hazards caused by nanoparticles at processing areas, and tracing the sources of these particles (Marra et al., 2010). For that reason, these nanotracers can be applied for the assessment of risks and hazards at every level of processing in meat chains.

4.6.3 Nanosensors

Various types of nanosensors have been utilized for the rapid monitoring and quantification of pathogenic microbes, spoilage causing agents and proteins responsible for allergic reactions (Silvestre et al., 2011; NSI, 2013). So, nanosensors possess the compatibility to cogently influence the way of processing in food systems (NSI, 2013). In general, these nanosensors are incorporated to the food packaging materials to monitor the deviations in internal as well as external environment of food. Mills (2005) employed oxygen indicators to the food packaging of raw bacon at elevated concentration of CO_2. Mills observed significant color changes in nanoindicators placed in the packaging material when food was exposed to oxygen.

Another advantage of using nanosensors is that these indicators have the ability to detect food spoilage in the early stages, thereby can be used to avoid different problems faced by food processors, retailers, and consumers. Nanosensors using CNTs along with antibodies are quite effective for the identification of food pathogens like salmonella on the surface of food products such as chicken fillets (Villamizar et al., 2008). These nanosensors can also be deployed for the detection of freshness in canned tuna with the help of several chemicals used as meat freshness indicator like xanthine and hypoxanthine (Çubukçua et al., 2007). Although devices based on use of nanotechnology are facing a lot of challenges in formulation, integration, and manufacturing of such devices at commercial scale but these sensors are anticipated to have a bright future for food industry (Evans, 2009).

5 Potential Risks of Using Nanotechnology in Food Processing

Use of nanotechnology in food processing shows potential benefits over conventional methods of processing but anxieties related to the risks associated with the use of nanomaterials for the processing and preservation of different food products have also been raised up. Therefore, a number of tests have been performed to assess the impact of using nanomaterials on human cells. Examples of some tests include estimation of oral and toxicity, acute and chronic tests, and estimation of the capability of nanomaterials to induce skin irritations and mutagenesis (Cushen et al., 2012). It has also been observed that many types of toxicities like cytotoxicity and genotoxicity, inflammation of cells and oxidative stress may be caused by using nanoparticles in food (Cockburn et al., 2012). Bioefficacy studies depicted that use of metallic nanoparticles such as titanium dioxide, copper, and zinc-induced acute oral toxicity in rodents at higher levels of dosage (Bouwmeester et al., 2009). Accordingly, toxic effects of using biopolymeric nanomaterials like PLGA were not found to be prominent but they have limitations such as reduced loading ability and elevated rupture (Danhier et al., 2012). However, it is critical to identify that variation in toxic effects instigating from free, bound, or biodegradable nanomaterials (Borm and Berube, 2008). Liposomes are also biodegradable compounds that are used in food processing as micro- and nanoparticles, and not reported to be toxic after consumption (Underwood and Eps, 2012; Akbarzadeh et al., 2013). Nevertheless, higher production cost, complicated manufacturing and processing methods and poor stability have impeded the use of liposomes in foods (Underwood and Eps, 2012).

Cockburn et al. (2012) postulated that use of engineered nanomaterials in food processing may have varied levels of uptake in comparison with macromaterials. Particularly, modified behavior interactions with the biological systems can also be anticipated when solubility of nanomaterials has been decreased. Contrarily, different features of particles to be used in processing such as size, shape, and charge may also lead to cell toxicity. As an example, pore formation in cellular membranes is often linked with the excessive use of nanomaterials (Lovric et al., 2005). Moreover, accrual of some nanoparticles in different cells also leads toward cell inflammation. In addition, accumulation of nanomaterials in body also induces oxidative stress, which ultimately leads toward different infections. For intense, elevated levels of reactive oxygen species (ROS) in cells can result in the reduction of glutathione production in the cell, which has significant antioxidant potential. These mechanistic approaches are generally concomitant with possible toxicity in humans. Likewise, sometimes nanoparticles interact with DNA that leads toward the onset of genotoxicity (Cockburn et al., 2012). Similarly, use of polymeric micelles and solid lipid nanoparticles also cause toxicity of low level. In the same way use of nanoemulsions can also cause significant health hazards (He et al., 2010; Underwood and Eps, 2012). To understand the mechanistic approach involved behind the toxic effects of some nanomaterials and to combat the hitch, case by case risk assessment would be required (Bouwmeester et al., 2009; Cockburn et al., 2012).

6 Problems Associated with the Use of Nanoparticles in Packaging

Food packaging is the most promising domain wherein nanotechnology has a significant impact (Chaudhry and Castle, 2011). A number of studies have revealed that use of nanoparticles in packaging materials is more preferable by consumers as compared to their direct incorporation in food due to several health concerns associated with their usage (Siegrist et al., 2007). However, many consumers have no awareness about the reality that nanosubstances present in the packaging materials are shifted to food and can be inhaled. They can also be transferred to humans through contact with skin (Carlson et al., 2008). Different groups of researchers showed that silver, titanium dioxide, and CNTs could enter the blood stream and accumulate in the organs due to their poor solubility (Gurr et al., 2005; Carrero-Sánchez et al., 2006; Kim et al., 2008; Rhim et al., 2013).

Avella et al. (2005) studied that particle migration is quite lower from nanocomposites to foods and the level of this relocation falls in the permissible level approved by European Commission (EC). Similarly, Panea et al. (2013) depicted that particle shifting of silver and zinc oxide nanocomposites was below the limits described by EC. However, it is also observed that zinc oxide nanoparticles have genotoxic effects even at very low concentration (Sharma et al., 2009). It is also confirmed through experimentation that the rate of particle migration has been significantly enhanced by decreasing the viscosity and size of nanoparticles (Simon et al., 2008). Another group of researchers, Huang et al. (2011), reported that time and temperature of storage are the two major factors that notably affect the rate of nanoparticle transfer from the packaging material to food. Besides, several environmental hazards are also related to the use of some nanocomposites because of their biodegradability. However, further environmental toxicity studies are needed to understand the science involved behind the toxicity induced by some nanoparticles, which would pointedly improve the consumer's acceptability toward use of foods prepared by the involvement of nanotechnology (Cushen et al., 2012). Although the release of these compounds from the packaging material is allowed in some countries, the long-term impacts of using nanomaterials are still unknown (Cushen et al., 2012; Rhim et al., 2013). Hence, proper risk assessment, maintenance, and management of nanoparticles, transfer of information regarding the food safety are among the big challenges faced by this sector (Bouwmeester and Marvin, 2010).

7 Regulations of Using Nanotechnology in Food Systems

The United States of America, China, Germany, and Japan are the leading countries that are currently dealing with the development of food products including meat products by using the knowledge of nanotechnology (Helmut Kaiser Consultancy, 2009). China and some other countries also provide open marketing of food products containing nanomaterials because of their poor legislation regarding the use of nanoparticles in food industry (Chaudhry and Castle, 2011). On the other hand, in developed countries such as the United States, several challenges exist in regulation of using nanotechnology due to complications in nanotechnology and are further worsened due to the unavailability of a uniform and comprehensive framework which ensures the safety of consumers (Corley et al., 2009).

Currently, no defined regulations are available regarding the use of nanotechnology in the food in the United States (FDA) and Europe (EFSA, 2009). The US Food and Drug Administration expresses that it manages "food products, but not technologies," and envisions that numerous results of nanotechnology will fall under the concentration of different centers of the FDA and will in this manner be controlled by the Office of Combination Products. The FDA claims that "particle sizing is not the issue" and stresses that new materials, no matter the technological know-how accustomed to develop all of them, are going to be subjected to the conventional battery regarding safety assessments. However, Institute of Food Sentients and Technologist (IFST)-UK based independent skill qualifying group consisting of food researchers and technologists consider that sizing is an issue and suggest that nano-particles will be handles as dangerous resource till confirmation of safety.

The impairments in the risk analysis originating from nano-products include incomplete information, inadequate models, and qualms from government agencies (Kuzma et al., 2008). Additionally, the pace of risk assessment programs is another major reason that needs the implementation of regulations for nanomaterials. Likewise, measuring the safety of new food packaging materials has also become compulsory for ensuring the food safety, due to the ability of nanomaterials to transfer from packaging to the food. Moreover, formation of food products containing nanoparticles cannot be kept up with the consumer's expectations regarding their safety status (Cushen et al., 2012). Even so, the worth of regulation in food systems depends on the extensiveness of definitions, product accountabilities, and applications involving use of nanoparticles possessing novel and modified features as well as the appropriate permissible concentrations refer to the nanomaterials (Chaudhry et al., 2010).

8 Public Acceptance of Nanoproducts

It is a well-known fact that public acceptance is authoritative for the successful marketing of a particular food product (Siegrist et al., 2007). Similar to the case of using genetically modified foods, acceptance is also hindered for products containing nanoparticles due to several health and environmental concerns (Gaskell et al., 1999). Generally, consumers are attracted to the products that are cost effective and offer additional benefits, but reluctance has also been seen toward the products manufactured with the help of nanotechnology due to health concerns

(Spence and Townsend, 2006; Siegrist et al., 2008). In a study proposed by Siegrist et al. (2007) a willing to buy (WTB) model was designed to elucidate the factors affecting the suitability of nanotechnology in the food industry. Results of the study depicted that affect heuristics can impose a vital influence on the perception of drawbacks and assistances associated with the use of novel products. Hence, WTB of these kinds of products is affected in a positive way by attainment of social trust and improvement in the perception of genuineness of nanoproducts (Siegrist et al., 2007). In another study, Cushen et al. (2012) claimed that most of the consumers in the United States of America are not aware of nanoproducts, but have an expectant insight. Similar findings were observed in Taiwan whereas less positive views were recorded in Europe regarding the use of nanoproducts (Chen et al., 2013; Gaskell et al., 2005). Therefore, public perception on the subject of using nanotechnology in food can be improved by refining the awareness among common people (Siegrist et al., 2007). So, processors and legislative authorities should take some measures for public understanding regarding the use of nanotechnology in food processing and preservation and safety status of these functional products (Chen et al., 2013).

9 Conclusions

Involvement of nanotechnology in the food systems has been gaining importance around the globe. A promising area for the application of nanotechnology is meat packaging that has potential benefits such as effectual barrier properties and antimicrobial possessions. Other areas of using nanoparticles include meat processing for the purpose of enhancement in the sensory characteristics of meat products and nanoencapsulation of bioactive ingredients. In addition, improved nutritional value and additional functional properties of meat products are also included in the objectives of using nanotechnology in the meat industry. However, many challenges have been faced by the researchers as well as producers while using nanotechnology in meat products. For instance, several problems like poor oxidative stability and short-term antimicrobial effects have been faced by using nanoemulsions in meat products. Other important challenges associated with the use of nanopowders are unavailability of suitable processing equipment and poor knowledge about the nature of the ingredients used. Furthermore, some nanoparticles have difficult production methodologies, lack of scalability, cost ineffectiveness, and poor performance efficiency due to their unique nature. A number of potential health-related risks are also concomitant with the consumption of some nanoparticles. These limitations depend on the specific characteristics of nanomaterials to be used and biokinetics of the human body in response to these materials. Nonetheless, with the advancement in research activities, progression of accompanying technologies, modified instruments, and improved methodologies, nanotechnology can expressively contribute to meat industries. Apart from filling the gaps in knowledge about the fabrication and safety of nanoparticles, improving consumers' acceptance, finances and rational regulation is also very imperative for the fruitful future applications of nanotechnology in the development of meat products and packaging.

References

Abdou, E.S., Osheba, A.S., Sorour, M.A., 2012. Effect of chitosan and chitosan-nanoparticles as active coating on microbiological characteristics of fish fingers. Int. J. Appl. Sci. Technol. 2, 158–169.

Akbarzadeh, A., Rezaei-Sadabady, R., Davaran, S., Joo, S.W., Zarghami, N., Hanifehpour, Y., Samiei, M., Kouhi, M., Nejati-Koshki, K., 2013. Liposome: classification, preparation, and applications. Nanoscale Res. Lett. 8, 102.

Althues, H., Henle, J., Kaskel, S., 2007. Functional inorganic nanofillers for transparent polymers. Chem. Soc. Rev. 36, 1454–1465.

Augustin, M.A., Sanguansri, P., 2009. Nanostructured materials in the food industry. Adv. Food Nutr. Res. 58, 183–213.

Avella, M., De Vlieger, J.J., Errico, M.E., Fischer, S., Vacca, P., Volpe, M.G., 2005. Biodegradable starch/clay nanocomposite films for food packaging applications. Food Chem. 93, 467–474.

Bilska, A., Rudzinska, M., Kowalski, R., Krysztofiak, K., 2009. The effect of soy hydrolysates on changes in cholesterol content and its oxidation products in fine-ground model sausages. Acta Sci. Pol. 8, 15–22.

Borm, P.J.A., Berube, D., 2008. A tale of opportunities, uncertainties, and risks. Nano Today 3, 56–59.

Borm, P.J.A., Robbins, D., Haubold, S., Kuhlbusch, T., Fissan, H., Donaldson, K., Schins, R., Stone, V., Kreyling, W., Lademann, J., Krutmann, J., Warheit, D., Oberdorster, E., 2006. The potential risks of nanomaterials: a review carried out for ECETOC part. Fibre Toxicol. 3, 11.

Bouwmeester, H., Marvin, H.J.P., 2010. Potential risks of nanofood to consumers. In: Chaudhry, Q.L., Castle, L., Watkins, R. (Eds.), Nanotechnologies in Food. Royal Society of Chemistry Publishers, Cambridge, UK, pp. 134–140.

Bouwmeester, H., Dekkers, S., Noordam, M.Y., Hagens, W.I., Bulder, A.S., De Heer, C., Voorde, T.S.E., Wijnhoven, S.W., Marvin, H.J., Sips, A.J., 2009. Review of health safety aspects of nanotechnologies in food production. Regul. Toxicol. Pharmacol. 53, 52–62.

Brody, A.L., Bugusu, B., Han, J.H., Sand, C.K., Mchugh, T.H., 2008. Innovative food packaging solutions. J. Food Sci. 73, 107–116.

Cagri, A., Ustunol, Z., Ryser, E.T., 2004. Antimicrobial edible films and coatings. J. Food Prot. 67, 833–848.

Carlson, C., Hussain, S.M., Schrand, A.M., Braydich-Stolle, L.K., Hess, K.L., Jones, R.L., Schlager, J.J., 2008. Unique cellular interaction of silver nanoparticles: size-dependent generation of reactive oxygen species. J. Phys. Chem. B 112, 13608–13619.

Carrero-Sánchez, J.C., Elías, A.L., Mancilla, R., Arrellín, G., Terrones, H., Laclette, J.P., Terrones, M., 2006. Biocompatibility and toxicological studies of carbon nanotubes doped with nitrogen. Nano Lett. 6, 1609–1616.

Chang, Y.C., Chen, D.G.H., 2005. Adsorption kinetics and thermodynamics of acid dyes on a carboxymethylated chitosan-conjugated magnetic nano-adsorbent. Macromol. Biosci. 5, 254–261.

Chaudhry, Q., Castle, L., 2011. Food applications of nanotechnologies: an overview of opportunities and challenges for developing countries. Trends Food Sci. Technol. 22, 595–603.

Chaudhry, Q., Scotter, M., Blackburn, J., Ross, B., Boxall, A., Castle, L., 2008. Applications and implications of nanotechnologies for the food sector. Food Addit. Contam. 25, 241–258.

Chaudhry, Q., Watkins, R., Castle, L., 2010. Knowns, unknowns and unknown unknowns. In: Chaudhry, Q.L., Castle, L., Watkins, R. (Eds.), Nanotechnologies in Food. Royal Society of Chemistry Publishers, Cambridge, UK, pp. 212–214.

Chen, H., Weiss, J., Shahidi, F., 2006. Nanotechnology in nutraceuticals and functional foods. Food Technol. 60, 30–36.

Chen, M.F., Lin, Y.P., Cheng, T.J., 2013. Public attitudes toward nanotechnology applications in Taiwan. Technovation 33, 88–96.

Cockburn, A., Bradford, R., Buck, N., Constable, A., Edwards, G., Haber, B., Hepburn, P., Howlett, J., Kampers, F., Klein, C., Radomski, M., Stamm, H., Wijnhoven, S., Wildeman, T., 2012. Approaches to the safety assessment of engineered nanomaterials (ENM) in food. Food Chem. Toxicol. 50, 2224–2242.

Corley, E., Scheufele, D.A., Hu, Q., 2009. Of risks and regulations: how leading US nanoscientists form policy stances about nanotechnology. Nanopart. Res. 11, 1573–1585.

Çubukçua, M., Timurb, S., Anik, U., 2007. Examination of performance of glassy carbon paste electrode modified with gold nanoparticle and xanthine oxidase for xanthine and hypoxanthine detection. Talanta 74, 434–439.

Cushen, M., Kerry, J., Morris, M., Cruz-Romero, M., Cummins, E., 2012. Nanotechnologies in the food industry—recent developments, risks and regulation. Trends Food Sci. Technol. 24, 30–46.

Danhier, F., Ansorena, E., Silva, J.M., Coco, R., Breton, A.L., Préat, V., 2012. PLGA-based nanoparticles: an overview of biomedical applications. J. Control. Release 161, 505–522.

de Azeredo, H.M.C., 2009. Nanocomposites for food packaging applications. Food Res. Int. 42, 1240–1253.

Diallo, M., Brinker, C.J., 2011. Nanotechnology for sustainability: environment, water, food, minerals, and climate. In: Roco, M.C., Mirkin, C.A., Hersam, M.C. (Eds.), Nanotechnology Research Directions for Societal Needs in 2020: Retrospective and Outlook. Springer, London, UK, p. 229.

Dias, M.V., Soares, N.F., Borges, S.V., de Sousa, M.M., Nunes, C.A., de Oliveira, I.R.N., Medeiros, E.A.A., 2013. Use of allyl isothiocyanate and carbon nanotubes in an antimicrobial film to package shredded, cooked chicken meat. Food Chem. 141, 3160–3166.

Duncan, T.V., 2011. Applications of nanotechnology in food packaging and food safety: barrier materials, antimicrobials and sensors. J. Colloid. Interface Sci. 363, 1–24.

EFSA, Scientific Opinion of the Scientific Committee, 2009. Request from the European Commission on the potential risks arising from nanoscience and nanotechnologies on food and feed safety. EFSA J. 958, 1–39.

Evans, H.M., 2009. Nanotechnology-enabled sensing. Report of the National Nanotechnology Initiative Workshop, pp. 27–34.

Fedotova, A.V., Snezhko, A.G., Sdobnikova, O.A., Samoilova, L.G., Smurova, T.A., Revina, A.A., Khailova, E.B., 2010. Packaging materials manufactured from natural polymers modified with silver nanoparticles. Int. Polym. Sci. Technol. 37, 59–64.

Flanagan, J., Singh, H., 2006. Microemulsions: a potential delivery system for bioactives in food. Crit. Rev. Food Sci. Nutr. 46, 221–237.

FSAI, 2008. The Relevance for Food Safety of Applications of Nanotechnology in the Food and Feed Industries. Food Safety Authority of Ireland, Dublin, Ireland, p. 11.

Garti, N., Benichou, A., 2001. Double emulsions for controlled-release applications: progress and trends. In: Sjoblom, J. (Ed.), Encyclopedic Handbook of Emulsion Technology. Marcel Dekker, New York, NY, pp. 377–407.

Garti, N., Benichou, A., 2004. Recent developments in double emulsions for food applications. In: Friberg, S., Larsson, K., Sjoblom, J. (Eds.), Food Emulsions. Marcel Dekker, New York, NY, pp. 353–412.

Garti, N., Spernath, A., Aserin, A., Lutz, R., 2005. Nano-sized self-assemblies of non-ionic surfactants as solubilization reservoirs and microreactors for food systems. Soft Matter 1, 206–218.

Gaskell, G., Bauer, M.W., Durant, J., Allum, N.C., 1999. Worlds apart? The reception of genetically modified foods in Europe and the US. Science 285, 384–387.

Gaskell, G., Eyck, T.T., Jackson, J., Veltri, G., 2005. Imagining nanotechnology: cultural support for technological innovation in Europe and the United States. Public Underst. Sci. 14, 81–90.

Golding, M., Sein, A., 2004. Surface rheology of aqueous casein-monoglyceride dispersions. Food Hydrocoll. 18, 451–461.

Graveland-Bikker, J.F., de Kruif, C.G., 2006. Unique milk protein based nanotubes: food and nanotechnology meet. Trends Food Sci. Technol. 17, 196–203.

Graveland-Bikker, J.F., Schaap, I.A.T., Schmidt, C.F., De Kruif, C.G., 2006. Structural and mechanical study of a self-assembling protein nanotube. Nano Lett. 6, 616–621.

Gruère, G.P., 2012. Implications of nanotechnology growth in food and agriculture in OECD countries. Food Policy 37, 191–198.

Gupta, A.K., Gupta, M., 2005. Synthesis and surface engineering of iron oxide nanoparticles for biomedical applications. Biomaterials 26, 3995–4021.

Gurr, J.R., Wang, A.S., Chen, C.H., Jan, K.Y., 2005. Ultrafine titanium dioxide particles in the absence of photoactivation can induce oxidative damage to human bronchial epithelial cells. Toxicology 213, 66–73.

Haruyama, T., 2003. Micro- and nanobiotechnology for biosensing cellular responses. Adv. Drug Deliv. Rev. 55, 393–401.

He, C.X., He, Z.G., Gao, J.Q., 2010. Microemulsions as drug delivery systems to improve the solubility and the bioavailability of poorly water-soluble drugs. Expert Opin. Drug. Deliv. 7, 445–460.

Helmut Kaiser Consultancy, 2009. Study: nanotechnology in food and food processing industry worldwide 2003-2006-2010-2015. Available from: http://www.hkc22.com/nanomarkets.html

HOL (House of Lords), 2010. House of Lords, Science and Technology Committee, 1st Report of Session 2009–2010, HL Paper 22-I. Stationery Office, London, USA. Nanotechnologies and Food, vol. I: Report; pp. 12, 51–52.

Huang, Y., Chen, S., Bing, X., Gao, C., Wang, T., Yuan, B., 2011. Nanosilver migrated into food-simulating solutions from commercially available food fresh containers. Packag. Technol. Sci. 24, 291–297.

IFIC (International Food Information Council), 2006. Functional foods. Available from: http://ific.org/nutrition/functional/indcx.cfm

Jasińska, M., Dmytrów, I., Mituniewicz-Małek, A., Wąsik, K., 2010. Cow feeding system versus milk utility for yoghurt manufacture. Acta Sci. Pol. 9, 189–199.

Kim, J.S., Kuk, E., Yu, K.N., Kim, J.H., Park, S.J., Lee, H.J., Kim, S.H., Park, Y.K., Park, Y.H., Hwang, C.Y., Kim, Y.K., Lee, Y.S., Jeong, D.H., Cho, M.H., 2007. Antimicrobial effects of silver nanoparticles. Nanomed. Nanotechnol. Biol. Med. 3, 95–101.

Kim, Y.S., Kim, J.S., Cho, H.S., Rha, D.S., Kim, J.M., Park, J.D., Choi, B.S., Lim, R., Chang, H.K., Chung, Y.H., Kwon, I.H., Jeong, J., Han, B.S., Yu, I.J., 2008. Twenty eight-day oral toxicity, genotoxicity, and gender-related tissue distribution of silver nanoparticles in Sprague-Dawley rats. Inhal. Toxicol. 20, 575–583.

Kotov, N.A., 2003. Layer-by-layer assembly of nanoparticles and nanocolloids: intermolecular interactions, structure and materials perspective. In: Decher, G., Schlenoff, J.B. (Eds.), Multilayer Thin Films: Sequential Assembly of Nanocomposite Materials. Wiley-VCH, Weinheim, Germany, pp. 207–243.

Kuzma, J., Romanchek, J., Kokotovich, A., 2008. Upstream oversight assessment for agrifood nanotechnology: a case studies approach. Risk Anal. 28, 1081–1098.

Lee, K.T., 2010. Quality and safety aspects of meat products as affected by various physical manipulations of packaging materials. Meat Sci. 86, 138–150.

Linton, J.D., Walsh, S.T., 2008. A theory of innovation for process-based innovations such as nanotechnology. Technol. Forecast Soc. Change 75, 583–594.

Lok, C.N., Ho, C.M., Chen, R., He, Q.Y., Yu, W.Y., Sun, H., Tam, P.K., Chiu, J.F., Che, C.M., 2007. Silver nanoparticles: partial oxidation and antibacterial activities. J. Biol. Inorg. Chem. 12, 527–534.

Lövenstam, G., Rauscher, H., Roebben, G., Sokull Klüttgen, B., Gibson, N., Putaud, J.P., Stamm, H., 2010. Considerations on a Definition of Nanomaterial for Regulatory Purposes Joint Research Center of the European Commission (JRC) Reference Reports. Publication Office of the European Union, Luxembourg, pp. 17–18.

Lovric, J., Bazzi, H.S., Cuie, Y., Fortin, G.R.A., Winnik, F.M., Maysinger, D., 2005. Differences in subcellular distribution and toxicity of green and red emitting CdTe quantum dots. J. Mol. Med. 83, 377–385.

Marra, J., Voetz, M., Kiesling, H.J., 2010. Monitor for detecting and assessing exposure to airborne nanoparticles. J. Nanopart. Res. 12, 21–37.

Maynard, A.D., Aitken, R.J., Butz, T., Colvin, V., Donaldson, K., Oberdörster, G., Philbert, M.A., Ryan, J., Seaton, A., Stone, V., Tinkle, S.S., Tran, L., Walker, N.J., Warheit, B., 2006. Safe handling of nanotechnology. Nature 444, 267–269.

McClements, D.J., Decker, E.A., 2000. Lipid oxidation in oil-in-water emulsions: impact of molecular environment on chemical reactions in heterogeneous food systems. J. Food Sci. 65, 1270–1282.

McClements, D.J., Rao, J., 2011. Food-grade nanoemulsions: formulation, fabrication, properties, performance, biological fate, and potential toxicity. Crit. Rev. Food Sci. Nutr. 51, 285–330.

Mills, A., 2005. Oxygen indicator and intelligent inks for packaging food. Chem. Soc. Rev. 34, 1003–1011.

Moraru, C.I., Panchapakesan, C.P., Huang, Q., Takhistov, P., Liu, S., Kokini, J.L., 2003. Nanotechnology: a new frontier in food science. Food Technol. 57, 24–29.

Moraru, C.I., Huang, Q., Takhistov, P., Dogan, H., Kokini, J.L., 2009. Food nanotechnology: current developments and future prospects. Global Issues Food Sci. Technol. 21, 369–399.

Morillon, V., Debeaufort, F., Blond, G., Capelle, M., Voilley, A., 2002. Factors affecting the moisture permeability of lipid-based edible films: a review. Crit. Rev. Food Sci. Nutr. 42, 67–89.

Morones, J.R., Elechiguerra, J.L., Camacho, A., Holt, K., Kouri, J.B., Ramírez, J.T., Yacaman, M.J., 2005. The bactericidal effect of silver nanoparticles. Nanotechnology 16, 2346–2353.

Neo, Y.P., Ray, S., Jin, J., Gizdavic-Nikolaidis, M., Nieuwoudt, M.K., Liu, D., Quek, S.Y., 2013. Encapsulation of food grade antioxidant in natural biopolymer by electrospinning technique: a physicochemical study based on zein–gallic acid system. Food Chem. 136, 1013–1021.

NSI, 2013. Nanotechnology for sensors and sensors for nanotechnology: improving and protecting health, safety, and the environment. Nanotechnology Signature Initiative. Available from: http://www.nano.gov/node/849

O'Brien, N., Cummins, E., 2010. Ranking initial environmental and human health risk resulting from environmentally relevant nanomaterials. J. Environ. Sci. Health 45, 992–1007.

Okutan, N., Terzi, P., Altay, F., 2014. Affecting parameters on electrospinning process and characterization of electrospun gelatin nanofibers. Food Hydrocoll. 39, 19–26.

Olmedilla-Alonsoa, B., Jiménez-Colmeneroa, F., Sánchez-Muniz, F.J., 2013. Development and assessment of healthy properties of meat and meat products designed as functional foods. Meat Sci. 95, 919–930.

Ozimek, L., Pospiech, E., Narine, S., 2010. Nanotechnology is food and meat processing. Acta. Sci. Pol. 9, 401–412.

Panea, B., Ripoll, G., González, J., Fernández-Cuello, A., Albertí, P., 2013. Effect of nanocomposite packaging containing different proportions of ZnO and Ag on chicken breast meat quality. J. Food Eng. 123, 104–112.

Philips Co. Inc., 2014. Ultrafine and nanoparticle monitors. Available from: http://www.aerasense.com/index.php?pageID=3

Picouet, P.A., Fernandez, A., Realini, C.E., Lloret, E., 2014. Influence of PA6 nanocomposite films on the stability of vacuum-aged beef loins during storage in modified atmospheres. Meat Sci. 96, 574–580.

Ravichandran, M., Hettiarachchy, N.S., Ganesh, V., Ricke, S.C., Surendra, S., 2011. Enhancement of antimicrobial activities of naturally occurring phenolic compounds by nanoscale delivery against *Listeria monocytogenes*, *Escherichia coli* O157:H7 and *Salmonella typhimurium* in broth and chicken meat system. J. Food Saf. 31, 462–471.

Rhim, J.W., Park, H.M., Ha, C.S., 2013. Bio-nanocomposites for food packaging applications. Prog. Polym. Sci. 38, 1629–1652.

Riley, T., Govender, T., Stolnik, S., Xiong, C.D., Garnett, M.C., Illum, L., Davis, S.S., 1999. Colloidal stability and drug incorporation aspects of micellar-like PLA-PEG nanoparticles. Colloids Surf. B 16, 147–159.

Ritzoulis, C., Scoutaris, N., Papademetriou, K., Stavroulias, S., Panayiotou, C., 2005. Milk proteinbased emulsion gels for bone tissue engineering. Food Hydrocoll. 19, 575–581.

Salminen, H., Herrmann, K., Weiss, J., 2013. Oil-in-water emulsions as a delivery system for n-3 fatty acids in meat products. Meat Sci. 93, 659–667.

Sharma, V., Shukla, R.K., Saxena, N., Parmar, D., Das, M., Dhawan, A., 2009. DNA damaging potential of zinc oxide nanoparticles in human epidermal cells. Toxicol. Lett. 185, 211–218.

Shibata, T., 2002. United States Patent US 6,416,803B1. Method for producing green tea in microfine powder.

Shimoni, E., 2009. Nanotechnology for foods: delivery systems. Global Issues in Food Science TechnologyIUFoST World Congress, pp. 411–424.

Siegrist, M., Cousin, M.E., Kastenholz, H., Wiek, A., 2007. Public acceptance of nanotechnology foods and food packaging: the influence of affect and trust. Appetite 49, 459–466.

Siegrist, M., Stampfli, N., Kastenholz, H., Keller, C., 2008. Perceived risks and perceived benefits of different nanotechnology foods and nanotechnology food packaging. Appetite 51, 283–290.

Silvestre, C., Duraccio, D., Cimmino, S., 2011. Food packaging based on polymer nanomaterials. Prog. Polym. Sci. 36, 1766–1782.
Simon, P., Chaudhry, Q., Bakos, D., 2008. Migration of engineered nanoparticles from polymer packaging to food—a physicochemical view. J. Food Nutr. Res. 47, 105–113.
Sozer, N., Kokini, J.L., 2009. Nanotechnology and its applications in the food sector. Trends Biotechnol. 27, 82–89.
Spence, A., Townsend, E., 2006. Examining consumer behavior toward genetically modified (GM) food in Britain. Risk Anal. 26, 657–670.
Troy, D.J., Kerry, J.P., 2010. Consumer perception and the role of science in the meat industry. Meat Sci. 86, 214–226.
Underwood, C., Eps, A.W.V., 2012. Nanomedicine and veterinary science: the reality and the practicality. Vet. J. 193, 12–23.
USDA, 2008. Food safety information: additives in meat and poultry products. United States Department of Agriculture. Available from: http://www.fsis.usda.gov/wps/portal/fsis/topics/food-safety-education/get-answers/food-safety-fact-sheets/food-labeling/additives-in-meat-and-poultry-products/additives-in-meat-and-poultry-products
Villamizar, R., Maroto, A., Xavier, R.F., Inza, I., Figueras, M., 2008. Fast detection of *Salmonella Infantis* with carbon analytical nanotechnology for food analysis 17 nanotube field effect transistors. Biosens. Bioelectron. 24, 279–283.
Weiss, J., Takhistov, P., McClements, J., 2006. Functional materials in food nanotechnology. J. Food Sci. 71, 107–116.
Weiss, J., Gibis, M., Schuh, V., Salminen, H., 2010. Advances in ingredient and processing systems for meat and meat products. Meat Sci. 86, 196–213.
Yam, K.L., Takhistov, P.T., Miltz, J., 2005. Intelligent packaging: concepts and applications. J. Food Sci. 70, 1–10.
Yang, M., Kostov, Y., Rasooly, A., 2008. Carbon nanotubes based optical immunodetection of *Staphylococcal enterotoxin* B (SEB) in food. Int. J. Food Microbiol. 127, 78–83.
Young, J.F., Therkildsen, M., Ekstrand, B., Che, B.N., Larsen, M.K., Oksbjerg, N., Stagsted, J., 2013. Novel aspects of health promoting compounds in meat. Meat Sci. 95, 904–911.
Yusop, S.M., O'Sullivan, M.G., Preuß, M., Weber, H., Kerry, J.F., Kerry, J.P., 2012. Assessment of nanoparticle paprika oleoresin on marinating performance and sensory acceptance of poultry meat. LWT Food Sci. Technol. 46, 349–355.
Zhang, W., Zhang, J., Jiang, Q., Xia, W., 2013. The hypolipidemic activity of chitosan nanopowder prepared by ultrafine milling. Carbohydr. Polym. 95, 487–491.
Zhao, X., Yang, Z., Gai, G., Yang, Y., 2009. Effect of superfine grinding on properties of ginger powder. J. Food Eng. 91, 217–222.
Zhu, K., Huang, S., Peng, W., Qian, H., Zhou, H., 2010. Effect of ultrafine grinding on hydration and antioxidant properties of wheat bran dietary fiber. Food Res. Int. 43, 943–948.

11

ANTIMICROBIAL NANOCOMPOSITES FOR FOOD PACKAGING APPLICATIONS: NOVEL APPROACHES

Alberto Jiménez, María Vargas, Amparo Chiralt
Universitat Politècnica de València, Research Institute of Food Engineering for Development, Valencia, Spain

1 Introduction

The field of nanoscience and nanotechnology is extending the applications of physics, chemistry, biology, engineering, and technology into previously unapproached infinitesimal length scales. The polymer–nanoparticles or nanocomposites have been the exponentially growing field of research for developing the materials in past few decades and have been mainly focusing on the structure–property relationships and their development. Along with the exponential momentum of the development in polymer nanocomposites (Alexandre and Dubois, 2000; Paul and Robeson, 2008; Pavlidou and Papaspyrides, 2008; Sinha Ray and Bousmina, 2005; Sinha Ray and Okamoto, 2003; Spitalsky et al., 2010), much attention has been focused on the use of nanosized fillers (at least one dimension in the nanometer range, that is, 1–100 nm) in improving the performance of and adding new functionalities to polymer-based materials. One of the additional functions is the antimicrobial properties exhibited by some nanocomposites on the basis of the antimicrobial activity of the added nanofillers (eg, silver or metal oxide nanoparticles). The controlled release of these active nanofillers or the attached compounds to their surface or nanostructure leads to a sustained antimicrobial activity of the nanocomposites, which is very useful in extending the food shelf life when these are used as packaging materials. Nevertheless, active properties of the nanocomposites are greatly affected by the molecular interactions in the system, which affect the level

Novel Approaches of Nanotechnology in Food. http://dx.doi.org/10.1016/B978-0-12-804308-0.00011-X

of nanoparticle aggregation and their release, which, in turn, seriously determine their active properties.

Many studies have been carried out into the use of nanomaterials as fillers of polymers in order to obtain active packaging materials for long-term food preservation, thus enhancing food safety in their commercialization in further markets and reducing food waste. Nanomaterials act against microbes by different mechanisms, which are generally related with changes in the permeability of the bacterial cytoplasmic membrane. Likewise, manufacturers fill polymers with particles in order to improve the stiffness and the toughness of the materials, to enhance their barrier properties, to enhance their resistance to fire and ignition or simply to reduce cost, although addition of particulate fillers sometimes imparts drawbacks to the resulting composites such as brittleness or opacity.

One important aspect from the food safety point of view concerns with the nanoparticle migration from packaging to the food product and their potential toxicity for humans. In this sense, regulations about their use are very restricted while deep studies into toxicity of these materials are required.

In this chapter, the most recent advances in nanocomposites for antimicrobial food packaging applications are discussed, considering both biodegradable and nonbiodegradable polymers. The production techniques, the different nanomaterials, and the last applications to food products are reviewed in a comprehensive way. In addition, regulatory aspects related to nanocomposites are discussed.

2 Nanoparticles and Nanomaterials Used in Antimicrobial Nanocomposites

During the past few years a great amount of nanoparticles and nanomaterials have been studied in relationship with their antimicrobial activity giving rise, in most of the cases, to promising results. Thus, metallic ions, metallic oxides, inorganic compounds, and biobased materials have been assessed in combination with both nonbiodegradable and biodegradable polymers to form antimicrobial nanocomposites (see Table 11.1).

2.1 Silver Nanoparticles

Silver nanoparticles may be considered the most used antimicrobial nanoparticles due to their application in different fields such as electrical appliances, medical equipment, and packaging.

Table 11.1 Different Nanoparticles and Nanomaterials Used in Antimicrobial Nanocomposites

Nanoparticles	Main Features	References
Silver nanoparticles	High thermal stability, versatile, and long term antimicrobial activity. Ag^+ acts by interaction with proteins and enzymes, cell wall and plasma membrane components and microbial DNA	López-Carballo et al., 2013; Wang et al., 2012
Celullose nanomaterials	Mainly used as reinforcing agents of polymeric composites improving their mechanical resistance and barrier properties. Synergistic antimicrobial effect in ternary systems in combination with chitosan.	Khan et al., 2012; Li et al., 2015; Salehudin et al., 2014; Slavutsky and Bertuzzi, 2014
Clay nanoparticles	The most used clays in nanoform are those derived from montmorillonite. Cloisite® 30B has been widely tested, showing antibacterial activity against gram positive strains due to the presence of quaternary ammonium groups.	Liu et al., 2014; Rhim and Wang, 2014; Sothornvit et al., 2009
Metal oxide nanoparticles	Nanoparticles of TiO_2, ZnO, MgO, CaO, CuO, Al_2O_3, AgO, and CeO_2 have been studied as components of antimicrobial composites. ZnO have shown good antimicrobial effect in both biodegradable and nonbiodegradable polymers. TiO_2 have also been used, although its antimicrobial activity is dependent on by the effect of UV and visible irradiation.	Kanmani and Rhim, 2014b; Manna, 2012; Muñoz-Bonilla et al., 2013; Tankhiwale and Bajpai, 2012; Wei et al., 2014
Copper nanoparticles	Copper nanoparticles may be obtained by physical or chemical methods by using heat and UV radiation or reducing agents, respectively. Proved antifungal and antibacterial activity when added to biodegradable and nonbiodegradable polymers.	Cioffi et al., 2005; Llorens et al., 2012; Shankar et al., 2014
Nanocarriers	Nanoliposomes, carbon nanotubes, and mesoporous materials have been assessed as carriers of antimicrobial compounds in different matrices. Although these nanocarriers do not present antimicrobial activity (in general) they play a relevant role in the stabilization and antimicrobial activity of the entrapped/encapsulated active compounds.	Bendahou et al., 2015; Imran et al., 2015; Jiménez et al., 2014a; Qi et al., 2012

(*Continued*)

Table 11.1 Different Nanoparticles and Nanomaterials Used in Antimicrobial Nanocomposites (*cont.*)

Nanoparticles	Main Features	References
Nanoemulsions	Techniques such as high-pressure homogenization and sonication may be considered to obtain nanoemulsions, since very low size droplets have been related with higher antimicrobial activity.	Abd-Elsalam and Khokhlov, 2015; Acevedo-Fani et al., 2015; Otoni et al., 2014a,b; Salvia-Trujillo et al., 2015
Other nanoparticles	Other nanoparticles obtained from gold, polyaniline, or chitosan have been evaluated as antimicrobials. Gold nanoparticles showed lower antimicrobial effect in comparison with silver. Polyaniline do not present a remarkable antimicrobial activity, but may be used to improve the activity of other substances.	Antoniou et al., 2015; Bogdanović et al., 2015; Guerra et al., 2013; Masadeh et al., 2015

In this way, silver is widely used to avoid the microbial growth in the internal surface of refrigerators, in surgery equipment and novel packaging materials. The applicability of these nanoparticles is related to their physicochemical properties, which include high thermal stability and versatile activity (including bacteria, fungus, and viruses) and long-term antimicrobial capacity (Wang et al., 2012).

Generally, the antibacterial effect of nanoparticles is explained in terms of the interaction between such particles and different cell components, mainly located in the plasma membrane and cell wall. Regarding silver nanoparticles, different mechanisms have been described as explained by Wang et al. (2012) and López-Carballo et al. (2013). Thus, antimicrobial activity of silver nanoparticles is related to different reactions: (1) between thiol groups of proteins and enzymes and silver ions (Ag^+), altering their structure and function, (2) between Ag^+ and peptide-glycan cell wall, provoking structural changes and cell lysis, and (3) between Ag^+ and bacterial DNA, avoiding its replication.

Composites containing silver nanoparticles have been obtained by different procedures. López-Carballo et al. (2013) and Martínez-Abad et al. (2012) added directly silver nitrate ($AgNO_3$) to obtain Ag^+, in chitosan and EVOH films and studied its release in different systems. However, this is not the more usual process used, but this salt is generally used as a precursor of silver nanoparticles by means of a reduction step, as considered by

Sadeghnejad et al. (2014) and Zhao et al. (2013). These authors obtained polyethylene and soy protein isolate-based nanocomposites, respectively. The reduction step is performed in different ways, by varying the conditions of processing (temperature, pH, stirring, etc.) and the reducing agents. Thus, compounds such as sodium borohydride (Abreu et al., 2015; Mahmoud, 2015) or fructose (Sadeghnejad et al., 2014) have been recently considered.

Silver nanoparticles are active as antimicrobials against a great amount of microorganisms and nanocomposites loaded with silver nanoparticles have been reported to be effective against *Listeria monocytogenes*, *Salmonella* spp., *Escherichia coli*, *Staphylococcus aureus*, *Candida albicans*, *Bacillus subtilis*, *Salmonella typhimurium*, *Bacillus cereus*, and *Aspergillus niger* when added to both natural or synthetic polymers (Jokar et al., 2012; Kanmani and Rhim, 2014a; Mahmoud, 2015; Martínez-Abad et al., 2012; Pinto et al., 2013; Zhang et al., 2010; Zhao et al., 2013). Likewise, it has been reported that the antimicrobial activity of silver nanoparticles differs depending on the type of microorganism. In fact, Fortunati et al. (2014) reported that *E. coli* was more affected by silver ions than *S. aureus*. This difference was related with the structural difference in the cell wall of gram positive and gram negative cells. These authors explained that the thick peptidoglycan layer of the staphylococcal cell wall prevents the penetration of the nanoparticles inside the cytoplasm, thus giving rise to a limited antimicrobial effect.

2.2 Cellulose Nanoparticles

Cellulose nanoparticles may be used to form antimicrobial composites, even though their main usefulness is related with their capacity to reinforce polymeric materials, increasing the mechanical resistance of the material and improving its barrier properties, as found by Khan et al. (2012) and Slavutsky and Bertuzzi (2014).

Cellulose nanomaterials may be divided in different types according to the TAPPI standard (WI-3021) that defines the different forms of cellulose nanoparticles. Table 11.2 shows the different cellulose derived nanomaterials as reported by Peponi et al. (2014).

Due to the fact that cellulose nanomaterials are mainly obtained from biowaste, these types of nanoparticles are frequently used to obtain biodegradable nanocomposites, in combination with polymers such as starch, poly(lactic acid) (PLA) or chitosan (Tomé et al., 2013; Salehudin et al., 2014; Salmieri et al., 2014). The antimicrobial activity of cellulose nanoparticles is not very clear

Table 11.2 Types of Cellulose-Derived Nanomaterials According to the TAPPI Standard WI-3021 (Peponi et al., 2014)

Type of Nanomaterial	Definition
Nanocellulose (synonymous of cellulose nanomaterials)	General term to refer to materials composed by cellulose nano and microcrystals and cellulose nano and microfibrils.
Cellulose nanocrystals (CNC)	Cellulose nanofiber totally crystalline with dimensions of 3–10 nm in width and aspect ratio[a] generally between 5 and 50. Obtained mainly by acid hydrolysis, oxidation, or other methods.
Cellulose nanowhiskers (CNW)	Aggregates or individual forms of cellulose nanocrystals obtained by acid hydrolysis.
Cellulose nanofibrils (CNF)	Cellulose nanofiber with both crystalline and amorphous regions with dimensions of 5–30 width and aspect ratio usually greater than 50. They are produced by bacteria (bacterial cellulose) or by mechanical treatment from vegetal tissues, usually containing hemicellulose.

[a] Aspect ratio: For a nanoparticle refers to the ratio of its sizes in different dimensions.

because contradictory results are reported by different authors. Thus, Li et al. (2015) and Salehudin et al. (2014) found that cellulose nanomaterials improved the inherent antimicrobial activity of chitosan when added at 1 and 2% (w/w), respectively. Nevertheless, contents of nanoparticles higher than 2% reduced the antimicrobial activity of the nanocomposites (Li et al., 2015), which is consistent with that reported by Tomé et al. (2013): a decrease in the antimicrobial activity of starch-chitosan composites when cellulose nanofibers content was 10%. Fortunati et al. (2013, 2014) did not find antimicrobial activity in poly(vinyl alcohol) (PVA) and PLA composites containing cellulose nanocrystals (CNCs).

From these results it seems that the ability of cellulose nanomaterials to avoid microbial growth should be further studied.

Li et al. (2015) reported an increase of the antimicrobial activity of PVA/chitosan films by the addition of cellulose nanowhiskers (CNW), which was related to a better availability of amino groups of chitosan to interact with the negatively charged cell membranes. When CNW were not present in the polymer blend, chitosan is homogeneously dispersed in the PVA matrix, which inhibits the availability of amino groups, thus reducing their effective concentration of the active groups, as compared to neat chitosan. However, the addition of CNW enhanced the polymer phase separation due to the new balance of molecular interactions, giving rise to high local amino concentrations, as compared to PVA/chitosan blends without CNW.

2.3 Clay Nanoparticles

Layered nanoclays have been widely used as reinforcing nanoparticles in biodegradable packaging materials due to their high-availability, low-cost, significant enhancements conferred to the polymeric matrix and relatively simple processability (Alexandre and Dubois, 2000). Generally, nanoclays increase the mechanical resistance of the materials due to their very high elastic modulus. Besides, barrier properties (water vapor and gases) are significantly improved by nanoclay addition in agreement with the increase in the pathway tortuosity for mass transfer of the polymeric matrix. Nevertheless, the properties of the clay-containing nanocomposites are highly dependent on the distribution of the nanoclays. Therefore, this type of nanoparticles may lead to micro- or nanocomposites, as can be observed in Fig. 11.1. If layered nanoclays aggregate inside the polymeric matrix, phase separation occurs, therefore giving rise to a microcomposite. On the contrary, if the nanomaterial is well dispersed, the final material will be a nanocomposite, whose properties could highly differ from those of microcomposites.

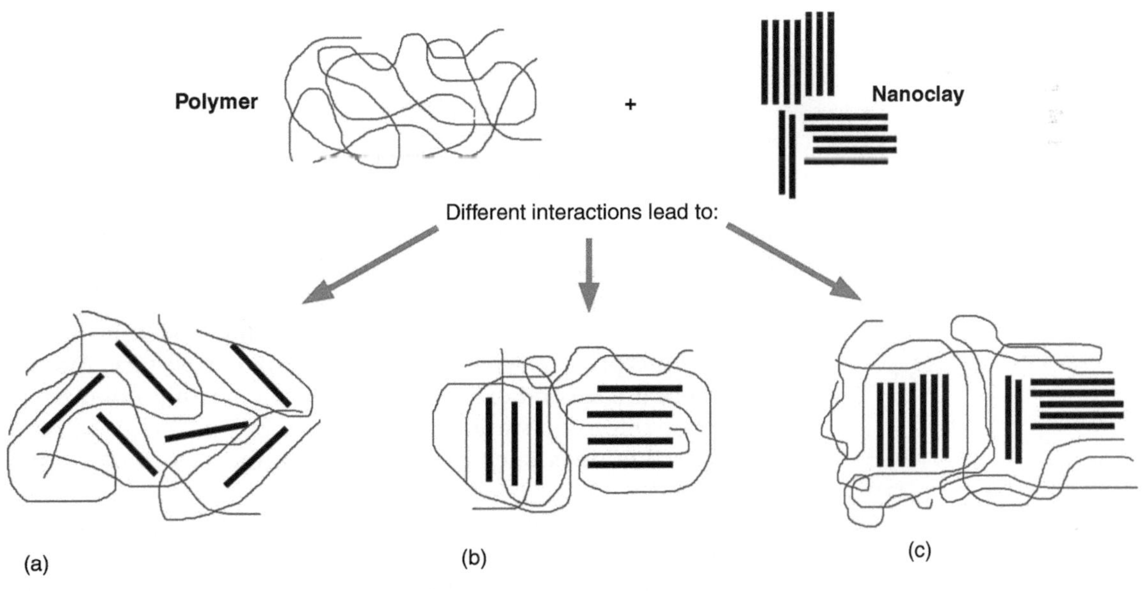

Figure 11.1. Different polymer–nanoclay composite structures. (a) Delamination of nanoclays (exfoliation). (b) Alternative layers of nanoclays and polymer. (c) Phase separation (microcomposite).

In addition to the studies about the influence on clays on mechanical, barrier, and other properties such as transparency or gloss of the nanocomposites, nanoclays have been assessed in polymeric matrices in terms of their antimicrobial or bacteriostatic effect. Thus, Sothornvit et al. (2009) evaluated the effect of three different clays (Cloisite® Na^+, Cloisite® 20A, and Cloisite® 30B) on the antimicrobial effect of whey protein isolate films. Among the three studied clays, only nanocomposites containing Cloisite® 30B showed bacteriostatic effect against gram positive *L. monocytogenes*. On the contrary gram negative *E. coli* was not affected by nanocomposites. This result was related by the authors with the quaternary ammonium groups in the silicate layer of Cloisite® 30B, which could disrupt bacterial cell membranes and cause cell lysis. Similar results were found by Liu et al. (2014) while assessing the effect of different montmorillonite-derived nanoclays on the antimicrobial activity of PVA films. Among all the evaluated samples, only the composites containing methyl, bis hydroxyethyl, octadecyl ammonium-montmorillonite showed antibacterial effect against gram positive bacteria. In this case, authors also related this observation to the presence of a quaternary ammonium group. The same effect was also found in films of κ-carrageenan (Rhim and Wang, 2014), PLA (Rhim et al., 2009), and linear low density polyethylene (Hong and Rhim, 2012) containing Cloisite® 30B.

2.4 Metal Oxide Nanoparticles

Metal oxides, in particular those derived from transition metals, have interesting applications in various fields due to their excellent optical, magnetic, electrical, and chemical properties (Manna, 2012). In nanoform, these oxides may be used to avoid the microbial growth in different fields such as biomedicine and food packaging (Espitia et al., 2012; Xing et al., 2012; Khan et al., 2015).

During the past years a wide variety of oxides, including TiO_2, ZnO, MgO, CaO, CuO, Al_2O_3, AgO, and CeO_2, have been tested for their antimicrobial activity under different conditions. Among them, there are several oxides that exhibited antimicrobial activity without photo-activation (eg, CaO, MgO, and ZnO), in contrast with other ones, such as TiO_2, that required photo-activation (Manna, 2012).

Considering all the mentioned compounds, ZnO is the oxide that has received much attention due to the fact that ZnO nanoparticles are considered nontoxic, biosafe, and biocompatible and have been found in many applications such as drug carriers, in cosmetics, and as fillings in medical materials or devices (Wang, 2004). These particles, can be obtained by "green synthesis" (from plant resources, Janaki et al., 2015) and have been recently used to obtain antimicrobial nanocomposites. In a recent work, Kanmani and Rhim (2014b) obtained different bionanocomposites by adding ZnO nanoparticles to carrageenan, agar, and carboxymethylcellulose (CMC). They found that nanoparticles inhibited the growth of *L. monocytogenes* and *E. coli* regardless the polymer. Nevertheless, the effect was greater when carrageenan or agar were considered, being *L. monocytogenes* more sensitive than *E. coli* to the ZnO action. These results, similar to those obtained by other authors (Nafchi et al., 2012), are related to the different structure of the bacterial cell wall in gram positive and gram negative bacteria. Surprisingly, Díez-Pascual and Díez-Vicente (2014a) found the opposite effect, when ZnO was added to poly(3-hydroxybutyrate-co-3-hydroxyvalerate) (PHBV); *E. coli* (Gram negative) was more affected than *S. aureus* (Gram positive) by ZnO nanoparticles. From these results it can be stated that the cell wall structure is not the only factor that determines the antimicrobial effect of ZnO nanoparticles. Thus, other factors such as the carrier (polymer), the specific bacteria, the concentration or other mechanisms of action are involved in the final response. In this way, Kanmani and Rhim (2014b) explained other possible mechanism related with the antimicrobial activity of ZnO nanoparticles. In this way, ZnO could promote the formation of reactive oxygen species (ROS), which would damage the proteins, lipids, and DNA or penetrate into the cell wall leading to cell death (in the case of hydrogen peroxide, Tankhiwale and Bajpai, 2012).

Besides their addition to biodegradable polymers, ZnO nanoparticles have also been added to synthetic matrices such as polyethylene (Tankhiwale and Bajpai, 2012) or polyurethane (El Saeed et al., 2015) acting effectively against *E. coli* and *B. subtilis*.

Other metal oxide that has received the attention of several authors is TiO_2. This inorganic compound has been tested in polymeric materials in order to demonstrate its antimicrobial activity with promising results. Xing et al. (2012) obtained antimicrobial materials based on polyethylene and TiO_2 nanoparticles and they found that the material effectively inhibited the growth of *E. coli* and *S. aureus*. Furthermore, the antimicrobial activity of the nanocomposites highly improved when they were irradiated under UV light for 60 min. In fact, nanocomposites containing TiO_2 are frequently doped with Ag or Zn to be able to spread out the absorption light into the visible region of the electromagnetic spectrum. Kubacka et al. (2009) obtained ethylene-vinyl alcohol (EVOH) copolymer composites containing Ag-TiO_2 nanoparticles. They found that Ag shows a very intense localized surface plasmon

(LSP) absorption band in the near-UV-visible region. Thus, Ag doping of TiO_2 improves the potential of these nanoparticles working under sunlight and/or diffuse artificial light, typical of human environments, without requiring UV excitation (Muñoz-Bonilla et al., 2013).

Muñoz-Bonilla et al. (2013) added TiO_2 nanoparticles to polycaprolactone films and found a remarkable biocidal effect of the nanocomposite against *E. coli* and *S. aureus* when samples were irradiated at 500 nm without adding a doping element. As these authors indicate, the observed results were extraordinary because they correspond with a test performed on visible-light excitation. Therefore, each application of TiO_2 should be carefully studied in order to determine the behavior of the nanocomposite upon UV and visible irradiation.

2.5 Copper Nanoparticles

Besides silver, copper nanoparticles have also been studied in agreement with their antimicrobial activity at low concentration (Mukherji et al., 2012). On the contrary, other metal-derived nanoparticles such as gold or iron oxide do not present activity when assessed at low concentration. Nevertheless, some antimicrobial activity has been reported for gold nanoparticles for concentrations above 240 μg/mL (Burygin et al., 2009).

Although some studies pointed out the antimicrobial activity of different nanoparticles before the beginning of the present century, the majority of the works studying the antimicrobial activity of copper nanoparticles incorporated in composites have been performed in the past decade. As explained by Cioffi et al. (2005), copper is a well-known germicidal whose activity has been considered since the 19th century. In fact, copper or copper oxide particles have been incorporated into organic matrices to be used as antifouling coatings by the paint industry, mainly for maritime applications. Furthermore, the use of copper as antifungal for plant pathogens has also been widely considered.

Cioffi et al. (2005) obtained nanocomposites by incorporating copper nanoparticles colloidal dispersion into polyvinylmethyl ketone (PVMK), poly(vinyl chloride) (PVC), and polyvinylidene-fluoride (PVDF). They found a significant antimicrobial activity against *S. cerevisiae* which slightly varied depending on the polymer. This difference correlated well with the lower copper loading found in the PVDF matrix (where the lower activity was found), which was assessed by XPS analysis (X-ray photoelectron spectroscopy). PVMK containing nanoparticles was also evaluated against other microorganisms, also showing antimicrobial activity against *E. coli*, *L. monocytogenes*, *S. aureus*, and molds.

More recently, copper nanoparticles have been used to obtain nanocomposites in combination with cellulose (Llorens et al., 2012) and agar (Shankar et al., 2014). Llorens et al. (2012) obtained their materials by considering two different methods. On one hand, a material with good interfacial adhesion was obtained by treating the composite with heat and UV radiation. On the other hand, when a chemical reduction with borohydride was performed, small, but partially aggregated (giving rise to microparticles) copper oxide nanoparticles were obtained. They evaluated the antimicrobial activity in vitro and in pineapple and melon juices and found that the studied materials may effectively be used to increase the shelf-life of food products. They pointed out the advantage of obtaining copper nanoparticles by physical methods, therefore allowing the control of the microbial spoilage without using chemical reagents. Shankar et al. (2014) also assessed the influence of different obtaining methods on different properties of the nanocomposites. In particular, they obtained nanoparticles from three different salts and two different reducing reagents, therefore obtaining particles with different shapes and sizes. Authors obtained a significant antimicrobial activity against *L. monocytogenes* and *E. coli*, without influence of the type of nanoparticles. Furthermore, these authors related the antimicrobial activity of copper nanoparticles with the inhibition of DNA replication and destruction of the cell wall of bacteria.

2.6 Nanocarriers (Nanoliposomes, Carbon Nanotubes, and Mesoporous Materials)

One of the most promising approaches from nanotechnology to food packaging or other relevant fields such as tissue engineering or anticancer drugs is the use of nanocarriers. These devices may contain active compounds and protect them from external agents until their release. In this way, nanocarriers may modulate the release of the active agents, therefore reducing the necessary amounts and avoiding their losses. In the field of food packaging different types of these devices have been proposed and assessed, including nanoliposomes, carbon nanotubes, or even mesoporous materials.

Lecithin nanoliposomes may be easily obtained by high-pressure homogenization or by sonication as explained by Imran et al. (2015) and Zhang et al. (2012), respectively. Due to the amphiphilic character of lecithins, these compounds may be used to encapsulate compounds with different polarity. Thus, compounds with well-known antimicrobial activity such as nisin or cinnamon essential oil have been effectively encapsulated (Imran et al., 2015; Wu et al., 2015).

In general, nanoliposomes have been assessed in composites obtained with natural-occuring polymers such as chitosan, starch, or gelatin (Liu et al., 2013; Jiménez et al., 2014a; Wu et al., 2015). However, PLA films have been also loaded with nanoliposomes. Thus, Makwana et al. (2014) obtained PLA films coated with cinnamaldehyde nanoliposomes, which were obtained by means of bilayers of polydiacetylene-N-hydroxysuccinimide.

Most of the studies performed with antimicrobial nanocomposites containing nanoliposomes have been assessed in their antilisterial activity because of the relevance of these bacteria in food-borne infections and their ability to grow properly at low temperature.

Carbon nanotubes and mesoporous materials, such as zeolites, can be also considered for obtaining antimicrobial nanocomposites for food packaging uses. Nevertheless, even though many studies pointed out the capacity of these materials, their applicability has been oriented to other fields such as tissue engineering, biomedicine, or water treatments.

Carbon nanotubes present weak antimicrobial activity when added directly to polymers (Qi et al., 2012). Nevertheless, they may carry or stabilize active compounds, as aforementioned, or even be functionalized. Rusen et al. (2014) obtained membranes based on polymer colloids/multiwall carbon nanotubes with silver nanoparticles. These authors found that the developed materials were effective in inhibiting the growth of both *S. aureus* and *E. coli*. Despite that the antimicrobial activity was related to silver release, multiwalled carbon nanotubes played a relevant role in the stabilization of silver nanoparticles. These type of nanotube was also used to immobilize cephalexin (a broad-spectrum antibiotic that inhibits the cell wall synthesis) by using poly(ethyleneglycol) (Qi et al., 2012). These authors found that the carbon nanotubes containing cephalexin were able to reduce the growth of both Gram-positive (*S. aureus* and *B. subtilis*) and Gram-negative bacteria (*E. coli* and *Pseudomonas aeruginosa*). Furthermore, the inmobilization of cephalexine by covalent linking with poly(ethyleneglycol) also presented superior antiadhesion ability.

Other approaches to use carbon nanotubes as antimicrobials is their functionalization as explained by Yadav et al. (2012). These authors functionalized carbon nanotubes by click-coupling with hyperbranched polyurethane. This procedure ("click chemistry") uses the Cu(I)-catalyzed [3 + 2] Huisgen cycloaddition reaction of azides and alkynes moieties. In this way, authors obtained a high-performance material with improved mechanical and antimicrobial properties. Thus, a significant antimicrobial activity of the nanocomposite was found, in agreement with the good dispersions of the functionalized nanotubes, as the authors explained.

As previously mentioned, mesoporous materials may also be considered to obtain antimicrobial nanocomposites. In this way, Bendahou et al. (2015) obtained PLA films with micro- and nano-zeolites ($NaAlO_2$, SiO_2). They found antibacterial activity (against *E. coli*) regardless the zeolite size. In other recent work, zeolite A was loaded with silver ions and silver nanoparticles (Jiraroj et al., 2014). In this case, authors found antimicrobial activity when zeolites contained silver ions and zeolites did not present any activity against the tested microorganisms.

For the formulation of antimicrobial composites, the choice of the carrier has to be carefully considered because the activity of antimicrobials may vary depending on their support and its physicochemical properties. This aspect was studied by Sohrabnezhad and Sadeghi (2015) by analyzing the influence of the carrier on the antimicrobial activity of Ag_2CO_3 nanoparticles. These authors compared MCM 41 (material with a neutral matrix with high surface area and high pore volume) with layered montmorillonite (Mt) and found that, even though both formulations showed antimicrobial activity against *E. coli*, MCM 41-Ag_2CO_3 was more effective. Authors correlated this result with the comparatively larger surface area and smaller particle size of MCM 41-Ag_2CO_3 in comparison with Mt- Ag_2CO_3, which makes possible a higher release of silver ions, at the same time that bacteria are more exposed to come in contact with the antimicrobial.

2.7 Nanoemulsions

During the past years, one of the most considered approaches to improve the properties of edible and biodegradable films has been the combination of hydrocolloids (polysaccharides and proteins) and lipids for obtaining composites. These materials present better properties than pure hydrocolloid films, mainly in terms of its water vapor permeability and, when lipids have antimicrobial activity (essential oils, terpens, polyphenols, etc.), the composites may be used as active packaging materials. Traditionally, the particle size of these composites is in the order of micrometers (Fabra et al., 2009; Jiménez et al., 2010). Nevertheless, techniques such as high-pressure homogenization or sonication lead to dispersed lipid nanodroplets, thus giving rise to nanoemulsions. Lipid nanodroplets incorporated in the polymeric materials yield high-performance nanocomposites (Acevedo-Fani et al., 2015; Otoni et al., 2014a).

Acevedo-Fani et al. (2015) obtained alginate films containing nanoemulsions of thyme (THEO), lemongrass (LGEO), and sage (SGEO) oil with particle sizes ranging from 20 to 6000 nm, by using high-pressure homogenization. They found a peak about 20 nm, which is related with the presence of surfactant micelles not adsorbed at the oil–water interface. Other main peaks also appeared at 190 nm, assigned to essential oil droplets. Highest particle sizes were attributed to larger oil droplets or alginate colloidal structures. As regards antimicrobial activity, authors found that the type of oil significantly affected the response. Meanwhile thyme oil provoked a dramatic decrease in the population of *E. coli*, the other tested oils did not affected its growth. Authors related these results with the great antimicrobial activity of thymol (main component of thyme oil). In the case of SGEO and LGEO, the absence of activity was related with their low antimicrobial activity or the evaporation of oil actives during the film drying.

Otoni et al. (2014a) obtained biodegradable pectin films added with papaya puree and cynnamaldehyde. These nanocomposite films showed antibacterial activity against *E. coli*, *Salmonella enterica*, *L. monocytogenes*, and *S. aureus*, with differences depending on the particle size. The highest inhibition corresponded with the smallest cynnamaldehyde droplets, pointing out the advantages of nanoemulsions in comparison with microemulsions. As droplet size decreases, the droplet specific surface exponentially increase, thus facilitating the diffusion of the active agent.

Antifungal activity of nanoemulsions also has been recently assessed. Abd-Elsalam and Khokhlov (2015) obtained eugenol nanoemulsions and studied their antifungal activity against *Fusarium oxysporum*. Authors observed a significant antifungal activity in both in vitro and in vivo analysis with a concentration of eugenol of 10%.

2.8 Other Nanoparticles

Besides the aforementioned, many other nanoparticles have been studied as antimicrobials for food packaging, surface and water treatments, and biomedical applications. Among them, the most widely assessed are gold, polyaniline, and chitosan nanoparticles.

Gold nanoparticles have been mainly evaluated for their application in the biomedical field. Hernández-Sierra et al. (2008) compared the effectiveness of gold, silver, and zinc oxide nanoparticles as antimicrobials against *Streptococcus mutans*, which is directly related to dental caries. Gold nanoparticles exhibited lower antimicrobial activity in comparison with silver, mainly at low concentrations. In a more recent work, gold was again compared with silver and assessed against *E. coli* and *S. typhi* (Guerra et al., 2013). In agreement with the previous work, silver nanoparticles were better biocide nanomaterials than gold nanoparticles, being that *E. coli* is more sensitive to the action of both nanoparticles. Nevertheless, even though gold nanoparticles were less effective, they showed a notable antimicrobial activity against the tested microorganisms.

Gold nanoparticles have been used to obtain high-performance nanocomposites in combination with chitosan (Regiel-Futyra et al., 2015). Films based on medium molecular weight chitosan and gold nanoparticles exhibited antimicrobial activity against biofilm forming strains of *S. aureus* and *P. aeruginosa*, producing a disruptive effect on the bacteria cell wall. In addition, obtained nanocomposites did not show cytotoxicity against mammalian somatic cells, which is positive in terms of the nanocomposite usefulness in the fields of food packaging or biomedicine. Other studies that use gold nanoparticles synthesized by a special strain of *B. subtilis*, incorporated in chitosan matrices (Youssef et al., 2014). These nanocomposites showed antimicrobial activity against *S. aureus*, *P. aeruginosa*, *A. niger*, and *C. albicans*, regardless of the well-known antimicrobial activity of chitosan.

Gold nanoparticles have been also added to quinoa starch matrices (Pagno et al., 2015). These authors observed that these nanoparticles inhibit at 99% the growth of *E. coli* and at 98% the growth of *S. aureus*, also improving the mechanical, optical, and morphological properties of the nanocomposites. This was obtained with contents of nanoparticles below 0.06% (w/w), therefore indicating the strong potential of this nanomaterial in this type of matrix. In fact, authors pointed out the usefulness of these nanocomposites for using as active packaging system to increase food shelf life during storage and distribution.

Polyaniline (PANI), obtained from the polymerization of aniline, has been also considered for obtaining antimicrobial nanocomposites. Thus, although there is a low number of reports, in comparison with those studying other types of nanoparticles, some advantages of this polymer were reported. Pandiselvi and Thambidurai (2015) obtained chitosan-zinc oxide with and without PANI nanocomposites. They observed that PANI addition improved the activity against Gram-positive and Gram-negative bacteria (*S. aureus* and *P. aeruginosa*). Furthermore, nanocomposites containing PANI were also effective against *C. albicans*. The greater antimicrobial activity was related to the fact that the composite attached to the cell membrane through amino groups (present in PANI and chitosan), thus causing cell death by disrupting microbial cell walls and membranes via electrostatic action. Additionally, PANI improved the dispersion of zinc oxide, thus enhancing the release of Zn^{2+}, which could also contribute to the increase of the antimicrobial action.

Bogdanović et al. (2015) obtained PANI-Cu nanocomposites by in situ polymerization. These authors found that the obtained material showed greater antimicrobial activity than PANI and Cu acting alone. On the contrary, Ghaffari-Moghaddam and Eslahi (2014) found that PVA-PANI composites did not exhibit antimicrobial activity, which was acquired when silver in nanoform was incorporated to the composite. Taking into account the different studies, it may be stated that PANI is not always as effective as antimicrobial for food packaging applications, although its use may be considered to improve the antimicrobial activity of other substances, because PANI improved the dispersion of nanoparticles, causing some synergistic effects, such as that observed in chitosan-ZnO composites.

Chitosan, whose antibacterial and antifungal activity is very well known, is also used to obtain antimicrobial nanoparticles. In this way, the effect of bulk chitosan and chitosan nanoparticles has been assessed in tara gum matrices (Antoniou et al., 2015). Authors found the addition of chitosan nanoparticles improved the majority of the composite properties in comparison with bulk chitosan. However, antimicrobial activity (against *E. coli* and *S. aureus*) was higher for composites containing bulk chitosan. Even though chitosan nanoparticles conferred antimicrobial activity to tara gum films, this property decreased when nanoparticles content increase from 10 to 15% (w/w), due to the aggregation of nanoparticles during film formation. This type of nanoparticle has been also employed to encapsulate active compounds, such as eugenol-loaded chitosan nanoparticles, as reported by Woranuch and Yoksan (2013). This approach led to high-performance nanoparticles with improved antimicrobial activity for food packaging applications.

Despite the high amount of antimicrobial nanoparticles that may be considered in formulation of nanocomposites, not all the nanomaterials must be considered for this purpose. Therefore, each specific application should be carefully assessed. In fact, Masadeh et al. (2015) have recently demonstrated that some nanoparticles may abolish the antimicrobial activity of the well-known ciprofloxacin. Authors studied the antimicrobial activity of CeO_2 and Fe_2O_3 nanoparticles and found that these nanomaterials not only did not show a better antimicrobial activity but they inhibited the action of the aforementioned antibiotic. This results in the need of assessment of each nanomaterial in each specific system, due to the different factors that may affect their response.

3 Polymers Used in Antimicrobial Nanocomposites

In the past few years, a great quantity of polymers has been assessed to obtain antimicrobial nanocomposites. In fact, nanoparticles (or other additives such as plasticizers, antioxidants) are

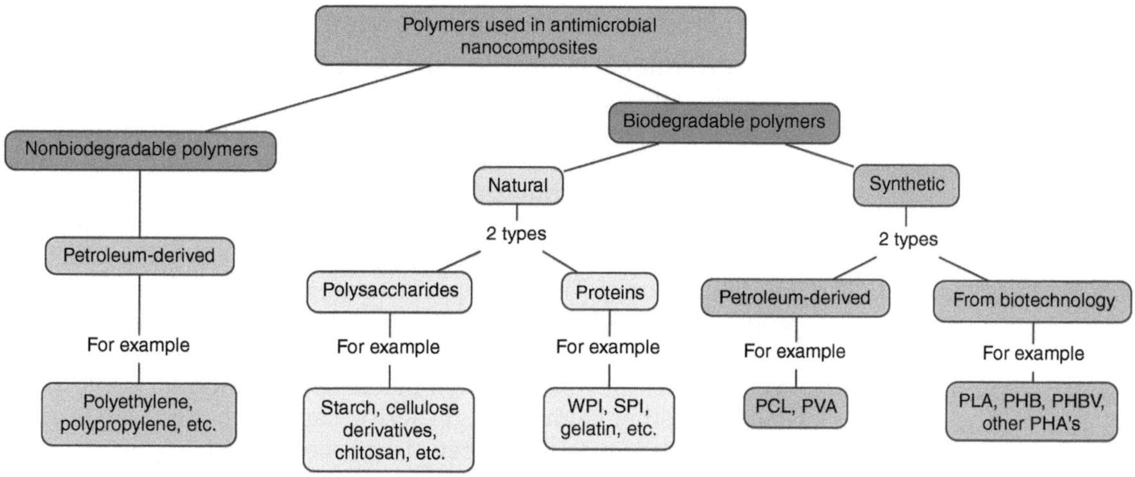

Figure 11.2. Polymers used in antimicrobial nanocomposites.

used to improve different properties of the polymers. This is due to the fact that these materials, when compared, present different advantages but also many drawbacks. As an example, polyolefins are very good barriers against moisture transfer and present good mechanical performance. On the contrary, their ability to avoid oxygen transfer is poorer in comparison with polysaccharide-based films.

It has to be taken into account that, when nanoparticles are added to polymers, not only their antimicrobial activity varies. Thus, in most cases, properties such as mechanical properties and barrier capacity varies with nanoparticles addition, as explained later.

Polymers used in antimicrobial nanocomposites may be classified in different ways. Therefore, they may be categorized as a function of their final use (food packaging, tissue engineering, surface treatment, etc.), their origin (natural, petroleum-derived, produced by microorganisms) or their biodegradability. In fact, in Fig. 11.2, polymers are classified in agreement with its biodegradability because most works published recently are performed with biodegradable polymers. This phenomenon occurs because, even though biodegradable polymers present more advantages from an environmental point of view, their properties are worse (mechanical performance and water vapor barrier) in comparison with traditional polymers. Therefore, nanoparticles incorporation is considered a promising approach to obtain materials with improved properties.

3.1 Nonbiodegradable Polymers

Conventional nonbiodegradable polymers, such as low density polyethylene or EVOH have been added with different types of nanoparticles in order to increase their functionality. As an example, polyethylene was used as matrix of antimicrobial nanocomposites in combination with ZnO and TiO_2 (Tankhiwale and Bajpai, 2012; Xing et al., 2012). In this work, ZnO nanoparticles were synthesized in situ on a starch-coated polyethylene surface. On the contrary, TiO_2 nanoparticles were produced in a previous step and then mixed and extruded with polyethylene, which demonstrates that antimicrobial nanocomposites for food packaging applications may be obtained by using conventional equipment. Polyether ether ketone, a high-performance polymer used in engineering applications, has been also processed by using an extruder, where this material was mixed, after the preparation and grafting with ZnO nanoparticles (Díez-Pascual and Díez-Vicente, 2014b).

Most of the nanocomposites with antimicrobial activity tested with nonbiodegradable polymers are conceived for nonfood contact materials applications, probably due to the lack of regulation in this way and the higher cost of some of them. Therefore, despite some works performed with polyethylene or EVOH (Tankhiwale and Bajpai, 2012; Xing et al., 2012; Kubacka et al., 2009), different researchers have considered other polymers such as poly(methyl methacrylate) (Rusen et al., 2014), polyurethane (El Saeed et al., 2015; Yadav et al., 2012), commercial epoxy resin (Wei et al., 2014), or polyvinylidenefluoride (Cioffi et al., 2005).

3.2 Biodegradable Polymers

3.2.1 Polymers from Natural Sources

As explained previously, there is a wide amount of published works in which biodegradable polymers are considered as matrices for antimicrobial nanocomposites. Among them, polysaccharides and proteins have been widely assessed due to their renewable character, low permeability against oxygen transfer and nontoxicity. These polymers are directly obtained from biomass products, such as plant tissues or seeds (eg, cellulose, starch and soy protein), or animal products (eg, gelatin and caseinate) (Jiménez et al., 2014b) and directly used after purification. Thus, no synthesis step is necessary for their obtaining.

Gelatin is one of the most studied proteins in the formulation of edible and biodegradable packaging materials for food uses, in agreement with its excellent film forming ability, low gelling and melting point, abundance and biodegradability (Wu et al., 2015).

This biopolymer is obtained from physical, chemical, or biochemical denaturation and hydrolysis of collagen and it has been studied to form nanocomposites in combination with different nanoparticles such as silver, zinc oxide, or nanoliposomes (Arfat et al., 2014; Kanmani and Rhim, 2014a; Wu et al., 2015). In all cases, obtained nanocomposites showed greater antimicrobial activity thus indicating the ability of gelatin to form high-performance biodegradable packaging materials. Besides gelatin, other proteins have been also used in nanocomposites formulation due to their amphiphilic nature, which allows to include both polar and nonpolar nanoparticles (Jiménez et al., 2014a; Sothornvit et al., 2009; Zhang et al., 2010; Zhao et al., 2013).

In an interesting work, Zhao et al. (2013) obtained nanocomposites based on silver nanoparticles and soy protein isolate. This polymer acted as a reducing and stabilizing agent for silver as well as constituted the nanocomposite matrix, without using any other reagents. In this way, the obtained material presented a great potential to be used as sustainable and active packaging material for food uses.

Likewise, polysaccharides are frequently employed in nanocomposites obtaining, being starch, cellulose (and its derivatives), and chitosan the most relevant.

Chitosan is widely used to develop novel materials for food packaging applications due to its inherent antimicrobial activity. Therefore, many studies have been focused on studying the effect of nanoparticles addition to chitosan matrices (Khan et al., 2012; López-Carballo et al., 2013; Regiel-Futyra et al., 2015). In some applications, nanoparticles, such as silver, have to be released from the matrix to inhibit microorganisms' growth. Thus, the polymers have to be able to modify their structure thus releasing the active agents. In this way, the hydrophilic nature of chitosan is of interest as a sustained releasing carrier when the release of the retained active compound is required in a moist environment (López-Carballo et al., 2013).

Starch, which is the second most abundant polysaccharide on earth, has also been widely used in nanocomposites formulation (Jiménez et al., 2014a; Nafchi et al., 2012; Pagno et al., 2015). Even though most studies consider the casting technique to obtain the nanocomposites, starch is a biopolymer that can be easily processed following a conventional method such as extrusion and/or compression molding. These process, known as dry processing (water is not present in excess) can be considered with those materials, which present thermoplastic properties; this means that they become soft (melted or rubbery) at a temperature lower than decomposition temperature and so, they can be molded into a determined shape when submitted to a thermal/mechanical process (Jiménez et al., 2012). Even though starch does not present this characteristic in its native state, it is capable of becoming a thermoplastic material if it is properly plasticized. Hietala et al. (2013) obtained starch nanocomposites containing cellulose nanofibers by twin-screw extrusion in one step, in which starch was gelatinized and mixed with the nanofibers powder. In another work, Salaberria et al. (2014) obtained starch nanocomposites containing chitin nanocrystals by melt-mixing and injection molding. These works point out that it is possible to obtain nanocomposites with a conventional process, as commented on earlier. However, these authors did not study the antimicrobial ability of the obtained materials. Therefore, further studies have to be performed to assess the developed materials for food packaging applications.

Due to its high availability and low cost if compared with other polymers, cellulose and its derivatives are widely considered to formulate nanocomposites and biodegradable packaging materials. As an example, Otoni et al. (2014b) obtained methylcellulose nanocomposites containing nanoemulsions of oregano and clove oil. Developed materials showed a significant antifungal activity when applied to bread slices, thus increasing the shelf life of the product. As mentioned in the previous section, cellulose is also used to obtain different nanomaterials (see Section 2 and Table 11.2), which may be considered to produce cheap nanocomposites with improved properties.

Besides chitosan, starch, and cellulose, other polysaccharides such as agar, carrageenan, or pullulan are considered in antimicrobial nanocomposites formulation in combination with silver, clay, or copper nanoparticles as reported by different authors (Rhim et al., 2013a; Rhim and Wang, 2014; Shankar et al., 2014). Thus, further information can be easily found in the published literature.

3.2.2 Polymers Obtained from Synthesis Processes

Different biodegradable polymers are obtained by means of synthesis processes. In this way, some of them are obtained from petroleum (nonrenewable) and other are synthesized by using natural sources or directly produced by bacteria (see Fig. 11.2).

Polycaprolactone is an aliphatic polyester obtained by chemical synthesis from crude oil or even from renewable resources, such as polysaccharides (Ortega-Toro et al., 2015). Poly(vinyl alcohol) is also obtained by a synthesis, in this case from poly(vinyl acetate), and widely used because of its biocompatibility and interesting physical properties, which are due to the presence of -OH groups and the hydrogen bond formation (Cano et al., 2015). Both polymers have been assessed as matrices of antimicrobial nanocomposites as reported by Muñoz-Bonilla et al. (2013) and Fortunati et al. (2013). TiO_2 and silver nanoparticles were used to confer antimicrobial activity to these polymers against *E. coli* and *S. aureus*.

Other polymers used in antimicrobial nanocomposites, such as polyhydroxyalcanoates and poly(lactic acid), are also obtained by synthesis processes from natural sources as introduced previously. Poly(3-hydroxybutyrate-co-3-hydroxyvalerate) the most common polyhydroxyalkanoate, was used by Díez-Pascual and Díez-Vicente (2014a) in combination with ZnO nanoparticles. SEM images showed very uniform nanofiller dispersion in the nanocomposite with the lowest ZnO content, without any agglomerates. On the contrary, when nanoparticles content reached 8% (w/w) agglomeration appeared. Fortunately, the best antimicrobial effect was found with 5% of nanoparticles, in agreement with the better dispersion in the polymeric matrix.

Regarding PLA, different nanomaterials including nanoclays, cellulose nanocrystals, and eugenol-loaded chitosan nanoparticles have been used in nanocomposites formulation (Rhim et al., 2009; Salmieri et al., 2014; Woranuch and Yoksan, 2013). In all cases, the developed materials presented improved properties, pointing out the ability of this biodegradable polymer to be used to obtain high-performance packaging.

4 Production Techniques: Electrospinning, Electrospraying, and Nanoencapsulation

In order to obtain active nanoparticles different chemical or physical methods have been used. For instance, to synthesize silver nanoparticles and silver-based nanocomposites different physical methods such as laser ablation in liquid media, ultrasonic-assisted reduction, photo-induced synthesis, microwave-assisted synthesis and irradiation reduction (Wang et al., 2012). Other methods to obtain silver-based nanocomposites are based on chemical approaches that include chemical and electrochemical reduction or microemulsion methods (Zhang et al., 2007). In addition, more classical methods such as acid hydrolysis are commonly used to obtain cellulose nanoparticles (Peponi et al., 2014).

Likewise, nanostructures as carriers of active compounds (nonexplained previously), have also been obtained by different techniques such as electrospinning and electrospraying (Anu Bhushani and Anandharamakrishnan, 2014).

Electrospinning is a promising fabrication technology based on high electric fields that can be used to develop antimicrobial

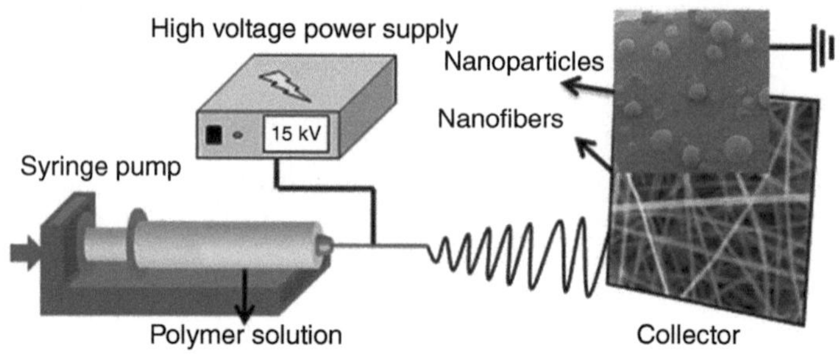

Figure 11.3. Typical electrospinning setup.

nanostructured packaging materials that are based on layer of nanofibers and other nanostructures (10–100 nm).

Electrospinning and electrospraying are electrohydrodynamic processes, where a polymer solution is spun or sprayed by the application of high potential electric field to obtain fibers or particles, respectively (Kessick et al., 2004).

The typical setup for electrospinning and electrospraying (Fig. 11.3) consists of these three main components: (1) a high voltage power supply (2) a syringe pump with a blunt ended stainless steel needle or capillary, and (3) a grounded collector.

Nanofiber and other nanoparticles formation via electrospinning and electrospraying requires dispersing the biopolymer in a liquid medium and thus both the process of electrospinning and the shape and size of the obtained materials depend on solvent-related properties like pH, conductivity and surface tension (Anu Bhushani and Anandharamakrishnan, 2014).

The most important factor that determines the size of electrospun materials is polymer concentration (Torres-Giner et al., 2008). In this sense, highly concentrated polymeric solutions lead to the production of high-diameter fibers whereas the reduction of the polymer content below a threshold value leads to the production of beads of mono-dispersed nanoparticles or extremely thin fibers. The latter is a variation of the electrospinning technology that is called "electrospraying" because the process is similar to an atomization procedure carried out by electrical forces.

Coaxial electrospinning and electrospraying setups are currently being developed for the production of nanomaterials with two miscible or immiscible core and wall components. In this method, a concentric or dual nozzle spinneret enables the production of coreeshell fibers or nanoparticles by incorporating the bioactive compound by an inner capillary and the wall material in an outer capillary (Zheng et al., 2012).

Electrospun nanofibers and nanoparticles carrying active ingredients can be incorporated to packaging materials by different methods such as melt-blending, extrusion, compression, or injection molding, and solvent casting to develop nanocomposites with antimicrobial performance.

In antimicrobial biodegradable food packaging concepts, the incorporation of bioactive molecules into a biopolymer matrix is a challenge because most natural antimicrobials are sensitive to heat. Electrospinning can be carried out at room temperature and thus it allows such molecules to be incorporated by nanoencapsulating them in nanofibers of proteins such as zein or biopolymers like PLA. In this sense, the antimicrobial activity of chitosan was used to produce zein/chitosan electrospun insoluble fiber materials with biocide properties (Torres-Giner et al., 2009).

Nanoencapsulation is based on the introduction of nonpolymeric antimicrobial agents (eg, enzymes, silver, or metal oxide nanoparticles) into the polymeric solution to electrospin that acts as a wall material. This leads to the formation of hybrid ultrathin fibers in which the antimicrobial particles remain packed into the polymer matrix as individual nanodroplets of liquid or solid nanoparticles. These fibers show improved stability and nanostructural distribution and allow for the controlled release of the encapsulated ingredients (Torres-Giner, 2011). For example, allyl isothiocyanate a natural antimicrobial component, was electrospun in fiber forming solutions of soy protein isolate or PLA to effectively control the release of the active ingredient (Vega-Lugo and Lim, 2009). In a similar way, PVA electrospun fibers containing nanosized dark tea powder showed antibacterial activity against *E. coli* (Sun et al., 2011).

5 Application of Nanocomposites for Food Packaging

Nanocomposite food packaging materials are generally produced from a mixture of polymers with at least one filler that has a dimension lower than 100 nm. Nanocomposite antimicrobial systems are particularly effective because of the high surface-to-volume ratio and enhanced surface reactivity of the nanosized antimicrobial agents, making them able to inactivate microorganisms more effectively than their micro- or macroscale counterparts (Damm et al., 2008). These nanomaterials can extend the product's shelf life by different mechanisms, depending on their barrier properties, their composition, and their interaction with the food. In this way, the protection function of the package can be carried out throughout direct contact between the active compounds of the material and the product or by adding to the package oxygen scavengers, water vapor absorbers, and ethylene removers, which create the protective modified atmosphere.

5.1 Food Contact Active Packaging Materials

Active packaging systems are those materials that incorporate into them active compounds such as antimicrobials and preservatives. The active components are assembled using nanotechnology to produce an active component carrier that can be integrated into a food package. The carrier can interact with internal and/or external factors thereby activating an intended action that can enhance the shelf life, food quality, and/or safety of the product (Mihindukulasuriya and Lim, 2014).

Metal nanoparticles such as silver, gold, zinc, or metal oxides have been evaluated in various active packaging applications for their antimicrobial properties (Bikiaris and Triantafyllidis, 2013). The efficiency of antimicrobial function of these antimicrobial polymeric nanocomposites is greatly influenced by various factors such as particle size, size distribution, degree of particle agglomeration, and interaction of the nanoparticle surface with the polymer matrix.

Silver nanoparticles have been proved to be effective against both gram positive and gram negative bacteria (Lewington, 2013; Rhim et al., 2014). PVC films coated with ZnO nanoparticles have shown improved antibacterial activity (Li et al., 2010).

Nanoclays can be functionalized to provide antimicrobial functions and thus nanocomposite films prepared with some organically modified montmorillonite (Cloisite® 30B), have shown antimicrobial activity (Rhim et al., 2006). Active packaging films based on EVA, functionalized nanoclays, and a natural extract obtained from brewery waste showed antibacterial properties against *E. coli* and *S. aureus* and extended the shelf life of beef meat products (Barbosa-Pereira et al., 2014).

Nanocomposite melt-processed films based on low-density polyethylene organically modified montmorillonite containing essential oil mixtures of carvacrol and thymol were effective in reducing *B. cinerea* infection in strawberry fruits (Campos-Requena et al., 2015). Following a similar approach, the shelf life of fresh fruit salads was extended by incorporating Ag-montmorillonite nanoparticles into PP packaging materials (Costa et al., 2011). The same particles were added to film-forming dispersions based on alginate, which were effective in extending the shelf life of cold-stored carrots (Costa et al., 2012). Soy lecithin nanoliposomes containing nisin were incorporated into HPMC-based films that inhibited *L. monocytogenes* growth (Imran et al., 2012).

5.2 Modified Atmosphere and Gas Scavenging Packaging Materials

The improvement in gas barrier properties of nanocomposites has been explored in the food packaging to obtain multilayer nanocomposites for rigid food packaging like thermoformed containers. Packaging of fresh food products in such containers results in a commodity-generated modified atmosphere (MA), which in turn depends on film permeability, respiration rate, gas diffusion properties, weight and surface area of the packaged products, free headspace volume and the atmospheric composition within the package.

Nanocomposite packaging materials modify the atmosphere of the packaged food products due to their reduced permeability to gases, especially to oxygen. The presence of oxygen inside the food package decreases food shelf life due to various degradation reactions such as growth of aerobic microorganisms, enzymatic browning, and oxidation reactions. Moreover, oxygen has a significant effect on the respiration rate and ethylene production in fruits and vegetables. In order to avoid these undesirable effects, oxygen within the package should be eliminated or reduced to a level acceptable for a given product. Other volatile compounds generated by fresh fruits and vegetables during storage can be undesirable. In particular, the formation of ethylene in climacteric fruits can accelerate the postharvest ripening thus leading to a reduced shelf life.

Oxygen scavenger systems have been widely used to eliminate the residual oxygen in the package. The combined effect of modified atmosphere packaging in trays with Ag-cellulose nanoparticles absorbent pads prepared led to reduction in the microbial load and delayed senescence in minimally processed melon samples (Fernández et al., 2010). Similarly, cellulose pads impregnated with silver nanocomposites promoted a reduction of total aerobic bacteria, yeast, and molds in kiwi fruit (Lloret et al., 2012).

Many packaging concepts have been applied for oxygen and ethylene scavenging. For instance, nanocrystalline TiO_2 deposited on polypropylene films was effective in removing ethylene in packaged horticultural products (Manikantan and Varadharaju, 2011).

5.3 Commercial Developments

Polymer nanocomposites are a promising alternative to conventional composites due to their improved performance in mechanical, thermal, barrier, optical, electrical, and other physical and chemical properties. Thus, many companies have invested in developing nanoclays and polymer nanocomposites and a great amount number of commercial products has become available. The first commercial nanocomposite was nylon 6 nanocomposite reported by Toyota Central Research Laboratories and Ube in 1990 (Zeng et al., 2005) and nylon composites have been the dominant commercial nanocomposites during decades. In fact, in the packaging industry, the improved oxygen and dioxide carbon barrier properties of the nylon nanocomposites have been used to produce PET multilayer bottles and films for food and beverages. For instance, Imperm®, a nanoclay has been used as a barrier in beer and carbonated drink bottles as well as in meat and cheese packaging. The addition of 5% of Imperm® in PET bottles increased the shelf life by 6 months and reduced the dioxide carbon loss to less than 10% (Anadão, 2012). Other commercial products are Durethan® KU2-2601, a nylon 6 nanocomposite for coating juice bottles as barrier films and AEGIS™ OX, which combines nanocomposite barrier and an oxygen scavenger in one system and is used for beer bottles.

At present, the nanofiller type that has received the most attention for commercial use in plastics is the organically modified montmorillonite clay. For example, patent-pending O2Block® technology is based on surface-modified phyllosilicate clay that is functionalized with active iron to create a naturally sourced and highly efficient oxygen scavenging product for maximum shelf life in barrier packaging. For antimicrobial application silver-based nanocomposites that can be added to packaging materials such as Biomaster®, Aglon ®, Irgaguard ®, and Bactiblock ® are available in the market. Agion® antimicrobial solutions incorporate silver, copper, and other elemental ions into a zeolite carrier. Irgaguard® is a special glass/zeolite design allow for controlled silver-ion release, whereas Bactiblock® is a silver-functionalized clay.

6 Physicochemical Properties of Antimicrobial Nanocomposites

Biopolymer-based biodegradable antimicrobial packaging materials are environmentally friendly and have shown the ability to keep food safety and quality. The use of nanoparticles or nanocomposites has been shown to improve some physical properties of biopolymer-based packaging materials (Mihindukulasuriya and Lim, 2014), in particular water barrier properties, gas permeability, thermal stability, and mechanical performance.

6.1 Water Vapor Permeability and Gas Transport Properties

The addition of nanofillers such as nanoplatelets with high surface to thickness ratio into the polymer leads to an improvement of the barrier properties of the packaging materials against water vapor and gases. These fillers are impermeable or less permeable to gases and water vapor than the polymer matrices, and thus their presence results in a longer and tortuous diffusion path taken by the permeant molecules, as compared to the pure polymer matrix.

The reduction in gas permeability of nanocomposites strongly depends on the compatibility between clay and polymer matrix, aspect ratio of clay platelets, and structure of the nanocomposites. In general, the best gas barrier properties would be obtained in polymer nanocomposites with fully exfoliated clay minerals with large aspect ratio (Choudalakis and Gotsis, 2009) (see Fig. 11.1). For instance, the oxygen barrier properties of a low density polyethylene (LDPE) improved by 7 times after incorporating organic montmorillonite (Xie et al., 2012). The addition of cellulose nanocrystals into a poly (ε-caprolactone) matrix led to a decreased in water vapor permeability with the increase in filler content (Follain et al., 2013). The oxygen permeability of EVOH/PLA films, LDPE, PP, and HDPE films was reduced by the addition of nanoclays (Lagarón et al., 2005; Arunvisut et al., 2007; Manikantan and Varadharaju, 2011; Horst et al., 2012).

6.2 Mechanical Performance

The addition of nanomaterials such as montmorllinonite (MMT) and metal oxides can have a positive effect on the reinforcement of material mechanical properties. The improvement in mechanical properties of polymer nanocomposites can be attributed to the high rigidity of nanoclay together with the good affinity through interfacial interaction between polymer matrix and dispersed nanoclay.

Formation of nanocomposite with organoclays has shown a significant improvement in the mechanical properties of various biopolymers even with low level of filler loading (Rhim et al., 2013b). Mechanical properties of polymer/clay nanocomposites are strongly dependent on filler content. In this sense, the addition of increased amounts of nanoclays to PET films obtained by extrusion led to an increase in the yield stress (Ghanbari et al., 2013). Tensile strength of PV-based films increased by 21.26% and 5.17% with the addition of 4% and 8% sodium MMT, respectively (Mondal et al., 2013). Tensile strength of starch/MMT nanocomposites prepared with various filler concentrations increased monotonously with increase in filler content up to 8%, then leveled off, while tensile strain decreased with increase in filler loading except for 8% loading (Huang and Yu, 2006). The tensile strength of soy protein/MMT nanocomposite sheets increased from 8.77 to 15.43 MPa as the MMT content increased up to 16%, while percent elongation at break of the nanocomposite materials decreased with increasing MMT content (Chen and Zhang, 2006).

7 Regulatory Aspects of Nanocomposites for Food Contact Applications

One of the most relevant drawbacks for nanocomposites application is the lack of regulation in some fields, mainly in those corresponding with food contact materials. This handicap creates difficulty for the commercialization of developments because each new material has to be tested in terms of toxicity for consumers and environmental concerns. The polymers used in nanocomposites formulation are well known and widely used. In fact, most of the studies are performed with biodegradable materials, which may be edible or even biocompatible. Therefore, the potential risk of polymers employed in nanocomposites is very low. On the contrary, conversion from bulk state to nanosize imparts new properties to the nanomaterials in regard to mechanical, electrical, optical, catalytic activity, and biological activity (Thomas et al., 2013), which have to be carefully studied because of their unknown effects.

Regarding the potential toxicity of the nanocomposites, the main risk of consumer exposure to nanoparticles from food packaging is likely to be through potential migration of nanoparticles into the foodstuffs (Han et al., 2011). In addition, the processing of the nanocomposites may lead to the formation of NIAS (non-intentionally added substances), which have to be studied. To this aim, analytical screenings have to be performed to identify hazardous substances. Besides, the exposure to these substances, if present, must be evaluated.

In this complex context it is difficult for researchers and developers to comply with existing regulation when no specific laws or norms are available. Nevertheless, some existing regulations apply for antimicrobial nanocomposites as food contact materials.

Any packaging material intended to be used in United States must comply with those requirements included in the Title 21 of the Code of Federal Regulations (CFR). This portion of the CFR governs food and drugs within the United States. Specifically, parts from 170 to 179 are directly related with food packaging substances and materials whereas other parts corresponding with good manufacturing practices have to be also considered.

Although this regulation does not specify any requirements directly for nanocomposites, each component has to be authorized for its use in food contact materials. Furthermore, to be authorized, any nanocomposite should be assessed previously to its commercialization by the Food Contact Substance Notification Program (FDA, 2015).

In the European Union, some regulations apply directly for all food contact materials. Regulation (EC) No. 1935/2004 on materials and articles intended to come into contact with food and Commission Regulation (EC) No. 2023/2006 on good manufacturing practice for materials and articles intended to come into contact with food apply for any packaging intended to be used in the European Union, including nanocomposites. Furthermore, Commission Regulation (EC) on active and intelligent materials and articles intended to come into contact with food applies also to antimicrobial nanocomposites since they are considered as active packaging.

Besides cited regulations, Commission Regulation (EU) No. 10/2011 on plastic materials and articles intended to come into contact with food establishes the conditions to determine the migration from packaging to foodstuffs and distinguish between overall and specific migration. Therefore, this regulation, in its annex 1, presents a positive list of substances authorized in the formulation of plastic packaging as well as the maximum migration limit for each substance.

The mentioned positive list includes some substances in nanoform that may be used in food contact materials and their conditions of use, as detailed in Table 11.3.

European regulations do not include the majority of developed nanoparticles for food packaging applications. Anyway, nanocomposites may be authorized for its use after positive safety evaluation by the European Safety Authority. Manufacturers have to submit a dossier with proper information about the developed material including its composition, toxicity, physical properties, and so on, that is assessed by a group of experts. Afterward, if the experts panel gives a positive opinion, the material may be authorized for food packaging uses and included in the corresponding regulation.

Due to its relevancy in the world commerce, China regulations should be taken into account. In this case no specific regulation for nanoparticles exists although there are some standards and regulations that have to be considered. Thus, in accordance with Chinese

Table 11.3 Authorized Nanoparticles for Food Contact Materials in the European Union (Commission Regulation (EU) No. 10/2011 on Plastic Materials and Articles Intended to Come into Contact with Food and Corresponding Amendments)

Nanoparticle	Restrictions and Specifications	Authorized Use
Titanium nitride	No migration of titanium nitride nanoparticles. In the PET, the agglomerates have a diameter of 100–500 nm consisting of primary titanium nitride nanoparticles; primary particles have a diameter of approximately 20 nm.	Only to be used in PET up to 20 mg/kg.
Carbon black	Primary particles of 10–300 nm, which are aggregated to a size of 100–1200 nm, which may form agglomerates within the size distribution of 300 nm–mm. Toluene extractables: maximum 0.1%, determined according to ISO method 6209. UV absorption of cyclohexane extract at 386 nm: < 0,02 AU for a 1 cm cell or < 0,1 AU for a 5 cm cell, determined according to a generally recognized method of analysis. Benzo(a)pyrene content: max 0,25 mg/kg carbon black.	Maximum use level of carbon black in the polymer: 2.5% w/w.
Silicon dioxide	For synthetic amorphous silicon dioxide: primary particles of 1–100 nm , which are aggregated to a size of 0.1–1 μm, which may form agglomerates within the size distribution of 0.3 μm to the mm size.	No specifications are included.
Kaolin	Particles can be thinner than 100 nm only if incorporated at a quantity of less than 12% w/w in an ethylene vinyl alcohol copolymer (EVOH) inner layer of a multilayer structure, in which the layer in direct contact with the food provides a functional barrier preventing migration of particles into the food.	Only to be used in EVOH, being this material the inner layer of a multilayer structure.
(Butadiene, ethyl acrylate, methyl methacrylate, styrene) copolymer crosslinked with divinylbenzene, in nanoform	Only to be used as particles in nonplasticized PVC up to 10% w/w in contact with all food types at room temperature or below including long-term storage. When used together with the substance with FCM No. 998 and/or the substance with FCM No. 1043, the restriction of 10% w/w applies to the sum of those substances. The diameter of particles shall be > 20 nm, and for at least 95% by number it shall be > 40 nm.	Only to be used in nonplasticized PVC.

Table 11.3 Authorized Nanoparticles for Food Contact Materials in the European Union (Commission Regulation (EU) No. 10/2011 on Plastic Materials and Articles Intended to Come into Contact with Food and Corresponding Amendments) (*cont.*)

Nanoparticle	Restrictions and Specifications	Authorized Use
(Butadiene, ethyl acrylate, methyl methacrylate, styrene) copolymer not cross-linked, in nanoform	Only to be used as particles in nonplasticized PVC up to 10% w/w in contact with all food types at room temperature or below including long-term storage. When used together with the substance with FCM No. 859 and/or the substance with FCM No. 1043, the restriction of 10% w/w applies to the sum of those substances. The diameter of particles shall be > 20 nm, and for at least 95% by number it shall be > 40 nm.	Only to be used in nonplasticized PVC.
(Butadiene, ethyl acrylate, methyl methacrylate, styrene) copolymer crosslinked with 1,3-butanediol dimethacrylate, in nanoform	Only to be used as particles in nonplasticized PVC up to 10% w/w in contact with all food types at room temperature or below including long-term storage. When used together with the substance with FCM No. 859 and/or the substance with FCM No. 998, the restriction of 10% w/w applies to the sum of those substances. The diameter of particles shall be > 20 nm, and for at least 95 % by number it shall be > 40 nm.	Only to be used in nonplasticized PVC.

FCM 859: (butadiene, ethyl acrylate, methyl methacrylate, styrene) copolymer crosslinked with divinylbenzene, in nanoform
FCM 998: (butadiene, ethyl acrylate, methyl methacrylate, styrene) copolymer not crosslinked, in nanoform
FCM 1043: (butadiene, ethyl acrylate, methyl methacrylate, styrene) copolymer crosslinked with 1.3-butanediol dimethacrylate, in nanoform

national standards on food contact materials, every product or raw material shall be compliant with corresponding National Standards (GB), which exists for specific polymers and materials. In addition, the GB 9685-2008 Hygienic Standard for Use of Additives in Food Containers and Packaging Materials is the most important food safety standard for food contact additive in China. This regulation (and its corresponding amendments) includes permitted food contact additives (including monomers) and resins. In most cases, specific migration levels included in this standard are similar to those specified in the European regulation. Nevertheless, this standard applies for all types of packaging (not only plastics) and includes, besides specific migration level, maximum levels of use.

Regardless these standards and regulations, it is possible to authorize new food contact additives not included in the positive lists. Similar to the aforementioned comment for European regulation, the authorization requires the following: (1) application form and physicochemical properties; (2) conditions of use and manufacturing; (3) safety assessment data; (4) migration results; (5) estimation of dietary exposure; and (6) status of the substance in other countries.

8 Conclusions

A considerable number of studies into antimicrobial nanocomposites for food packaging applications has been carried out. The effect of the incorporation of antimicrobial nanomaterials (eg, silver or metal oxide nanoparticles) on the functional properties of different biodegradable and nonbiodegradable polymer film, as well as on the antimicrobial functions has been analyzed.

The controlled release of these active nanofillers or the attached compounds to their surface or nanostructure leads to a sustained antimicrobial activity of the nanocomposites, which is very useful to extend the food shelf life. Antimicrobial activity of some nanomaterials is generally based on the changes provoked in the permeability of the cytoplasmic membrane of microorganisms. Nevertheless, active properties of the nanocomposites are greatly affected by the molecular interactions in the system, and particularly, by nanoparticle aggregation and their specific release capacity to different food systems.

Likewise, particulate fillers improve the stiffness and the toughness of the materials and enhance their barrier properties, although they sometimes impart drawbacks to the resulting composites such as brittleness or opacity.

From the food safety point of view the nanoparticles migration from packaging to the food product and their potential toxicity for humans must be seriously considered, through regulations about their use, based on studies into the potential toxicity of these materials.

References

Abd-Elsalam, K.A., Khokhlov, A.R., 2015. Eugenol oil nanoemulsion: antifungal activity against *Fusarium oxysporum* f. sp. vasinfectum and phytotoxicity on cottonseeds. Appl. Nanosci. 5, 255–265.

Abreu, A.S., Oliveira, M., de Sá, A., Rodrigues, R.M., Cerqueira, M.A., Vicente, A.A., Machado, A.V., 2015. Antimicrobial nanostructured starch based films for packaging. Carbohydr. Polym. 129, 127–134.

Acevedo-Fani, A., Salvia-Trujillo, L., Rojas-Graü, M.A., Martín-Belloso, O., 2015. Edible films from essential-oil-loaded nanoemulsions: physicochemical characterization and antimicrobial properties. Food Hydrocoll. 47, 168–177.
Alexandre, M., Dubois, P., 2000. Polymer-layered silicate nanocomposites: preparation, properties and uses of a new class of materials. Mater. Sci. Eng. R Rep. 28, 1–63.
Anadão, P., 2012. Polymer/clay nanocomposites: concepts, researches, applications and trends for the future. In: Ebrahimi, F. (Ed.), Nanocomposites—New Trends and Developments. InTech, pp. 1–16.
Antoniou, J., Liu, F., Majeed, H., Zhong, F., 2015. Characterization of tara gum edible films incorporated with bulk chitosan and chitosan nanoparticles: a comparative study. Food Hydrocoll. 44, 309–319.
Anu Bhushani, J., Anandharamakrishnan, C., 2014. Electrospinning and electrospraying techniques: potential food based applications. Trends Food Sci. Technol. 38, 21–33.
Arfat, Y.A., Benjakul, S., Prodpran, T., Sumpavapol, P., Songtipya, P., 2014. Properties and antimicrobial activity of fish protein isolate/fish skin gelatin film containing basil leaf essential oil and zinc oxide nanoparticles. Food Hydrocoll. 41, 265–273.
Arunvisut, S., Phummanee, S., Somwangthanaroj, A., 2007. Effect of clay on mechanical and gas barrier properties of blown film LDPE/clay nanocomposites. J. Appl. Polym. Sci. 106, 2210–2217.
Barbosa-Pereira, L., Angulo, I., Lagarón, J.M., Paseiro-Losada, P., Cruz, J.M., 2014. Development of new active packaging films containing bioactive nanocomposites. Innov. Food Sci. Emerg. Technol. 26, 310–318.
Bendahou, D., Bendahou, A., Grohens, Y., Kaddami, H., 2015. New nanocomposite design from zeolite and poly(lactic acid). Ind. Crops Prod. 72, 107–118.
Bikiaris, D.N., Triantafyllidis, K.S., 2013. HDPE/Cu-nanofiber nanocomposites with enhanced antibacterial and oxygen barrier properties appropriate for food packaging applications. Mater. Lett. 93, 1–4.
Bogdanović, U., Vodnik, V., Mitrić, M., Dimitrijević, S., Škapin, S.D., Žunič, V., Budimir, M., Stoiljković, M., 2015. Nanomaterial with high antimicrobial efficacy--copper/polyaniline nanocomposite. ACS Appl. Mater. Inter. 7, 1955–1966.
Burygin, G.L., Khlebtsov, B.N., Shantrokha, A.N., Dykman, L.A., Bogatyrev, V.A., Khlebtsov, N.G., 2009. On the enhanced antibacterial activity of antibiotics mixed with gold nanoparticles. Nanoscale Res. Lett. 4, 794–801.
Campos-Requena, V.H., Rivas, B.L., Pérez, M.A., Figueroa, C.R., Sanfuentes, E.A., 2015. The synergistic antimicrobial effect of carvacrol and thymol in clay/polymer nanocomposite films over strawberry gray mold. LWT—Food Sci. Technol. 64, 390–396.
Cano, A., Fortunati, E., Cháfer, M., Kenny, J.M., Chiralt, A., González-Martínez, C., 2015. Properties and ageing behaviour of pea starch films as affected by blend with poly(vinyl alcohol). Food Hydrocoll. 48, 84–93.
Chen, P., Zhang, L., 2006. Interaction and properties of highly exfoliated soy protein/montmorillonite nanocomposites. Biomacromolecules 7, 1700–1706.
Choudalakis, G., Gotsis, A.D., 2009. Permeability of polymer/clay nanocomposites: a review. Eur. Polym. J. 45, 967–984.
Cioffi, N., Torsi, L., Ditaranto, N., Tantillo, G., Ghibelli, L., Sabbatini, L., Bleve-Zacheo, T., D'Alessio, M., Zambonin, P.G., Traversa, E., 2005. Copper nanoparticle/polymer composites with antifungal and bacteriostatic properties. Chem. Mater. 17, 5255–5262.
Costa, C., Conte, A., Buonocore, G.G., Del Nobile, M.A., 2011. Antimicrobial silver-montmorillonite nanoparticles to prolong the shelf life of fresh fruit salad. Int. J. Food Microbiol. 148, 164–167.

Costa, C., Conte, A., Buonocore, G.G., Lavorgna, M., Del Nobile, M.A., 2012. Calcium-alginate coating loaded with silver-montmorillonite nanoparticles to prolong the shelf-life of fresh-cut carrots. Food Res. Int. 48, 164–169.

Damm, C., Münstedt, H., Rösch, A., 2008. The antimicrobial efficacy of polyamide 6/silver-nano- and microcomposites. Mater. Chem. Phys. 108, 61–66.

Díez-Pascual, A.M., Díez-Vicente, A.L., 2014a. ZnO-reinforced poly(3-hydroxybutyrate-co-3-hydroxyvalerate) bionanocomposites with antimicrobial function for food packaging. ACS Appl. Mater. Inter. 6, 9822–9834.

Díez-Pascual, A.M., Díez-Vicente, A.L., 2014b. Development of nanocomposites reinforced with carboxylated poly(ether ether ketone) grafted to zinc oxide with superior antibacterial properties. ACS Appl. Mater. Inter. 6, 3729–3741.

El Saeed, A.M., El- Fattah, M.A., Azzam, A.M., 2015. Synthesis of ZnO nanoparticles and studying its influence on the antimicrobial, anticorrosion and mechanical behavior of polyurethane composite for surface coating. Dye Pigment 121, 282–289.

Espitia, P.J.P., Soares, N., de, F.F., Coimbra, J.S., dos, R., de Andrade, N.J., Cruz, R.S., Medeiros, E.A.A., 2012. Zinc oxide nanoparticles: synthesis, antimicrobial activity and food packaging applications. Food Bioprocess Technol. 5, 1447–1464.

Fabra, M.J., Jiménez, A., Atarés, L., Talens, P., Chiralt, A., 2009. Effect of fatty acids and beeswax addition on properties of sodium caseinate dispersions and films. Biomacromolecules 10, 1500–1507.

Fernández, A., Picouet, P., Lloret, E., 2010. Cellulose-silver nanoparticle hybrid materials to control spoilage-related microflora in absorbent pads located in trays of fresh-cut melon. Int. J. Food Microbiol. 142, 222–228.

Follain, N., Belbekhouche, S., Bras, J., Siqueira, G., Marais, S., Dufresne, A., 2013. Water transport properties of bio-nanocomposites reinforced by *Luffa cylindrica* cellulose nanocrystals. J. Memb. Sci. 427, 218–229.

Food and Drug Administration, 2015. Packaging & food contact substances (FCS). Available from: http://www.fda.gov/Food/IngredientsPackagingLabeling/PackagingFCS/ucm2006853.htm.

Fortunati, E., Luzi, F., Puglia, D., Terenzi, A., Vercellino, M., Visai, L., Santulli, C., Torre, L., Kenny, J.M., 2013. Ternary PVA nanocomposites containing cellulose nanocrystals from different sources and silver particles: part II. Carbohydr. Polym. 97, 837–848.

Fortunati, E., Rinaldi, S., Peltzer, M., Bloise, N., Visai, L., Armentano, I., Jiménez, A., Latterini, L., Kenny, J.M., 2014. Nano-biocomposite films with modified cellulose nanocrystals and synthesized silver nanoparticles. Carbohydr. Polym. 101, 1122–1133.

Ghaffari-Moghaddam, M., Eslahi, H., 2014. Synthesis, characterization and antibacterial properties of a novel nanocomposite based on polyaniline/polyvinyl alcohol/Ag. Arab. J. Chem. 7, 846–855.

Ghanbari, A., Heuzey, M.-C., Carreau, P.J., Ton-That, M.-T., 2013. Morphological and rheological properties of PET/clay nanocomposites. Rheol. Acta 52, 59–74.

Guerra, R., Lima, E., Guzmán, A., 2013. Antimicrobial supported nanoparticles: gold versus silver for the cases of *Escherichia coli* and *Salmonella typhi*. Micropor. Mesopor. Mater. 170, 62–66.

Han, W., Yu, Y., Li, N., Wang, L., 2011. Application and safety assessment for nano-composite materials in food packaging. Chinese Sci. Bull. 56, 1216–1225.

Hernández-Sierra, J.F., Ruiz, F., Pena, D.C.C., Martínez-Gutiérrez, F., Martínez, A.E., Guillén, A.de J.P., Tapia-Pérez, H., Castañón, G.M., 2008. The antimicrobial sensitivity of *Streptococcus mutans* to nanoparticles of silver, zinc oxide, and gold. Nanomedicine 4, 237–240.

Hietala, M., Mathew, A.P., Oksman, K., 2013. Bionanocomposites of thermoplastic starch and cellulose nanofibers manufactured using twin-screw extrusion. Eur. Polym. J. 49, 950–956.
Hong, S.-I., Rhim, J.-W., 2012. Preparation and properties of melt-intercalated linear low density polyethylene/clay nanocomposite films prepared by blow extrusion. LWT—Food Sci. Technol. 48, 43–51.
Horst, M.F., Quinzani, L.M., Failla, M.D., 2012. Rheological and barrier properties of nanocomposites of HDPE and exfoliated montmorillonite. J. Thermoplast. Compos. Mater. 27, 106–125.
Huang, M., Yu, J., 2006. Structure and properties of thermoplastic corn starch/montmorillonite biodegradable composites. J. Appl. Polym. Sci. 99, 170–176.
Imran, M., Revol-Junelles, A.-M., René, N., Jamshidian, M., Akhtar, M.J., Arab-Tehrany, E., Jacquot, M., Desobry, S., 2012. Microstructure and physico-chemical evaluation of nano-emulsion-based antimicrobial peptides embedded in bioactive packaging films. Food Hydrocoll. 29, 407–419.
Imran, M., Revol-Junelles, A.-M., Paris, C., Guedon, E., Linder, M., Desobry, S., 2015. Liposomal nanodelivery systems using soy and marine lecithin to encapsulate food biopreservative nisin. LWT—Food Sci. Technol. 62, 341–349.
Janaki, A.C., Sailatha, E., Gunasekaran, S., 2015. Synthesis, characteristics and antimicrobial activity of ZnO nanoparticles. Spectrochim. Acta. A. Mol. Biomol. Spectrosc. 144, 17–22.
Jiménez, A., Fabra, M.J., Talens, P., Chiralt, A., 2010. Effect of lipid self-association on the microstructure and physical properties of hydroxypropyl-methylcellulose edible films containing fatty acids. Carbohydr. Polym. 82, 585–593.
Jiménez, A., Fabra, M.J., Talens, P., Chiralt, A., 2012. Edible and niodegradable starch films: a review. Food Bioprocess Technol. 5, 2058–2076.
Jiménez, A., Sánchez-González, L., Desobry, S., Chiralt, A., Arab-Tehrany, E., 2014a. Influence of nanoliposomes incorporation on properties of film forming dispersions and films based on corn starch and sodium caseinate. Food Hydrocoll. 35, 159–169.
Jiménez, A., Arab-Tehrany, E., Sánchez-González, L., 2014b. Progress in biodegradable packaging materials. In: Rhim, J.W. (Ed.), Progress in Nanomaterials for Food Packaging. Future Science Ltd., London, pp. 50–65.
Jiraroj, D., Tungasmita, S., Tungasmita, D.N., 2014. Silver ions and silver nanoparticles in zeolite A composites for antibacterial activity. Powder Technol. 264, 418–422.
Jokar, M., Abdul Rahman, R., Ibrahim, N.A., Abdullah, L.C., Tan, C.P., 2012. Melt production and antimicrobial efficiency of low-density polyethylene (LDPE)-silver nanocomposite film. Food Bioprocess Technol. 5, 719–728.
Kanmani, P., Rhim, J.-W., 2014a. Physicochemical properties of gelatin/silver nanoparticle antimicrobial composite films. Food Chem. 148, 162–169.
Kanmani, P., Rhim, J.-W., 2014b. Properties and characterization of bionanocomposite films prepared with various biopolymers and ZnO nanoparticles. Carbohydr. Polym. 106, 190–199.
Kessick, R., Fenn, J., Tepper, G., 2004. The use of AC potentials in electrospraying and electrospinning processes. Polymer 45, 2981–2984.
Khan, A., Khan, R.A., Salmieri, S., Le Tien, C., Riedl, B., Bouchard, J., Chauve, G., Tan, V., Kamal, M.R., Lacroix, M., 2012. Mechanical and barrier properties of nanocrystalline cellulose reinforced chitosan based nanocomposite films. Carbohydr. Polym. 90, 1601–1608.
Khan, S.T., Al-Khedhairy, A.A., Musarrat, J., 2015. ZnO and TiO_2 nanoparticles as novel antimicrobial agents for oral hygiene: a review. J. Nanoparticle Res. 17, 276.
Kubacka, A., Cerrada, M.L., Serrano, C., Fernández-García, M., Ferrer, M., Fernández-Garcia, M., 2009. Plasmonic nanoparticle/polymer

nanocomposites with enhanced photocatalytic antimicrobial properties. J. Phys. Chem. C 113, 9182–9190.

Lagarón, J.M., Cabedo, L., Cava, D., Feijoo, J.L., Gavara, R., Gimenez, E., 2005. Improving packaged food quality and safety. Part 2: nanocomposites. Food Addit. Contam. 22, 994–998.

Lewington, S., 2013. Antimicrobial nanostructures in food packaging. Food New Zealand 13, 40–41.

Li, X.H., Xing, Y.G., Li, W.L., Jiang, Y.H., Ding, Y.L., 2010. Antibacterial and physical properties of poly(vinyl chloride)-based film coated with ZnO nanoparticles. Food Sci. Technol. Int. 16, 225–232.

Li, H.-Z., Chen, S.-C., Wang, Y.-Z., 2015. Preparation and characterization of nanocomposites of polyvinyl alcohol/cellulose nanowhiskers/chitosan. Compos. Sci. Technol. 115, 60–65.

Liu, W., Liu, J., Liu, W., Li, T., Liu, C., 2013. Improved physical and in vitro digestion stability of a polyelectrolyte delivery system based on layer-by-layer self-assembly alginate-chitosan-coated nanoliposomes. J. Agric. Food Chem. 61, 4133–4144.

Liu, G., Song, Y., Wang, J., Zhuang, H., Ma, L., Li, C., Liu, Y., Zhang, J., 2014. Effects of nanoclay type on the physical and antimicrobial properties of PVOH-based nanocomposite films. LWT—Food Sci. Technol. 57, 562–568.

Llorens, A., Lloret, E., Picouet, P., Fernandez, A., 2012. Study of the antifungal potential of novel cellulose/copper composites as absorbent materials for fruit juices. Int. J. Food Microbiol. 158, 113–119.

Lloret, E., Picouet, P., Fernández, A., 2012. Matrix effects on the antimicrobial capacity of silver based nanocomposite absorbing materials. LWT—Food Sci. Technol. 49, 333–338.

López-Carballo, G., Higueras, L., Gavara, R., Hernández-Muñoz, P., 2013. Silver ions release from antibacterial chitosan films containing in situ generated silver nanoparticles. J. Agric. Food Chem. 61, 260–267.

Mahmoud, K.H., 2015. Synthesis, characterization, optical and antimicrobial studies of polyvinyl alcohol-silver nanocomposites. Spectrochim. Acta. A. Mol. Biomol. Spectrosc. 138, 434–440.

Makwana, S., Choudhary, R., Dogra, N., Kohli, P., Haddock, J., 2014. Nanoencapsulation and immobilization of cinnamaldehyde for developing antimicrobial food packaging material. LWT—Food Sci. Technol. 57, 470–476.

Manikantan, M.R., Varadharaju, N., 2011. Preparation and properties of polypropylene-based nanocomposite films for food packaging. Packag. Technol. Sci. 24, 191–209.

Manna, A.C., 2012. Synthesis, characterization, and antimicrobial activity of zinc oxide nanoparticles. In: Cioffi, N., Rai, M. (Eds.), Nano-Antimicrobials. Springer, Berlin Heidelberg, pp. 151–180.

Martínez-Abad, A., Lagaron, J.M., Ocio, M.J., 2012. Development and characterization of silver-based antimicrobial ethylene-vinyl alcohol copolymer (EVOH) films for food-packaging applications. J. Agric. Food Chem. 60, 5350–5359.

Masadeh, M.M., Karasneh, G.A., Al-Akhras, M.A., Albiss, B.A., Aljarah, K.M., Al-Azzam, S.I., Alzoubi, K.H., 2015. Cerium oxide and iron oxide nanoparticles abolish the antibacterial activity of ciprofloxacin against gram positive and gram negative biofilm bacteria. Cytotechnology 67, 427–435.

Mihindukulasuriya, S.D.F., Lim, L.-T., 2014. Nanotechnology development in food packaging: a review. Trends Food Sci. Technol. 40, 149–167.

Mondal, D., Mollick, M.M.R., Bhowmick, B., Maity, D., Bain, M.K., Rana, D., Mukhopadhyay, A., Dana, K., Chattopadhyay, D., 2013. Effect of poly(vinyl pyrrolidone) on the morphology and physical properties of poly(vinyl alcohol)/sodium montmorillonite nanocomposite films. Prog. Nat. Sci. Mater. Int. 23, 579–587.

Mukherji, S., Ruparelia, J., Agnihotri, S., 2012. Antimicrobial activity of silver and copper nanoparticles: variation in sensitivity across various strains of bacteria and fungi. In: Cioffi, N., Rai, M. (Eds.), Nano-Antimicrobials. Springer, Berlin Heidelberg, pp. 225–251.

Muñoz-Bonilla, A., Cerrada, M.L., Fernández-García, M., Kubacka, A., Ferrer, M., Fernández-García, M., 2013. Biodegradable polycaprolactone-titania nanocomposites: preparation, characterization and antimicrobial properties. Int. J. Mol. Sci. 14, 9249–9266.

Nafchi, A.M., Alias, A.K., Mahmud, S., Robal, M., 2012. Antimicrobial, rheological, and physicochemical properties of sago starch films filled with nanorod-rich zinc oxide. J. Food Eng. 113, 511–519.

Ortega-Toro, R., Contreras, J., Talens, P., Chiralt, A., 2015. Physical and structural properties and thermal behaviour of starch-poly(ε-caprolactone) blend films for food packaging. Food Packag. Shelf Life 5, 10–20.

Otoni, C.G., de Moura, M.R., Aouada, F.A., Camilloto, G.P., Cruz, R.S., Lorevice, M.V., de Soares, N.F.F., Mattoso, L.H.C., 2014a. Antimicrobial and physical-mechanical properties of pectin/papaya puree/cinnamaldehyde nanoemulsion edible composite films. Food Hydrocoll. 41, 188–194.

Otoni, C.G., Pontes, S.F.O., Medeiros, E.A.A., Soares, N.de.F.F., 2014b. Edible films from methylcellulose and nanoemulsions of clove bud (*Syzygium aromaticum*) and oregano (*Origanum vulgare*) essential oils as shelf life extenders for sliced bread. J. Agric. Food Chem. 62, 5214–5219.

Pagno, C.H., Costa, T.M.H., de Menezes, E.W., Benvenutti, E.V., Hertz, P.F., Matte, C.R., Tosati, J.V., Monteiro, A.R., Rios, A.O., Flôres, S.H., 2015. Development of active biofilms of quinoa (*Chenopodium quinoa* W.) starch containing gold nanoparticles and evaluation of antimicrobial activity. Food Chem. 173, 755–762.

Pandiselvi, K., Thambidurai, S., 2015. Synthesis, characterization, and antimicrobial activity of chitosan–zinc oxide/polyaniline composites. Mater. Sci. Semicond. Process. 31, 573–581.

Paul, D.R., Robeson, L.M., 2008. Polymer nanotechnology: nanocomposites. Polymer 49, 3187–3204.

Pavlidou, S., Papaspyrides, C.D., 2008. A review on polymer–layered silicate nanocomposites. Prog. Polym. Sci. 33, 1119–1198.

Peponi, L., Puglia, D., Torre, L., Valentini, L., Kenny, J.M., 2014. Processing of nanostructured polymers and advanced polymeric based nanocomposites. Mater. Sci. Eng. R Rep. 85, 1–46.

Pinto, R.J.B., Almeida, A., Fernandes, S.C.M., Freire, C.S.R., Silvestre, A.J.D., Neto, C.P., Trindade, T., 2013. Antifungal activity of transparent nanocomposite thin films of pullulan and silver against *Aspergillus niger*. Colloids Surf. B Biointerfaces 103, 143–148.

Qi, X., Gunawan, P., Xu, R., Chang, M.W., 2012. Cefalexin-immobilized multi-walled carbon nanotubes show strong antimicrobial and anti-adhesion properties. Chem. Eng. Sci. 84, 552–556.

Regiel-Futyra, A., Kus-Liśkiewicz, M., Sebastian, V., Irusta, S., Arruebo, M., Stochel, G., Kyzioł, A., 2015. Development of noncytotoxic chitosan-gold nanocomposites as efficient antibacterial materials. ACS Appl. Mater. Interfaces 7, 1087–1099.

Rhim, J.-W., Wang, L.-F., 2014. Preparation and characterization of carrageenan-based nanocomposite films reinforced with clay mineral and silver nanoparticles. Appl. Clay Sci. 97-98, 174–181.

Rhim, J.-W., Hong, S.-I., Park, H.-M., Ng, P.K.W., 2006. Preparation and characterization of chitosan-based nanocomposite films with antimicrobial activity. J. Agric. Food Chem. 54, 5814–5822.
Rhim, J.-W., Hong, S.-I., Ha, C.-S., 2009. Tensile, water vapor barrier and antimicrobial properties of PLA/nanoclay composite films. LWT—Food Sci. Technol. 42, 612–617.
Rhim, J.W., Wang, L.F., Hong, S.I., 2013a. Preparation and characterization of agar/silver nanoparticles composite films with antimicrobial activity. Food Hydrocoll. 33, 327–335.
Rhim, J.-W., Park, H.-M., Ha, C.-S., 2013b. Bio-nanocomposites for food packaging applications. Prog. Polym. Sci. 38, 1629–1652.
Rhim, J.-W., Wang, L.-F., Lee, Y., Hong, S.-I., 2014. Preparation and characterization of bio-nanocomposite films of agar and silver nanoparticles: laser ablation method. Carbohydr. Polym. 103, 456–465.
Rusen, E., Mocanu, A., Nistor, L.C., Dinescu, A., Călinescu, I., Mustăţea, G., Voicu, Ş.I., Andronescu, C., Diacon, A., 2014. Design of antimicrobial membrane based on polymer colloids/multiwall carbon nanotubes hybrid material with silver nanoparticles. ACS Appl. Mater. Interfaces 6, 17384–17393.
Sadeghnejad, A., Aroujalian, A., Raisi, A., Fazel, S., 2014. Antibacterial nano silver coating on the surface of polyethylene films using corona discharge. Surf. Coatings Technol. 245, 1–8.
Salaberria, A.M., Labidi, J., Fernandes, S.C.M., 2014. Chitin nanocrystals and nanofibers as nano-sized fillers into thermoplastic starch-based biocomposites processed by melt-mixing. Chem. Eng. J. 256, 356–364.
Salehudin, M.H., Salleh, E., Mamat, S.N.H., Muhamad, I.I., 2014. Starch based active packaging film reinforced with empty fruit bunch (EFB) cellulose nanofiber. Procedia Chem. 9, 23–33.
Salmieri, S., Islam, F., Khan, R.A., Hossain, F.M., Ibrahim, H.M.M., Miao, C., Hamad, W.Y., Lacroix, M., 2014. Antimicrobial nanocomposite films made of poly(lactic acid)–cellulose nanocrystals (PLA–CNC) in food applications—part B: effect of oregano essential oil release on the inactivation of *Listeria monocytogenes* in mixed vegetables. Cellulose 21, 4271–4285.
Salvia-Trujillo, L., Rojas-Graü, M.A., Soliva-Fortuny, R., Martín-Belloso, O., 2015. Use of antimicrobial nanoemulsions as edible coatings: impact on safety and quality attributes of fresh-cut Fuji apples. Postharvest Biol. Technol. 105, 8–16.
Shankar, S., Teng, X., Rhim, J.-W., 2014. Properties and characterization of agar/CuNP bionanocomposite films prepared with different copper salts and reducing agents. Carbohydr. Polym. 114, 484–492.
Sinha Ray, S., Bousmina, M., 2005. Biodegradable polymers and their layered silicate nanocomposites: in greening the 21st century materials world. Prog. Mater. Sci. 50, 962–1079.
Sinha Ray, S., Okamoto, M., 2003. Polymer/layered silicate nanocomposites: a review from preparation to processing. Prog. Polym. Sci. 28, 1539–1641.
Slavutsky, A.M., Bertuzzi, M.A., 2014. Water barrier properties of starch films reinforced with cellulose nanocrystals obtained from sugarcane bagasse. Carbohydr. Polym. 110, 53–61.
Sohrabnezhad, S., Sadeghi, A., 2015. Matrix effect of montmorillonite and MCM-41 matrices on the antibacterial activity of Ag_2CO_3 nanoparticles. Appl. Clay Sci. 105-106, 217–224.
Sothornvit, R., Rhim, J.-W., Hong, S.-I., 2009. Effect of nano-clay type on the physical and antimicrobial properties of whey protein isolate/clay composite films. J. Food Eng. 91, 468–473.
Spitalsky, Z., Tasis, D., Papagelis, K., Galiotis, C., 2010. Carbon nanotube–polymer composites: chemistry, processing, mechanical and electrical properties. Prog. Polym. Sci. 35, 357–401.

Sun, L., Zhang, C., Li, P., 2011. Characterization, antimicrobial activity, and mechanism of a high-performance (-)-epigallocatechin-3-gallate (EGCG)-CuII/polyvinyl alcohol (PVA) nanofibrous membrane. J. Agric. Food Chem. 59, 5087–5092.

TAPPI Standard WI-3021: standard terms and their definition for cellulose nanomaterials, draft. <http://www.tappi.org/content/hide/draft3.pdf>

Tankhiwale, R., Bajpai, S.K., 2012. Preparation, characterization and antibacterial applications of ZnO-nanoparticles coated polyethylene films for food packaging. Colloids Surf. B. Biointerfaces 90, 16–20.

Thomas, S.P., Al-Mutairi, E.M., De, S.K., 2013. Impact of nanomaterials on health and environment. Arab. J. Sci. Eng. 38, 457–477.

Tomé, L.C., Fernandes, S.C.M., Perez, D.S., Sadocco, P., Silvestre, A.J.D., Neto, C.P., Marrucho, I.M., Freire, C.S.R., 2013. The role of nanocellulose fibers, starch and chitosan on multipolysaccharide based films. Cellulose 20, 1807–1818.

Torres-Giner, S., 2011. Electrospun nanofibers for food packaging applications. In: Lagarón, J.M. (Ed.), Multifunctional and Nanoreinforced Polymer for Food Packaging. Woodhead, Cambridge, pp. 108–125.

Torres-Giner, S., Gimenez, E., Lagaron, J.M., 2008. Characterization of the morphology and thermal properties of zein prolamine nanostructures obtained by electrospinning. Food Hydrocoll. 22, 601–614.

Torres-Giner, S., Ocio, M.J., Lagaron, J.M., 2009. Novel antimicrobial ultrathin structures of zein/chitosan blends obtained by electrospinning. Carbohydr. Polym. 77, 261–266.

Vega-Lugo, A.-C., Lim, L.-T., 2009. Controlled release of allyl isothiocyanate using soy protein and poly(lactic acid) electrospun fibers. Food Res. Int. 42, 933–940.

Wang, Z.L., 2004. Zinc oxide nanostructures: growth, properties and applications. J. Phys. Condens. Matter 16, R829–R858.

Wang, D., An, J., Luo, Q., Li, X., Yan, L., 2012. Synthesis, characterization and application of silver-based antimicrobial nanocomposites. In: Cioffi, N., Rai, M. (Eds.), Nano-Antimicrobials. Springer, Berlin Heidelberg, pp. 47–84.

Wei, X., Yang, Z., Tay, S.L., Gao, W., 2014. Photocatalytic TiO_2 nanoparticles enhanced polymer antimicrobial coating. Appl. Surf. Sci. 290, 274–279.

Woranuch, S., Yoksan, R., 2013. Eugenol-loaded chitosan nanoparticles: II. application in bio-based plastics for active packaging. Carbohydr. Polym. 96, 586–592.

Wu, J., Liu, H., Ge, S., Wang, S., Qin, Z., Chen, L., Zheng, Q., Liu, Q., Zhang, Q., 2015. The preparation, characterization, antimicrobial stability and in vitro release evaluation of fish gelatin films incorporated with cinnamon essential oil nanoliposomes. Food Hydrocoll. 43, 427–435.

Xie, L., Lv, X.-Y., Han, Z.-J., Ci, J.-H., Fang, C.-Q., Ren, P.-G., 2012. Preparation and performance of high-barrier low density polyethylene/organic montmorillonite nanocomposite. Polym. Plast. Technol. Eng. 51, 1251–1257.

Xing, Y., Li, X., Zhang, L., Xu, Q., Che, Z., Li, W., Bai, Y., Li, K., 2012. Effect of TiO_2 nanoparticles on the antibacterial and physical properties of polyethylene-based film. Prog. Org. Coatings 73, 219–224.

Yadav, S.K., Mahapatra, S.S., Cho, J.W., 2012. Synthesis of mechanically robust antimicrobial nanocomposites by click coupling of hyperbranched polyurethane and carbon nanotubes. Polymer 53, 2023–2031.

Youssef, A.M., Abdel-Aziz, M.S., El-Sayed, S.M., 2014. Chitosan nanocomposite films based on Ag-NP and Au-NP biosynthesis by *Bacillus subtilis* as packaging materials. Int. J. Biol. Macromol. 69, 185–191.

Zeng, Q.H., Yu, A.B., Lu, G.Q., Paul, D.R., 2005. Clay-based polymer nanocomposites: research and commercial development. J. Nanosci. Nanotechnol. 5, 1574–1592.

Zhang, W., Qiao, X., Chen, J., 2007. Synthesis of silver nanoparticles—effects of concerned parameters in water/oil microemulsion. Mater. Sci. Eng. B 142, 1–15.

Zhang, B., Luo, Y., Wang, Q., 2010. Development of silver-zein composites as a promising antimicrobial agent. Biomacromolecules 11, 2366–2375.

Zhang, H.Y., Arab Tehrany, E., Kahn, C.J.F., Ponçot, M., Linder, M., Cleymand, F., 2012. Effects of nanoliposomes based on soya, rapeseed and fish lecithins on chitosan thin films designed for tissue engineering. Carbohydr. Polym. 88, 618–627.

Zhao, S., Yao, J., Fei, X., Shao, Z., Chen, X., 2013. An antimicrobial film by embedding in situ synthesized silver nanoparticles in soy protein isolate. Mater. Lett. 95, 142–144.

Zheng, J., Zhang, H., Zhao, Z., Han, C.C., 2012. Construction of hierarchical structures by electrospinning or electrospraying. Polymer 53, 546–554.

12

NANOMATERIAL-BASED SENSORS FOR MYCOTOXIN ANALYSIS IN FOOD

Bing Huei Chen, Baskaran Stephen Inbaraj
Fu Jen University, Department of Food Science, Taipei, Taiwan

1 Introduction

Mycotoxins are organic compounds produced as secondary metabolites by fungi. Filamentous fungi produce thousands of toxic compounds, but the more important mycotoxins belong to species of *Aspergillus, Fusarium,* and *Penicillium* (Vidal et al., 2013). From the toxicological and legislative point of view, the mycotoxins of interest include aflatoxins, ochratoxins, fumonisins, deoxynivalenol, T-2, HT-2, zearalenone, patulin, citrinin, and ergot alkaloids (Bennett and Klich, 2003). The presence of mycotoxins in foods and feeds not only retards the international trade but also represents a health hazard for humans and animals (Shephard et al., 2012; Vidal et al., 2013). Storage of food and feed under high temperature and humidity conditions favor the growth of filamentous fungi (Campas et al., 2012). In addition, agricultural products, particularly a wide variety of grains in the field and during storage after harvest, are the main source of mycotoxin production in the food chain (Prieto-Simon et al., 2007). Mycotoxins affect a broad range of agricultural products including cereals, cereal-based foods, corn, rice, rye, wheat, barley, oats, sorghum, soybeans, buckwheat, peanuts, malt, beer, dried fruits, wine, milk, baby and infant foods, coffee beans, cocoa, bakery, and meat products, all of which play a vital role for economy in many developing countries (Prieto-Simon et al., 2007; Shephard et al., 2012). Only since 5 years mycotoxins attracted worldwide interest as a result of serious economic losses due to deleterious impact on human health, animal production, domestic and international commerce (Bhat et al., 2010). However, their existence has been as long as crops have grown on earth. Mycotoxins can produce acute and chronic effects causing serious impact on

Novel Approaches of Nanotechnology in Food. http://dx.doi.org/10.1016/B978-0-12-804308-0.00012-1

central nervous, pulmonary, and cardiovascular systems (Bhat et al., 2010). Consequently, to protect consumers and maintain sustainability of agriculture, the maximum permissible limits (MPLs) in several matrices as well as detection limits of analytical methods have been established by several organizations including World Health Organization (WHO), United States Food and Drug Administration (USFDA), European Commission (EU), and Food and Agriculture Organization (FAO) (Campas et al., 2012; Vidal et al., 2013). The EU has recommended the MPL for fusarium mycotoxins in maize products to be in the range from 50 to 2000 μg/kg (Vidal et al., 2013). However, the MPLs may vary depending on the toxicity and variety of food. For example, aflatoxin B_1 is regulated by EU legislation with MPL at 2 μg/kg in foods for human consumption, while aflatoxin M_1, which is not as toxic as aflatoxin B_1, is regulated with MPL at 0.05 μg/kg owing to the high consumption of milk by infants and children (Vidal et al., 2013). Thus, it is imperative to develop analytical methods for rapid and sensitive determination of mycotoxins with capability for multiplexing and on-site analysis.

Analyses of mycotoxins have been difficult because of their heterogeneous distribution in agricultural products and complexity of mycotoxin-containing food matrices (Koppen et al., 2010). Accordingly, the selection of detection method strongly depends on the mycotoxin and the matrix to be analyzed (Campas et al., 2012). Moreover, the MPLs of mycotoxins set by the legislation have necessitated development of analytical methods for qualitative and quantitative determination of mycotoxins in both food and feed samples (Valdes et al., 2009). While thin-layer chromatography (TLC) has been widely used to detect and identify the presence of mycotoxins, some other methods with superior analytical performance are also currently used for quantitative determination of mycotoxins (Turner et al., 2009). For instance, most of the official methods are based on high-performance liquid chromatography (HPLC) coupled with ultraviolet (UV), fluorescence (FL), or mass spectrometry (MS) detectors (Krska et al., 2005; Shephard, 2008). Detection methods based on gas chromatography (GC) coupled with MS detectors are not widely used due to interference drawbacks (Koppen et al., 2010). Of the various methods, liquid chromatography-tandem MS (LC-MS/MS) have been used as a valuable analytical technique for simultaneous multitoxin determination. Though these techniques provide very low detection limits, their cost and need of skilled personnel along with the incapability to perform in situ analysis have considerably reduced their application in many parts of the developing countries (Campas et al., 2012). Routine screening of mycotoxins is commonly performed by enzyme-linked immunosorbent assays (ELISA), which is capable of analyzing a large number of samples in a single assay (Tothill, 2011; Malhotra et al., 2014). Yet, the disadvantages are laborious and tedious procedures, as well as cross-reactivity resulting in false positive results especially for single mycotoxin determination. Extract cleanup and/or dilution and addition of detergents are some of the pretreatment steps carried out to minimize the matrix effects (Shephard, 2008). For rapid analysis of mycotoxins, both quantitative and semiquantitative ELISA kits as well as immunochromatography test kits are commercially available (Campas et al., 2012). Emerging methods for mycotoxin analysis include fluorescence polarization assays and Fourier transform near-infrared spectroscopy (FT-NIR) (Cruz-Aguado and Penner, 2008a; Chun et al., 2009; Tripathi and Mishra, 2009). In addition, new receptors such as recombinant fragments of antibodies, DNA aptamers or molecularly imprinted polymers (MIPs) have been proposed for incorporation into high-throughput analysis (Romanazzo et al., 2010; Cruz-Aguado and Penner, 2008b; Navarro-Villoslada et al., 2007; Maragos, 2009). Over the past two decades, many approaches for developing new affinity biosensors for mycotoxin analysis have been evaluated and some excellent reviews have been published on analytical methods (Cigic and Prosen, 2009; Koppen et al., 2010; Shephard et al., 2012), general biosensors (Logrieco et al., 2005; Maragos and Busman, 2010; Prieto-Simon et al., 2007) and electrochemical biosensors (Palchetti and Mascini, 2008; Laschi et al., 2011) for mycotoxins.

Biosensors are valuable devices for detection of many different target toxins including mycotoxins in a wide range of matrices. Among different sensors such as enzyme sensors, electrochemical sensors, optical immunosensors, tissue biosensors, quartz crystal sensors, and surface plasmon resonance (SPR) biosensors, electrochemical sensors have been used for a long time, which consists of a transducer and recognition receptor (Brett and Oliveira-Brett, 2011). Several transducers widely used in conventional electrochemical sensors include amperometric, voltammetric, impedimetric, potentiometric, and conductometric transducers (Tothill, 2011; Vidal et al., 2013). For recognition receptors, antibodies, DNA, antibody fragments, aptamers, and molecularly imprinted polymers are commonly employed (Prieto-Simon et al., 2007; Tothill, 2011). The application of nanotechnology in biosensors can range from the transducer device to the recognition ligand, the label and the running systems (Campas et al., 2012; Malhotra et al., 2014). The application of nanomaterials in sensor development provides excellent advantages such as miniaturization of devices and signal enhancements (Tothill, 2011). Moreover, the use of nanoparticles as labels can not only result in high precision and accuracy as well as signal amplification, but also the high surface-to-volume ratio of nanomaterials can enhance sensitivity and allow single molecule detection of toxins (Malhotra et al., 2014). The small size of nanomaterials also shorten the time of assay, decrease the distance of electron travel, lower both power and voltage, improve component density, and enhance chip functionality for multiplexing capability (Tothill, 2011; Malhotra et al., 2014). This chapter provides an overview on various nanomaterial-based sensors used for detection of mycotoxins in food and/or feed. Fig. 12.1 shows the chemical structures of various mycotoxins reviewed in this chapter.

2 Aflatoxins

Aflatoxins are a group of secondary fungal metabolites mainly produced by *Aspergillus flavus* and *A. parasiticus* under certain optimum conditions (Vidal et al., 2013). They are the most widely spread group of toxins resulting in contamination of agricultural and food products (Ye et al., 2010). *Aspergillus* molds grow mostly on crops such as grains and nuts as well as in milk, cheese, corn, peanuts, cottonseed, almonds, figs, spices, and a variety of other foods and feeds (Malhotra et al., 2014). Besides, milk, eggs, and meat products are sometimes contaminated due to animal consumption of aflatoxin-contaminated feed. The *Aspergillus* species can grow well under warm temperature and moisture level ≥7% (Malhotra et al., 2014). Of more than 20 aflatoxins identified, the major ones are AFB_1, AFB_2, AFG_1, and AFG_2, with AFB_1 (2,3,6a,9a-tetrahydro-4-methoxycyclopenta[c]furo-[3′,2′:4,5]furo[2,3-h][1]-benzopyran-1,11-dione) being the most toxic one, followed by AFG_1, AFB_2, and AFG_2 (Fig. 12.1) (Delmulle et al., 2005). The International Agency for Research on Cancer (IARC) has categorized AFB_1 to be under group 1 carcinogen, which can cause liver cancer in animals (Vidal et al., 2013). The EU has regulated its level at 2.0 μg/kg in foods for direct human consumption and 5 μg/kg for all kinds of animals (Delmulle et al., 2005). Therefore, there is an urgent need to develop sensitive and selective analytical methods for rapid detection of aflatoxins in food and feed samples.

2.1 Detection of Aflatoxin B_1

An immunoassay based on bioelectrocatalytic reaction was developed by Liu et al. (2006), who fabricated a microcomb electrode by self-assembling horseradish peroxidase (HRP) and AFB_1 antibody molecules on gold nanoparticles (GNPs). A good

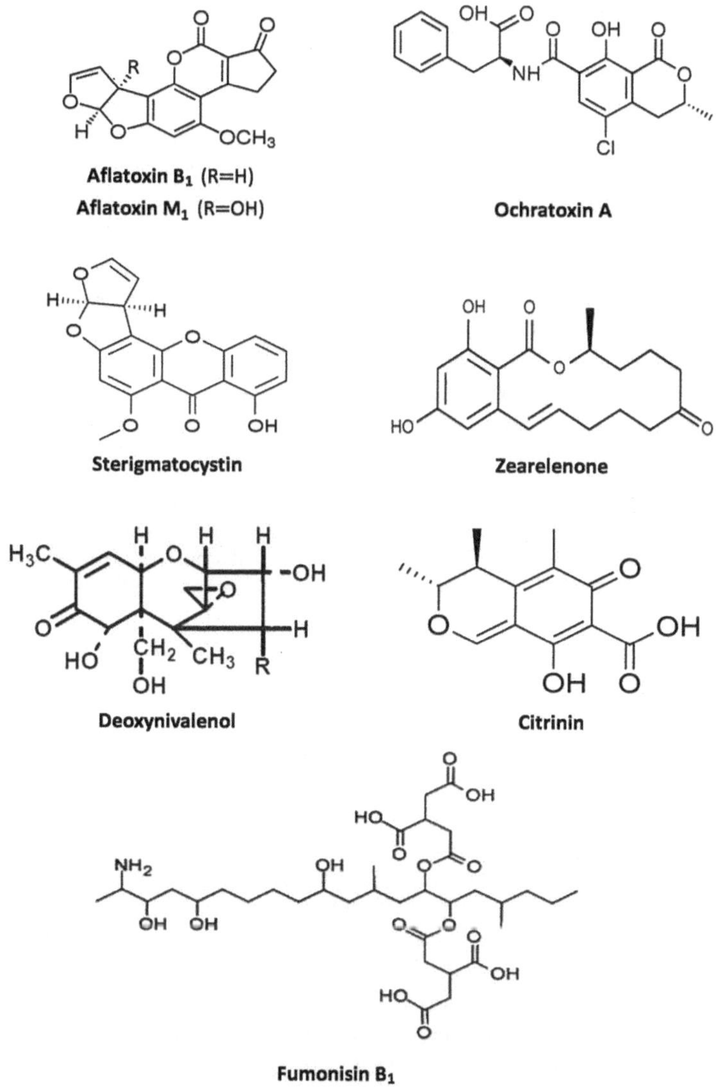

Figure 12.1. Chemical structures of some mycotoxins.

conductometric response relative to AFB_1 concentration was shown in a linear range from 0.5 to 1.0 ng/mL with a low detection limit of 0.1 ng/mL at 3δ. The developed immunoassay shows high accuracy and is comparable with that obtained by ELISA, with the interassay coefficient of variation being 7.1 and 4.6% for 1.0 and 8.0 ng/mL AFB_1, respectively. The presence of GNPs provides a congenial environment for the immobilized biomolecules and decreases the electron transfer impedance leading to a direct electrochemical behavior of the immobilized HRP. In principle, a barrier of direct electrical communication generated between the immobilized HRP and the electrode surface through the formation of antibody–antigen reaction of immobilized anti-AFB_1 and AFB_1 in samples cause variation in local conductivity, which is measured by the HRP bioelectrocatalytic reaction in 0.02 M phosphate buffer solution (pH 7.0) containing 80 μM hydrogen peroxide, 0.05 M potassium iodide, and 0.15 M sodium chloride. However, a two-fold decrease in detection limit (0.05 ng/mL) with a wide linear dynamic range (1.0–12 ng/mL) was attained in a later study (Sun et al., 2008), in which ionic liquid 1-ethyl-3-methyl imidazolium tetrafluoroborate was immobilized on the surface of a glassy carbon electrode (GCE) through titania sol and Nafion film, followed by GNPs adsorption onto titania surface and HRP-labeled anti-AFB_1 antibodies on GNPs surface. Likewise, Owino et al. (2008) developed an electrochemical immunosensor by immobilizing AFB_1-bovine serum albumin (BSA) on a polythionine/GNPs conjugate-modified glass carbon electrode, with the surface of AFB_1-BSA being covered with HRP for prevention of nonspecific binding of ions onto the sensor. The detection was based on competition between free AFB_1 and the immobilized AFB_1-BSA conjugate for the binding sites of free anti-AFB_1 antibody. The response by differential pulse voltammetry dropped following an increase in the level of AFB_1 in the linear working range of 0.6–2.4 ng/mL with a detection limit of 0.07 ng/mL being attained.

Application of CNTs is becoming increasingly important for construction of biosensors because of their unique physical and chemical characteristics, especially the high specific superficial area can increase the amount of enzymes immobilized, enlarge the reaction areas between enzyme and substrate, elevate the electrical conductivity and enhance the biosensor response. Li et al. (2011) constructed an amperometric biosensor by embedding aflatoxin-oxidase (AFO) in sol–gel and linking to multiwalled carbon nanotubes MWCNTs-modified Pt electrode and demonstrated that the covalent linkage between AFO and MWCNTs could retain enzyme activity for sensitive detection of AFB_1 oxidation. The biosensor exhibited good affinity for AFB_1 as shown by the apparent Michaelis–Menten constant value of 7.03 μmol/L. Moreover, the sensor was sensitive over the wide linear range from 3.2 to 721 μmol/L (1–225 ng/mL) with the limit of detection of 1.6 nmol/L and average response time of 44 s. A high activation energy (18.8 kJ/mol) obtained for this method demonstrates significant catalyzation of AFO for oxidation of AFB_1. However, in a recent study dealing with evaluating carboxylated MWCNTs (c-MWCNTs)-based biosensor, Singh et al. (2013) attained a much lower detection limit (0.08 ng/mL) in the linear dynamic range of 0.25–1.375 ng/mL for AFB_1 by electrophoretically depositing c-MWCNTs onto indium tin oxide (ITO) glass, followed by functionalizing with monoclonal AFB_1 antibody and analyzing using an electrochemical technique. The method showed high sensitivity (95.2 μA/ng mL/cm² and improved detection limit (0.08 ng/mL), with low association constant (0.0915 ng/mL) indicating high affinity of the immunoelectrode toward AFB_1.

2.2 Novel On-Site Immunological Techniques for Detection of Aflatoxins AFB_1 and AFB_2

Due to consumers' demand for food safety, a one-step assay is necessary for determination of mycotoxin residues. After researching on many techniques, scientists have developed a novel concept of immunochromatography (IC), which has several benefits including user-friendly format, short assay time, long-term stability over a wide range of climates and cost effectiveness (Delmulle et al., 2005; Shim et al., 2007). This IC technique, also known as lateral-flow assay, utilizes antigen, and antibody properties for the rapid detection of toxin residues (Xiulan et al., 2006). For fast screening of AFB_1, Xiulan et al. (2005) developed an IC technique by using a nanogold-labeled polyclonal antibody probe and a detection limit of 2.5 ng/mL was attained. Compared to the conventional ELISA method, a 6- to 10-fold reduction in assay time and two-fold decrease of detection limit was achieved by this IC method. The entire analysis can be completed within 10 min and the technique is simple for easy on-site application even by an untrained personnel. By adopting a similar IC technique, the same research group employed colloidal gold-labeled polyclonal antibody specific to AFB_1 as marker, AFB_1-BSA conjugate as competitive antigen and goat-anti-rabbit antibody as control for AFB_1 detection in 67 naturally contaminated food samples (Xiulan et al., 2006). This IC method shows a comparable detection limit of 2.5 ppb as the above method by visual observation while a much lower value of 0.05–0.1 ppb by photometric strip reader. However, the detection limit in food extract matrix increased to 2 ng/mL, with the recovery data of AFB_1-spiked rice, corn, and wheat ranging from 80.79 to 110.56%. A high correlation ($R^2 = 0.93$) between IC and ELISA kit test was shown for all the 67 food samples. Likewise, Delmulle et al. (2005) employed a membrane coated with two capture reagents, namely, AFB_1-BSA conjugate and rabbit antimouse antibodies, conjugate pad saturated with GNPs as detection reagent and absorbent pad for rapid detection of AFB_1 in pig feed. The AFB_1 in the sample competes with AFB_1 immobilized on the membrane for binding to limited amounts of antibodies in the detection reagent with line color intensity of positive AFB_1 being visually distinguishable from that of AFB_1-negative sample. The whole analysis can be completed within 10 min, but the visual detection limit (VDL) of 5 ng/mL is high when compared to the two aforementioned methods. However, in a later study, Shim et al. (2007) reported a much lower VDL of 0.5 ng/mL by using an IC strip test with nanocolloidal gold-monoclonal antibody as a probe. Application of both IC and HPLC methods to 172 grain and feed samples showed comparable results for detection of AFB_1. However, a cross reactivity to AFB_2, AFG_1, and AFG_2 can occur for IC assay, which should be overcome in the future study.

Immunoassay-based lateral flow dipstick (LFIDS) method is also gaining considerable interest in recent years for on-site detection of mycotoxins in food and feed. Moreover, besides employment of one kind of nanoparticle, composite nanoparticles are also used as nanomaterial supports in sensors based on an immunoassay technique. For example, Tang et al. (2009) developed a novel membrane-based LFIDS assay for fast screening of AFB_2 in foods by using magnetic nanogold microspheres (nano-Fe_2O_3 as core and GNPs as shell) biofunctionalized with monoclonal anti-AFB_2 antibodies. A VDL (0.9 ng/mL) three-fold lower than the GNPS-based conventional immunostick test was obtained. Qualitative results can be obtained within 15 min with no false negative results and without need of any expensive equipment. An excellent correlation of results was shown as HPLC for AFB_2-spiked food samples such as peanuts, hazelnuts, pistacia, and almonds. Likewise, Liao and Li (2010) used a monoclonal antibody immobilized nanoparticles with a silver core and a gold shell (Ag/Au) in a LFIDS assay. The Ag/Au dipstick exhibited a lower VDL of 0.1 ng/mL for AFB_1, with sensitivity, reproducibility, and stability being higher than that for pure GNPs. Also, the assay was validated for application to naturally contaminated samples including rice, wheat, sunflower, cotton, chilies, and almonds.

Similar to IC and LFIDS, dot-immunogold filtration (DIGF) assay is a new technique, which uses nitrocellulose membrane as solid phase support and GNPs as label. Ye et al. (2010) demonstrated rapid detection (15 min) of AFB_1 in 45 different food samples with a VDL of 2 ng/mL and high specificity for AFG_1, AFG_2, and AFM_1 could be attained. A high correlation of results was obtained as HPLC for all the samples. However, a slight cross reactivity with AFB_2 is claimed to be a pitfall of this method. Overall, these immunological techniques (IC, LFIDS, and DIGF) have the potential to be used as user-friendly, rapid, and cost-effective screening tools, especially for preliminary screening of mycotoxin detection in food and agricultural products within a very short time. In addition, these techniques enable on-site application by an untrained personnel providing long-term stability over a wide range of climates.

2.3 Simultaneous Determination of AFB_1, AFB_2, AFG_1, and AFG_2

Usually aflatoxins of more than one variety exist together in the contaminated foods (Zhang et al., 2009). Although more than 20 aflatoxins have been identified, the major aflatoxins of great concern include AFB_1, AFB_2, AFG_1, and AFG_2 owing to their classification under group 1 human carcinogens by the International Agency for Research on Cancer (Cavaliere et al., 2007; Ventura et al., 2006). Moreover, the current maximum level set by the EU is 4 ng/g for total aflatoxins ($B_1 + B_2 + G_1 + G_2$) in groundnuts, nuts, dried fruits, and cereals (Kolosava et al., 2006). To meet the growing demand in rapid monitoring of total aflatoxins in agroproducts, Zhang et al. (2011) developed an ultrasensitive immunochromatographic assay (ICA) for simultaneous detection of 4 aflatoxins AFB_1, AFB_2, AFG_1, and AFG_2 in peanuts. The ICA is based on a competitive format whose sensitivity is improved by using a novel criterion to screen the optimal amount of MAb labeled onto nanogold particles. Initially, antiaflatoxin MAb was conjugated with nanogold solution for preparation of nanogold probe, followed by preparation of IC strip containing a sample pad, conjugate release pad, absorbent pad, and nitrocellulose membrane. AFB_1-BSA conjugate and rabbit antimouse IgG were used for test and control lines, respectively, and coated onto the nitrocellulose membrane using the BioDot XYZ3050 platform, nanogold probe on conjugate pad and no treatment on absorbent pad. Then, the coated membrane, conjugate pad, sample pad, and absorbent pad were laminated, pasted to a plastic scaleboard and the assembled scaleboard was cut vertically for dividing into strips and storing in a desiccator at 4°C. After dipping the test strip in sample solution, the released nanogold probe moved along the nitrocellulose membrane chromatographically due to capillary action and the different shades of red color depending on aflatoxin concentration formed after 15 min were visually compared. In the peanut matrix, all the four aflatoxins could be detected in the linear working range of 0.03–2.0 ng/mL for AFB_1 and AFB_2 as well as 0.06–4.0 ng/mL for AFG_1 and AFG_2, with the VDL being 0.03, 0.06, 0.12, and 0.25 ng/mL, respectively. The results were also shown to be in good agreement with those obtained by HPLC for analysis of aflatoxins in peanuts, demonstrating practical application of ICA in real samples. It is worth pointing out that ICA possesses several advantages such as procedure simplicity, rapid operation, immediate results, low cost and no requirement of skilled technicians or expensive equipment (Cui et al., 2008; Zhang et al., 2011). Owing to these characteristics, ICA is suitable for online, in-field, and immediate applications. However, until now, only a few studies have focused on application of ICA-based technique in analysis of total aflatoxins.

2.4 Detection of Aflatoxin M_1

Aflatoxin M_1 (AFM_1) is the hydroxylated metabolite of AFB_1 and the mammals ingesting AFB_1 contaminated diets excrete AFM_1 into milk, resulting in its presence in a large variety of dairy products (Fig. 12.1) (Stoloff, 1980; Radoi et al., 2008). It is relatively stable during pasteurization, storage, and preparation of various dairy products, thereby raising significant threats to human health especially to children as they are the major consumers of milk (Stubblefield and Shannon, 1974; Wang et al., 2011). Several studies have demonstrated the toxic and carcinogenic effects of AFM_1, which led WHO-IARC to change its classification from group 2 to group 1 for carcinogens (IARC, 2002). Also, the European Community Legislation has limited the concentration of AFM_1 in milk, dried and processed milk products intended for adults and infants to be at 0.050 and 0.025 ppb, respectively (ECR, 2001, 2004).

A superparamagnetic nanoparticle-based enzyme-linked immunosorbent assay was developed for AFM_1 detection by Radoi et al. (2008), who tested five different clones of antibodies (1C6, 3G11, 6G4, ATX9, and ATX2) in a direct competitive ELISA format and compared with the classical indirect and direct competitive ELISA formats aiming to reduce the time consumed during coating, blocking, and competition steps. Of the five antibody clones, the best IC_{50} value of 15 ng/L in buffer and 41 ng/L in milk was shown by the clone 3G11 in the linear working range of 4–250 ng/L with recovery ranging from 93 to 98% depending on the spiked AFM_1 concentration (30–120 ng/L). In addition, the precision data for replicate measurements showed a relative standard deviation ranging from 3 to 7%. The conjugation of superparamagnetic iron oxide nanoparticles (SPIONs) with protein G and antimouse IgG reduced the coating time from 12–14 h (in conventional ELISA methods) to 20 min. Also, the competition time was reduced with complete skipping of blocking step. The application of SPIONs enabled fast, reliable, and easy detection of AFM_1 in milk with their separation from the bulk solution being accomplished by using a simple magnet.

In a later study, Vig et al. (2009) developed an impedimetric immunosensor based on colloidal gold and silver electrodeposition for detection of AFM_1. An indirect competitive ELISA assay performed on screen-printed electrodes (SPE) in the presence of anti-AFM_1 gold-labeled antibodies and chronoamperometrically electrodeposited silver provided good response signal with the calculated charge transfer resistance correlating well with AFM_1 concentration. The linear working range achieved by this method was from 15 to 1000 ng/L with a detection limit of 15 ng/L. However, in milk sample matrix, a wide range of linearity (25–125 ng/L) and higher detection limit (25 ng/L) was shown, which may be attributed to absorption of protein compounds on gold-labeled AFM_1. A high mean recovery of 116% was achieved for milk samples spiked with different concentrations of AFM_1 (25, 50, and 100 ng/L). Dinckaya et al. (2011) developed an impedimetric AFM_1 biosensor based on DNA probe and GNPs. A self-assembled monolayer of cysteamine and GNPs were prepared layer-by-layer on gold electrodes by monitoring using EIS and CV techniques and then a thiol-modified single stranded DNA probe was immobilized to specifically bind AFM_1. The biosensor used $K_3[Fe(CN)_6]/K_4[Fe(CN)_6]$ as a redox probe for electrochemical measurements providing a linear response over the concentration range of 1–14 ng/mL and detection limit of 0.4 ng/mL. Both cysteamine and GNPs enhanced the charge transfer of the redox probe through the electrode surface. Overall, the biosensor exhibited good analytical characteristics such as high affinity of AFM_1 toward the DNA probe, low quantification limit (0.4 ng/mL), high recovery (107%), reproducibility (R^2 ranging from 0.9244 to 0.9913), and repeatability (coefficient of variation = 6.2%).

Immunochromatographic strip method using GNPs and electrochemical immunochip sensor is becoming popular for detection of low molecular weight toxins (Liu et al. 2008; Paek et al., 2000). This technique involves comigration of test samples and antibody–gold nanoparticle conjugates along membrane strips for binding interactions to occur and comparison of color intensity visually provides efficient and simple on-site detection in less than 10 min without requirement of any skilled personnel or sophisticated instruments (Parker et al., 2009). Wang et al. (2011) developed a gold nanoparticle immunochromatographic strip method for effective on-site detection of AFM_1 providing linear response in the concentration range of 0–10 ng/mL in milk samples with detection limit at 0.5–1.0 ng/mL, which is close to the Taiwan FDA and United States FDA regulatory limit of 0.5 ng/mL. Also, of the 144 milk samples analyzed in Taiwan, only 1 sample (0.054 ng/mL) was found to slightly exceed the regulatory limit of the European Union, suggesting that the milk products sold in Taiwan are safe for human consumption.

3 Ochratoxin A

Ochratoxin A (OTA) is a ubiquitous mycotoxin produced by fungi of improperly stored food products. Of at least nine ochratoxins identified so far, only three (OTA, OTB, and OTC) contribute mainly to overall proportion with OTA being the most prevalent and toxic (Bonel et al., 2011). Structurally, OTA is 7-(L-β-phenylalanylcarbonyl)-carboxyl-5-chloro-8-hydroxy-3, 4-dihydro-3R-methyl-isocumarin (Fig. 12.1) (Kaushik et al., 2009a). It is commonly found in cereals, barley, wheat, corn, rye, coffee, wines, dried fruits, nuts, baby foods, and animal feeds as well as in blood of animals and humans (Bonel et al., 2011). Also OTA is found in tissues and organs of animals and is known to produce nephrotoxic, teratogenic, carcinogenic, and immunotoxic effects, mainly associated with the Balkan endemic nephropathy and urinary tract tumors in human (Pfohl-Leszkowicz and Manderville, 2007). The IARC has classified OTA as Group 2B human carcinogen and the mode of action is possibly by induction of oxidative DNA damage (Kaushik et al., 2009a; Vidal et al., 2013). Pfohl-Leszkowicz and Manderville (2007) have presented an overview on OTA's toxicity and carcinogenicity in animals and humans. Consequently, the EU has imposed a strict permissible limit of 3–5 μg/kg in cereals and cereal products (Bonel et al., 2011). Thus, the development of analytical methods for detection of OTA with improved sensitivity is important.

For detection of OTA using GNPs, Bonel et al. (2010) developed competitive electrochemical immunosensors based on disposable SPE through physical adsorption of OTA onto BSA, followed by binding of OTA-BSA on GNPs-modified working electrode surface and detecting by differential-pulse voltammetry using α-naphtyl-phosphate as substrate. The secondary IgG antibody labeled with alkaline phosphatase enhances the detection signal. The response signal changes linearly in the concentration range from 0.3 to 8.5 ng/mL with the detection limit being 0.20 ng/mL, which is about four-fold lower than that obtained by OTA-BSA based immunosensor without GNPs. Application of this sensor to OTA-spiked wheat sample provides a high recovery of 104–107%, implying no matrix interference on OTA detection. In a later study, compared to the previous method, a further reduction in detection limit to one-half (0.10 ng/mL) was achieved by an improved competitive immunosensor employing biotinylated monoclonal antibody capture probe and GNPs. The differential pulse voltammetric measurements show linear response with OTA concentration in the range of 0.15–9.94 ng/mL. A high recovery (101–105%) of OTA from OTA-spiked wheat sample demonstrates high accuracy of the proposed immunosensor method. Through covalent immobilization via amine coupling to carboxymethylated dextran hydrogel-modified gold working electrode, Heurich et al. (2011) reduced further the detection limit to 0.05 ng/mL, which remained unaltered in the wine matrix, while a linear dynamic response in the wide concentration range from 0.01 to 100 ng/mL was obtained. Electrochemical detection was performed using 3,3′,5,5′-tetramethylbenzidine and hydrogen peroxide with HRP as the enzyme label. The generated signal was detected by chronoamperometry at −150 mV using onboard screen-printed Ag-AgCl as pseudoreference electrode. Earlier to the previously mentioned methods, Liu et al. (2009) constructed a more sensitive immunosensor by loading OTA-ovalbumin conjugate on self-assembled monolayer of 1,6-hexanedithiol assembly on GNPs and a much lower detection limit of 8.2×10^{-3} ng/mL in the linear working range between 0.01 and 100 ng/mL was obtained. After competition of the limited anti-OTA mouse monoclonal antibody between immobilized hapten and OTA in sample, alkaline phosphatase-labeled horse antimouse IgG antibody was selectively bound on electrode surface, thereby providing response signal inversely proportional to OTA concentration in the sample. For OTA determination in corn samples, negligible matrix effect and good recovery (85–105%) were obtained, revealing feasibility of the proposed method for application to real sample matrix.

Instead of antibodies, recently aptamer-based biosensors are developed for detection of OTA. Aptamers are functional nucleic acids selected from combinatorial oligonucleotide libraries through in vitro selection to bind to specific target molecules such as drugs, proteins, peptides, vitamins, and other organic/inorganic compounds. Aptamers have several advantages over antibodies due to their small size. Also, they can provide greater surface density for receptors and bind a wide range of targets including nonimmunogenic targets. They are thermally stable, reusable, and easily modified to facilitate immobilization of reporter molecules and functional groups. By making use of SPIONs, Barthelmebs et al. (2011) developed a novel electrochemical aptasensor on disposable SPE for sensitive detection of OTA. More specifically, the competition between free OTA and labeled alkaline phosphatase (ALP)-OTA for binding to the DNA aptamer immobilized on magnetic beads generated signals from the ALP enzyme for measurement by differential pulse voltammetry (DVP). The OTA can be detected at a level as low as 0.11 ng/mL in buffer and 0.15 ng/mL in wine samples, with a high recovery of spiked OTA (0.5–10 ng/mL) ranging from 94 to 98%. In another study, a much lower detection limit (0.07 ng/mL) in the linear range from 0.78 to 8.74 ng/mL can be attained by using superparamagnetic beads functionalized with an DNA biotinylated aptamer specific to OTA, which can compete with OTA-HRP conjugate and free OTA (Bonel et al., 2011). The modified-magnetic beads are then localized on disposable SPE under a magnetic field for OTA detection by DPV. Spiking of OTA in wheat samples yielded average recoveries ranging from 102 to 104% with the relative standard deviation (RSD) being <8%. Fabricated aptamer-based electrochemical "signal-off" sensor was developed by Kuang et al. (2010), who immobilized three single-stranded DNA molecules (including aptamer) on the surface of an electrode and electrochemically measured the change in redox current of methylene blue produced by binding of OTA target to the aptamer. A detection limit as low as 0.03 ng/mL in the linear working range of 0.1–20 ng/mL is attained due to sensitive detection of aptamer by OTA and signal enhancement caused by GNPs-functionalized DNA molecules. A highly sensitive and selective fluorescent aptamer sensor after modification with carboxyfluorescein was constructed by Guo et al. (2011) using single-walled carbon nanotubes (SWCNTs) as fluorescence quencher. The fluorescence measurements provide a linear response over the OTA concentration range from 0.025 to 0.2 ng/mL, while the lowest detection limit was 0.024 ng/mL. This method has been successfully applied to real sample such as OTA-spiked beer sample. Thus, the integration of nanostructures with the use of specific recognition molecules such as nucleic acids (aptamers) to recognize and bind target analytes in the sample matrix is promising for application to a wide range of toxins and pathogens in food and feed.

Of the various metal oxides, zinc oxide nanoparticles (ZNPs) has been of interest for immunosensor applications due to their unique properties such as high isoelectric point (9.5) and biocompatibility facilitating immobilization of an enzyme and protein having low isoelectric point via electrostatic interactions. Moreover, ZNPs not only provide a friendly microenvironment for immobilizing negatively charged rabbit antibodies (r-IgGs) to retain its bioactivity, but also accelerate electron transfer communication between protein and the electrode to a large extent. For detection of OTA, Ansari et al. (2010) deposited ZNPs onto indium-tin-oxide (ITO) glass plate and coimmobilized with r-IgGs and BSA. The electrochemical impedimetric response of BSA/r-IgGs/ZNPs/ITO immune electrode obtained as a function of OTA concentrations showed a linear response ranging from 6×10^{-6} to 1×10^{-5} ng/mL, detection limit of 6×10^{-6} ng/mL and response time of 25 s. In an earlier study, Kaushik et al. (2009a) also employed a similar immunoelectrode (BSA/r-IgGs/CNPs/ITO) using cerium oxide nanoparticles (CNPs) instead of ZNPs and developed an immunosensor for detection of

OTA with a detection limit of 2.5×10^{-5} ng/mL in the linear working range of 5×10^{-5}–6×10^{-4} ng/mL. In addition, a high affinity of the electrode toward OTA is evident as shown by a high association constant (0.9×10^{11} L/mol). Similar to ZNPs, CNPs also possess unique properties of high mechanical strength, oxygen ion conductivity, high isoelectric point, biocompatibility, high absorption, and high oxygen storage capacity. Several advantages of using CNPs in biosensors are low temperature processing, tunable physical parameters, optical transparency, chemical inertness, thermal stability, and insignificant swelling in both aqueous and nonaqueous solutions for immobilization of enzymes.

The surface functionalization of nanoparticles with organic moieties can result in controlled release and improved molecular recognition capabilities of nanoparticles for biosensing application. Kaushik et al. (2009b) utilized silica nanoparticles (SiNPs) after functionalizing with chitosan (CH) and coimmobilized with r-IgGs and BSA (BSA/r-IgGs/CH-SiNPs/ITO) for detection of OTA. A rapid response (25 s) with a detection limit of 3×10^{-5} ng/mL in the linear dynamic range of 5×10^{-5}–6×10^{-4} ng/mL was obtained. It is worth pointing out that SiNPs with controlled particle size, morphology, porosity, and surface area, along with its chemical, thermal, and easy functionalization properties make SiNPs suitable for several technological applications including biosensing. Besides, their biocompatibility, nontoxic nature, high ionic conductivity and surface-to-volume ratio are also suitable for construction of biosensors. In another study, Khan and Dhayal (2008) also conjugated chitosan with titanium oxide nanoparticle (CH/TNPs) and immobilized with r-IgGs and proteins for detection of OTA with response signal by electrochemical impedance spectroscopy increasing linearly to 10 ng/mL. The sensitivity of CH/TNPs conjugate was shown to be four-fold higher than that by using only CH.

3.1 Simultaneous Determination of AFB_1 and OTA

Analyzing more than one mycotoxin using the same method at the same time is important as multiplexing capability reduces both time and cost. Shim et al. (2009) developed a one-step simultaneous immunochromatographic assay for rapid analysis of AFB_1 and OTA in feed samples by using colloidal–monoclonal antibody conjugate and a VDL of 0.5 ng/mL for AFB_1 and 2.5 ng/mL for OTA was obtained. The matrix interference from the feed extracts was significantly reduced by appropriate dilution with buffer and the detection limit for the feed spiked with different AFB_1/OTA mixtures were 10 and 50 μg/kg for AFB_1 and OTA, respectively. In a later study,

Wu et al. (2011) further lowered the detection limit of both AFB_1 and OTA by 50-fold by employing antigen-modified magnetic nanoparticles as immunosensing probes and antibody functionalized upconversion nanoparticles (UCNPs) as signal probes for simultaneous detection of AFB_1 and OTA. The signal probes were prepared by functionalizing $NaY_{0.78}F_4$:$Yb_{0.2}$, $Tm_{0.02}$, and $NaY_{0.28}F_4$:$Yb_{0.7}$,$Er_{0.02}$ UCNPs with immobilized anti-AFB_1 and anti-OTA antibodies. The detection limit of both AFB_1 and OTA was 0.01 ng/mL in the linear range from 0.01 to 10 ng/mL. Both high sensitivity and selectivity obtained by this method can be attributed to the magnetic separation and concentration effect of magnetic particles, the high sensitivity of UCNPs and the different emission lines of Yb/Tm and Yb/Er doped $NaYF_4$ UCNPs by 980 nm laser. These two methods have been successfully applied for simultaneous determination of AFB_1 and OTA in naturally contaminated feed samples or maize samples, with the results being comparable with those obtained by conventional ELISA and HPLC methods.

4 Sterigmatocystin

Sterigmatocystin (SMC) with a molecular formula ($C_{18}H_{12}O_6$) is a mycotoxin with toxicity second to AFB_1 (Versilovskis and De Saeger, 2010; Georgianna and Payne, 2009). Being a biogenic precursor of AFB_1, both SMC and AFB_1 possess a similar structure consisting of a xanthone nucleus attached with a bifuran (Fig. 12.1) (Versilovskis and De Saeger, 2010). Several *Aspergillus* species responsible for the production of SMC include *A. versicolor, A. nidulans, A. chevalieri, A. ruber, A. amstelodami, A. aureolatus, A. quadrilineatus,* and *A. sydowi* (Versilovskis and De Saeger, 2010). Some other fungal species such as *Penicillium, Bipolaris, Chaetomium, Emiricella* are also capable of producing this mycotoxin. The severity of SMC's toxicity is mainly due to its xerophilic nature, that is, the tendency to grow even at low water activity (<0.8) (Versilovskis and De Saeger, 2010). In addition, it can grow in a wide range of temperature between 4 and 40°C. It commonly contaminates foods and crops such as wheat, corn, peanut, barley, pecan nuts, soybeans, coffee beans, ham, and cheese (Table 12.1) (Prieto-Simon et al., 2007). Like aflatoxins, SMC is carcinogenic to humans and has been classified as group 2B by the IARC (Chen et al., 2010). SMC is acutely toxic to the liver and kidney and closely associated with the occurrence of hepatocellular, gastric and esophagus carcinoma (Purchase and van der Watt, 1970). Moreover, SMC is found to be one of the predominant contaminating mycotoxins in highly incident areas of malignant tumors in China (Tian and Liu, 2004). Although many countries do not have regulations

Table 12.1 Source, Harmful Effects as Well as Food Products and Animals Affected by Certain Mycotoxins

Mycotoxins	Sources	Contaminated Food Products	Affected Animals	Harmful Effects
Aflatoxins (B_1, B_2, G_1, G_2, M_1, M_2)	*Aspergillus flavus*, *Aspergillus nominus*, *Aspergillus parasiticus*	Corn, rice, pasta, Brazil nuts, peanuts, peanut butter, pistachios, cassava, tobacco, cottonseed cake, oilseeds, figs, milk, cheese, yoghurt, spices	Fish, rabbits, dogs, birds, turkeys, ducks, primates, cows	Humans: hepatomegaly, jaundice, hepatitis, cirrhosis, liver cancer, Reye's syndrome, kwashiorkor, immunosuppression Animals: weight loss, reduced reproductive function, teratogenesis, immunosuppression, mutagenesis, kidney, cancer, liver cancer, death
Ochratoxin A	*Aspergillus alliaceus*, *Aspergillus auricomus*, *Aspergillus carbonarius*, *Aspergillus glaucus*, *Aspergillus melleus*, *Aspergillus niger*, *Aspergillus ochraceus*, *Penicillium cyclopium*, *Penicillium verrucosum*, *Penicillium viridicatum*	Corn, rice, rye, wheat, buckwheat, barley, millet, oats, grapes, raisins, currants, nuts, coffee, cocoa, spices, wine, beer, pork, cheese	Mice, hamsters, rats, chickens, dogs	Humans: nephropathy (Balkan endemic nephropathy) Animals: reduced reproductive function, teratogenesis, nephrotoxicity, immunosuppression, genotoxicity, mutagenesis, cancer
Sterigmato-cystin	*Aspergillus nidulans*, *Aspergillus versicolor*	Corn, wheat, barley, peanuts, pecan nuts, soya beans, green coffee beans, ham, cheese	Cows, mice, monkeys, rats, zebra fish	Animals: diarrhea, lactation loss, teratogenesis, pulmonary tumors, liver lesions, hepatomas, liver tumors, liver cancer, kidney lesions, skin tumors, mutagenesis, cancer, death
Zearlenone	*Fusarium culmorum*, *Fusarium crookwellense*, *Fusarium equiseti*, *Fusarium graminearum*, *Fusarium sporotrichioide*	Corn, wheat, wheat flour, bread, breakfast cereals, noodles, rice, barley, oats, sorghum, walnuts, milk, corn beer, meat, animal-feed products	Cows, mice, pigs, sheep	Humans: cervical cancer Animals: vaginitis, sterility, infertility, abortion, neonatal mortality, reduced litter size and weight, hyperestrogenism, earlier sexual maturity, prolonged estrus, reduced lactation, genotoxicity

(*Continued*)

Table 12.1 Source, Harmful Effects as Well as Food Products and Animals Affected by Certain Mycotoxins (cont.)

Mycotoxins	Sources	Contaminated Food Products	Affected Animals	Harmful Effects
Deoxynivalenol	*F. culmorum, F. graminearum*	Corn, popcorn, rice, rye, wheat, wheat flour, bread, buckwheat, barley, barley products, oats, sorghum, triticale, breakfast cereals, noodles, baby and infant foods, malt, beer	Pigs, horses	Humans: throat irritation, gastrointestinal and abdominal pains, nausea, vomiting, diarrhea, hemorrhages, dizziness, fever, headache Animals: appetite loss, weight loss, vomiting, diarrhea, necrosis, teratogenesis, immunosuppression
Citrinin	*Aspergillus carneus, Aspergillus niveus, Aspergillus terreus, Penicillium citrinum, P. verrucosum, Penicillium expansum*	Corn, rice, rye, wheat, barley, oats, peanuts, fruits, sausages	Birds, mice, pigs, guinea, pigs, rabbits, rats	Humans: fatal kidney disease, Balkan endemic nephropathy, Shoshin-kakke beriberi Animals: fetotoxicity, abortion, teratogenesis, kidney damage, vasodilatation, bronchial constriction, liver damage, increased muscular tone, nephropathy, necrosis
Fumonisins (B_1, B_2)	*Aspergillus alternate, Fusarium anthophilum, Fusarium dlamini, Fusarium moniliforme, Fusarium naphiformel, Fusarium nygama, Fusarium proliferatum, Fusarium verticillioides*	Corn, polenta, corn-based breakfast cereals, snack products, rice, sorghum, mung beans, navy beans, asparagus beer	Cows, horses, pigs, rats	Humans: esophageal cancer Animals: appetite loss, reduced litter weight, low bone development in fetus, fetal mortality, respiratory problems, porcine pulmonary edema, lethargy, hydrothorax, hepatic lesions, fibrosis, hepatocellular carcinoma, hypercholesterolemia, neurotoxicity, equine leukoencephalomalacia, immunosuppression Plants: cell-membrane damage, reduced chlorophyll synthesis, disrupted sphingolipids synthesis

on SMC, Czech Republic and Slovakia have set 5 μg/kg for rice, vegetables, potatoes, flour, poultry, meat, and milk as well as 20 μg/kg for other foods (Stroka et al., 2004). For humans, the California Department of Health Services has recommended no significant risk level of 8 μg/kg bw per day for a 70 kg adult (Versilovskis and De Saeger, 2010). Consequently, the EU-based European Food Safety Authority (EFSA) has suggested that analytical methods for SMC determination in foods should attain a limit of quantification of <1.5 μg/kg (EFSA, 2013) (Table 12.2).

Novel electrochemical biosensors were designed incorporating SWCNTs and MWCNTs and enzymes. Ever since MWCNTs came into use in 1991, the manufacture of enzyme biosensors have increased substantially as MWCNTs can facilitate large quantities of enzymes to be immobilized, thereby enlarging the reactive area between enzyme and substrate, elevating the electrical conductivity and enhancing the biosensor response (Yao et al., 2006). Yao et al. (2004) conducted a preliminary study on SMC detection using a triplet electrode enzyme-biosensor system, in which Ag/AgCl is used as reference electrode, while Pt and Au deposited on MWCNTs and aflatoxin-detoxifizyme (ADTZ) are used as counter and working electrode, respectively. The response signals are shown to be substrate correlative with linear working range and detection limit being 8.32×10^{-5}–66.56×10^{-5} mg/mL and 8.32×10^{-5} mg/mL (0.268 μM) respectively. However, the narrow linear range and low sensitivity are claimed to be the disadvantages of this method. In a later study, the same research group improved the sensitivity and widened the linear working range of this method by saturating the ADTZ with SMC prior to covalently immobilizing on the surface of MWCNT-modified electrode (Yao et al., 2006). Initially, the MWCNTs are oxidized by refluxing with aqua regia and the MWCNTs-modified electrode are prepared by dropping 10 μL of MWCNTs suspension on the gold electrode and evaporating the solvent in a 60°C oven. Then, the SMC-saturated ADTZ is immobilized onto MWCNTs-modified electrode through covalent bonding. More specifically, upon oxidation of MWCNTs, its surface area is enhanced and the hydrophobic surface of MWCNTs are converted into hydrophilic ones favoring attachment with hydrophilic sites on ADTZ enzyme and thereby the attachment of hydrophobic sites on ADTZ with the hydrophobic biosensor substrate (SMC) remains undisturbed. Differential pulse voltammetric measurements revealed the linear working range of 0.134–4.29 μM and a detection limit of 0.134 μM could be achieved with two-fold improvement in sensitivity compared to the aforementioned method using randomly absorbed ADTZ-MWCNTs electrode. The improvement in sensitivity may

Table 12.2 Some Nanoparticle-Based Biosensors and Their Analytical Performance for Detection of Certain Mycotoxins

Nanomaterial Support	Mycotoxin	Biorecognition/ Electrochemical Technique	LOD/Linear Dynamic Range	Sample	References
GNPs	AFB_1	Anti-AFB_1 Ab; conductometry	0.1 ng/mL; 0.5–10 ng/mL	Buffer	Liu et al. (2006)
	OTA	OTA-(CMD); amperometry	0.05 ng/mL; 0.01–100 ng/mL	Buffer, wine	Heurich et al. (2011)
	OTA	Aptamer to OTA; CV	0.03 ng/mL; 0.1–20 ng/mL	Buffer	Kuang et al. (2010)
	OTA	Anti-OTA MAb; EI	0.008 ng/mL; 0.01–100 ng/mL	Corn	Liu et al. (2009)
	AFB_1 and OTA	MAb; ICA	AFB_1 = 0.5 ng/mL; OTA = 2.5 ng/mL	Feed sample	Shim et al. (2009)
	AFB_1, AFB_2, AFG_1, AFG_2	MAb; ICA	AFB_1 = 0.03 ng/mL; AFB_2 = 0.06 ng/mL; AFG_1 = 0.12 ng/mL; AFG_2 = 0.25 ng/mL	Peanuts	Zhang et al. (2011)
	AFB_1	Anti-AFB_1 MAb; EI	0.05 ng/mL; 0.1–12 ng/mL	Naturally contaminated samples	Sun et al. (2008)
	AFM_1	Thiol-modified ssDNA; EIS and CV	1–14 ng/mL	Milk	Dinckaya et al. (2011)
	AFM_1	Anti-AFM_1 Ab; EIS and LSV	0.015 ng/mL; 0.015–1 ng/mL	Buffer and milk	Vig et al. (2009)
	OTA	Anti-OTA MAb; DPV	0.15–9.94 ng/mL; 0.10 ng/mL	Cereals	Vidal et al. (2011)
	AFB_1	PI; ICIA	0.01 ng/mL; 0.1–10 ng/mL	Milk	Jin et al. (2009)
	CIT	Anti-CIT IgG MAb; amperometry	0.1 ng/mL; 0.5–50 ng/mL	Rice	Arevalo et al. (2011)
	AFB_1	Anti-AFB_1 Ab; LFIDS	5 ng/mL; 2–10 ng/mL	Pig feed	Delmulle et al. (2005)
	AFB_1	Anti-AFB_1 Ab; DPV and CV	0.07 ng/mL; 0.6–2.4 ng/mL	Buffer	Owino et al. (2008)

Table 12.2 Some Nanoparticle-Based Biosensors and Their Analytical Performance for Detection of Certain Mycotoxins (*cont.*)

Nanomaterial Support	Mycotoxin	Biorecognition/ Electrochemical Technique	LOD/Linear Dynamic Range	Sample	References
	OTA	Anti-OTA PAb; DPV	0.86 ng/mL; 0.3–8.5 ng/mL	Wheat	Bonel et al. (2010)
	AFM_1	Anti-AFM_1 PAb; ELISA-ICA	1.0 ng/mL; 0.01–10 ng/mL	Milk	Wang et al. (2011)
	AFB_1	Anti-AFB_1 PAb; ICA	2.5 ng/mL; 0.5–5 ng/mL	Buffer	Xiulan et al. (2005)
	AFB_1	Anti-AFB_1 PAb; ICA	2 ng/mL; 2–50 ng/mL	Rice, corn, wheat	Xiulan et al. (2006)
	AFB_1	Anti-AFB_1 MAb; DIGFA	2 ng/mL; 0.5 – 10 ng/mL	45 Food samples	Ye et al. (2010)
	AFB_1	Anti-AFB_1 MAb; ICA	0.5 ng/mL; 0.1–10 ng/mL	172 Grain and feed samples	Shim et al. (2007)
	FMN	Anti-FMN MAb; SPGE	5 ng/mL; 1–1000 ng/mL	Corn	Kadir and Tothill (2010)
CNTs	SMC	Aflatoxin-detoxifizyme electrode; EIS and CV	0.13 ng/mL; 0.13–4.29 ng/mL	Buffer	Yao et al. (2006)
	ZRL	Anti-ZRL Ab; EI	0.77 ppb; 0–500 ppb	Corn, food and feed samples	Panini et al. (2010)
	AFB_1	Anti-AFB_1; EI	0.08 ng/mL; 0.25–1.38 ng/mL	Buffer	Singh et al. (2013)
	AFB_1	Aflatoxin-oxidase; amperometry	1.6 ng/mL; 1–225 ng/mL	Buffer	Li et al. (2011)
	SMC	Aflatoxin-oxidase; CV	3 ng/mL; 10–1480 ng/mL	Buffer	Chen et al. (2010)
	OTA	Aptamer-carboxyfluorescein conjugate; FS	0.024 ng/mL; 0.025–0.2 ng/mL	Beer	Guo et al. (2011)

(*Continued*)

Table 12.2 Some Nanoparticle-Based Biosensors and Their Analytical Performance for Detection of Certain Mycotoxins (*cont.*)

Nano-material Support	Myco-toxin	Biorecognition/ Electrochemical Technique	LOD/Linear Dynamic Range	Sample	References
MNPs	AFB_1 and OTA	Anti-AFB_1 Ab and Anti-OTA Ab	0.01 ng/mL; 0.01–10 ng/mL for AFB_1 and OTA	Maize	Wu et al. (2011)
	AFM_1	Anti-AFM_1 Ab; ELISA	0.004 ng/mL; 0.004–0.25 ng/mL	Milk	Radoi et al. (2008)
	OTA	Alkaline phosphatase-aptamer; DPV	0.11 ng/mL; 0.11–15 ng/mL	Wine	Barthelmebs et al. (2011)
MBs	ZRL	r-IgG MAb; amperometry	0.011 ng/mL	Maize, baby food	Hervas et al. (2009)
	ZRL	EI using carbon SPE; EIS	0.007 ng/mL	Baby foods	Hervas et al. 2010)
	ZRL	MAb; ELISA and EIS	0.4 ng/mL	Solid and liquid samples	Hervas et al. (2011)
	OTA	Aptamer to OTA; DPV	0.07 ng/mL; 0.78–8.4 ng/mL	Wheat	Bonel et al. (2011)
	DON	Anti-DON Ab; amperometry	63 ng/mL; 100–4500 ng/mL	Wheat, cereal, and baby food	Romanazzo et al. (2010)
CNPs	OTA	r-IgG Ab; CV and EIS	0.0025 ng/mL; 0.005–0.06 ng/mL	Buffer	Kaushik et al. (2009a)
ZNPs	OTA	r-IgG; EIS	6×10^{-6} ng/mL; 6×10^{-6}– 1×10^{-5} ng/mL	Buffer	Ansari et al. (2010)
TNPs	OTA	r-IgG Ab; EIS	0–10 ng/mL	Buffer	Khan and Dhayal (2008)
SiNPs	OTA	r-IgG Ab; CV	0.003 ng/mL; 0.005–0.06 ng/mL	Buffer	Kaushik et al. (2009b)

Table 12.2 Some Nanoparticle-Based Biosensors and Their Analytical Performance for Detection of Certain Mycotoxins (*cont.*)

Nano-material Support	Myco-toxin	Biorecognition/ Electrochemical Technique	LOD/Linear Dynamic Range	Sample	References
GNPs@ MNPs	AFB_2	Anti-AFB_2 MAb; LFIDS	0.9 ng/mL	Peanuts, hazlenuts, pistachio, almonds	Tang et al. (2009)
GNPs@ SNPs	AFB_1	Anti-AFB_1 MAb; LFIDS	0.1 ng/mL	Rice, wheat, sunflower, cotton, chilies, almonds	Liao and Li (2010)

LOD, limit of detection; AFB_1, aflatoxin B_1; AFB_2, aflatoxin B_2; AFG_1, aflatoxin G_1; AFG_2, aflatoxin G_2; OTA, ochratoxin A; FMN, fumonisin; CIT, citrinin; SMC, sterigmatocystin; ZRL, zearalenone; DON, deoxynivalenol; GNPs, gold nanoparticles; CNTs, carbon nanotubes; MNPs, superparamagnetic iron oxide nanoparticles; SNPs, silver nanoparticles; CNPs, cerium oxide nanoparticles; ZNPs, zinc oxide nanoparticles; TNPs, titanium oxide nanoparticles; QDs, quantum dots; SiNPs, silica nanoparticles; GNPs@MNPs, gold nanoparticles coated over magnetic iron oxide nanoparticles; GNPs@SNPs, gold nanoparticles coated over silver nanoparticles; Ab, antibody; MAb, monoclonal antibody; PAb, polyclonal antibody; r-IgG, rabbit immunoglobulin G; CMD, carboxymethylated dextran; ssDNA, single-stranded DNA; EI, electrochemical immunosensor; FS, fluorescence spectroscopy; PI, piezoelectric immunosensor; EIS, electrochemical impedance spectroscopy; DPV, differential pulse voltammetry; LSV, linear sweep voltammetry; ICIA, indirect competitive immunoassay; LFIDS, lateral-flow immunodipstick; CV, cyclic voltammetry; ELISA, enzyme-linked immunosorbent assay; ICA, immunochromatographic assay; DIGFA, dot-immunogold filtration assay; SPGE, screen printed gold electrode; SPCE, screen printed carbon electrode; EMC, electrochemical microfluidic chips; and ITA, immunoturbidimetric assay.

be attributed to the protection of ADTZ's active centers and its appropriate orientation provided by saturation of ADTZ and direct attachment of ADTZ to MWCNTs. However, several parameters such as sample temperature, ion intensity, and acidity are still required to be optimized for further improvement in sensitivity and applicability.

Recently, Chen et al. (2010) developed a simple, rapid, and highly sensitive electrochemical biosensor for SMC based on a new enzyme aflatoxin-oxidase (AFO), which is cloned from *Armillariella tabescens* and expressed in *Pichia pastoris*. The chitosan-coated SWCNTs are prepared first by dispersing acid-functionalized SWCNTs in 0.2% CH solution by weight. Then, for preparation of CH-AFO-SWCNTs/Au electrode, the purified AFO is ultrasonicated with CH-SWCNTs suspension to obtain black-color CH-AFO-SWCNTs solution containing 5 mg/mL AFO, followed by dropping 5 μL of CH-AFO-SWCNTs on the gold electrode surface and drying at 4°C for 24 h. The AFO enzyme is immobilized onto CH-SWCNTs-modified Au electrode through electrostatic and hydrophobic interactions. The CH-SWCNTs nanocomposite matrix could provide a favorable microenvironment for AFO to retain its native conformation and accelerate direct electron transfer between protein and underlying electrode surface. Accordingly, the CH-AFO-SWCNTs/Au electrode exhibiting excellent electrocatalytic response, long-term stability, and high reproducibility was successfully prepared for detection of SMC. An SMC response of 95% is reached within 10 s with the linear concentration range from 10 to 1480 ng/mL and the detection limit (3 ng/mL) obtained is comparable to that obtained by the conventional GC-MS and LC-MS methods. The kinetic parameters such as charge transfer coefficient, electron transfer rate constant, the apparent Michaelis–Menten constant and activation energy obtained for this method are 0.4, 0.9 ± 0.02/s, 7.8 μM, and 38.8 kJ/mol, respectively. The biosensor also showed high operational stability, reproducibility, and long-term stability, which can be attributed to three major aspects: (1) electrostatic interaction between positively charged CH-SWCNTs and negatively charged AFO can favor AFO immobilization; (2) the 3-D-porous structure of CH-SWCNTs biofunctional film can prevent leaching of the enzyme; and (3) the biopolymer chitosan can provide desirable environment around the enzyme for maintenance of AFO's bioactivity. In addition, the 3-D-porous structure of composite film can also accelerate the direct electron transfer between protein and electrode surface.

5 Zearalenone

Zearalenone (ZRL), also known as F-2 toxin, is a nonsteroidal estrogenic mycotoxin biosynthesized through a polyketide pathway by several *Fusarium* species including *F. graminearum*, *F. culmorum*, *F. sporotrichioide*, *F. cerealis*, *F. equiseti*, *F. crookwellense*, and *F. semitectum*, all of which are common contaminants of cereal crops worldwide in both temperate and warm regions (Zinedine et al., 2007; Bennett and Klich, 2003). It belongs to a phenolic resorcyclic acid lactone mycotoxin chemically described as 6-[10-hydroxy-6-oxo-*trans*-1-undecenyl]-B-resorcyclic acid lactone (Fig. 12.1) (Panini et al., 2010). Under the moist and cool field conditions, ZRL can contaminate several food products such as maize, corn, wheat, bread, cereals, noodles, rice, oats, barley, sorghum, walnuts, milk, meat, rye, and animal-feed products (Table 12.1) (Prieto-Simon et al., 2007). More specifically, the highest concentrations of ZRL are reported to be found in wheat bran, corn and corn products, grains and grain-based foods as well as vegetable oil, all of which contribute largely to the overall ZRL exposure (EFSA, 2011). The worldwide contamination of food and feed products is estimated to range from a minimum of 180 ng/kg in Qatar and maximum of 6×10^8 ng/kg in India (Zinedine et al., 2007). Although steroidal estrogens and ZRL are structurally dissimilar, several derivatives (mainly ZRL and zearalenol) exhibit estrogenic activity comparable to diethylstilbestrol through appropriate folding to orient hydroxyl groups and facilitate attachment to estrogen-binding tissue receptors (Zinedine et al., 2007; Panini et al., 2010). It is frequently associated with reproductive disorders of farm animals and hyperestrogenic syndromes in humans (Diekman and Green, 1992; Farnworth and Trenholm, 1983; Green et al., 1990). The biotransformation of ZRL in animals involves formation of two ZRL metabolites, namely, α-zearalenol and β-zearalenol, both of which conjugate with glucuronic acid (Long and Diekman, 1986). Zinedine et al. (2007) have presented an overview of acute, subacute, and chronic toxicity along with reproductive and developmental toxicity, carcinogenicity, genotoxicity, and immunotoxicity of ZRL and its metabolites. The mean dietary intake for ZRL is 20 ng/kg bw per day for people in Canada, Denmark, and Norway, while a higher intake of 30 ng/kg bw per day is estimated among Americans (Zinedine et al., 2007). The joint FAO/WHO Expert Committee on Food Additives (JECFA) established a maximum tolerable daily intake of 0.5 μg/kg bw for ZRL (JECFA, 2000).

For rapid detection of ZRL in corn silage samples, an electrochemical-based method was developed by Panini et al. (2010) by coupling an immunosensor with GCE modified with MWCNTs and integrating with a continuous-flow system. The MWCNTs-modified GCE was prepared by dropping a 20 μL aliquot of MWCNTs dispersed in 0.1% Nafion solution onto the GC electrode surface and drying under infrared lamp for solvent evaporation. Next, the immunosensor was designed by immobilizing mouse monoclonal anti-ZRL antibody to ZRL on 3-aminopropyl-modified controlled-pore glass, which was previously spread smoothly on a double-coated tape affixed to the disk surface. The immobilized antibodies were finally washed with phosphate buffer (pH 7) and stored at 5°C until further use. The ZRL in corn sample was then allowed to compete immunologically with HRP-bound ZRL for the immobilized antibodies. HRP in the presence of hydrogen peroxide catalyzes the oxidation of 4-tert-butylcatechol, followed by measuring the back electrochemical reduction on MWCNTs-modified glass carbon electrode at −0.15 V by an amperometric technique. The developed electrochemical immunosensor is rapid (15 min) and enables detection of ZRL in the linear working range of 0–500 ppb with detection limit at 0.77 ppb, recovery from 93 to 106.5%, coefficient of variation for within-assay precision at <2.42% and between-assay precision at <4.36%. Compared to the conventional ELISA method, a higher sensitivity could be achieved by 3.2-fold reduction in detection limit. The main advantages of this method are enhancement of electrochemical responses by MWCNTs, minimization of expensive reagents, physical and chemical stability of immunosensor, low background noise, wide working potential range and accuracy.

Application of magnetic beads in electrochemical immunoassay for detection of ZRL was explored by Hervas et al. (2009, 2010, 2011). In an amperometric analysis based on direct competitive immunoassay involving antibody-coated magnetic beads as immobilization support and HRP as enzymatic label, ZRL could be detected in baby food samples with low detection limit (0.011 μg/L) and high recovery (95–108%) (Hervas et al., 2009). However, in a later study, a remarkably lower detection limit (0.007 μg/L) and a good recovery (101–111%) are achieved by an electrochemical immunosensor consisting of magnetic beads coupled with disposable carbon SPE (Hervas et al., 2010). In addition, a very low systematic error (<4%) and excellent reproducibility (RSD = 6%) are obtained by this method. After the immunochemical reaction, the modified paramagnetic beads were confined by a magnet on the surface of SPE where the electrochemical detection is accomplished using hydrogen peroxide substrate and hydroquinone mediator for the HRP enzyme.

Microfluidic technology has now become a novel sensing platform in which different analytical steps, biological recognition materials and suitable transducers are cleverly integrated to yield a generation of efficient sensors for analytical applications. Hervas et al. (2011) reported a novel "lab-on-chip" strategy integrating an electrokinetic magnetic bead-based electrochemical immunosensor on a microfluidic chip for reliable control of permitted levels of ZRL in infant foods. Immunological reaction is performed using a competitive enzyme-linked immunosorbent assay where the ZRL and enzyme-labeled derivative compete for the binding sites of the specific monoclonal antibody immobilized onto protein G coupled with magnetic beads. The back electrochemical reduction obtained by hydrogen peroxide–mediated HRP catalysis of hydroquinone oxidation into benzoquinone was measured at +0.1 V. This method was performed in less than 15 min attaining a limit of detection of 0.4 μg/L well below the legislative requirements, low systematic error of 2%, and good recovery of 101% for solid samples and 103% for liquid samples. This novel microfluidic technology involved creative use of simple channel layout of double-T microchip for sequential performance of immunointeraction and enzymatic reaction through application of electric fields suitably connected to reservoirs for driving the fluidics to different chambers for specific reactions. In addition, the nonspecific adsorption is avoided in a simple and elegant way by operating both zones through application of magnetic field.

6 Deoxynivalenol

Deoxynivalenol (DON) is a naturally occurring trichothecene-type B mycotoxin mainly produced by two fungi *F. graminearum* and *F. culmorum* (Kushiro, 2008; Sobrova et al., 2010). DON is also named as "vomitoxin" because of its strong emetic effects after consumption and was first diagnosed among Japanese men in 1972 who consumed moldy barley containing *Fusarium* fungi (Ueno, 1988; Sobrova et al., 2010). It is commonly found in cereals, posing serious threat for the safety of cereal-based foods and feedstuffs (Romanazzo et al., 2010). Besides, DON is also present in food products such as corn, popcorn, rice, rye, wheat, buckwheat, barley, oats, sorghum, noodles, malt, beer, baby and infant foods (Table 12.1) (Prieto-Simon et al., 2007). Structurally, it is a polar compound with the chemical name 12,13-epoxy-3,7, 15-trihydroxytrichothec-9-*en*-8-on and the presence of three free hydroxy groups are responsible for toxicity of DON (Fig. 12.1), and the possible detoxification route of DON in human is cytochrome P450 (Nagy et al., 2005). The most critical physicochemical property of DON lies its ability to withstand high temperatures (up to 350°C), which intensifies its risk of occurrence in foods/food products (Manthey et al., 2004; Sobrova et al., 2010). However, owing to its high solubility in water, its level is considerably reduced during food processing (Sugita-Konishi et al., 2006; Visconti et al., 2004). Consequently, the European Union has recommended a maximum permissible level (MPL) of DON to range from 50 to 2000 μg/kg depending on the type of cereal and cereal products (Romanazzo et al., 2010). However, most of the countries have set the MPL to be 750 μg/kg, with the provisional tolerable daily intake recommended by JECFA being 1 μg/kg bw per day (FAO, 2003; Chen et al., 2012). Thus, it is important to develop reliable analytical methods for analysis of DON in food products.

For detection of DON in food samples, a cost-effective enzyme-linked immunomagnetic electrochemical assay (ELIME) was developed by Romanazzo et al. (2010), who employed a novel anti-DON fragment antibody (Fab) and immunomagnetic beads combined with 8 magnetized SPE as electrochemical transducers.

The entire ELIME assay is performed in three major steps with first step involving washing of magnetic beads coating with DON conjugate and stored at 4°C until use. In the second step, the coated particles are subjected to immunological chain, followed by pipetting particles on the surface of eight magnetized SPE for electrochemical measurement. Based on the cross-reactivity of antigen-binding fragment toward different trichothecene mycotoxins, a good selectivity was shown for DON with the exception of 3-acetyldeoxynivalenol in the working range between 100 and 4500 ng/mL with an EC_{50} (amount of DON required to produce 50% decrease in the response signal) of 380 ng/mL. Application to real food samples showed precision and recovery values of 9–24% and 82–110% for breakfast cereals, respectively, and 10–33% and 97–108% for baby foods. This ELIME assay is claimed to be extremely robust and insensitive to sample matrix interference as compared to some other techniques such as SPR, fluorescent array biosensor and electronic olfactory system. Additionally, a simple water extraction was used for wheat, while acetonitrile-water extraction coupled with a drying step can be employed for breakfast cereal and baby food.

7 Citrinin

Citrinin (CIT) or (3R,4S)-4,6-dihydro-8-hydroxy-3,4,5-trimethyl-6-oxo-3H-2- benzopyran-7-carboxylic acid is a secondary metabolite produced by the three fungal genera *Aspergillus* (*A. carneus*, *A. niveus*, and *A. terreus*), *Penicillium* (*P. citrinum*, *P. verrucosum*, and *P. expansum*), and *Monoscus* (*M. ruber*) (Fig. 12.1) (Flajs and Peraica, 2009; Bragulat et al., 2008). It usually contaminates maize, wheat, rye, rice, corn, barley, oat, peanut, fruit, and sausage (Table 12.1) (Prieto-Simon et al., 2007). In addition, CIT contamination has been reported in numerous agricultural commodities, foods, feedstuffs, and biological fluids from a wide variety of geographical origins (Flajs and Peraica, 2009; Arevalo et al., 2011). Though CIT possesses antibiotic property against Gram-positive bacteria, it is seldom used as a drug due to its high nephrotoxicity (Gupta et al., 1983). Besides kidney, the other target organs of CIT include liver, and bone marrow (Flajs and Peraica, 2009). The decomposition products (CIT-H_1 and CIT-H_2) obtained by heating CIT with water at 140–150°C are more toxic than CIT as the decomposition temperature declines from 175 to 140°C in the presence of water (Kitabatake et al., 1991; Xu et al., 2006). Since CIT is commonly found along with OA, it was shown to increase the toxicity of the latter either additively or synergistically causing a kidney disease called “Balkan endemic nephropathy” in human (Speijers and

Speijers, 2004; Klaric et al., 2013). The IARC (1986) classified CIT in Group III of carcinogens due to limited evidence in animals and no evidence in human. The scientific opinion published by EU on behalf of EFSA (2012) has recommended no-observed-adverse-effect level (NOAEL) of 20 μg/kg bw per day for both rats and pigs, while the NOAEL between 9 and 53 μg/kg was suggested for high consumers as well as 19 and 100 μg/kg for average consumers.

Recently, Arevalo et al. (2011) have developed an electrochemical immunosensor incorporated in a microfluidic cell for quantification of CIT in rice samples. The assay involved competition of monoclonal anti-CIT (mAb-CIT) IgG antibody in solution with CIT in rice samples as well as CIT immobilized on a gold surface, which is previously electrodeposited over a GCE and modified with a cysteamine self-assembled monolayer. An excess of rabbit IgG (H + L)-labeled HRP added reacts with mAb-CIT on the electrode surface. The HRP catalyzes the oxidation of catechol to benzoquinone in the presence of hydrogen peroxide and its electrochemical reduction back to catechol is measured on the GCE at −0.15 V versus Ag/AgCl by amperometry. The measured current is directly proportional to the enzyme activity and inversely proportional to the amount of CIT present in rice samples. Detection of CIT by this assay is fast (2 min for detection and 45 min for total assay) and sensitive in the linear working range between 0.5 and 50 ng/mL, with the LOD and LOQ being 0.1 and 0.5 ng/mL, respectively, while the coefficient of variation for both intraday and interday assays was <6%. The main advantages of this immunosensor assay are higher sensitivity and shorter analysis time compared to other conventional chromatographic methods.

8 Fumonisins

Fumonisins (FMNs) are naturally occurring mold toxins produced mainly by *Fusarium verticillioides, F. proliferatum,* and *Gibberella fujikuroi* (Leslie, 1996; Ghali et al., 2009). Though there are seven structurally similar FMNs, only FMN-B_1, FMN-B_2, and FMN-B_3 are the most abundant ones contaminating food and feed worldwide with FMN-B_1 (molecular formula $C_{34}H_{59}NO_{15}$) contributing about 75% of the total FMs content (Fig. 12.1) (Branham and Plattner, 1993). It is reported to occur in several human foods including maize, maize-based products, barley, wheat, rice, and sorghum (Table 12.1) (Jimenez et al., 1997; Mateo and Jimenez, 2000; Prieto-Simon et al., 2007). FMNs are potentially hazardous to humans and animals causing various diseases such as immunosuppression, neurotoxicity, carcinogenicity, liver and kidney toxicity (Branham and Plattner, 1993; Gelderblom et al., 1994). Moreover, the exposure to FMNs has been closely associated with disruption of sphingolipids metabolism causing leukoencephalomalacia in equines and pulmonary edema in pigs (Voss et al., 1990; Colvin and Harrison, 1992; Ross et al., 1993). In humans, FMNs are suspected to cause esophageal cancer among certain South African and Chinese people (Pitt, 2000), and the IARC (1993) has classified FMN-B_1 as 2B carcinogen. The USFDA (2001) has enforced permissible limits on total FMNs to be ranged from 0.2 to 4 ppm for direct human consumption of food products, while a level of 5–100 ppm recommended for animal feed.

An electrochemical affinity sensor for detection of FMNs in foods was developed by Kadir and Tothill (2010), who employed a gold SPE modified with monoclonal antibody and silver–silver chloride pseudoreference electrode. Initially, a direct ELISA assay was developed, followed by transferring the platform to the surface of a gold SPE, monitoring the reaction of 3,3',5,5'-tetramethylbenzidine dihydrochloride (mediator) and hydrogen peroxide (substrate) catalyzed by HRP at −100 mV and detecting FMNs by chronoamperometry. Compared to conventional ELISA method, this immunosensor assay attained a 20-fold reduction in detection limit of 5 μg/L in the linear working range of 1–1000 μg/L, which is lower than the detection limit (2–4 mg/L) of FMNs required by European Union legislation. The sensor also showed >1% cross reactivity in the presence of some other mycotoxins like aflatoxin B_1 and ochratoxin A, implying that a good selectivity could be achieved by this method. Also the application to ground corn samples spiked with four different concentrations (50, 250, 500, and 2500 μg/kg) of FMN-B_1 and FMN-B_2 mixture showed a recovery of 101.7, 76.04, 112.6, and 104.64%, respectively. The advantages of this electrochemical assay include low matrix interference, high sensitivity and reproducibility, as well as rapid detection for on-site application.

9 Conclusions and Future Perspective

Recent advancements in the field of nanotechnology has enabled improved sensitivity, on-site analysis, and multiplexing capability of electrochemical sensors for detection of mycotoxins. This chapter has reviewed some applications of nanomaterial-based sensors for detection of aflatoxins (B_1, B_2, G_1, G_2, and M_1), ochratoxin A, sterigmatocystin, zearalenone, deoxynivalenol, citrinin, and fumonisins in food/feed samples. However, more research still needs to be dedicated to the application of other nanomaterials including metal and metal-oxide nanoparticles as analytical sensors for mycotoxin detection. In addition, the nanomaterial-based detection should be extended to mycotoxins other than aflatoxin and ochratoxin A. More studies should also focus on the application to real samples in the future.

References

Ansari, A.A., Kaushik, A., Solanki, P.R., Malhotra, B.D., 2010. Nanostructured zinc oxide platform for mycotoxin detection. Bioelectrochemistry 77, 75–81.

Arevalo, F.J., Granero, A.M., Fernandez, H., Raba, J., Zon, M.A., 2011. Citrinin (CIT) determination in rice samples using a micro fluidic electrochemical immunosensor. Talanta 83, 966–973.

Barthelmebs, L., Hayat, A., Limiadi, A.W., Marty, J.L., Noguer, T., 2011. Electrochemical DNA aptamer-based biosensor for OTA detection, using superparamagnetic nanoparticles. Sens. Actuat. B 156, 932–937.

Bennett, J.W., Klich, M., 2003. Mycotoxins. Clin. Microbiol. Rev. 16, 497–516.

Bhat, R., Rai, R.V., Karim, A.A., 2010. Mycotoxins in food and feed: present status and future concerns. Comp. Rev. Food Sci. Food Saf. 9, 57–81.

Bonel, L., Vidal, J.C., Duato, P., Castillo, J.R., 2010. Ochratoxin A nanostructured electrochemical immunosensors based on polyclonal antibodies and gold nanoparticles coupled to the antigen. Anal. Methods 2, 335–341.

Bonel, L., Vidal, J.C., Duata, P., Castillo, J.R., 2011. An electrochemical competitive biosensor for ochratoxin A based on a DNA biotinylated aptamer. Biosens. Bioelectron. 26, 3254–3259.

Bragulat, M.R., Martinez, E., Castella, G., Cabanes, F.J., 2008. Ochratoxin A and citrinin producing species of the genus *Penicillium* from feedstuffs. Int. J. Food Microbiol. 126, 43–48.

Branham, B.E., Plattner, R.D., 1993. Alanine is a precursor in the biosynthesis of fumonisin B1 by *Fusarium moniliforme*. Mycopathologia 124, 99–104.

Brett, C.M.A., Oliveira-Brett, A.M., 2011. Electrochemical sensing in solution-origins, applications and future perspectives. J. Solid State Electrochem. 15, 1487–1494.

Campas, M., Garibo, D., Prieto-Simon, B., 2012. Novel nanobiotechnological concepts in electrochemical biosensors for the analysis of toxins. Analyst 137, 1055–1067.

Cavaliere, C., Foglia, P., Guarino, C., Nazzari, M., Samperi, R., Laganà, A., 2007. Determination of aflatoxins in olive oil by liquid chromatography–tandem mass spectrometry. Anal. Chim. Acta 596, 141–148.

Chen, J., Liu, D., Li, S., Yao, D.S., 2010. Development of an amperometric enzyme electrode biosensor for sterimatocystin detection. Enzyme Microb. Technol. 47, 119–126.

Chen, Y.C., Su, W.T., Liao, C.D., Lin, H.Y., Chiueh, L.C., Shih, D.Y.C., 2012. A survey of deoxynivalenol and zearalenone in cereal products. Ann. Rep. Food Drug Res. 3, 159–164, (in Chinese).

Chun, H.S., Choi, E.H., Chang, H.J., Choi, S.W., Eremin, S.A., 2009. A fluorescence polarization immunoassay for the detection of zearalenone in corn. Anal. Chim. Acta 639, 83–89.

Cigic, I.K., Prosen, H., 2009. An overview of conventional and emerging analytical methods for the determination of mycotoxins. Int. J. Mol. Sci. 10, 62–115.

Colvin, B.M., Harrison, L.R., 1992. Fumonisins induced pulmonary oedema ad hydrothorax in swine. Mycopathologia 117, 79–82.

Cruz-Aguado, J.A., Penner, G., 2008a. Fluorescence polarization based displacement assay for the determination of small molecules with aptamers. Anal. Chem. 80, 8853–8855.

Cruz-Aguado, J.A., Penner, G., 2008b. Detemination of ochratoxin A with a DNA aptamer. J. Agric. Food Chem. 56, 10456–10461.
Cui, S.J., Chen, C.M., Tong, G.Z., 2008. A simple and rapid immunochromatographic strip test for monitoring antibodies to H5 subtype Avian Influenza Virus. J. Virol. Methods 152, 102–105.
Delmulle, B.S., De Saeger, S.M.D.G., Sibanda, L., Barna-Vetro, I., Peteghem, V., 2005. Development of an immunoassay-based lateral flow dipstick for the rapid detection of aflatoxin B_1 in pig feed. J. Agric. Food Chem. 53, 3364–3368.
Diekman, M.A., Green, M.L., 1992. Mycotoxins and reproduction in domestic livestock. J. Anim. Sci. 70, 1615–1627.
Dinckaya, E., Kinik, O., Sezginturk, M.K., Altug, C., Akkoca, A., 2011. Development of an impedimetric aflatoxin M1 biosensor based on a DNA probe and gold nanoparticles. Biosens. Bioelectron. 26, 3806–3811.
ECR, 2001. European Commission Regulation (ECR) No. 466/2001/EC. Setting maximum levels for certain contaminants in food stuffs. Official Journal of European Communities L077, pp. 1–13.
ECR, 2004. European Commission Regulation (ECR) No. 683/2004/EC. Amending regulation No. 466/2001 as regards aflatoxins and ochratoxin A in foods for infants and young children. Official Journal of European Communities L106, pp. 3–5.
European Food Safety Authority (EFSA), 2011. Scientific opinion on the risks for public health related to the presence of zearalenone in food. EFSA J. 9, 2197.
European Food Safety Authority (EFSA), 2012. Scientific opinion on the risks for public and animal health related to the presence of citrinin in food and feed. EFSA J. 10, 2605.
European Food Safety Authority (EFSA), 2013. Scientific opinion on the risks for public health related to the presence of sterigmatocystin in food and feed. EFSA J. 11, 3254.
FAO, 2003. Mycotoxin regulations in 2003 and current developments. Available from: http://www.fao.org/docrep/007/y5499e/y5499e07.htm
Farnworth, E.R., Trenholm, H.L., 1983. The metabolism of the mycotoxin zearalenone and its effects on the reproductive tracts of young male and female pigs. Can. J. Anim. Sci. 63, 967–975.
Flajs, D., Peraica, M., 2009. Toxicological properties of citrinin. Arh. Hig. Rada. Toksikol. 60, 457–464.
Gelderblom, W.C., Cawood, M.E., Snyman, S.D., Marasas, W.F., 1994. Fumonisin B1 dosimetry in relation to cancer initiation in rat liver. Carcinogenesis 15, 209–214.
Georgianna, D.R., Payne, G.A., 2009. Genetic regulation of aflatoxin biosynthesis: from gene to genome. Fungal Genet. Biol. 109, 243–248.
Ghali, R., Ghorbel, H., Hedilli, A., 2009. Fumonisin determination of Tunisian foods and feeds ELISA and HPLC methods comparison. J. Agric. Food Chem. 57, 3955–3960.
Green, M.L., Diekman, M.A., Malayer, J.R., Scheidt, A.B., Long, G.G., 1990. Effect of prepubertal consumption of zearalenone on puberty and subsequent reproduction of gilts. J. Anim. Sci. 68, 171–178.
Guo, Z., Ren, J., Wang, J., Wang, E., 2011. Single-walled carbon nanotubes based quenching of free FAM-aptamer for selective determination of ochratoxin A. Talanta 85, 2517–2521.
Gupta, M., Sasmal, D., Bandyopadhyay, S., Bagchi, G., Chatterjee, T., Dey, S., 1983. Hematological changes produced in mice by ochratoxin A and citrinin. Toxicology 26, 55–62.
Hervas, M., Lopez, M.A., Escarpa, A., 2009. Electrochemical immunoassay using magnetic beads for the determination of zearalenone in baby food. an anticipated analytical tool for food safety. Anal. Chim. Acta 653, 167–172.

Hervas, M., Lopez, M.A., Escarpa, A., 2010. Simplified calibration and analysis on screen-printed disposable platforms for electrochemical magnetic bead-based immunosensing of zearalenone in baby food samples. Biosens. Bioelectron. 25, 1755–1760.
Hervas, M., Lopez, M.A., Escarpa, A., 2011. Integrated electrokinetic magnetic bead-based electrochemical immunoassay on microfluidic chips for reliable control of permitted levels of zearalenone in infant foods. Analyst 136, 2131–2138.
Heurich, M., Kadir, M.K.A., Tothill, I.E., 2011. An electrochemical sensor based on carboxymethylated dextran modified gold surface for ochratoxin A analysis. Sens. Actuat. B 156, 162–168.
International Agency for Research of Cancer (IARC), 1986. Some naturally occurring and synthetic food components, furocumarins and ultraviolet radiation. IARC monographs on the evaluation of the carcinogenic risk of chemicals to humans, vol. 40. IARC, Lyon, pp. 67–82.
International Agency for Research on Cancer (IARC), 1993. Ochratoxin A and some naturally occurring substances: food items and constituents, heterocyclic aromatic amines and mycotoxins. Monographs on the evaluation of carcinogenic risk to humans, vol. 56. IARC, Lyon, pp. 489–452.
International Agency for Research on Cancer (IARC), 2002. Monograph on the evaluation of carcinogenic risk to humans, vol. 82. World Health Organization, Lyon, p. 171.
Joint FAO/WHO Expert Committee on Food Additives (JECFA), 2000. Zearalenone. In: Joint FAO/WHO Expert Committee on Food Additives (Ed.), Safety Evaluation of Certain Food Additives and Contaminants. WHO/FAO Food Additives Series, vol. 44. IPCS, International Programme on Chemical Safety, WHO, Geneva.
Jimenez, M., Huerta, T., Mateo, R., 1997. Secondary metabolites produced by *Fusarium*. Appl. Environ. Microbiol. 63, 364–372.
Jin, X., Jin, X., Chen, L., Jiang, J., Shen, G., Yu, R., 2009. Piezoelectric immunosensor with gold nanoparticles enhanced competitive immunoreactions technique for quantification of aflatoxin B1. Biosens. Bioelectron. 24, 2580–2585.
Kadir, M.K.A., Tothill, I.E., 2010. Development of an electrochemical immunosensor for fumonisins detection in foods. Toxins 2, 382–398.
Kaushik, A., Solanki, P.R., Ansari, A.A., Ahmad, S., Malhotra, B.D., 2009a. Nanostructured cerium oxide film-based immunosensor for mycotoxin detection. Nanotechnology 20, 055105.
Kaushik, A., Solanki, P.R., Sood, K.N., Ahmad, S., Malhotra, B.D., 2009b. Fumed silica nanoparticles-chitosan nanobiocomposite for ochratoxin-A detection. Electrochem. Commun. 11, 1919–1923.
Khan, R., Dhayal, M., 2008. Nanocrystalline bioactive TiO_2-chitosan impedimetric immunosensor for ochratoxin-A. Electrochem. Commun. 10, 492–495.
Kitabatake, N., Trivedi, B.A., Doi, E., 1991. Thermal decomposition and detoxification of citrinin under various moisture conditions. J. Agric. Food Chem. 39, 2240–2244.
Klaric, M.S., Rasic, D., Peraica, M., 2013. Deleterious effects of mycotoxin combinations involving ochratoxin A. Toxins 5, 1965–1987.
Kolosava, A.Y., Shim, W.B., Yang, Z.Y., Eremin, S.A., Chung, D.H., 2006. Direct competitive ELISA based on a monoclonal antibody for detection of aflatoxin B1 stabilization of ELISA kit components and application to grain samples. Anal. Bioanal. Chem. 384, 286–294.
Koppen, R., Koch, M., Siegel, D., Merkel, S., Maul, R., Nehls, I., 2010. Determination of mycotoxins in foods: current state of analytical methods and limitation. Appl. Microbiol. Biotechnol. 86, 1595–1612.

Krska, R., Welzig, E., Berthiller, F., Molinelli, A., Mizaikoff, B., 2005. Advances in the analysis of mycotoxins and its quality assurance. Food Addit. Contam. 22, 345–353.

Kuang, H., Chen, W., Xu, D., Liguang, X., Zhu, Y., Liu, L., Chu, H., Peng, C., Xu, C., Zhu, S., 2010. Fabricated aptamer-based electrochemical "signal-off" sensor of ochratoxin A. Biosen. Bioelectron. 26, 710–716.

Kushiro, M., 2008. Effects of milling and cooking processes on the deoxynivalenol content in wheat. Int. J. Mol. Sci. 9, 2127–2145.

Laschi, S., Centi, S., Mascini, M., 2011. Electrochemical arrays coupled with magnetic separators for immunochemistry. Bioanal. Rev. 3, 11–25.

Leslie, J.F., 1996. Introductory biology of *Fusarium moniliforme*. Adv. Exp. Med. Biol. 392, 153–164.

Li, S.C., Chen, J.H., Cao, H., Yao, D.S., Liu, D.L., 2011. Amperometric biosensor for aflatoxin B_1 based on aflatoxin-oxidase immobilized on multiwalled carbon nanotubes. Food Control 22, 43–49.

Liao, J.Y., Li, H., 2010. Lateral flow immunodipstick for visual detection of aflatoxin B1 in food using immune-nanoparticles composed of a silver core and a gold shell. Microchim. Acta 171, 289–295.

Liu, Y., Qin, Z., Wu, X., Jiang, H., 2006. Immune-biosensor for aflatoxin B1 based bio-electrocatalytic reaction on micro-comb electrode. Biochem. Eng. J. 32, 211–217.

Liu, B.H., Tsao, Z.J., Wang, J.J., Yu, F.Y., 2008. Development of a monoclonal antibody against ochratoxin and its application in enzyme-linked immunesorbent assay and gold nanoparticle immunochromatographic strip. Anal. Chem. 80, 7029–7035.

Liu, X.P., Deng, Y.J., Jin, X.Y., Chen, L.G., Jiang, J.H., Shen, G.L., Yu, R.Q., 2009. Ultrasensitive electrochemical immunosensor for ochratoxin A using gold colloid-mediated hapten immobilization. Anal. Biochem. 389, 63–68.

Logrieco, A., Arrigan, D.W.M., Brengel-Pesce, K., Siciliano, P., Tothill, I., 2005. DNA arrays, electronic noses and tongues, biosensors and receptors for rapid detection of toxigenic fungi and mycotoxins: a review. Food Addit. Contam. 22, 335–344.

Long, G.G., Diekman, M.A., 1986. Characterization of effects of zearalenone in swine during early pregnancy. Am. J. Vet. Res. 47, 184–187.

Malhotra, B.D., Srivastava, S., Ali, M.A., Chandan, S., 2014. Nanomaterial-based biosensors for food toxin detection. Appl. Biochem. Biotechnol. 174, 880–896.

Manthey, F.A., Wolf-Hall, C.E., Yalla, S., Vijayakumar, C., Carlson, D., 2004. Microbial loads, mycotoxins, and quality of durum wheat from the 2001 harvest of the northern plains region of the United States. J. Food Prot. 67, 772–780.

Maragos, C.M., 2009. Recent advances in the development of novel materials for mycotoxin analysis. Anal. Bioanal. Chem. 395, 1205–1213.

Maragos, C.M., Busman, M., 2010. Rapid and advanced tools for mycotoxin analysis: a review. Food Addit. Contam. 27, 688–700.

Mateo, J.J., Jimenez, M., 2000. Trichothecenes and fumonisins produced in autoclaved tiger nuts by strains of *Fusarium sporotrichioides* and *Fusarium moniliforme*. Food Microbiol. 17, 167–176.

Nagy, C.M., Fejer, S.N., Berek, L., Molnar, J., Viskolcz, B., 2005. Hydrogen bondings in deoxynivalenol (DON) conformations—a density functional study. J. Mol. Struct. 726, 55–59.

Navarro-Villoslada, F., Urraca, J.L., Moreno-Bondi, M.C., Orellana, G., 2007. Zearalenone sensing with molecularly imprinted polymers and tailored fluorescent probes. Sens. Actuat. B 121, 67–73.

Owino, J.H.O., Arotiba, O.A., Hendricks, N., Songa, E.A., Jahed, N., Waryo, T.T., Ngece, R.F., Baker, P.G.L., Iwuoha, E.I., 2008. Electrochemical immunosensor based on polythionine/gold nanoparticles for the determination of aflatoxin B_1. Sensors 8, 8262–8274.

Paek, S.H., Lee, S.H., Cho, J.H., Kim, Y.S., 2000. Development of rapid one-step immunochromatographic assay. Methods 22, 53–60.

Palchetti, I., Mascini, M., 2008. Electroanalytical biosensors and their potential for food pathogen and toxin detection. Anal. Bioanal. Chem. 391, 455–471.

Panini, N.V., Bertolino, F.A., Salinas, E., Messina, G.A., Raba, J., 2010. Zearalenone determination in corn silage samples using an immunosensor in a continuous-flow/stopped-flow systems. Biochem. Eng. J. 51, 7–13.

Parker, C.O., Lanyon, Y.H., Manning, M., Arrigan, D.W., Tothill, I.E., 2009. Electrochemical immunochip sensor for aflatoxin M1 detection. Anal. Chem. 81, 5291–5298.

Pfohl-Leszkowicz, A., Manderville, R.A., 2007. Ochratoxin A: an overview on toxicity ad carcinogenicity in animals and humans. Mol. Nutr. Food Res. 51, 61–99.

Pitt, J.I., 2000. Toxigenic fungi and mycotoxins. Br. Med. Bull. 56, 184–192.

Prieto-Simon, B., Noguer, T., Campas, M., 2007. Emerging biotools for assessment of mycotoxins in the past decade. Trends Anal. Chem. 26, 689–702.

Purchase, I.F.H., van der Watt, J.J., 1970. Carcinogenecity of sterigmatocystin. Food Cosmet. Toxicol. 8, 289–295.

Radoi, A., Targa, M., Prieto-Simon, Marty, J.L., 2008. Enzyme-linked immunosorbent assay (ELISA) based on superparamagnetic nanoparticles for aflatoxin M1 detection. Talanta 77, 138–143.

Romanazzo, D., Ricci, F., Volpe, G., Elliott, C.T., Vesco, S., Kroeger, K., Moscone, D., Stroka, J., Egmond, H.V., Vehniainen, M., Palleschi, G., 2010. Development of a recombinant Fab-fragment based electrochemical immunosensor for deoxynivalenol detection in food samples. Biosens. Bioelectron. 25, 2615–2621.

Ross, P.F., Ledet, A.E., Owens, D.L., Rice, L.G., Nelson, H.A., Osweiler, G.D., Wilson, T.M., 1993. Experimental equine leukoencephalomalacia, toxic hepatosis, and encephalopathy caused by corn naturally contaminated with fumonisins. J. Vet. Diagn. Invest. 5, 69–74.

Shephard, G.S., 2008. Determination of mycotoxins in human foods. Chem. Soc. Rev. 37, 2468–2477.

Shephard, G.S., Berthiller, F., Burdaspal, P.A., Crews, C., Jonker, M.A., Krska, R., MacDonald, S., Malone, R.J., Maragos, C., Sabino, M., Solfrizzo, M., Van Egmond, H.P., Whitaker, T.B., 2012. Developments in mycotoxin analysis: an update for 2010–2011. World Mycotoxin J. 5, 3–30.

Shim, W.B., Yang, Z.Y., Kim, J.S., Kim, J.Y., Kang, S.J., Woo, G.J., Chung, Y.C., Eremin, S.A., Chung, D.H., 2007. Development of immunochromatography strip-test using nanocolloidal gold-antibody probe for the rapid detection of aflatoxin B1 in grain and feed samples. J. Microbiol. Biotechnol. 17, 1629–1637.

Shim, W.B., Kim, G., Ryu, H.J., Nam, M., Chung, D.H., 2009. Development of one-step simultaneous immunochromatographic assay for rapid analysis of aflatoxin B1 and ochratoxin A. Food Sci. Biotechnol. 18, 641–648.

Singh, C., Srivastava, S., Ali, M.A., Gupta, T.K., Sumana, G., Srivastava, A., Mathur, R.B., Malhotra, B.D., 2013. Carboxylated multiwalled carbon nanotubes based biosensor for aflatoxin detection. Sens. Actuat. B 185, 258–264.

Sobrova, P., Adam, V., Vasatkova, A., Beklova, M., Zeman, L., Kizek, R., 2010. Deoxynivalenol and its toxicity. Interdisc. Toxicol. 3, 94–99.

Speijers, G.J.A., Speijers, M.H.M., 2004. Combined toxic effects of mycotoxins. Toxicol. Lett. 153, 91–98.

Stoloff, L., 1980. Aflatoxin M1 in perspective. J. Food Prot. 43, 226–230.
Stroka, J., Dasko, L., Spangenberg, B., Anklam, E., 2004. Determination of the mycotoxin, sterigmatocystin, by thin-layer chromatography and reagent-free derivatisation. J. Liq. Chromatogr. Relat. Technol. 27, 2101–2111.
Stubblefield, R.D., Shannon, G.M., 1974. Aflatoxin M1 analysis in dairy products and distribution in dairy foods made from artificially contaminated milk. J. Assoc. Off. Anal. Chem. 57, 847–851.
Sugita-Konishi, Y., Park, B.J., Kobayashi-Hattori, K., Tanaka, T., Chonan, T., Yoshikawa, K., Kumagai, S., 2006. Effect of cooking process on the deoxynivalenol content and its subsequent cytotoxicity in wheat products. Biosci. Biotechnol. Biochem. 70, 1764–1768.
Sun, A.L., Qi, Q.A., Dong, Z.L., Liang, K.Z., 2008. An electrochemical enzyme immunoassay for aflatoxin B1 based on bio-electrocatalytic reaction with room-temperature ionic liquid and nanoparticle-modifed electrodes. Sens. Instrum. Food Qual. Saf. 2, 43–50.
Tang, D., Sauceda, J.C., Lin, Z., Ott, S., Basova, E., Goryacheva, I., Biselli, S., Lin, J., Niessner, R., Knopp, D., 2009. Magnetic nanogold microspheres-based lateral-flow immunodipstick for rapid detection of aflatoxin B2 in food. Biosens. Bioelectron. 25, 514–518.
Tian, H.J., Liu, X.M., 2004. Survey and analysis on sterigmatocystin contaminated in grains in China. Wei Sheng Yan Jiu 33, 606–608.
Tothill, I.E., 2011. Biosensors and nanomaterials and their application for mycotoxin determination. World Mycotoxin J. 4, 361–374.
Tripathi, S., Mishra, H.N., 2009. A rapid FT-IR method for estimation of AFB1 in red chili powder. Food Control 20, 840–846.
Turner, N.W., Subrahmanyam, S., Piletsky, S.A., 2009. Analytical methods for determination of mycotoxins: a review. Anal. Chim. Acta 632, 168–180.
Ueno, Y., 1988. Toxicology of trichothecene mycotoxins. ISI Atlas Sci-Pharm. 2, 121–124.
United States Food and Drug Administration (USFDA), 2001. Center for food safety and applied nutrition. Fumonisin levels in human food and animal feeds. Guidance for Industry, Final guidance, Rockville, MD. Available from: http://www.cfsab.fda.gov/dms/fumongu2.html
Valdes, M.G., Gonzalez, A.C.V., Calzon, J.A.G., Diaz-Garcia, M.E., 2009. Analytical nontechnology for food analysis. Microchim. Acta 166, 1–19.
Ventura, M., Guillen, D., Anaya, I., Broto-Puig, F., Lliberia, J.L., Agut, M., Comellas, L., 2006. Ultra-performance liquid chromatography/tandem mass spectrometry for the simultaneous analysis of aflatoxins B1, G1, B2, G2 and ochratoxin A in beer. Rapid Commun. Mass Spectrom. 20, 3199–3204.
Versilovskis, A., De Saeger, S., 2010. Sterigmatocystin: occurrence in foodstuffs and analytical methods—an overview. Mol. Nutr. Food Res. 54, 136–147.
Vidal, J.C., Bonel, L., Duato, P., Castillo, J.R., 2011. Improved electrochemical competitive immunosensor for ochratoxin A with a biotinylated monoclonal antibody capture probe and colloidal gold nanostructuring. Anal. Methods 3, 977.
Vidal, J.C., Bonel, L., Ezquerra, A., Hernandez, S., Bertolin, J.R., Cubel, C., Castillo, J.R., 2013. Electrochemical affinity biosensors for detection of mycotoxins: a review. Biosens. Bioelectron. 49, 146–158.
Vig, A., Radoi, A., Munoz-Berbel, X., Gyemant, G., Marty, J.L., 2009. Impedimetric aflatoxin M_1 immunosensor based on colloidal gold and silver electrodeposition. Sens. Actuat. B 138, 214–220.
Visconti, A., Haidukowshi, E.M., Pascale, M., Silvestri, M., 2004. Reduction of deoxynivalenol during durum wheat processing and spaghetti cooking. Toxicol. Lett. 153, 181–189.

Voss, K.A., Plattner, R.D., Bacon, C.W., Norred, W.P., 1990. Comparative studies of hepatotoxicity and fumonisin B1 and B2 content of water and chloroform/methanol extracts of *Fusarium moniliforme* strain MRC 826 culture material. Mycopathologia 112, 81–92.

Wang, J.J., Liu, B.H., Hsu, Y.T., Yu, F., 2011. Sensitive competitive direct enzyme-linked immunosorbent assay and gold nanoparticle immunochromatographic strip for detecting aflatoxin M1 in milk. Food Control 22, 964–969.

Wu, S., Duan, N., Zhu, C., Ma, X., Wang, M., Wang, Z., 2011. Magnetic nanobead-based immunoassay for the simultaneous detection of aflatoxin B1 and ochratoxin A using upconversion nanoparticles as multicolor labels. Biosens. Bioelectron. 30, 35–42.

Xiulan, S., Xiaolian, Z., Jian, T., Zhou, T., Chu, F.S., 2005. Preparation of gold-labeled antibody probe and its use in immunochromatography assay for detection of aflatoxin B1. Int. J. Food Microbiol. 99, 185–194.

Xiulan, S., Xiaolian, Z., Jian, T., Xiaohong, G., Jun, Z., Chu, F.S., 2006. Development of an immunocromatographic assay for detection of aflatoxin B1 in foods. Food Control 17, 256–262.

Xu, B.J., Jia, X.Q., Gu, L.J., Sung, C.K., 2006. Review on the qualitative and quantitative analysis of the mycotoxin citrinin. Food Control 17, 271–285.

Yao, D.S., Wen, S.M., Liu, D.L., Xie, C.F., Bai, Y., Ran, Y.H., 2004. The primary study on the detection of sterigmatocystin by biologic enzyme electrode modified with the multiwall carbon nanotubes. Chin. J. Biotechnol. 20, 601–606.

Yao, D.S., Cao, H., Wen, S., Liu, D.L., Bai, Y., Zheng, W.J., 2006. A novel biosensor for sterimatocystin constructed by multi-walled carbon nanotubes (MWCNT) modified with aflatoxin-detoxifizyme (ADTZ). Bioelectrochemistry 68, 126–133.

Ye, Y., Zhou, Y., Mo, Z., Cheng, W., Yang, S., Wang, X., Chen, G., 2010. Rapid detection of aflatoxin B1 on membrane by dot-immunogold filtration assay. Talanta 81, 792–798.

Zhang, D., Li, P., Zhang, Q., Zhang, W., Huang, Y., Ding, X., Jiang, J., 2009. Production of ultrasensitive generic monoclonal antibodies against major aflatoxins using a modified two-step screening procedure. Anal. Chim. Acta 636, 63–69.

Zhang, D., Li, P., Zhang, Q., Zhang, W., 2011. Ultrasensitive nanogold prove-based immunochromatographic assay for simultaneous detection of total aflatoxins in peanuts. Biosens. Bioelectron. 26, 2877–2882.

Zinedine, A., Soriano, J.M., Molto, J.C., Manes, J., 2007. Review on the toxicity, occurrence, metabolism, detoxification, regulations and intake of zearalenone: an oestrogenic mycotoxin. Food Chem. Toxicol. 25, 1–18.

13

APPLICATIONS OF CHITOSAN AS A FUNCTIONAL FOOD

Monica Gallo*, Daniele Naviglio, Arturo Armone Caruso†, Lydia Ferrara‡**

**University of Naples Federico II, Department of Molecular Medicine and Medical Biotechnology, Naples, Italy; **University of Naples Federico II, Department of Chemical Sciences, Naples, Italy; †ENT Department, AIAS Structure of Afragola, Contrada Leutrec snc, Naples, Italy; ‡University of Naples Federico II, Department of Pharmacy, Naples, Italy*

Where nature finishes to produce its species, man begins, using species of nature, and in harmony with nature, to create an infinite number of species.

—Leonardo da Vinci

1 Introduction

Chitosan is a by-product of the fish industry that has no cost and its reuse will reduce its environmental impact. The disposal of waste materials from shellfish, shrimp, and fish has become a serious problem in coastal areas around the world, in addition to the disposal of plant biomass. Chitosan is a natural biopolymer from the endless applications, even if actually the variability of the molecular weight and degree of acetylation does not allow the production of a compound always reproducible, with experimental results sometimes contradictory. Its application has increased over the past 20 years especially as biomaterial in nanotechnology. The selective isolation of a bioactive material such as chitin or other polymers from industrial waste is the easiest way to not only decrease environmental pollution from the presence of these unused biomaterials but also to increase the production of bioinnovative products. Moreover, because of its unique chemical characteristics, chitosan can be used in industrial activities that require specific purification, resistance, and antibacterial activity; it is remarkable in its use as an antimicrobial and antifungal agent, both in agriculture and food preservation due to its

Novel Approaches of Nanotechnology in Food. http://dx.doi.org/10.1016/B978-0-12-804308-0.00013-3

biodegradability and ability to form coating films. The primary mechanism of action of chitosan resides in its ability to bind to fatty and oily substances due to its chemical composition, favoring their elimination, and thus it can be used for different chemical–pharmaceutical preparations: chitosan is used in slimming diets, for its ability to reduce the absorption of fat. In fact, chitosan is widely exploited for the preparation of supplements that are designed to promote weight loss. According to many industry experts, taking the drug orally in special supplements reduces the absorption of fat that is ingested in the diet. In fact, chitosan binds to fats and then wraps them in a sort of gelatinous casing until they become partially indigestible and are eliminated with the feces. The accumulation of adipose tissue is consequently limited, which is very useful in fighting obesity. In addition, the use of supplements containing this substance can also decrease feelings of hunger. There is no scientific confirmation of the slimming function of chitosan, and many scientists seriously question this idea. However, because these studies are still ongoing, the scientific community has not arrived at an unambiguous judgment, but it is believed that this type of substance derived from chitin is effective only if it is accompanied by a low-calorie diet and regular physical activity. Although many effects of its use are controversial and/or yet to be verified, chitosan has been shown to be a multifunctional substance with applications in many areas.

1.1 Precursor of Chitosan: Chitin. Chemical and Physical Characteristics

Chitosan is a natural polymer derived from the deacetylation of chitin, which is the primary component of the exoskeleton of insects, the shells of crustaceans and the cell walls of fungi. Chitin, the most abundant natural biopolymer after cellulose, was discovered in fungi by the French scholar Henry Braconnot (1811) after conducting several reactions on raw materials that had been isolated from the following different fungal species: *Agaricus volvaceus*, *A. acris*, *Cantharellus agaricus*, *A. piperatus*, *Hydnum repandum*, *H. hybridum*, and *Boletus viscidus*. Fungal material can be partially purified by boiling it with diluted potassium hydroxide to remove its proteins and pigments and obtain a chitin-glucan compound. Ammonia was developed from the distillation of the alkaline solution containing fungal material to demonstrate the presence of nitrogen in the molecule; the addition of sulfuric acid then freed acetic acid, revealing the presence of acetylated groups. A few years later, chitin was also discovered in the armor of some insects. Chitin is a polymer consisting

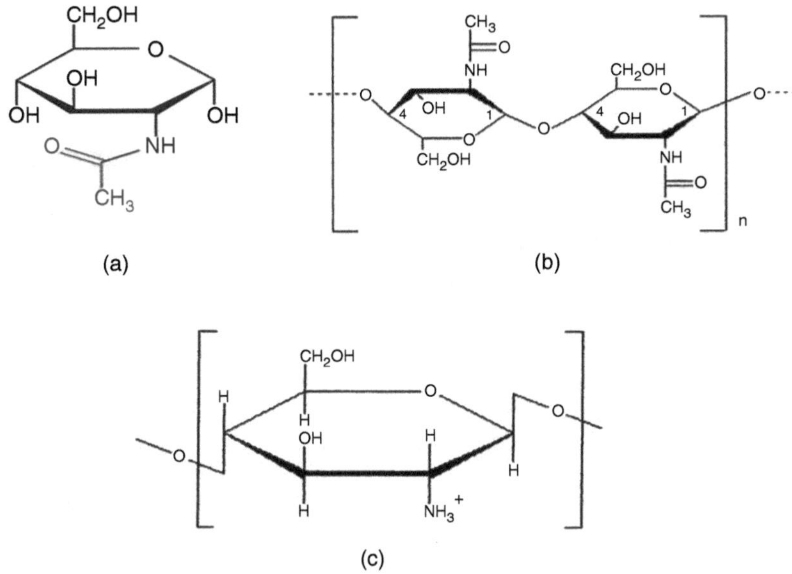

Figure 13.1. (a) The principal unit of chitin: b-D-N-acetilglucosammina (NAG); (b) structure of chitin; (c) structure of chitosan.

of D-N-acetylglucosamine units that are linked by β-1,4 bonds (Fig. 13.1).

Ledderhose (1878) determined the structure of chitin and demonstrated that it could be synthesized from glucosamine and acetic acid molecules. In various units, chitin forms long chains that aggregate into sheets through hydrogen bonding; the various ways in which these sheets aggregate relative to one another determine the three known crystalline forms of chitin: α, β, and γ. The hydrogen bonds that form between the adjacent polymer chains guarantee a remarkable hardness and flexibility; these characteristics, together with its degradability by endogenous enzymes, make chitin an excellent material for the production of wires for surgical sutures, bandages, and synthetic leather; it also has an accelerated healing property when applied to superficial wounds in humans. Chitin is now obtained primarily from the shells of shellfish, especially shrimp, amounting to 50,000 tons per year, and because its biodegradation is very slow, its disposal was a serious problem for a long time in the fishing industry until it could be converted into products with high added value, such as chitosan, and for research on its possible uses. In this way, chitin was transformed from a waste product into an inexpensive secondary raw material and thus became a product with great utility (Shahidi et al., 1999).

2 Chitosan, a Natural and Extraordinary Polymer

Chitosan was discovered by chance by Professor Rouget (1859) while trying to produce a natural soap by boiling chitin in alkali; this boiling led to the deacetylation of the amino groups, producing chitosan. Chitosan is a linear polymer with the chemical formula $(C_6H_{11}O_4N)n$, and it possesses a 3-dimensional alpha helix configuration that is stabilized by intramolecular hydrogen bonds. Depending on the length of the chain, chitosan presents different molecular weights (between 50 kDa and 200 kDa) and different degrees of deacetylation, from 40% to 98%. To obtain chitosan, the waste products of shrimp are subjected to demineralization with 0.25 N HCl at room temperature in a 1:40 w/v solution for approximately 15 min. This step is followed by deproteinization with N NaOH at 70°C for 24 h, after which chitin is obtained and subjected to deacetylation at elevated temperatures from 100 to 120°C in a solution of 60% (p/v) NaOH (Juang et al., 2002) (Fig. 13.2).

The resulting crude chitosan product obtained in the manner described above must undergo complex purification treatments, for which traditional chemical extraction methods are not very economical and also have a high environmental impact, to remove substances present in the starting material, namely proteins, lipids,

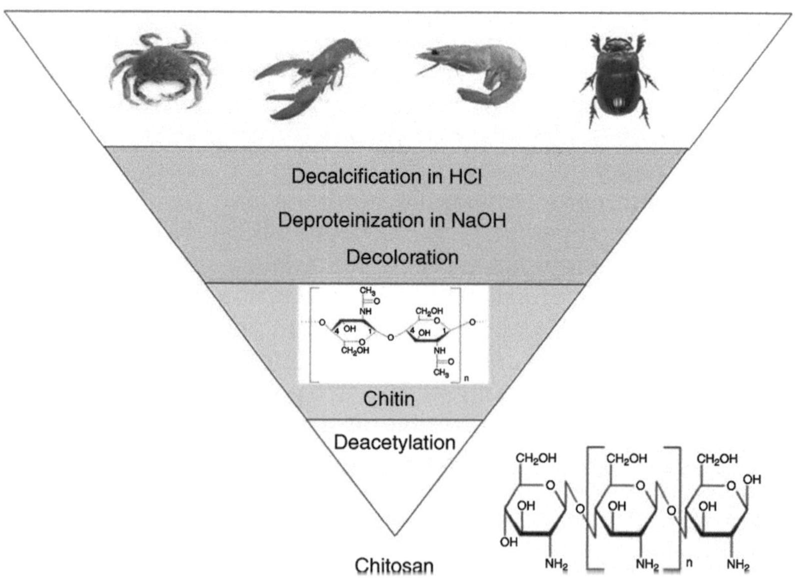

Figure 13.2. Formation of chitosan.

mineral salts (especially calcium carbonate), and pigments. Alternatively, a method of fermentation with lactic acid bacteria was proposed (Rao et al., 2000) and optimized by considering various factors, such as the choice of the most active microorganism, the culture medium, the addition of nutrients to the culture medium (Khanafari et al., 2008) to obtain high product yields, cost effectiveness, and a low environmental impact. *Lactobacillus plantarum* (PTTC 1058) was inoculated into MRS broth that was mixed with agar to solidify the culture medium, which was then incubated at 35–37°C in the presence of CO_2 for 48–72 h. This species was shown to be the most efficient among the lactobacilli that were considered for extracting chitosan from shrimp, with a yield greater than that obtained by chemical methods and without harming the environment (Khanafari et al., 2008). In addition, the use of organic acids for demineralization led to a significant change in the purification process, giving way to numerous advantages. Lactic acid is produced at a low cost by the bacteria themselves, and the resulting salts can be employed as antifreeze products or food preservatives; protein substances are precipitated and recovered for use as feed.

The physical characteristics of the chitosan obtained by following the two methods described earlier do not differ appreciably. The resulting products come in the form of fine powder or flakes ranging in color from yellow to white and are tasteless with a slight characteristic odor. Chitosan is insoluble in water at neutral and alkaline pH and in some organic solvents, and it is soluble in acids or as a consequence of the protonation of the amino group, leading to the formation of salt with the anion of the given acid (glutamic, hydrochloric, acetic, or lactic acid). The most commonly used salified form contains glutamic acid or hydrochloric acid, and the salts are water soluble, although the degree of solubilization depends on the deacetylation degree of the chitosan and the pH of the solution. If the pH range of the aqueous systems is natural (pH = 6–8), then the amine groups of the chitosan are present in their protonated forms and the compound assumes the characteristics of a Lewis acid, which can then interact with basic group binders. Chitosan that has a low degree of deacetylation of approximately 40% is also soluble at a pH below 9; chitosan with a deacetylation degree of approximately 85% is soluble only at a pH below 5 or 6. The degree of deacetylation also influences the viscosity of the solution and determines chitosan's configuration, from an elongated, very flexible form with a high degree of deacetylation to an alpha-helix form at a lower degree of deacetylation. Finally, the solubility of chitosan is affected by the presence of salts in solution: the greater the ionic strength of the solution, the lower its solubility. The behavior of chitosan as a polycation in acid explains its ability to form complexes with negatively charged molecules, such as lipids and proteins that are present in solution or in a colloidal state, causing its flocculation (Muzzarelli, 1985; Shahidi et al., 1999) and consequent removal from the solution.

3 Industrial Applications of Chitosan

At present, chitosan is produced in various countries primarily by the chemical method, and the primary interest in this product stems from its biological characteristics affording its use in a wide range of applications. It is also used in the food industry because it has very low toxicity; in mice, its LD_{50} was assessed at greater than 16 g/day/kg body weight. As mentioned earlier, the difficulties related to extraction that, although not complex, require accurate control of the layers to obtain a final product with specific purity and functionality characteristics have prevented increased use of at the industrial level. Chitosan has been recognized as an important biomaterial only in recent decades, but it is widely used in environmental applications such as wastewater treatment to remove heavy metals or in water purification, in the textile industry as an antibacterial agent or a natural fiber, in the cosmetics industry as an additive, in the pharmaceutical industry in drug delivery processes, and in food supplements as a weight loss additive.

3.1 Chitosan as a Biomaterial

A biomaterial is a substance that presents affinity for biological systems and is used to treat, improve, or replace parts of organs or tissues and improve their functions. Biocompatibility is critical for the use of the selected material (Ratner et al., 1996) for medical and pharmaceutical applications. The interactions between a material and a living organism occur at different levels, namely physical, chemical, and molecular, and they induce a state of "action-reaction"; a biomaterial may cause a biological response in the organism, which can react by degrading or rejecting the biomaterial. The materials used in medicine can be of biological or synthetic origin and can constitute the primary components of prostheses. In ancient times, these materials were used to replace damaged parts of the human body. In ancient Egyptian tombs, mummies with implants made of various metals, primarily iron, have been discovered; among the Phoenicians, it was a common practice to tie artificial teeth to natural ones with gold thread. Hieronymus Fabricius D'Acquapendente (1628) was a professor of anatomy at Padua, and in his *Opera surgery*, specifically in the chapter *De eruto and ammisso oculus*, he recounts a replacement that was performed with an artificial eye made of various materials, including glass, stone, and silver, to correct the deformity resulting from the loss of an eye. A mummy dating back to the XVIII dynasty (approximately 3000 years ago) found in Luxor that belonged to a woman had a big toe made of wood and leather on her right foot with visible signs of wear as evidence of its actual functional use. Wood was the first biomaterial used for the replacement of artificial limbs and other body parts and was later replaced by metal. During the American Civil War (1861–65), carpenter's nails were employed on the battlefield to block femur fractures in wounded soldiers. The tendency of metals such as iron and lead to corrode brought about their replacement with alloys, especially platinum and titanium, which both have great chemical stability and high mechanical strength and can be applied in an endless variety of formats. The introduction of plastic materials has considerably improved prostheses due to their light weight. Polymethylmethacrylate was used as a material for intraocular lenses; without this material, patients had reactions to the foreign body. Because rejection is a frequent phenomenon, it becomes necessary to purify the possible polymers that are already available, minimizing contamination, and to research and synthesize new and innovative biomaterials for specific applications that solve the problems that arise from biomaterial contact with different parts of the humans or animals. New biomaterials have been well developed for use in bone implants due to their osteointegrative capacity, in an aortocoronary stent coated with materials that prevents the narrowing of systems during the controlled release of drugs in the bloodstream, and in the construction of new skin by combining polymers with patient cells to rebuild tissues that have suffered burns. Today, tissue engineering is an emerging interdisciplinary field that applies the principles of biological, chemical, and engineering sciences using biomaterials alone or in combination to restore, maintain, or improve tissue function (Langer and Vacanti, 1993).

The principle of this discipline consists of the isolation of cells from a healthy individual, followed by their expansion in vitro; these expanded cells are subsequently seeded onto a scaffold that is made from a 3-dimensional biomaterial that acts as both a support and a growth factor and degrades slowly over time; the biomaterial is ultimately replaced by the fabric formed by seeded cells. The materials used for these applications are of different types: natural polymers, such as chitosan, collagen, gelatin, and hyaluronic acid, or synthetic polymers, namely derivatives of lactic acid, polyurethanes, polyvinyl acetate ethylene, and polycaprolate that are reduced by electrospinning with nanofibers to construct scaffolding with different degrees of porosity (Fig. 13.3).

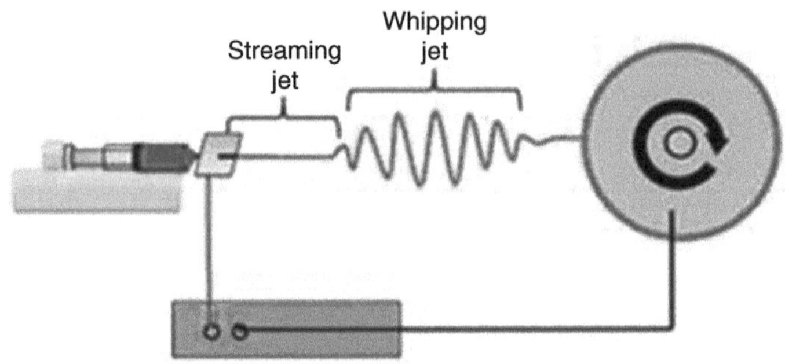

Figure 13.3. In the polymer industry, electrospinning is a production process that allows to obtain continuous filaments of synthetic material of the diameter extraordinarily small, less than one micron.

The possibility of obtaining electrospun chitosan of different thicknesses and of obtaining tissue has led to further chitosan applications not only in the medical field but also in the food, agricultural, pharmaceutical, and cosmetics industries.

4 Antibacterial and Antifungal Activity of Chitosan

In recent years, chitosan and its derivatives have received a great deal of attention as antimicrobial agents against fungi, bacteria, and viruses and as activators of plants' biochemical defense systems (Chirkov, 2002; Didenko et al., 2005; Je and Kim, 2006). Chitosan has been employed in applications within many sectors, from agriculture and food to pharmaceuticals and cosmetics, including commodities, due to its strong antimicrobial activity. A large amount of research has been conducted to understand how this polymer exerts its antibacterial activity and how its mechanism of action works. One of the most well-regarded hypotheses is linked to its structure. The cationic nature of chitosan when amminic nitrogen (N^+) is present in the glucosamine molecule is fundamental in chitosan's interactions with a negatively charged microbial cell surface that block bacterial life. Some researchers have explored chitosan's characteristics, such as its ability to bind water, chelate trace metals, and interact with DNA, to explain its mechanism of antimicrobial action. Chitosan's antimicrobial activity is influenced by several factors established by treating *Staphylococcus simulans* 22 cells with chitosan, namely the polymer molecular weight of chitosan and its degree of acetylation. Unfortunately, the inhibition of bacterial growth is dose-dependent; the permeabilization of the cell membrane coupled with a significant depolarization of the membrane constitutes a complex system

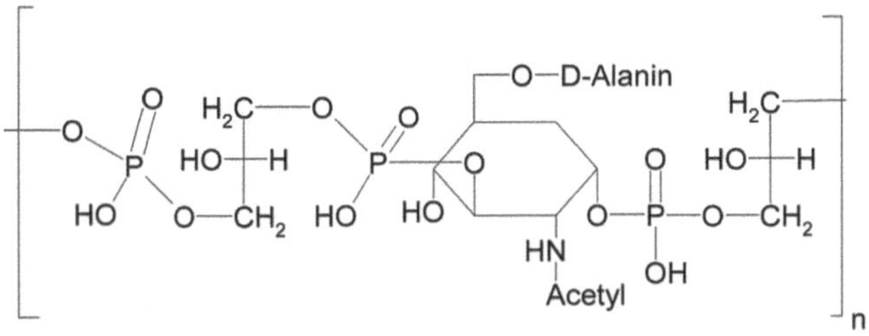

Figure 13.4. Structure of teichoic acids.

that must be carefully assessed on a case-by-case basis to achieve its antibacterial effect. Chitosan treatment was also found to involve numerous changes in the genetic profile of *Staphylococcus aureus* SG511 involved in the regulation of stress conditions and in autolysis, as well as genes associated with energy metabolism.

4.1 Chemical Mechanism of Chitosan Action in Relation to Antimicrobial Activity

The experimental results reported in the literature clearly show that antimicrobial activity cannot focus on a single target; however, the mechanism of action is not yet clear, and we are inclined to place greater emphasis on the primary characteristic of chitosan, namely its electropositivity. In Gram-positive bacteria, teichoic acids are present, which are polymers of polyhydric alcohols linked by phosphate groups that contain alanine residues. These structures are important components of the cytoplasmic membrane (Fig. 13.4).

Because of their negative charge, teichoic acids are partly responsible for the overall negative charge of the cell surface and may contribute to the passage of ions through the cell wall. Teichoic acids also bind divalent cations, such as Ca^{2+} and Mg^{2+}, carrying a part of them inside the cell. In some Gram-positive bacteria, teichoic acids that contain glycerol are closely related to lipid membrane-forming lipoteichoic acids. The positively charged chitosan would be attracted to the negative charge of the bacterial cytoplasmic membrane, binding to lipoteichoic acids, and the extraction of the membrane lipids would cause bacterial cell death. There are significant electrostatic and hydrophobic interactions, and the hydrogen bonding that occurs between lipids and chitosan support this theory (Wydro et al., 2007; Raafat et al., 2008). According to other authors, chitosan or some of its derivatives could be absorbed by bacterial cells because they are positively charged and interact with DNA by inhibiting the transcription of not only

DNA but also RNA and thus protein synthesis (Tarsi et al., 1997; Liu et al., 2001). The antimicrobial potential of chitosan is of great importance in the design of antimicrobial systems in the industrial sector in food preservation, pharmaceuticals, and the protection of drugs (Roller, 2003). Chitosans with a low average molecular weight (<10 kDa) have been shown to have greater antimicrobial activity than high molecular weight chitosan; this difference could be related to the former's greater solubility in aqueous medium, an important condition for being able to interact with microorganisms. To increase the antibacterial activity, it is also necessary to have a polymer that is made of at least 7 units of glucosamine, and the degree of deacetylation should be high to increase the charge density. Overall, the solubility and charge density are two important factors in the adhesion of chitosan to a bacterial cell. Synergistic effects were observed in both a strongly acidic environment and at an operating temperature of 37°C (Tsai and Su, 1999) due to the effect of temperature on the viscosity of the chitosan solution, which decreases with increasing temperatures. The results regarding the effect of the ionic strength on chitosan activity are mixed. Although Chung et al. (2003) propose that increased ionic strength could improve the solubility of chitosan and thus increase its antibacterial action regardless of the test strain, Tsai and Su (1999) suggest that the presence of sodium ions at a 100 mM concentration would reduce its activity against *Escherichia coli*. The addition of bivalent metal ions rendered chitosan completely inactive through the formation of complexes, as shown in soybean cells (Young et al., 1982). According to other researchers (Sudarshan et al., 1992; Tokura et al., 1997), the inhibition of bacterial growth could occur by blocking the nutrient flow to deposited chitosan layers on the bacterial surface, thereby preventing the mass transfer and suppression of the metabolic activity in bacteria.

5 Chemistry of Chitosan and Training Films

Chitosan possesses three reactive functional groups: an amino group at position C-2 of each deacetylated unit and an OH group at the C-6 and C-3 primary and secondary positions of each repeating unit (Fig. 13.1). These reactive groups can be derivatized to react with certain drugs, or the amino group can be neutralized with either organic acids or hydrochloric acid to obtain soluble chitosan salts (Henriksen et al., 1996). Soluble chitosan forms solutions with viscosities that are higher or lower in relation to the degree of deacetylation, the molecular weight, the concentration, the ionic strength of the solution, the solvent type, and the temperature, and these forms may act as thickeners, stabilizers, or suspension agents (Singla and Chawla, 2001). The solubility factor is very important because it not only affects the antifungal and antimicrobial activity of chitosan, as seen previously, but also allows for the production of films for food protection when other substances have been added to enhance its antimicrobial activity. Chitosan-based films have modest ability as an oxygen barrier and good ability as a carbon dioxide barrier, even if they show a high permeability to water vapor. To increase their water barrier properties, compounds such as essential oils are typically incorporated into the films of hydrophobic chitosan compounds. Since ancient times, essential oils have been known for their fragrance and their medicinal properties, including antiseptic, bactericidal, virucidal, and fungicidal properties; they were used both in the preservation of foods and in medicinal remedies such as antiinflammatories, spasmolytics, and local anesthetics. Essential oils consist of numerous components that are present at different concentrations, but they are characterized by having only two or three primary components at rather high concentrations compared with the other trace components. The primary group of compounds are the terpenes, characterized by low molecular weights and high volatility. These odorous substances are found in aromatic plants and are contained in oil-bearing cells distributed in various parts of the plant. The in vitro efficacy of essential oils against food-borne pathogens and their mechanisms of action have been extensively studied. The primary oils that have demonstrated antimicrobial properties against spoilage microorganisms are coriander, oregano, rosemary, sage, clove, and thyme. Rosemary oil exhibits strong antimicrobial action against spoilage and pathogenic microorganisms and also demonstrates antioxidant properties. Recently, there has been increasing interest in using these molecules as active components in edible films or coatings to preserve and extend the shelf life of foods such as meat and fish and minimally processed fruit. The only limitation in using edible films or coatings as support materials is that the incorporated antimicrobial agents must be selected from edible compounds. For films based on chitosan and essential oils, possible interactions between the antimicrobial agent and chitosan, which has its own antimicrobial activity, must always be considered because they could increase the response of the antimicrobial film. The intensity of these interactions depends on the type of essential oil used; the incorporation of cinnamon essential oil into chitosan films decreases the compactness of the film, which is not acceptable because the films must have good mechanical properties, an appropriate appearance in terms of gloss and transparency and good barrier properties against oxygen, carbon dioxide, and water.

Edible films with these desired properties have been created by adding bergamot essential oil to chitosan (Sánchez-González et al., 2011), which can then be used to prevent microbial growth without interfering with the flavor of the food. These composite films showed significant growth inhibition against *Penicillium italicum*, the fungus that is a major cause of citrus fruit deterioration with the appearance of so-called "blue mold." This antimycotic activity is directly proportional to the bergamot oil concentration and reaches a maximum at a bergamot oil/chitosan weight ratio equal to 3:1, with total inhibition of fungal growth in 5 days at 20°C. In addition, the incorporation of bergamot essential oil into chitosan films decreases water vapor permeability that stabilizes the internal humidity in the packaged product.

6 Chitosan and Yeast

Although the antibacterial activity of chitosan toward microorganisms is well documented, a few experiments have been performed on yeast spoilage that causes swollen packaging and unwanted flavors in food products due to the production of CO_2 during fermentation. In recent years, there has been an increased need to identify natural products that can inhibit the growth of undesirable yeasts, and chitosan has been preferred for its antimicrobial potential that antagonizes the action of yeast in consumer products such as fruit juices, increasing safety and lengthening the retention period. Three yeast species, *Zygosaccharomyces bailii*, *Saccharomyces cerevisiae*, and *Candida krusei*, have been tested in samples of apple juice purified by pasteurization and the addition of 1% acetic acid (Elosiebo, 2004). After adding chitosan solutions at concentrations of 0.005%, 0.01%, and 0.05% (w/v) with degrees of polymerization of 15 and molecular weights of 450 kDa, 0.005% chitosan significantly inhibited the growth of spoilage yeasts, but the inhibitory effect at 0.05% was more effective against yeast growth. Unlike other coating materials, chitosan has greater antifungal than antibacterial activity, which acts against various fungi, including *Botrytis cinerea* Pers. and *Rhizopus stolonifer* Ehr that have deleterious effects on fresh fruits and vegetables (El Ghaouth et al., 1992).

7 Environmental Applications of Chitosan

The first modern use of chitosan was for the purification of seawater contaminated with oil due to its ability to capture oils and fats. Its properties purified the water by attracting the biohazardous substances in the body of liquid to form a powerful foam containing the potentially harmful substances, which can then be easily removed by removing the foam. Its absorbent capacity was subsequently enhanced to use chitosan in a pharmaceutical as a supplement to ward off fat from the body. One of the most interesting prospects for the use of chitosan and its derivatives involves its applications in the environmental sector. Dyes are significant water pollutants that are generally present in effluents from various industries, namely the textile, leather, paper, and dye industries. Some of these dyes reportedly cause allergies, dermatitis, skin irritation, and cancer in humans. It is important to remove dyes from effluents before they are mixed into natural water bodies. Several studies on the use of chitin and chitosan to remove dyes from wastewater have been conducted over several years (Jain et al., 2003; Crini, 2005). When comparing various purification techniques and the use of different materials, bioadsorbents based on chitosan materials were cheap and efficient, and their absorption was more effective than other methods, demonstrating a high affinity for many dye classes. Chitosan containing hydrophobic groups in the molecule is particularly efficient in removing dyes from various wastewaters, and it is proposed for the removal of dyes from effluents resulting from textile fiber processing (Kim et al., 1997), being especially efficient in the removal of acid dyes and azo-dyes (Joseph, 1994; Shimizu et al., 1995).

Furthermore, chitosan is used to clarify water that contains protein substances derived from the processing of fruit, meat, fish, and milk (Juang et al., 2002); these substances are precipitated by flocculation and then eliminated. Chitosan has long been used in the beverage industry not only for its antimicrobial activity but also for the clarification of fruit juices (Shahidi et al., 1999). One application of this polysaccharide has also been suggested in the oenological field (Spagna et al., 1996), demonstrating that chitosan has great affinity for phenolic compounds implicated in white wine browning and that it can be used for sedimentation, which would be compatible with the techniques and equipment currently used in winemaking. In fact, chitosan could potentially be used as a stabilizer of white wine characteristics, although the possible side effects of this polysaccharide could be numerous. The proteins in grape juice have considerable technological importance, especially for their effects on the stability and clarity of wine (Waters et al., 1991), and chitosan acts by binding these proteins, causing their precipitation and thereby preventing the turbidity that may occur over time. In addition to its antibacterial activity, chitosan also acts on the microflora of white wines by preventing the growth of the bacteria that are responsible for malolactic fermentation, which is another factor in the instability of white wines (Radler, 1990). In this respect, chitosan could also reduce the concentration of sulfur dioxide that is normally used as an antibacterial agent in winemaking.

Another important application of chitosan in the purification of drinking water is addition to sand as part of the clarification process to remove both phosphorus and heavy metals (including arsenic), which are particularly toxic to organisms at high concentrations. Chitosan is a very versatile substance that can be used in various forms, for example, crystalline, powder, film, fiber, beads, and fabric; choosing the correct formulation depends on the use for which it is intended. Pearls of chitosan-coated nanoparticles made from titanium dioxide and zinc oxide were very efficient at removing arsenites and arsenates, with an added benefit over nanopowders in that there is no need for filtration after treatment. During experiments, the process of arsenate ion adsorption to the surface of the chitosan pearls occurred within the first 30 min. After a subsequent stationary phase, during which the arsenate concentration in solution does not vary excessively, the desorption process is initiated, which restores the initial concentrations after approximately 72 h. This experiment highlights chitosan's effectiveness in arsenate ion absorption within a short period of time, which could be exploited in columns arranged in a series to improve the process performance and economics. The adsorption process, with maximum removal efficiency in the first 20 min, could alternate with desorption for successive periods of 30 min, regenerating the adsorbent material at practically no cost (Consoli, 2011).

Many diseases are transmitted by water contaminated with pathogenic microorganisms, particularly *E. coli*, and decontamination of this water is essential. To this end, the use of chitosan as a cotton gauze coating has been proposed by means of polymerization with UV rays to use as a filter for disinfecting both Gram-negative and Gram-positive bacteria (Periolatto et al., 2014). The chitosan-treated gauze exhibited good antibacterial activity under conditions of both static and dynamic responses, even after a few seconds of contact, and thus it can be used as a biological filter. Moreover, gauze that was functionalized with chitosan has been tested as an absorbent medium for the removal of dyes from wastewater with good continuous flow performance (Periolatto and Ferrero, 2013).

7.1 Water Purification to Remove Heavy Metals Using Chitosan

Many studies have demonstrated the absorption capacity of chitosan for heavy metals that are present in galvanic waste, such as silver, zinc, lead, copper, nickel, cobalt, cadmium, iron, and chromium (Udyabhaskar et al., 1990). Various ions of toxic heavy metals are discharged into the environment through a number of industrial activities, constituting a major source of environmental pollution. McKay et al. (1989) investigated the adsorption of several metal ions by chitosan, noting that its adsorption capacity for Hg (II), Cu (II), Ni (II), and Zn (II) were 815 mg, 222 mg, 164 mg, and 75 mg/g, respectively. The complexation of metals by chitosan and its derivatives has been described by Varma et al. (2004). Chitosan possesses ion exchange characteristics that allow it to chelate with a higher capacity than other polymers of natural origin. Wan Ngah et al. (2002) showed that chitosan's adsorption capacity is significantly greater for the recovery of copper compared with that of commercial synthetic resins, and this superiority in removing heavy metals from aqueous solutions is linked to its high percentage of nitrogen, present as amino groups responsible for binding with metal ions through chelation mechanisms. Even the hydroxyl groups contribute to adsorption, especially at the C-3 position (Varma et al., 2004). The kinetics of metal adsorption are very rapid during the first few minutes and reach a maximum after 5–10 min at 25°C, as reported by (Salehizadeh and Shojaosadati, 2003). The interaction of chitosan with metallic species is strongly dependent on the pH; the optimum pH for complexation is 3–5, above which most metals precipitate, making them unavailable to chelating groups. Aside from pH, chitosan's shape is also important in increasing the absorption of metals; evaluation of powder, flakes, membranes, fibers, and beads has demonstrated that most absorbent capacity is related to the surface and porosity. To increase the surface area involved in complexation, chitosan gel beads were coated with a glutaraldehyde pattern, giving them an advantage of larger area and resistance to a low pH for a greater scope of action (Jeon and Holl, 2003). The possibility of regeneration makes chitosan reusable with high efficiency in industrial processes; regeneration of the adsorbent is important in reducing the process costs and in recovering the extracted pollutant from the solution. This biopolymer has attracted particular interest for its high complexing value in addition to its low cost and for its amino and hydroxyl functional groups that can adsorb a wide range of molecules, such as phenolic compounds, dyes, and metal ions (Crini, 2005). Resins based on chitosan have demonstrated high efficiency in the absorption of noble metals and are used for their recovery. In the case of gold, an efficiency of 11.7 mmoles of metal per gram of resin was achieved (Caihua and Yuwu, 1996). Promising results have also been obtained for the adsorption of uranium ions that are adsorbed and exchanged through the carboxylic and amine groups of chitosan (Jansson-Charrier et al., 1995), and, a technique based on this interesting complexation has been developed to remove uranium present in groundwater (Lenhart et al., 1997).

7.2 Chitosan in Aquaculture

Macrobrachium rosenbergii is known as the giant river prawn, a species that lives in the fresh waters of tropical regions in northern Australia and Southeast Asia and is commercially important for its value as a food source (Delves-Broughton and Poupard, 1976). In India, the rearing of *M. rosenbergii* larvae in fresh water is an important industrial sector, but many drawbacks have been observed with regard to the presence of different *Vibrio* spp. strains that reduce larval survival below a profitable level (Sindermann, 1977). These bacteria grow on both the eggs and larvae of shrimp, increasing significantly during the culture stages; the use of antibiotics has been documented (Karunasagar et al., 1994), but the development of antibiotic resistance is feared with prolonged use. Tendencia and de la Pena (2001) tested the effectiveness of chitosan as a possible prophylactic agent. They employed chitosan solutions at various concentrations and noted an increase in antimicrobial activity with increasing concentrations of chitosan. In this case, optimized chitosan use was observed under the following conditions: 1% w/v concentration of chitosan in solution, a molecular weight of <180 kDa, and a degree of deacetylation of 80%. The use of chitosan in aquaculture farming systems has been shown to inhibit the growth of *Vibrio* bacteria and thus lead to a decline in its growth, which may help prevent larval vibriosis (Anas et al., 2005).

8 Chitosan in Cosmetics

Chitin and chitosan have been used as conditioners and moisturizers in the preparation of cosmetic creams in place of other compounds such as hyaluronic acid. In particular, chitosan, which is nontoxic, can be used at any desired percentage in cosmetic and pharmaceutical applications. Its film-forming ability can impart a pleasant feeling of smoothness to the skin, protecting it from both adverse environmental conditions and the consequences of detergent use, with better results than that of hyaluronic acid in terms of lasting moisturizing effects on the skin (Muzzarelli, 1988; Chen et al., 2002).

In particular, a derivative of chitosan called *N*-carboxymethyl-chitosan has antimicrobial and reparative capacities in epidermal tissue, and it was introduced in cosmetic products for functional odonto-stomatological applications and oral hygiene or in the form of emulsions for milk formulations after sun exposure and as moisturizers for the body. The functional characteristics of *N*-carboxymethyl-chitosan include hemostatic properties, reconstruction of torn tissues, antiinflammatory and soothing properties, and bacteriostatic and bactericidal action against *candidiasis*, with full biocompatibility, biodegradability, and mucoadhesiveness being very important qualities in dermatology. Furthermore, chitosan is indicated for the preparation of toothpastes, mouthwashes, and especially in the treatment of pyorrhea, for which it significantly improves the health of the gums and oral mucosa. *Streptococcus mutans*, which is frequently present in the oral cavity, is considered the most cariogenic agent, and its action is exerted through the metabolism of sucrose to lactic acid (Li et al., 2001). The acidic environment that arises in the dental area as a result of this process leads to high demineralization, which makes the teeth susceptible to caries; *S. mutans* is also one of the microorganisms with the greatest receptors for adhesion to the tooth surface. Various substances are used for the control of plaque, such as alkaloids of plant origin, biguanides, essential oils, fluorinated amines, and triclosan, but the frequent use of these substances can lead to negative changes in the oral microflora, favoring the selection of resistant strains. Chlorhexidine is the best known antiplaque agent in use; however, it is not free from side effects. Low molecular weight chitosan inhibits the absorption of *S. mutans*, and hydroxyapatite appears to have bioadhesive properties that increase the retention time on the surface of oral mucosa. Decker et al. (2005) reported the synergistic antiplaque combination of chlorhexidine/chitosan on the adhesive property of chitosan to the antibiotic. Fujiwara et al. (2004) showed *S. mutans* growth inhibition using soluble chitosan solutions. The influence of different chitosan concentrations on bacterial biofilms was studied, and the adhesion step was completely inhibited by all chitosan concentrations; the subsequent accumulation stages were instead inhibited only at the highest concentrations and longest exposure times. These results support the efficacy of chitosan for the prevention of biofilm formation by *S. mutans*.

Chitosan has achieved good results in the treatment of onychomycosis, a common nail disease that affects 2% to 13% of the population, especially in old age; the topical application of a lacquer based on hydroxypropyl-chitosan, a biopolymer film-forming water-soluble formulation, penetrated more deeply into the nail structure due to its high affinity for keratin (Monti et al., 2005, 2010). Soluble hydroxypropyl-chitosan forms a hydrolipidic film that is invisible, odorless, and permeable to air and water and that acts as a vehicle to ensure high penetration of the curative into the nail without being removed by adherent surfaces such as socks or shoes. The hydroxypropyl-chitosan film also makes the surface of the nail smoother and creates a barrier against the external environment and pathogens, accelerating the physiological processes underlying nail growth and allowing for optimum treatment.

9 Applications of Chitosan in the Textile Industry

In recent years, nanotechnology has gained great importance in various fields, including the textile industry; nanotechnology has improved the performance of textile materials and is being used for new functions in tissues. In this field, research has been primarily focused on two technical aspects, that is, the insertion of nanoscale particles in the spinning stage or as a coating, and on the use products or technical specifications for the construction of nanostructures to compose the surface of textile products. Application of these new techniques can produce "multifunctional" cloths with novel properties for a large number of applications. Nanotechnology is an interdisciplinary science that allows for the organization of matter at the nanoscale and sometimes at the atomic scale, influencing the structure and behavior of the material itself. In various studies on the creation of functional surfaces, the sol-gel technique was used to construct materials with chemical, physical, and mechanical properties that could be modulated at the molecular level to obtain products with specific characteristics related to their composition, microstructure, and targeted applications. Creation of the sol phase requires available precursors, which are generally metal-organic compounds that are transformed by hydrolysis into hydroxides that are very unstable at high concentrations, which are subsequently used in the deposition step on the textile substrates. The deposition of the sol onto the substrate can occur within conventional systems that are already used in textile finishes, namely impregnation, coating, or spraying. A thermal treatment is required to induce condensation of the hydrolyzed precursor and its crosslinking on a treated material. The sol that is condensed and deposited on the substrate can be modified with additives, generally bulky molecules such as polymers, pigments, dyes, or enzymes, to increase the desired functionality. Sol-gel technology is particularly suitable for surface treatment because it takes place at ambient pressure and temperature, and crosslinking occurs at over

100°C. The crosslinked products confer new properties to the textile substrates, influencing the physical–mechanical (stability, conductivity, adhesion, and hydrophilicity), optical (color and luminescence), and bioactive (biocompatibility, antibacterial activity, release of aromas, and drug delivery) characteristics, transforming a textile material into a new product that is more aesthetically pleasing, comforting, protected and resistant against external stress, and long-lived and that can be used in most diversified sectors as "functional coatings."

Chitosan can be employed in the textile industry, although few studies have characterized tissues based on chitosan and their properties. Chitosan acts in cloth as a color-leveling agent; it facilitates the exhaustion of the bath, favors the use of a lesser amount of auxiliary textile products and considerably reduces costs in the dyeing process. The primary reason for these phenomena is the affinity between the chitosan and dyes, primarily of the anionic type, with which there are established interactions between the protonated amine groups of chitosan and the acidic groups of sulfonated anionic dyes. Chitosan can improve the properties of printed cloth; a press is necessary for use with suitable conventional pretreatment with a chemical agent to ensure a good print on the cloth by limiting the excessive diffusion and migration of the dye and proper penetration into the fibers, with improved sharpness. Chitosan is an alternative to the pretreatment agents currently in use because the protonated amine groups of the biopolymer form electrostatic interactions with the groups of the anionic dye that consequently favor greater absorption of the pigment on the cloth compared with an untreated cloth. The antibacterial properties imparted to the manufactured article are related to the release of amino deoxy-soluble sugars as a result of enzymatic hydrolysis and their binding properties towards the walls of bacterial cells; both bacteriostatic and bactericidal properties are connected to increases in the chitosan amounts employed during finishing. Moreover, chitosan application to wool fibers leads to an antifelting effect that is better than that of wool treated with chlorine or other chemicals with high environmental impacts.

A new fiber was obtained from a mixture of chitosan and viscose by reusing food industry waste, a new use for substances that would otherwise be discarded. The chemical structure of chitosan fibers imparts strong antibacterial properties; chitosan is also very similar to cellulose, and when matched with viscose and cotton, it makes extremely versatile fibers. The demonstrated antibacterial and antimicrobial properties represent a characteristic that is not temporary but is durable over time; in fact, chitosan and viscose are structurally bound in a fiber whose antibacterial function remains unchanged for over long periods of time, even in the face of frequent washing or abrasion (Takai, 2002). This cloth is suitable for the packaging of various textile products, from carpets to underwear, and for its ability to absorb moisture, preventing the skin from drying out, keeping it hydrated, and providing comfort; chitosan-treated fabric is hypoallergenic and suitable for sensitive skin; it has excellent dyeability and complete biodegradability (Varesano et al., 2011). Given these characteristics, chitosan can also be used in the medical field for treating dermatitis and wounds, as an adjuvant in tissue regeneration and as an antihemorrhagic agent.

10 Chitosan in the Preservation of Food Products

In recent years, there has been a growing demand for a more rational use of added chemicals in food, with an increasing preference for natural substances with antimicrobial activity and preservative properties. In this regard, many studies have focused on the safety and efficacy of chitosan and its use as a natural preservative in food preparations, similar to its uses in the pharmaceutical field (Rhoades and Roller, 2000; Helander et al., 2001; No et al., 2002). Several researchers have increased their studies on the antimicrobial potential of chitosan and its derivatives. Jumaa et al. (2002) suggested the use of chitosan as an antimicrobial preservative in emulsions that will be applied to mucous membranes and in parenteral applications. Sagoo et al. (2002) proposed its use as an adjuvant for enhancing the efficacy of other antimicrobial compounds used as preservatives, such as benzoates and phenylmercuric alcohol derivatives. Furthermore, chitosan exhibited the same antimicrobial activity against several bacterial strains (Raafat et al., 2008). The possibility of obtaining soluble low molecular weight chitosan with a high degree of deacetylation allows its transformation into a variety of useful forms, such as gels, membranes, sponges, films, fibers, and beads, for use in a wide variety of applications. Chitosan-based films and gels are also permeable to humidity, oxygen, and functional substances, such as flavoring additives, antioxidants, vitamins, and coloring agents, especially in the food industry, and they can improve the organoleptic characteristics of food, as well as its preservation (Hirano et al., 1991). When antimicrobial agents, such as benzoic acid, sorbic acid, propionic acid, lactic acid, nisin, and lysozyme, were incorporated into edible films, they delayed the development of bacteria, yeasts, molds, and pathogenic microorganisms such as

Listeria monocytogenes on the surface of many foods, including meats and cheese; this bacteria may contaminate the surface of cooked food that is ready to be eaten after preparation. Gómez-Estaca et al. (2010) studied chitosan films to which essential oils had been added to observe their antimicrobial activity against different bacterial species responsible for alterations of and toxicity in food, noting that clove essential oil showed the greatest inhibition. To evaluate the usefulness of essential oils as food preservatives, they studied fish extracts, noting that the films with added clove and thyme oils were the most effective. A further confirmation of the bactericidal effect of chitosan film with clove oil was observed during the refrigerated storage of fish. The growth of gram-negative bacteria was dramatically reduced, enterobacteria in particular, and lactic bacteria remained unaltered for most of the storage period.

An emulsion of chitosan in mineral oil was used as a cover film to maintain the internal quality of eggs during a storage period of 5 weeks at 25°C and 20 weeks at 4°C compared with eggs without a cover (Torrico et al., 2011). Although the cooling temperature is the most important parameter for preserving eggs by delaying the deterioration of their organoleptic properties, the use of a chitosan emulsion in mineral oil has considerable advantages, prolonging the shelf life of the eggs by well over 3 weeks compared with the control samples. The synergistic effects between the temperature and the coating decreased the weight of the eggs by $<2\%$ under storage at 25°C relative to that at 4°C, and in the latter case, the albumen of the egg yolk was maintained for 5 weeks more than that of the unprotected eggs.

10.1 Chitosan in the Preservation of Fruits and Vegetables

Our busy modern life does not allow people to devote much time to preparing meals, favoring the consumption of products that are ready to use, especially vegetables and fruits that have been suitably cut and washed (products in the IV range). However, these products have a limited retention time due to oxidation and browning; in fact, a minimum treatment involves alterations to the cut parts of the fresh products, such as variations in color, softening of the tissues, and increased microbial deterioration, reducing product preservation and integrity (Erturk and Picha 2006; Lin and Zhao, 2007). Therefore, fruits and vegetables in the IV range must have an expiration date that is no later than the 17th day after treatment. The most important characteristics for determining the quality of fresh, minimally processed produce are the following:

appearance, color, texture, taste, nutritional value, and microbial safety. To preserve the quality and extend the storage time of fresh produce, several measures have been employed, such as treatment with antimicrobial agents, edible film coatings, and modified atmosphere packaging of fresh produce (Banks, 1985). Edible films are made from biodegradable polymers that act as a barrier against moisture, gases, and volatile compounds to prevent moisture loss, controlling the respiratory activity of the plant to minimize the loss of volatile aromatic compounds and natural color. In addition, edible films can serve as a vehicle for substances and for the addition of other functional ingredients such as antibiotics, antifungal agents, and antioxidants (Lin and Zhao, 2007; Marsh and Bugusu, 2007). Chitosan is a biocompatible, biodegradable, biofunctional, and nontoxic polymer that has been widely used as a coating material for fresh products because of its ability to form films, its compatibility with other substances such as vitamins and minerals (Park and Zhao, 2004; Durango et al., 2006; Ribeiro et al., 2007) and its antimicrobial activity (Sudarshan et al., 1992; No et al., 2002). Chitosan coatings have been applied to a large variety of food products and clearly improve the quality and extend the life of various foods.

Xihong Li (2012) presented a paper at the 244th National Meeting & Exposition of the American Chemical Society meeting in Philadelphia on spraying green bananas with a chitosan aerogel that maintained the bananas' freshness for up to 12 days. When bananas begin to ripen quickly, they turn yellow in color, and the pulp later softens and rots. Protecting bananas with a chitosan coating kept the bananas green for a longer period of time and inhibited the rapid maturation that occurs following the breathing process of this fruit. Unlike many other fruits, respiration in bananas does not slow down after harvest, and bananas do not ripen slowly. Banana pulp releases ethylene, which increases breathing, accelerating the production of the sugar that imparts a sweet flavor to the fruit. However, as bananas continue to breathe, the process accelerates, and the banana pulp becomes unpleasant, sweet, and soft, with increasing bacterial growth on the skin that causes the deterioration of the banana. Xihong Li showed that the chitosan hydrogel coating slowed down respiration and killed bacteria via decomposition, keeping bananas fresh for almost 2 weeks. A derivative of chitosan called carboxymethylchitosan reduced the respiration and production rate of ethylene and was employed as an edible film coating for bananas and other fruits (Davis et al., 1989; Banks, 1984). This coating can also be used by consumers at home, in supermarkets, or during the shipment of bananas, given chitosan's nontoxicity.

Figure 13.5. Strawberry infected by *Botrytis cinerea.*

The primary cause of tomato deterioration after harvest is fungal lesions in the skin through physiological processes. To avoid wasting products, different methods of conservation have been used, namely the use of moderately low temperatures and high humidity (Morris, 1982), the use of modified atmospheres (low O_2, high CO_2) (Kader et al., 1989), and coating with waxes or sucrose esters with fatty acids (El Ghaouth et al., 1992). Given that a coating should extend the life of preserved fresh fruits by reducing decay without creating anaerobic conditions, these methods are unsatisfactory. A chitosan solution and its carboxymethyl-chitosan derivative were used at a concentration of 2% (w/v) to coat freshly picked green tomatoes with a semi-permeable film, maintained at 20°C. After 4 weeks of storage, the chitosan coating was observed for its antibacterial and antifungal properties, and it reduced the development of ethylene gas, delayed the production of red color, and increased the production of CO_2 compared with O_2; the pulp had greater consistency than that of the control, even if the total acidity was slightly increased. The chitosan coating also reduced the incidence of decay caused primarily by *B. cinerea*. In the tomato control, the first signs of fungal growth appeared after two days of storage, and signs of infection were visible only after 10 days of storage at 20°C on tomatoes covered with chitosan. The chitosan film cover effectively inhibits the growth of certain fungal species, such as *Alternaria alternata*, *B. cinerea*, and *Penicillium expansum*, responsible for the rot of some fruits such as mangoes (Prusky et al., 2002), strawberries (El Ghaouth et al., 1991, 1992; Han et al., 2004) (Fig. 13.5), raspberries (Tezotto-Uliana et al., 2014), apples (Morales et al., 2010), pears (Meheriuk and Lau, 1988), and peaches (Zhang et al., 2007).

Microscopic observations have shown that chitosan exerts its effects by causing mycelia aggregation and reducing the cell wall and the size of the hyphae, making it difficult for nutrients to enter fungal cells. The fungistatic activity was once again related to pH, particularly at low values, to the degree of deacetylation and to the cationic charge. Polycationic chitosan is able to interact with components of cell membrane possessing a negative charge, such as proteins and phospholipids, by interfering with the normal growth and metabolism of fungal cells. Roller and Covill (1999) reported that amino group-substituted chitosan can interact with the anionic groups present on the surface of yeast cells and form an impermeable layer due to chitosan's tendency to form films. Consequently, this barrier prevents the development and growth of pathogens.

Several applications involve the use of chitosan to prolong the storage time of plants after harvest. During storage, fruits and vegetables are exposed to bacterial attack and to deterioration that is promoted by enzymatic reactions. Because of its breathing rate, green asparagus deteriorates significantly during postharvest storage; it is also easily attacked by *Fusarium* species that cause rotting of the shoot. Qiu et al. (2014) evaluated the effectiveness of chitosan against *Fusarium concentricum* by immersing an asparagus sample in highly concentrated solutions of low molecular weight chitosan and a solution of carboxymethylchitosan for 10 min. After treatment, the asparagus were dried for 2 h at room temperature and stored at 2°C and in a relative humidity of 95% for 35 days. Concentrations of low molecular weight chitosan solutions ranging from highly concentrated to 0.05 mg/mL effectively inhibited the development of *F. concentricum* and blocked spore germination; treated asparagus showed no signs of phytotoxicity and were well preserved for up to 28 days without wilting or degradation of external quality.

Chitosan can be used to form antimicrobial coatings that are semipermeable, restricting gas exchange and decreasing the loss of water during fruit storage. This use is of particular relevance in fruit export from one country to another for the duration under transport conditions. For example, papaya is a highly perishable tropical fruit subject to weight loss by dehydration, loss of firmness and attack by microorganisms. It is generally consumed as a fresh fruit or in whole pieces for fruit salad and is appreciated by consumers for its high antioxidants and its anticancer properties. Chitosan coatings with different molecular weights and at different concentrations were evaluated with respect to the quality of freshly cut papaya cubes stored at 5°C. Covering the papaya cubes with a medium molecular weight chitosan film at a 0.02 mg/L concentration prolonged their preservation time beyond 12 days while maintaining the color and firmness of the fruit. In addition, the film inhibited the growth of mold and pathogens, with decreased activity of the polygalacturonase enzyme and a reduction in degeneration and pectin (Gonzalez-Aguilar et al., 2009).

Apple juice and cider are popular drinks consumed by people of all ages for their organoleptic and nutritional qualities. These drinks are low in sodium, cholesterol, and fat and high in vitamin C, polyphenols, flavonoids, and antioxidants (Lee et al., 2003; Leontowicz et al., 2003). Cider is a liquid obtained by crushing apples, and it is clarified by removing the suspended solids by filtration, resulting in a transparent product. Heat treatment is the most widely used method to preserve apple juice and apple cider for its effectiveness in reducing microorganisms, such as *Salmonella, E. coli* K-12 or *Cryptosporidium parvum*, responsible for infections in different foods, but it affects the sensory qualities and nutritional value (Kumar et al., 2009). Many studies have been performed on the development of new nonthermal preservation methods for the inactivation of microorganisms with minimal changes to the chemical and physicochemical quality of food products. Among these methods is high pressure homogenization (Lacroix et al., 2005; Diels and Michiels, 2006), a physical process that involves subjecting a fluid to a combination of pressure, shear, turbulence, and cavitation, promoting the division of particles or droplets of micrometer dimensions and creating a stable dispersion or emulsion combined with chitosan for further processing to destroy *E. coli* K-12 in apple juice and cider. The homogenization pressure and chitosan concentration have exhibited synergistic effects to inhibit the development of *E. coli* K-12, with more evidence in juice than in cider (Paster et al., 1995).

In recent years, the food industry has recognized that consumers demand fresh, healthy produce that is free of synthetic preservatives. For this reason, there is a global trend toward exploring new alternatives based on natural compounds that can control pathogenic diseases postharvest, activating methods that reduce the incidence of disease and prevent adverse effects on human health from the excessive application of synthetic fungicides. Chitosan has been recognized for its antimicrobial properties against Gram-negative and Gram-positive bacteria, molds, and yeasts, its structural, film-forming and adhesive properties and its biodegradability, and it can thus be considered a viable and safe alternative to chemicals.

10.2 Chitosan in Agriculture

In agriculture, chitosan is used to increase the innate ability of plants to defend against fungal infections and to improve productivity. Chitosan action occurs in the plant cell membrane and nuclear chromatin. When applied to the leaves of plants or to the soil, chitosan increases photosynthesis, promotes and enhances plant growth, and stimulates the absorption of nutrients by increasing the vigor of the plant (Kim, 1998). When used in seed treatments, chitosan imparts a sort of immunity in the development of roots that react by destroying parasitic nematode cysts, which prevent the development and growth of the future plant (Lee et al., 1999; Choi et al., 2000). No et al. (2003) showed that processing soya beans with chitosan increases both the total amount of soy sprouts and the concentration of vitamin C by approximately 10% compared to a reference sample (Kim et al., 1990). Chitosan treatment of sunflower seeds led to higher bud yields, faster germination, and a higher concentration of amino acids, melatonin, and polyphenols. Furthermore, the antioxidant capacity was increased because of a higher concentration of total polyphenols, particularly of the isoflavone daidzein. These ameliorative effects are influenced by the following parameters: the molecular weight of chitosan, the concentrations of the chitosan solutions, the type of solvent, and the soaking time of the seeds (Cho et al., 2008).

Since 1989, chitosan salt solutions have been applied to crops for frost protection or to improve seed germination, receiving EPA (US Environmental Protection Agency) approval for labeling as a *biopesticide*.

Chitosan has the ability to exercise natural biocontrol, awakening innate defense reactions in plants against insects, pathogens, and fungal diseases when applied via foliar or soil routes. Chitosan stimulates photosynthesis, promotes plant growth, increases the absorption of nutrients, revitalizes the plant, and boosts the immunity level against radicle parasitic cyst nematodes without destroying useful microorganisms (Rabea et al., 2003).

Chitosan applications in agronomy can reduce the environmental stress caused by drought or soil deficiencies, strengthening the seed viability, increasing yields, and reducing the decay of fruits and vegetables. The use of chitosan in seed treatments reduces weight loss during storage and prevents senescence (Devlieghere et al., 2004); its use in addition to fertilizers to control this release was explored by Sukwattanasinitt et al. (2001), and, in some cultures, it was highlighted for increasing the yield (Chandrkrachang, 2002; Nwe et al., 2004). Chitosańs stimulating effects on plant growth were also assessed (Bautista-Baños et al., 2003) along with the differentiation and growth of the roots of various plants (Lay Nge et al., 2006). In plants, chitosan acts as a growth regulator that induces resistance by protecting plants from late blight, powdery mildew, and botrytis without leading to cytotoxicity problems.

The power of chitosan to act as an inducer of resistance in wheat plants (*Triticum durum* Desf.) was evaluated by treating seeds to improve plant defenses during the early stages of growth against the fungus *Fusarium graminearum* (Schwabe), a leading causative agent of eyespot in cereals. Chitosan is a component of the fungal cell wall and, by mimicking the presence of fungus; it warns the plant to begin activating its baseline defense system.

Treatment was performed by immersing the kernels in an aqueous 0.5% chitosan-based solution (w/v); the kernels were initially evaluated with three different concentrations of the chitosan solution, and the final selection of 0.5% (w/v) was made by analyzing the seed quality and symptoms caused by the fungal inoculation of seedlings (cv. Simeto and Croesus). The experimental design included the following treatments: (1) treating seeds with chitosan by immersion, (2) treating seeds with chitosan by dipping and adding fungal inoculum to the soil next to the first leaf stadium, (3) untreated seeds and subsequent fungal inoculum in the ground at the first leaf stadium, (4) and untreated or inoculated seeds (control). The effect of the treatment on the seeds was evaluated according to agronomic, biochemical, and enzymatic parameters. First, we analyzed the enzymatic activities of important defense mechanism enzymes: guaiacol peroxidase (G-POD) and ascorbate peroxidase (APX), involved in the deactivation of reactive oxygen species and other metabolic processes; polyphenol oxidase (PPO), which catalyzes the oxidative reaction of polyphenols; and phenylalanine ammonia lyase (PAL), a key enzyme in phenylpropanoic acid metabolism that leads to the synthesis of important secondary metabolites.

Plant resistance to diseases can be obtained by genetic means or can be induced in the plant through various forms of assistance. Induced resistance develops after *preinoculation* of plants with various biotic agents or their *treatment* with chemical and physical agents (various proteins, lipoproteins, polysaccharides, RNA, polyanions, polyacrylic acids, salicylic acid, etc.). *Induced drag* is a mechanism that is present in the plant, but not active, that is induced from the outside; it is a man-made attempt to ensure that the plant is protected from pathogen attacks, putting the plant in a state of alert. The plant is treated with substances or pathogens that do not cause the disease, but that only activate a defense response without sickening the plant. Induced resistance can be defined as the set of mechanisms that induce a defense response in the plant; the postinfection responses are activated even in the absence of the pathogen. This resistance is activated by biotic or abiotic stimuli (ie, chemicals such as acetylsalicylic acid or aspirin). Treatment of a plant with an inactivated virus or with a nonpathogenic strain induces the plant's defense mechanisms, but the plant does not become sick; it is then able to defend itself from a pathogen attack that may infect the plant later. This mechanism therefore prepares the plant for pathogen attack.

Weeds and microorganisms in the soil that attack the plants at their roots, preventing development, are a significant problem in organic agriculture; the prohibited use of herbicides and the lack of effective natural products pose serious economic problems for companies handling crops for human consumption and require specific solutions. Soil mulching is a rapidly expanding technique and has many economic advantages; it consists of covering the soil with a layer of impermeable material to prevent the growth of weeds, maintain soil moisture, protect the soil from erosion by diverting rain, decrease the compaction of the soil and raise the soil temperature (Sun et al., 2014; Shiukhy et al., 2014). Traditional plastic mulching films made of polyethylene or ethylene copolymers-vinyl acetate have a drawback in that they must be removed and transported to disposal sites for "special waste" at the end of a culture, with substantial additional costs. In recent years, biodegradable plastic films produced for mulching consist of polymer mixtures with various additives (Brodhagen et al., 2015) that do not need to be removed at the end of culture; they are stirred into the ground and completely decompose within a month. Biodegradable films have been applied to several fruit crops, in some cases involving premature aging and in other cases improved organoleptic characteristics compared with traditional mulch (Qiu et al., 2014).

Biodegradable polymers constitute a very important material for mulching; they may be natural polymers such as polysaccharides of terrestrial or marine origin, which are solubilized, or they may include various additives to solidify into a transparent membrane after the evaporation of the solvent. These membranes have different thicknesses depending on the needs of the farmer and can be colored or enriched with fertilizers and herbicides.

The development of new strategies for the implementation of effective and safe agricultural chemicals for crop protection has become more important in recent years. The micro/nanoencapsulation of pesticides in a controlled release matrix is one of the most promising areas of research for the future (Cota-Arriola et al., 2013). Microencapsulation has been found to increase the efficiency of pesticides, further reducing their volatilization along with decreased toxicity and environmental contamination. The choice of the polymer used as a matrix for releasing the pesticide over a prolonged period of time depends on various factors, including the matrix preparation method, the type of pesticide and the experimental conditions of the process. One important aspect is that the polymer is biodegradable, and a classification system was created to recognize this property (Wilkins, 1990); the polymer should also not react with the given pesticide and should be inexpensive.

Chitosan is currently one of the most promising natural polymers for use in micro- and nanotechnology, and it is very effective for use in agriculture when combined with natural functional compounds to eliminate the problems associated with the misuse of harmful chemicals. Many studies have highlighted its potential to fight agricultural infections caused by viruses, bacteria, and fungi, and micro- and nanoparticles based on chitosan have improved this activity even more. The study of micro- and nanoparticles based on chitosan with particular respect to the release of natural active ingredients for the control of bacteria, viruses, and fungi in food and agriculture has not been widely explored at present, but it is one of the most promising areas of future research to find new strategies for the use of active compounds in chitosan matrices that meet the requirements for food and agricultural applications.

11 Chitosan as a Dietary Supplement

Food supplements are foodstuffs that are used to supplement the common diet, and they are concentrated sources of nutrients, such as vitamins and minerals, or other substances with nutritional or physiological effects. Food supplements primarily serve to compensate for an inadequate intake of essential nutrients by certain people or specific population groups, or, in some cases, to increase the intake of these nutrients. However, food supplements cannot be used to replace a healthy, balanced diet, nor can they be proposed for the treatment or cure of diseases.

In this context, one of the most well-known applications of chitosan is in the dietetic field for its ability to act as dietary fiber. It is not digested, and it intervenes in the absorption of ingested fats before they are metabolized. In this sense, chitosan prevents the absorption and accumulation of excess calories and is then removed from the body together with lipid substances. Chitosan fiber comes in contact with water and swells in the stomach to form a jelly, which gives a sense of fullness and reduces the sense of hunger; it binds tightly to lipid substances because of the electrostatic force between the positive charge present in the molecule and the negative charge present in the fats. The beneficial effects, in addition to weight loss from an increased sense of satiety, are highlighted by decreased levels of cholesterol and triglycerides in the blood (Muzzarelli, 1999). The safety of using chitosan in humans at various doses confirms that this dietary fiber is also well-tolerated, without causing deficiencies in fat-soluble vitamins or minerals. The cholesterol-lowering action of chitosan is caused by the minor gastrointestinal absorption of exogenous cholesterol. A concentration of 2 g/day in the form of dietary supplements can be used for a long periods of time without causing damage to the body.

In a study by Cornelli et al. (2008), a dough sample containing 2% (w/w) chitosan was tested in overweight and obese subjects to verify the effectiveness of the product. The polymer was standardized for both its molecular weight and its degree of deacetylation, important factors for the reproducibility of the results. The samples were administered in a double-blind study to volunteers aged between 25 and 70 years who were compared with control patients who received pasta without chitosan and took part in a reduced the caloric intake regimen with reduced portions/week of cereals and increased servings of fruit per week. After two months of treatment, a slight weight loss was noted in the control group that was attributed to the decreased caloric intake, and patients who had consumed the pasta with chitosan exhibited not only a decrease in their weight but also a reduction in their serum cholesterol and triglycerides (Cornelli et al., 2008). The effect of chitosan is likely related to its fiber content, which may reduce the glycemic index of the pasta. Therefore, the dough with added chitosan is active in the control of body weight without the need for any caloric restriction and can be taken regularly without major restrictions. However, the data in the literature on the efficacy of chitosan as a "slimming agents" are discordant, and they appear to indicate that the polymer is completely ineffective, although they cannot deny the positive results for the reduction of cholesterol and triglycerides. However, the various trials have not employed the same chitosan samples, the effectiveness of which is dependent on numerous factors, namely the molecular weight, degree of deacetylation, concentration, addition of other components (eg, vitamin C and guar gum), trial duration, clinical examination of patients, and quality of life. Chitosan certainly cannot be considered a cure, especially for obese people with very complex diseases, but it can help prevent and regulate lipid absorption in the gastrointestinal tract (Bokura and Kobayashi, 2003; Wuolijoki et al., 1999).

Additionally, chitosan in the form of soluble fiber has shown a specific action as a prebiotic that restores and enhances the development of the autochthonous intestinal microflora, which is damaged as a result of inflammatory diseases of the intestinal tract, or the ingestion of certain drugs such as antibiotics. An experiment conducted on rats revealed the bifidogenic effect of chitosan and chito-oligosaccharides with short chains from 2 to 10 units of D-glucosamine at a concentration 0.1–0.5%, with increases in bifidobacteria and lactobacilli concentrations and reduced concentrations of *Enterococcus* and enterobacteria.

Chitosan's protective action in the liver is less obvious, where it acts not only to protect this organ from poisoning by heavy metals and organochlorine compounds but also as an antioxidant that fights the formation of free radicals.

Natural polymeric materials have recently attracted the attention of researchers for use as biodegradable substances to carry drugs. The use of these materials provides better drug availability and effectiveness, reducing undesired pharmacological effects. The most important characteristic of chitosan is its good chemical reactivity; it possesses reactive atoms at positions 2, 3, and 6 of the molecule (Fig. 13.1) that allow reactions of ethers/esterifications that lead to chemical changes, such as hydrolysis, oxidation, and enzymatic degradation, to obtain polysaccharide derivatives for specific applications. Chitosan and its derivatives have attracted great interest because of their biocompatibility, biodegradability, and bioactivity, while remaining nontoxic.

Chitosan has applications in the pharmaceutical industry as an excipient that is used directly in the preparation of tablets, as a disintegrant to improve the dissolution of drugs, and for the production of solid forms of dosage-controlled release. Compared with traditional excipients, chitosan has better characteristics and an overall greater flexibility of use. This biopolymer is used as a form of deposit for the release of components to improve the permeability of the cell membrane and act as a specific agent for a particular objective. In fact, it has been employed for the production of implant systems with controlled release for the administration of hormones for prolonged periods of time; it was also used for the transmucosal absorption of polar drugs via nasal or oral proteins and also in vaccines for its ability to form mucoadhesive films. The antimicrobial activity of chitosan suggests its use in a variety of different formulations, for example, in dressing tapes, toothpaste, or artificial tears; chitosan nanoparticles are used as vehicles for drugs that are targeted to neoplastic diseases or disorders of the central nervous system (Xu et al., 2011; Yamaguchi et al., 2003; Haastert-Talini et al., 2013) (Fig. 13.6).

In conclusion, chitosan is a natural biopolymer from the endless applications, even if the variability of the molecular weight and degree of acetylation not allow the production of a compound always reproducible, with experimental results sometimes contradictory. Its application has increased over the past

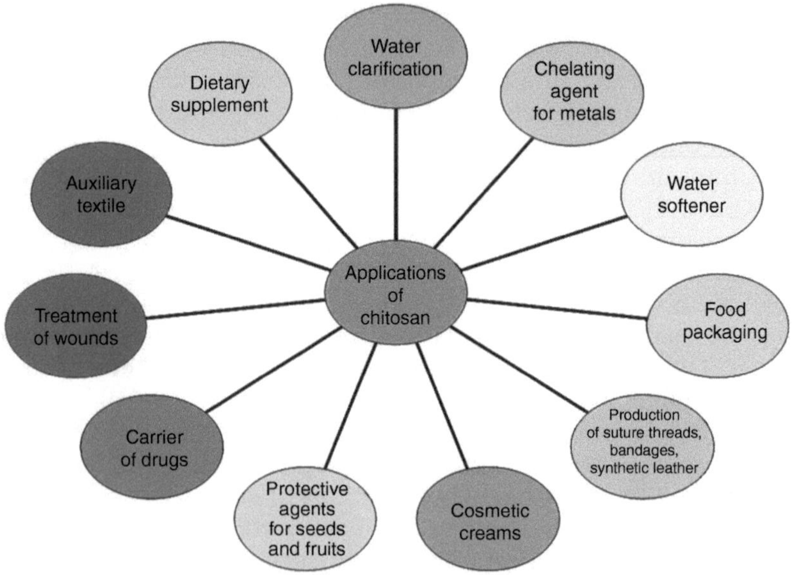

Figure 13.6. The numerous applications of chitosan: food, agriculture, packaging, environment, drug, textile, and cosmetic.

20 years especially as biomaterial in nanotechnology chitosan is a by-product of fish that has no cost and its reuse will reduce its environmental impact. The disposal of waste materials from shellfish, shrimp, and fish has become a serious problem in coastal areas around the world, in addition to the disposal of plant biomass. The selective isolation of a bioactive material such as chitin or other polymers from industrial waste is the easiest way to not only decrease environmental pollution from the presence of these unused biomaterials but also to increase the production of bioinnovative products. Moreover, because of its unique chemical characteristics, chitosan can be used in industrial activities that require specific purification, resistance, and antibacterial activity; it is remarkable in its use as an antimicrobial and antifungal agent, both in agriculture and food preservation due to its biodegradability and ability to form coating films. The primary mechanism of action of chitosan resides in its ability to bind to fatty and oily substances due to its chemical composition, favoring their elimination, and thus it can be used for different chemical–pharmaceutical preparations: chitosan is used in slimming diets, for its ability to reduce the absorption of fat. In fact, chitosan is widely exploited for the preparation of supplements that are designed to promote weight loss. According to many industry experts, taking

the drug orally in special supplements reduces the absorption of fat that is ingested in the diet. In fact, chitosan binds to fats and then wraps them in a sort of gelatinous casing until they become partially indigestible and are eliminated with the feces. The accumulation of adipose tissue is consequently limited, which is very useful in fighting obesity. In addition, the use of supplements containing this substance can also decrease feelings of hunger. There is no scientific confirmation of the slimming function of chitosan, and many scientists seriously question this idea. However, because these studies are still ongoing, the scientific community has not arrived at an unambiguous judgment, but it is believed that this type of substance derived from chitin is effective only if it is accompanied by a low-calorie diet and regular physical activity. Although many effects of its use are controversial and/or yet to be verified, chitosan has been shown to be a multifunctional substance with applications in many areas.

12 Conclusions

Chitosan is the most common natural polymer after natural cellulose. It is obtained by deacetylating chitin and has many versatile uses because of its solubility in weak acids; it can be used to produce biodegradable adhesive films, fibers of different thicknesses obtained by means of electrospinning, functional hydrogels, nanoemulsions, and nanoparticles for functions such as the packaging of active substances for controlled release administration, and other derivatives with great potential, in addition to its use in mineral complexing. The range of possible applications is very broad, from food to the environment, agriculture, pharmaceuticals, medical products, textiles, and cosmetics, and it is used not only for its chemical and physical characteristics but also for its antimicrobial, antibacterial, wound-healing, lipid-lowering, antiinflammatory, antiatherosclerosis, and anticancer properties. Moreover, its nontoxicity allows chitosan to be used long term as a soluble dietary fiber to improve digestion and aid the development of intestinal microflora with significant benefits to the health of the organism.

On the other hand, in addition to the positive aspects, we must also consider the downside: the intake of supplements of chitosan, in fact, may cause some side effects. Among them we can cite: nutritional deficiencies: chitosan can interfere with the absorption of vitamins, particularly fat-soluble vitamins (A, E, K, and D), essential fatty acids such as omega-3 and omega-6, and some minerals, causing nutritional deficiencies and malabsorption of macro- and micronutrients.

In conclusion, although many effects of its uses are controversial and/or yet to be verified, the chitosan has proved to be a multifunctional substance, with applications in many areas. In the near future all the innovative products, such as chitosan, if introduced on the international market, will profoundly modify the characteristics, also changing consumer behavior.

References

Anas, A., Paul, S., Jayaprakash, N.S., Philip, R., Bright Singh, I.S., 2005. Antimicrobial activity of chitosan against vibrios from freshwater prawn *Macrobrachium rosenbergii* larval rearing. Dis. Aquat. Org. 67, 177–179.

Banks, N.H., 1984. Some effects of TAL pro-long coating on ripening bananas. J. Exp. Bot. 35, 127–137.

Banks, N.H., 1985. Internal atmosphere modification in pro-long coated apples. Acta Hort. 517, 105–112.

Bautista-Baños, S., Hernandez-Lopez, M., Bosquez-Molina, E., Wilson, C.L., 2003. Effects of chitosan and plant extracts on growth of *Colletotrichum gloeosporioides*, anthracnose levels and quality of papaya fruit. Crop Prot. 22, 1087–1092.

Bokura, H., Kobayashi, S., 2003. Chitosan decreases total cholesterol in women: a randomized, double-blind, placebo-controlled trial. Eur. J. Clin. Nutr. 57, 721–725.

Braconnot, H., 1811. Sur la nature des champignons. Ann. Chim. Phys. 79, 265–304.

Brodhagen, M., Peyron, M., Miles, C., Inglis, D.A., 2015. Biodegradable plastic agricultural mulches and key features of microbial degradation. Appl. Microbiol. Biotechnol. 99 (3), 1039–1056.

Caihua, N., Yuwu, X., 1996. Studies on synthesis and properties of chelating resins based on chitosan. J. Appl. Polym. Sci. 59, 499–504.

Chandrkrachang, S., 2002. The applications of chitin and chitosan in agriculture in Thailand. In: Suchiva, K., Chandrkrachang, S., Methacanon, P., Peter, M.G. (Eds.), Advances in Chitin Science, Vol. 5, Bangkok, pp. 458–462.

Chen, L.Y., Du, Y.M., Wu, H.Q., Xiao, L., 2002. Relationship between molecular structure and moisture-retention ability of carboxymethyl chitin and chitosan. J. Appl. Polym. Sci. 83, 1233–1241.

Chirkov, S.N., 2002. The antiviral activity of chitosan. Appl. Biochem. Microbiol. 38, 1–8 (in Russian).

Cho, M.H., No, H.K., Prinyawtwatkul, W., 2008. Chitosan treatments affect growth and selected quality of sunflower sprouts. J. Food Sci. 73 (1), S70–S77.

Choi, H.D., Kim, S.S., Kim, K.T., Lee, J.Y., Park, W.M., 2000. Effect of presoaking treatments on growth and rot of soybean sprouts. Korean J. Food Sci. Technol. 32, 584–589.

Chung, Y.C., Wang, H.L., Chen, Y.M., Li, S.L., 2003. Effect of abiotic factors on the antibacterial activity of chitosan against waterborne pathogens. Bioresour. Technol. 88, 179–184.

Consoli, M., 2011. Survey on the applicability of chitosan in the removal of arsenate ions from water intended for human consumption. Thesis.

Cornelli, U., Belcaro, G., Cesarone, M.R., Cornelli, M., 2008. Use of polyglucosamine and physical activity to reduce body weight and dyslipidemia in moderately overweight subjects. Minerva Cardioangiol. 56 (5 Suppl.), 71–78.

Cota-Arriola, O., Onofre, M., Cortez-Rocha, M.O., Burgos-Hernandez, A., Ezquerra-Brauer, J.M., Plascencia-Jatomea, M., 2013. Controlled release matrices and micro/nanoparticles of chitosan with antimicrobial potential: development of new strategies for microbial control in agriculture. J. Sci. Food Agric. 93, 1525–1536.

Crini, G., 2005. Recent developments in polysaccharide-based materials used as adsorbents in wastewater treatment. Prog. Polym. Sci. 30, 38–70.

D'Acquapendente, H.F., 1628. Opera chirurgica ex officina Joannis Pillehotte. Università di Ghent, p. 83.

Davis, D.H., Elson, C.M., Hayes, E.R., 1989. N,O-carboxymethyl chitosan, a new water soluble chitin derivative. In: SkjakBraek, G., Anthonsen, T., Sandford, P. (Eds.), Chitin and Chitosan: Sources, Chemistry, Biochemistry, Physical Properties and Applications. Elsevier, New York, pp. 467–472.

Decker, E.M., von Ohle, C., Weiger, R., Wiech, I., Brecx, M., 2005. A synergistic chlorhexidine/chitosan combination for improved antiplaque strategies. J. Periodontal Res. 40, 373–377.

Delves-Broughton, J., Poupard, C.W., 1976. Disease problems of prawn in recirculation systems in UK. Aquaculture 7, 201–217.

Devlieghere, F., Vermeulen, A., Debevere, J., 2004. Chitosan: antimicrobial activity, interactions with food components and applicability as a coating on fruit and vegetables. Food Microbiol. 21 (6), 703–714.

Didenko, L.V., Gerasimenko, D.V., Konstantinova, N.D., Silkina, T.A., Avdienko, I.D., Bannikova, G.E., Varlamov, V.P., 2005. Ultrastructural study of chitosan effects on Klebsiella and staphylococci. Bull. Exp. Biol. Med. 140, 356–360.

Diels, A.M., Michiels, C.W., 2006. High-pressure homogenization as a non thermal technique for inactivation of microorganisms. Crit. Rev. Microbiol. 32 (4), 201–216.

Durango, A.M., Soares, N.F., Andrade, N.J., 2006. Microbiological evaluation of an edible antimicrobial coating on minimally processed carrots. Food Control 17, 336–341.

El Ghaouth, A., Arul, J., Ponnampalam, R., Boulet, M., 1991. Chitosan coating effect on storability and quality of fresh strawberries. J. Food Sci. 56, 1618–1620.

El Ghaouth, A., Ponnampalam, R., Castaigne, F., Arul, J., 1992. Chitosan coating to extend the storage life of tomatoes. Hort. Sci. 27 (9), 1016–1018.

El Ghaouth, A., Arul, J., Grenier, J., Asselin, A., 1992. Antifungal activity of chitosan on two postharvest pathogens of strawberry fruits. Phytopathology 82, 398–402.

Elosiebo Mechelle, E., 2004. Antimicrobial activity of chitosan against *Candida krusei*, *Saccharomyces cerevisiae*, and *Zygosaccharomyces bailii*. University of Tennessee honors thesis projects. <http://trace.tennessee.edu/utk_chanhonoproj/733>

Erturk, E., Picha, D.H., 2006. Microbiological quality of fresh-cut sweet potatoes. Int. J. Food Sci. Technol. 41, 366–374.

Fujiwara, M., Hayashi, Y., Ohara, N., 2004. Inhibitory effect of water-soluble chitosan on growth of *Streptococcus mutans*. New Microbiol. 27, 83–86.

Gómez-Estaca, J., López de Lacey, A., López-Caballero, M.E., Gómez-Guillén, M.C., Montero, P., 2010. Biodegradable gelatin-chitosan films incorporated with essential oils as antimicrobial agents for fish preservation. Food Microbiol. 27 (7), 889–896.

Gonzalez-Aguilar, G., Valenzuela-Soto, E., Lizardi-Mendoza, J., Goycoolea, F., Martinez-Tellez, M., Villegas-Ochoa, M., Monroy-Garcia, I., Ayala-Zavala, F., 2009. Effect of chitosan coating in preventing deterioration and preserving the quality of fresh-cut papaya "Maradol". J. Sci. Food Agric. 89, 15–23.

Haastert-Talini, K., Geuna, S., Dahlin, L.B., Meyer, C., Stenberg, L., Freier, T., Heimann, C., Barwig, C., Pinto Luis, F.V., Raimondo, S., Gambarotta, G., Samy, S.R., Sousa, N., Salgano, A.J., Ratzka, A., Wrobel, S., Groth, C., 2013. Chitosan tubes of varying degrees of acetylation for bridging peripheral nerve defects. Biomaterials 34 (38), 9886–9904.

Han, C., Zhao, Y., Lionard, S.W., Traber, M., 2004. Edible coatings to improve storability and enhance nutritional value offresh and frozen strawberries (*Fragaria ananassa*) and raspberries (*Rubus ideaus*). Postharvest Biol. Technol. 33, 67–78.

Helander, I.M., Nurmiaho-Lasdsila, E.L., Ahvenainen, R., Rhoades, J., Roller, S., 2001. Chitosan disrupt the barrier properties of the outer membrane of Gram–negative bacteria. Int. J. Food Microbiol. 71, 235–244.

Henriksen, I., Green, K.L., Smart, J.D., Smistad, G., Karlsen, J., 1996. Bioadhesion of hydrated chitosans: an in vitro and in vivo study. Int. J. Pharm. 145, 231–240.

Hirano, S., Hirochi, K., Hayashi, K.I., Mikami, T., Tachibana, H., 1991. In: Gebelein, C.G., Cheng, T.C., Yang, V.C. (Eds.), Cosmetic and Pharmaceutical Uses of Chitin and Chitosan. Plenum Press, New York, pp. 95–104.

Jain, A.K., Gupta, V.K., Bhatnagar, A., Suhas, 2003. Utilization of industrial waste products as adsorbents for the removal of dyes. J. Hazard. Mater. B101, 31–42.

Jansson-Charrier, M., Saucedo, I., Guibal, E., Le Cloirec, P., 1995. Approach on uranium sorption mechanisms on chitosan and glutamate glucan by IR and 13C-NMR analysis. React. Funct. Polym. 27, 209–221.

Je, J.Y., Kim, S.K., 2006. Chitosan derivatives killed bacteria by disrupting the outer and inner membrane. J. Agric. Food Chem. 54, 6629–6633.

Jeon, C., Holl, W.H.H., 2003. Chemical modification of chitosan and equilibrium study for mercury ions removal. Water Res. 37, 4770–4780.

Joseph, A.L., 1994. Removing acid dyes from textile waste water using biomass for decolourisation. Am. Dyestuff Rep. 83, 5–13.

Juang, R.S., Wu, F.C., Tseng, R.L., 2002. Use of chemically modified chitosan beads for sorption and enzyme immobilization. Adv. Environ. Res. 6 (2), 171–177.

Jumaa, M., Furkert, F.H., Muller, B.W., 2002. A new lipid emulsion formulation with high antimicrobial efficacy using chitosan. Eur. J. Pharm. Biopharm. 53, 115–123.

Kader, A., Zagory, D., Kerbel, E., 1989. Modified atmosphere packaging of fruits and vegetables. CRC Crit. Rev. Food Nutr. 28, 1–30.

Khanafari, A., Marandi, R., Sanatei, S., 2008. Recovery of chitin and chitosan from shrimp waste by chemical and microbial methods Iran. J. Environ. Health Sci. Eng. 5 (1), 19–24.

Karunasagar, I., Pai, R., Malathi, G.R., Karunasagar, I., 1994. Mass mortality of *Penaeus monodon* larvae due to antibiotic resistant *Vibrio harveyi* infection. Aquaculture 128, 203–209.

Kim, D.H., Choi, H.S., Kim, W.J., 1990. Comparison study of germination and cooking rate of several soybean varieties. Korean J. Food Sci. Technol. 22, 94–98.

Kim, C.Y., Choi, H.M., Cho, H.T., 1997. Effect of acetylation on sorption of dyes from chromium on chitin. J. Appl. Polym. Sci. 63, 725–736.

Kim, S.K., 1998. Applications of chitin and chitosan in agriculture. Korean J. Chitin Chitosan 3, 327–342.

Kumar, S., Thippareddi, H., Subbiah, J., Zivanovic, S., Davidson, P.M., Harte, F., 2009. Inactivation of *Escherichia coli* K-12 in apple juice using combination of high-pressure homogenization and chitosan. J. Food Sci. 74 (1), M8–M14.

Lacroix, N., Fliss, I., Makhlouf, J., 2005. Inactivation of pectin methylesterase and stabilization of opalescence in orange juice by dynamic high pressure. Food Res. Int. 38 (5), 569–576.

Langer, R., Vacanti, J.P., 1993. Tissue engineering. Science 260, 920–926.

Lay Nge, K., New, N., Chandrkrachang, S., Willem, F., Stevens, W.F., 2006. Chitosan as a growth stimulator in orchid tissue culture. Plant Sci. 170, 1185–1190.
Ledderhose, G., 1878. Ueber chitin und seine Spaltungprodukte Z. Physiol. Chem. 2, 213.
Lee, Y.S., Park, R.D., Rhee, C.O., 1999. Effect of chitosan treatment on growing characteristics of soybean sprouts. Korean J. Food Sci. Technol. 31, 153–157.
Lee, K.W., Kim, Y., Kim, D.O., Lee, H.J., Lee, C.Y., 2003. Major phenolics in apple and their contribution to the total antioxidant capacity. J. Agric. Food Chem. 51 (22), 6516–6520.
Lenhart, J.J., Figueroa, L.A., Honeyman, B.D., Kaneko, D., 1997. Modelling the adsorption of U(VI) onto animal chitin using coupled mass transfer and surface complexation. Colloids Surf. 120 (1–3), 243–254.
Leontowicz, M., Gorinstein, S., Leontowicz, H., Krzeminski, R., Lojek, A., Katrick, E., Cizm, M., Martin-Bellooso, M., Soliva-Fortuny, R., Haruenkit, R., Trakhtenberg, S., 2003. Apple and pear peel and pulp and their influence on plasma lipid and antioxidant potentials in rats fed cholesterol-containing diets. J. Agric. Food Chem. 51 (19), 5780–5785.
Li, Y.H., Hanna, M.N., Svensater, G., Ellen, R.P., Cvitkovitch, D.G., 2001. Cell density modulates acid adaptation in *Streptococcus mutans*: implication for survival in biofilm. J. Bacteriol. 183, 6875–6884.
Lin, D., Zhao, Y., 2007. Innovations in the development and application of edible coatings for fresh and minimally processed fruits and vegetables. Comp. Rev. Food Sci. Food Saf. 6 (3), 60–75.
Liu, X.F., Guan, Y.L., Yang, D.Z., Li, Z., De Yao, K., 2001. Antibacterial action of chitosan and carboxymethylated chitosan. J. Appl. Polym. Sci. 79, 1324–1335.
McKay, G., Blair, H.S., Findon, A., 1989. Equilibrium studies for the sorption of metal ions onto chitosan. Ind. J. Chem. 28A, 356–360.
Marsh, K., Bugusu, B., 2007. Food packaging—roles, materials, and environmental issues. J. Food Sci. 72 (3), R39–R55.
Meheriuk, M., Lau, O.L., 1988. Effect of two polymeric coatings on fruit quality of "Bartlett" and "d'Anjou" nears. J. Amer. Soc. Hort. Sci. 113, 222–226.
Monti, D., Saccomani, L., Chetoni, P., Burgalassi, S., Saettone, M.F., Mailland, F., 2005. In vitro transungual permeation of ciclopirox from a hydroxypropyl chitosan-based, water-soluble nail lacquer. Drug Dev. Ind. Pharm. 31 (1), 11–17.
Monti, D., Saccomani, L., Chetoni, P., Burgalassi, S., Senesi, S., Ghepardi, E., Mailland, F., 2010. Hydrosoluble medicated nail lacquers: in vitro drug permeation and corresponding antimycotic activity. Br. J. Dermatol. 62 (2), 311–317.
Morales, H., Marín, S., Ramos, A.J., Sanchis, V., 2010. Influence of post-harvest technologies applied during cold storage of apples in *Penicillium expansum* growth and patulin accumulation: a review. Food Control. 21, 953–962.
Morris, L.L., 1982. Chilline injury of horticultural crops: an overview. Hort. Sci. 17, 161–162.
Muzzarelli, R.A.A., 1985. Chitin. In: The Polysaccharides, vol. 3, G.O. Aspinall (Ed.). Academic Press, New York, pp. 417–450.
Muzzarelli, R.A.A., 1988. Carboxymethyl chitins and chitosans. Carbohydr. Polym. 8, 1–21.
Muzzarelli, R.A.A., 1999. Clinical and biochemical evaluation of chitosan for hypercholesterolemia and overweight control. In: Jollès, P., Muzzarelli, R.A. (Eds.), Chitin and Chitinases. Birkhauser Verlag, Basel.
No, H.K., Park, N.Y., Lee, S.H., Meyers, S.P., 2002. Antibacterial activity of chitosans and chitosan oligomers with different molecular weights. Int. J. Food Microbiol. 74, 65–72.

No, H.K., Lee, K.S., Kim, I.D., Park, M.J., Kim, S.D., Meyers, S.P., 2003. Chitosan treatment affects yield, ascorbic acid content, and hardness of soybean sprouts. J. Food Sci. 68, 680–685.

Nwe, N., Chandrkrachang, S., Stevens, W.F., 2004. Application of chitosan in Myanmar's agriculture sector. In: Proceedings of the Sixth Asia Pacific Chitin and Chitosan Symposium, May 23–26, the National University of Singapore, pp. 27–32.

Park, S.I., Zhao, Y., 2004. Incorporation of a high concentration of mineral or vitamin into chitosan-based films. J. Agric. Food Chem. 52, 1933–1939.

Paster, N., Huppert, D., Barkai-Golan, R., 1995. Production of patulin by different strains of *Penicillium expansum* in pear and apple cultivars stored at different temperatures and modified atmospheres. Food Addit. Contam. 12 (1), 51–58.

Periolatto, M., Ferrero, F., 2013. Cotton filter fabrics functionalization by chitosan UV-grafting for removal of dyes. Chem. Engin. Trans. 32, 85–90.

Periolatto, M., Ferrero, F., Vineis, C., Varesano, A., 2014. Antibacterial water filtration by cationized or chitosan coated cotton gauze antibacterial water filtration by cationized or chitosan coated cotton gauze. Chem. Engin. Trans. 38, 235–240.

Prusky, D., Shalom, Y., Kobiler, I., Akerman, M., Fuchs, Y., 2002. The level of quiescent infection of *Alternaria alternata* in mango fruits at harvest determines the postharvest treatment applied for the control of rots during storage. Postharvest Biol.Technol. 42, 134–141.

Qiu, M., Wu, C., Ren, G., Liang, X., Wang, X., Huang, J., 2014. Effect of chitosan and its derivatives as antifungal and preservative agents on postharvest green asparagus. Food Chem. 155, 105–111.

Raafat, D., von Bargen, K., Haas, A., Sahl, H.G., 2008. Insights into the mode of action of chitosan as an antibacterial compound. Appl. Environ. Microbiol. 74, 3764–3773.

Rabea, E.I., Badawy, M.E.T., Stevens, C.V., Smagghe, G., Steurbaut, W., 2003. Chitosan as antimicrobial agent: applications and mode of action. Biomacromolecules 4 (6), 1457–1465.

Radler, F., 1990. Action of nisin against lactic-acid bacteria and wine yeasts in solid and liquid-media. Am. J. Enol. Viticult. 41 (1), 1–6.

Rao, M.S., Muñoz, J., Stevens, W.F., 2000. Critical factors in chitin production from biomaterials of black tiger shrimp (*Penaeus monodon*) by fermentation. Appl. Microbiol. Biotechnol. 54, 808–813.

Ratner, B., Hoffmann, A.S., Schoen, F.J., Lemons, J.E., 1996. Biomaterials Science. An Introduction to Materials in Medicine. Academic Press, San Diego.

Ribeiro, C., Vicente, A.A., Teixeira, J.A., Miranda, C., 2007. Optimization of edible coating composition to retard strawberry fruit senescence. Postharvest Biol. Technol. 44, 63–70.

Rhoades, J., Roller, S., 2000. Antimicrobial actions of degraded and native chitosan against spoilage organisms in laboratory media and foods. Appl. Environ. Microbiol. 66, 80–86.

Roller, S., 2003. Chitosan: new food preservative or laboratory curiosity? In: Roller, S. (Ed.), Natural Antimicrobials for the Minimal Processing of Foods. Woodhead Publishing, Cambridge, UK, pp. 158–175.

Roller, S., Covill, N., 1999. The antifungal properties of chitosan in laboratory media and apple juice. Int. J. Food Microbiol. 47 (1–2), 67–77.

Rouget, M.C., 1859. Des substances amylacées dans les tissus des animaux, spécialement des Articulés (chitine). Comp. Rend. 48, 792–795.

Sagoo, S.K., Board, R., Roller, S., 2002. Chitosan potentiates the antimicrobial action of sodium benzoate on spoilage yeasts. Lett. Appl. Microbiol. 34, 168–172.

Salehizadeh, H., Shojaosadati, S.A., 2003. Removal of metal ions from aqueous solution by polysaccharide produced from *Bacillus firmus*. Water Res. 37, 4231–4235.

Sánchez-González, L., Cháfer, M., González-Martínez, C., Chiralt, A., Desobry, S., 2011. Study of the release of limonene present in chitosan films enriched with bergamot oil in food simulants. J. Food Eng. 105 (1), 138–143.
Shahidi, F., Arachchi, J.K.V., Jeon, Y.J., 1999. Food applications of chitin and chitosan. Trends Food Sci. Tech. 10 (2), 37–51.
Shimizu, Y., Kono, K., Kim, I.S., Takagishi, T., 1995. Effects of added metal ions on the interaction of chitin and partially deacetylated chitin with an azo dye carrying hydroxyl grouphs. J. Appl. Biomater. 5, 245–254.
Shiukhy, S., Raeini-Sarjaz, M., Chalavi, V., 2014. Colored plastic mulch microclimates affect strawberry fruit yield and quality. Int. J. Biometeorol., 1–6.
Sindermann, C.J., 1977. Disease Diagnosis and Control in North Marine Aquaculture. Elsevier, Amsterdam.
Singla, A.K., Chawla, M., 2001. Chitosan: some pharmaceutical and biological aspects—an update. J. Pharm. Pharmacol. 53, 1047–1067.
Spagna, G., Pifferi, P.G., Rangoni, C., Mattivi, F., Nicolini, G., Palmonari, R., 1996. The stabilization of white wines by adsorption of phenolic compounds on chitin and chitosan. Food Res. Int. 29 (3–4), 241–248.
Sudarshan, N.R., Hoover, D.G., Knorr, D., 1992. Antibacterial action of chitosan. Food Biotechnol. 6, 257–272.
Sukwattanasinitt, M., Klaikherd, A., Skulnee, K., Aiba, S., 2001. Chitosan as a releasing device for 2,4-D herbicide. In: Uragami, T., Kurita, K., Fukamizo, T. (Eds.), Chitin and Chitosan in Life Science. Yamaguchi, Japan, pp. 142–143.
Sun, H.X., Wang, W., Zhang, P., Wang, Z.W., Jia, Z.K., Yang, B.P., Han, Q.F., 2014. Effects of straw mulching on soil moisture and watermelon yield in dryland. Ying Yong Sheng Tai Xue Bao 25 (7), 2004–2010.
Takai, K., 2002. Antibacterial properties of antimicrobial-finished textile products. Microbiol. Immunol. 46 (2), 75–81.
Tarsi, R., Muzzarelli, R.A.A., Guzman, C.A., Pruzzo, C., 1997. Inhibition of *Streptococcus mutans* adsorption to hydroxyapatite by low-molecular-weight chitosans. J. Dent. Res. 76, 665–672.
Tendencia, E.A., de la Pena, L.D., 2001. Antibiotic resistance of bacteria from shrimp ponds. Aquaculture 195, 193–204.
Tezotto-Uliana, J.V., Fargoni, G.P., Geerdink, G.M., Kluge, R.A., 2014. Chitosan applications pre-or postharvest prolong raspberry shelf-life quality. Postharvest Biol. Technol. 91, 72–77.
Tokura, S., Ueno, K., Miyazaki, S., Nishi, N., 1997. Molecular weight dependent antimicrobial activity by chitosan. Macromol. Symp. 120, 1–9.
Torrico, D.D., No, H.K., Prinyawiwatkul, W., Janes, M., Corredor José, A.H., Osorio Luis, F., 2011. Mineral oil-chitosan emulsion coatings affect quality and shelf-life of coated eggs during refrigerated and room temperature storage. J. Food Sci. 76 (4), S262–S268.
Tsai, G.J., Su, W.H., 1999. Antibacterial activity of shrimp chitosan against *Escherichia coli*. J. Food. Prot. 6, 239–243.
Udyabhaskar, P., Leda, I., Prabhakara, A.V.S., 1990. Hexavalent chromium interaction with chitosan. J. Appl. Polym. Sci. 39, 739–747.
Varesano, A., Vineis, C., Aluigi, A., Rombaldoni, F., 2011. Antimicrobial polymers for textile products. In: Mendez-Vilas, A. (Ed.), Science Against Microbial Pathogens: Communicating Current Research and Technological Advance, Vol. 1, 3rd ed. Formatex Microbiology Book Series, Badajoz, Spain, pp. 99–110.
Varma, A.J., Desphande, S.V., Kennedy, J.F., 2004. Metal complexation by chitosan and its derivative. A review. Carbohydr. Polym. 55, 77–93.
Wan Ngah, W.S., Endud, C.S., Mayanar, R., 2002. Removal of copper(II) ions from aqueous solution onto chitosan and cross-linked chitosan beads. React. Funct. Polym. 50, 181–190.

Waters, E.J., Wallace, W., Williams, P.J., 1991. Heat haze characteristics of fractionated wine proteins. Am. J. Enol. Viticult. 42 (2), 123–127.

Wilkins, R.M., 1990. Controlled Delivery of Crop Protection Agents. Taylor and Francis, London.

Wuolijoki, E., Hirvela, T., Ylitalo, P., 1999. Decrease in serum LDL cholesterol with microcrystalline chitosan. Meth. Find. Exp. Clin. Pharmac. 21, 357–361.

Wydro, P., Krajewska, B., Hac-Wydro, K., 2007. Chitosan as a lipid binder: a Langmuir monolayer study of chitosan lipide. Biomacromolecules 8 (8), 2611–2617.

Xihong L., 2012. American Chemical Society (ACS). Good news for banana lovers: Help may be on the way to slow that rapid over-ripening. ScienceDaily. <www.sciencedaily.com/releases/2012/08/120822100813.htm>.

Xu, H., Yan, Y., Li, S., 2011. PDLLA/chondroitin sulfate/chitosan/NGF conduits for peripheral nerve regeneration. Biomaterials 32, 4506–4516.

Yamaguchi, I., Itoh, S., Suzuki, M., Osaka, A., Tanaka, J., 2003. The chitosan prepared from crab tendons: II. the chitosan/apatite composites and their application to nerve regeneration. Biomaterials 24, 3285–3292.

Young, D.Y., Kohle, H., Kauss, H., 1982. Effect of chitosan on membrane permeability of suspension-cultured *Glycine max* and *Phaseolus vulgaris* cells. Plant Physiol. 70, 1449–1454.

Zhang, H., Wang, L., Zheng, X., Dong, Y., 2007. Effect of yeast antagonist in combination with heat treatment on postharvest blue mold decay and *Rhizopus* decay of peaches. Int. J. Food Microbiol. 115, 53–58.

14

ACTIVE FOOD PACKAGING FROM CHITOSAN INCORPORATED WITH PLANT POLYPHENOLS

Ubonrat Siripatrawan
Chulalongkorn University, Department of Food Technology, Faculty of Science, Bangkok, Thailand

1 Introduction

Biobased active packaging is the main focus of current food packaging research and developments. Among various biobased materials, chitosan takes precedence due to its intrinsic antimicrobial and antioxidant properties. Chitosan, a functional biopolymer, is the new generation of active food packaging performing some functions in the preservation of foods. Due to consumers' concerns about food-related health risks and environmental problems, current research in active packaging has focused on biobased packaging materials incorporated with natural active ingredients in order to control undesirable changes of food quality and to provide particular functions that do not exist in conventional packaging (Dutta et al., 2009; Fernandez-Saiz et al., 2009; Leerahawong et al., 2008; Salgado et al., 2011; Siripatrawan and Harte, 2010).

Incorporation of additives or active ingredients into packaging materials to inhibit microbial growth and lipid oxidation, which are the major problems affecting food quality, has become extensive. This approach can reduce the addition of large amounts of food additives that are usually incorporated into food. The most common additives used in food industry include organic acids and their derivatives (ie, benzoic acids, parabens, and propionic acid), fungicides, alcohols, and antibiotics. Although these synthetic additives have been practically used in active food packaging, there are critical concerns related to their toxicological consequences (Gómez-Estaca et al., 2009; Siripatrawan and Noipha, 2012). Therefore, extensive research has pointed in the direction

Novel Approaches of Nanotechnology in Food. http://dx.doi.org/10.1016/B978-0-12-804308-0.00014-5

of employing alternative natural additives or active substances (Jongjareonrak et al., 2008; Schreiber et al., 2013; Siripatrawan and Vitchayakitti, 2014; Yen et al., 2008).

The interest in the development and application of natural antimicrobials and antioxidants to be used as additives in packaging materials has increased apparently due to their safety advantages over synthetic compounds for food applications. Many research studies have demonstrated that plant polyphenols are potential natural alternatives for synthetic antimicrobials and antioxidant agents (Altiok et al., 2010; Bodini et al., 2013; Xi and Shouqin, 2007). Various plant polyphenols have been proposed to be used as active ingredients in biodegradable films, such as green tea (Siripatrawan and Harte, 2010; Wang et al., 2013), yerba mate (Reis et al., 2015), propolis (Bodini et al., 2013; Siripatrawan and Vitchayakitti, 2014), and spices and herbs (Altiok et al., 2010; Jiarpinijnun et al., 2013; Gómez-Estaca et al., 2010; Ojagh et al., 2010a; Martucci et al., 2015; Shen and Kamdem, 2015). Among them, polyphenols from green tea and cinnamon essential oil have proved to act as effective antimicrobials and antioxidants and preferentially be incorporated into biodegradable films (Elsabee and Abdou, 2013; Leerahawong et al., 2008; Siripatrawan and Harte, 2010).

2 Chitosan Films

Chitosan, a deacetylated product of chitin, is a linear polysaccharide of randomly distributed β-(1–4)-linked D-glucosamine (deacetylated unit) and N-acetyl-D-glucosamine (acetylated unit) (Fig. 14.1). Chitosan is considered a biobased environmentally friendly material (Bégin and Van Calsteren, 1999). Chitosan has received extensive attention as active films for food preservation due to its biodegradability, biocompatibility, and antioxidant and antimicrobial activities (Dutta et al., 2009; Fernandez-Saiz et al., 2009).

2.1 Physical Properties of Chitosan Films

Chitosan is of interest as active packaging because of its good film-forming ability, good oxygen and carbon dioxide barrier, and intrinsic antimicrobial and antioxidant properties (Fan et al., 2009; Hernández-Muñoz et al., 2008). Physical and functional properties of chitosan films and coatings rely on numerous factors, such as molecular weight and degree of deacetylation (DD) of chitosan, type and concentration of acid solutions and plasticizers, and environmental factors (eg, pH, temperature) of film preparation

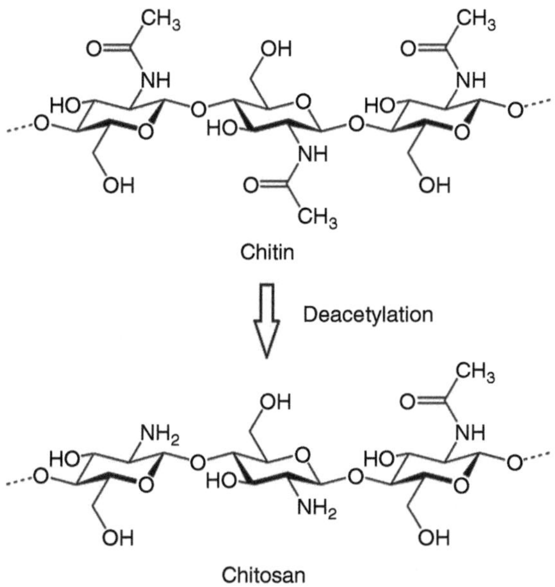

Figure 14.1. Structure of chitin and chitosan.

(Kong et al., 2010). Chitosan films have a selective permeability to gasses (CO_2 and O_2) and good mechanical properties (eg, tensile strength and elongation at break). However, the fact that chitosan films are highly permeable to water vapor limits their use since an effective control of moisture transfer is a desirable property for most foods (Elsabee and Abdou, 2013). Therefore, several strategies have been applied to improve the functional and physical properties of chitosan based films.

Nunthanid et al. (2001) characterized physicochemical properties of chitosan films prepared from chitosan with different molecular weights and degrees of deacetylation. They found that the tensile strength of the chitosan films prepared from chitosan with 90% DD increased with increasing molecular weight of chitosan. Chitosan films prepared from 100% DD chitosan had a higher mechanical strength than those from 80 to 85% DD. The authors explained that the higher mechanical strength of chitosan with higher degree of deacetylation arises from the denser packing of the chitosan polymer chains substituted with small amino groups, leading to higher interchain interactions. However, an increase in the degree of deacetylation of chitosan decreases the elongation at break (the ability of a film to stretch) and moisture absorption of the films. The authors also reported that an increase in molecular weight of chitosan would increase the tensile strength and the elongation as well as the moisture absorption of the films due to chain entanglement network forming in high molecular weight chitosan.

Plasticizers are usually added to polymer to modify certain physical properties. Plasticizers make film more flexible and lower film brittleness by reducing the intermolecular bonds between polymer chains and thus facilitate elongation of the film. However, this sometimes compromises gas and moisture barrier properties of the film. Leerahawong et al. (2007) reported that water vapor permeability of chitosan films increased with the addition of plasticizer due to the structural modifications of the chitosan network. Incorporation of plasticizers increases the free volume and the molecular mobility of the polymer network. The network becomes less dense and, as a consequence, more permeable. Incorporation of hydrophilic plasticizers has been reported to increase the water vapor permeability of hydrocolloid-based films and this can be related to the hydrophilicity of plasticizer molecules (Srinivasa et al., 2007; Talja et al., 2007). For instance, chitosan films plasticized with sorbitol had lower water vapor permeability than those plasticized with glycerol at similar concentration due to the fact that sorbitol has less ability to bind water than glycerol (Leerahawong et al., 2007).

Chitosan films and coatings have been reported to provide some mechanical and barrier protection for foods and, consequently, maintain quality and prolong shelf life of foods. Although chitosan film has low moisture barrier, this deficiency can be improved by incorporating hydrophobic substances. The control of gas exchanges, particularly of oxygen, allows better control of the ripening of fruits or reduces the oxidation of oxygen-sensitive foods and the rancidity of polyunsaturated fats. Moreover, chitosan films can function as carriers for various active substances to provide specific properties, to enhance protective functions, or to improve quality and shelf life of the packaged food and agricultural products.

2.2 Antimicrobial Activity of Chitosan Films

Chitosan has intrinsic antimicrobial activity against a wide variety of microorganisms including bacteria, filamentous fungi, and yeasts. The antimicrobial activity of chitosan has been reported to be due to its positively charged amino group that would interact with negatively charged microbial cell membrane, causing the leakage of proteinaceous and other intracellular constituents of the microorganisms (Dutta et al., 2009; Pereda et al., 2011; Shahidi et al., 1999; Younes et al., 2014).

No et al. (2002) demonstrated that chitosan has stronger antibacterial efficacy on Gram-positive than Gram-negative bacteria. This is probably because for Gram-positive bacteria, the major constituent of their outermost thick cell wall structure is peptidoglycan (up to 90%) and a little amount of protein, whereas the cell wall of Gram-negative bacteria is thinner than that of Gram-positive bacteria, but more complex and contains various polysaccharides, proteins, and lipids, besides peptidoglycan (Nikaido, 2009). An outer lipopolysaccharide membrane lies outside of the thin peptidoglycan layer of Gram-negative bacteria. The outer membrane comprises a double layer of phospholipids that is linked to inner membrane by lipopolysaccharides. The peptidoglycan layer is covered by an outer membrane that contains various proteins and lipopolysaccharides. The peptidoglycan and the outer membrane are tightly linked by Braun's lipoprotein, which is bound by a covalent bond to the peptidoglycan and is embedded in the outer membrane (Fig. 14.2) (Department of Biology, The Pennsylvania State University, 2015).

On the contrary to No et al. (2002); Devlieghere et al. (2004) and Li et al. (2010) have demonstrated that Gram-negative bacteria are

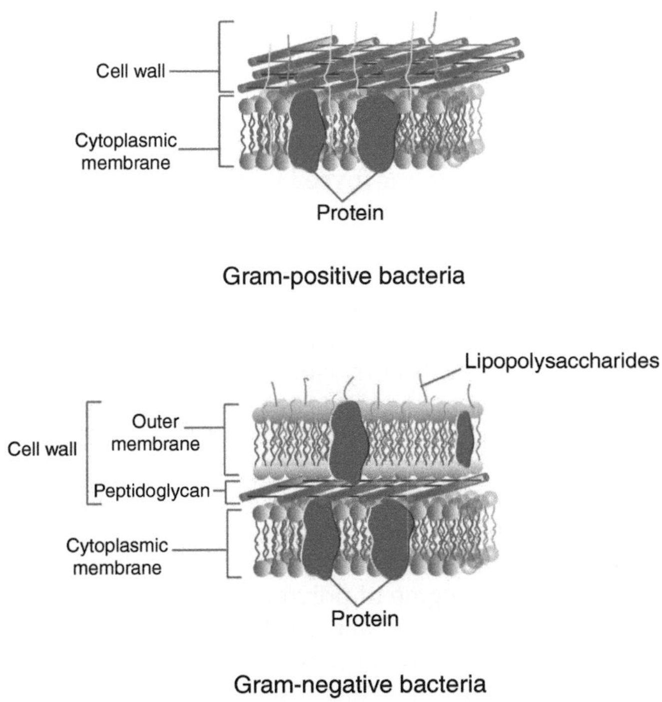

Figure 14.2. Cell membrane of Gram-positive and Gram-negative bacteria.

more sensitive to chitosan. The antibacterial mechanism involves the interactions of negative charge, hydrophilicity of bacterial cell surface, and the adsorption of chitosan onto the bacterial cell. Gram-negative bacteria have higher cell wall hydrophilicity and negative charge of cell surface, and thus have a greater interaction with positively charged chitosan than Gram-positive bacteria. This exposition has been supported by Chung et al. (2004) and Younes et al. (2014) who demonstrated that there is a high value of correlation coefficient between adsorbed chitosan and inhibition efficiency.

Younes et al. (2014) studied antibacterial activities of different degrees of acetylation (DA) and molecular weights of chitosan (0.1% w/v) against Gram-negative bacteria (*Escherichia coli, Pseudomonas aeruginosa, Klebsiella pneumoniae*, and *Salmonella typhi*) and Gram-positive bacteria (*Staphylococcus aureus, Bacillus cereus, Enterococcus faecalis*, and *Micrococcus luteus*), determined by colony counting of viable bacteria (log CFU/mL) after incubation for 24 h at 37°C. It was found that chitosan showed stronger antibacterial effects for Gram-negative than for Gram-positive bacteria. Chitosan samples inhibited the growth of Gram-negative bacteria by 3–6 log cycles for *S. typhi* and by 5–8 log cycles for *E. coli*. For Gram-positive bacteria, bacterial growth was reduced by 1–5 log cycles for *M. luteus* and by 4–6 log cycles for *B. cereus*. These results are in line with other studies (Devlieghere et al., 2004; Li et al., 2010), which showed higher antibacterial activity for Gram-negative bacteria.

For antifungal activities, chitosan has been reported to suppress fungal infection by inhibiting mycelial growth, sporangial production, release of zoospores, and spore germination (Atia et al., 2005; Wojdyła, 2004). Kong et al. (2010) proposed three different mechanisms for the fungal inhibition by chitosan. First, interaction of positively charged chitosan and negatively charged phospholipid components of fungal membrane results in increasing membrane permeability and consequently causes a leakage of cellular contents. Second, chitosan, which acts as a chelating agent by binding to trace elements, causes the essential nutrients unavailable for fungal growth. Last, chitosan can penetrate the cell wall of bacteria and interact with DNA, which may interfere with the production of essential proteins and enzymes.

Chitosan antimicrobial activity varies with chitosan characteristics (eg, average degree of polymerization, average degree of acetylation (DA), and molecular weight), target microorganisms, and conditions of the environmental media (eg, pH and ionic strength). Moreover, the presence of solutes susceptible to electrostatic interaction and/or covalent bonding with chitosan, which block the reaction of the active amino groups also affects the antimicrobial efficacy of chitosan (Aider, 2010; Younes et al., 2014). In addition, the inhibitory effect differs with regard to the acetylation degree of chitosan. It has been demonstrated that for both Gram-negative and Gram-positive bacteria, antibacterial effect strengthens as the acetylation degree decreases. The role of the acetylation degree has been described to support the charge density on antibacterial activity. Chung and Chen (2008) studied the antibacterial activity of chitosan with different degrees of acetylation (5 and 25% DA) against *E. coli* and *S. aureus* and found that 5% DA chitosan has stronger antimicrobial effect than 25% DA chitosan because chitosan with a low DA has higher positive charge density, which confers stronger antibacterial activity against bacteria (Kong et al., 2010). Moreover, the antimicrobial effect is also dependent on the molecular weight of chitosan. For Gram-negative bacteria, chitosan with a lower molecular weight enters the cell through permeation, and antibacterial activity appears to increase with decreasing chitosan molecular weight. However, this trend was not evident for Gram-positive bacteria (Younes et al., 2014). The difference in the inhibitory effect is probably because for Gram-positive bacteria (*S. aureus*) the chitosan adsorbed on the cell surface can form a polymeric membrane which inhibits nutrients from entering the cell (Zheng and Zhu, 2003).

Cao and Sun (2009) and Ganan et al. (2009) demonstrated that antibacterial activity of chitosan increased with decreasing molecular weight. On the contrary, Qin et al. (2006) suggested that antibacterial activity of chitosan increased with increasing molecular weight. However, the study by Chang et al. (2015) revealed that antibacterial activity of chitosan was affected not only by the molecular weight, but also by pH of the mixture. They described that pH of chitosan solution affected the correlation between the antibacterial activity (against *E. coli* and *S. aureus*) and the chitosan molecular weights (300, 156, 72.1, 29.2, 7.1, and 3.3 kDa). For instance, in acidic solution (pH 5.0 and 6.0) the antibacterial activity increased as the molecular weight increased. On the contrary, at pH 7.0 the chitosan having molecular weight higher than 29.2 kDa (ie, 300, 156, and 72.1 kDa) lost their activity, whereas for small chitosan molecular weight (29.2, 7.7, and 3.3 kDa) the antibacterial activity increased with decreasing molecular weight. These pH effects are probably due to the discrepancies in the water solubility and zeta potential (the magnitude of the electrostatic repulsion between particles) of chitosan with various molecular weights when the pH changed from acidic to neutral (Chang et al., 2015). However, Zivanovic et al. (2005) indicated that chitosan with low DA (<15% DA) and molecular weight in a range of 28–1671 kDa showed strong antibacterial effect in aqueous solutions regardless of the type of acid used for solubilization.

Ouattara et al. (2000) reported that highly deacetylated chitosan showed stronger antibacterial activity than chitosan with higher proportion of acetylated amino groups. However, for antifungal activity there is no general trend related to increasing or decreasing molecular weight or DA, but more likely to depend on the specific fungal strains (Younes et al., 2014). Further investigation will be needed to explain these differences which could be based on specific mechanism.

2.3 Antioxidant Activity of Chitosan Films

Chitosan films not only have low oxygen permeability but have also antioxidant effect. Many research studies have documented that the in vitro antioxidant activity of chitosan is related to its scavenging ability on hydroxyl radials and chelating ability on ferrous ions (Park et al., 2004; Yen et al., 2007, 2008). Rhazi et al. (2002) and Schreiber et al. (2013) also suggested that antioxidant activity of chitosan is generally attributed to its chelation efficiency since chitosan, acting as a secondary antioxidant, prevents initiation of lipid oxidation by binding with metal ions. However, some research works have demonstrated that chitosan has low or no antioxidant activity, and the activity may increase significantly with proper chemical modifications of the chitosan polymer (Alexandrova et al., 1999; Casettari et al., 2012; Schreiber et al., 2013; Xie et al., 2001; Xing et al., 2005).

Yen et al. (2008) studied the effects of chitosan prepared from different degrees of deacetylation of crab chitosan, including C60, C90, and C120 having different deacetylation degrees of 83.3, 88.4, and 93.3%, respectively, on the antioxidant activity based on scavenging ability on hydroxyl radicals and chelating ability on ferrous ions. They found that with respect to the antioxidant properties assayed, the efficacy of crab chitosan C60, C90, and C120 correlated with their degree of deacetylation, with chitosan C120 being the most effective and antioxidant activity was in the descending order of chitosan C120 > C90 > C60. It is obvious that chitosan with higher degree of deacetylation has more amino groups to enhance its antioxidant activity. The prolonged deacetylation results in chitosan with more effective antioxidant activity. Park et al. (2004) also reported the antioxidant activity increased with degree of deacetylation. Moreover, due to their low oxygen permeability characteristics, chitosan films and coatings can lower the rate of oxidation of packaged foods simply by preventing their contact with oxygen (Georgantelis et al., 2007; Siripatrawan and Vitchayakitti, 2014).

3 Chitosan Films Incorporated With Plant Polyphenols

The potential of chitosan of natural origin to act as active food packaging to maintain qualities and prolong shelf life of foods has been widely reported. More recently, the addition of natural antimicrobials and antioxidants derived from plant polyphenols for enhancing functional properties and improving physical properties of the chitosan films has become increasingly popular. Along these lines, polyphenols from green tea and cinnamon essential oil have proved to be effective antimicrobials and antioxidants and can be practically incorporated into chitosan-based films.

3.1 Chitosan Films Incorporated With Green Tea

3.1.1 Green Tea

Green tea (*Camellia sinensis*) is a good source of polyphenolic compounds including catechins, theaflavins, and thearubigenes (Gramzaa et al., 2006). Green tea catechins, including (+)-catechin, (−)-epicatechin, (−)-epigallocatechin, (−)-epicatechin gallate, and (−)-epigallocatechin gallate (Fig. 14.3), are proved to exhibit antimicrobial activity against some bacteria and fungi and have good antioxidant activity (Rice-Evans et al., 1997; Rohn et al., 2004).

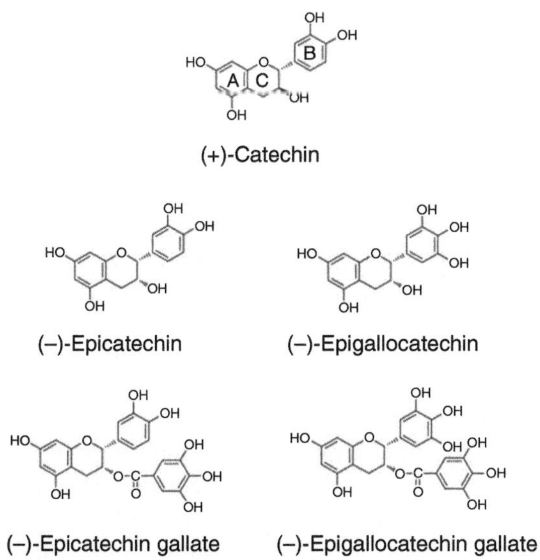

Figure 14.3. Structure of green tea catechins.

Polyphenolic compounds in green tea have been reported to exhibit strong antioxidant activity (Von Gadow et al., 1997). Green tea polyphenols can act as antioxidants by donation of a hydrogen atom and interrupting chain oxidation reactions or by chelating metals (Gramzaa et al., 2006). Among green tea catechins, catechin, and epicatechin have shown to have powerful antioxidant capacities to approximately the same extents (Rice-Evans et al., 1997; Von Gadow et al., 1997). The scavenging ability of epigallocatechin and gallocatechin has been proved to be stronger than those of epicatechin and catechin, suggesting that the presence of a hydroxyl group (at the 5′ position) also plays an important role in their ability to scavenge free radicals. Epigallocatechin gallate and epigallocatechin have been reported as the most effective antioxidant in green tea due to the presence of a gallate group at 3′ position of epigallocatechin gallate, and a hydroxyl group at the 5′ position of epigallocatechin gallate and epigallocatechin (Komes et al., 2010).

Green tea extract has shown inhibitory effect against foodborne pathogens including *Listeria monocytogenes, E. coli* O157:H7, *S. typhimurium, Campylobacter jejuni, S. aureus, Salmonella* Enteriditis, *Shigella flexneri, Shigella dysenteriae,* and *Vibrio cholera* (An et al., 2004; Hamilton-Miller, 1995; Perumalla and Hettiarachchy, 2011). Similarly, Noipha and Siripatrawan (2009) demonstrated inhibitory effect of green tea aqueous extract against *S. aureus, E. coli, Pseudomonas fluorescens,* and *S.* Enteritidis with *S. aureus* being the most susceptible to green tea polyphenols. Their result is in line with those of Hamilton-Miller (1995) who reported tea polyphenols inhibited *E. coli, S. typhimurium, L. monocytogenes, S. aureus,* and *C. jejuni* in foods, Wu et al. (2007) who found epicatechin and caffeine in tea inhibited *S. aureus, Bacillus subtilis,* and *E. coli,* and Cushnie and Lamb (2005) who reported green tea inhibited *S. aureus, Staphylococci,* and *Yersinia enterocolitica.*

Antimicrobial activity of green tea extract has been reported to be attributed to the interaction between green tea polyphenols and phospholipid bilayer and protein in the outer cell membrane of bacteria, leading to the loss of cell structure and function and, consequently, causing the cell death (Perumalla and Hettiarachchy, 2011; Shimamura et al., 2007). Catechins have deteriorating effect on the phospholipid bilayer membrane due to their galloyl and gallic moieties. Among the catechins present in green tea extract, epigallocatechin gallate and epicatechin gallate are the most potent in exhibiting antimicrobial effect due to the galloyl moiety present in their structures (Shimamura et al., 2007). Functional hydroxyl groups and conjugated double bonds in the reactive groups of green tea may be accountable to their binding to the cell wall components (usually proteins) of bacteria (Ikigai et al., 1993; Perumalla and Hettiarachchy, 2011; Yam et al., 1997).

The cytoplasmic membrane of a bacterium is crucially composed of a phospholipid bilayer and proteins and is the major site of interaction with functional hydroxyl groups and conjugated double bonds in the reactive groups of green tea polyphenols. Deteriorative effect to this vital membrane is ascribed to (1) physical disturbance of the cell membrane (Shimamura et al., 2007); (2) depletion of the proton motive force (PMF); and (3) inhibition of membrane-associated enzyme activity (Juven et al., 1994; Perumalla and Hettiarachchy, 2011). Consequently, the cell membrane damage causes the death of bacterial cells (Shimamura et al., 2007). Ikigai et al. (1993) investigated the antimicrobial activity of green tea catechins against Gram-positive and Gram-negative bacteria by using liposomes as model bacterial membranes. They found that catechins appeared to have higher antimicrobial effect against Gram-positive than Gram-negative bacteria. The authors suggested two theories of how catechins cause bacterial cell membrane damage. First, catechins may discompose the phospholipid bilayers of bacterial cell membrane by directly penetrating them and disturbing their barrier function. Alternatively, catechins may cause membrane fusion, a process that results in a leakage of intramembranous materials. The low susceptibility of Gram-negative bacteria to catechins may be partially ascribed to the lipopolysaccharide of bacterial cell membrane acting as a barrier for the bacterial cell.

3.1.2 Interaction of Chitosan and Green Tea Polyphenols

Investigation of interactions between green tea polyphenols and functional groups of chitosan is necessary to explain the effects of the incorporated green tea polyphenols on the mechanical, water vapor barrier, and antioxidant properties of the resulting films. The interactions can be investigated using a Fourier transform infrared (FTIR) spectrometry.

Siripatrawan and Harte (2010) prepared and characterized chitosan films incorporated with green tea. Chitosan film forming solution (2% v/v) was prepared from chitosan with 95% DD in 1% acetic acid. Glycerol, as plasticizer, was added to the film-forming solution at the concentration of 30% w/w of chitosan powder. Green tea aqueous extract was prepared by mixing ground tea powder in distilled water (1:5 w/w) controlled at 90°C. The green tea extract was incorporated into chitosan film forming solution to obtain the final green tea concentrations of 0, 5, 10, and 20% (w/v).

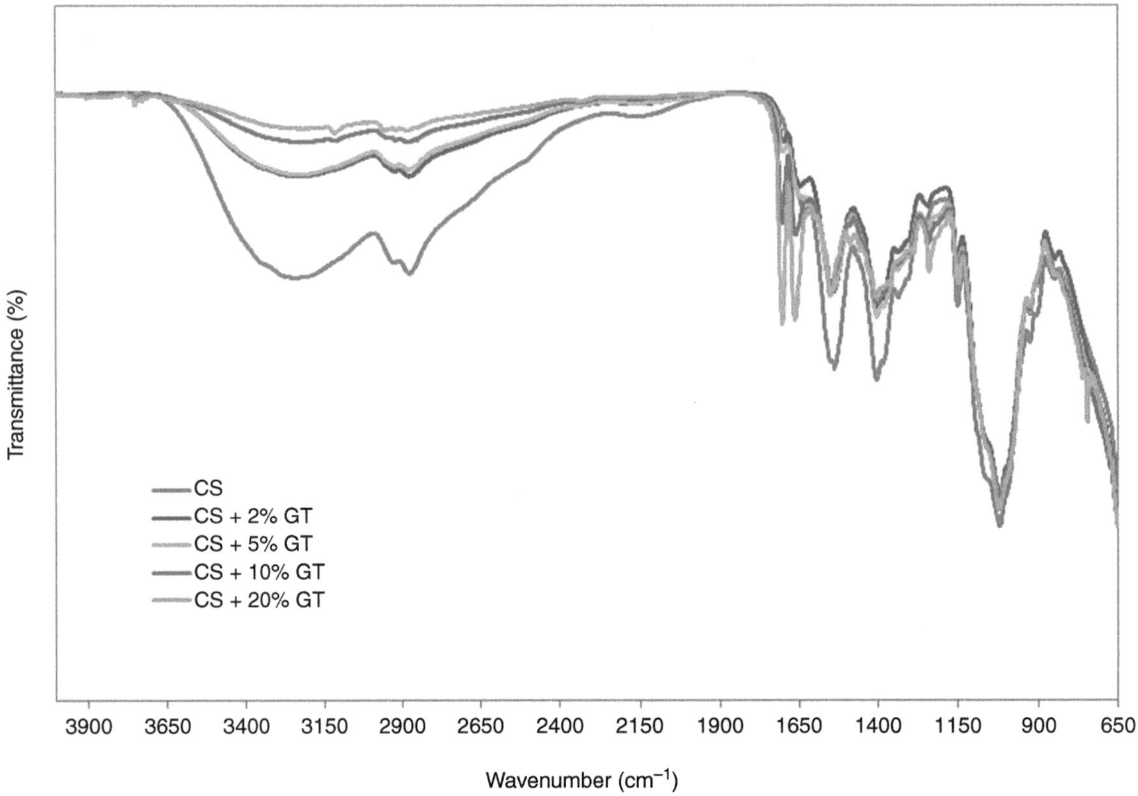

Figure 14.4. FTIR spectra of chitosan films incorporated with green tea extract (CS + GT) at 0, 2, 5, 10, and 20% (w/v) (Siripatrawan and Harte, 2010).

The FTIR spectra of the chitosan films incorporated with 0, 5, 10, and 20% green tea extract were recorded from 4000 to 650 cm^{-1} (Fig. 14.4) using a FTIR spectrometer to investigate the structural interactions of chitosan and green tea polyphenols. The peaks between 3500 and 3000 cm^{-1}, corresponding to stretching vibration of free hydroxyl groups and to asymmetric and symmetric stretching of the N—H bonds in amino group, respectively, are stronger in control chitosan film. Two strong water bands at 1530 and 1400 cm^{-1}, associated to OH in-plane bending, are less noticeable in the films incorporated with green tea extract. These bands become more flattened with increasing green tea extract concentrations. When green tea extract was incorporated into film, new peaks at 1700 cm^{-1}, corresponding to carbon-to-oxygen (C═O) stretching within the carboxylic group, and at 1660 cm^{-1}, correlated to carbon-to-carbon (C═C) stretching within the aromatic ring, which indicate functional groups of phenolic compounds are apparent. These bands are more discernible with increasing green tea extract concentration (Siripatrawan and Harte, 2010).

Changes in the FTIR spectra supported the assumption that there could be a particular arrangement in the films due to the interactions of green tea polyphenolic compounds with hydroxyl and amino groups in chitosan matrix (Abdollahi et al., 2012). This result is in good agreement with the study of Zhang and Kosaraju (2007) who investigated physicochemical interaction between chitosan and catechin and found that amine functional groups of the chitosan decreased when catechin was added.

3.1.3 Physical Properties of Chitosan Films Incorporated With Green Tea

Changes in physical properties of chitosan-based films as a result of plant polyphenols infused in chitosan have been reported by many researchers (Siripatrawan and Harte, 2010; Wang et al., 2013).

Siripatrawan and Harte (2010) studied the effect of green tea extract on the physical properties, including tensile strength, elongation at break, and water vapor permeability coefficient of chitosan film. Tensile strength, which is the strength of a film to withstand against applied tensile stress, and elongation at break, which is the ability of a film to stretch, were measured with an Instron Universal Testing Machine following the ASTM Standard Test Method D 882-91 (ASTM, 2010). The result suggested that green tea extract affected the mechanical resistance and extensibility of the chitosan-based films. Tensile strength and percentage elongation at break increased when green tea extract concentration increased. The improved mechanical properties of the films incorporated with green tea extract may be attributed to the interaction between reactive groups of chitosan matrix and polyphenolic compounds from green tea extract.

Water vapor permeability coefficient of a film is a constant value for permeation of the water vapor at a given temperature. Siripatrawan and Harte (2010) and Wang et al. (2013) reported that incorporation of green tea polyphenols into chitosan film caused the resulting films to become less water vapor permeability possibly because polyphenolic compounds may be able to fit into chitosan matrix and establish interactions such as hydrogen or covalent bonding with reactive groups of chitosan. The hydrogen and covalent interactions between chitosan network and polyphenolic compounds limit the availability of hydrogen groups to form hydrophilic bonding with water, and consequently lead to a decrease in the affinity of chitosan film toward water (Rohn et al., 2004). It is suggested that the barrier characteristic of the film is improved when green tea extract is incorporated.

3.1.4 Antimicrobial Activity of Chitosan Films Incorporated With Green Tea

Green tea has been reported to be able to improve antimicrobial activity of chitosan-based films (Noipha and Siripatrawan, 2009; Wang et al., 2013). Noipha and Siripatrawan (2009) investigated the antimicrobial efficacy of chitosan films containing green tea extract (0, 2, 5, 10, and 20% w/v) against *P. fluorescens*, *S. aureus*, *E. coli*, and *S.* Enteritidis using an agar diffusion method and compared with pure green tea extract of similar concentrations contained in sterilized paper discs. The test bacterial strains were grown in nutrient broth to give a final density of 10^7 CFU/mL. The film discs and paper discs were placed on nutrient agar plates, which had been previously seeded with inoculum containing the target microorganisms (10^7 CFU/mL). The diameter of the inhibitory zone (clear zone) surrounding the film discs was measured. If there was no clear zone surrounding the film disc, it was marked as no inhibitory zone. The contact area of the film disc with the agar surface was also observed to evaluate the growth inhibition underneath the film in direct contact with the target microorganisms on the agar.

The chitosan films without incorporation of green tea extract (control film) showed no inhibitory zone for all tested bacteria, but could inhibit bacterial growth underneath the films (Fig. 14.5). This could be due to the intrinsic antimicrobial characteristic of chitosan (Darmadji and Izumimoto, 1994). However, the antimicrobial effect of chitosan occurred without migration of active agents. As chitosan is in a solid form, therefore only organisms in direct contact with the active sites of chitosan is inhibited. It has been reported that chitosan increases the permeability of the outer and inner membranes and finally disrupts bacterial cell membrane due to the release of cellular contents. This damage is likely caused by the electrostatic interaction between ammonium groups of chitosan and phosphoryl groups of phospholipid components of cell membrane (Liu et al., 2004).

When compared with the antimicrobial effect of green tea extract contained in the sterilized paper discs, the size of inhibition zone formed due to the effect of green tea extract incorporated in the chitosan film was smaller. This is probably because the positive charge of amino group in chitosan film can react with negative charge from tea phenolic acids (eg, epigallocatechin gallate, epicatechin gallate, epigallocatechin, and epicatechin). The green tea polyphenols may be partially trapped in the chitosan network retarding the diffusion of these active compounds on the adjacent culture media. This result is comparable to that of Altiok et al. (2010) who studied the antimicrobial activity of film containing thyme oil and found no clear zone of *E. coli*,

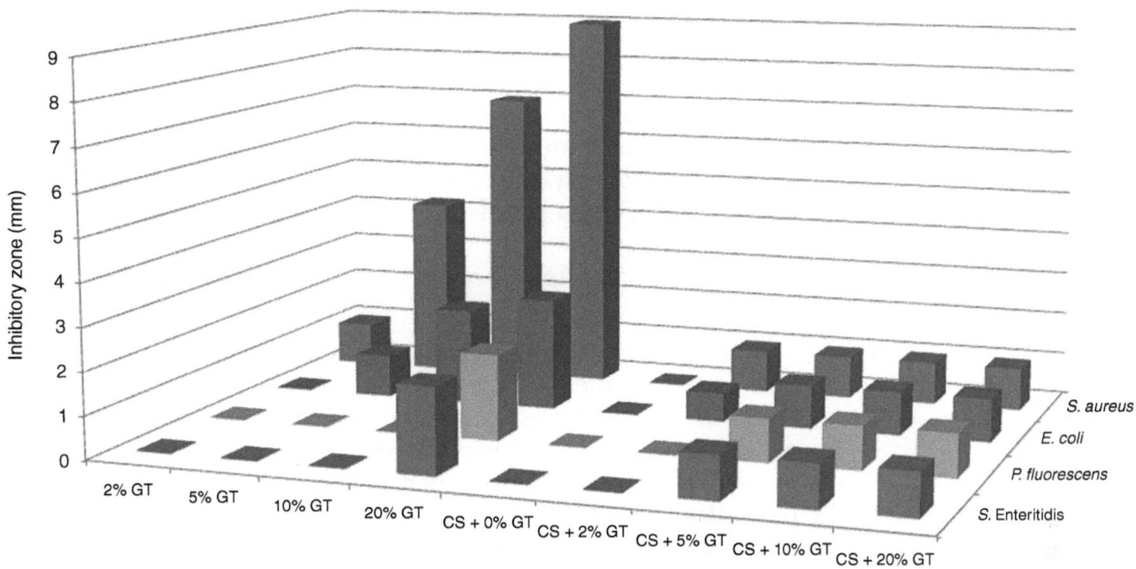

Figure 14.5. Inhibitory zone surrounding paper discs containing green tea extract (GT) at 2, 5, 10, and 20% GT in comparison with chitosan films incorporated with green tea extract (CS + GT) at 0, 2, 5, 10, and 20% (w/v) against *P. fluorescens*, *S. aureus*, *E. coli*, and *S.* Enteritidis. The inhibitory zone is a diameter measured in millimeter, and calculated by subtracting overall clear zone by the diameter of the paper or film disc (Noipha and Siripatrawan, 2009).

P. aeruginosa, and *S. aureus*, but the film can inhibit bacteria underneath the films.

Noipha and Siripatrawan (2009) demonstrated that the antimicrobial activity of chitosan films containing green tea extract is influenced by the microbial strain and the concentration of the green tea extract. Films containing green tea extract had antimicrobial activity against *P. fluorescens*, *S. aureus*, *E. coli*, and *S.* Enteritidis higher than chitosan films without green tea extract and the clear zone was larger with increasing green tea concentration. When compared among the four bacteria studied, the strain *S. aureus* was the most sensitive, showing the largest inhibition diameter in the presence of the green tea extract. The second most sensitive strain was *E. coli*, while *P. fluorescens* and *S.* Enteritidis were inhibited at high green tea concentration (20% w/v). The authors suggested that green tea could inhibit Gram-positive bacteria (*S. aureus*) better than Gram-negative (*E. coli*, *P. fluorescens*, and *S.* Enteritidis) probably because cell wall of Gram-negative bacteria is more complicated than that of Gram-positive bacteria. Gram-negative bacteria, apart from the cell membrane, possess an additional outer layer membrane which consists of phospholipids, proteins, and lipopolysaccharides, and this outer membrane is impermeable to most molecules (Inouye, 1986). This result is in line with those of Pilar Almajano et al. (2008) who studied antimicrobial activity of tea infusions and found that Gram-negative bacteria were more resistant to polyphenols than Gram-positive bacteria. Siripatrawan et al. (2013) also reported that propolis polyphenols were more effective against Gram-positive bacteria (*S. aureus*) than Gram-negative bacteria (*E. coli*, *P. aeruginosa*, and *S.* Enteritidis) probably due to the structural differences of the bacterial cell wall of Gram-positive and Gram-negative bacteria.

3.1.5 Antioxidant Activity of Chitosan Films Incorporated With Green Tea Extract

In the study of Siripatrawan and Harte (2010), the correlation between total phenolic content, expressed as gallic acid equivalents (GAE), and antioxidant activity of chitosan films incorporated with green tea extract was investigated. Folin–Ciocalteu phenol reagent was used to estimate the amount of phenolic groups present in the chitosan films. The total phenolic content of the chitosan films was found to be increased with increasing green tea concentration (Fig. 14.6). The phenolic content can be used as an indicator of the antioxidant capacity, which has been used as a preliminary screen for any substance intended to use as a natural source of antioxidants in functional foods.

The antioxidant activity of the films was evaluated using 2,2-diphenyl-1-picrylhydrazyl (DPPH) free radical scavenging assay. The DPPH scavenging activity of the film increased with increasing green tea extract concentration and correlated with the total phenolic content. Similarly, Wang et al. (2013) found that the incorporation

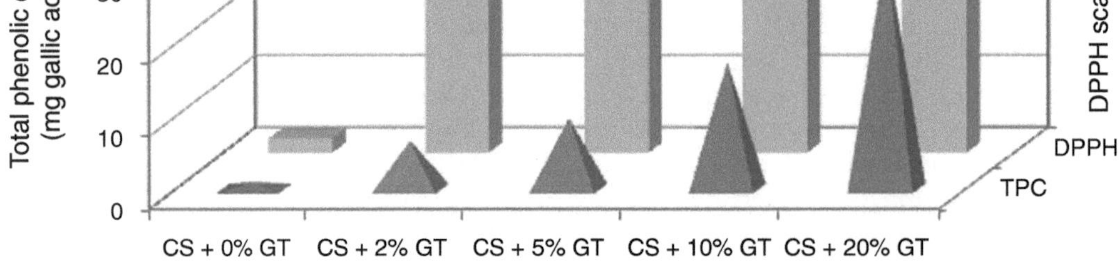

Figure 14.6. Total polyphenolic content (TPC) and DPPH scavenging activity of chitosan films incorporated with green tea extract (CS + GT) at 0, 2, 5, 10, and 20% (w/v) (Siripatrawan and Harte, 2010).

of tea polyphenols enhanced the total phenolic content and increased the antioxidant activity of active chitosan-based films. A high correlation between free radical scavenging and the phenolic contents has also been reported for chitosan film containing propolis extract (Siripatrawan and Vitchayakitti, 2014). Giménez et al. (2013) also indicated a positive correlation between phenolic compounds released from agar and agar–gelatin films with the antioxidant activity measured by two spectrophotometric methods including the ferric reducing antioxidant power (FRAP) method and the 2,2'-azinobis-3-ethylbenzotiazoline-6-sulfonic acid (ABTS) method.

Siripatrawan and Harte (2010) documented that chitosan films without green tea extract exhibited a scavenging activity on DPPH. The scavenging ability of the chitosan films has been reported to be contributed to the presence of the amino groups in the chitosan polymer which can react with free radicals to form stable macromolecule radicals. (Park et al., 2004; Yen et al., 2008). Schreiber et al. (2013) suggested that the antioxidant activity of chitosan is generally attributed to its metal chelating efficiency. The amino groups in chitosan molecules induce the ability of chitosan, acting as a secondary antioxidant, to chelate metal and prevent initiation of lipid oxidation.

Phenolic compounds, classified as primary antioxidants, can scavenge free radicals by donating a hydrogen atom to free radicals, and the resulting phenoxyl radicals are stabilized by delocalization of unpaired electrons around the aromatic ring, leading to the inhibition of lipid peroxidation (Schreiber et al., 2013; Sforcin and Bankova, 2011). When compared to chitosan-alone film, the chitosan film incorporated with green tea extract exhibited higher antioxidant activity and the activity increased with green tea concentration. Therefore, incorporating phenolic compounds to chitosan films appears to be an efficient way of introducing the primary antioxidant capability to the chitosan biopolymer.

3.2 Chitosan Films Incorporated With Cinnamon Essential Oil

Essential oils are aromatic oily liquid extracted from plant materials and commonly utilized as flavoring in foodstuffs. Various essential oils from spices and herbs, such as rosemary, thyme, clove, and cinnamon, have been reported to possess antioxidant and antimicrobial activities (Perdones et al., 2014). Their antimicrobial activities against various pathogenic microorganisms involved in foodborne illness have been demonstrated. However, different results are observed to be depending on test conditions, microorganisms, and the sources of the antimicrobial compounds.

The composition, structure, as well as functional groups of essential oils (eg, phenolic groups) in spices and herbs play important roles in determining their antioxidant and antimicrobial activities (Alboofetileh et al., 2014; Oriani et al., 2014).

The food industry is considering essential oils as potential preservatives of natural origin. Nevertheless, their incorporation into foods is often limited by flavor considerations since effective antimicrobial doses may exceed consumers' organoleptic acceptance (Acevedo-Fani et al., 2015; Elsabee and Abdou, 2013). The addition of essential oils into films may have supplementary applications (eg, antimicrobial and antioxidant activities) to food packaging. Incorporation of plant essential oils into a food packaging material as a form of active packaging, instead of directly adding into foods, may be an alternative way to preserve quality of the packaged foods. Varieties of spices and herbs and their essential oils have been widely studied for their antioxidant and antimicrobial properties. Among them, cinnamon has proved to exhibit effective antimicrobial and antioxidant activities.

3.2.1 Cinnamon Essential Oil

Cinnamon (*Cinnamomum zeylanicum* and *Cinnamon cassia*) is one of the most important spices used in various food products. Cinnamon primarily contains essential oils and other derivatives, such as cinnamaldehyde, eugenol, cinnamic acid, and cinnamate. The main compound of cinnamon essential oil is cinnamaldehyde (Fig. 14.7). Cinnamaldehyde (*trans*-3-phenyl-2-propenal) or cinnamic aldehyde occurs naturally in the bark of cinnamon trees and other species of the genus *Cinnamomum*. The essential oil of cinnamon bark contains about 90% cinnamaldehyde. Cinnamaldehyde is the chemical compound that gives a specific spicy aroma and flavor of cinnamon. Besides cinnamaldehyde, cinnamon oil contains a great variety of chemical compounds (eg, cinnamic acid, coumaric acid, cinnamyl alcohol, and eugenol) that present antioxidant and antimicrobial activities. Cinnamon essential oil has shown not only antifungal and antibacterial properties against a broad spectrum of food spoilage and pathogenic microorganisms but also antioxidant activity (Jiarpinijnun et al., 2013; Leerahawong et al., 2008; Perdones et al., 2014). Moreover, it has

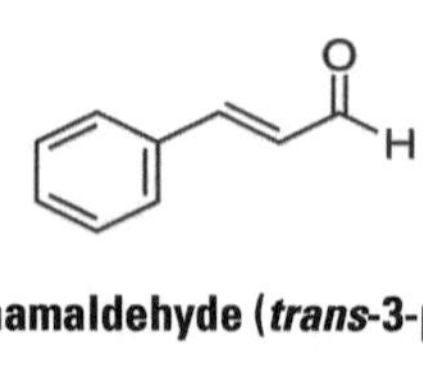

Figure 14.7. Structure of cinnamaldehyde (*trans*-3-phenyl-2-propenal).

antiinflammatory, antioxidant, antimicrobial, antidiabetic, antitumor, antimutagenic, memory enhancer, and many other activities (Mathew and Abraham, 2006).

Mathew and Abraham (2006) studied the antioxidant activity of essential oil from cinnamon (*Cinnamomum verum*) bark extract. The cinnamon bark extract yields an essential oil containing cinnamaldehyde and eugenol. The evaluation of antioxidant activities of the methanolic extract of the cinnamon bark revealed that cinnamon bark extract, due to its hydrogen donating ability, possesses antioxidant activity, which is concentration dependent. Cinnamon bark extract exhibits reducing power and is found to be potent in free radical scavenging activity against DPPH radicals and ABTS radical cations. Moreover, although cinnamon bark extract is weak chelators of metal ions, its free radical scavenging capacity is comparable to that of synthetic antioxidants such as butylated hydroxyanisole (BHA).

Cinnamaldehyde has been reported to be among the most active components against Gram-positive and Gram-negative bacteria. Sanla-Ead et al. (2012) studied antimicrobial activity of cinnamaldehyde (50 μL/mL) using a modified agar-well diffusion. The inhibition zone of cinnamaldehyde was observed with descendent from *L. monocytogenes*, *S. aureus*, *E. faecalis*, and *B. cereus*, respectively, for Gram-positive bacteria, and *E. coli*, *S.* Enteritidis, *E. coli* O157:H7, and *P. aeruginosa*, respectively for Gram-negative bacteria. Zhang et al. (2016) reported that cinnamon essential oil exhibited effective antibacterial activity against foodborne pathogenic bacteria in model systems using *E. coli* and *S. aureus*. The minimum inhibition concentration (MIC) of cinnamon oil was similar (1.0 mg/mL) for both bacteria, while the minimum bactericide concentration (MBC) was 4.0 mg/mL for *E. coli* and 2.0 mg/mL for *S. aureus*.

The mechanism of antimicrobial effect of cinnamon oil has been demonstrated by Zhang et al. (2016) that cinnamon essential oil impairs the permeability and integrity of cell membrane, and alters morphology of bacterial cell membrane, as observed using a scanning electron microscope (SEM). As described by Diao et al. (2014), the bacterial cytoplasmic membrane provides a barrier to the access of small ions such as K^+, Na^+, and H^+. The membrane impermeability to small ions is maintained and managed by structure and chemical compositions of the membrane. Maintenance of ion homeostasis is crucial for bacterial cell membrane functions to maintain the energy status of the cell for solute transport, control enzyme activity and normal metabolism, and manage turgor pressure and motility. Small variations to the membrane structure can lead to the loss of inner cell materials, the disruption of cell metabolism, and, consequently, the bacterial cell death. This assumption was evaluated by measuring the relative electrical conductivity of *E. coli* and *S. aureus*, which is correlated to the absorbance (at 260 nm) of nucleic acids released through the malfunctioning membrane. The relative electrical conductivity is found to be increased with increasing concentration of cinnamon essential oil, suggesting that there is a leakage of electrolytes, due to disruption of cell permeability in respond to the addition of cinnamon essential oil. Consequently, loss of electrolytes causes the death of *E. coli* and *S. aureus* (Zhang et al., 2016).

Zhang et al. (2016) also found that *E. coli* exerted more resistant to cinnamon oil than *S. aureus*, probably due to structural differences in the outer membrane of bacteria. Gram-negative bacteria have a layer of outer lipopolysaccharide membrane covering the cell wall. This cell membrane structure of Gram-negative bacteria has shown to be more resistant to the hydrophobic substance of essential oils when compared with the Gram-positive *S. aureus*, which possess single peptidoglycan layer structure (Hsouna et al., 2011; Zhang et al., 2016). Literature has reported that essential oils of cinnamon and other spices (eg, thyme, origanum, mint, salvia, and clove) can affect bacterial cell membrane structure, as the antibacterial compounds in essential oils can penetrate through cell wall and destroy the cytoplasmic membrane, and consequently cause the cell death (Weerakody et al., 2010; Zhang et al., 2016).

3.2.2 Interaction of Chitosan and Cinnamon Essential Oil

When cinnamon essential oil was incorporated into chitosan, the interaction between functional groups of chitosan and cinnamon bark oil can be observed using a FTIR spectrometer (Jiarpinijnun et al., 2013). The spectra of chitosan film and chitosan film incorporated with cinnamon oil (100 μL/g of chitosan) were compared (Fig. 14.8). Both spectra show similar pattern with the peak around 3500 cm^{-1}. Absorption in this area indicates stretching vibration of hydroxyl group (OH) and indicates intermolecular hydrogen bonding of chitosan molecules (Leerahawong et al., 2008). They are also overlapped in the same region of an NH stretching (Nunthanid et al., 2001). The carbonyl (C=O), amine (NH_2), and ammonium (NH_3^+) bands are situated in the region between 1700 and 1400 cm^{-1}. A change in the band at 1649 cm^{-1} is evident, indicating some interactions between amine group of chitosan and carbonyl group of cinnamon oil. In addition, imine (C=N) absorption band around 1600 cm^{-1} appeared when cinnamon oil was incorporated. The spectral

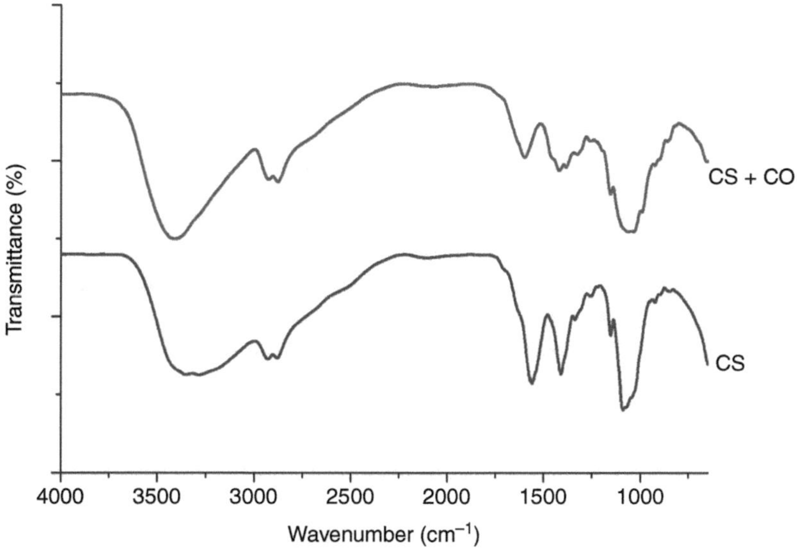

Figure 14.8. FTIR spectra of control chitosan film (CS) and chitosan film incorporated with cinnamon oil (CS + CO) at 100 μL/g.

results indicated that the possible interaction between functional groups of chitosan and cinnamon oil may be due to imine, or Schiff base, formation. An increase in the absorption bands at 1700 and 1660 cm^{-1} in the FTIR spectra is coincidental with the decrease in peaks at 3500–3000, 1530, and 1440 cm^{-1}. Moreover, the bands at 2830–2695 cm^{-1}, associated to aldehyde (H—C═O) stretching, at 1500–1400 cm^{-1}, ascribed to C—C stretching, and at 900–629 cm^{-1}, correlated to C—H stretching within aromatic ring, are observed for film containing cinnamon oil.

The FTIR spectra suggested that polyphenols from cinnamon oil can form hydrogen bonding and covalent bonding with the functional groups of chitosan matrix, and consequently lower the free hydroxyl groups, which can form hydrophilic bonding with water. As a result, improved mechanical and water barrier properties and enhanced antioxidant activity are evident in the chitosan films incorporated with cinnamon oil (Jiarpinijnun et al., 2013). Azevedo and Kumar (2012) demonstrated that cinnamaldehyde (the main component of the cinnamon bark oil) can react with the amino groups of the chitosan structure, inducing the formation of a Schiff base. The Schiff base subsequently induces crosslinking of the chitosan polymer. In addition, other authors also reported that cinnamon oil can act as a crosslinking agent with chitosan (Ojagh et al., 2010a; Higueras et al., 2015; López-Mata et al., 2015).

3.2.3 Physical Properties of Chitosan Films Incorporated With Cinnamon Essential Oil

Jiarpinijnun et al. (2013) studied the effects of cinnamon bark oil with different concentrations (0, 50, 100, and 150 μL/g chitosan) on the physical and functional properties of chitosan films. The chitosan film-forming solution (1% w/v) was prepared by dissolving chitosan powder (95% DD) in 1% acetic acid and plasticized with glycerol at 30% w/w of chitosan powder. The chitosan solution was casted, dried, and conditioned at 25°C and 50% RH. The result suggested that when the concentration of cinnamon bark oil increased, tensile strength of the film increased but elongation at break decreased (Fig. 14.9). Tensile strength and elongation at break are usually related to the film network microstructure and the intermolecular force (Atarés et al., 2010). The changes of both tensile strength and elongation at break of the films may be caused by the interchain interactions between functional groups of chitosan and some compounds in cinnamon bark oil (Jiarpinijnun et al., 2013). Shen and Kamdem (2015) found that the addition of different essential oils including citronella essential oil and cedarwood oil in the considered concentration range (10, 20, and 30% w/w) significantly affected the mechanical properties of the chitosan-based films. Their finding suggested that an increase in the concentration of the citronella essential oil significantly increased the elongation at break. However, the elongation at break showed an opposite trend in chitosan films containing 30% citronella essential oil, exhibiting a reduction of 36% compared to the chitosan films without citronella essential oil. Their result is in agreement with that of Ojagh et al. (2010a) who reported that the tensile strength increased when introducing cinnamon essential oil into chitosan films, but the elongation decreased. Poster et al. (2010) reported that the reduction of elongation at break of hydroxypropylmethylcellulose films when incorporated with propolis extract may be attributed to the crystalline formation of flavonoids and phenolic acid or their esters, which reduces the film flexibility.

The mechanical properties obtained for chitosan films incorporated with cinnamon essential oil (Jiarpinijnun et al., 2013) were different from those found by some investigators (Leerahawong et al., 2008; Shen and Kamdem, 2015) possibly due to the difference in reactivity in binding or interacting with the polymer network. Leerahawong et al. (2008) reported that both tensile strength and elongation at break increased when the chitosan film was incorporated with cinnamaldehyde. On the other hand, Sánchez-González et al. (2010) reported that the tensile strength and elongation at break of chitosan film decreased with the

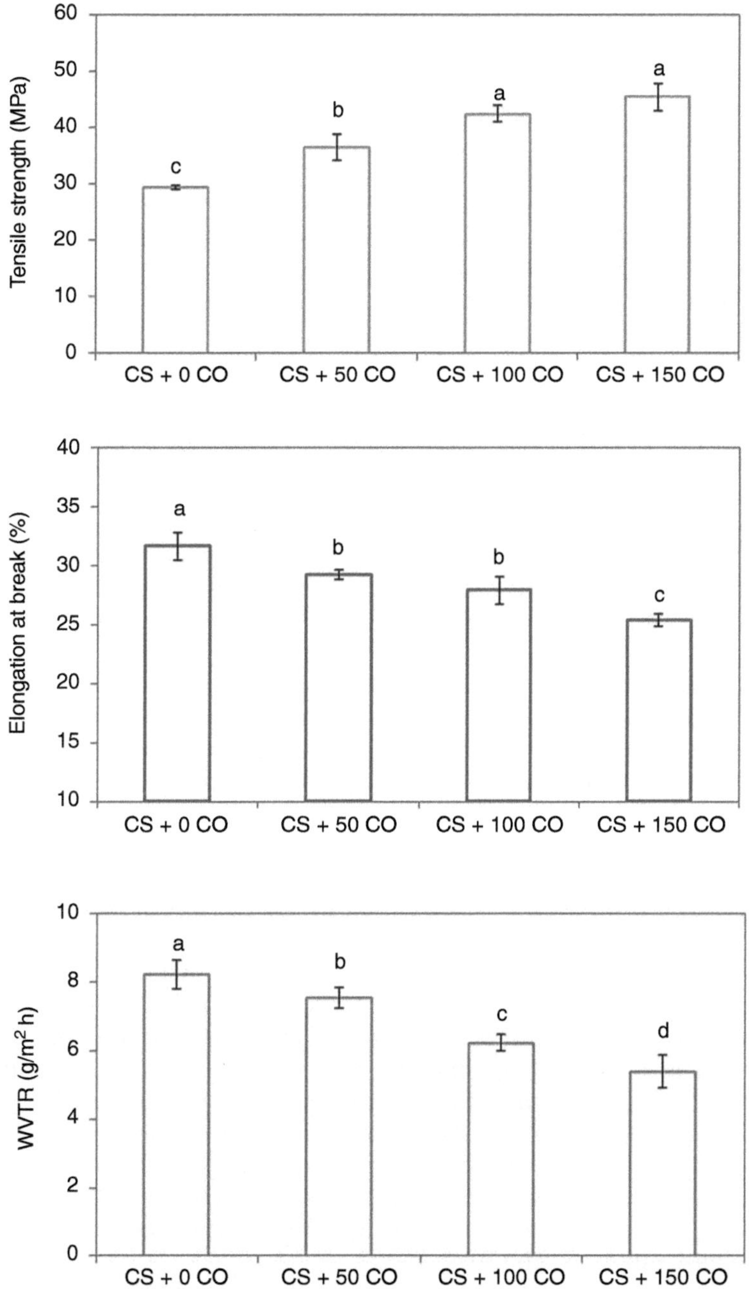

Figure 14.9. Tensile strength, elongation at break, and water vapor transmission rate (WVTR) of chitosan films incorporated with cinnamon oil (CS + CO) at 0, 50, 100, and 150 μL/g chitosan. Values are given as mean ± standard deviation. Different letters indicate significantly different values ($p \leq 0.05$) (Jiarpinijnun et al., 2013).

addition of bergamot essential oil. Peng and Li (2014) and Shen and Kamdem (2015) suggested that different essential oils affected the mechanical properties of the films differently because the interactions in the matrix are varied with types of essential oils. Moreover, chitosan compositions, plasticizers, emulsifiers, and experimental conditions (eg, pH, concentration, and relative humidity used during equilibration of the film) also affect the mechanical properties of the films (Leerahawong et al., 2008; Peng and Li, 2014; Shen and Kamdem, 2015).

In the study of Jiarpinijnun et al. (2013), the effect of different cinnamon bark oil concentrations (0, 50, 100, and 150 μL/g chitosan) on water vapor permeability of the films was investigated. The result (Fig. 14.9) indicated that increasing the concentration of cinnamon bark oil significantly decreased the water vapor transmission rate of the film due to the hydrophobic properties of cinnamon oil and an increase in concentration of cinnamon oil lessened the hydrophilic-hydrophobic ratio, and consequently reduced water vapor permeability of the chitosan film. The water vapor transfer generally occurs through the hydrophilic portion of the film and depends on the hydrophilic-hydrophobic ratio of the film components. Essential oils, although being complex mixtures, are highly hydrophobic and the increase in hydrophobicity of the film matrix should lower the water absorption. Similar result was also observed by López-Mata et al. (2015) who investigated the effect of cinnamon oil on the water vapor permeability of chitosan film and found that the permeability decreased when cinnamon oil was added. This result is in good agreement with those of Moradi et al. (2012) who found that adding grape seed oil improved moisture barrier property of chitosan film and Siripatrawan and Vitchayakitti (2014) who observed an improved moisture barrier of chitosan film when propolis (a chemically complex resinous bee product consisting of waxes, essential oils, and various organic compounds) was added. Leerahawong et al. (2008) reported that water vapor permeability decreased significantly upon addition of cinnamaldehyde (the main component of cinnamon oil). The water vapor permeability decreased with increasing cinnamaldehyde concentration. Incorporation of cinnamaldehyde also showed the same effect on water vapor permeability of alginate–apple puree edible films (Rojas-Graü et al., 2007).

Many studies have demonstrated that an addition of cinnamon oil can improve mechanical and barrier properties of chitosan films because there is a structural modification of the chitosan network caused by the interactions between the functional groups of chitosan and the cinnamon oil polyphenols. Tensile strength and moisture barrier of the chitosan films improved, whereas elongation at break may increase or decrease upon the incorporation of cinnamon oil. The hydrophobic nature of cinnamon oil can make chitosan films a better moisture barrier against humidity, and in turn the chitosan matrix can preserve the properties of cinnamon oil, which improves the physical and antioxidant characteristics of chitosan films (López-Mata et al., 2015; Senanayake and Wijesekera, 2004). However, this incorporation has disadvantage in terms of surface color and transparency. The films become more yellowish and more opaque with addition of cinnamon oil (Jiarpinijnun et al., 2013; Leerahawong et al., 2008).

3.2.4 Antimicrobial Activity of Chitosan Films Incorporated With Cinnamon Essential Oil

Incorporation of cinnamon essential oil into chitosan films has been reported to enhance antimicrobial activity of the films (Jiarpinijnun et al., 2013; López-Mata et al., 2015; Ojagh et al., 2010b). Chitosan (85% DD) films containing different concentrations of cinnamaldehyde (50, 100, and 150 μL/g) were tested for their antimicrobial activity against the selected Gram-positive bacteria (*S. aureus* and *Bacillus licheniformis*) and Gram-negative bacteria (*E. coli* and *Shewanella putrefaciens*) using an agar diffusion method (Leerahawong et al., 2008). The film discs (8.6 mm diameter) were placed on the solid media of tryptic soy agar which was previously inoculated with the selected bacterial cultures set at c. 10^5–10^6 CFU/mL and incubated at 30°C for 24 h. The Petri dishes were examined for inhibitory zone or clear zone. The inhibitory zone is a diameter measured in millimeter, and calculated by subtracting overall clear zone by the diameter of the film disc (8.6 mm). No growth underneath the film disc indicated inhibitory effect on the contact surface. The test was also determined using sterilized paper discs (with similar diameter to the film disc) containing the same concentrations of cinnamaldehyde as those incorporated into the chitosan films.

The result demonstrated that chitosan films (control films) without cinnamaldehyde did not show inhibition against the tested bacteria (Fig. 14.10). Although chitosan has been reported to have intrinsic antimicrobial activities against both Gram-positive and Gram-negative bacteria, it is possible that when chitosan is in a form of insoluble films the antimicrobial activity may become negligible due to lack of interactions between functional groups in chitosan network and bacterial cell wall. Torlak and Sert (2013) also reported that chitosan films did not exhibit antibacterial activity by an agar diffusion method because chitosan

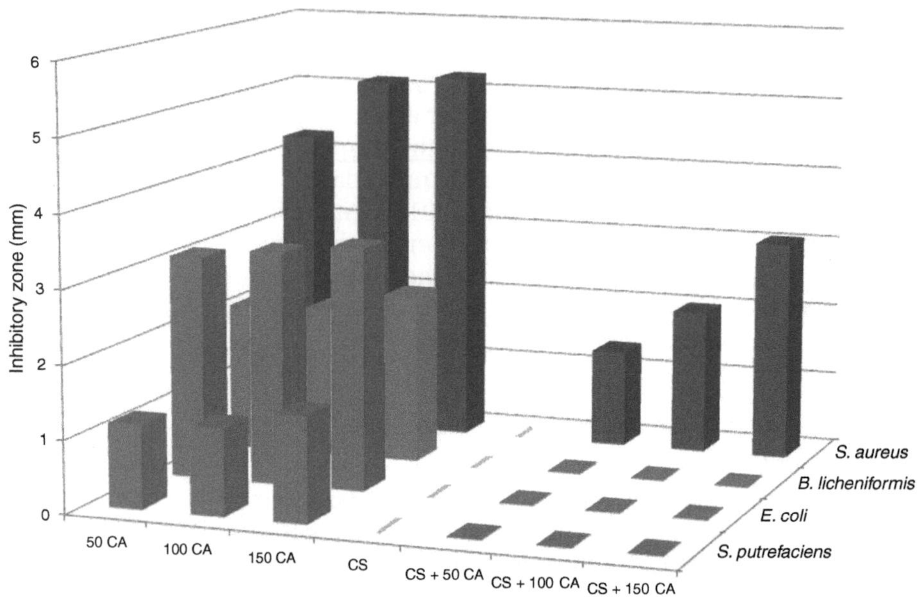

Figure 14.10. Inhibitory zone surrounding paper discs containing cinnamaldehyde (CA) at 50, 100, 150 μL/g in comparison with chitosan film (CS, control) and chitosan films incorporated with cinnamaldehyde at 50, 100, 150 μL/g chitosan against Gram-positive bacteria (*S. aureus* and *B. licheniformis*) and Gram-negative bacteria (*E. coli* and *S. putrefaciens*). The inhibitory zone is a diameter measured in millimeter, and calculated by subtracting overall clear zone by the diameter of the paper or film disc (Leerahawong et al., 2008).

in a film form is unable to diffuse through the adjacent agar media. Conversely, another research reported that chitosan film exhibit antibacterial activity when tested against *P. fluorescens*, *S. aureus*, *E. coli*, and *S.* Enteritidis using agar diffusion method (Noipha and Siripatrawan, 2009). It has been demonstrated that antibacterial activity of chitosan in solid state is affected by various factors, including chitosan characteristics, target microorganisms, and environmental media (Kong et al., 2010; Younes et al., 2014).

When cinnamaldehyde was incorporated into the film, inhibitory effect on contact surface underneath the chitosan films was observed against all tested bacteria with markedly against *S. aureus* probably because cinnamaldehyde could act on the cytoplasmic membrane by impairing the integrity of membrane and consequently causing the death of bacteria (Zhang et al., 2016). However, when compared with paper discs containing cinnamaldehyde at the same concentrations, the chitosan films incorporated with cinnamaldehyde showed lower inhibitory effect probably because cinnamaldehyde could not properly release from chitosan matrix due to the interaction between cinnamaldehyde and functional groups of chitosan. This effect was also observed in chitosan films containing green tea polyphenols (Noipha and Siripatrawan, 2009). Incorporation of cinnamaldehyde into chitosan edible films improves antimicrobial efficacy of the films especially on the contact surface of the films. Therefore, chitosan films may be used as antimicrobial wrapping films, coatings or inner layers of multilayer packaging films for food products to inhibit microbial growth on the contact surface.

3.2.5 Antioxidant Activity of Chitosan Films Incorporated With Cinnamon Essential Oil

In the study of Jiarpinijnun et al. (2013), the effect of cinnamon bark oil with different concentrations (0, 50, 100, and 150 μL/g chitosan) on the antioxidant activity of chitosan (95% DD) film was investigated using DPPH method. The result suggested that the antioxidant activity of the chitosan films increased with increasing cinnamon oil concentration. When compared with chitosan films containing cinnamon oil, control films without cinnamon oil exhibited minimal DPPH radical scavenging activity and, thus, have potential as a free radical scavenger. Other studies also reported that chitosan possesses DPPH radical scavenging activity and its activity depends on the degree of deacetylation of chitosan (Moradi et al., 2012; Ruiz-Navajas et al., 2013). Chitosan with higher degree of deacetylation shows higher radical scavenging effect on free radicals (Park et al., 2004; Xie et al., 2001).

As reported by Jiarpinijnun et al. (2013), an addition of cinnamon oil enhanced the antioxidant activity of chitosan films. Cinnamon oil is considered a good source of various phenolic compounds and flavonoids. Phenolic compounds are considered the most active metabolites in plants. The phenolic compounds are capable of donating hydrogen or electrons to form stable radical intermediates. Therefore, there is a close correlation between the content of phenolic compounds and antioxidant activity (Moradi et al., 2012).

López-Mata et al. (2015) investigated the physicochemical and antioxidant properties of chitosan films incorporated with cinnamon oil, prepared by an emulsion method. The chitosan films were prepared by dissolving chitosan (34% DA with an average molecular weight of 128 kDa) in acetic acid (1% v/v) solution and glycerol was added as plasticizer. Films with different concentrations (0, 0.25, 0.5, and 1.0% v/v) of cinnamon oil were prepared and investigated. The antioxidant activity of the films was measured using DPPH and ABTS. A significant increase in the antioxidant capacity of the films containing cinnamon oil at 0.5% (6.01 times) and 1.0% (14.5 times) was observed, when compared to the control. This activity may be attributed to the presence of cinnamon oil antioxidant, and may also be due to a synergetic effect among chitosan, eugenol, and cinnamaldehyde (Mathew and Abraham, 2006).

4 Active Chitosan Films and Coatings for Food Shelf Life Extension

Biobased active films and coatings have potential to meet consumer demand for safe and high quality foods without chemical preservatives. Chitosan films incorporated with plant polyphenols have been proved to be able to provide effective gas barrier, mechanical strength, and competent functional properties beneficial to maintain quality and prolong shelf life of fruits and vegetables (Suseno et al., 2014; Vu et al., 2011), processed meat products (Siripatrawan and Noipha, 2012; Yingyuad et al., 2006), and fish and seafood (Ojagh et al., 2010b; Yuan et al., 2016).

4.1 Application to Meat Products

Yingyuad et al. (2006) applied chitosan coatings to maintain quality and extend shelf life of vacuum packaged grilled pork stored at refrigerated temperature (2°C). Chitosan dipping solutions were prepared by dissolving chitosan (87% DD) in aqueous solutions of 1% w/v acetic acid to a final concentration of 2.0 and 2.5% w/v. Grilled pork samples prepared from boneless pork loins were dipped into 0% (control), 2.0%, and 2.5% chitosan solutions for 1 min and air-dried. Each pork sample was individually packaged in a 10 × 15 cm pouch of 60 μm polyvinylidene chloride coated with polyamide (PVDC/nylon) having oxygen permeation of 7 cm^3/m^2 per atm per day at 25°C, and water vapor transmission rate of 4 g/m^2 per day at 25°C and heat-sealed under vacuum condition.

The result showed that lipid oxidation, expressed as peroxide values, in grilled pork could be minimized by the use of chitosan coating probably due to its low oxygen permeability characteristic and antioxidant activity. The antioxidant efficacy of chitosan, acting as a secondary antioxidant, is due to its metal chelating activity. When compared with control samples, samples coated with chitosan showed a slow microbial growth and pH changes throughout the entire storage period probably because chitosan could inhibit the growth of lactic acid and other bacteria which are predominant on cooked and cured meat products (Samelis et al., 2000). Changes in color of chitosan coated grilled pork were lower than those of uncoated samples. Since lipid oxidation and microbial growth are important factors causing the discoloration of the products, inhibiting this factor would be beneficial in delaying product deterioration. The antimicrobial effect of chitosan is thought to be related to the electrostatic interaction between positively charged chitosan molecules and negatively charged microbial cell membranes, causing the leakage of intracellular constituents of bacteria (Xu et al., 2005). This result is in close agreement with that of Darmadji and Izumimoto (1994) who reported a reduction of microbial counts by an average of 2 log CFU/g in beef patties coated with 1.0% chitosan and stored at 4°C. Their result is also supported by the study of Devlieghere et al. (2004) who found that chitosan coatings can act as an oxygen barrier and can inhibit the growth of aerobic bacteria. However, the authors reported that chitosan coating did not protect the samples against moisture loss probably because although chitosan film is highly impermeable to oxygen, it has poor water vapor barrier property (Xu et al., 2005). The poor moisture barrier property of chitosan coating may be attributed to the large volume of NH_2 groups in chitosan matrix having strong affinity to water molecules (Wu et al., 2004).

Sensory attributes of grilled pork with and without chitosan coating were evaluated by an experienced 10-member panel, using a 5-point scale to determine color, odor, and overall acceptability of the samples. Shelf life criteria assumed that rejection would occur when the sensory attributes declined below 3.0. The sensory evaluation result (Fig. 14.11) appeared to be correlated to color index, peroxide values, and microbiological analyses. Samples without chitosan coating were unacceptable after 14 days, while the samples coated with chitosan maintained acceptable sensory qualities throughout the storage period up to 28 days. However, increasing the concentration of chitosan from 2.0 to 2.5% did not enhance the beneficial effects. Chitosan coating delayed changes in sensory characteristics and allowed longer shelf life of grilled pork by minimizing lipid oxidation and inhibiting microbial growth.

Siripatrawan and Noipha (2012) investigated the efficacy of chitosan film incorporated with green tea for shelf life extension of cooked pork sausages. The chitosan film was prepared by dissolving chitosan powder (2% w/v) with a molecular weight of 100 kDa and 95% DD into 1% acetic acid solution and glycerol was used as plasticizer. Green tea aqueous extract was prepared by mixing ground tea in distilled water (1:5 w/w) at 90°C and was then mixed with chitosan solution to obtain the final concentration of 20%

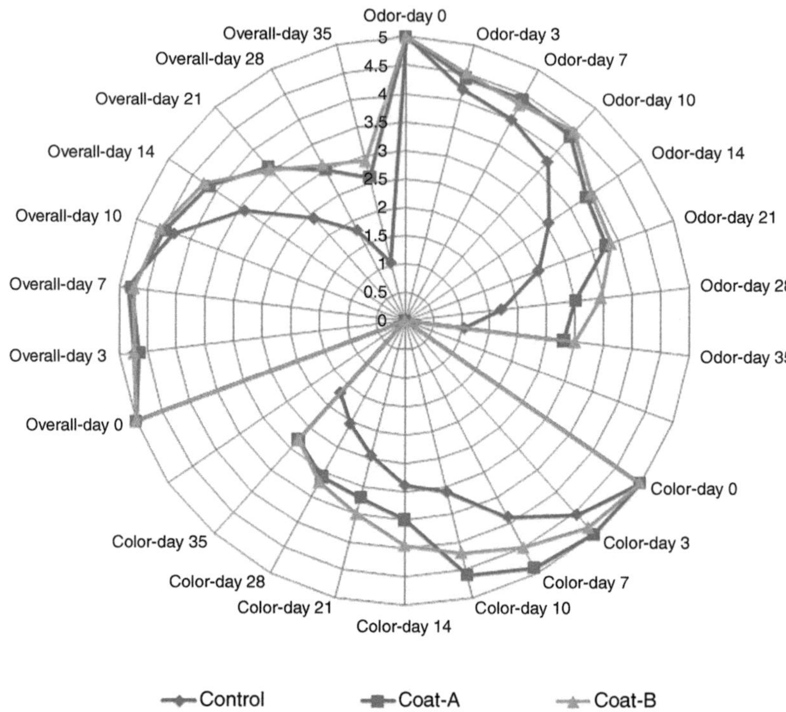

Figure 14.11. Sensory attribute scores (odor, discoloration, and overall acceptance) of grilled pork without chitosan coating (Control), coated with 2% chitosan (Coat-A), and coated with 2.5% chitosan (Coat-B) and stored at 2 ± 1°C. The sensory evaluation was based on a five-point scale to determine color discoloration (5 = no discoloration; 1 = extremely discoloration), odor (5 = extremely desirable; 1 = extremely unacceptable/off-odors), and overall acceptability (5 = extremely desirable; 1 = extremely unacceptable) of the samples. Shelf life criteria assumed that rejection would occur when the sensory attributes declined below 3.0 (Yingyuad et al., 2006).

(w/v) of green tea in the chitosan film-forming solution. The films were casted and conditioned at 25°C and 50% RH for 48 h before used. Cooked pork sausages containing 61.43% moisture, 18.94% fat, and 16.26% protein were wrapped with chitosan films and chitosan films containing green tea extract. The samples without chitosan film wrapping were used as control. Each sample was individually packaged into a 8 × 12 cm pouch of 85 μm linear low density polyethylene coated with polyamide plastic bag, heat-sealed under atmospheric condition, and stored at 4°C.

When compared with control samples, samples wrapped with chitosan films had lower variability of total aerobic counts, yeasts, and molds, and lactic acid bacteria. Their result demonstrated that chitosan films could inhibit microbial growth on the sausages. Moreover, incorporation of chitosan films with green tea extract enhanced the antimicrobial property of the films, as total aerobic counts, yeasts, and molds on the sausages wrapped with chitosan films containing green tea extract were lower than those wrapped with chitosan films.

The antimicrobial effect of chitosan was described to be related to the electrostatic interaction between a positive charge on the NH_3^+ groups of glucosamine monomer in chitosan molecules and negative charge of microbial cell membrane that leads to the leakage of intracellular constituents (Dutta et al., 2009). The enhanced antimicrobial activity of chitosan film containing green tea extract is potentially attributed to green tea polyphenols, which are capable of bactericidal activity by inhibiting cytoplasmic membrane function of bacteria, which cause a leakage of intramembranous materials, suppressing DNA and RNA synthesis of bacterial cells, and interfering with energy metabolisms of bacteria (Mori et al., 1987; Yilmaz, 2006).

Moreover, their result suggested that lipid oxidation in pork sausages, determined using thiobarbituric acid (TBA) measurement and expressed as milligrams of malonaldehyde per kg of sample, was minimized by the use of chitosan film probably due to the antioxidant activity of chitosan as well as its low oxygen permeability characteristic. The antioxidant ability of chitosan film is thought to be due to chelation of free ion, which is released from hemoproteins of meat during storage (Xu et al., 2005; Shahidi et al., 1999). Addition of green tea into chitosan film improved the antioxidant activity of the film, as TBA values of samples wrapped with chitosan film containing green tea extract were lower than those wrapped with chitosan film. Green tea polyphenols, including catechin, epicatechin, epigallocatechin, epicatechin gallate, epigallocatechin gallate, and gallocatechin gallate, are effective free radical-scavengers (Chan et al., 2007; Farhoosh et al., 2007). The antioxidant activity of the inferred green tea polyphenols are attributed to assorted mechanisms, including prevention of radical chain initiation, decomposition of peroxides, and interaction with the free radicals to inhibit lipid oxidation (Noipha and Siripatrawan, 2009; Perumalla and Hettiarachchy, 2011).

Four quality attributes including odor, discoloration, slime formation, and overall acceptance of the samples were evaluated by an experienced 10-member panel using quantitative descriptive analysis. The sample rejection was set at score ≤5. Off-odor occurred in control samples that could be detected faster than those wrapped with chitosan film and chitosan film containing green tea extract, respectively. Unacceptable odor scores (score ≤5) were observed in control samples and those wrapped with chitosan films by days 12 and 20, respectively. The changes in color quality were probably caused by microbial growth and lipid oxidation during the storage since the color scores were in good relation to microbial counts and TBA values. Slime formation on the sausage surface, caused by yeasts and molds (Ambrosiadis et al., 2004), was observed at day 16 for control samples and at day 20 for chitosan film wrapped samples, but it was not observed on the samples wrapped with chitosan films containing green tea extract. The overall acceptance of control samples was significantly lower than those wrapped with chitosan films and chitosan films containing green tea extract, respectively, throughout the storage. Samples wrapped with chitosan film containing green tea extract could maintain quality attributes longer than those wrapped with chitosan film and control samples, respectively, suggesting that chitosan film could inhibit the growth of microorganisms and lipid oxidation, and that its antimicrobial and antioxidant activities could be enhanced when green tea extract was incorporated into the film.

4.2 Application to Fish and Seafood

Ojagh et al. (2010b) studied the effect of chitosan coating and chitosan enriched with cinnamon oil on the quality of refrigerated rainbow trouts. The chitosan coating solution was prepared using 2% w/v medium molecular weight chitosan (450 kDa) in 1% v/v acetic acid and cinnamon essential oil was added. Fresh water rainbow trouts with an average weight of 550–600 g were coated with the prepared chitosan containing cinnamon oil and stored at refrigerated temperature (4°C). The samples were evaluated based on a five-point scale to determine texture, discoloration, odor, and overall acceptability by a six-member trained panel. The shelf life criteria assumed that rejection would occur when the sensory attributes declined below 4.0.

The total volatile base nitrogen (TVB-N) content of samples coated with chitosan enriched with cinnamon oil was lower than those coated with chitosan. This can be attributed to either a more rapidly reduced bacterial population or decreased capacity of bacteria for oxidative deamination of nonprotein nitrogen compounds or both, due to the effect of cinnamon oil on fish fillets. The antibacterial effect of chitosan containing cinnamon oil coating was observed since cinnamon oil contains high concentrations of *trans*-cinnamaldehyde (a well-known antimicrobial compound), linalool, eugenol, and other phenolic compounds. Other studies have also identified *trans*-cinnamaldehyde as the major antibacterial constituent of cinnamon oil

(Ouattara et al., 1997; Shan et al., 2007). The antimicrobial effects of cinnamon oil polyphenols have been described to involve a sensitization of the phospholipid bilayer of the bacterial cell membrane, causing an increase in permeability and leakage of vital intracellular constituents, or impairment of bacterial enzyme systems. Cinnamaldehyde also acts by inhibiting the amino acid decarboxylase in target bacteria (Ouattara et al., 1997).

Moreover, the authors also demonstrated that cinnamon oil added into the chitosan coating exhibited antioxidant ability to break the free radical chain by donating a hydrogen atom, resulting in effectively retarding the production of peroxide value in trout fillets stored at refrigerated temperature (4°C). Chitosan containing cinnamon oil could maintain trout fillet shelf life throughout the end of the storage period (16 days) without any significant loss of quality attributes and without significant microbial growth, while control samples had a shelf life of 12 days.

Yuan et al. (2016) investigated the effect of chitosan coating combined with pomegranate peel extract (PPE) on the melanosis and quality of Pacific white shrimps (*Litopenaeus vannamei*) during iced storage for 10 days. PPE is considered as rich sources of polyphenolic compounds including punicalagin A, punicalagin B, gallic acid, ellagic acid, chlorogenic acid, caffeic acid, catechin, epicatechin, rutin, quercetin, and galangal, which show antioxidant and antimicrobial effects (Akhtar et al., 2015; Li et al., 2015). In their study, chitosan/acetic acid solution (1% w/w) was prepared from chitosan having 90% DD and the average molecular weight of 25 kDa. The 1.5% PPE dipping solution was prepared by dissolving PPE in distilled water. Pacific white shrimps with the size of 50–55 shrimps/kg were randomly assigned into different groups, including control (uncoated) group, PPE group (dipping in 1.5% PPE aqueous solution), chitosan coating group (dipping in 1% chitosan solution), and chitosan coating + PPE group (dipping in 1.5% PPE at 4°C for 30 min and then in 1% chitosan solutions at 4°C for 30 min). The samples were drained, covered in plastic bags, and stored in ice.

Their results indicated that chitosan coating, PPE, and chitosan coating + PPE retarded the melanosis and the changes of total color difference (ΔE values), and maintained the texture parameters and the sensory scores of the shrimp samples. The increase of TVB-N value of shrimps treated by chitosan coating + PPE was significantly less than those treated by PPE or chitosan coating alone. Moreover, the increase in total viable counts was highest in uncoated shrimps, but it was lower in shrimps treated by PPE, chitosan coating, and chitosan coating + PPE, respectively. They described that chitosan coating could inhibit the microbial growth by the chelation of nutrients and essential metals as well as the penetration of the nuclei of the microorganisms, which leads to the interference with protein synthesis by binding with DNA. Moreover, the efficacy of chitosan coating to retard microbial growth was increased when chitosan coating was used in combination with PPE, probably due to the synergism in antimicrobial activity of chitosan and PPE. Chitosan coating combined with pomegranate peel extract could be used as an effective natural alternative to synthetic antimelanosic agents to inhibit postmortem melanosis and maintain the shrimp quality during iced storage. Huang et al. (2012); Wang et al. (2014); and Wu (2014) have also reported the use of chitosan coating to maintain the refrigerated shrimp quality.

4.3 Application to Fresh Fruits and Vegetables

Chitosan coatings have been effectively used for shelf life extension of fresh fruits and vegetables (Suseno et al., 2014; Vu et al., 2011). Suseno et al. (2014) studied the effect of chitosan coating on the shelf life of Cavendish bananas. The chitosan coating solutions were prepared by dissolving chitosan with different degrees of deacetylation (70% DD and 80% DD) at various concentrations (1, 1.5, 2% w/w) in 0.6% w/w acetic acid solution. Fresh bananas with uniform size, shape, and color was dipped into chitosan coating solutions of different concentrations and degrees of deacetylation for 1 min, and then stored at room temperature of 30°C. All samples were analyzed for weight loss, vitamin C, and sensory attributes. The effect of different degrees of deacetylation of chitosan in various chitosan concentrations on weight loss and vitamin C loss were investigated. Sensory evaluation was conducted to monitor the changes in color, texture, and aroma. Their results revealed that chitosan coated bananas showed delayed ripening processes when compared to the uncoated bananas. This result was correlated to the reduction in weight loss and vitamin C loss in comparison to the uncoated bananas. The result also revealed that weight loss and vitamin C loss decreased with increasing chitosan concentration and degree of deacetylation. The chitosan coating prepared with 2% chitosan of 80% DD was proved to be the most effective for reducing the weight loss, retention of vitamin C, and maintaining desirable sensory quality. Similarly, Tattiyakul and Siripatrawan (2015) also demonstrated that chitosan coating effectively delayed the ripening process, maintained good commodity quality attributes and prolong shelf life of cultivated bananas (*Musa acuminata* spp. *Siamea*).

Vu et al. (2011) developed an active edible coating from modified chitosan containing limonene (LIM) or peppermint (PM), as bioactive compounds, to increase shelf life of fresh strawberries. The modified chitosan polymer was prepared by dissolving 5 g of chitosan in 600 mL aqueous acetic acid (0.12 M, pH 4). Thereafter, the pH of the solution was adjusted to 7.5 by adding 0.1 M NaOH. The acylation reactions were carried out by adding palmitoyl chloride (d = 0.907 g/mL) to chitosan solution in a ratio of 1:2 (w/w), at 55°C. The mixture was neutralized (pH 6.8–7.0) and precipitated with acetone and the precipitate was washed at 50°C with an excess of methanol. The modified product was then dried to obtain the corresponding functionalized polymer powder.

The developed modified chitosan coating solution was prepared by solubilizing 2% of modified chitosan polymer into 3% acetic acid solution and the bioactive compounds (LIM or PM) in the presence of 0.4% Tween-80 was added to obtain a final concentration of bioactive compounds at 0.02%, w/v. The strawberries were separately coated with (1) modified chitosan + LIM; (2) modified chitosan + PM; and (3) uncoated fruits (control). Visual decay of strawberries was evaluated and the percentage of mold contamination was calculated during storage at 4°C. The result showed that modified chitosan + LIM coating was more effective since its related percentage of fruit decay was lower than that of modified chitosan + PM coating. The visual inspection of the fruits also revealed that uncoated strawberries were dehydrated and largely contaminated by molds at day 10 whereas modified chitosan + LIM coated strawberries kept a good hydrated, red-color appearance even at day 21. The authors concluded that the edible bioactive coating based on modified chitosan containing LIM could effectively extend the shelf life of strawberries.

The effect of chitosan containing cinnamon oil coating on qualitative properties of sweet peppers (*Capsicum annuum* L.) stored at 8°C for 35 days was investigated by Xing et al. (2011). Chitosan solution (1.0%) with 95% DD was prepared using 0.5% acetic acid solution as a medium to dissolve chitosan and 0.75% glycerol (plasticizer) and the cinnamon oil (0 and 0.25%) mixed with Tween-80 was then added to the chitosan solution. Sweet peppers at 80% maturity (green stage) were dipped in the chitosan solution for 5 min and dried before wrapped with a polypropylene film and stored at 8°C. Samples dipped in distilled water were used as control.

The chitosan containing cinnamon oil coating treatment provided the best control of sample decay (below 5%). At the end of 35 day storage, samples coated with chitosan containing cinnamon oil maintained good sensory quality, whereas the sensory quality of control samples was no longer acceptable. The authors suggested that higher antioxidant activities of oxyradical detoxification enzymes, including superoxide dismutase (SOD), peroxidase (POD), and catalase (CAT), in treated sweet peppers were contributed to the chitosan containing cinnamon oil coating. The increase in activities of SOD, POD, and CAT improved the disease resistance of sweet peppers during storage. Changes in malondialdehyde (MDA) and electrolyte leakage contents of sweet peppers coated with chitosan containing cinnamon oil were lower than those of control samples. The result was verified using atomic force microscopic images, which showed that the surface of sweet pepper without coating treatment was rougher than that of sweet peppers treated with chitosan containing cinnamon oil. The authors concluded that chitosan containing cinnamon oil coated could maintain the quality of sweet peppers.

5 Conclusions

Chitosan edible films and coatings have been developed as active food packaging to maintain quality, improve safety, and prolong shelf life of food products. Chitosan films and coatings can serve as carriers of active compounds such as antimicrobials and antioxidants. Researchers have attempted to improve functional properties of chitosan by introducing phenolic groups into chitosan network so that the chitosan films and coatings can be effectively exploited. Different sources of plant polyphenols affected the functional properties of the chitosan-based film due to different active chemical substances. The enhancement of antimicrobial and antioxidant activities and the improvement of physical properties of chitosan films have been achieved by incorporation with polyphenols derived from green tea extract and cinnamon essential oil. Many studies have demonstrated that chitosan films and coatings incorporated with plant polyphenols can maintain qualities and prolong shelf life of various food and agricultural products in a healthy way since they can eliminate the use of synthetic additives and, on the other, they will also be beneficial for the environment given that they will reduce the use of synthetic polymeric materials.

References

Abdollahi, M., Rezaei, M., Farzi, G., 2012. Improvement of active chitosan film properties with rosemary essential oil for food packaging. Int. J. Food Sci. Technol. 47, 847–853.

Acevedo-Fani, A., Salvia-Trujillo, L., Alejandra Rojas-Graü, M., Martín-Belloso, O., 2015. Edible films from essential-oil-loaded nanoemulsions: physicochemical characterization and antimicrobial properties. Food Hydrocoll. 47, 168–177.

Aider, M., 2010. Chitosan application for active bio-based films production and potential in the food industry: review. LWT Food Sci. Technol. 43, 837–842.
Akhtar, S., Ismail, T., Fraternale, D., Sestili, P., 2015. Pomegranate peel and peel extracts: chemistry and food features. Food Chem. 174, 417–425.
Alboofetileh, M., Rezaei, M., Hosseini, H., Abdollahi, M., 2014. Antimicrobial activity of alginate/clay nanocomposite films enriched with essential oils against three common foodborne pathogens. Food Control 36, 1–7.
Alexandrova, V.A., Obukhova, G.V., Domnina, N.S., Topchiev, D.A., 1999. Modifications of chitosan for construction of efficient antioxidant biodegradable macromolecular systems. Macromol. Symp. 144, 413–422.
Altiok, D., Altiok, E., Tihminlioglu, F., 2010. Physical, antibacterial and antioxidant properties of chitosan films incorporated with thyme oil for potential wound healing applications. J. Mater. Sci. Mater. Med. 21, 2227–2236.
Ambrosiadis, J., Soultos, N., Abrahim, A., Bloukas, J.G., 2004. Physicochemical, microbiological and sensory attributes for the characterization of Greek traditional sausages. Meat Sci. 66, 279–287.
An, B.J., Kwak, J.H., Son, J.H., Park, J.M., Lee, J.Y., Jo, C., et al., 2004. Biological and antimicrobial activity of irradiated green tea polyphenols. Food Chem. 88, 447–451.
ASTM, 2010. Annual Book of ASTM Standards. American Society for Testing and Materials, Pennsylvania.
Atarés, L., De Jesús, C., Talens, P., Chiralt, A., 2010. Characterization of SPI-based edible films incorporated with cinnamon or ginger essential oils. J. Food Eng. 99, 384–391.
Atia, M.M.M., Buchenauer, H., Aly, A.Z., Abou-Zaid, M.I., 2005. Antifungal activity of chitosan against *Phytophthora infestans* and activation of defence mechanisms in tomato to late blight. Biol. Agric. Hortic. 23, 175–197.
Azevedo, E.P., Kumar, V., 2012. Rheological, water uptake and controlled release properties of a novel self-gelling aldehyde functionalized chitosan. Carbohydr. Polym. 90, 894–900.
Bégin, A., Van Calsteren, M.R., 1999. Antimicrobial films produced from chitosan. Int. J. Biol. Macromol. 26, 63–67.
Bodini, R.B., Sobral, P.J.A., Favaro-Trindade, C.S., Carvalho, R.A., 2013. Properties of gelatin-based films with added ethanol–propolis extract. LWT Food Sci. Technol. 51, 104–110.
Cao, Z.B., Sun, Y.Y., 2009. Chitosan-based rechargeable long-term antimicrobial and biofilm-controlling systems. J. Biomed. Mater. Res. A 89 (4), 960–967.
Casettari, L., Gennari, L., Angelino, D., Ninfali, P., Castagnino, E., 2012. ORAC of chitosan and its derivatives. Food Hydrocoll. 28, 243–247.
Chan, E.W.C., Lim, Y.Y., Chew, Y.L., 2007. Antioxidant activity of Camellia sinensis leaves and tea from a lowland plantation in Malaysia. Food Chem. 102, 1214–1222.
Chang, S.-H., Lin, H-T.V., Wu, G.-J., Tsai, G.J., 2015. pH Effects on solubility, zeta potential, and correlation between antibacterial activity and molecular weight of chitosan. Carbohydr. Polym. 134, 74–81.
Chung, Y.C., Chen, C.Y., 2008. Antibacterial characteristics and activity of acid-soluble chitosan. Bioresource Technol. 99, 2806–2814.
Chung, Y.C., Su, Y.P., Chen, C.C., Jia, G., Wang, H.L., Wu, J.G., 2004. Relationship between antibacterial activity of chitosan and surface characteristics of cell wall. Acta Pharm. Sin. B 25, 932–936.
Cushnie, T.P.T., Lamb, A.J., 2005. Antimicrobial activity of flavonoids. Int. J. Antimicrob. Agents 26, 343–356.
Darmadji, P., Izumimoto, M., 1994. Effect of chitosan in meat preservation. Meat Sci. 38, 243–254.

Department of Biology, The Pennsylvania State University, 2015. Prokaryotes II—structure and function. Available from: https://online.science.psu.edu/biol110_sandbox_8862/node/8971

Devlieghere, F., Vermeulen, A., Debevere, J., 2004. Chitosan: antimicrobial activity, interactions with food components and applicability as a coating on fruit and vegetables. Food Microbiol. 21, 703–714.

Diao, W.R., Hu, Q.P., Zhang, H., Xu, J.G., 2014. Chemical composition, antibacterial activity and mechanism of action of essential oil from seeds of fennel (*Foeniculum vulgare* Mill.). Food Control 35, 109–116.

Dutta, P.K., Tripathi, S., Mehrotra, G.K., Dutta, J., 2009. Perspectives for chitosan based antimicrobial films in food applications. Food Chem. 114, 1173–1182.

Elsabee, M.Z., Abdou, E.S., 2013. Chitosan based edible films and coatings. Mater. Sci. Eng. C 33, 1819–1841.

Fan, W., Sun, J., Chen, Y., Qiu, J., Zhang, Y., Chi, Y., 2009. Effects of chitosan coating on quality and shelf life of silver carp during frozen storage. Food Chem. 115, 66–70.

Farhoosh, R., Golmovahhed, G.A., Khodaparast, M.H.H., 2007. Antioxidant activity of various extracts of old tea leaves and black tea wastes (*Camellia sinensis* L.). Food Chem. 100, 231–236.

Fernandez-Saiz, P., Lagaron, J.M., Ocio, M.J., 2009. Optimization of the biocide properties of chitosan for its application in the design of active films of interest in the food area. Food Hydrocoll. 23, 913–921.

Ganan, M., Carrascosa, A.V., Martinez-Rodriguez, A.J., 2009. Antimicrobial activity of chitosan against *Campylobacter* spp. and other microorganisms and its mechanism of action. J. Food Prot. 72 (8), 1735–1738.

Georgantelis, D., Ambrosiadis, I., Katikou, P., Blekas, G., Georgakis, S.A., 2007. Effect of rosemary extract, chitosan and α-tocopherol on microbiological parameters and lipid oxidation of fresh pork sausages stored at 4°C. Meat Sci. 76, 172–181.

Giménez, B., López de Lacey, A., Pérez-Santín, E., López-Caballero, M.E., Montero, P., 2013. Release of active compounds from agar and agar–gelatin films with green tea extract. Food Hydrocoll. 30, 264–271.

Gómez-Estaca, J., Gimnez, B., Montero, P., Gómez-Guillén, M.C., 2009. Incorporation of antioxidant borage extract into edible films based on sole skin gelatin or a commercial fish gelatin. J. Food Eng. 92, 78–85.

Gómez-Estaca, J., López de Lacey, A., López-Caballero, M.E., Gómez-Guillén, M.C., Montero, P., 2010. Biodegradable gelatin-chitosan films incorporated with essential oils as antimicrobial agents for fish preservation. Food Microbiol. 27, 889–896.

Gramzaa, A., Khokharb, S., Yokob, S., Gliszczynska-Swigloc, A., Hesa, M., Korczaka, J., 2006. Antioxidant activity of tea extracts in lipids and correlation with polyphenol content. Eur. J. Lipid Sci. Technol. 108, 351–362.

Hamilton-Miller, J.M., 1995. Antimicrobial properties of tea (*Camellia sinensis* L.). Antimicrob. Agents Chemother. 39, 2375–2377.

Hernández-Muñoz, P., Almenar, E., Valle, V.D., Del Velez, V., Gavara, R., 2008. Effect of chitosan coating combined with postharvest calcium treatment on strawberry (Fragaria × ananassa) quality during refrigerated storage. Food Chem. 110, 428–435.

Higueras, L., López-Carballo, G., Gavara, R., Hernández-Muñoz, P., 2015. Reversible covalent immobilization of cinnamaldehyde on chitosan films via schiff base formation and their application in active food packaging. Food Bioprocess Technol. 8, 526–538.

Hsouna, A.B., Trigui, M., Mansour, R.B., Jarraya, R.M., Damak, M., Jaoua, S., 2011. Chemical composition, cytotoxicity effect and antimicrobial activity of *Ceratonia siliqua* essential oil with preservative effects against Listeria inoculated in minced beef meat. Int. J. Food Microbiol. 148, 66–72.

Huang, Y.J., Chen, Q.C., Qiu, M., Li, S.Q., 2012. Chitosan-based edible coatings for quality preservation of postharvest whiteleg shrimp (*Litopenaeus vannamei*). J. Sci. Food Agric. 77, C491–C496.

Ikigai, H., Nakae, T., Hara, Y., Shimamura, T., 1993. Bactericidal catechins damage the lipid bilayer. Biochim. Biophys. Acta 1147, 132–137.

Inouye, M., 1986. Bacterial Outer Membranes as Model Systems. Wiley, New York, NY.

Jiarpinijnun, A., Pomchaksin, P., Makino, Y., Siripatrawan, U., 2013. Chitosan-based film containing cinnamon bark oil as an active film for food application.In: The 22nd Meeting. The Society of Packaging Science & Technology, Tokyo, Japan.

Jongjareonrak, A., Benjakul, S., Visessaguan, W., Tanaka, M., 2008. Antioxidant activity and properties of fish skin gelatin films incorporated with BHT and α-tocopherol. Food Hydrocoll. 22, 449–458.

Juven, B.J., Kanner, J., Schved, F., Weisslowicz, H., 1994. Factors that interact with the antibacterial action of thyme essential oil and its active constituents. J. Appl. Bacteriol. 76, 626–631.

Komes, D., Horžić, D., Belščak, A., Ganić, K.K., Vulić, I., 2010. Green tea preparation and its influence on the content of bioactive compounds. Food Res. Int. 43, 167–176.

Kong, M., Chen, X.G., Xing, K., Park, H.J., 2010. Antimicrobial properties of chitosan and mode of action: a state of the art review. Int. J. Food Microbiol. 144, 51–63.

Leerahawong, A., Siripatrawan, U., Sanguandeekul, U., Tanaka, M., 2007. Effect of acids and plasticizers on physical properties of chitosan films. The Ninth National Graduate Research Conference, March 14–15, 2007, Chonbury, Thailand.

Leerahawong, A., Siripatrawan, U., Sanguandeekul, R., Tanaka, M., 2008. Antimicrobial property of chitosan-based film containing cinnamaldehyde. The Society for Industrial Microbiology (SIM) Annual Meeting and Exhibition 2008, August 10–14, 2008, San Diego, CA.

Li, X.F., Feng, X.Q., Yang, S., 2010. A mechanism of antibacterial activity of chitosan against Gram-negative bacteria. Food Sci. 13, 37.

Li, J., He, X., Li, M., Zhao, W., Liu, L., Kong, X., 2015. Chemical fingerprint and quantitative analysis for quality control of polyphenols extracted from pomegranate peel by HPLC. Food Chem. 176, 7–11.

Liu, H., Du, Y., Wang, X., Sun, L., 2004. Chitosan kills bacteria through cell membrane damage. Int. J. Food Microbiol. 95, 147–155.

López-Mata, M.A., Ruiz-Cruz, S., Silva-Beltrán, N.P., Ornelas-Paz, J.e.J., Ocaño-Higuera, V.M., Rodríguez-Félix, F., Cira-Chávez, L.A., Del-Toro-Sánchez, C.L., Shirai, K., 2015. Physicochemical and antioxidant properties of chitosan films incorporated with cinnamon oil. Int. J. Polym. Sci. 2015, 1–8.

Martucci, J.F., Gende, L.B., Neira, L.M., Ruseckaite, R.A., 2015. Oregano and lavender essential oils as antioxidant and antimicrobial additives of biogenic gelatin films. Ind. Crops Prod. 71, 205–213.

Mathew, S., Abraham, T.E., 2006. Studies on the antioxidant activities of cinnamon (*Cinnamomum verum*) bark extracts, through various in vitro models. Food Chem. 94, 520–528.

Moradi, M., Tajik, H., Rohani, S.M.R., Oromiehie, A.R., Malekinejad, H., Aliakbarlu, J., Hadian, M., 2012. Characterization of antioxidant chitosan film incorporated with *Zataria multiflora* Boiss essential oil and grape seed extract. LWT Food Sci. Technol. 46, 477–484.

Mori, A., Nishino, C., Enoki, N., Tawata, S., 1987. Antibacterial activity and mode of action of plant flavonoids against *Proteus vulgaris* and *Staphylococcus aureus*. Phytochemistry 26, 2231–2234.

Nikaido, H., 2009. Outer membrane, Gram-negative bacteria. In: Schaechter, M. (Ed.), Encyclopedia of Microbiology. third ed. Academic Press, San Diego, CA.

No, H.K., Park, N.Y., Lee, S.H., Meyers, S.P., 2002. Antibacterial activity of chitosans and chitosan oligomers with different molecular weights. Int. J. Food Microbiol. 74, 65–72.

Noipha, S., Siripatrawan, U., 2009. Antioxidant and antimicrobial activity of green tea. In: Proceeding of the 35th Congress on Science and Technology of Thailand, October 15–17, 2009, Chonbury, Thailand.

Nunthanid, J., Puttipipatkhachorn, S., Yamamoto, K., Peck, G.E., 2001. Physical properties and molecular behavior of chitosan films. Drug Dev. Ind. Pharm. 27, 43–57.

Ojagh, S.M., Rezaei, M., Razavi, S.H., Hosseini, S.M.H., 2010a. Development and evaluation of a novel biodegradable film made from chitosan and cinnamon essential oil with low affinity toward water. Food Chem. 122, 161–166.

Ojagh, S.M., Rezaei, M., Razavi, S.H., Hosseini, S.M.H., 2010b. Effect of chitosan coatings enriched with cinnamon oil on the quality of refrigerated rainbow trout. Food Chem. 120, 193–198.

Oriani, V.B., Molina, G., Chiumarelli, M., Pastore, G.M., Hubinger, M.D., 2014. Properties of cassava starch-based edible coating containing essential oils. J. Food Sci. 79, E189–E194.

Ouattara, B., Simard, R.E., Holley, R.A., Piette, G.J.P., Bégin, A., 1997. Antibacterial activity of selected fatty acids and essential oils against six meat spoilage organisms. Int. J. Food Microbiol. 37, 155–162.

Ouattara, B., Simard, R., Piette, G., Begin, A., Holley, R., 2000. Diffusion of acetic and propionic acids from chitosan-based antimicrobial packaging films. J. Food Sci. 65, 768–773.

Park, P., Je, J., Kim, S., 2004. Free radical scavenging activities of differently deacetylated chitosans using an ESR spectrometer. Carbohydr. Polym. 55, 17–22.

Peng, Y., Li, Y., 2014. Combined effects of two kinds of essential oils on physical, mechanical and structural properties of chitosan films. Food Hydrocoll. 36, 287–293.

Perdones, T., Vargas, M., Atarés, L., Chiralt, A., 2014. Physical, antioxidant and antimicrobial properties of chitosan-cinnamon leaf oil films as affected by oleic acid. Food Hydrocoll. 36, 256–264.

Pereda, M., Ponce, A.G., Marcovich, N.E., Ruseckaite, R.A., Martucci, J.F., 2011. Chitosan-gelatin composites and bi-layer films with potential antimicrobial activity. Food Hydrocoll. 25, 1372–1381.

Perumalla, A.V.S., Hettiarachchy, N.S., 2011. Green tea and grape seed extracts—potential applications in food safety and quality. Food Res. Int. 44, 827–839.

Pilar Almajano, M., Carbó, R., Angel López Jiménez, J., Gordon, M.H., 2008. Antioxidant and antimicrobial activities of tea infusions. Food Chem. 108, 55–63.

Qin, C.Q., Li, H.R., Xiao, Q., Liu, Y., Zhu, J.C., Du, Y.M., 2006. Water-solubility of chitosan and its antimicrobial activity. Carbohydr. Polym. 63 (3), 367–374.

Reis, L.C.B., Oliveira de Souza, C., Alves da Silva, J.B., Martins, A.C., Nunes, I.L., Druzian, J.I., 2015. Active biocomposites of cassava starch: the effect of yerba mate extract and mango pulp as antioxidant additives on the properties and the stability of a packaged product. Food Bioprod. Process. 94, 382–391.

Rhazi, M., Desbrières, J., Tolaimate, A., Rinaudo, M., Vottero, P., Alagui, A., El Meray, M., 2002. Influence of the nature of the metal ions on the complexation with chitosan: application to the treatment of liquid waste. Eur. Polym. J. 38, 1523–1530.

Rice-Evans, C., Miller, N., Paganga, G., 1997. Antioxidant properties of phenolic compounds. Trends Plant Sci. 2, 152–159.
Rohn, S., Rawel, H.M., Kroll, J., 2004. Antioxidant activity of protein-bound quercetin. J. Agric. Food Chem. 52, 4725–4729.
Rojas-Grau, M.A., Avena-Bustillos, R.J., Olsen, C., Friedman, M., Henika, P.R., Martın-Belloso, O., Pan, Z., McHugh, T.H., 2007. Effects of plant essential oils and oil compounds on mechanical, barrier and antimicrobial properties of alginate–apple puree edible films. J. Food Eng. 81, 634–641.
Ruiz-Navajas, Y., Viuda-Martos, M., Sendra, E., Perez-Alvarez, J.A., Fernández-López, J., 2013. In vitro antibacterial and antioxidant properties of chitosan edible films incorporated with *Thymus moroderi* or *Thymus piperella* essential oils. Food Control 30, 386–392.
Salgado, P.R., Fernández, G.B., Drago, S.R., Mauri, A.N., 2011. Addition of bovine plasma hydrolysates improves the antioxidant properties of soybean and sunflower protein-based films. Food Hydrocoll. 25, 1433–1440.
Samelis, J., Kakouri, A., Rememtzis, J., 2000. Selective effect of the product type and the packaging conditions on the species of lactic acid bacteria dominating the spoilage microbial association of cooked meats at 4°C. Food Microbiol. 17, 329–340.
Sánchez-González, L., González-Martínez, C., Chiralt, A., Cháfer, M., 2010. Physical and antimicrobial properties of chitosan–tea tree essential oil composite films. J. Food Eng. 98, 443–452.
Sanla-Ead, N., Jangchud, A., Chonhenchob, V., Suppakul, P., 2012. Antimicrobial activity of cinnamaldehyde and eugenol and their activity after incorporation into cellulose-based packaging films. Packag. Technol. Sci. 25, 7–17.
Schreiber, S.B., Bozell, J.J., Hayes, D.G., Zivanovic, S., 2013. Introduction of primary antioxidant activity to chitosan for application as a multifunctional food packaging material. Food Hydrocoll. 33, 207–214.
Senanayake, U.M., Wijesekera, R.O.B., 2004. Cinnamon and cassia: the future vision. In: Ravindran, P.N., Nirmal-Babu, K., Shylaja, M. (Eds.), Cinnamon and Cassia: The Genus Cinnamomum. CRC Press, Boca Raton, FL.
Sforcin, J.M., Bankova, V., 2011. Propolis: is there a potential for the development of new drugs? J. Ethnopharmacol. 133, 253–260.
Shahidi, F., Arachchi, J.K.V., Jeon, Y.-J., 1999. Food applications of chitin and chitosans. Trends Food Sci. Technol. 10, 37–51.
Shan, B., Cai, Y., Brooks, J.D., Corke, H., 2007. Antibacterial properties and major bioactive components of cinnamon stick (*Cinnamomum burmannii*): activity against foodborne pathogenic bacteria. J. Agric. Food Chem. 55, 5484–5490.
Shen, Z., Kamdem, D.P., 2015. Development and characterization of biodegradable chitosan films containing two essential oils. Int. J. Biol. Macromol. 74, 289–296.
Shimamura, T., Zhao, W.H., Hu, Z., 2007. Mechanism of action and potential for use of tea catechin as an anti-infectiveagent. Antiinfect. Agents Med. Chem. 6, 57–62.
Siripatrawan, U., Harte, B.R., 2010. Physical properties and antioxidant activity of an active film from chitosan incorporated with green tea extract. Food Hydrocoll. 24, 770–775.
Siripatrawan, U., Noipha, S., 2012. Active film from chitosan incorporating green tea extract for shelf life extension of pork sausages. Food Hydrocoll. 27, 102–108.
Siripatrawan, U., Vitchayakitti, W., 2014. Effect of propolis extract on physical and antioxidant properties of chitosan film. Responsible Packaging for a Global Market. In: Sek, M.A., Rouillard, V., Brigger, S.W. (Eds.), The 19th IAPRI World Conference on Packaging, June 15–19, 2014, Melbourne, Australia.

Siripatrawan, U., Vitchayakitti, W., Sanguandeekul, R., 2013. Antioxidant and antimicrobial properties of Thai propolis extracted using ethanol aqueous solution. Int. J. Food Sci. Technol. 48, 22–27.

Srinivasa, P.C., Ramesh, M.N., Tharanathan, R.N., 2007. Effect of plasticizers and fatty acids on mechanical and permeability characteristics of chitosan films. Food Hydrocoll. 21, 1113–1122.

Suseno, N., Savitri, E., Sapei, L., Padmawijaya, K.S., 2014. Improving shelf-life of Cavendish banana using chitosan edible coating. Procedia Chem. 9, 113–120.

Talja, R.A., Helén, H., Roos, Y.H., Jouppila, K., 2007. Effect of various polyols and polyol contents on physical and mechanical properties of potato starch-based films. Carbohydr. Polym. 67, 288–295.

Tattiyakul, T., Siripatrawan, U., 2015. Shelf life extension of banana fruit using chitosan containing photocatalyst. The Third ISHS Southeast Asia Symposium on Quality Management in Postharvest Systems (SEAsia2015), Siem Reap, Cambodia.

Torlak, E., Sert, D., 2013. Antibacterial effectiveness of chitosan–propolis coated polypropylene films against foodborne pathogens. Int. J. Biol. Macromol. 60, 52–55.

Von Gadow, A.V., Joubert, E., Hansmann, C.F., 1997. Comparison of the antioxidant activity of rooibos tea (*Aspalathus linearis*) with green, oolong and black tea. Food Chem. 60, 73–77.

Vu, K.D., Hollingsworth, R.G., Leroux, E., Salmieri, S., Lacroix, M., 2011. Development of edible bioactive coating based on modified chitosan for increasing the shelf life of strawberries. Food Res. Int. 44, 198–203.

Wang, L., Dong, Y., Men, H., Tong, J., Zhou, J., 2013. Preparation and characterization of active films based on chitosan incorporated tea polyphenols. Food Hydrocoll. 32, 35–41.

Wang, Y., Liu, L., Zhou, J., Ruan, X., Lin, J., Fu, L., 2014. Effect of chitosan nanoparticle coatings on the quality changes of postharvest whiteleg shrimp, *Litopenaeus vannamei*, during storage at 4°C. Food Bioprocess Technol. 8, 907–915.

Weerakody, N.S., Caffin, N., Turner, M.S., Dykes, G.A., 2010. In vitro antimicrobial activity of less-utilized spice and herb extracts against selected food-borne bacteria. Food Control 21, 1408–1414.

Wojdyła, A.T., 2004. Chitosan (biochikol 020 PC) in the control of some ornamental foliage diseases. Commun. Agric. Appl. Biol. Sci. 69, 705.

Wu, S.J., 2014. Effect of chitosan-based edible coating on preservation of white shrimp during partially frozen storage. Int. J. Biol. Macromol. 65, 325–328.

Wu, Y.B., Yu, S.H., Mi, F.L., Wu, C.W., Shyu, S.S., Peng, C.K., Chao, A.C., 2004. Preparation and characterization on mechanical and antibacterial properties of chitosan/cellulose blends. Carbohydr. Polym. 57, 435–440.

Wu, S.C., Yen, G.C., Wang, B.S., Chiu, C.K., Yen, W.J., Chang, L.W., Duh, P.D., 2007. Antimutagenic and antimicrobial activities of pu-erh tea. LWT Food Sci. Technol. 40, 506–512.

Xi, J., Shouqin, Z., 2007. Antioxidant activity of ethanolic extracts of propolis by high hydrostatic pressure extraction. Int. J. Food Sci. Technol. 42, 1350–1356.

Xie, W., Xu, P., Liu, Q., 2001. Antioxidant activity of water-soluble chitosan derivatives. Bioorg. Med. Chem. Lett. 11, 1699–1701.

Xing, R., Liu, S., Guo, Z., Yu, H., Wang, P., Li, C., Li, Z., Li, P., 2005. Relevance of molecular weight of chitosan and its derivatives and their antioxidant activities in vitro. Bioorg. Med. Chem. 13, 1573–1577.

Xing, Y., Li, X., Xu, Q., Yun, J., Lu, Y., Tang, Y., 2011. Effects of chitosan coating enriched with cinnamon oil on qualitative properties of sweet pepper (*Capsicum annuum* L.). Food Chem. 124, 1443–1450.

Xu, Y.S., Kim, K.M., Hanna, M.A., Nag, D., 2005. Chitosan-starch composite film: preparation and characterization. Ind. Crops Prod. 21, 185–192.
Yam, T.S., Shah, S., Hamilton-Miller, J.M.T., 1997. Microbiological activity of whole and fractionated crude extracts of tea (*Camellia sinensis*), and of tea components. FEMS Microbiol. Lett. 152, 169–174.
Yen, M.T., Tseng, Y.H., Li, R.C., Mau, J.L., 2007. Antioxidant properties of fungal chitosan from shiitake stipes. LWT Food Sci. Technol. 40, 255–261.
Yen, M.T., Yang, J.H., Mau, J.L., 2008. Antioxidant properties of chitosan from crab shells. Carbohydr. Polym. 74, 840–844.
Yilmaz, Y., 2006. Novel uses of catechins in foods. Trends Food Sci. Technol. 17, 64–71.
Yingyuad, S., Ruamsin, S., Leekprokok, T., Douglas, S., Pongamphai, S., Siripatrawan, U., 2006. Effect of chitosan coating and vacuum packaging on the quality of refrigerated grilled pork. Packag. Technol. Sci. 19, 149–157.
Younes, I., Sellimi, S., Rinaudo, M., Jellouli, K., Nasri, M., 2014. Influence of acetylation degree and molecular weight of homogeneous chitosans on antibacterial and antifungal activities. Int. J. Food Microbiol. 185, 57–63.
Yuan, G., Lv, H., Tang, W., Zhang, X., Sun, H., 2016. Effect of chitosan coating combined with pomegranate peel extract on the quality of Pacific white shrimp during iced storage. Food Control 59, 818–823.
Zhang, L., Kosaraju, S.L., 2007. Biopolymeric delivery system for controlled release of polyphenolic antioxidants. Eur. Polym. J. 43, 2956–2966.
Zhang, Y., Liu, X., Wang, Y., Jiang, P., Quek, S.Y., 2016. Antibacterial activity and mechanism of cinnamon essential oil against *Escherichia coli* and *Staphylococcus aureus*. Food Control 59, 282–289.
Zheng, L.-Y., Zhu, J.-F., 2003. Study on antimicrobial activity of chitosan with different molecular weights. Carbohydr. Polym. 54, 527–530.
Zivanovic, S., Chi, S., Draughon, A.F., 2005. Antimicrobial activity of chitosan films enriched with essential oils. J. Food Sci. 70, M45–M51.

15

APPLICATIONS OF NANOTECHNOLOGY IN NUTRITION: POTENTIAL AND SAFETY ISSUES

Aditya S. Kulkarni*, Padmini S. Ghugre, Shobha A. Udipi****

**Aten Porus Lifesciences, Bengaluru, Karnataka, India; **SNDT Women's University, Department of Food Science and Nutrition, Mumbai, Maharashtra, India*

Nanotechnology is one of the major technologies that could help address numerous human needs to improve health, well-being, and quality of life. Extensive research is being carried out for development of drug delivery systems, including their handling and fate in biological systems. However, most work so far has been devoted to drugs and bioactive phytochemicals whereas their application in the area of food science, technology, and nutrition is still limited. The promise that nanotechnology holds is being tested in the field of nutrition, although several areas still need to be understood and explored such as biocompatibility, bioavailability, stability, and delivery. This chapter attempts to summarize the work carried out in the area of nanotechnological approaches for food science and nutrition.

1 Introduction

Micronutrient deficiencies are prevalent in several countries. There are concerns about meeting requirements of essential nutrients such as folic acid, vitamin D, trace elements, and calcium. In developing countries, the problems are worse with micronutrient malnutrition extending to iodine, iron, zinc, vitamin A, and in many cases selenium and vitamin B12. The deficiency of these essential nutrients are likely to add to public nutrition and health concerns. Food fortification provides a strategy for covering large segments of the population on a sustainable basis in contrast to

Novel Approaches of Nanotechnology in Food. http://dx.doi.org/10.1016/B978-0-12-804308-0.00015-7

supplementation, which is a therapeutic approach with significantly limited outreach.

Fortification of both liquid and solid foods is possible. In liquid products, either the soluble form of the nutrient or bioactive needs to be added or a dispersion of an insoluble substance has to be used. Additives that help in solubilizing an insoluble nutrient or bioactive make its absorption easier and the compound more bioavailable. However, the additives can impart their own flavor or taste to a liquid product and there is an additional associated risk of unwanted physical or chemical interactions of the additives with bioactives or nutrients, which in turn can lead to coalescence, sedimentation/precipitation, and changes in texture and consistency as well as impacting the stability of the food product. Similar problems may be encountered with insoluble fortificants. Nanotechnology can be used in the development of nanoformulations that can be designed to deliver nutrients through effective encapsulation that would not negatively impact the flavor or stability of the nutrient or bioactive. The advantage of these systems is that hydrophobic substances can be rendered water soluble and vice versa so that they are compatible with the food vehicle into which they are incorporated.

2 Carrier Materials

Many colloidal systems require the use of carrier materials and, for food applications, emphasis has to be put on the nature of materials that can be used to encapsulate functional ingredients such as vitamins (McCrae et al., 2009; Wandrey et al., 2010). Some of the biopolymeric carrier materials that can be used in developing colloidal delivery systems for foods include polysaccharides from different sources, food proteins, chitosan (Table 15.1). Because many of the materials are hydrophilic, delivery systems are designed based on protein–polysaccharide interactions, bioactive molecule-biopolymer binding, self-assembling, pH-dependent solubility differences, crosslinking, and hydrophobic modifications (Tiyaboonchai, 2003; Benichou et al., 2004, 2007; Neirynck et al., 2007; McCrae et al., 2009; Wandrey et al., 2010; O'Regan and Mulvihill, 2010; Perrechil and Cunha, 2012; Patel and Bhandari, 2014); whereas, for inherently hydrophobic materials such as lipids, waxes, zein, and shellac, colloidal particles can be prepared by simple alteration of solubility (Patel et al., 2010a, 2010b, 2011; Wandrey et al., 2010; Patel, 2012) or use of emulsion-based techniques.

Table 15.1 Carrier Materials Suited for Microencapsulation in the Food Industry

Type	Typical Examples
Carbohydrates	Starch, maltodextrin, corn syrup, cyclodextrin, sucrose, alginates
Proteins	Albumin, casein, gelatin, whey protein, soy protein, gluten
Cellulose derivatives	Ethylcellulose, methylcellulose, carboxymethylcellulose
Gums	Guar gum, gum Arabic, locust bean gum, xanthan gum
Lipids	Lecithin, stearic acid, tristearin, glycerides, fatty acids/alcohols
Waxes	Beeswax, candelilla wax, carnuaba wax, paraffin wax
Inorganic materials	Silicon oxide, aluminum oxide, tripolyphosphate
Miscellaneous	Chitosan, polyvinyl pyrollidone, shellac, zein

3 Challenges for Nutrient Nanoencapsulation

The use of nanoformulations in fortification necessitates a suitable vehicle wherein the stability of the nutrient is guaranteed, while it does not interact with the food vehicle and its components and also does not cause any sensory changes that may reduce consumer acceptance. Fortification with vitamins is challenged by their sensitivity to and consequent instability in the presence of light, oxygen, heat, humidity, and pH alterations beyond the optimum. Vitamin A and D are sensitive to light, oxidizing agents, heat, and acid; vitamin D is also sensitive to alkaline pH; and vitamin E is sensitive to light, oxidizing agents, and heat but its sensitivity is less compared to A and D. The high susceptibility of vitamin C and folic acid to oxidation is well known. Folic acid is also sensitive to reducing agents. Riboflavin, vitamin B12 and B6 are sensitive to light. B12 and riboflavin are also sensitive to alkali or high pH (Kohlmeier, 2003). Fig. 15.1 indicates various factors that influence stability and bioavailability of nutrients.

Nutrients such as iron can cause degradation of vitamins and lipid oxidation in foods resulting in rancidity. In high protein foods, proteins may be destabilized by the addition of calcium and magnesium. On the other hand, addition of some amino acids to high protein foods could result in competition between amino acids during transport and reduce absorption. In milk and milk

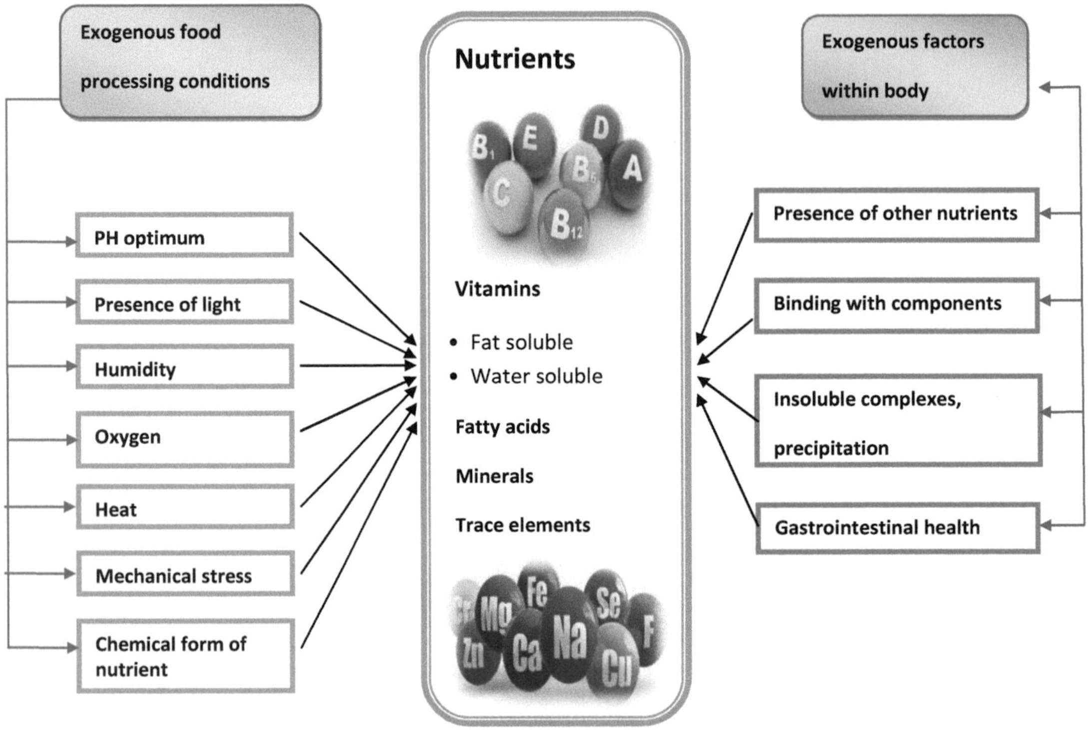

Figure 15.1. Factors affecting stability and bioavailability of nutrients.

products, calcium could bind to the nutraceutical added, form an insoluble complex and reduce its bioavailability. At the same time, addition of some nutrients like vitamin C may well protect bioactive phytochemicals like lutein or vitamin E/vitamin A from oxidation. Thus, the physical and chemical properties of nutrients pose challenges because nutrients and other components in food can alter the digestion and bioavailability of the nanoparticles (NPs) and the cargo (Huynh, 2012).

4 Applications

In the global market, several nanoscale additives are available, the fat-soluble vitamins—A, E, D, and K, carotenoids, polyunsaturated fatty acids (PUFAs), and co-enzyme Q_{10}. Carotenoids have been focused on for their color, strength, and bioavailability. Solid NPs have been available since the 1960s. In the case of vitamins A, D, E, the focus has been on improving their stability through encapsulation. For the same reasons, PUFAs have been encapsulated and are available since the 1990s (End, 2006). Magnuson et al. (2011) reported that the Woodrow Wilson Project on Emerging Nanotechnologies listed 97 food and food-related products. All these are not food items, some are vitamin or dietary supplements, some are cookware and others food storage containers. But, these products have been included primarily based on claims made by manufacturers and it does not indicate that such claims can be entirely relied on, as they may not have been validated.

Möller et al. (2009) reported that there may be several companies producing carotenoids. Several of the carotenoid additives are water soluble and are sold under various trade or brand names. BASF manufactures β-carotene that can be added to beverages, and lycopene for food supplements. A mixture of β-carotene and lycopene is also available. A supplement containing Q10, different trace elements and vitamins is marketed by NovaConcept N.E. GmBH. Nanocalcium and magnesium is also available (Magnuson et al., 2011). A formulation "SoluTM E 200 Clear" containing vitamin E is manufactured by BASF. Bio Delivery Sciences International has developed nanocochleates from soy and calcium to carry and deliver nutrients such as vitamins, lycopenes, and omega fatty acids directly to cells. The company claims nanocochleates deliver omega-3 fatty acids to cakes and muffins without affecting their color and taste. Yada et al. (2014) and Ranjan et al. (2014) noted that several companies manufacture and market foods containing nanoformulations of different nutrients (Table 15.2).

Fortitech Strategic Nutrition (2011) has projected numerous possible applications. For example, sports persons may need

Table 15.2 Examples of Nanoencapsulated Nutrients Claimed to be Delivered Through Food Vehicles

Nutrient	Food Vehicles
β-carotene	Beverages
ω-3 fatty acids	Cakes, muffins
Calcium, iron	Flavored milk
Calcium	Milk powder, biscuits, chocolates, infant milk, fish past, peanut drink
Coenzyme Q10	Flavored beverages, orange juice
Polyglycerol fatty acid esters	Oil
Phytosterol	Cooking oil

specific amino acids to promote recovery but some are bitter in taste. Similarly, omega-3 fatty acids have a typical fishy odor. Such unpleasant/unacceptable tastes could well be masked in nanoformulations in order to provide the nutritional benefits. Nanoformulations can be added to a variety of beverages ranging from dairy products to juices and enhanced waters (Grebler et al., 2010).

Beverages are sold in the form of shots, that is, small, portable, single servings providing 8 or 12 ounces or even 2–3 ounces. In these shots, several ingredients like caffeine, taurine, D-ribose, B vitamins are added in addition to sweetener and flavoring. Shots have been developed to promote gastrointestinal health, boost immunity, or enhance mood, to enable consumers to relax. Stick packs that contain powders are used to deliver different beverages like tea, ready-to-drink tea, enhanced waters, and energy drinks. Many stick packs contain nutrients and offer a vehicle for providing a mix of nutrients in a product that is easy to use in different settings. There are several products that claim to be based on nanotechnology formulations containing flavorings or nutrients, for example, diet milk shakes (Möller et al., 2009).

Ready-to-drink teas provide another vehicle for delivery of nutrients. These products may be worthwhile vehicles, as research has shown that the polyphenolic compounds in tea have several health benefits. Milk or dairy-based beverages can be developed in which fatty acids, antioxidants, and different functional ingredients could be added, which are suitable for various health conditions. Similarly enhanced waters can be developed not only for fulfilling the thirst needs of the consumer but also to provide nutraceuticals for specific needs such as enhancing immunity and providing different nutrients. Wellness and sports drinks are popular and may be available in different flavors. Some products claim to contain micelles that encapsulate vitamins and Q10 have a label "MICELLE Q10 INSIDE" (Möller et al., 2009).

It has been pointed out that researchers are looking to develop formulations containing nanocapsules and nanosensors that will release nutrient(s) or other bioactives in the body on demand as per need or deficits. However, the question is whether such a product can be termed a food or a therapeutic. Such products may require approval as therapeutics or medicines from regulatory authorities because depositing nanocapsules or nanosensors in the body come under the category of medical intervention. Although nanoformulated food additives or supplements are available, for example, regulatory peptides from plants, nanodroplets/-clusters, with claims of increased bioavailability, yet valid studies on relative bioavailability comparing these products with similar nonnanoformulated products are lacking and the validity of the claimed effects for such formulations need verification (Bouwmeester et al., 2009).

Increasing prevalence of obesity and noncommunicable diseases such as diabetes mellitus, heart disease, cancer, and osteoporosis add to challenges faced by health-care systems. Several of these problems are linked to inflammatory processes and oxidative stress in the human body. Hence, there is considerable interest in dietary bioactive compounds/phytochemicals food components with beneficial effects, that is, anticarcinogenic, antiangiogenic, antiproliferative properties. These compounds include carotenoids, flavonoids, dithiolthiones, phenolic compounds, glucosinolates, saponins, indoles, inositol hexaphosphate, isothiocyanates, coumarins, phenols, plant sterols, isoflavones/lignans, saponins, allium compounds, and limonene (Slavin and Lloyd, 2012; Wang et al., 2014). Many exogenous bioactives can be used either as supplements or added to a variety of vehicles that can be functionalized in order to provide consumers with affordable yet efficient remedies (Anand et al., 2010).

Several bioactive compounds like curcuminoids, carotenoids, phytosterols, and phenolic compounds are lipophilic (McClements and Xiao, 2012). Carotenoids, plant sterols are easily oxidized as well as photosensitive (Bohn, 2012). Also, many interact with other food components like phospholipids, proteins. This reduces their availability and the in vivo activity of the molecule. Besides this, the addition of some fortificants like ferrous sulfate can adversely affect the color and taste of foods and thus reduce acceptability. For such compounds, nanoformulations can help improve stability as well as enhance availability.

5 Digestion, Absorption, Metabolism, and Excretion

The challenge in delivering nanformulations-nanoliposomes, nanoemulsions, solid lipid nanoparticles (SLNs) is that they must withstand the mechanical pressure, varying pH, and enzymes in different parts of the gastrointestinal tract and other parts of the body. Several physicochemical properties such as particle size and surface charge, hydrophobicity, or hydrophilicity influence the fate of NPs and are important considerations that influence the cargo's bioactivity (Naahidi et al., 2013).

Human studies with administration of nanoformulations are very limited; hence, several researchers have attempted to study the fate of nanoparticles using in vitro models of human digestive processes. However, such models have some limitations because the action of different enzymes, minerals, bile salts, and even the mechanical forces can be simulated, they generally do not include several vital elements that play crucial roles in digestion and absorption, such as feedback mechanisms, the role of hormones, gut microflora, the gut immune system. Also there is an absence of mucus that lines the gastrointestinal tract. In static models the absorption process is not included and the dynamic mechanical pressure, peristalsis, changes in pH, flow rates of different secretions within the gastrointestinal tract, are not simulated. In the dynamic model, the physical forces and temporal alterations in the lumen are included, thus making these closer to in vivo conditions (Wickham et al., 2009). Fig. 15.2 shows the factors influencing digestion, absorption and excretion of NPs.

5.1 Digestion

McClements and Xiao (2012) and Fröhlich and Roblegg (2012) have delineated the digestion and absorption of NPs. Like any other food, formulations/delivery systems mix with saliva. In the oral cavity, NPs may be subjected to amylase besides physical exposure, "flow/force" and several properties such as pH, ionic strength, temperature may be altered. There may be coalescence resulting in altered particle size or interfacial properties due to displacement of emulsifiers or being coated with salivary mucin. Upon passage into the stomach, NPs would be exposed to highly acid pH, gastric proteases besides the mechanical pressure. Consequently, there may be displacement and changes in the original surface active materials leading to changes in interfacial characteristics such as composition, thickness, charge, and permeability. Alterations in pH and ionic composition may further encourage flocculation or aggregation such that the original size of the particles (<100 nm) may not be maintained. Lipid digestion can be initiated by gastric lipases. Similarly, hydrocolloids will be degraded by the gastric proteases.

Transfer of NPs from stomach to duodenum exposes them to a near neutral pH (6–7.5), proteolytic enzymes, pancreatic lipase, phospholipases, bile salts, phospholipids, bicarbonate (McClements and Li, 2010a,b). In the small intestine, there may be competition between surface active compounds with those that are at the lipid droplet surface. Bile salts are required to facilitate passage of products of digestion such as lipids. Pancreatic lipases can hydrolyze triglycerides, diglycerides, and phospholipids resulting in the formation of monoglycerides and free fatty acids that are incorporated into micelles as are bioactive lipophilic compounds. Even at this stage, the interfacial characteristics could

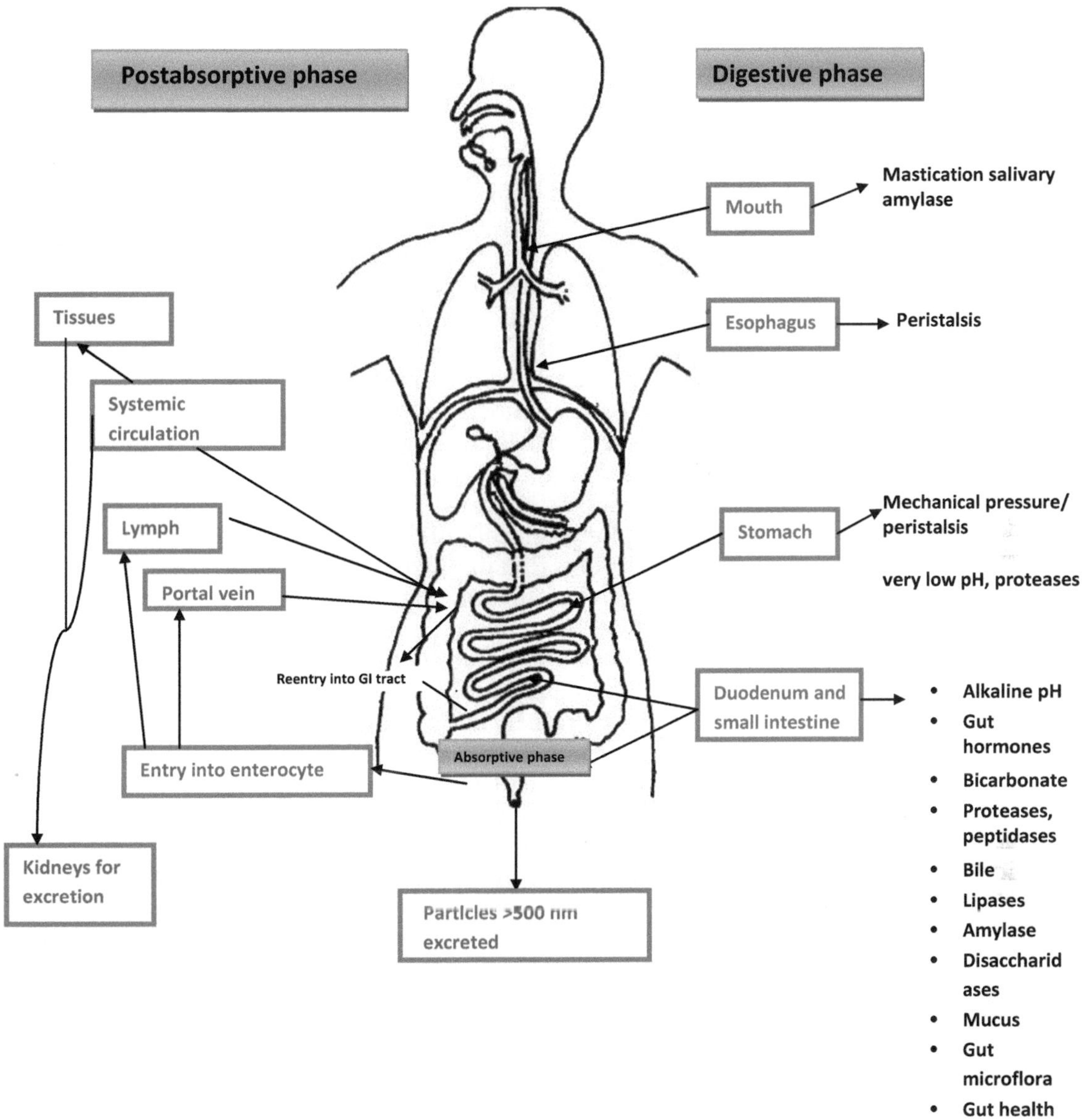

Figure 15.2. Digestion, absorptive fate of nanoparticles.

alter with changes in particle size. Proteins and carbohydrates that are used to encapsulate the cargo will be degraded in the digestive tract by proteolytic or carbohydrate-digesting enzymes. Pancreatic proteases would hydrolyze proteins with formation of peptides and amino acids and larger carbohydrate molecules would be hydrolyzed to smaller ones.

5.2 Absorption

NPs remain in the small intestines for a long enough period, (3–6 h), where the NP and/or the bioactive cargo that has been released after digestion are absorbed. While NPs have been shown to enhance absorption of phytochemicals, there is much less work on nutrients delivered in a similar fashion. NPs' stability in the gastrointestinal tract should be taken into account since various factors such as pH, ions, digestive enzymes, and mucus layer, impact the properties of NP delivery system. Future work needs to focus on encapsulated NPs designed for oral administration, better gastrointestinal stability, mucus penetrating function, and intestinal epithelial cell targeting properties.

Absorption of nanoparticles may occur through (1) small or solid lipid particles via the tight junctions between the intestinal cells; (2) cellular uptake (transcellular mechanism) involving entry of the nanocarrier into the cell (Bouwmeester et al., 2009; Dube et al., 2010). Between cells of the gastrointestinal epithelium there are tight junctions that may expand and become permeable under the influence of some polymers and allow large particles to cross this barrier. Metallic iron (Fe) particles may cross into circulation through paracellular uptake through tight junctions.

In the transcellular route, NPs will be taken up by the epithelial cell at its apical side by endocytosis. The particles are transported through M cells in Peyer's patches or also in enterocytes and then released from the basolateral side of the epithelium (Bouwmeester et al., 2009). Physicochemical characteristics (surface charge, particle size, attachment of ligands, or coating with surfactants) play an important role in transcellular uptake (Yoo et al., 2010b). The path that NPs traverse and the site of their localization within the cell determine the functionality of the nutrient or bioactive phytochemical cargo. If the coating remains undegraded locally within the intestines, the likelihood of the encapsulated nutrient or phytoactive remaining intact and reaching either blood or lymphatic circulation is higher.

5.2.1 *Size*

Uptake of NPs is determined by their size. Size also affects their cytotoxicity, distribution, and how they are cleared from the system (Chithrani et al., 2006; De Jong et al., 2008; Sonavane et al., 2008; Hirn et al., 2011). Small-sized molecules may enter cells through endocytosis. The translocation of the molecule that is trapped inside the endosome is determined by the type of endocytosis (Sahay et al., 2010). Generally, the size of vesicles formed by endocytosis, ranges from 60 to 1000 nm. However, particles exceeding 500 nm in size are not taken up by enterocytes via endocytosis, and larger particles may be eliminated by cells in the mononuclear phagocyte system before they reach the blood stream (Yoo et al., 2010a; Choi et al., 2011). Length of stay in the enterocyte is influenced by particle size, smaller particles –1~20 nm in size reside longer than larger particles. Particles that exceed 500 nm in size are generally excreted through feces before they enter the blood stream, as they are not absorbed or absorbed poorly (Naahidi et al., 2013).

5.2.2 Shape

Yoo et al. (2010b) compared poly(lactic-*co*-glycolic) acid (PLGA) spherical and elliptical particles. Particle shape influenced endocytosis and the intracellular distribution of the particles in endothelial cells, with the elliptical disks being endocytosed slowly compared to the spherical ones, despite the volume of disks of both shapes being similar.

5.2.3 Mucus

Uptake may also be influenced by mucus, which protects the underlying epithelium while permitting nutrients, ions, and some proteins to be absorbed (Mackie et al., 2012). Mucus is a viscoelastic gel made up of glycoproteins, lipids, protein, and salts. The mucus layer facilitates digestion by immobilizing enzymes that are located close to the epithelial surface, as well as adsorption of nutrients (Mahler et al., 2009). The extent of uptake of bioactive compounds is different in cells with and without mucus secreting ability (Anand et al., 2010; Guri et al., 2013). Presence of the mucus layer has been found to influence transport and uptake of drugs (Lai et al., 2009; D'Agostino et al., 2012).

The mucus barrier uses a combination of porosity and charge to keep out large particles, those that do not have sufficient charge and invading microorganisms while permitting nutrients/other compounds to pass. Pore size is an important consideration for delivery of NPs through the mucus layer especially with reference to endocytosis. NPs ranging from 0.5 to 2 μm have been reported to be able to diffuse through the mucus layer (Macierzanka et al., 2011). The optimal size for mucus transfer is <100 nm, as this is generally the pore size in the mucus (Lai et al., 2007, 2009). Particles ~200 nm have thrice the diffusion coefficient and hence are preferred over smaller particles <100 nm in size (Acosta, 2009). Particles greater than 500 nm do not undergo mucosal transfer (Norris et al., 1998). M cells in the Peyer's patches can pick up particles that are about 1 μm, although smaller sized particles, ~100 nm, may also be taken up. Particles that are between 60 and 500 nm in size can be endocytosed.

Mucus may be important for the bioavailability of some nutrients. Mucin binds with iron, forms mucin-iron complexes, thus preventing precipitation of iron at the neutral pH in the small intestines (Conrad et al., 1994). Mahler et al. (2009) demonstrated that mucus has an important role in iron absorption using an in vitro digestion model.

5.2.4 Ionic Charge

Uptake is also dependent on ionic charge. Negatively charged particles may be able to penetrate the mucus layer better, in contrast, diffusion of neutral or positively charged particles may be limited (Macierzanka et al., 2011). Negative charges on the particle surface can be increased by adsorption of bile salts on the particle surface, or micelle formation (Macierzanka et al., 2011, 2012; Maldonado-Valderrama et al., 2011). Endocytosis can be augmented by adding positive charges to the surface of the NPs, as cations may be absorbed more readily than anions. The coating used therefore may be important as use of cationic chitosan or ions was found to protect the NPs from being degraded in the enterocyte. Partially digested lipid droplets are likely to attach to the mucus layer (Mackie et al., 2012). Once they enter the intestinal cell, NPs may be acted upon by cellular enzymes and the digestive products may be absorbed, or they cross the cell intact into either blood or the lymph.

5.2.5 Hydrophobicity

Hydrophobicity also determines cellular uptake, distribution, recognition by immune cells, clearance and interaction with plasma proteins (Naahidi et al., 2013). Hydrophilic particles can cross through the mucus faster than hydrophobic particles (Lai et al., 2009). They have a large activity coefficient; hence, they diffuse less easily into epithelial cells (Choi et al., 2011). Hydrophilic particles are less likely to be recognized as foreign in blood (Lai et al., 2009). Macrophages have been reported to be able to recognize hydrophobic particles as foreign, whereas the risk is less with hydrophilic particles (Naahidi et al., 2013). Lipophilicity was found to be an important characteristic for diffusion of drugs through purified gastrointestinal mucus compared to size and charge (Karlsson et al., 1993; Larhed et al., 1977).

5.2.6 Type of Material

The material used for coating can alter hydrophobic and/or electrostatic interactions. Specifically, coating NPs with polyethylene glycol reduced their hydrophobicity or electrostatic interaction greatly, which was associated with less mucoadhesion and faster penetration through human mucus (Cu and Saltzman, 2009). Liposomes modified with nonionic polymers were stable in simulated gastric and intestinal fluid, were able to penetrate the mucus layer and reach the epithelial surface (Chen et al., 2013). Since pectin is mucoadhesive, it has been studied when present on the surface of delivery systems. High methoxyl-pectin coated liposomes were found to have the highest interaction with mucin, reflected by particle size and turbidity (Klemetsrud et al., 2013). Alginate results in the formation of weak viscoelastic gels and promotes mucin-mucin interaction (Taylor et al., 2005). In contrast, low-molecular-weight glucaronic acid oligomers extracted from alginate interfered with mucus interactions leading to rheological changes, less crosslink density, and resistance of mucus gel deformation (Nordgård and Draget, 2011). Thus, different biopolymers can interact differently with mucus, interact with it in either direction, that is, either enhance or reduce its viscoelastic properties and thereby influence absorption.

Salvia-Trujillo et al. (2013) reported that with decrease in the diameter of emulsion droplets containing β-carotene, the rate, amount of lipid digested as well as bioaccessibility of β-carotene increased. They attributed this to greater exposure of surface area to lipases.

Liu et al. (2012) used in vitro digestion to examine the influence of the interfacial structure on the bioaccessibility and microstructural changes in β-carotene emulsions, prepared with different emulsifiers—whey protein isolate, soybean soluble polysaccharides, and decaglycerol monolaurate. These emulsifiers significantly influenced particle size, electric charge, and the microstructural changes that occurred during the digestion of β-carotene. Decaglycerol monolaurate was associated with the highest release rate in gastric fluids, and the emulsion made with whey protein concentrate exhibited large amounts of changes in droplets and rate of release of the provitamin.

The lipid structure may also be important. Long chain triglycerides were digested less as compared to medium-chain or short chain triglycerides and they tended to accumulate at the oil/water interface, which prevented exposure of the undigested lipids to hydrolysis by lipase. This was attributed to lower water dispersibility of the former. Bioaccessibility of curcumin was lowest with short chain triglycerides and highest with medium chain triglycerides, possibly because mixed micelles that solubilized the curcumin were formed when the short chain or medium chain triglycerides were used in contrast to long chain triglycerides. Salvia-Trujillo et al. (2015) investigated the bioavailability of fucoxanthin from nanoemulsions using long chain, medium chain triglycerides, and orange/mineral oil as carrier oils. The nanoemulsion containing mineral oil was not digested whereas those containing the long chain and medium chain triglycerides were completely digested, the former showing better solubility of cargo in mixed micelles. In animals fed fucoxanthin nanoliposomes, it was absorbed with no difference in the serum concentrations with either the long chain or medium chain triglycerides.

Because phospholipids are emulsifiers, milk protein may be suitable for use in liposomes and such liposomes had higher stability during in vitro digestion than did soy phospholipids (Thompson and Singh 2006; Livney, 2010; Farhang, 2013).

5.3 Metabolism and Excretion

Generally nutrients and drugs are distributed via the vascular and lymphatic systems (Faraji and Wipf, 2009). After absorption, highly hydrophobic NPs will be transported to lymph. If they have been taken up by gut-associated lymphoid tissue, they will enter lymphoid circulation (Bouwmeester et al., 2009). The other route is for the particles to enter the portal vein, be transported to the liver where they are processed, and then enter systemic circulation (Chen et al., 2011). However, some NPs can exit out of the basolateral membrane of the intestinal cells. All NPs that enter the portal vein or lymphatic circulation are not necessarily metabolized and may accumulate in some tissues (Dube et al., 2010). Lu et al. (2010) reported that NPs may return to the intestines from the liver. NPs that enter the lymphatic system then enter systemic circulation (Lu et al., 2010) and will ultimately be metabolized in the liver (Moghimi et al., 2005).

Particles <20 nm in size and those with size between 30 and 100 nm will remain longer in circulation as they are able to avoid being cleared by macrophages (Faraji and Wipf, 2009). In the bloodstream, NPs can interact with red blood cells, white blood cells, and platelets as well as plasma proteins. In biological fluids, they can interact with proteins, glycans, and ions, depending on their properties and have the potential to change the activity and distribution of such molecules (Pan et al., 2012). NPs are transported to different organs—heart, lungs, spleen, kidney, liver, and perhaps the brain—with the physicochemical properties of the NPs influencing their transport through cell membranes and cell organelles. Those organs that have higher blood flow will naturally be exposed to higher concentrations (Faraji and Wipf, 2009). In the organ, the kind, rate, and extent of interactions will determine the length of stay within the cell (Sahay et al., 2010). If the nanoparticle localizes within the lysosome and there is lysosomal release, small sized hydrophobic molecules may escape the endosome or lysosomal membranes in contrast to larger molecules that remain trapped within the lysosome. The cargo will be metabolized depending on their hydrophilicity or hydrophobicity and their physicochemical properties (Sahay et al., 2010).

Particles less than 100 nm in size will remain in systemic circulation longer, hence various tissues and organs will be exposed to them for a longer period of time. Luo et al. (2012) reported that chemical stability of nutrients like α-tocopherol and β-carotene was better with NPs formulated with chitosan or with SLNs. Luo et al. (2010) formulated selenite-loaded chitosan with zein coating to develop a selenite supplement. The antioxidant properties of selenite-loaded NPs were better in an in vitro system as compared to pure selenite. Hu et al. (2008) fabricated chitosan NP carriers for delivering tea catechins. They reported from observations of in vitro experiments that there was controlled release of the catechins.

5.4 Excretion

The liver, kidneys, and colon are primarily responsible for excretion of NPs and their metabolites (Bouwmeester et al., 2009; Bertrand and Leroux, 2012). Depending on their particle size, NPs may be excreted via the gastrointestinal tract or through the kidney. Particles with size ≥500 nm are not likely to be absorbed in the intestines and are excreted in the feces.

Excretion from the kidney may also depend on particle size. Pore size in the glomerular epithelium is about 80–100 nm in diameter and in the glomerular basement membrane it is about 3–8 nm. This filters small molecules based on their size and charge. Underlying the basement membrane are podocytes, shaped such that there are filtration slits about 32 nm wide. Overall, the kidney filters out molecules that are less than 10 nm in diameter, hence if NPs are smaller than 5–10 nm in diameter, they will be rapidly cleared by the kidney (Choi et al., 2011; Naahidi et al., 2013). A higher hydrodynamic diameter has been shown to delay renal clearance. (Choi et al., 2011). Larger NPs would have longer residence time in the kidneys. The longer the NPs stay in the kidney, the higher the risk of toxicity. Choi et al. (2007, 2011) showed that with excessive uptake of NPs by the kidney, there is cell shrinkage. These authors also noted that NPs submicron in size could diffuse through and accumulate in the mesangial cells after exiting the glomerulus.

Several researchers have developed NPs for delivery of selected nutrients such as β-carotene, vitamin A, vitamin D, PUFA, and minerals like iron, selenium, and calcium. However, there is scope for preparing nanoformulations for vitamin B_{12}, folic acid, vitamin K, and amino acids like branched chain amino acids/ conditionally essential amino acids.

6 Nanoformulations of Selected Nutrients

6.1 β-Carotene

Carotenoids and terpenoids together constitute the major proportion of secondary metabolites in plants. Besides acting as antioxidants, carotenoids have been shown to be antiproliferative, be involved in apoptosis, as well as cell–cell communication. Carotenoid compounds include α- and β-carotene, α- and β-cryptoxanthin, zeaxanthin, lycopene, lutein. β-carotene is well known for its provitamin A activity. Carotenoids have low solubility in water and during digestion they are generally incorporated into mixed micelles with phospholipids and fatty acids. Xanthophllys like β-cryptoxanthin that are relatively more hydrophilic will be located in the outer part of the micelles whereas the more hydrophobic molecules like β-carotene are present in the inner part of the micelle. Incorporation into micelles is higher for more hydrophilic molecules (Bohn, 2012). High concentrations of divalent minerals in the gastrointestinal tract could bind free fatty acids and bile acids forming insoluble soaps, and in turn reduce the amount of carotenoids incorporated into micelles (Biehler et al., 2011).

β-carotene is only slightly soluble in oil at room temperature. It is highly sensitive to light and oxygen. This vulnerability to being degraded under general conditions of food processing poses challenges for delivery of this provitamin. Also, because it is crystalline, its bioavailability is fairly low (Yuan et al., 2008; Qian et al., 2012), and presents challenges because crystalline materials are not easily incorporated into liquid food materials, as they tend to aggregate and may sediment, and adversely affect the texture and mouthfeel of food products (Xia, 2014).

Nik et al. (2012) observed that SLN dispersions of β-carotene stabilized with the non-ionic surfactants Poloxamer 188 and Tween-20 were stable when they were exposed to gastric conditions, that is, pepsin, acid, and high ionic strength and in the duodenal phase, they were more stable compared to a canola oil-in-water emulsion. In contrast, when emulsions were stabilized with proteins like whey protein isolates/β-lactoglobulin, proteolysis of the protein interfacial layers by pepsin occurred and the surface charge of the emulsion was lost.

Xia (2014) compared the in vitro bioaccessibility of β-carotene crystals with that of solubilized β-carotene using a simulated gastrointestinal digestion model, with three systems: predissolved β-carotene nanoemulsion ($d < 200$ nm), with a corn oil emulsion ($d < 200$ nm) wherein the β-carotene crystals were added prior to digestion and with phosphate buffer saline to which β-carotene was added before digestion. The phosphate buffer saline model showed the poorest performance in terms of rate of digestion and release of free fatty acids.

Bioavailability was highest with the predissolved nanoemulsion, which performed in between the phosphate buffer saline and the corn oil emulsion for rate of digestion and release of free fatty acids. Oil concentrations were also compared: high fat (20%) with low fat (4%) in order to determine whether the digestion is influenced by the amount of fat used. Bioaccessibility was lower with the high fat group, which was attributed to insufficient fat digestion (Huynh, 2012).

6.2 Vitamin D

Vitamin D is a fat soluble vitamin that promotes intestinal absorption of calcium and its retention. It plays a role in gene regulation, is important for bone growth, and connective tissue as well as having a protective role against cancer (Kohlmeier, 2003). As it is fat soluble, it cannot be dissolved in aqueous solutions and is sensitive to oxidation. Therefore, nanoencapsulation offers a solution for providing a supplement for fortifying foods because vitamin D is generally not available from foods. Several researchers have formulated NPs to deliver this vitamin.

Domingues (2013) reported that PLGA/vitamin D NPs 170–180 nm in size were less cytotoxic compared to those without the cargo in a human pancreatic cell line. Patel and San Martin-Gonzalez (2012) studied the properties of ergocalciferol (ranging from 0 to 20% w/w/of lipid) encapsulated in tripalmitin SLNs stabilized by polysorbate 20. With higher ergocalciferol load (concentrations, 5–20%), turbidity of the SLN dispersions was less, clarity and stability were better, and capacity of the SLN for incorporating lipid was greater; suggesting that clear beverages or milk, margarine could be fortified with ergocalciferol NPs.

Mohammadi et al. (2014) developed vitamin D nanoliposomes by thin-film hydration-sonication method. There was no interaction between the cargo and the liposome constituents, the encapsulation efficiency being 93%. They suggested such a nanoformulation could be used for fortifying beverages with vitamin D.

Proteins show promise for encapsulating vitamin D. β-Lactoglobulin helped to make the complex soluble in aqueous solutions. Pure casein micelles have been used, although there are issues of cost and less efficacy compared to emulsion based techniques (Cherian, 2009). Almouazen et al. (2012) examined the efficiency of calcidiol/calcitriol loaded NPs at inhibiting growth of human breast adenocarcinoma cells. The loaded NPs were as effective as nonencapsulated metabolites of vitamin D on the 4th day, with inhibition being better after a week. Abbasi et al. (2014) used whey protein isolate to encapsulate vitamin D3. With calcium in the NP, the structure was more compact and degradation was less during storage. Other investigators have also used casein (Semo et al., 2007; Haham et al., 2012). Haham et al. (2012) found that the bioavailability of a dose of 50,000 IU of vitamin D3 encapsulated in casein micelles and added to 1% fat milk, was high, indicating the potential of using such NPs for fortification.

6.3 Vitamin A

Vitamin A and its metabolites have diverse roles including vision and ocular health, growth and development, and neurological function. It is essential for effective immune responses as it is important for maintaining the integrity of mucosal surfaces in the respiratory tract, digestive tract, and urinary tract. Deficiency is associated with increased rates of invasive respiratory, gastrointestinal, and ocular infections because of the loss of integrity of the mucosal barriers (Kohlmeier, 2003). Vitamin A deficiency is a public health problem in several countries. Therefore, besides supplementation programs, food fortification has been explored as a viable alternative. Nanoformulation is challenging, as this vitamin does not disperse easily in aqueous systems, it is highly labile under ambient conditions, and losses can occur during processing, transportation and storage (Dary and Mora, 2002). Exposure to light especially ultraviolet light (<415 nm) also results in degradation (Garcia-Fuentes et al., 2003).

Cho (2012) studied the thermal stability of retinol encapsulated in polyester NPs. Different polyesters were used, poly(ethylene adipate), poly(butylene adipate), poly(hexamethylene adipate), and three polycaprolactones with molecular weights of approximately 10, 40, and 80K. Stability was monitored in aqueous solution at 25°C and 40°C for 4 weeks. The thermal stability was related to the melting point of the polyesters and the NPs manufactured therefrom. Jee et al. (2006) also studied the stability of solid lipid NPs containing all transretinol using Tween-80. The NPs were subjected to irradiation with a 60-W bulb for testing their photostability. Addition of antioxidant in the formulation improved the stability of the SLN as well as photostability. However, stability was influenced by the type of antioxidant used and BHT–BHA were found to be effective.

Retinoic acid is important for commitment, maturation, and survival of neural cells and there is interest in determining whether retinoic acid can be used for treatment of neurodegenerative diseases. Esteves et al. (2015) recently reported on the neuroprotection of dopaminergic neurons by retinoic acid NPs. There was significantly lower loss of these neurons in the substantia nigra and the neuronal fibers. Also,expression of transcription factors Pitx3 and Nurr1 were higher, indicating that retinoic acid containing NPs can offer a means of preventing the progression of neurological disorders, for example, Parkinson's disease. However, caution is required if nanoformulations of vitamin A are used because vitamin A accumulates in adipose tissue and high levels of this vitamin are known to be toxic.

6.4 Vitamin E/α–Tocopherol

Vitamin E collectively refers to eight related compounds-tocopherols and tocotrienols, of which α-tocopherol has the greatest bioavailability. Vitamin E is a well-known antioxidant but it is also anti-inflammatory, neuroprotective, and antithrombotic as well as being protective against cancer (Neophytou and Constantinou, 2015). Like other vitamins, vitamin E is also susceptible to oxidation, including prolonged exposure to light and heat. Several investigators have developed NPs of α-tocopherol (Duclairoir et al., 2002; Zigoneanu et al., 2008; Anais et al., 2009). Luo (2012) observed that coating chitosan NPs for selenite improved the physicochemical properties. Zein NPs with chitosan or carboxymethyl chitosan coating also were developed to encapsulate hydrophobic nutraceuticals, including vitamin E, vitamin D3, indole-3-carbinol, and diindolylmethane Luo (2012). Zigoneanu et al. (2008) synthesized spherical PGLA NPs with α-tocopherol as the cargo and two surfactants: sodium dodecyl sulfate and polyvinyl alcohol. With SDS, the size decreased when α-tocopherol was loaded in the PGLA matrix, in contrast to the polyvinyl, containing NPs.

6.5 Vitamin C

Vitamin C is a water-soluble vitamin required for the metabolism of tyrosine, tryptophan, folic acid, and cholesterol. It is also required for the synthesis of carnitine, catecholamines, and degradation of histamine. It plays an important role in improving iron bioavailability. An important role is in the synthesis of collagen and for neutralization of free radicals along with vitamin E and other nutrients (Iqbal et al., 2004). It is easily oxidized at room temperature and hence there can be considerable oxidative losses in food systems. It is also susceptible to light and its stability is affected by pH and presence of metal ions. During processing, its destruction often results in discoloration. It can interact with other food ingredients and result in undesirable changes in taste and/or color. Hence, nanoencapsulation has been exploited (perhaps more than other nutrients) to ensure stability during food processing (Patel and Bhandari, 2014). Kirby et al. (1991) and Farhang et al. (2012) reported that ascorbic acid in NPs is stable.

Several attempts have been made to encapsulate vitamin C sustained release carriers (Alishahi et al., 2011a; Zhao et al., 2011; Cho et al., 2012; de Britto et al., 2014; Jiménez-Fernández et al., 2014). The polysaccharide chitosan has been used to encapsulate active compounds because of its nontoxicity, biodegradability, biocompatibility, and immunity system induction (Tiyaboonchai, 2003). Vitamin C has been encapsulated for sustained release using crosslinked chitosan as a delivery material. Some preparations have controlled-release properties by varying manufacturing parameters like inlet temperature, flow rate, and using different molecular weights and concentrations of chitosan. Vitamin C-encapsulated chitosan microcapsules had a spherical shape and their surface was smooth. The heat stability of vitamin C-loaded chitosan NPs has been examined with aqueous solutions. Stability was dependent on temperature and that internal stability was better than their surface stability.

Resistant starch-pectin has been used to encapsulate vitamin C, although with increasing amount of starch, the size of the particles increased. Release of ascorbic acid was different at a pH of 1.2 versus pH 7.0 as well as the type of starch used (Liu, 2014).

Liposomes have also been made with ascorbic acid and α-tocopherol to obtain better (synergistic) antioxidant function and as a possible means for fortification (Marsanasco et al., 2011). When vitamin C was incorporated into the inner core of liposome, decomposition due to copper, ascorbate oxidase, and lysine was prevented (Liu and Park, 2010; Suntres, 2011). In rainbow fish, chitosan NPs showed controlled release up to 48 h with increase in lysozyme and complement upon exposure, indicating that the immune system was stimulated (Alishahi et al., 2011b).

6.6 Folic Acid

Folic acid is a water-soluble vitamin. Folate plays an important role in transfer of one-carbon atoms and thus is required for synthesis of purine nucleotides, carnitine, creatine, lipids, and hormones as well as for several specific biochemical reactions, for example, the conversion of homocysteine to methionine, metabolism of specific anticonvulsant drugs. Folate deficiency increases the risk of neural tube defects and increases homocysteine levels. Hyperhomocysteinemia has been associated with increased risk of cardiovascular disease and neural tube defects (Kohlmeier, 2003). Naturally occurring folate is not stable during storage and preparation of food as it is highly sensitive to oxidation. 5-methyl tetrahydrofolate is sensitive to oxygen, heat, light, rapidly degraded during cooking and food processing (Kitts and Liu, 2015). It has a half life of of 20 min at 50°C at pH 6.8 and 80 min at pH 4.0 in buffer solutions. (Zhao et al., 2011). In freshly made bread only 53% of the folate was retained. After storage at room temperature for 3 days, only 21% of the original vitamin amount remained and after 7 days only 10% of the vitamin was retained.

Given the implications of folate deficiency for fetal development and the risk of neural tube defects, folate fortification has been made mandatory in as many as 53 countries (Crider et al., 2011). Folic acid the stable form of the vitamin is generally used for supplementation and food fortification. Zhao et al. (2011) reported that 5-methyl tetrahydrofolate has been nanoencapsulated with stearate emulsion, pectin, and skimmed milk powder and modified starch. The folate in the nanoemulsion was unstable. With pectin there was leakage, whereas the skimmed milk powder and modified starch have shown better results. Further work is warranted for delivery of folate with NPs.

6.7 Iron

Iron deficiency results in anemia, a public health problem affecting millions of individuals worldwide (World Health Organization, 2008). Besides causing anemia, iron deficiency has adverse effects on growth and development, cognitive functions resulting in developmental abnormalities. This calls for measures to tackle the problem. There are several iron compounds (ferrous and ferric), for example, ferrous sulfate, ferrous fumarate, ferrous gluconate, ferric pyrophosphate, ferric ammonium citrate), Fe(III) chelates (eg, sodium ferric EDTA, ferric citrate) that can be used to reach iron to the target group through food vehicles. Besides this, elemental iron in the form of electrolytic iron, carbonyl iron can be used. Several strategies have been developed such as fortifying staple foods like rice or salt. However, most iron compounds have gastrointestinal side effects resulting in poor compliance. Also the iron can generate reactive oxygen species (ROS) in the gastrointestinal tract through the Fenton reaction. Soluble Fe (III) chelates have also been implicated in colorectal carcinogenesis (Pereira et al., 2014a). Sadeghiani et al. (2005) reported that uncoated iron oxide nanoparticles had low solubility, which would increase risk of precipitation and in the biological system could possibly aggregate and impede blood flow in blood vessels.

Xia and Xu (2005) prepared liposomes for encapsulating ferrous sulfate using different methods: thin-film hydration, thin-film sonication, reverse-phase evaporation, freeze-thawing. They observed that capacity for encapsulating the ferrous ions was influenced by the concentration of the salt and hydrating medium. With reverse-phase evaporation method, encapsulation efficiency was best, followed by thin-film hydration and thin-film sonication. Fortification of milk with the ferrous sulfate liposomes increased the iron concentration to 15 mg/L. The liposomes were stable to heat treatment at 100°C for 30 min and over a period of 1 week when stored at 4°C.

Hilty et al. (2010) developed nanoparticles of iron and zinc compounds having a surface area of about 190 m^2/g. The change in color of the food to which the nanoparticles were added was less compared to other iron compounds that are generally used for fortification. Also the bioavailability of iron in these nanoparticles when tested in rats was similar to ferrous sulfate. Also addition of zinc to $FePO_4$ enhanced the absorption of iron. There was no detectable iron in the gut wall or in other tissues including lymphatics in the gastrointestinal tract. Based on this the investigators concluded that there were no adverse effects.

Jahna et al. (2011) prepared iron (III)-hydroxide NPs with a polymaltose shell and studied their uptake in Caco-2 cells. Uptake of nanoparticles involved carrier mediated transport, possibly involving receptor-mediated endocytosis. Pereira et al. (2013) used conditions to mimic the in vivo environment using Caco-2 cells to examine the uptake of iron from NPs. The NPs dissolved easily in mild acid environment that contained citrate, a mimic of lysosomal pH. They also observed that iron may be taken up by endocytosis. Formation of ferritin was observed indicating that the iron from the NPs was utilized by the cells. Subsequently, Pereira et al. (2014a) examined five nanoformulations containing Fe(III) oxo-hydroxide. Synthetic ferrihydrite was modified by incorporating small organic ligands. The particles were small and had a mean hydrodynamic diameter of 10 nm. When the NPs were added to the cell culture medium, they were fully dispersed and were acid soluble in simulated gastric conditions. Pereira et al. (2014b) developed a nanoparticulate tartrate-modified ferrihydrite ligand-(Fe(III) polyoxo-hydroxide) that is similar to the ferrihydrite core of ferritin, with the aim of examining whether this nanoformulation can be used for treatment of iron deficiency. They demonstrated that the efficacy of iron from the formulation was similar to ferrous sulfate in improving hemoglobin levels in a mouse model using CD1 mice.

Zhao et al. (2011) reported the development of NPs of synthetic Fe(III) poly oxo-hydroxide structure combined with dietary ligands of low molecular weight such as tartaric acid and succinic acid. Absorption efficiency of the iron from these formulations was about 80% of that of ferrous sulfate and 3 times more compared to ferric chloride salt. More than solubility in gastric acid, adhesion to the epithelial cells or uptake of the intact NP by the epithelial cells was an important determining factor. Results of toxicity studies showed that the iron in the NPs will have adverse cellular effects only at levels that are far above those used for iron supplementation. Soluble ferrous iron has been shown to be toxic at 1 mM concentration but the level of iron in the nanoformulation was about 7 times lower.

Rohner et al. (2007) investigated the bioavailability of ferric phosphate ($FePO_4$) from amorphous NP powder in rats. Solubility was studied in vitro, at pH 1 and 2. The hemoglobin repletion method was used to determine relative bioavailability by comparison with $FeSO_4$. Also, oxidative stress was assessed by measuring thiobarbituric acid reactive substances (TBARS). Solubility of the NPs was approximately 80% of $FeSO_4$ and relative bioavailability was between 70 and 96%. Using TBARS and histological examination as indicators, there appeared to be no evidence of toxicity.

It is possible that small particles between 5 and 10 mm have good relative bioavailability. Thus, it may be worthwhile to use small particle sizes for iron compounds with low solubility when the NPs are to be used for food fortification. The sensory properties of various food products, with incorporation of such NPs warrants study (Jani et al.,1990).

In Japan, a company Taiyo Kagaku manufactures a compound SunActive Fe™, that contains ferric pyrophosphate particles with average size of 0.5 μm. Although the particle size is larger than what is the size for NPs in general, it has been added to rice and in table salt and some beverage mixes.

Pereira et al. (2015) admininstered soluble $FeSO_4$ to severely anemic rats with diarrhea and high prevalence of the pathogen *Escherichia fergusonii*. When iron stores were repleted, there was an accompanying shift of the microbiota with bacteroides dominating. In contrast, when repletion of iron stores was achieved with a NP Fe(III) compound (tartrate-modifiedferrihydrite) the gut microbiota shifted toward the genus *Lactobacillus*. The authors suggested that nano Fe(III)) is readily bioavailable to the host, is acquired endocytically by the enterocyte and undergoes chelator-induced dissolution in the cell lysosome, but it is not easily available to bacteria. In mice weaned on an iron-deficient diet and maintained on it for 2 months, the gut microbial diversity was low whereas with nanoferric iron supplementation, it was better, although not to the same extent as was observed with $FeSO_4$. The authors suggested that availability of nanoiron to bacteria in the gastrointestinal tract may be less. This would be advantageous if nanoparticules of oral iron are given for repletion in gastrointestinal disease and enteropathogenic infections.

Nel et al. (2006) prepared magnetite and folate coated magnetite, the ratio of iron to folate being 7:1. The nanoiron and nanoiron with folate at three different doses (nanoiron, 87,174, and 348 mg/kg diet, and nanoiron plus folate, 87.3, 174.5, and 349 mg/kg diet) was administered to adult female rats. The doses chosen were equivalent to half, the requirement per se and twice the requirement of standard diets. These animals were compared with a group given ferrous sulfate. Weekly estimations of erythrocyte count, hemoglobin, mean corpuscular volume, mean corpuscular hemoglobin, mean corpuscular hemoglobin concentration, platelet count, and total leucocyte count were done for 8 weeks. The group receiving the dose supplying half the iron requirement achieved normal hemoglobin and hematocrit within 4 weeks whereas those in the other two groups reached it within 3 weeks but there was a decline thereafter. Histological examination of the livers of rats treated with magnetite showed cytoplasmic vacuolization of hepatocytes, infiltration of centrolobular hepatocytes with leucocytic inflammatory cells, sinusoidal leukocytosis, and portal infiltration with leucocytes. At the higher doses there was Kupffer cells activation, focal hepatic necrosis associated with leucocytic cell infiltration and apoptosis of hepatocytes. Similar histopathological alterations were seen in the epithelial lining of renal tubules and glomerular tufts in the kidneys.

Hafeli et al. (2009) also reported that nanoparticles may have toxic effects like inflammation, impairment of mitochondrial function, membrane leakage of lactate dehydrogenase and formation of ROS.

6.8 Zinc

Zinc is an important micronutrient, present in all cells and plays catalytic, structural, and regulatory roles in the body. It is an essential component of more than 300 enzymes including those involved in synthesis and breakdown of carbohydrates, proteins, lipids, nucleic acids. It stabilizes the structure of cellular components and membrane and plays an important role in genetic expression. It is also important for cellular and humoral immunity (Kohlmeier, 2003; Roohani et al., 2013). Studies on zinc NPs have been undertaken in applications other than nutrition such as cosmetics, paints, fillings in medical materials. The cargo loaded has been generally zinc oxide. The antibacterial activity of zinc oxide has been found to increase inversely in relation to particle size (Vaseem et al., 2009; Premanathan et al., 2011).

Rasmussen et al. (2010) delineated use of zinc NPs in cancer applications. Treatment with zinc NPs resulted in changes in metabolism of zinc, chaperonin proteins, and protein folding genes but the NPs did not have a pro-inflammatory effect. Cytotoxicity for the human colon carcinoma cells was influenced by whether there was direct contact between the particles and the cells but not the zinc concentration in the medium. Several effects were observed when LoVo human colon carcinoma cells were exposed to zinc oxide nanoparticles (size 50–70 nm) leading to decreased viability, oxidative stress, depolarization of the inner mitochondrial membrane, apoptosis, and IL-8 release. The cytotoxic effects were size dependent (De Berardis et al., 2010; Guan et al., 2012). Studies show that zinc oxide NPs have cytotoxic effects (Ostrovsky et al., 2009; Namvar et al., 2015).

Studies conducted with mice showed that zinc from the NPs was absorbed and concentrations in blood exhibited a dose response relationship. The NPs were concentrated in the liver, spleen, and kidney, but was different from animals administered zinc oxide microparticles. However, there were transient pathological changes in the liver with oral administration of the NPs but not with the microparticles. Damage was also observed in lung, liver, and kidney tissue in the animals administered the NPs. Also, there was significant increase in the levels of the two enzymes indicative of hepatic injury, that is, alanine and aspartate aminotransferases. Administration of zinc oxide NPs 100 nm in size was associated with accumulation of the NPs in the liver (dose response curves were seen), spleen, lung, bone, muscle, and kidney. Reduction in body weight has also been observed when animals were exposed to a high dose (Cho et al., 2013). The gender of the animals and size of the nanoparticles did not appear to play a role.

6.9 Calcium

Besides being a major constituent of bone, calcium plays an important role in basic regulatory functions that include contraction and relaxation of muscle, coagulation of blood, transmission of nerve impulses, activation of enzyme reactions, stimulation of hormone secretions, and integrity of intracellular cement substance (Kohlmeier, 2003). Various calcium supplements containing calcium carbonate, calcium citrate, calcium phosphate, or calcium gluconate are available. Among these, calcium carbonate is the cheaper form although its bioavailability is very low. Hence, there is potential for enhancing its bioavailability through nanoformulations. Kim et al. (2014) developed layered particles coated with enteric coating polymer, consisting of methacryalate and metharcylic acid. The polymer can be soluble at pH ≥ 6.0 that is compatible with intestinal pH. The particles did not exhibit cytotoxicity with human epithelial colorectal CaCo-2 cells, up to concentration of 1000 μg/mL. The absorption efficiency was approximately 8% that is almost 2 times the absorption efficiency of food grade calcium carbonate. Vijayakumar and Balakrishnan (2014) examined the effect of supplemention with calcium phosphate NPs on the growth of male broiler chicks. They reported that cumulative feed intake was significantly higher in chicks fed the calcium phosphate NPs and their weight gain was higher after 4 weeks of the feeding trial compared to birds fed conventionally by incorporating dicalcium phosphate into the broiler diet.

6.10 Selenium

Another mineral that has received attention is selenium. Selenium is a constituent of selenoproteins and is well known as an antioxidant. It is also a catalyst for synthesis of thyroxin and is required for immune functioning and sperm motility (Rayman, 2000). It has been shown to protect against inflammation and cancer.

When selenium NPs were administered in different doses to deficient mice, there was a dose-dependent increase in selenium-dependent enzymes as well as tissue selenium. There was a five- to sixfold increase in hepatic glutathione peroxidase activity with unheated selenium NPs and a threefold increase with heated selenium NPs. Similarly, hepatic thioredoxin reductase activity was almost 2 times higher (Zhang et al., 2012).

Proteins such as bovine serum albumin have been used and the particle size has been altered by using different concentrations of the protein. An inverse association was observed between the concentration of bovine serum albumin and the size of the selenium NPs. The type of nanoformulation may be important as it has been reported that with nanorods, the bioavailability was reduced. Similarly heat treatment would also have an adverse effect, with both temperature and duration of heat processing being important factors for conserving the bioactivity of the selenium nanoparticles. Thermostability of the NPs depends on their size, the smaller the particles the more resistant they are. The time temperature combination used for heat treatment of milk (62°C for 30 min) did not alter the particle size (Zhang et al., 2012).

6.11 Fatty Acids

Fatty acids are chemically diverse lipophilic molecules that can occur in free form or as part of triglycerides, phospholipids, glycolipids, lipoproteins, lipopolysaccharides, alkaloids, and so on. They perform important roles being structural components of biological membranes and as biologically active molecules in addition to being important dietary sources of energy. Polyunsaturated fatty acids both *n*-6 and *n*-3 have been shown to lower cholesterol levels. The polyunsaturated fatty acids as eicosapentaenoic acid (EPA) and docosahexaenoic acid (DHA) have important functions such as being the substrates for synthesis of molecules with anti-inflammatory action and being important for cognitive function, as they play a central role in normal development and functioning of the brain and central nervous system. EPA, DHA, and arachidonic acid are involved in neuronal processes, membrane fluidity as well as regulation of gene expression (Schuchardt et al., 2009). However, these fatty acids are lipophilic in nature and are highly susceptible to oxidation because they are unsaturated. Also EPA and DHA possess a typical fishy odor, thus limiting their acceptability. Therefore, NPs offer a means of overcoming this problem (Pereira et al., 2014c).

Medium chain fatty acids are nutritionally important because they are hydrolyzed by pancreatic lipase faster than are long chain fatty acids and after digestion are transported to the liver where they are preferentially metabolized by β-oxidation, through a carnitine independent pathway. Also they inhibit formation of body fat. However, like other fatty acids they are lipophilic, hence their inclusion in parenteral and oral formulas poses a challenge (Yu et al., 2013). Attempts have been made to use proteins like β-lactoglobulin to protect *n*-3 fatty acids. Zimet and Livney (2009) used this protein as it is a natural carrier for hydrophobic molecules. They prepared stable nanocomplexes of the protein combined with low methoxyl pectin. Use of excess pectin resulted in formation of negatively charged particles. Average particle size was approximately 100 nm and the DHA was protected against degradation. NPs subjected to 40°C for 100 h exhibited only 5–10% loss, in contrast to 80% for the unprotected DHA. Yang et al. (2013) formulated nanoliposomes encapsulating vitamin C and medium chain fatty acids with sucrose. The freeze-dried nanoliposomes were stable at 4°C for 60 days, without a significant increase in their particle size and no change in their spherical shape.

Liu et al. (2011) used nanoliposomes to encapsulate medium-chain fatty acids and evaluated the reduction in body fat for 2 weeks and 6 weeks in mice. Compared to the control animals, mice given the nanoliposomes had lower body fat,total cholesterol and triglyceride levels. The authors suggested that nanoliposomes carrying medium chain fatty acids may be useful to prevent accumulation of body fat.

Several researchers have worked on multilayered nanoformulations to encapsulate ω-3 fatty acids such as spray dried tuna oil-in-water emulsion droplets coated with chitosan and lecithin multilayer systems that were found to be stable and did not get rapidly oxidized (Zhao et al., 2011). The in vivo digestibility of lipids encapsulated with chitosan was not compromised, but such results need to be confirmed in humans. Chitosan microcapsules by themselves were not found to be as effective as dispersed emulsions in protecting the cargo EPA and DHA against oxidation. Hadian et al. (2014) prepared nanoliposomes using several methods such as extrusion, ultrasonic irradiation, bath sonication, probe sonication, and a combination of probe and bath sonication. The liposomes were spherical in shape and the structural integrity was good.

Omega-3 fatty acids have also been encapsulated with whey protein in order to develop fats with granular texture and are stable so that they can be used in processed foods like bread and meat and enhance their nutritional value while masking the fishy odor.

Conjugated linoleic acid (CLA) is another fatty acid shown to have the potential to help reduce body weight. This is attributed to reduction in lipid accumulation in adipocyte, decreased fat mass due to induction of apoptosis in adipocytes. In mice it has been shown to increase muscle mass as well promote use of fat for energy. Kim et al. (2013) developed very small nano CLA micelles. They compared their formulation with CLA and Orlistat, a drug for treating obesity. Orlistat inhibits lipase activity in the gastrointestinal tract and thus reduces absorption of dietary fat. However, it has side effects such as fatty/oily stools, fecal urgency, and absorption of fat soluble vitamins may also be reduced. There was reduction in leptin secretion by nano CLA, but less than Orlistat or CLA. The nano CLA showed significant lipolytic activity and degradation of triglycerides. The authors stated that the nano CLA did not exhibit cytotoxic effects on the adipocytes. There was significant inhibition of weight gain and the feed efficiency ratio for the animals was low. Overall the authors concluded that the efficiency of nano CLA was better than CLA in terms of reducing liver triglyceride and cholesterol, possibly by suppressing appetite concomitant with augmenting energy expenditure. It was absorbed into cells and organs indicating that it is bioavailable.

NICE DB (2013) has suggested that fatty acids can be nanoencapsulated with whey protein to prepare granular fats that may be used in food products such as breads, processed meat, sweet foods, and beverages. The encapsulation would mask the flavor and odor of the fatty acids harvested from fish or algae.

7 Other Applications

Azevedo et al. (2014) developed alginate/chitosan NPs for controlled release of riboflavin. Polymeric relaxation was found to be important for release of the vitamin. The NPs were stable for 5 months.

7.1 Probiotics

Probiotics improve gut health and have been reported to have other health promoting effects including immune performance, antioxidant effect, facilitating mineral absorption, reduction of allergic symptoms, lowering cholesterol, and so on (Ooi and Liong, 2010). For these benefits to be accrued probiotic viability and activity are critical. Studies have been undertaken for encapsulating probiotics, alginate beads coated with chitosan or spraying of particles with chitosan. These render the formulation shelf stable during refrigerated storage with the bacterial cells being viable. When different coating materials were compared, alginate beads with chitosan coating performed better.

8 Importance of Nanoparticle Characteristics and Barriers to Nanoparticle-Mediated Delivery

Several studies of nanoparticle synthesis, production, and formulation have been reported elucidating the different types and their varied applications. Nanoparticles of various types have been studied including inorganic nanoparticles such as gold, titanium oxide, and silver; polymeric nanoparticles produced by emulsion-based or nanofabrication techniques; self-assembled; hydrogel-based; supramolecular; lipid-based such as solid lipid or liposomal; peptide or protein based. However, despite their varied nature, all these nanoparticles need to have optimized characteristics to ensure efficient delivery or the nutrient or bioactive to the target site. These characteristics include:

1. *Size*: Nanoparticles that are too small, that is, lower than 4–8 nm, will undergo rapid renal filtration hence requiring large doses or repeated administrations to ensure efficacy. On the other hand, nanoparticles that are too large, that is, >700–800 nm will not be taken up by cells and also have the risk of bioaccumulation in key organs such as lungs, kidney, spleen, and so on.
2. *Shape*: Various reports, especially studies, conducted by Discher and coworkers (Rodriguez et al., 2013) and Champion and Mitragotri (2009) indicate that spherical nanoparticles have a lower circulation time when administered intravenously, as compared to worm-like or rod-like particles. This is attributed to the fact that spherical particles undergo rapid macrophage uptake, which hence lowers the circulation time compared to worm-like or rod-like particles that can evade macrophage uptake.
3. *Surface charge*: The particles need to be either neutral charged or slightly negatively charged to avoid opsonization by serum proteins or uptake by macrophages or the reticulo-endothelial system.

Hence, several barriers exist that can affect the efficient delivery of nutrients or bioactives mediated by nanoparticles. Extracellular barriers that nanoparticles usually face are serum stability when administered intravenously, stability in GI tract when administered orally, and appropriate size, surface charge, and surface characteristics to avoid renal filtration. The nanoparticles can be administered either locally, intravenously (IV), or orally. If administered locally, the vectors need to penetrate into the tissue and enter the target cells. When administered by IV, the vectors need to have serum stability and have long enough circulation times in the body so that they can effectively extravasate from the circulation into the target site (Fig. 7.3). When administered orally, the nanoparticle carrier needs to be stable in the GI tract in harsh conditions such as low pH and presence of several degradative enzymes. Also, the particles need to then diffuse across the GI tract to the target site to achieve efficacy. Oral delivery is generally preferred for administration of nutrients or nutraceutical supplements as well as drugs. This is because of patient compliance, lower production costs, and ease of administration (Renukuntla et al., 2013).

Additionally, after administration of the nanoparticles, to achieve high target specificity and limit off-target effects, a targeting strategy needs to be used. The nanoparticles can either be targeted passively or actively. In passive targeting, no targeting ligand or antibody is employed, and merely the surface characteristics and physical characteristics of the nanoparticles would be exploited to ensure tissue uptake. Herein, the nanoparticles can be designed so that they are neutrally charged, has high serum/GI tract stability and the appropriate size to avoid renal filtration or macrophage uptake.

Active targeting to cells can be achieved by attachment of certain targeting ligands such as small molecules, peptides, proteins, or antibodies. The ligands are able to bind to receptors that are overexpressed on the surface of the cell of interest hence ensuring specificity of the targeted vector toward these cells over normal cells (Fig. 15.3).

Once the nanoparticles are able to circumvent the extracellular barriers and penetrate into the target tissue, they are able to attach to the target cell and internalize. On internalization into the cell, the nanoparticles need to face a host of intracellular barriers before the nutrient can be released and its therapeutic effect can be

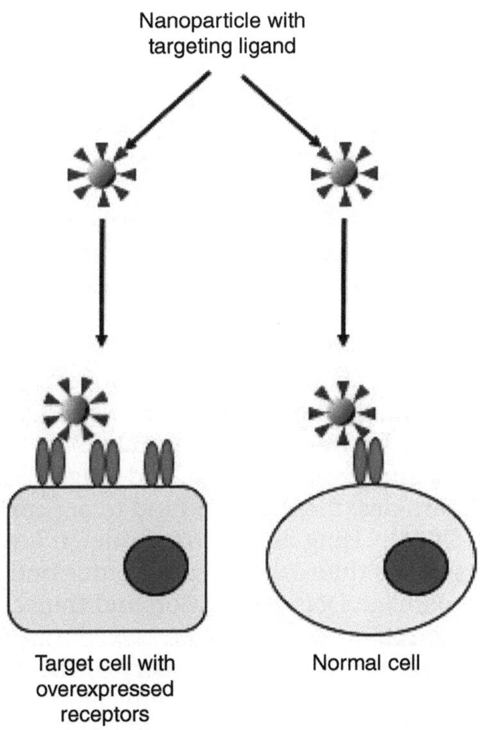

Figure 15.3. Multivalent binding of NPs to targeting ligands on cell surfaces with overexpressed receptors and normal cells.

achieved. On endocytosis, the nanoparticles need to escape from the endosome so that the harsh endolysosomal conditions do not degrade the nutrient cargo. The nanoparticles then need to unpackage and release the cargo to achieve the desired effect.

9 Safety and Toxicity Implications of Nanotechnology in Food Science

Although nanotechnology has a number of potential benefits, the use of nanoparticles can also lead to health complications. Currently, knowledge regarding the biological outcomes of consumption of nanoparticle fortified foods and their possible side effects in humans is still limited and relatively unexplored. There is a need to consider risks posed by use of nanotechnology and establish protocols for their risk assessment in order to ensure safe utilization of this promising technology (Magnuson et al., 2013).

9.1 Toxic Potential of Nanoparticles

The unique characteristics of NPs which make them useful for varied application can also be the cause for toxicity and safety issues. Some of the unique physicochemical properties of NPs are their small size, high surface area, chemical composition, surface structure, surface charge, and shape. All these properties make them more reactive than their bulk forms (Nel et al., 2006; Amini et al., 2014). Among these various characteristics, the size and surface area of NPs are important for the potential side effects. As the size decreases, the surface area increases, as a result greater proportion of atoms are displayed on the surface rather than in the interior. This could cause alterations in the electronic properties, which could then change reactive groups on the surface (Nel et al., 2006). Additionally, the small size allows NPs to pass through different biological barriers and accumulate in various tissues such as the liver, kidney, spleen, brain, and lungs. When mice were fed 4 nm gold particles, these were found to accumulate in these organs (FOEA, 2014). Using an in vitro model, it has been shown that particles smaller than 70 nm could enter human epithelial cell nuclei and damage DNA replication and transcription (Chen and Von Mickecz, 2005).

Besides the size, the surface chemistry of the NPs is another characteristic that can cause toxic side effects. For example, coating of NPs with polyethylene glycol could reduce the side effects (Amini et al., 2014) by giving it stealth properties that lower the rates of serum opsonization and macrophage uptake. On the other hand, surface active sites may participate in the generation of ROS and may thus induce oxidative stress (Nel et al., 2006). ROS generation can stimulate inflammatory or immune responses, cytotoxicity, genotoxicity, and can also have adverse effects on respiratory, cardiovascular, and nervous systems. In fact, generation of ROS is one of the important mechanisms implicated in NP toxicity (Martirosyan et al., 2014).

Surface charge of NPs also plays an important role in their targeting and uptake properties and most importantly their toxic side effects. Positively charged NPs undergo nonspecific uptake in various cells and tissues in the body and can then cause cytotoxicity and other side effects. On the other hand neutrally or mildly negatively charged particles do not undergo nonspecific uptake and hence have a lower risk of toxicity.

Solubility plays a vital role wherein soluble NPs have been shown to be more toxic than insoluble NPs due to their ease of uptake in various tissues. Soluble titanium oxide NPs have been shown to be more toxic than the insoluble particles (Amini et al., 2014).

Naahidi et al. (2013) pointed out that NPs have adjuvant properties and can influence the immune system in different ways such as immune stimulation and inflammation as well as immune suppression. This and particle clearance are considerations when a particle is to be considered for specific applications. The immune cells recognize the NP as foreign depending on the surface charge. Particles with a positive charge provoke immune response more than anionic or neutral particles. Tan et al. (1999) observed that cationic but not anionic particles were associated with cytokine secretion. Diwan et al. (2004) observed that exposure to oligonucleotides bound to PLGA-based nanoparticles resulted in large amount of cytokine secretion and greater proliferation of T-cells in contrast to naked oligonucleotides. Thus, NPs will differ in their toxicological effects based on their composition and physicochemical properties.

Among the various NPs, metal/metal oxide NPs have many applications in food industry. Toxicological effects of these nanomaterials have been investigated using in vitro studies and in vivo studies using experimental animals and the results have been delineated in recent reviews (Shi et al., 2013; Martirosyan and Schneider, 2014). It is reported that a high dose of silver NPs and titanium oxide have the potential to cross the blood brain barrier and cause neurotoxicity in rats and mice. NPs of silver, titanium oxide, and zinc oxide accumulate in the liver, kidneys, spleen, and can have localized toxic effects. Acute inhalation of zinc oxide at a dose of 2.5 mg/kg body weight in rats have been shown to damage liver and lung tissues.

Hong et al. (2014) studied the effect of zinc oxide NPs during pregnancy. These researchers administered zinc oxide NPs to rats from day 5 through 19 days of pregnancy at dose levels of 0,100, 200, and 400 mg/kg per day. The results showed that repeated oral dosing with zinc oxide NPs for 15 days during pregnancy resulted in maternal toxicity at 400 mg/kg per day, however, this dose did not cause serious teratogenic toxicity.

Although NPs have the potential for toxicity and have been shown to have adverse consequences, most of these effects were observed using in vitro models or in vivo studies with experimental animals. Moreover, the doses administered in these studies were very high. Therefore, these results cannot directly be extrapolated to humans in real-life situations. Also, many of these effects have been studied on exposure through inhalation. The effects of nanoparticles after oral consumption in humans is not known. To our knowledge there exist no studies to study the effects of chronic exposure of humans to nanoparticles in the food science sector.

However, in view of large number of nanofoods and nanofood packaging already in the market, which increase consumer exposure to NPs and the fact that NPs can cross cellular barriers and induce oxidative stress, safety aspects of nanotechnology in foods should be considered.

9.2 Risk Assessment: Issues and Knowledge Gaps

In order to address the safety of NPs for humans, there is a need to understand the risks associated with NPs in foods following oral exposure. Currently there are gaps in knowledge regarding human risk assessment of NPs in food. Some of these aspects are:

1. NPs are used in food packaging materials and as sensors. Very little information is available on the possible migration of these components. A study conducted to evaluate food contact materials had tested the migration of NPs from nanoclay and nanosilver. Migration of silver occurred at very low level while nanoclay did not migrate at all. Also presence of these NPs did not affect migration of other particles (Science and Technology Committee, 2010). However, migration pattern may vary for other NPs and also may vary from food to food (depending on the physical and chemical properties) for the same NP.
2. In order to assess the risk, accurate information is required on the characteristics of size, shape, structure of NPs used in various foods. Further, it is not clear whether NPs behave differently in different food matrices. Therefore, in view of the complexity of food, the characterization should be done in the matrix containing NPs.

3. In order to assess the potential risk posed by NPs information on the dose used is required with respect to NPs. Size, surface area, surface characteristics, surface charge, and shape are all important to describe the dose. Probably, the parameters used to describe dose may vary from food to food. There is a need to undertake research to describe dose metrics.
4. Risk assessment is possible if relevant information is available on the fate of ingested NPs in the gastrointestinal tract. As seen in Fig. 15.4, the ingested materials may remain in the gastrointestinal tract and exert local effects or cross the gut wall and may have systemic effects (Powell et al., 2010). Uptake from the gastrointestinal tract depends on the diffusion of NPs through mucus, which is determined by a number of factors including the charge on the particles. After diffusion, uptake depends on the translocation of NPs through gut epithelium. This can occur either by the paracellular route or transcellular route. It is not known whether NPs can pass through the paracellular pathway; however, this possibility cannot be ruled out. Additionally, transcellular uptake may be influenced by specific characteristics of NPs. Once NPs cross the GI epithelium, it is not fully clear as to how these are distributed, metabolized, and excreted. The physicochemical properties, which may determine the fate of ingested NPs, depend on the interaction of NPs with food matrix as well as the contents of the gastrointestinal tract and its condition. It remains to be elucidated how

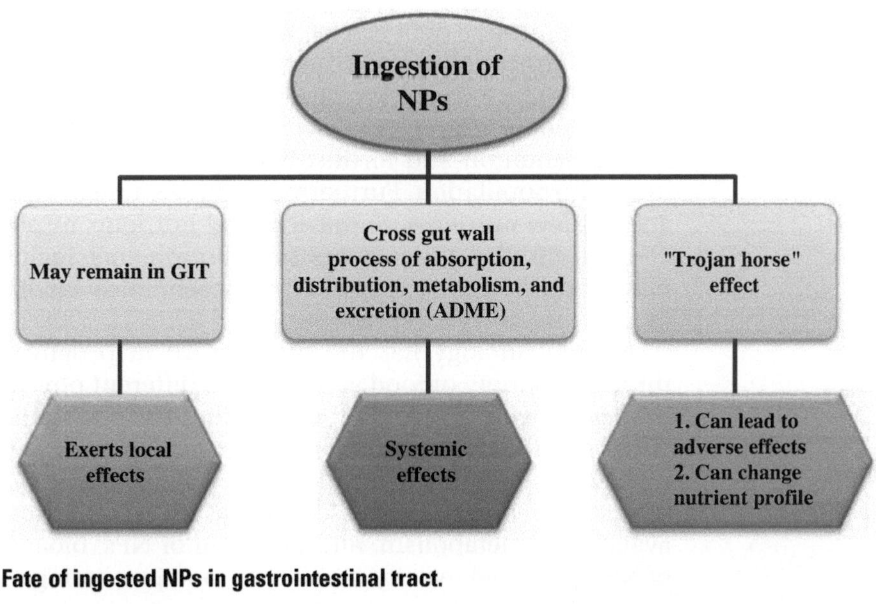

Figure 15.4. Fate of ingested NPs in gastrointestinal tract.

diseased gut will handle NPs and whether NPs can be given to persons with gastrointestinal, hepatic, or renal problems. This raises questions about fortified foods that are available to the masses. Whether the individual ingests NPs in the fed or fasting state may also be an important factor (McClements and Li, 2000).

5. As seen in Fig. 15.4, NPs added to the food to enhance nutrient delivery could also simultaneously increase uptake of unintended foreign materials, which can lead to adverse effects, known as the "Trojan Horse" effect.

Dwyer et al. (2015) stated that factors such as high aspect ratio and biopersistence need to be considered while assessing safety and biocompatibility of NPs used to deliver nutrients. Particulate NPs may accumulate in the reticuloendothelial system, which may adversely affect the immune system. Also, risk assessment should consider whether the exposure to NPs is acute or chronic vis-à-vis the dose.

All the above mentioned issues (Bouwmeester et al., 2009; Hwang et al., 2012; Yada et al., 2014) clearly point out the dire need for fundamental research to elucidate the fate of NPs vis-à-vis their interaction with food matrices and their contents, before regulatory standards are in place. There is also a need to examine the influence of long-term storage of foods fortified with NPs.

10 Conclusions

Micronutrient deficiencies are prevalent in India and other developing economies and their adverse consequences have a significant impact on the socioeconomic development of the country. Currently, strategies to overcome these deficiencies include supplementation and fortification, however, these do not cover the entire population. Furthermore, these strategies are limited to a select few nutrients. Supplying vital nutrients of public health concern through foods poses several problems, including instability of nutrients in the typical physicochemical environments of foods.

Nanotechnology has the potential to help deliver nutrients through a variety of food vehicles with different physicochemical properties while ensuring their stability. Various studies have attempted nanoencapsulation techniques of some of the important nutrients, however, the technology is still in its infancy. Several issues still need to be addressed such as digestion,absorption, bioavailability, metabolism, and excretion of NPs, bioaccumulation of NPs, potential toxicity, stability to food processing techniques/

methods, influence of storage at varied temperatures and risk assessment. Detailed studies to specifically understand the efficacy of nanoparticle mediated delivery of nutrients, and the chronic exposure to nanoparticles still need to be carried out before the technology can be fully exploited.

References

Abbasi, A., Emam-Djomeh, Z., Ebrahimzadeh, M.A., Mousavi, M.A., Davoodi, D., 2014. Stability of vitamin D3 encapsulated in nanoparticles of whey protein isolate. Food Chem. 143, 379–383.

Acosta, E., 2009. Bioavailability of nanoparticles in nutrient and nutraceutical delivery. Curr. Opin. Colloid Interface Sci. 14, 3–15.

Alishahi, A., Mirvaghefi, A., Tehrani, M.R., Farahmand, H., Shojaosadati, S.A., Dorkoosh, F., 2011a. Shelf life and delivery enhancement of vitamin C using chitosan nanoparticles. Food Chem. 126 (3), 935–940.

Alishahi, A., Mirvaghefia, A., Tehrani, M.R., Farahmanda, H., Koshio, S., Dorkoosh, F.A., Elsabee, M.Z., 2011b. Chitosan nanoparticle to carry vitamin C through the gastrointestinal tract and induce the non-specific immunity system of rainbow trout (*Oncorhynchus mykiss*). Carbohydr. Polym. 86, 142–146.

Almouazen, E., Bourgeois, S., Boussaïd, A., Valot, P., Malleval, C., Fessi, H., Nataf, S., Briançon, S., 2012. Development of a nanoparticle-based system for the delivery of retinoic acid into macrophages. Int. J. Pharm. 430 (1–2), 207–215.

Amini, M., Gilaki, M., Karchani, M., 2014. Safety of nanotechnology in food industries. Electron. Physician 6 (4), 962–968.

Anais, J.P., Razzouq, N., Carvalho, M., Fernandez, C., Astier, A., Paul, M., Astier, A., Fessi, H., Lorino, A.M., 2009. Stabilization of all-trans retinol by loading lipophilic antioxidants in solid lipid nanoparticles. Drug Dev. Ind. Pharm. 35 (2), 216–223.

Anand, P., Nair, H.B., Sung, B., Kunnumakkara, A.B., Yadav, V.R., Tekmal, R.R., Aggarwal, B.B., 2010. Design of curcumin-loaded PLGA nanoparticles formulation with enhanced cellular uptake, and increased bioactivity in vitro and superior bioavailability in vivo. Biochem. Pharmacol. 79 (3), 330–338.

Astete, C.E., Sabliov, C.M., 2006. Synthesis and characterization of PLGA nanoparticles. J. Biomater. Sci. Polym. Ed. 17 (3), 247–289.

Azevedo, M.A., Bourbon, A.I., Vicente, A.A., Cerqueira, M.A., 2014. Alginate/ chitosan nanoparticles for encapsulation and controlled release of vitamin B2. Int. J. Biol. Macromol. 71, 141–146.

Benichou, A., Aserin, A., Garti, N., 2004. Double emulsions stabilized with hybrids of natural polymers for entrapment and slow release of active matters. Adv. Colloid Interface Sci. 108–109, 29–41.

Benichou, A., Aserin, A.A., Garti, N., 2007. O/W/O double emulsions stabilized with WPI–polysaccharide complexes. Colloids Surf. A. 294 (1–3), , 20–32, 211–220.

Bertrand, N., Leroux, J.C., 2012. The journey of a drug-carrier in the body: an anatomophysiological perspective. J. Control. Release 161 (2), 152–163.

Biehler, E., Kaulmann, A., Krause, E., Hoffman, L., Bohn, T., 2011. Dietary and host-related factors influencing carotenoid accessibility from spinach (*Spinacia olearacea*). Food Chem. 125, 1328–1334.

Bohn, T., 2012. Provitamin A carotenoids: occurrence, intake and bioavailability. In: Preedy, V. (Ed.), Vitamin A and Carotenoids. Chemistry, Analysis, Function and Effects. RSC Publishing, Cambridge, UK, p. 142 (Chapter 10).

Bouwmeester, H., Dekkers, S., Noordam, M.Y., Hagens, W.I., Bulder, A.S., de Heer, C., ten Voorde, S.E.C.G., Wijnhoven, S.W.P., Marvin, H.J.P., Sips, A.J.A.M., 2009. Review of health safety aspects of nanotechnologies in food production. Reg. Toxicol. Pharmacol. 53, 52–62.

Champion, J.A., Mitragotri, S., 2009. Shape induced inhibition of phagocytosis of polymer particles. Pharm. Res. 26, 244–249.

Chen, M., Von Mickecz, A., 2005. Formation of nucleoplasmic proteins. Exp. Cell Res. 305, 51–62.

Chen, M.C., Sonaje, K., Chen, K.J., Sung, H.W., 2011. A review of the prospects for polymeric nanoparticle platforms in oral insulin delivery. Biomaterials 32 (36), 9826–9838.

Chen, D., Xia, D., Li, X., Zhu, Q., Yu, H., Zhu, C., Gan, Y., 2013. Comparative study of pluronic F127-modified liposomes and chitosan-modified liposomes for mucus penetration and oral absorption of cyclosporine A in rats. Int. J. Pharm. 449 (1–2), 1–9.

Cherian, A.M., 2009. Feasibility of soluble leaf proteins as a carrier for vitamin D. Master's thesis submitted to the Graduate School of the University of Maryland, College Park, MD.

Chithrani, B.D., Ghazani, A.A., Chan, W.C., 2006. Determining the size and shape dependence of gold nanoparticles uptake into mammalian cells. Nano Lett. 6 (4), 662–668.

Cho, E.C., 2012. Effect of polymer characteristics on the thermal stability of retinol encapsulated in aliphatic polyester nanoparticles. Bull. Korean Chem. Soc. 33 (8), 2560–2566.

Cho, Y., Kim, J.T., Park, H.J., 2012. Size-controlled self-aggregated N-acyl chitosan nanoparticles as a vitamin C carrier. Carbohydr. Polym. 88, 1087–1092.

Cho, W.-S., Kang, B.-C., Lee, J.K., Jeong, J., Che, J.-H., Seok, S.H., 2013. Comparative absorption, distribution, and excretion of titanium dioxide and zinc oxide nanoparticles after repeated oral administration. Part. Fibre Toxicol. 10, 9–18.

Choi, H.S., Liu, W., Misra, P., Tanaka, E., Zimmer, J.P., Ipe, B.I., Bawendi, M.G., Frangioni, J.V., 2007. Renal clearance of nanoparticles. Nat. Biotechnol. 25 (10), 1165–1170.

Choi, C.H.J., Zuckerman, J.E., Webster, P., Davis, M.E., 2011. Targeting kidney mesangium by nanoparticles of defined size. Proc. Natl. Acad. Sci. USA 108 (16), 6656–6661.

Conrad, M.E., Umbreit, J.N., Moore, E.G., Uzel, C., Berry, M.R., 1994. Alternate iron transport pathway—mobilferrin and integrin in K562 cells. J. Biol. Chem. 269 (10), 7169–7173.

Crider, K.S., Bailey, L.B., Berry, R.J., 2011. Folic acid food fortification—its history, effect, concerns, and future directions. Nutrients 3 (3), 370–384.

Cu, Y., Saltzman, W.M., 2009. Controlled surface modification with poly(ethylene) glycol enhances diffusion of PLGA nanoparticles in human cervical mucus. Mol. Pharm. 6 (1), 173–181.

D'Agostino, E.M., Rossetti, D., Atkins, D., Ferdinando, D., Yakubov, G.E., 2012. Interaction of tea polyphenols and food constituents with model gut epithelia: the protective role of the mucus gel layer. J. Agric. Food. Chem. 60, 3318–3328.

Dary, O., Mora, J.O., 2002. Food fortification to reduce vitamin A deficiency: International Vitamin A Consultative Group recommendations. J. Nutr. 132 (9), 2927S–2933S.

De Berardis, B., Civitelli, G., Condello, M., Lista, P., Pozzi, R., Arancia, G., Meschini, S., 2010. Exposure to ZnO nanoparticles induces oxidative stress and cytotoxicity in human colon carcinoma cells. Toxicol. Appl. Pharmacol. 246 (3), 116–127.

de Britto, D., de Moura, M.R., Aouada, F.A., Pinola, F.G., Lundstedt, L.M., Assis, O.B.G., Mattoso, L.H.C., 2014. Entrapment characteristics of hydrosoluble vitamins loaded into chitosan and *N,N,N*-trimethyl chitosan nanoparticles. Macromol. Res. 22 (12), 1261–1267.

De Jong, W.H., Hagens, W.I., Krystek, P., Burger, M.C., Sips, A.J., Geertsma, R.E., 2008. Particle-size dependent organ distribution of gold nanoparticles after intravenous administration. Biomaterials 29 (12), 1912–1919.

Diwan, M., Elamanchili, P., Cao, M., Samuel, J., 2004. Dose sparing of CpG oligodeoxynucleotide vaccine adjuvants by nanoparticle delivery. Curr. Drug Deliv. 1, 405–412.

Domingues, N.J.A., 2013. Carrier systems for vitamin D. Master's dissertation submitted to Universidade de Porto.

Dube, A., Nicolazzo, J.A., Larson, I., 2010. Chitosan nanoparticles enhance the intestinal absorption of the green tea catechins (+)-catechin and (−)-epigallocatechin gallate. Eur. J. Pharm. Sci. 41, 219–225.

Duclairoir, C., Orecchioni, A.M., Depraetere, P., Nakache, E., 2002. Alpha-tocopherol encapsulation and in vitro release from wheat gliadin nanoparticles. J. Microencapsul. 19 (1), 53–60.

Dwyer, J.T., Wiemer, K.L., Dary, O., Keen, C.L., King, J.C., Miller, K.B., Martin, A., Philbert, M.A., Tarasuk, V., Taylor, C.L., Gaine, P.C., Jarvis, A.B., Bailey, R.L., 2015. Fortification and health: challenges and opportunities. Technological issues with use of nanomaterials for fortification. Adv. Nutr. 6, 124–131.

End, L., 2006. Nanoscale formulations of health ingredients. Fine Chemicals Division R&D Formulation-BASF, Bethesda, MD.

Esteves, M., Cristóvão, A.C., Saraiva, T., Rocha, S.M., Baltazar, G., Ferreira, L., Bernardino, L., 2015. Retinoic acid-loaded polymeric nanoparticles induce neuroprotection in a mouse model for Parkinson's disease. Front. Aging Neurosci. 7, 20–39.

Faraji, A.H., Wipf, P., 2009. Nanoparticles in cellular drug delivery. Bioorg. Med. Chem. 17, 2950–2962.

Farhang, B., 2013. Encapsulation of bioactive compounds in liposomes prepared with milk fat globule membrane-derived phospholipids. MSc thesis submitted to University of Guelph.

Farhang, B., Kakuda, Y., Corredig, M., 2012. Encapsulation of ascorbic acid in liposomes prepared with milk fat globule membrane-derived phospholipids. Dairy Sci. Technol. 92, 353–366.

FOEA, Friends of Earth Australia, 2014. Way too little. Our government's failure to regulate nanomaterials in food and agriculture, May.

Fortitech Strategic Nutrition, 2011. New innovations in functional beverages.

Fröhlich, E., Roblegg, E., 2012. Models for oral uptake of nanoparticles in consumer products. Toxicology 291 (1–3), 10–17.

Garcia-Fuentes, M., Torres, D., Alonso, M.J., 2003. Design of lipid nanoparticles for the oral delivery of hydrophilic molecules. Colloids Surf. B 27 (2–3), 159–168.

Grebler, S., Gazso, A., Simko, M., Fiedeler, U., Nentwich, M., 2010. Nanotechnology in cosmetics. Nano Trust Dossiers 8, 1–6.

Guan, R., Kang, T., Lu, F., Zhang, Z., Shen, H., Liu, M., 2012. Cytotoxicity, oxidative stress, and genotoxicity in human hepatocyte and embryonic kidney cells exposed to ZnO nanoparticles. Nanoscale Res. Lett. 7, 602–609.

Guri, A., Gülseren, I., Corredig, M., 2013. Utilization of solid lipid nanoparticles for enhanced delivery of curcumin in cocultures of HT29-MTX and Caco-2 cells. Food Funct. 4, 1410–1419.

Hadian, Z., Sahari, M.A., Moghimi, H.R., Barzegar, M., 2014. Formulation, characterization and optimization of liposomes containing eicosapentaenoic and docosahexaenoic acids; a methodology approach. Iran. J. Pharm. Res. 13 (2), 393–404.

Hafeli, U.O., Riffle, J.S., Harris-Shekhawat, L., Carmichael- Baranauskas, A., Mark, F., Dailey, J.P., Bardenstein, D., 2009. Cell uptake and in vitro toxicity of magnetic nanoparticles suitable for drug delivery. Mol. Pharm. 6, 1417–1428.

Haham, M., Ish-Shalom, S., Nodelman, M., Duek, I., Segal, E., Kustanovich, M., Livney, Y.D., 2012. Stability and bioavailability of vitamin D nanoencapsulated in casein micelles. Food Funct. 3 (7), 737–744.

Hilty, F.M., Arnold, M., Hilbe, M., Teleki, A., Knijnenburg, J.T., Ehrensperger, F., Hurrell, R.F., Pratsinis, S.E., Langhans, W., Zimmerman, M.B., 2010. Iron from nanocompounds containing iron and zinc is highly bioavailable in rats without tissue accumulation. Nat. Nanotechnol. 5 (5), 374–380.

Hirn, S., Semmier-Behenke, M., Schel, C., WEnk, A., Lipka, J., Schaffler, M., Takenaka, S., Möller, W., Schmid, G., Simon, U., Kreyling, W.G., 2011. Particle size dependent and surface charge dependent biodistribution of gold particles after intravenous administration. Eur. J. Pharm. Biopharm. 77 (3), 407–416.

Hong, J.S., Park, M.K., Kim, M.S., Lim, J.H., Park, G.J., Maeng, E.H., Shin, J.H., Kim, M.K., Jeong, J., Park, J.A., Kim, J.C., Shin, H.C., 2014. Prenatal development toxicity study of zinc oxide nanoparticles in rats. Int. J. Nanomed. 9 (Suppl. 2), 159–171.

Hu, B., Pan, C., Sun, Y., Hou, Z., Ye, H., Hu, B., Zeng, X.X., 2008. Optimization of fabrication parameters to produce chitosan-tripolyphosphate nanoparticles for delivery of tea catechins. J. Agric. Food Chem. 56, 7451–7458.

Huynh, P.T., 2012. Solvent-free beta-carotene nanoparticles for food fortification. PhD dissertation submitted to Rutgers, State University of New Jersey, New Brunswick.

Hwang, M., Lee, E., Kweon, S., Park, M.S., Jeong, J.Y., Um, J.H., Kim, S.A., Han, B.S., Lee, K.H., Yoon, H.J., 2012. Risk assessment principle for engineered nanotechnology. Food Drug Toxicol. Res. 28 (2), 73–79.

Iqbal, K., Khan, A., Khattak, M.A.K., 2004. Biological significance of ascorbic acid (vitamin C) in human health: a review. Pakistan J. Nutr. 3 (1), 5–13.

Jahna, M.R., Shukoorb, I., Tremelb, W., Wolfrumc, U., Kolbd, U., Nawrotha, T., Peter Langgutha, P., 2011. Hemin-coupled iron(III)-hydroxide nanoparticles show increased uptake in Caco-2 cells. J. Pharm. Pharmacol. 63, 1522–1530.

Jani, P.J., Halbert, G.W., Langridge, J., Florence, A.T., 1990. Nanoparticle uptake by the rat gastrointestinal mucosa: quantitation and particle size dependency. J. Pharm. Pharmacol. 42, 821–826.

Jee, J.-P., Soo-Jeong Lim, S.-J., Jeong-Sook Park, J.-S., Kim, C.-K., 2006. Stabilisation of all-trans retinol by loading lipophilic antioxidants in solid nanoparticles. Eur. J. Pharm. Biopharm. 63, 134–139.

Jiménez-Fernández, E., Ruyra, A., Roher, N., Zuastia, E., Infantea, C., Fernández-Díaz, C., 2014. Nanoparticles as a novel delivery system for vitamin C administration in aquaculture. AQUA-631071. Available from: http://dx.doi.org/10.1016/j.aquaculture2014. 03.006

Karlsson, J., Wikman, A., Artursson, P., 1993. The mucus layer as a barrier to drug absorption in monolayers of human intestinal epithelial HT 29-H goblet cells. Int. J. Pharm. 99, 209–218.

Kim, D., Park, J.-H., Kweon, D.-J., Han, D.G., 2013. Bioavailability of nanoemulsified conjugated linoleic acid for an antiobesity effect. Intern. J. Nanomed. 8, 451–459.

Kim, T.-H., Lee, J.-A., Choi, S.-J., Oh, J.-M., 2014. Polymer coated CaAl-layered double hydroxide nanomaterials for potential calcium supplement. Int. J. Mol. Sci. 15, 22563–22579.
Kirby, C.J., Whittle, C.J., Rigby, N., Coxon, D.T., Law, B.A., 1991. Stabilization of ascorbic acid by microencapsulation in liposomes. Int. J. Food Sci. Technol. 26, 437–449.
Kitts, D.D., Liu, Y., 2015. Encapsulation strategies to stabilize a natural folate, L-5-methyltetrahydrofolic acid, for food fortification practices. In: Sabliov, C., Chen, H., Yada, R. (Eds.), Nanotechnology and Functional Foods: Effective Delivery of Bioactive Ingredients. John Wiley and Sons, UK, pp. 142–157.
Klemetsrud, T., Jonassen, H., Hiorth, M., Kjøniksen, A.L., Smistad, G., 2013. Studies on pectin-coated liposomes and their interaction with mucin. Colloids Surf. B 103, 158–165.
Kohlmeier, M., 2003. Nutrient Metabolism. Academic Press, Elsevier, London.
Lai, S.K., O'Hanlon, D.E., Harrold, S., Man, S.T., Wang, Y.Y., Cone, R., Hanes, J., 2007. Rapid transport to large polymeric nanoparticles in fresh undiluted human mucus. Proc. Natl. Acad. Sci. USA 104 (5), 1482–1487.
Lai, S.K., Wang, Y.-Y., Hanes, J., 2009. Mucus-penetrating nanoparticles for drug and gene delivery to mucosal tissues. Adv. Drug Deliv. Rev. 61 (2), 158–171.
Larhed, A.W., Artursson, P., Gräsjö, J., Björk, E., 1977. Diffusion of drugs in native and purified gastrointestinal mucus. J. Pharm. Sci. 86, 660–665.
Liu, Y., 2014. Starch-pectin matrices for encapsulation of ascorbic acid. Master's thesis submitted to the University of Nebraska.
Liu, N., Park, H.J., 2010. Factors effect on the loading efficiency of Vitamin C loaded chitosan-coated nanoliposomes. Colloids Surf. B 76 (1), 16–19.
Liu, W.L., Liu, W., Liu, C.M., Yang, S.B., Liu, J.H., Zheng, H.J., Su, K.M., 2011. Medium-chain fatty acid nanoliposomes suppress body fat accumulation in mice. Br. J. Nutr. 106 (9), 1330–1336.
Liu, Y., Hou, Z., Lei, F., Chang, Y., Gao, Y., 2012. Investigation into the bioaccessibility and microstructure changes of β-carotene emulsions during in vitro digestion. Innov. Food Sci. Emerg. Technol. 15, 86–95.
Livney, Y.D., 2010. Milk proteins as vehicles for bioactives. Curr. Opin. Colloid Interface Sci. 15 (1–2), 73–83.
Lu, Y., Feskens, E.J., Boer, J.M., Muller, 2010. The potential influence of genetic variants in genes along bile and bile metabolic pathway on blood cholesterol levels in the population. Atherosclerosis 210 (1), 14–27.
Luo, Y., 2012. Development of food polymer-based colloidal delivery systems for nutraceuticals. PhD dissertation submitted to the faculty of the graduate school of the University of Maryland.
Luo, Y., Zhang, B., Cheng, W.H., Wang, Q., 2010. Preparation, characterization and evaluation of selenite-loaded chitosan/TPP nanoparticles with or without zein coating. Carbohydr. Polym. 82, 942–951.
Macierzanka, A., Rigby, N.M., Corfield, A.P., Wellner, N., Bottger, F., 2011. Adsorption of bile salts to particles allows penetration of intestinal mucus. Soft Matter 7, 8077–8084.
Macierzanka, A., Böttger, F., Rigby, N.M., Lille, M., Poutanen, K., Mills, E.N.C., 2012. Enzymatically structured emulsions in simulated gastrointestinal environment: impact on interfacial proteolysis and diffusion in intestinal mucus. Langmuir 28 (50), 17349–17362.
Mackie, A.R., Round, A.N., Rigby, N.M., Macierzanka, A., 2012. The role of the mucus barrier in digestion. Food Dig. 3, 8–15.
Magnuson, B.A., Jonaitis, T.A., Card, J.W., 2011. A brief review of the occurrence, use, and safety of food-related nanomaterials. J. Food Sci. 76, R126–R133.

Magnuson, B., Munro, I., Baldwin, N., Lopez-Garcia, R., Ly, K., McGirr, L., Roberts, A., Socolovsky, S., 2013. Review of regulation and safety assessment of food substances in various countries and juristication. Food Addit. Contam. 30 (7), 1147–1220.

Mahler, G.J., Shuler, M.L., Glahn, R.P., 2009. Characterization of Caco-2 and HT29-MTX cocultures in an in vitro digestion/cell culture model used to predict iron bioavailability. J. Nutr. Biochem. 20 (7), 494–502.

Maldonado-Valderrama, J., Wilde, P., Macierzanka, A., Mackie, 2011. The role of bile salt in digestion. Adv. Colloid Interface Sci. 165 (1), 36–46.

Marsanasco, M., Márquez, M., Wagner, J.R., Alsonso, S.D.V., Chiaramoni, N.S., 2011. Liposomes as vehicles for vitamins E and C: an alternative to fortify orange juice and offer vitamin C protection after heat treatment. Food Res. Int. 44 (9), 3039–3046.

Martirosyan, A., Schneider, Y.J., 2014. Engineered nanomaterials in food: implications for food safety and consumer health. Int. J. Environ. Res. Public Health 11, 5720–5750.

McClements, D.J., Li, Y., 2010a. Structured emulsion-based delivery systems controlling the digestion and release of lipophilic food components. Adv. Colloid Interface Sci. 159 (2), 213–218.

McClements, D.J., Li, Y., 2010b. Review of in vitro digestion models for rapid screening of emulsion-based systems. Food Funct 1 (1), 32–59.

McClements, D.J., Xiao, H., 2012. Potential biological fate of ingested nanoemulsions: influence of particle characteristics. Food Funct. 3, 202–220.

McCrae, C., Debon, S., Guthrie, B., Heigis, J., Mondro, G., Shieh, W., 2009. Materials for encapsulation of food ingredients: understanding the properties to find practical solutions. Micro/Nanoencapsulation of Active Food Ingredients. American Chemical Society, Washington, DC, 213–232.

Moghimi, S.M., Hunter, A.C., Murray, J.C., 2005. Nanomedicine: current status and future prospects. FASEB J. 19, 311–330.

Mohammadi, M., Ghanbarzadeh, B., Hamishehkar, H., 2014. Formulation of nanoliposomal vitamin D3 for potential application in beverage fortification. Adv. Pharm. Bull. 4 (Suppl. 2), 569–575.

Möller, M., Eberle, U., Hermann, A., Moch, K., Stratmann, B., 2009. Nanotechnology in the food sector. Science and technology options assessment. Directorate General for Internal Policies Policy Department A: Economic and Scientific Policy. European Parliament IP/A/STOA/2009-09 PE 424.75.

Naahidi, S., Jafari, M., Edalat, F., Raymond, K., Khademhosseini, A., Chen, P., 2013. Biocompatibility of engineered nanoparticles for drug delivery. J. Control. Release 166 (2), 182–194.

Namvar, R., Rahman, H.S., Mohamad, R., Azizi, S., Tahir, P.M., Chartrand, M.S., Yeap, S.W., 2015. Cytotoxic effects of biosynthesized zinc oxide nanoparticles on murine cell lines. Evid.Based Complement. Alternat. Med. 2015, Article ID 593014, 11 pages.

Neirynck, M., Van Lent, K., Dewettinck, K., Van der Meeren, P., 2007. Influence of pH and biopolymer ratio on sodium caseinate guar gum interactions in aqueous solutions and in O/W solutions. Food Hydrocoll. 21, 862–869.

Nel, A., Xia, T., Madler, L., Li, N., 2006. Toxic potential of materials at the nano level. Science 311 (3), 622–627.

Neophytou, C.M., Constantinou, A.I., 2015. Drug delivery innovations for enhancing the anticancer potential of vitamin E isoforms and their derivatives. Bio. Med. Res. Int. 2015, Article ID 584862, 16 pages.

NICE DB for Nanotechnology in Society, 2013. Nanoencapsulation of omega fatty acids in processed food products. Arizona State University. Available from: https://nice.asu.edu

Nik, A.M., Langmaid, S., Wright, A.J., 2012. Digestibility and β-carotene release from lipid nanodispersions depend on dispersed phase crystallinity and interfacial properties. Food Funct. 3, 234–245.
Nordgård, C.T., Draget, K.I., 2011. Oligosaccharides as modulators of rheology in complex mucous systems. Biomacromolecules 12 (8), 3084–3090.
Norris, G., Puri, N., Sinko, P.J., 1998. The effect of physical barriers and properties on the oral absorption of particulates. Adv. Drug Deliv. Rev. 34, 135–154.
O'Regan, J., Mulvihill, D.M., 2010. Sodium caseinate-maltodextrin conjugate stabilized double emulsions: encapsulation and stability. Food Res. Int. 43, 224–231.
Ooi, L.G., Liong, M.T., 2010. Cholesterol-lowering effects of probiotics and prebiotics: a review of in vivo and in vitro findings. Int. J. Mol. Sci. 11, 2499–2522.
Ostrovsky, S., Kazimirsky, G., Gedanken, A., Chaya Brodie, C., 2009. Selective cytotoxic effect of ZnO nanoparticles on glioma cells. Nano Res. 2, 882–890.
Pan, Y., Bartneck, M., Jahnen-Dechent, W., 2012. Cytotoxicity of gold nanoparticles. In: Dũzgunes, N. (Ed.), Nanomedicine: Infectious Diseases, Immunotherapy, Diagnostics, Antifibrotics, Toxicology and Gene Medicine. Academic Press, San Diego, CA, pp. 225–239.
Patel, A.R., 2012. Colloidal delivery systems for phytochemicals. In: Sebastia, M., Ninan, H., Haghi, A. (Eds.), Nanomedicine and Drug Delivery. Apple Academic Press, Vancouver, BC, pp. 17–32.
Patel, A.R., Bhandari, B., 2014. Nano- and microencapsulation of vitamins. In: Kwak, H.-S. (Ed.), Nano- and Microencapsulation for Foods. John Wiley and Sons, Ltd, UK, pp. 225–250 (Chapter 9).
Patel, M.R., San Martin-Gonzalez, M.F., 2012. Characterization of ergocalciferol loaded solid lipid nanoparticles. J. Food Sci. 77, N8–N13.
Patel, A.R., Bouwens, E.C.B., Velikov, K.P., 2010a. Sodium caseinate stabilized zein colloidal particles. J. Agri. Food Chem. 58, 12497–12503.
Patel, A.R., Hu, Y., Tiwari, J.K., Velikov, K.P., 2010b. Synthesis and characterization of zein-curcumin colloidal particles. Soft Matter 6, 6192–6199.
Patel, A.R., Heussen, P., Hazekamp, J., Velikov, K.P., 2011. Stabilisation and controlled release of silibinin from pH responsive shellac colloidal particles. Soft Matter 7, 8549–8555.
Pereira, D.I., Mergler, B.I., Faria, N., Bruggraber, S.F., Aslam, M.F., Poots, L.K., Prassmayer, L., Lonnerdal, B., Brown, A.P., Powell, J.J., 2013. Caco-2 cell acquisition of dietary iron(III) invokes a nanoparticulate endocytic pathway. PLoS One 8 (11), e81250.
Pereira, D.I.A., Bruggraber, S.F.A., Faria, N., Poots, L.K., Tagmount, M.A., Aslam, M.F., Frazer, D.M., Vulpe, C.D., Anderson, G.J., Powell, J.J., 2014a. Nanoparticulate iron(III) oxo-hydroxide delivers safe iron that is well absorbed and utilised in humans. Nanomedicine 10, 1877–1886.
Pereira, D.I.A., Tempest, B., Ilyas, H., Flynn, A.C., Aslam, M.F., Simpson, R.J., Powell, J.J., O Latunde-Dada, G.O., 2014b. A nanoparticulate ferritin-core mimetic is well taken up by HuTu 80 duodenal cells and its absorption in mice is regulated by body iron. J. Nutr. 144, 1896–1902.
Pereira, D.M., Valentão, P., Andrade, P.B., 2014c. Nano- and microdelivery systems for marine bioactive lipids. Mar. Drugs 12, 6014–6027.
Pereira, D.I.A., Aslam, M.F., Frazer, D.M., Schmidt, A., Walton, G.E., McCartney, A.L., Gibson, G.R., Anderson, G.J., Powell, J.J., 2015. Dietary iron depletion at weaning imprints low microbiome diversity and this is not recovered with oral nano Fe(III). Microbiol. Open 4 (1), 12–27.
Perrechil, F.A., Cunha, R.L., 2012. Development of multiple emulsions based on the repulsive interaction between sodium caseinate and LBG. Food Hydrocoll. 26, 126–134.

Powell, J.J., Faria, N., Thomas-McKay, E., Pele, L.C., 2010. Origin and fate of dietary nanoparticles and microparticles in the gastrointestinal tract. J. Autoimmun. 34, 226–233.

Premanathan, M., Karthikeyan, K., Jeyasubramanian, K., Manivannan, G., 2011. Selective toxicity of ZnO nanoparticles toward gram-positive bacteria and cancer cells by apoptosis through lipid peroxidation. Nanomedicine 7, 184–192.

Qian, C., Decker, E.A., Xiao, H., McClements, D.J., 2012. Physical and chemical stability of β-carotene-enriched nanoemulsions: influence of pH, ionic strength, temperature, and emulsifier type. Food Chem. 132, 1221–1229.

Ranjan, S., Dasgupta, N., Chakraborty, A.K., Samuel, S.M., Chidambaram, R., Shanker, R., Kumar, A., 2014. Nanoscience and nanotechnologies in food industries: opportunities and research trends. J. Nanopart. Res. 16, 1–23.

Rasmussen, J.W., Martinez, E., Louka, P., Wingett, D.G., 2010. Zinc oxide nanoparticles for selective destruction of tumor cells and potential for drug delivery applications. Exp. Opin. Drug Deliv. 7 (9), 1063–1077.

Rayman, M.P., 2000. The importance of selenium to human health. Lancet 356 (9225), 233–241.

Renukuntla, J., Vadlapudi, A.D., Patel, A., Boddu, S.H.S., Mitra, A.K., 2013. Approaches for enhancing oral bioavailability of peptides and proteins. Int. J. Pharm. 447 (1–2), 75–93.

Rodriguez, P., Harada, T., Christian, D.A., Pantano, D.A., Tsai, R.K., Discher, D.E., 2013. Minimal "Self" peptides that inhibit phagocytic clearance and enhance delivery of nanoparticles. Science 339 (6122), 971–975.

Rohner, F., Ernst, F.O., Arnold, M., Hilbe, M., Biebinger, R., Ehrensperger, F., Pratsinis, S.E., Langhans, W., Hurrell, R.F., Zimmermann, M.B., 2007. Synthesis, characterization, and bioavailability in rats of ferric phosphate nanoparticles. J. Nutr. 137, 614–619.

Roohani, N., Hurrell, R., Kelishadi, R., Schulin, R., 2013. Zinc and its importance for human health: an integrative review. J. Res. Med. Sci. 18 (2), 144–157.

Sadeghiani, N., Barbosa, L.S., Silva, L.P., Azevedo, R.B., Morais, P.C., Lacava, Z.G.M., 2005. Genotoxicity and inflammatory investigation in mice treated with magnetite nanoparticles surface coated with polyaspartic acid. J. Magn. Magn. Mat. 289, 466–468.

Sahay, G., Alakhova, D.Y., Kabanov, A., 2010. Endocytosis of nanomedicines. J. Control. Release 145 (3), 182–195.

Salvia-Trujillo, L., Qian, C., Martín-Belloso, O., McClements, D.J., 2013. Influence of particle size on lipid digestion and β-carotene bioaccessibility in emulsions and nanoemulsions. Food Chem. 141 (2), 1472–1480.

Salvia-Trujillo, L., Sun, Q., Um, B.H., Park, Y., McClements, D.J., 2015. In vitro and in vivo study of fucoxanthin bioavailability from nanoemulsion-based delivery systems: impact of lipid carrier type. J. Funct. Foods 17, 293–304.

Schuchardt, J.P., Huss, M., Stauss-Grabo, M., Hahn, A., 2009. Significance of long-chain polyunsaturated fatty acids (PUFAs) for the development and behaviour of children. Eur. J. Pediatr. 169 (2), 149–164.

Science and Technology Committee, 2010. 1st Report of Session 2009–10. Nanotechnologies and Food. Authority of House of Lords, UK.

Semo, E., Kesselman, E., Danino, D., Livney, Y.D., 2007. Casein micelle as a natural nano-capsular vehicle for nutraceuticals. Food Hydrocoll. 21, 936–942.

Shi, H., Magaye, R., Castranova, V., Zao, J., 2013. Titanium dioxide nano particles: a review of current toxicological data. Part. Fibre Toxicol. 10, 10–15.

Slavin, J.L., Lloyd, B., 2012. Health benefits of fruits and vegetables. Adv. Nutr. 3, 506–516.

Sonavane, G., Tomoda, K., Makino, K., 2008. Biodistribution of colloidal gold particles after intravenous administration: effect of particle size. Colloids Surf. B 66, 274–280.
Suntres, Z.E., 2011. Liposomal antioxidants for protection against oxidant-induced damage. J. Toxicol. 2011, 152474.
Tan, Y., Li, S., Pitt, B.R., Huang, L., 1999. The inhibitory role of CpG immunostimulatory motifs in cationic lipid vector-mediated transgene expression in vivo. Hum. Gene Ther. 10, 2153–2161.
Taylor, V.K., Ussher, A.L., Burrell, R.E., 2005. Impact of heat on nanocrystalline silver dressings. Part I: chemical and biological properties. Biomaterials 26 (35), 7221–7229.
Thompson, A.K., Singh, H., 2006. Preparation of liposomes from milk fat globule membrane phospholipids using a microfluidizer. J. Dairy Sci. 89, 410–419.
Tiyaboonchai, W., 2003. Chitosan nanoparticles: a promising system for drug delivery. Naresuan Univ. J. 11 (3), 51–66.
Vaseem, M., Umar, A., Yoon-Bong Hahn, Y.-B., 2009. ZnO Nanoparticles: growth, properties, and applications. Metal Oxide Nanostructures and Their Applications, vol. 5, American Scientific Publishers, Valencia, CA, pp. 1–36 (Chapter 4).
Vijayakumar, M.P., Balakrishnan, V., 2014. Effect of calcium phosphate nanoparticles supplementation on growth performance of broiler chicken. Indian J. Sci. Technol. 7 (8), 1149–1154.
Wandrey, C., Bartkowiak, A., Harding, S., 2010. Materials for encapsulation. In: Zuidam, N.J., Nedovic, V. (Eds.), Encapsulation Technologies for Active Food Ingredients and Food Processing. Springer, New York, NY, pp. 31–100.
Wang, S., Moustaid-Moussa, N., Chen, L., Mo, H., Shastri, A., Su, R., Bapat, P., Kwun, I., Shen, C.-L., 2014. Novel insights of dietary polyphenols and obesity. J. Nutr. Biochem. 25 (1), 1–18.
Wang, S., Su, R., Nie, R., Sun, M., Zhang, J., Wu, D., Moustaid-Moussa, N., 2014. Application of nanotechnology in improving bioavailability and bioactivity of diet-derived phytochemicals. J. Nutr. Biochem. 25 (4), 363–376.
Wickham, M., Faulkas, R., Mills, C., 2009. In vitro digestion methods for assessing the effect of food structure on allergen breakdown. Mol. Nutr. Food Res. 53 (8), 952–958.
World Health Organization, 2008. Worldwide prevalence of anaemia 1993–2005. In: de Benoist, B., McLean, E., Egli, I., Cogswell, M. (Eds.), WHO Global Database on Anaemia. WHO, Geneva.
Xia, A., 2014. Influence of physical states of bioactive components (crystalized versus solubilized) and oil composition on bioaccessibility and bioavailability. MSc thesis submitted to the University of Massachusetts, Amherst.
Xia, Y., 2014. Are we entering the nano era? Angew. Chem. Int. 53 (46), 12268–12271.
Xia, S., Xu, S., 2005. Ferrous sulfate liposomes: preparation, stability and application in fluid milk. Food Res. Int. 38, 289–296.
Yada, R.Y., Buck, N., Canady, R., DeMerlis, C., Duncan, T., Janer, G., Juneja, L., Lin, M., McClements, D.J., Noonan, G., Oxley, J., Sabliov, C., Tstsikova, L., Vazquez-Campos, S., Yourick, J., Zhong, Q., Thurmond, S., 2014. Engineered nano scale food ingredients: evaluation of current knowledge on material characteristics relevant to uptake from the gastointestinal tract. Comp. Rev. Food Sci. Food Saf. 13, 730–744.
Yang, S., Liu, C., Liu, W., Yu, H., Zheng, W.Z., Hu, Y., 2013. Preparation and characterization of nanoliposomes entrapping medium-chain fatty acids and vitamin C by lyophilization. Int. J. Mol. Sci. 14, 19763–19773.
Yoo, J.W., Chambers, E., Mitragotri, S., 2010a. Factors that control the circulation time of nanoparticles in blood: challenges, solutions and future prospects. Curr. Pharm. Des. 16 (21), 2298–2307.

Yoo, J.W., Doshi, N., Mitragotri, S., 2010b. Endocytosis and intracellular distribution of PLGA particles in endothelial cells: effect of particle geometry. Macromol. Rapid Commun. 31 (2), 142–148.

Yu, Y., Li, Y., Wang, W., Jin, M., Du, Z., Sun, Z., 2013. Acute toxicity of amorphous silica nanoparticles in intravenously exposed ICR mice. PLoS One 8, e61346.

Yuan, Y., Gao, Y., Zhao, J., Mao, L., 2008. Characterization and stability evaluation of β-carotene nanoemulsions prepared by high pressure homogenization under various emulsifying conditions. Food Res. Int. 41, 61–68.

Zhang, J., Taylor, E.W., Wan, X., Peng, D., 2012. Impact of heat treatment on size, structure, and bioactivity of elemental selenium nanoparticles. Int. J. Nanomed. 7, 815–825.

Zhao, L.M., Shi, L.E., Zhang, Z.L., Chen, J.M., Shi, D.D., Yang, J., Tang, Z.-X., 2011. Preparation and application of chitosan nanoparticles and nanofibers. Brazilian J. Chem. Eng. 28 (03), 353–362.

Zigoneanu, I.G., Astete, C.E., Cristina Mirela Sabliov, C.M., 2008. Nanoparticles with entrapped α-tocopherol: synthesis, characterization, and controlled release. Nanotechnology 19, 105606.

Zimet, P., Livney, Y.D., 2009. Beta-lacto globulin and its nancomplexes with pectin as vehicles for omega-3 polyunsaturated fatty ac ids. Food Hydrocoll. 23, 1120–1126.

16

NANOENCAPSULATION OF GREEN TEA CATECHINS AND ITS EFFICACY

Gene Chi Wai Man*, Kai On Chu†, Chi Chiu Wang**

**The Chinese University of Hong Kong, Department of Orthopaedics and Traumatology, Faculty of Medicine, Prince of Wales Hospital, Shatin, New Territories, Hong Kong; **The Chinese University of Hong Kong, Department of Obstetrics and Gynaecology, Faculty of Medicine, Prince of Wales Hospital, Shatin, New Territories, Hong Kong; †The Chinese University of Hong Kong, Department of Ophthalmology and Visual Sciences, Hong Kong Eye Hospital, Kowloon, Hong Kong*

1 Introduction

Tea is one of the most popular beverages being consumed worldwide. Apart from water, it records a massive consumption of 120 mL/day per capita (Graham, 1992). History of tea is long and complex, spreading across multiple cultures over thousands of years. This history can be dated back as far as the 3rd century AD in ancient China. With different varieties of tea available in the market, the most commonly found are white, green, oolong, and black tea. And over the past few decades, green tea has been subjected to many scientific and medical studies on its potential beneficial effects to human health (Kris-Etherton and Keen, 2002; Mukhtar and Ahmad, 2000; Shankar et al., 2007; Trevisanato and Kim, 2000; Schneider and Segre, 2009).

Green tea is produced from nonfermented leaves of the plant *Camellia sinensis* (Graham, 1992). To produce green tea, freshly harvested tea leaves are steamed to prevent fermentation, yielding a dry stable product. On brewing the green tea leaves with hot water, the main substances found in green tea include polyphenolic catechins (25–35%), protein (15%), lignin (6.5%), amino acids (4%), caffeine (2–4%), organic acids (1.5%), and chlorophyll (0.5%) (Nagle et al., 2006). These phenolic catechins have shown a great potential and efficacy against multiple types of cancer (Carlson et al., 2007; Spinella et al., 2006; Katiyar and Mukhtar, 1996) and

Novel Approaches of Nanotechnology in Food. http://dx.doi.org/10.1016/B978-0-12-804308-0.00016-9

vascular diseases (Stangl et al., 2006; Riegsecker et al., 2013). The catechins, a type of natural phenol and antioxidant, present in high content in green tea are the major components responsible for the biological functions of green tea (Graham, 1992). These catechins account for 30–40% of the tea's dry weight (Balentine et al., 1997; Graham, 1992). Green tea contains higher catechins concentration than black tea or oolong tea, due to minimal oxidation during processing. The compositions of green tea catechins include (−)-epicatechin (EC), (−)-epigallocatechin (EGC), (−)-epigallocatechingallate (EGCG), (−)-epicathecin gallate (ECG), (−)-gallocatechingallate (GCG), (−)-catechingallate (CG), and (+)-catechin (C) (Fig. 16.1) (Miura et al., 1997). Among them, EGCG is the most abundant and the most extensively studied catechin, accounting for 50–80% of the total catechins in green tea (Yang et al., 2002). Moreover, it is considered as one of the most potent molecules known for their antioxidant properties among the catechins (Higdon and Frei, 2003).

It is widely accepted that consumption of green tea has beneficial effects on health. It has been believed that it can improve blood flow, eliminate alcohol and toxins, enhance resistance to disease, relieve joint pain, and improve urination (Balentine et al., 1997). Green tea extract is also able to decrease the lipid absorption, inhibit angiogenesis, and enhance cardioprotective effects (Sueoka et al., 2001). Epidemiology studies have illustrated that tea consumption can reduce cancers risks (Blot et al., 1996; Gao et al., 1994; Nakachi et al., 2000; Sun et al., 2006). Strong antioxidant and free radical scavenging properties of EGCG contribute to health promoting effects. However, the low bioavailability by oral administration results diminishes the effectiveness. Therefore, there is a need to modify the pharmaceutical formulation to enhance the efficacy and reduced toxicity. One of the solutions is

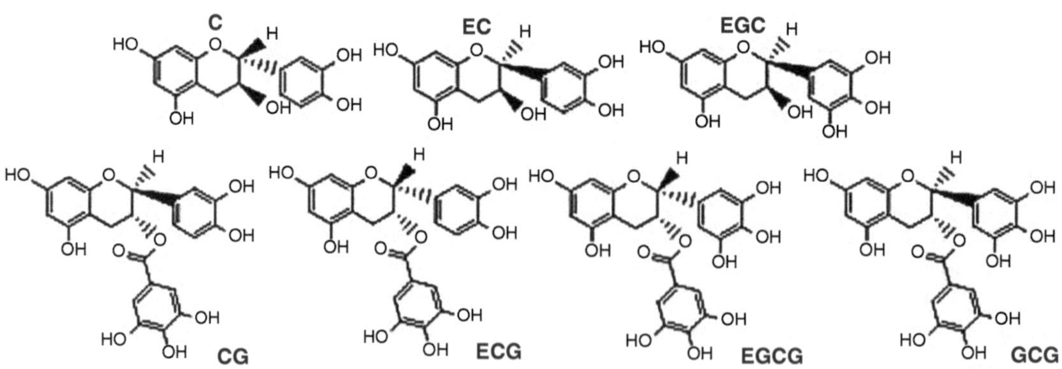

Figure 16.1. Seven types of catechins found in green tea.

to use nanotechnology. The aim of this chapter is to discuss the problem of using EGCG alone as a therapeutic agent, and how nanotechnology can help to overcome bioavailability limitation.

2 The Potential Health Benefits of EGCG

In literatures, EGCG has been well recognized for its cancer-preventive effects. These are widely supported by results from epidemiological, cell culture, animal, and clinical studies. It has been reported that EGCG suppresses the inflammatory processes that lead to transformation, hyperproliferation, and initiation of carcinogenesis (Thawonsuwan et al., 2010). Studies conducted in animals showed the potential of EGCG on inhibiting tumor incidence and growth in different organs, such as the prostate gland, stomach, mammary gland, liver, skin, and colon (Yang et al., 2002; Jankun et al., 1997; Mukhtar and Ahmad, 1999; Sato, 1999; Su and Arab, 2002; Zhang et al., 2002). In vitro cell studies showed that EGCG can strongly induce apoptosis and promote cell growth arrest by modifying cell cycle regulatory proteins expression, caspases activation, and suppressing nuclear factor-kappaB (NFκB) expression (Gupta et al., 2004). Another study showed that the reduction of breast cancer cells proliferation was inhibited in vitro and breast tumor growth was retarded in rodents (Rosengren, 2003). Although the anticarcinogenic properties on human studies remained limited, there were a few studies showing its prevention in human cancer. Regular green tea drinkers showed a significantly reduced risk of breast cancer when compared to women who did not drink green tea regularly (Wu et al., 2003). Two studies on Japanese women with breast cancer indicated green tea consumption is inversely associated with the rate of recurrence, especially in the early stages of breast cancer (Nakachi et al., 1998; Inoue et al., 2001). In addition, an overview of studies to find green tea consumption could reduce the risk of developing mouth cancers (Ide et al., 2007; Shin, 2009). Other studies in women consuming green tea in their diet suggested that it could reduce the risk of lung cancer (Kuroda and Hara, 1999), protect the bladder from cancer development (Laurie et al., 2005; You and Bergman, 1998), and reduce the risk of developing tracheal cancer (Inoue et al., 1998; Nasseri-Moghaddam, 2009).

Besides its anticancer potential, evidence from epidemiologic studies supports a potential role of EGCG in metabolic syndrome (Kim, 2008; Lee et al., 2009; Shimada et al., 2007). Obesity is the consequence associated with a cluster of metabolic and vascular disorders that have been termed the metabolic syndrome. This syndrome causes cardiovascular-related diseases, which is a major cause of mortality worldwide. EGCG have reported modulating dietary lipid absorption. In an in vivo study, the high fat diet supplemented with EGCG resulted in antiobesity in obese mice (Grove et al., 2012). This study reported a decrease in body weight gained in high fat-fed obese C57bl/6J mice treated with 0.32% EGCG for 6 weeks when compared with vehicle control. EGCG can decrease food digestibility to affect substrate metabolism of intestinal mucosal and liver, which leads to the increase of postprandial fat oxidation and reduces incorporation of dietary lipids into the murine tissues.

Also, EGCG can prevent endothelial dysfunction (Legeay et al., 2015). Endothelial dysfunction related to several pathological conditions, including anticoagulant and antiinflammatory properties, vascular growth impairment, and vascular dysregulation, leading to cardiovascular diseases. In addition, EGCG intake was associated with lower risk of cardiovascular diseases (Wang et al., 2014). Prevention of endothelial dysfunction may be due to averting oxidation of low-density lipoproteins (LDL) (Warnakulasuriya et al., 2014), platelet aggregation and adhesion (Wu et al., 2014), and smooth muscle cell migration and proliferation (Ahmad et al., 2013). Also, in another study, the beneficial properties of EGCG have been examined on spontaneously hypertensive rats (Potenza et al., 2007). In this study, rats fed 200 mg/kg per day EGCG showed a significant decrease of blood pressure when compared with control. This study also demonstrated that EGCG stimulated nitric oxide production from endothelium through a PI-3-kinase pathway to suppress hypertension (Potenza et al., 2007).

Another beneficial effect of EGCG is its protective effect on autoimmune diseases. The immune-modulating effect was related to alteration of lymphocyte proliferation, such as inhibiting both B- and T-cell proliferation, after in vitro supplementation with tea extracts (Hu et al., 1992). EGCG improved symptoms and reduced pathological changes in animal models of autoimmune diseases. A study conducted by Bayer et al. (2004) showed prolonged skin graft survival and decreased frequency of donor reactive caused by interferon-c-producing T-cell reactivity in mice after feeding with 0.3% green tea polyphenol solution, containing 50–60% of EGCG. In arthritis, oral green tea extract administered (200 mg/kg per day) was shown to ameliorate adjuvant-induced arthritis in rats (Marotte et al., 2010). It was found that a decrease in MCP-1/CCL2 and GROa/CXCL1 levels and an increase in CCR-1, -2, -5, and CXCR1 receptor expression in the joints of rats fed with green tea extract. Chemokines are well-established regulators of gene transcription, cell proliferation, and leucocyte trafficking to normal and inflamed tissues. The result indicated that the chemokine receptor overexpression with reduced chemokine production by EGCG may be one potential mechanism to limit the overall inflammation and joint destruction in the arthritis model. However, the effect may mediate with TH17 and regulatory T cells. In short, with the many studies showing the potential benefit of EGCG on health, it encourages investigators to further elucidate their essential molecular targets for possible translational therapy.

3 Biological Activities

3.1 Antioxidant Activity

Free-radical damage has been postulated to contribute to the etiology of aging, and many chronic health problems such as cardiovascular, inflammatory diseases, and cancer (Rice-Evans and Burdon, 1993; Spiteller, 2001). Reactive oxygen species (ROS) can modify biomolecules structure and function, leading to aging and chronic diseases development (McCall and Frei, 1999). The antioxidative property can also lead to antiproliferation. EGCG significantly reduces oxidative damage to both lipids and DNA (Rietveld and Wiseman, 2003) and inhibiting lipoxygenase, cyclooxygenase, and xanthine oxidase enzymes activities involving oxidative damage in tissues (Frei and Higdon, 2003).

Recently, a study showed regular green tea consumption in our balanced diet can promote defenses against oxidative damage (Erba et al., 2005). Green tea protects the human body from oxidative stress diseases (Coimbra et al., 2006). Total antioxidant capacity of plasma after intake of 300 mL green tea significantly increased 7.0% after 60 min, and 6.2% after 120 min. After consuming 450 mL of green tea, there was an increase significantly to 12.0% after 60 min, and 12.7% after 120 min over baseline value (Sung et al., 2000).

Green tea can increase the activity of SOD (superoxide dismutase) and the expression of catalase against reactive oxygen species (Negishi et al., 2004). This action, combined with direct antioxidative action on reactive nitrogen species, can decrease nitric oxide level in plasma. Oxidative stress biomarkers, like malondialdehyde, also decrease after green tea intake (Yokozawa et al., 1999).

Tumor is associated with the oxidative damage in tissues (Cutler, 2005). Animal studies strongly supported green tea in preventing oxidative DNA damages induced by carcinogens. Apart from the number of lung adenomas decrease, EGCG also lower 8-hydroxydeoxyguanosine (8-OHdG) level in lungs induced by tobacco carcinogens (Xu et al., 1992). In rats, green tea extract protects DNA lesion as indicated by lowering 8-OHdG level in colon mucosa induced by the colon carcinogen, 1,2-dimethylhydrazine (DMH) (Inagake et al., 1995). In addition, catechins in tea beverage can inhibit 8-OHdG increases in rat livers induced by the hepatic carcinogen, 2-nitropropane (2NP) (Hasegawa et al., 1995; Sai et al., 1998).

In hamsters, catechins can inhibit thiobarbituric acid-reacting substances (TBARS) and nuclear 8-OHdG level increase in pancreas induced by carcinogen, *N*-nitrobis (2-oxopropyl) amine (Takabayashi et al., 1997). However, damage to the DNA can still be afflicted by the presence of Cu(II) and Fe(III) complexes (Furukawa et al., 2003; Elbling et al., 2005). EGCG can also promote DNA damage in vitro in the human esophageal squamous cell carcinoma cell lines KYSE 510 and 150 (Fang et al., 2003) and the leukemia cell line HL-60 (Furukawa et al., 2003; Elbling et al., 2005).

EGCG inhibits transcription factor-mediated gene activation, such as NFκB and activator protein 1 (AP-1) (Ahmad et al., 2000). Inhibition of NF-κB and AP-1-mediated gene activation is the central phenomenon that explains the convergence in the antioxidant activity of the green tea catechins and their effects on specific molecular targets. NF-κB, in response to ROS, activates transcription of many proinflammatory and antiapoptotic genes (Schoonbroodt and Piette, 2000). The ROS-scavenging activity of green tea catechins inhibits NF-κB nuclear translocation and binding activity, leading to inhibition of expression of these proinflammatory and survival genes (Levites et al., 2002). Green tea also represses the catalytic activity of several cytochrome P-450 enzymes, which may involve in cancer initiation (Stoner and Mukhtar, 1995). This activity is brought upon by EGCG inhibiting the rise of ROS from the cytochrome P-450 system initiating futile cycling (Wang et al., 1988). Furthermore, green tea blocks nitrosamines and inhibits estrogen receptor interaction in breast cancer cell lines through inhibiting specific P-450 enzymes (Komori et al., 1993).

The effects of green tea catechins on cancer prevention, especially in antioxidation, appear very promising in animal models, but data on oxidative stress biomarkers in humans are limited. It is still not clear if the cancer prevention effect is due to its antioxidant activity. It may be due to the antioxidative effect on the cancer prediction, which is difficult to establish (Moyers and Kumar, 2004). ECG and EGCG displayed better antioxidant activity than EC and EGC to prevent lipid peroxidation (Guo et al., 1996). Catechins have been shown to protect or regenerate α-tocopherol that protects low-density lipoprotein (LDL) from oxidation (Zhu et al., 1999). Consumption of 300 mg of green tea polyphenol extract twice daily for 1 week can delay the oxidation of human LDL ex vivo (Miura et al., 2000).

3.2 Antimiotic

In the complicated multiple processes of carcinogenesis, adhesion and invasion of cancer cells to the host cell layer and the basement membrane are the most critical steps (Fidler, 1991; Nicolson, 1989; Liotta et al., 1991). Cancer cells are known to produce large amounts of ROS to promote invasion and metastasis (Nonaka et al., 1993). These free oxygen radicals can also serve as a key initiator in downstream signaling pathways to ultimately lead to the induction of invasion genes expression. Studies showed EGCG can prevent formation of hydroxyl radicals by binding to the transition metal ions from the exogenous ROS (Zhang et al., 2000). In other words, EGCG could inhibit tumor cell invasion by scavenging oxygen radicals. EGCG inhibits the invasion and adhesion by suppressing urokinase (Jankun et al., 1997), and matrix metalloproteinases (MMPs) activities (Garbisa et al., 2001), and by removing oxygen radicals (Zhang et al., 2000).

Urokinase (uPA), a hydrolase implicated in tumor cell invasion (Huang et al., 2000), facilitates degradation of the basement membrane and extracellular matrix. EGCG directly impairs the activity of urokinase, interferes with its enzymatic activity, and improves the degradation of extracellular matrix (Jankun et al., 1997). In molecular modeling, EGCG binds to uPA, blocks the histidine 57, serine 195, and extends toward arginine 35 from a positively charged loop to prevent uPA recognizing its substrates and thereby inhibit its enzymatic activity.

MMP-2 and MMP-9 are the most commonly expressed in cancer and in activated endothelial cells. They have been shown to degrade the basement membrane to facilitate cell invasion (Liotta et al., 1980). Studies showed that EGCG effectively inhibits MMP-2 and MMP-9 in vitro and can reduce tumor cell invasion by 50% (Garbisa et al., 2001). This inhibition activity was reported to be partially related to EGCG ability to chelate zinc, a necessary element for MMP enzymatic activity (Maeda-Yamamoto et al., 1999).

EGCG suppresses Kaposi's sarcoma cells and normal umbilical vein endothelial cells migration in response to fibroblast-conditioned media (Fassina et al., 2004). Endothelial cell invasion is also inhibited (Fassina et al., 2004). Catechins suppress rat ascites hepatoma cell line, AH109A (Miura et al., 1997; Zhang et al., 1999). It also reported that green tea and EGCG inhibited mouse lung carcinoma cells (Taniguchi et al., 1992). Investigators have also reported that increased consumption of green tea was associated with decreased numbers of auxiliary lymph node metastases among premenopausal women with stage I and II breast cancer and with increased expression of progesterone and estrogen receptors among postmenopausal women (Nakachi et al., 1998). However, the underlying mechanisms remain to be elucidated.

3.3 Apoptosis and Cell Cycle Regulation

Cancerous cells can evade apoptosis. EGCG causes cancer cells to growth inhibition, regulates cell cycle progression, and induces apoptosis (Ahmad and Mukhtar, 1999; Morre et al., 2000; Gupta et al., 2000). Apoptosis is regulated by a complex cascade of molecular events. Green tea is associated with suppressor protein p53 levels increase and apoptosis induction (Chung et al., 2003). EGCG is thought to activate the p53 protein by inducing phosphorylation of critical serine residues and activating arrest at the G1 phase (Hastak et al., 2003). In addition, apoptosis can be induced by inhibiting the activity of fatty acid synthase (FAS), a metabolic enzyme involved in lipid synthesis in the tumor cells, by EGCG perhaps through accumulation of toxic precursors, such as malonyl-coenzyme A, in the process of lipid synthesis (Brusselmans et al., 2003).

Remarkably, a report shown in vitro treatment with EGCG induced more apoptosis in transformed cancer cells (50%) than in normal cells (1%) (Chen et al., 1998). Others have provided further evidences that EGCG induced apoptosis specifically in abnormal or cancerous cells only (Ahn et al., 2003; Brusselmans et al., 2003). Using DNA microarray technology, EGCG down-regulated 16 genes and upregulated 4 genes related to cellular metabolism and cell cycle regulation (Ahn et al., 2003). EGCG induced pro-apoptotic clustering expression and led to cell death in cancer cells (Weinreb et al., 2003). In an in vitro study, cervical cancer cells exposed to EGCG triggers apoptosis (Ahn et al., 2003). EGCG reduced the proliferation rate and induced apoptosis of lung cancer cells (Yang et al., 1998). It also reported that EGCG inhibited prostate cancer adenoma cells growth and induced apoptosis (Yu et al., 2004).

Recent research showed that EGCG protects against bladder cancer. EGCG caused 100% cell death in the AY-27 cell line, and strong banding on the DNA ladder assay was seen with the mouse leukemia cell line. After treating for 3 weeks, EGCG prevented 64% animals from developing tumors (Kemberling et al., 2003). In another system, using NOD/SCID mice transplanted with human non-Hodgkin's lymphoma (NHL) cell lines, green tea inhibited 50% of tumor growth by inducing apoptosis and endothelial cell death (Bertolini et al., 2000).

3.4 Antiinflammatory

EGCG has been shown to possess antiinflammatory properties in vivo and in vitro (Mandel et al., 2005; Navarro-Peran et al., 2008; Peng et al., 2006; Tokuda et al., 2007). There is substantial evidence that the antiinflammatory effect of EGCG is partially due to inhibition of inducible nitric-oxide synthase (iNOS) (Lin and Lin, 1997). In this regard, in vitro studies have shown that EGCG inhibited the induction of iNOS mRNA after treatment with tumor necrosis factor-α (TNF-α) and interleukin-1 (IL-1) (Suganuma et al., 2000; Wheeler et al., 2004). Moreover, other studies showed that EGCG can inhibit iNOS activity and expression following brain damage (Sutherland et al., 2005; Suzuki et al., 2004). Also, tea polyphenols appear to modulate at different antiinflammatory targets relating to arachidonic acid-dependent pathways, like cyclooxygenase (COX). Within the arachidonic acid-independent pathways, nitric-oxide synthase (NOS) and NF-κB are targets of polyphenols (Miles et al., 2005). The released phytochemicals from the tea polyphenols inhibit cyclooxygenase-2 (COX-2) and iNOS expression by blocking NF-κB activation. EGCG also suppresses NF-κB activation by repression of degradation of inhibitory, IκBκ, which hampers formation of functionally active subunit of NF-κB and subsequently translocation to nucleus (Navarro-Peran et al., 2008).

3.5 Antiangiogenic

The growth of almost all types of tumors is dependent on angiogenesis. The growth and metastasis of tumors is disrupted when angiogenesis is suppressed (Folkman, 1995). At the initial phase of the tumor growth, nutrients and oxygen can be gained by diffusion alone. However, in the long-term supply of these nutrients and oxygen, the tumor must create a network of blood vessels to maintain its growth when it is larger than 0.5 mm in diameter (Bamias and Dimopoulos, 2003).

It has been shown that green tea can inhibit angiogenesis in in vitro proliferation studies (Laschke et al., 2008; Park et al., 2006; Slivova et al., 2005) and in vivo angiogenesis assays (Laschke et al., 2008; Xu et al., 2009). EGCG may inhibit angiogenesis at different stages of tumor growth. The initiation step of tumor growth involves multiple growth factors, proteases, and adhesion molecules secreted by the endothelial cells, as well as supporting cells from the surrounding (Carmeliet and Jain, 2000). EGCG does not only inhibit further vascularization of existing tumors, but prevents them from progressing toward invasive cancer (Tosetti et al., 2002). Cultures of human microvascular endothelial cells on Matrigel induced formation of capillary-like tubes, which was suppressed by EGCG (Fassina et al., 2004; Bertolini et al., 2000). EGCG suppresses endothelial cell growth in vitro and the formation of new blood vessels in chick chorioallantoic. It inhibited endothelial cell growth in a dose-dependent manner of the inhibition appeared to be specific to endothelial cells (Cao et al., 1996). EGCG reduced vascularization of highly angiogenic tumors, and reduced 30% tumor vessels in colon cancer mice (Jung et al., 2001). Further evidence to support the antiangiogenic activity of EGCG is that EGCG inhibits in vivo growth and angiogenesis of tumors by blocking the induction of vascular endothelial growth factor (VEGF) (Jung et al., 2001). Green tea catechins inhibited VEGF induced growth of umbilical vein endothelial cells, reduced their migration and size of endothelial tubes (Fassina et al., 2004). EGCG also acts as a potent inhibitor of the chymotrypsin-like activity in proteasome, which further contributes to anticancer prevention (Nam et al., 2001). Interleukin-12 (IL-12) is an antiangiogenic molecule that activates antiangiogenic programs in mouse spleens and human peripheral blood mononuclear cells (Strasly et al., 2001) to block neovascularization in human Burkitt's lymphoma, colon carcinoma, and ovarian carcinoma (Yao et al., 2000). EGCG can induce IL-12 level elevation in draining lymph nodes of mice (Katiyar et al., 1999).

EGCG selectively inhibits receptor tyrosine kinases that play a regulatory role in angiogenesis, including EGF-R (Liang et al., 1997), PDGF-receptor beta, and phosphatidylinositol-3 (PI-3) kinase, and the downstream p44/p42 MAP kinase pathway in vascular smooth muscle cells (Ahn et al., 1999). TNF-α and MMPs were also suppressed by EGCG (Annabi et al., 2007; Yang et al., 1998). These antiangiogenic effects may be clinically relevant (Kabbinavar et al., 2003). However, no clinical trials have been reported as medication for antiangiogenesis therapy.

3.5.1 *Vascular Endothelial Growth Factor*

VEGF plays a central role in angiogenesis. EGCG potentially acts as an antiangiogenic agent by suppressing the expression of VEGF (Jung et al., 2001). It inhibits angiogenesis by blocking VEGF binding to 67 Lamin receptor to prevent the receptor associating with integrin. In addition, EGCG also directly binds to the VEGF receptor to decrease VEGF receptor phosphorylation (Smith, 1997).

3.5.2 Matrix Metalloproteinases

MMPs play crucial roles in a variety of physiological angiogenic (eg, blood vessel formation, bone development, wound healing) and pathophysiological angiogenic (eg, diabetic retinography, cancer) processes. This is not only due to their ability to degrade the surrounding extracellular matrix (ECM), but also because MMPs function to reveal cryptic matrix binding sites, release matrix-bound growth factors inherent to these processes, and activate a variety of cell surface molecules. EGCG inhibits MMP-2 and MMP-9 activities in endothelial cell. EGCG inhibits MMP-2 activity even at physiological concentrations. EGCG binds to premature as well as mature forms of MMP-2, leading to reversible conformational changes in MMP-2 molecules and masks its catalytic region, which is essential for proteolytic activity (Annabi et al., 2002). Inhibition of MMP-2 and MMP-9 levels was also observed in oral cancer cells after treatment with EGCG (Kato et al., 2008).

3.5.3 Cadherin

Vascular endothelial (VE)-cadherin/-catenin complex is associated with VEGFR-2 (Flk-1/KDR)/PI3-kinase during VEGF-induced angiogenesis to activate the survival signal Akt (phospho-kinase B) protein, to induce Bcl-2 in vascular endothelial cells and to stimulate endothelial cell proliferation (Carmeliet et al., 1999; Gerber et al., 1998). VEGF-induced tube formation is inhibited by anti-VE-cadherin antibody and dose-dependently by green tea catechins (Naasani et al., 2003). It also demonstrated that inhibition of tube formation by EGCG, which is partially mediated through suppression of VE-cadherin tyrosine phosphorylation and inhibition of Akt activation during the tube formation (Naasani et al., 2003). These findings indicated that VE-cadherin and Akt, known as downstream proteins of VEGFR-2-mediated angiogenesis cascade, are the new targeting proteins of catechins.

3.5.4 Hypoxia-Inducible Factor-1

Hypoxia-inducible factor-1 (HIF-1) belongs to the basic helix-loop-helix-periodic acid-Schiff domain transcription factor family and consists of two subunits, HIF-1α and HIF-1β. Only the expression and activation of HIF-1α is tightly regulated by cellular oxygen concentration (Wang and Semenza, 1995). EGCG significantly inhibit hypoxia- and serum-induced HIF-1α protein accumulation in human cervical carcinoma and hepatoma cells without affecting HIF-1α mRNA expression. HIF-1α protein suppressed by green tea extract and EGCG also resulted in a drastic decrease in

VEGF expression at both mRNA and protein levels. The inhibitory effect of EGCG on serum-induced HIF-1α protein and VEGF expression is related to the phosphatidylinositol 3-kinase/Akt of rapamycin signaling pathways, which play a crucial role in the protein translational machinery cascade (Zhang et al., 2006).

3.5.5 *Nuclear Factor-KappaB and Activator Protein 1*

NF-κB and AP-1 are transcriptional factors regulating expression of multiple genes involved in tumor growth, metastasis, and angiogenesis (Shukla et al., 2005; Sweeney et al., 2004). NFκB plays a critical role in regulating genes related to cell survival, proliferation, and apoptosis. EGCG exhibits antiinflammatory properties by inhibiting components of the cell signaling pathways, especially those that converge on redox-sensitive transcription factors, including NFκB and AP-1, which have also been implicated in carcinogenesis (Gupta et al., 2004).

3.5.6 *Mitogen-Activated Protein Kinase*

The mitogen-activated protein kinase (MAPK) signaling pathways play essential roles in cell proliferation and differentiation. Activation of MAPK signaling pathways involve in tumorigenesis, metastasis, and angiogenesis of multiple human malignancies (Depeille et al., 2007; Johnson and Lapadat, 2002). EGCG inhibits endothelial cell proliferation and differentiation by blocking MAPK. It is a powerful inhibitor of the intracellular cleavage of MAPKK-2 and MAPKK-3 induced by anthrax lethal factor. It inhibits the phosphorylation of extracellular signal-regulated kinases 1 and 2 (ERK1/2), which are the members of an MAPK family necessary for MMP upregulation (Kim et al., 2004b).

3.5.7 *Cyclooxygenase-2*

Cyclooxygenase-2 (COX-2), a target of NSAIDs, is upregulated in many cancers and has been associated with increased VEGF production and angiogenesis (Toomey et al., 2009). EGCG inhibits upregulation of COX-2 and thus reduces inflammation (Ahmed et al., 2002) without affecting COX-1 expression (Hussain et al., 2005). Previous studies described that COX-2 induction and PGE2 level are markedly reduced by pretreatment with EGCG.

3.5.8 *Insulin-Like Growth Factor*

Insulin-like growth factor (IGF) is known to induce VEGF, and it has been proposed that IGF is involved in the angiogenic switch leading to neovascularization (Kaplan et al., 1999). EGCG reduces IGF levels, which can induce DNA synthesis and cell proliferation via activation of PI3K/Akt and Ras/MAPK pathways (Adhami et al., 2004).

3.5.9 Urokinase-Type Plasminogen Activator

uPA is a multifunctional protein that is the crucial protein for neointimal growth and vascular remodeling (Clowes et al., 1990). EGCG also inhibits uPA, which converts zymogen plasminogen to active plasmin. Plasmin can prevent apoptosis to stimulate angiogenesis, mitogenesis, cell migration, and cell adhesion. EGCG impairs adhesion of various types of cells to extracellular matrix proteins by binding to fibronectin and laminin and by impairing interaction between the cells and integrin beta 1 (Kim et al., 2004a).

4 Toxicity and Side Effects of EGCG

Although most studies used the regimen of greater administration of just adding EGCG or green tea extract with EGCG as the major component to drinking water, the underlying potential toxicity related to such a regimen has not been fully investigated. Most literatures demonstrated that highly consumed EGCG is a nontoxic natural compound when metabolized in our gastrointestinal tract, or these studies simply disregard the possible toxicity induced from the EGCG used. Until recently, an association from an Italian surveillance system of natural health has demonstrated a possible link between the consumption of green tea extract in humans with liver damage (Mazzanti et al., 2009). Histological evaluation on these patients' livers revealed pathologies of inflammatory reactions, cholestasis, occasional steatosis, and necrosis. Also, the study showed a positive dechallenge was reported in 29 cases studies, and a positive rechallenge occurred in 7 out of 34 cases studies (Mazzanti et al., 2009). This study concluded that there is a safety concern on green tea consumption relating to hepatotoxicity (Mazzanti et al., 2009). In addition, animal studies based on intervention with EGCG administration by oral or intraperitoneal route also demonstrated an adverse effect. When mice were fed orally with a high dose of EGCG at 1% for 6 weeks, it promoted an increase in splenocyte and macrophage proinflammatory markers, such as tumor necrosis factor-α, interleukin-6, interleukin-1β, and prostaglandin E2, and in overall disturbed immune cell population (Pae et al., 2011). Another study demonstrated a single interperitoneal injection of 100 mg/kg EGCG to mice can generate hepatotoxicity, whereas 150 mg/kg results in 100% mortality within 24 h (Galati et al., 2006). On the other hand, there are some studies showing the lack of adverse effect when EGCG was given in clinical trials. A study also showed daily doses of 800 mg caffeine-free EGCG for 4 weeks was shown to be safe and well tolerated in healthy human subjects (Chow et al., 2003). In a Phase I trial, it showed EGCG can be safely administered for a relatively long period without any adverse effect (Shanafelt et al., 2009).

The cause of hepatotoxicity has been ascribed to the prooxidant action of EGCG or its metabolites, which can induce oxidative stress in the liver. However, the underlying reason for mentioned toxicity in these studies could also be due to a variety of reasons. Such reasons include contaminated tea containing pesticide residues resulting to adverse effects. Endosulfan, a potential hepatotoxic organochlorine pesticide, has been used extensively in some green tea plantations due to its low price (Jia et al., 2009). In addition, the reports showing adverse effects of EGCG in experimental animals were obtained by using purified EGCG. The steps in the purification process may account for the contaminant residue being reflected in the experimental animals treated. Another known side effect reported from drinking green tea is to cause insomnia, increased heart rate, and nausea in selected individuals, mainly due to the presence of caffeine. An average cup of tea (10 g of tea leaves in 1 L water) contains around 300 mg of crude solids with 30–42% catechins and 3–6% caffeine (Khan et al., 2007). However, the amount of caffeine is 3 times less than that of drinking coffee. Thus, this makes drinking green tea a prospective and very safe treatment for clinical use. Hence, the reference on dose levels for different diseases, daily administration frequency and side effects in long-term therapy should be studied prior to using EGCG as a therapeutical agent.

5 Low Bioavailability of EGCG

The controversial degree of responsiveness to EGCG mentioned may also depend on its plasma and tissue bioavailability. Although green tea, green tea extracts, or pure EGCG may be largely consumed in the diet and high concentrations may exist in the intestine after oral administration, the concentration of EGCG in the blood circulation is likely to be low. It is undeniable that the plasma/tissue concentrations of this unstable compound are lower than their effective therapeutic doses. In order to achieve biological active level, increased EGCG levels could result in side-effect generation (Fig. 16.2).

Typically, the peak plasma concentrations of EGCG can be reached in 1–2 h of a healthy individual after intake of a single oral

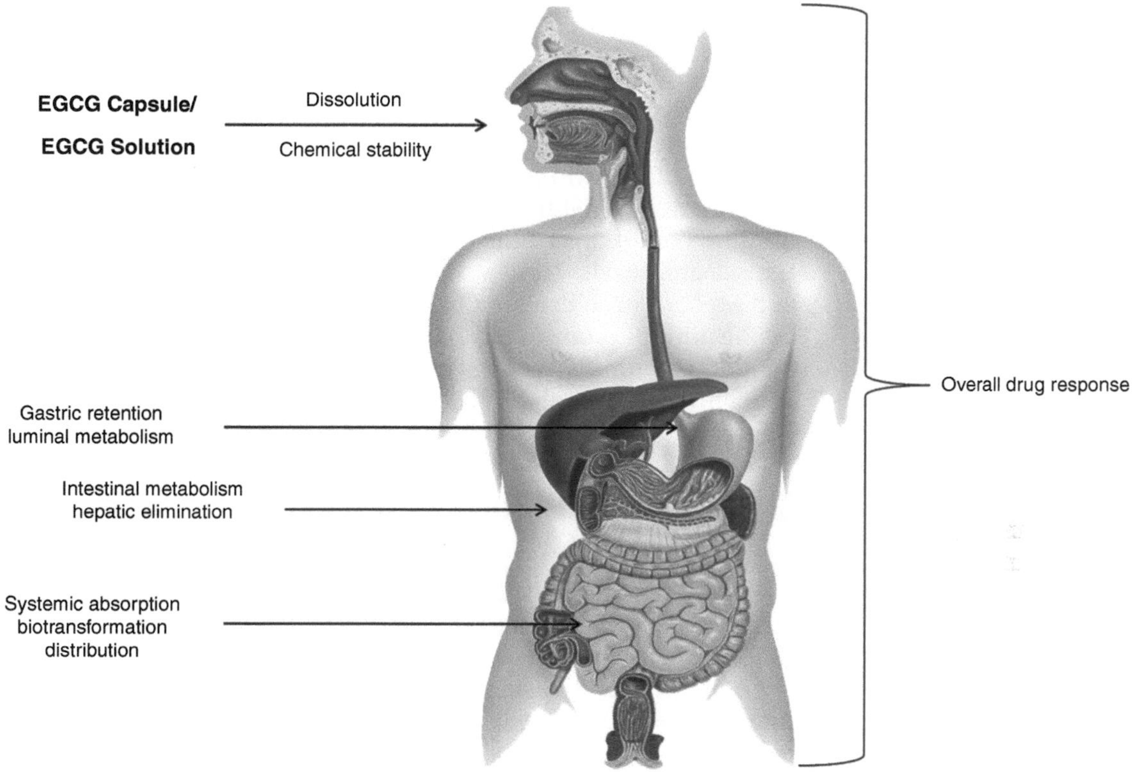

Figure 16.2. Things to be considered for EGCG to be used as a drug therapy.

administration in the morning, with overnight fasting period. This level will decrease to an undetectable level gradually within 24 h. The elimination of half-life of EGCG was reported to be 3.4 ± 0.3 h (Lee et al., 2002). Evolution has fashioned the intestinal wall as the first line of defense against damage from foreign substances. Hence, absorption would mainly takes place in the small gut, yet substantial quantities would also pass to the intestines for further degradation by the local microbiota (Roowi et al., 2010; Stalmach et al., 2009). EGCG needs to pass through a number of hurdles before reaching to the tissue target. The hurdles include solubility, permeability, transportation, metabolism, elimination, and absorption.

Similar to most flavonoids, EGCG are too hydrophilic to penetrate through the intestinal wall. After oral absorption, tea catechins undergo extensive methylation, glucuronidation, and sulfation. Rapid methylation of EGCG would usually be catalyzed by liver enzyme, catechol-O-methyltransferase, to decrease the hydrophilicity. Further glucuronidation and sulfation would only be needed to effectively eliminate the methylation product from the body (Lu et al., 2003). However, as EGCG possesses strong antioxidative activity, this results in polymerization and decomposition before reaching the site (Kadoma et al., 2006).

Human serum albumin facilitates transportation and stabilization of EGCG and protects it from oxidation through a reversible interaction (Ishii et al., 2011). Hence, low serum albumin in alkaline conditions of the blood may decrease EGCG stability in the plasma levels. Hard water with high concentrations of Ca^{2+} and Mg^{2+}, or even drinking milk together with EGCG can result to inactivation (Lorenz et al., 2007). Moreover, several other factors, including temperature, oxygen concentration, and antioxidant concentration, can also affect the stability of EGCG (Li et al., 2011).

6 Nanotechnologies in Cancer Therapy

Nanotechnology refers to the objects typically in the range of 1–100 nm (Odom and Pileni, 2008). The advantage of such parameter in drug formulations is that it enables the protection from the destructive action of external media, as they are biologically inert (Winterhalter and Lasic, 1993). Also, it is now well documented that drugs encapsulated in nanoparticles exhibit distinct pharmacokinetic and pharmacodynamic profiles. Nanospheres can help dramatically to increase blood circulation times and reduce liver accumulation in mice, as compared with the nonencapsulated free drugs (Gref et al., 1994).

Over the past two decades, the rapid developments in nanotechnology altered the incorporation of multiple therapeutic, sensing, and targeting agents into nanoparticle for various applications. The incorporation of nanotechnology can offer many advantages as drug carrier systems in improving the solubility of poorly water-soluble drugs, modifying pharmacokinetics, increasing drug half-life, reducing immunogenicity, improving bioavailability, and diminishing drug metabolism. The releasing rate of therapeutic compounds can be controlled and two or more drugs can be simultaneously delivered.

Liposomes represent versatile and advanced nanodelivery systems for a wide range of biologically active compounds (Hofheinz et al., 2005). Toward the clinical usage, liposomal incorporation of doxorubicin and daunorubicin was considered most advanced and widely used (Swenson et al., 2003; Gill et al., 1995). This incorporation of nanoparticles increased the plasma concentration, reduced clearance rate and volume of distribution, which increased the bioavailability of the drug. In addition, it showed a decrease on the cardial-related toxicity in liposome-coated doxorubicin than standard doxorubicin (Batist et al., 2001). These demonstrate an improvement in the safety and pharmacokinetics of liposome-coated drugs. With polymeric nanoparticles, the therapeutic window of conventional drugs can be widened. Albumin nanoparticle-based formulation has been used for paclitaxel to facilitate endothelial transcytosis to improve drug accumulation in breast tumors (Miele et al., 2009). It also improves drug tolerance compared to traditional paclitaxel (Ibrahim et al., 2002), as indicated from lowering the side effects of neutropenia and hypersensitivity (Gradishar et al., 2005).

Recent RNA interference technologies have also shown effectiveness in curing diseases, including cancer. This new approach could open another field of therapeutics when traditional small molecules are unsuccessful. With the many challenges associated with the delivery of short interfering RNA (siRNA) systemically, nanoparticle platforms can help overcome these hurdles. Nanoliposomes can improve delivery (approximately 10-fold) of siRNA and enhance antitumor effects systemically (Landen et al., 2005).

Another potential of nanoparticles used in therapy is immunotherapy. One of the main difficulties facing immunotherapy is cytokines and antigens delivery to the target region. Nanotechnology delivery can slow down degradation of cytokines, reduce proteins nonspecific binding, lower excretion rate, and minimize undesired toxic effects. By incorporating poly-lactic-*co*-glycolic acid in nanoparticles, it acts as intracellular antigen reservoirs for dendritic cells so as to retard the degradation of antigens and need of repeated dosing (Shen et al., 2006). In addition, studies with interleukin-2 and interleukin-12 have also shown that nanoparticle incorporation increased plasma retention time (Shimizu et al., 2008; Yao et al., 2011).

Likewise, novel nanoparticles have been used to increase the specificity and sensitivity in detecting small tumor nodules, while shortening imaging time, and lowering toxic effects. Ferumoxtran-10, an iron oxide nanoparticle, has been applied in conventional MRI scans (Harisinghani et al., 2003). The result showed significantly higher sensitivity (90.5 vs 35.4%) in detecting lymph node metastasis as compared with conventional MRI scans. Additionally, the use of nanoparticle can reduce the toxicity commonly found in conventional contrast agent. Nanoparticle incorporation to gadolinium also prevented nephrogenic systemic fibrosis (Liu et al., 2011; Zhang et al., 2014).

However, only a few formulations have been approved for human therapy. The full potential of nanotechnology applied in clinical application has not been implemented. In addition, the toxicity is still an issue needed for further evaluation. Nevertheless, with the advanced nanotechnology, new hopes may begin in cancer therapy.

7 In Vivo Efficacy of Nanoencapsulated EGCG

There have been reports on EGCG showing excellent potential to be used as effective agents in preventing and treating many cancers in animal models. However, the limited bioavailability could be the main reason to prevent it from proceeding to human translation study. To overcome this, the need to be small enough to diffuse through the mucus layer and avoid elimination by mucociliary clearance would be the main issue. Hence, the use of encapsulate useful bioactive food components by nanotechnology might be the key to sustain a slow release for higher efficacy of EGCG to the targeted site (Fig. 16.3).

EGCG encapsulated in lipid nanocapsules is stable in the aqueous phase over 4 weeks, whereas free EGCG totally degraded within

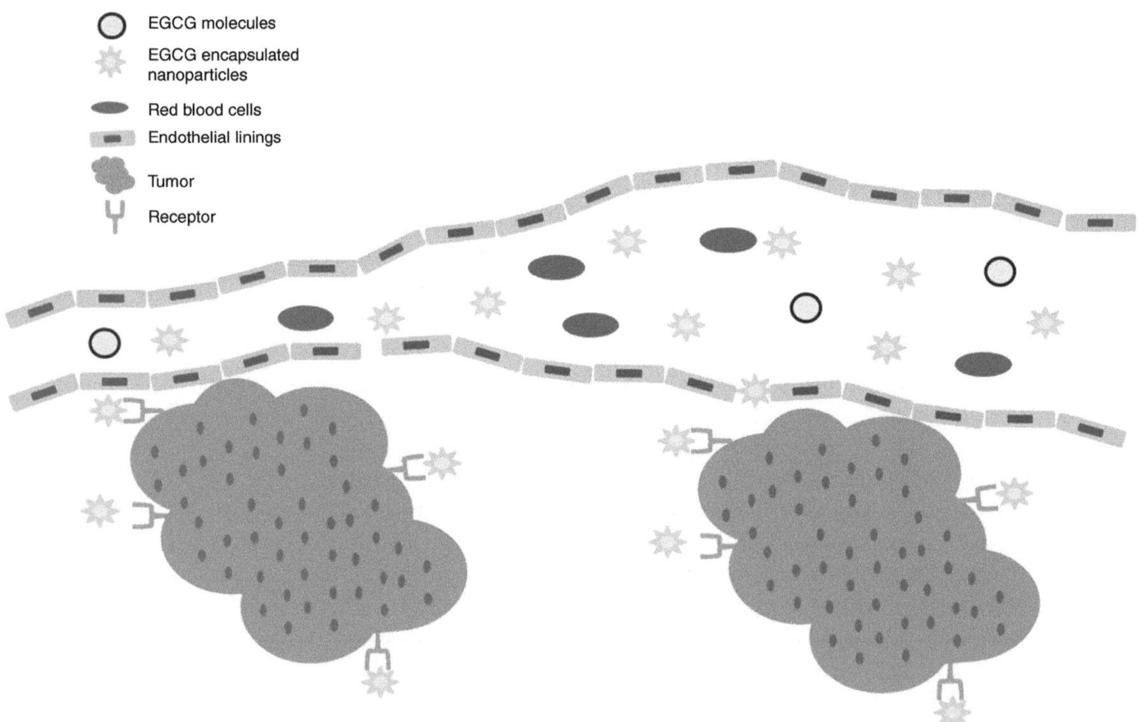

Figure 16.3. Delivery of nanoparticle through vascularization of tumor.

4 h (Barras et al., 2009). Also, the bioavailability of EGCG increased in encapsulated with chitosan (Dube et al., 2010, 2011). Intestinal absorption increased by a factor of 1.8 and plasma exposure of total EGCG by a factor of 1.5, when compared with nonencapsulated EGCG, in oral administration (Dube et al., 2010, 2011). EGCG (Chit-nano EGCG) effectively inhibits prostate cancer cell growth in a xenograft model in chitosan-based nanoformulation (Khan et al., 2014). The tumor growth and secreted prostate-specific antigen levels were greatly reduced compared to natural EGCG and control groups. The excised tumors from the Chit-nano EGCG treated mice showed an increase of poly (ADP-ribose) polymerases cleavage and caspases activity, while cell proliferation decreased, when compared with vehicle control. Similarly, Siddiqui et al. (2014) showed the encapsulated EGCG increased eightfold dose enhancement compared to native EGCG. The encapsulated EGCG also inhibited proliferation and induced apoptosis in Mel 928 tumor xenograft from the treated mice.

The main mechanism for the survival of cancer cells is their ability to escape from death. This is achieved by the disruption of spatiotemporal behavior of signaling cascades in cancer cells leading to downregulate tumor suppressor genes and overexpress oncogenes. Proapoptotic effect of EGCG can overcome the cancer cells' advantages. Nanotechnology can improve the bioavailability of EGCG to enhance proapoptotic effect. Polylactic acid–polyethylene glycol (PLA–PEG) encapsulated nano-EGCG give more degradation resistant in blood. It produces superior chemopreventive effects due to more than a 10-fold dose to induce apoptosis and inhibit angiogenesis and tumor growth than nonencapsulated EGCG (Siddiqui et al., 2009). Also, EGCG encapsulated in nanoparticles (5.48 μM) induces apoptosis more efficiently than nonencapsulated EGCG (40 μM) in a prostate cell line due to significant increase of proapoptotic Bax levels and decrease of antiapoptotic Bcl-2 level, tilting the Bax/Bcl-2 ratio in favor of apoptosis (Siddiqui and Mukhtar, 2010).

Besides inhibiting cell proliferation and inducing apoptosis, EGCG is an efficient inhibitor of angiogenesis as shown in an inhibition of FGF-induced angiogenesis with nano-EGCG treated ex ovo chick chorioallantoic membrane (CAM) (Siddiqui and Mukhtar, 2010). It also significantly inhibits mean branch formation as well as suppresses tumor weight of neuroblastoma-induced angiogenesis in CAM. Nano-EGCG (10 μM) can also significantly reduce triglyceride accumulation compared to nonencapsulated EGCG (Siddiqui et al., 2014). Most importantly, these results demonstrate that nano-EGCG retained its characteristics properties as EGCG at 1/10th dose. Moreover, these studies reveal the potential of nanoparticle-mediated delivery of EGCG that could serve as a basis for enhancing the bioavailability of EGCG in future clinical therapy.

8 Conclusions and Future Prospectives

Throughout the history of modern medicine, natural products are the source of new drug development. Considering that the current anticancer drugs are very expensive, toxic, and unsatisfactory, green tea extracts can be an alternative candidate. With extensive researches on EGCG, evidence shows that EGCG can be a potent and effective medicine for cancer prevention and treatment. However, the low bioavailability of EGCG, resulting in subtherapeutic concentrations causes failure to show the clinical efficacy. Therefore, with the help of nanotechnology, the concentration of EGCG can achieve therapeutic levels in the targeted site. Nanotechnology can also reduce the adverse effects found in the crude EGCG. Moreover, the concept of nanochemoprevention may help to open a new medical horizon in the development and discovery of novel chemopreventive agents, with efficacious improvement and toxicity reduction.

References

Adhami, V.M., Siddiqui, I.A., Ahmad, N., Gupta, S., Mukhtar, H., 2004. Oral consumption of green tea polyphenols inhibits insulin-like growth factor-I-induced signaling in an autochthonous mouse model of prostate cancer. Cancer Res. 64, 8715–8722.

Ahmad, N., Mukhtar, H., 1999. Green tea polyphenols and cancer: biologic mechanisms and practical implications. Nutr. Rev. 57, 78–83.

Ahmad, N., Cheng, P., Mukhtar, H., 2000. Cell cycle dysregulation by green tea polyphenol epigallocatechin-3-gallate. Biochem. Biophys. Res. Commun. 275, 328–334.

Ahmad, A., Ahmad, S., Malcolm, K.C., Miller, S.M., Hendry-Hofer, T., Schaack, J.B., White, C.W., 2013. Differential regulation of pulmonary vascular cell growth by hypoxia-inducible transcription factor-1alpha and hypoxia-inducible transcription factor-2alpha. Am. J. Respir. Cell Mol. Biol. 49, 78–85.

Ahmed, S., Rahman, A., Hasnain, A., Lalonde, M., Goldberg, V.M., Haqqi, T.M., 2002. Green tea polyphenol epigallocatechin-3-gallate inhibits the IL-1 beta-induced activity and expression of cyclooxygenase-2 and nitric oxide synthase-2 in human chondrocytes. Free Radic. Biol. Med. 33, 1097–1105.

Ahn, H.Y., Hadizadeh, K.R., Seul, C., Yun, Y.P., Vetter, H., Sachinidis, A., 1999. Epigallocathechin-3 gallate selectively inhibits the PDGF-BB-induced intracellular signaling transduction pathway in vascular smooth muscle cells and inhibits transformation of sis-transfected NIH 3T3 fibroblasts and human glioblastoma cells (A172). Mol. Biol. Cell 10, 1093–1104.

Ahn, W.S., Huh, S.W., Bae, S.M., Lee, I.P., Lee, J.M., Namkoong, S.E., Kim, C.K., Sin, J.I., 2003. A major constituent of green tea, EGCG, inhibits the growth of a human cervical cancer cell line, CaSki cells, through apoptosis, G(1) arrest, and regulation of gene expression. DNA Cell Biol. 22, 217–224.

Annabi, B., Lachambre, M.P., Bousquet-Gagnon, N., Page, M., Gingras, D., Beliveau, R., 2002. Green tea polyphenol (−)-epigallocatechin 3-gallate inhibits MMP-2 secretion and MT1-MMP-driven migration in glioblastoma cells. Biochim. Biophys. Acta 1542, 209–220.

Annabi, B., Currie, J.C., Moghrabi, A., Beliveau, R., 2007. Inhibition of HuR and MMP-9 expression in macrophage-differentiated HL-60 myeloid leukemia cells by green tea polyphenol EGCg. Leuk. Res. 31, 1277–1284.

Balentine, D.A., Wiseman, S.A., Bouwens, L.C., 1997. The chemistry of tea flavonoids. Crit. Rev. Food Sci. Nutr. 37, 693–704.

Bamias, A., Dimopoulos, M.A., 2003. Angiogenesis in human cancer: implications in cancer therapy. Eur. J. Intern. Med. 14, 459–469.

Barras, A., Mezzetti, A., Richard, A., Lazzaroni, S., Roux, S., Melnyk, P., Betbeder, D., Monfilliette-Dupont, N., 2009. Formulation and characterization of polyphenol-loaded lipid nanocapsules. Int. J. Pharm. 379, 270–277.

Batist, G., Ramakrishnan, G., Rao, C.S., Chandrasekharan, A., Gutheil, J., Guthrie, T., Shah, P., Khojasteh, A., Nair, M.K., Hoelzer, K., Tkaczuk, K., Park, Y.C., Lee, L.W., 2001. Reduced cardiotoxicity and preserved antitumor efficacy of liposome-encapsulated doxorubicin and cyclophosphamide compared with conventional doxorubicin and cyclophosphamide in a randomized, multicenter trial of metastatic breast cancer. J. Clin. Oncol. 19, 1444–1454.

Bayer, J., Gomer, A., Demir, Y., Amano, H., Kish, D.D., Fairchild, R., Heeger, P.S., 2004. Effects of green tea polyphenols on murine transplant-reactive T cell immunity. Clin. Immunol. 110, 100–108.

Bertolini, F., Fusetti, L., Rabascio, C., Cinieri, S., Martinelli, G., Pruneri, G., 2000. Inhibition of angiogenesis and induction of endothelial and tumor cell apoptosis by green tea in animal models of human high-grade non-Hodgkin's lymphoma. Leukemia 14, 1477–1482.

Blot, W.J., Chow, W.H., Mclaughlin, J.K., 1996. Tea and cancer: a review of the epidemiological evidence. Eur. J. Cancer Prev. 5, 425–438.

Brusselmans, K., De Schrijver, E., Heyns, W., Verhoeven, G., Swinnen, J.V., 2003. Epigallocatechin-3-gallate is a potent natural inhibitor of fatty acid synthase in intact cells and selectively induces apoptosis in prostate cancer cells. Int. J. Cancer 106, 856–862.

Cao, J., Xu, Y., Chen, J., Klaunig, J.E., 1996. Chemopreventive effects of green and black tea on pulmonary and hepatic carcinogenesis. Fundam. Appl. Toxicol. 29, 244–250.

Carlson, J.R., Bauer, B.A., Vincent, A., Limburg, P.J., Wilson, T., 2007. Reading the tea leaves: anticarcinogenic properties of (−)-epigallocatechin-3-gallate. Mayo Clin. Proc. 82, 725–732.

Carmeliet, P., Jain, R.K., 2000. Angiogenesis in cancer and other diseases. Nature 407, 249–257.

Carmeliet, P., Lampugnani, M.G., Moons, L., Breviario, F., Compernolle, V., Bono, F., Balconi, G., Spagnuolo, R., Oosthuyse, B., Dewerchin, M., Zanetti, A., Angellilo, A., Mattot, V., Nuyens, D., Lutgens, E., Clotman, F., De Ruiter, M.C., Gittenberger-De Groot, A., Poelmann, R., Lupu, F., Herbert, J.M., Collen, D., Dejana, E., 1999. Targeted deficiency or cytosolic truncation of the VE-cadherin gene in mice impairs VEGF-mediated endothelial survival and angiogenesis. Cell 98, 147–157.

Chen, Z.P., Schell, J.B., Ho, C.T., Chen, K.Y., 1998. Green tea epigallocatechin gallate shows a pronounced growth inhibitory effect on cancerous cells but not on their normal counterparts. Cancer Lett. 129, 173–179.

Chow, H.H., Cai, Y., Hakim, I.A., Crowell, J.A., Shahi, F., Brooks, C.A., Dorr, R.T., Hara, Y., Alberts, D.S., 2003. Pharmacokinetics and safety of green tea polyphenols after multiple-dose administration of epigallocatechin gallate and polyphenon E in healthy individuals. Clin. Cancer Res. 9, 3312–3319.

Chung, J.H., Han, J.H., Hwang, E.J., Seo, J.Y., Cho, K.H., Kim, K.H., Youn, J.I., Eun, H.C., 2003. Dual mechanisms of green tea extract (EGCG)-induced cell survival in human epidermal keratinocytes. FASEB J. 17, 1913–1915.

Clowes, A.W., Clowes, M.M., Au, Y.P., Reidy, M.A., Belin, D., 1990. Smooth muscle cells express urokinase during mitogenesis and tissue-type plasminogen activator during migration in injured rat carotid artery. Circ. Res. 67, 61–67.

Coimbra, S., Castro, E., Rocha-Pereira, P., Rebelo, I., Rocha, S., Santos-Silva, A., 2006. The effect of green tea in oxidative stress. Clin. Nutr. 25, 790–796.

Cutler, R.G., 2005. Oxidative stress profiling: part I. Its potential importance in the optimization of human health. Ann. NY Acad. Sci. 1055, 93–135.

Depeille, P.E., Ding, Y., Bromberg-White, J.L., Duesbery, N.S., 2007. MKK signaling and vascularization. Oncogene 26, 1290–1296.

Dube, A., Nicolazzo, J.A., Larson, I., 2010. Chitosan nanoparticles enhance the intestinal absorption of the green tea catechins (+)-catechin and (−)-epigallocatechin gallate. Eur. J. Pharm. Sci. 41, 219–225.

Dube, A., Nicolazzo, J.A., Larson, I., 2011. Chitosan nanoparticles enhance the plasma exposure of (−)-epigallocatechin gallate in mice through an enhancement in intestinal stability. Eur. J. Pharm. Sci. 44, 422–426.

Elbling, L., Weiss, R.M., Teufelhofer, O., Uhl, M., Knasmueller, S., Schulte-Hermann, R., Berger, W., Micksche, M., 2005. Green tea extract and (−)-epigallocatechin-3-gallate, the major tea catechin, exert oxidant but lack antioxidant activities. FASEB J. 19, 807–809.

Erba, D., Riso, P., Bordoni, A., Foti, P., Biagi, P.L., Testolin, G., 2005. Effectiveness of moderate green tea consumption on antioxidative status and plasma lipid profile in humans. J. Nutr. Biochem. 16, 144–149.

Fang, M.Z., Wang, Y., Ai, N., Hou, Z., Sun, Y., Lu, H., Welsh, W., Yang, C.S., 2003. Tea polyphenol (−)-epigallocatechin-3-gallate inhibits DNA methyltransferase and reactivates methylation-silenced genes in cancer cell lines. Cancer Res. 63, 7563–7570.

Fassina, G., Vene, R., Morini, M., Minghelli, S., Benelli, R., Noonan, D.M., Albini, A., 2004. Mechanisms of inhibition of tumor angiogenesis and vascular tumor growth by epigallocatechin-3-gallate. Clin. Cancer Res. 10, 4865–4873.

Fidler, I.J., 1991. Cancer metastasis. Br. Med. Bull. 47, 157–177.

Folkman, J., 1995. Angiogenesis in cancer, vascular, rheumatoid and other disease. Nat. Med. 1, 27–31.

Frei, B., Higdon, J.V., 2003. Antioxidant activity of tea polyphenols in vivo: evidence from animal studies. J. Nutr. 133, 3275S–3284S.

Furukawa, A., Oikawa, S., Murata, M., Hiraku, Y., Kawanishi, S., 2003. (−)-Epigallocatechin gallate causes oxidative damage to isolated and cellular DNA. Biochem. Pharmacol. 66, 1769–1778.

Galati, G., Lin, A., Sultan, A.M., O'brien, P.J., 2006. Cellular and in vivo hepatotoxicity caused by green tea phenolic acids and catechins. Free Radic. Biol. Med. 40, 570–580.

Gao, Y.T., Mclaughlin, J.K., Blot, W.J., Ji, B.T., Dai, Q., Fraumeni, Jr., J.F., 1994. Reduced risk of esophageal cancer associated with green tea consumption. J. Natl. Cancer Inst. 86, 855–858.

Garbisa, S., Sartor, L., Biggin, S., Salvato, B., Benelli, R., Albini, A., 2001. Tumor gelatinases and invasion inhibited by the green tea flavanol epigallocatechin-3-gallate. Cancer 91, 822–832.

Gerber, H.P., Dixit, V., Ferrara, N., 1998. Vascular endothelial growth factor induces expression of the antiapoptotic proteins Bcl-2 and A1 in vascular endothelial cells. J. Biol. Chem. 273, 13313–13316.

Gill, P.S., Espina, B.M., Muggia, F., Cabriales, S., Tulpule, A., Esplin, J.A., Liebman, H.A., Forssen, E., Ross, M.E., Levine, A.M., 1995. Phase I/II clinical and pharmacokinetic evaluation of liposomal daunorubicin. J. Clin. Oncol. 13, 996–1003.

Gradishar, W.J., Tjulandin, S., Davidson, N., Shaw, H., Desai, N., Bhar, P., Hawkins, M., O'shaughnessy, J., 2005. Phase III trial of nanoparticle albumin-bound paclitaxel compared with polyethylated castor oil-based paclitaxel in women with breast cancer. J. Clin. Oncol. 23, 7794–7803.

Graham, H.N., 1992. Green tea composition, consumption, and polyphenol chemistry. Prev. Med. 21, 334–350.

Gref, R., Minamitake, Y., Peracchia, M.T., Trubetskoy, V., Torchilin, V., Langer, R., 1994. Biodegradable long-circulating polymeric nanospheres. Science 263, 1600–1603.

Grove, K.A., Sae-Tan, S., Kennett, M.J., Lambert, J.D., 2012. (−)-Epigallocatechin-3-gallate inhibits pancreatic lipase and reduces body weight gain in high fat-fed obese mice. Obesity 20, 2311–2313.

Guo, Q., Zhao, B., Li, M., Shen, S., Xin, W., 1996. Studies on protective mechanisms of four components of green tea polyphenols against lipid peroxidation in synaptosomes. Biochim. Biophys. Acta 1304, 210–222.

Gupta, S., Ahmad, N., Nieminen, A.L., Mukhtar, H., 2000. Growth inhibition, cell-cycle dysregulation, and induction of apoptosis by green tea constituent (−)-epigallocatechin-3-gallate in androgen-sensitive and androgen-insensitive human prostate carcinoma cells. Toxicol. Appl. Pharmacol. 164, 82–90.

Gupta, S., Hastak, K., Afaq, F., Ahmad, N., Mukhtar, H., 2004. Essential role of caspases in epigallocatechin-3-gallate-mediated inhibition of nuclear factor kappa B and induction of apoptosis. Oncogene 23, 2507–2522.

Harisinghani, M.G., Barentsz, J., Hahn, P.F., Deserno, W.M., Tabatabaei, S., Van De Kaa, C.H., De La Rosette, J., Weissleder, R., 2003. Noninvasive detection of clinically occult lymph-node metastases in prostate cancer. N. Engl. J. Med. 348, 2491–2499.

Hasegawa, R., Chujo, T., Sai-Kato, K., Umemura, T., Tanimura, A., Kurokawa, Y., 1995. Preventive effects of green tea against liver oxidative DNA damage and hepatotoxicity in rats treated with 2-nitropropane. Food Chem. Toxicol. 33, 961–970.

Hastak, K., Gupta, S., Ahmad, N., Agarwal, M.K., Agarwal, M.L., Mukhtar, H., 2003. Role of p53 and NF-kappaB in epigallocatechin-3-gallate-induced apoptosis of LNCaP cells. Oncogene 22, 4851–4859.

Higdon, J.V., Frei, B., 2003. Tea catechins and polyphenols: health effects, metabolism, and antioxidant functions. Crit. Rev. Food Sci. Nutr. 43, 89–143.

Hofheinz, R.D., Gnad-Vogt, S.U., Beyer, U., Hochhaus, A., 2005. Liposomal encapsulated anti-cancer drugs. Anticancer Drugs 16, 691–707.

Hu, Z.Q., Toda, M., Okubo, S., Hara, Y., Shimamura, T., 1992. Mitogenic activity of (−)epigallocatechin gallate on B-cells and investigation of its structure-function relationship. Int. J. Immunopharmacol. 14, 1399–1407.

Huang, S., New, L., Pan, Z., Han, J., Nemerow, G.R., 2000. Urokinase plasminogen activator/urokinase-specific surface receptor expression and matrix invasion by breast cancer cells requires constitutive p38alpha mitogen-activated protein kinase activity. J. Biol. Chem. 275, 12266–12272.

Hussain, T., Gupta, S., Adhami, V.M., Mukhtar, H., 2005. Green tea constituent epigallocatechin-3-gallate selectively inhibits COX-2 without affecting COX-1 expression in human prostate carcinoma cells. Int. J. Cancer 113, 660–669.

Ibrahim, N.K., Desai, N., Legha, S., Soon-Shiong, P., Theriault, R.L., Rivera, E., Esmaeli, B., Ring, S.E., Bedikian, A., Hortobagyi, G.N., Ellerhorst, J.A., 2002. Phase I and pharmacokinetic study of ABI-007, a Cremophor-free, protein-stabilized, nanoparticle formulation of paclitaxel. Clin. Cancer Res. 8, 1038–1044.

Ide, R., Fujino, Y., Hoshiyama, Y., Mizoue, T., Kubo, T., Pham, T.M., Shirane, K., Tokui, N., Sakata, K., Tamakoshi, A., Yoshimura, T., 2007. A prospective study of green tea consumption and oral cancer incidence in Japan. Ann. Epidemiol. 17, 821–826.

Inagake, M., Yamane, T., Kitao, Y., Oya, K., Matsumoto, H., Kikuoka, N., Nakatani, H., Takahashi, T., Nishimura, H., Iwashima, A., 1995. Inhibition of 1,2-dimethylhydrazine-induced oxidative DNA damage by green tea extract in rat. Jpn. J. Cancer Res. 86, 1106–1111.

Inoue, M., Tajima, K., Hirose, K., Hamajima, N., Takezaki, T., Kuroishi, T., Tominaga, S., 1998. Tea and coffee consumption and the risk of digestive tract cancers: data from a comparative case-referent study in Japan. Cancer Causes Control 9, 209–216.

Inoue, M., Tajima, K., Mizutani, M., Iwata, H., Iwase, T., Miura, S., Hirose, K., Hamajima, N., Tominaga, S., 2001. Regular consumption of green tea and the risk of breast cancer recurrence: follow-up study from the Hospital-based Epidemiologic Research Program at Aichi Cancer Center (HERPACC) Japan. Cancer Lett. 167, 175–182.

Ishii, T., Ichikawa, T., Minoda, K., Kusaka, K., Ito, S., Suzuki, Y., Akagawa, M., Mochizuki, K., Goda, T., Nakayama, T., 2011. Human serum albumin as an antioxidant in the oxidation of (−)-epigallocatechin gallate: participation of reversible covalent binding for interaction and stabilization. Biosci. Biotechnol. Biochem. 75, 100–106.

Jankun, J., Selman, S.H., Swiercz, R., Skrzypczak-Jankun, E., 1997. Why drinking green tea could prevent cancer. Nature 387, 561.

Jia, H., Li, Y.F., Wang, D., Cai, D., Yang, M., Ma, J., Hu, J., 2009. Endosulfan in China 1-gridded usage inventories. Environ. Sci. Pollut. Res. Int. 16, 295–301.

Johnson, G.L., Lapadat, R., 2002. Mitogen-activated protein kinase pathways mediated by ERK, JNK, and p38 protein kinases. Science 298, 1911–1912.

Jung, Y.D., Kim, M.S., Shin, B.A., Chay, K.O., Ahn, B.W., Liu, W., Bucana, C.D., Gallick, G.E., Ellis, L.M., 2001. EGCG, a major component of green tea, inhibits tumour growth by inhibiting VEGF induction in human colon carcinoma cells. Br. J. Cancer 84, 844–850.

Kabbinavar, F., Hurwitz, H.I., Fehrenbacher, L., Meropol, N.J., Novotny, W.F., Lieberman, G., Griffing, S., Bergsland, E., 2003. Phase II, randomized trial comparing bevacizumab plus fluorouracil (FU)/leucovorin (LV) with FU/LV alone in patients with metastatic colorectal cancer. J. Clin. Oncol. 21, 60–65.

Kadoma, Y., Ishihara, M., Okada, N., Fujisawa, S., 2006. Free radical interaction between vitamin E (alpha-, beta-, gamma- and delta-tocopherol), ascorbate and flavonoids. In Vivo 20, 823–827.

Kaplan, P.J., Mohan, S., Cohen, P., Foster, B.A., Greenberg, N.M., 1999. The insulin-like growth factor axis and prostate cancer: lessons from the transgenic adenocarcinoma of mouse prostate (TRAMP) model. Cancer Res. 59, 2203–2209.

Katiyar, S., Mukhtar, H., 1996. Tea in chemoprevention of cancer. Int. J. Oncol. 8, 221–238.

Katiyar, S.K., Challa, A., Mccormick, T.S., Cooper, K.D., Mukhtar, H., 1999. Prevention of UVB-induced immunosuppression in mice by the green tea polyphenol (−)-epigallocatechin-3-gallate may be associated with alterations in IL-10 and IL-12 production. Carcinogenesis 20, 2117–2124.

Kato, K., Long, N.K., Makita, H., Toida, M., Yamashita, T., Hatakeyama, D., Hara, A., Mori, H., Shibata, T., 2008. Effects of green tea polyphenol on methylation status of RECK gene and cancer cell invasion in oral squamous cell carcinoma cells. Br. J. Cancer 99, 647–654.

Kemberling, J.K., Hampton, J.A., Keck, R.W., Gomez, M.A., Selman, S.H., 2003. Inhibition of bladder tumor growth by the green tea derivative epigallocatechin-3-gallate. J. Urol. 170, 773–776.

Khan, S.A., Priyamvada, S., Arivarasu, N.A., Khan, S., Yusufi, A.N., 2007. Influence of green tea on enzymes of carbohydrate metabolism, antioxidant defense, and plasma membrane in rat tissues. Nutrition 23, 687–695.

Khan, N., Bharali, D.J., Adhami, V.M., Siddiqui, I.A., Cui, H., Shabana, S.M., Mousa, S.A., Mukhtar, H., 2014. Oral administration of naturally occurring chitosan-based nanoformulated green tea polyphenol EGCG effectively inhibits prostate cancer cell growth in a xenograft model. Carcinogenesis 35, 415–423.

Kim, J.A., 2008. Mechanisms underlying beneficial health effects of tea catechins to improve insulin resistance and endothelial dysfunction. Endocr. Metab. Immune Disord. Drug Targets 8, 82–88.

Kim, M.H., Jung, M.A., Hwang, Y.S., Jeong, M., Kim, S.M., Ahn, S.J., Shin, B.A., Ahn, B.W., Jung, Y.D., 2004a. Regulation of urokinase plasminogen activator by epigallocatechin-3-gallate in human fibrosarcoma cells. Eur. J. Pharmacol. 487, 1–6.

Kim, H.S., Kim, M.H., Jeong, M., Hwang, Y.S., Lim, S.H., Shin, B.A., Ahn, B.W., Jung, Y.D., 2004b. EGCG blocks tumor promoter-induced MMP-9 expression via suppression of MAPK and AP-1 activation in human gastric AGS cells. Anticancer Res. 24, 747–753.

Komori, A., Yatsunami, J., Okabe, S., Abe, S., Hara, K., Suganuma, M., Kim, S.J., Fujiki, H., 1993. Anticarcinogenic activity of green tea polyphenols. Jpn. J. Clin. Oncol. 23, 186–190.

Kris-Etherton, P.M., Keen, C.L., 2002. Evidence that the antioxidant flavonoids in tea and cocoa are beneficial for cardiovascular health. Curr. Opin. Lipidol. 13, 41–49.

Kuroda, Y., Hara, Y., 1999. Antimutagenic and anticarcinogenic activity of tea polyphenols. Mutat. Res. 436, 69–97.

Landen, Jr., C.N., Chavez-Reyes, A., Bucana, C., Schmandt, R., Deavers, M.T., Lopez-Berestein, G., Sood, A.K., 2005. Therapeutic EphA2 gene targeting in vivo using neutral liposomal small interfering RNA delivery. Cancer Res. 65, 6910–6918.

Laschke, M.W., Schwender, C., Scheuer, C., Vollmar, B., Menger, M.D., 2008. Epigallocatechin-3-gallate inhibits estrogen-induced activation of endometrial cells in vitro and causes regression of endometriotic lesions in vivo. Hum. Reprod. 23, 2308–2318.

Laurie, S.A., Miller, V.A., Grant, S.C., Kris, M.G., Ng, K.K., 2005. Phase I study of green tea extract in patients with advanced lung cancer. Cancer Chemother. Pharmacol. 55, 33–38.

Lee, M.J., Maliakal, P., Chen, L., Meng, X., Bondoc, F.Y., Prabhu, S., Lambert, G., Mohr, S., Yang, C.S., 2002. Pharmacokinetics of tea catechins after ingestion of green tea and (−)-epigallocatechin-3-gallate by humans: formation of different metabolites and individual variability. Cancer Epidemiol. Biomarkers Prev. 11, 1025–1032.

Lee, J.W., Lee, Y.K., Ban, J.O., Ha, T.Y., Yun, Y.P., Han, S.B., Oh, K.W., Hong, J.T., 2009. Green tea (−)-epigallocatechin-3-gallate inhibits beta-amyloid-induced cognitive dysfunction through modification of secretase activity via inhibition of ERK and NF-kappaB pathways in mice. J. Nutr. 139, 1987–1993.

Legeay, S., Rodier, M., Fillon, L., Faure, S., Clere, N., 2015. Epigallocatechin gallate: a review of its beneficial properties to prevent metabolic syndrome. Nutrients 7, 5443–5468.

Levites, Y., Youdim, M.B., Maor, G., Mandel, S., 2002. Attenuation of 6-hydroxydopamine (6-OHDA)-induced nuclear factor-kappaB (NF-kappaB) activation and cell death by tea extracts in neuronal cultures. Biochem. Pharmacol. 63, 21–29.

Li, N., Taylor, L.S., Mauer, L.J., 2011. Degradation kinetics of catechins in green tea powder: effects of temperature and relative humidity. J. Agric. Food Chem. 59, 6082–6090.

Liang, Y.C., Lin-Shiau, S.Y., Chen, C.F., Lin, J.K., 1997. Suppression of extracellular signals and cell proliferation through EGF receptor binding by (−)-epigallocatechin gallate in human A431 epidermoid carcinoma cells. J. Cell Biochem. 67, 55–65.

Lin, Y.L., Lin, J.K., 1997. (−)-Epigallocatechin-3-gallate blocks the induction of nitric oxide synthase by down-regulating lipopolysaccharide-induced activity of transcription factor nuclear factor-kappaB. Mol. Pharmacol. 52, 465–472.

Liotta, L.A., Tryggvason, K., Garbisa, S., Hart, I., Foltz, C.M., Shafie, S., 1980. Metastatic potential correlates with enzymatic degradation of basement membrane collagen. Nature 284, 67–68.

Liotta, L.A., Stracke, M.L., Aznavoorian, S.A., Beckner, M.E., Schiffmann, E., 1991. Tumor cell motility. Semin. Cancer Biol. 2, 111–114.

Liu, Y., Chen, Z., Liu, C., Yu, D., Lu, Z., Zhang, N., 2011. Gadolinium-loaded polymeric nanoparticles modified with anti-VEGF as multifunctional MRI contrast agents for the diagnosis of liver cancer. Biomaterials 32, 5167–5176.

Lorenz, M., Jochmann, N., Von Krosigk, A., Martus, P., Baumann, G., Stangl, K., Stangl, V., 2007. Addition of milk prevents vascular protective effects of tea. Eur. Heart J. 28, 219–223.

Lu, H., Meng, X., Yang, C.S., 2003. Enzymology of methylation of tea catechins and inhibition of catechol-O-methyltransferase by (−)-epigallocatechin gallate. Drug Metab. Dispos. 31, 572–579.

Maeda-Yamamoto, M., Kawahara, H., Tahara, N., Tsuji, K., Hara, Y., Isemura, M., 1999. Effects of tea polyphenols on the invasion and matrix metalloproteinases activities of human fibrosarcoma HT1080 cells. J. Agric. Food Chem. 47, 2350–2354.

Mandel, S.A., Avramovich-Tirosh, Y., Reznichenko, L., Zheng, H., Weinreb, O., Amit, T., Youdim, M.B., 2005. Multifunctional activities of green tea catechins in neuroprotection modulation of cell survival genes, iron-dependent oxidative stress and PKC signaling pathway. Neurosignals 14, 46–60.

Marotte, H., Ruth, J.H., Campbell, P.L., Koch, A.E., Ahmed, S., 2010. Green tea extract inhibits chemokine production, but upregulates chemokine receptor expression, in rheumatoid arthritis synovial fibroblasts and rat adjuvant-induced arthritis. Rheumatology 49, 467–479.

Mazzanti, G., Menniti-Ippolito, F., Moro, P.A., Cassetti, F., Raschetti, R., Santuccio, C., Mastrangelo, S., 2009. Hepatotoxicity from green tea: a review of the literature and two unpublished cases. Eur. J. Clin. Pharmacol. 65, 331–341.

McCall, M.R., Frei, B., 1999. Can antioxidant vitamins materially reduce oxidative damage in humans? Free Radic. Biol. Med. 26, 1034–1053.

Miele, E., Spinelli, G.P., Miele, E., Tomao, F., Tomao, S., 2009. Albumin-bound formulation of paclitaxel (Abraxane ABI-007) in the treatment of breast cancer. Int. J. Nanomed. 4, 99–105.

Miles, E.A., Zoubouli, P., Calder, P.C., 2005. Effects of polyphenols on human Th1 and Th2 cytokine production. Clin. Nutr. 24, 780–784.

Miura, Y., Shiomi, H., Sakai, F., Yagasaki, K., 1997. Assay systems for screening food components that have anti-proliferative and anti-invasive activity to rat ascites hepatoma cells: in vitro and ex vivo effects of green tea extract. Cytotechnology 23, 127–132.

Miura, Y., Chiba, T., Miura, S., Tomita, I., Umegaki, K., Ikeda, M., Tomita, T., 2000. Green tea polyphenols (flavan 3-ols) prevent oxidative modification of low density lipoproteins: an ex vivo study in humans. J. Nutr. Biochem. 11, 216–222.

Morre, D.J., Bridge, A., Wu, L.Y., Morre, D.M., 2000. Preferential inhibition by (−)-epigallocatechin-3-gallate of the cell surface NADH oxidase and growth of transformed cells in culture. Biochem. Pharmacol. 60, 937–946.

Moyers, S.B., Kumar, N.B., 2004. Green tea polyphenols and cancer chemoprevention: multiple mechanisms and endpoints for phase II trials. Nutr. Rev. 62, 204–211.

Mukhtar, H., Ahmad, N., 1999. Green tea in chemoprevention of cancer. Toxicol. Sci. 52, 111–117.

Mukhtar, H., Ahmad, N., 2000. Tea polyphenols: prevention of cancer and optimizing health. Am. J. Clin. Nutr. 71, 1698S–1702S, discussion 1703S–1704S.

Naasani, I., Oh-Hashi, F., Oh-Hara, T., Feng, W.Y., Johnston, J., Chan, K., Tsuruo, T., 2003. Blocking telomerase by dietary polyphenols is a major mechanism for limiting the growth of human cancer cells in vitro and in vivo. Cancer Res. 63, 824–830.

Nagle, D.G., Ferreira, D., Zhou, Y.D., 2006. Epigallocatechin-3-gallate (EGCG): chemical and biomedical perspectives. Phytochemistry 67, 1849–1855.

Nakachi, K., Suemasu, K., Suga, K., Takeo, T., Imai, K., Higashi, Y., 1998. Influence of drinking green tea on breast cancer malignancy among Japanese patients. Jpn. J. Cancer Res. 89, 254–261.

Nakachi, K., Matsuyama, S., Miyake, S., Suganuma, M., Imai, K., 2000. Preventive effects of drinking green tea on cancer and cardiovascular disease: epidemiological evidence for multiple targeting prevention. Biofactors 13, 49–54.

Nam, S., Smith, D.M., Dou, Q.P., 2001. Ester bond-containing tea polyphenols potently inhibit proteasome activity in vitro and in vivo. J. Biol. Chem. 276, 13322–13330.

Nasseri-Moghaddam, S., 2009. Tea drinking habits and oesophagial cancer in a high-risk area in northern Iran: population based case-control study. Arch. Iran. Med. 12, 330–332.

Navarro-Peran, E., Cabezas-Herrera, J., Sanchez-Del-Campo, L., Garcia-Canovas, F., Rodriguez-Lopez, J.N., 2008. The anti-inflammatory and anti-cancer properties of epigallocatechin-3-gallate are mediated by folate cycle disruption, adenosine release and NF-kappaB suppression. Inflamm. Res. 57, 472–478.

Negishi, H., Xu, J.W., Ikeda, K., Njelekela, M., Nara, Y., Yamori, Y., 2004. Black and green tea polyphenols attenuate blood pressure increases in stroke-prone spontaneously hypertensive rats. J. Nutr. 134, 38–42.

Nicolson, G.L., 1989. Metastatic tumor cell interactions with endothelium, basement membrane and tissue. Curr. Opin. Cell Biol. 1, 1009–1019.

Nonaka, Y., Iwagaki, H., Kimura, T., Fuchimoto, S., Orita, K., 1993. Effect of reactive oxygen intermediates on the in vitro invasive capacity of tumor cells and liver metastasis in mice. Int. J. Cancer 54, 983–986.

Odom, T.W., Pileni, M.P., 2008. Nanoscience. Acc. Chem. Res. 41, 1565.

Pae, M., Ren, Z., Meydani, M., Shang, F., Smith, D., Meydani, S.N., Wu, D., 2011. Dietary supplementation with high dose of epigallocatechin-3-gallate promotes inflammatory response in mice. J. Nutr. Biochem. 23, 526–531.

Park, J.S., Kim, M.H., Chang, H.J., Kim, K.M., Kim, S.M., Shin, B.A., Ahn, B.W., Jung, Y.D., 2006. Epigallocatechin-3-gallate inhibits the PDGF-induced VEGF expression in human vascular smooth muscle cells via blocking PDGF receptor and Erk-1/2. Int. J. Oncol. 29, 1247–1252.
Peng, G., Dixon, D.A., Muga, S.J., Smith, T.J., Wargovich, M.J., 2006. Green tea polyphenol (−)-epigallocatechin-3-gallate inhibits cyclooxygenase-2 expression in colon carcinogenesis. Mol. Carcinog. 45, 309–319.
Potenza, M.A., Marasciulo, F.L., Tarquinio, M., Tiravanti, E., Colantuono, G., Federici, A., Kim, J.A., Quon, M.J., Montagnani, M., 2007. EGCG, a green tea polyphenol, improves endothelial function and insulin sensitivity, reduces blood pressure, and protects against myocardial I/R injury in SHR. Am. J. Physiol. Endocrinol. Metab. 292, E1378–E1387.
Rice-Evans, C., Burdon, R., 1993. Free radical-lipid interactions and their pathological consequences. Prog. Lipid Res. 32, 71–110.
Riegsecker, S., Wiczynski, D., Kaplan, M.J., Ahmed, S., 2013. Potential benefits of green tea polyphenol EGCG in the prevention and treatment of vascular inflammation in rheumatoid arthritis. Life Sci. 93, 307–312.
Rietveld, A., Wiseman, S., 2003. Antioxidant effects of tea: evidence from human clinical trials. J. Nutr. 133, 3285S–3292S.
Roowi, S., Stalmach, A., Mullen, W., Lean, M.E., Edwards, C.A., Crozier, A., 2010. Green tea flavan-3-ols: colonic degradation and urinary excretion of catabolites by humans. J. Agric. Food Chem. 58, 1296–1304.
Rosengren, R.J., 2003. Catechins and the treatment of breast cancer: possible utility and mechanistic targets. IDrugs 6, 1073–1078.
Sai, K., Kai, S., Umemura, T., Tanimura, A., Hasegawa, R., Inoue, T., Kurokawa, Y., 1998. Protective effects of green tea on hepatotoxicity, oxidative DNA damage and cell proliferation in the rat liver induced by repeated oral administration of 2-nitropropane. Food Chem. Toxicol. 36, 1043–1051.
Sato, D., 1999. Inhibition of urinary bladder tumors induced by *N*-butyl-*N*-(4-hydroxybutyl)-nitrosamine in rats by green tea. Int. J. Urol. 6, 93–99.
Schneider, C., Segre, T., 2009. Green tea: potential health benefits. Am. Fam. Physician 79, 591–594.
Schoonbroodt, S., Piette, J., 2000. Oxidative stress interference with the nuclear factor-kappa B activation pathways. Biochem. Pharmacol. 60, 1075–1083.
Shanafelt, T.D., Call, T.G., Zent, C.S., Laplant, B., Bowen, D.A., Roos, M., Secreto, C.R., Ghosh, A.K., Kabat, B.F., Lee, M.J., Yang, C.S., Jelinek, D.F., Erlichman, C., Kay, N.E., 2009. Phase I trial of daily oral Polyphenon E in patients with asymptomatic Rai stage 0 to II chronic lymphocytic leukemia. J. Clin. Oncol. 27, 3808–3814.
Shankar, S., Ganapathy, S., Srivastava, R.K., 2007. Green tea polyphenols: biology and therapeutic implications in cancer. Front Biosci. 12, 4881–4899.
Shen, H., Ackerman, A.L., Cody, V., Giodini, A., Hinson, E.R., Cresswell, P., Edelson, R.L., Saltzman, W.M., Hanlon, D.J., 2006. Enhanced and prolonged cross-presentation following endosomal escape of exogenous antigens encapsulated in biodegradable nanoparticles. Immunology 117, 78–88.
Shimada, M., Mochizuki, K., Sakurai, N., Goda, T., 2007. Dietary supplementation with epigallocatechin gallate elevates levels of circulating adiponectin in non-obese type-2 diabetic Goto-Kakizaki rats. Biosci. Biotechnol. Biochem. 71, 2079–2082.
Shimizu, T., Kishida, T., Hasegawa, U., Ueda, Y., Imanishi, J., Yamagishi, H., Akiyoshi, K., Otsuji, E., Mazda, O., 2008. Nanogel DDS enables sustained release of IL-12 for tumor immunotherapy. Biochem. Biophys. Res. Commun. 367, 330–335.

Shin, D.M., 2009. Oral cancer prevention advances with a translational trial of green tea. Cancer Prev. Res. 2, 919–921.
Shukla, S., Maclennan, G.T., Marengo, S.R., Resnick, M.I., Gupta, S., 2005. Constitutive activation of P I3 K-Akt and NF-kappaB during prostate cancer progression in autochthonous transgenic mouse model. Prostate 64, 224–239.
Siddiqui, I.A., Mukhtar, H., 2010. Nanochemoprevention by bioactive food components: a perspective. Pharm. Res. 27, 1054–1060.
Siddiqui, I.A., Adhami, V.M., Bharali, D.J., Hafeez, B.B., Asim, M., Khwaja, S.I., Ahmad, N., Cui, H., Mousa, S.A., Mukhtar, H., 2009. Introducing nanochemoprevention as a novel approach for cancer control: proof of principle with green tea polyphenol epigallocatechin-3-gallate. Cancer Res. 69, 1712–1716.
Siddiqui, I.A., Bharali, D.J., Nihal, M., Adhami, V.M., Khan, N., Chamcheu, J.C., Khan, M.I., Shabana, S., Mousa, S.A., Mukhtar, H., 2014. Excellent anti-proliferative and pro-apoptotic effects of (−)-epigallocatechin-3-gallate encapsulated in chitosan nanoparticles on human melanoma cell growth both in vitro and in vivo. Nanomedicine 10, 1619–1626.
Slivova, V., Zaloga, G., Demichele, S.J., Mukerji, P., Huang, Y.S., Siddiqui, R., Harvey, K., Valachovicova, T., Sliva, D., 2005. Green tea polyphenols modulate secretion of urokinase plasminogen activator (uPA) and inhibit invasive behavior of breast cancer cells. Nutr. Cancer 52, 66–73.
Smith, S.K., 1997. Angiogenesis. Semin. Reprod. Endocrinol. 15, 221–227.
Spinella, F., Rosano, L., Decandia, S., Di Castro, V., Albini, A., Elia, G., Natali, P.G., Bagnato, A., 2006. Antitumor effect of green tea polyphenol epigallocatechin-3-gallate in ovarian carcinoma cells: evidence for the endothelin-1 as a potential target. Exp. Biol. Med. 231, 1123–1127.
Spiteller, G., 2001. Peroxidation of linoleic acid and its relation to aging and age dependent diseases. Mech. Ageing Dev. 122, 617–657.
Stalmach, A., Troufflard, S., Serafini, M., Crozier, A., 2009. Absorption, metabolism and excretion of Choladi green tea flavan-3-ols by humans. Mol. Nutr. Food Res. 53 (Suppl. 1), S44–S53.
Stangl, V., Lorenz, M., Stangl, K., 2006. The role of tea and tea flavonoids in cardiovascular health. Mol. Nutr. Food Res. 50, 218–228.
Stoner, G.D., Mukhtar, H., 1995. Polyphenols as cancer chemopreventive agents. J. Cell Biochem. Suppl. 22, 169–180.
Strasly, M., Cavallo, F., Geuna, M., Mitola, S., Colombo, M.P., Forni, G., Bussolino, F., 2001. IL-12 inhibition of endothelial cell functions and angiogenesis depends on lymphocyte-endothelial cell cross-talk. J. Immunol. 166, 3890–3899.
Su, L.J., Arab, L., 2002. Tea consumption and the reduced risk of colon cancer—results from a national prospective cohort study. Public Health Nutr. 5, 419–425.
Sueoka, N., Suganuma, M., Sueoka, E., Okabe, S., Matsuyama, S., Imai, K., Nakachi, K., Fujiki, H., 2001. A new function of green tea: prevention of lifestyle-related diseases. Ann. NY Acad. Sci. 928, 274–280.
Suganuma, M., Sueoka, E., Sueoka, N., Okabe, S., Fujiki, H., 2000. Mechanisms of cancer prevention by tea polyphenols based on inhibition of TNF-alpha expression. Biofactors 13, 67–72.
Sun, C.L., Yuan, J.M., Koh, W.P., Yu, M.C., 2006. Green tea, black tea and colorectal cancer risk: a meta-analysis of epidemiologic studies. Carcinogenesis 27, 1301–1319.
Sung, H., Nah, J., Chun, S., Park, H., Yang, S.E., Min, W.K., 2000. In vivo antioxidant effect of green tea. Eur. J. Clin. Nutr. 54, 527–529.

Sutherland, B.A., Shaw, O.M., Clarkson, A.N., Jackson, D.N., Sammut, I.A., Appleton, I., 2005. Neuroprotective effects of (−)-epigallocatechin gallate following hypoxia-ischemia-induced brain damage: novel mechanisms of action. FASEB J. 19, 258–260.

Suzuki, M., Tabuchi, M., Ikeda, M., Umegaki, K., Tomita, T., 2004. Protective effects of green tea catechins on cerebral ischemic damage. Med. Sci. Monit. 10, BR166–BR174.

Sweeney, C., Li, L., Shanmugam, R., Bhat-Nakshatri, P., Jayaprakasan, V., Baldridge, L.A., Gardner, T., Smith, M., Nakshatri, H., Cheng, L., 2004. Nuclear factor-kappaB is constitutively activated in prostate cancer in vitro and is overexpressed in prostatic intraepithelial neoplasia and adenocarcinoma of the prostate. Clin. Cancer Res. 10, 5501–5507.

Swenson, C.E., Bolcsak, L.E., Batist, G., Guthrie, Jr., T.H., Tkaczuk, K.H., Boxenbaum, H., Welles, L., Chow, S.C., Bhamra, R., Chaikin, P., 2003. Pharmacokinetics of doxorubicin administered i.v. as Myocet (TLC D-99; liposome-encapsulated doxorubicin citrate) compared with conventional doxorubicin when given in combination with cyclophosphamide in patients with metastatic breast cancer. Anticancer Drugs 14, 239–246.

Takabayashi, F., Harada, N., Tahara, S., Kaneko, T., Hara, Y., 1997. Effect of green tea catechins on the amount of 8-hydroxydeoxyguanosine (8-OHdG) in pancreatic and hepatic DNA after a single administration of *N*-nitrosobis(2-oxopropyl)amine (BOP). Pancreas 15, 109–112.

Taniguchi, S., Fujiki, H., Kobayashi, H., Go, H., Miyado, K., Sadano, H., Shimokawa, R., 1992. Effect of (−)-epigallocatechin gallate, the main constituent of green tea, on lung metastasis with mouse B16 melanoma cell lines. Cancer Lett. 65, 51–54.

Thawonsuwan, J., Kiron, V., Satoh, S., Panigrahi, A., Verlhac, V., 2010. Epigallocatechin-3-gallate (EGCG) affects the antioxidant and immune defense of the rainbow trout *Oncorhynchus mykiss*. Fish Physiol. Biochem. 36, 687–697.

Tokuda, H., Takai, S., Hanai, Y., Matsushima-Nishiwaki, R., Hosoi, T., Harada, A., Ohta, T., Kozawa, O., 2007. (−)-Epigallocatechin gallate suppresses endothelin-1-induced interleukin-6 synthesis in osteoblasts: inhibition of p44/p42 MAP kinase activation. FEBS Lett. 581, 1311–1316.

Toomey, D.P., Murphy, J.F., Conlon, K.C., 2009. COX-2 VEGF and tumour angiogenesis. Surgeon 7, 174–180.

Tosetti, F., Ferrari, N., De Flora, S., Albini, A., 2002. Angioprevention: angiogenesis is a common and key target for cancer chemopreventive agents. FASEB J. 16, 2–14.

Trevisanato, S.I., Kim, Y.I., 2000. Tea and health. Nutr. Rev. 58, 1–10.

Wang, G.L., Semenza, G.L., 1995. Purification and characterization of hypoxia-inducible factor 1. J. Biol. Chem. 270, 1230–1237.

Wang, Z.Y., Das, M., Bickers, D.R., Mukhtar, H., 1988. Interaction of epicatechins derived from green tea with rat hepatic cytochrome P-450. Drug Metab. Dispos. 16, 98–103.

Wang, X., Zhang, J., Lei, F., Liang, C., Yuan, F., Gao, Y., 2014. Covalent complexation and functional evaluation of (−)-epigallocatechin gallate and alpha-lactalbumin. Food Chem. 150, 341–347.

Warnakulasuriya, S.N., Ziaullah, Rupasinghe, H.P., 2014. Long chain fatty acid acylated derivatives of quercetin-3-o-glucoside as antioxidants to prevent lipid oxidation. Biomolecules 4, 980–993.

Weinreb, O., Mandel, S., Youdim, M.B., 2003. Gene and protein expression profiles of anti- and pro-apoptotic actions of dopamine, R-apomorphine, green tea

polyphenol (−)-epigallocatechine-3-gallate, and melatonin. Ann. NY Acad. Sci. 993, 351–361, discussion 387–393.
Wheeler, D.S., Catravas, J.D., Odoms, K., Denenberg, A., Malhotra, V., Wong, H.R., 2004. Epigallocatechin-3-gallate, a green tea-derived polyphenol, inhibits IL-1 beta-dependent proinflammatory signal transduction in cultured respiratory epithelial cells. J. Nutr. 134, 1039–1044.
Winterhalter, M., Lasic, D.D., 1993. Liposome stability and formation: experimental parameters and theories on the size distribution. Chem. Phys. Lipids 64, 35–43.
Wu, A.H., Yu, M.C., Tseng, C.C., Hankin, J., Pike, M.C., 2003. Green tea and risk of breast cancer in Asian Americans. Int. J. Cancer 106, 574–579.
Wu, C.M., Lin, K.W., Teng, C.H., Huang, A.M., Chen, Y.C., Yen, M.H., Wu, W.B., Pu, Y.S., Lin, C.N., 2014. Chalcone derivatives inhibit human platelet aggregation and inhibit growth in human bladder cancer cells. Biol. Pharm. Bull. 37, 1191–1198.
Xu, Y., Ho, C.T., Amin, S.G., Han, C., Chung, F.L., 1992. Inhibition of tobacco-specific nitrosamine-induced lung tumorigenesis in A/J mice by green tea and its major polyphenol as antioxidants. Cancer Res. 52, 3875–3879.
Xu, H., Lui, W.T., Chu, C.Y., Ng, P.S., Wang, C.C., Rogers, M.S., 2009. Anti-angiogenic effects of green tea catechin on an experimental endometriosis mouse model. Hum. Reprod. 24, 608–618.
Yang, G.Y., Liao, J., Kim, K., Yurkow, E.J., Yang, C.S., 1998. Inhibition of growth and induction of apoptosis in human cancer cell lines by tea polyphenols. Carcinogenesis 19, 611–616.
Yang, C.S., Maliakal, P., Meng, X., 2002. Inhibition of carcinogenesis by tea. Annu. Rev. Pharmacol. Toxicol. 42, 25–54.
Yao, L., Pike, S.E., Setsuda, J., Parekh, J., Gupta, G., Raffeld, M., Jaffe, E.S., Tosato, G., 2000. Effective targeting of tumor vasculature by the angiogenesis inhibitors vasostatin and interleukin-12. Blood 96, 1900–1905.
Yao, H.J., Ju, R.J., Wang, X.X., Zhang, Y., Li, R.J., Yu, Y., Zhang, L., Lu, W.L., 2011. The antitumor efficacy of functional paclitaxel nanomicelles in treating resistant breast cancers by oral delivery. Biomaterials 32, 3285–3302.
Yokozawa, T., Nakagawa, T., Lee, K.I., Cho, E.J., Terasawa, K., Takeuchi, S., 1999. Effects of green tea tannin on cisplatin-induced nephropathy in LLC-PK1 cells and rats. J. Pharm. Pharmacol. 51, 1325–1331.
You, M., Bergman, G., 1998. Preclinical and clinical models of lung cancer chemoprevention. Hematol. Oncol. Clin. North Am. 12, 1037–1053.
Yu, H.N., Yin, J.J., Shen, S.R., 2004. Growth inhibition of prostate cancer cells by epigallocatechin gallate in the presence of Cu2+. J. Agric. Food Chem. 52, 462–466.
Zhang, G., Miura, Y., Yagasaki, K., 1999. Effects of green, oolong and black teas and related components on the proliferation and invasion of hepatoma cells in culture. Cytotechnology 31, 37–44.
Zhang, G., Miura, Y., Yagasaki, K., 2000. Suppression of adhesion and invasion of hepatoma cells in culture by tea compounds through antioxidative activity. Cancer Lett. 159, 169–173.
Zhang, M., Binns, C.W., Lee, A.H., 2002. Tea consumption and ovarian cancer risk: a case-control study in China. Cancer Epidemiol. Biomarkers Prev. 11, 713–718.
Zhang, Q., Tang, X., Lu, Q., Zhang, Z., Rao, J., Le, A.D., 2006. Green tea extract and (−)-epigallocatechin-3-gallate inhibit hypoxia- and serum-induced HIF-1alpha protein accumulation and VEGF expression in human cervical carcinoma and hepatoma cells. Mol. Cancer Ther. 5, 1227–1238.

Zhang, H.W., Wang, L.Q., Xiang, Q.F., Zhong, Q., Chen, L.M., Xu, C.X., Xiang, X.H., Xu, B., Meng, F., Wan, Y.Q., Deng, D.Y., 2014. Specific lipase-responsive polymer-coated gadolinium nanoparticles for MR imaging of early acute pancreatitis. Biomaterials 35, 356–367.

Zhu, Q.Y., Huang, Y., Tsang, D., Chen, Z.Y., 1999. Regeneration of alpha-tocopherol in human low-density lipoprotein by green tea catechin. J. Agric. Food Chem. 47, 2020–2025.

17

POTENTIAL OF NANOMATERIALS IN FOOD PACKAGING

Konstantinos G. Gatos
R&D Center, Megaplast S.A., Athens, Greece

1 Introduction

The importance of packaging in the food sector is not only related to the preservation of food but also to its transportation and selling. It has been reported that the main four functions of packaging are containment, protection, convenience, and communication (Robertson, 2013). For example, on one hand a faulty containment may lead to leakage of the content through closures or seals, while insufficient protection would make food realize external conditions like humidity, odors, or compressive forces. On the other hand, an inconvenient packaging could raise great concerns on how to open, handle, and store a food product, while improper communication might discourage a customer from buying it.

In this view, primary packaging appears to be the most crucial type because it is in direct contact with food and it is highly involved in all these functions. Secondary or tertiary or quaternary packaging is less probable to be in contact with food. For example, secondary packaging could be a carton box, which contains a number of primary packages. Likewise, tertiary packaging could be a film-wrapped pallet with stacked carton boxes, while quaternary packaging could be a container with a controlled environment, which carries pallets. Nevertheless, all types of packaging should sustain safe transportation, adequate protection, and selling of food.

Currently, both industry and academia put a lot of effort on research and development mainly of primary packaging in order to optimize the aforementioned four functions. For this purpose, nanomaterials, which have shown notable performances as fillers in polymeric matrices, are examined as possible candidates in food packaging applications (Ayhan, 2013). In several cases these fillers are characterized as engineered nanomaterials in order to indicate that they are processed or manufactured (Morris, 2011).

Novel Approaches of Nanotechnology in Food. http://dx.doi.org/10.1016/B978-0-12-804308-0.00017-0

It is now widely accepted that a material is characterized as nanomaterial when at least one of its dimensions is at the range of 1–100 nm. According to the dimensionality of their elementary units, nanomaterials can be classified as 0-D, 1-D, and 2-D, relating, for example, to nanoparticles, nanotubes, and nanoplatelets, respectively (Gatos and Leong, 2016). The added value that nanomaterials bring to the food packaging sector is worth the undertaking in each application. In this respect, undesirable changes of important food quality characteristics like texture, flavor, color, appearance, and nutritive value should be prolonged, which in turn will extend significantly the shelf life of the relevant product (Robertson, 2013). Regarding terminology, Robertson has defined as “active” packaging when a subsidiary constituent has been deliberately included in or on the packaging material in order to enhance the performance of the system and as “intelligent” packaging when a packaging contains an internal indicator to provide information about the history of the package and/or the quality of the food (Robertson, 2013).

In this chapter, food packaging involving nanomaterials is described as “passive,” “active,” or “intelligent.” As illustrated on Fig. 17.1, the basic category relates to passive role of nanomaterials in food packaging, while passing to active or intelligent role the functionality of the related nanomaterials rises, accordingly. More

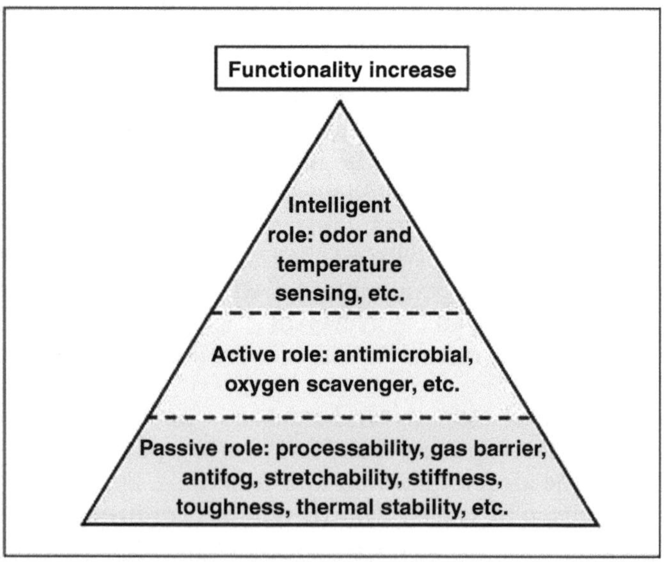

Figure 17.1. The three categories regarding the role of nanomaterials in food packaging wherein the functionality of the respective nanomaterials increases from bottom to top.

specifically, the first category relates to nanomaterials that alter passively the properties of food packaging due to their presence. In literature this category has been also characterized as "improved" (Silvestre et al., 2011). The second category relates to nanomaterials that play an active role in food preservation. Finally, the third category relates to nanomaterials that provide information about the quality of the food.

2 Passive Role of Nanomaterials in Food Packaging

In general, optimization of the packaging cost is a major concern for packaging manufacturers and food producers especially for fast-moving food packaging applications (eg, consumer packed goods). Therefore, there should be a strong reason for exploiting nanomaterials. Typical food packaging materials include polyethylene (PE), polypropylene (PP), polyvinylchloride (PVC), polyamide (PA), polyethylene terephthalate (PET), ethylene vinyl alcohol (EVOH), poly(lactic) acid (PLA), and polystyrene (PS). The initial selection of one or other material for each application by the engineers is based on the inherent properties of the material itself like rheological, thermal, mechanical, barrier, optical, printability, sealability, and recyclability. Fig. 17.2 presents the water vapor and oxygen permeability of some materials used for packaging applications. For example, polyamide 6 is less oxygen permeable than bioriented polypropylene (BOPP) or low density polyethylene (PE-LD), however, it performs worse as far as the water vapor permeability is concerned (Silvestre et al., 2013).

Usually, nanomaterials are considered as expensive additives, which complicate the manufacturing process. In practice, use of nanomaterials might be favored if the cost of incorporation is compensated by the property improvement compared to the conventional package. This improvement on one hand might relate to less overall packaging material cost (eg, by down gauging, by eliminating an expensive component) and on the other hand to increased productivity (eg, easier melt-flow, lower processing temperatures, faster sealing). Additionally, the desirable option of removing a nonrecyclable component from a multilayer or a laminated package due to the use of nanomaterials could make the handling of the scrap easier, optimizing further the packaging cost.

2.1 0-D Nanomaterials

In case of 0-D nanomaterials, silica with a primary particle size of 15nm has been shown to improve stiffness and toughness of

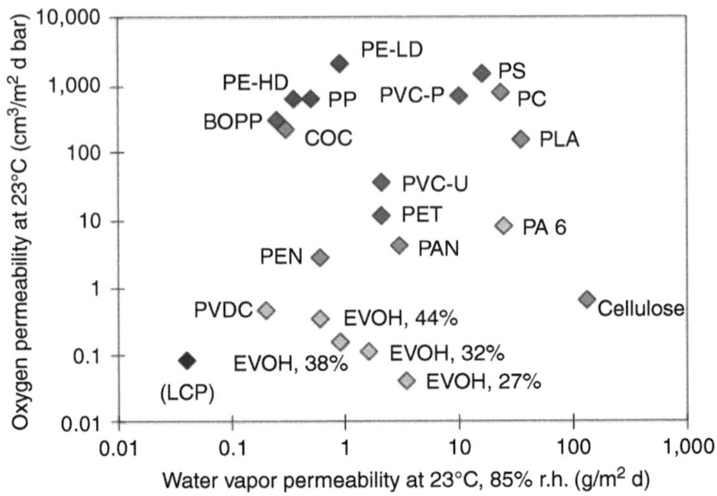

Figure 17.2. Permeability for water vapor and oxygen at 23°C and 100 μm film thickness of commonly used thermoplastics and special packaging type polymers. Reproduced from Silvestre, C., Pezzuto, M., Cimmino, S., Duraccio, D., (Eds.), 2013. Polymer nanomaterials for food packaging: current issues and future trends. In: Silvestre, C., Cimmino, S. (Eds.), Ecosustainable Polymer Nanomaterials for Food Packaging, CRC Press/Taylor & Francis, Boca Raton, pp. 1–25, with permission of Taylor & Francis Books, UK.

a PP matrix at a filler content of 0.5 vol.% (Wu et al., 2002). Liu et al. presented that the surface modification of nanosilica was able to provide better mechanical performance than the neat particles and assist the blending of the PP matrix with an ethylene-octene copolymer (Liu and Kontopoulou, 2006). As shown in Fig. 17.3, the trimethoxyoctylsilane modification of the silica nanoparticles ($mSiO_2$) improved the dispersion of silica particles within the polypropylene continuous phase, which also contained 10 wt.% of PP grafted with maleic anhydride (PP-g-MA). At the same time, the size of the discontinuous phase of the ethylene-octene copolymer polyolefin elastomer (POE) was decreased. Similarly, a prominent droplet size reduction has been reported also for PP/PA and PP/PC blends when 5 wt.% of silica nanoparticles treated with hexamethyl disilazane were mixed with each mixture. In these immiscible blends, the apolar PP (at 80 wt.%) was the continuous phase while the polar PA or PC (at 20 wt.%) represented the discontinuous one. In both cases, the silica nanoparticles were detected on the one hand to be preferentially located at the interfaces and on the other hand to be dispersed within the PP matrix (Laoutid et al., 2013).

Moreover, grafting polymers onto the silica surface has proved to be an efficient way to improve the interface between PP and

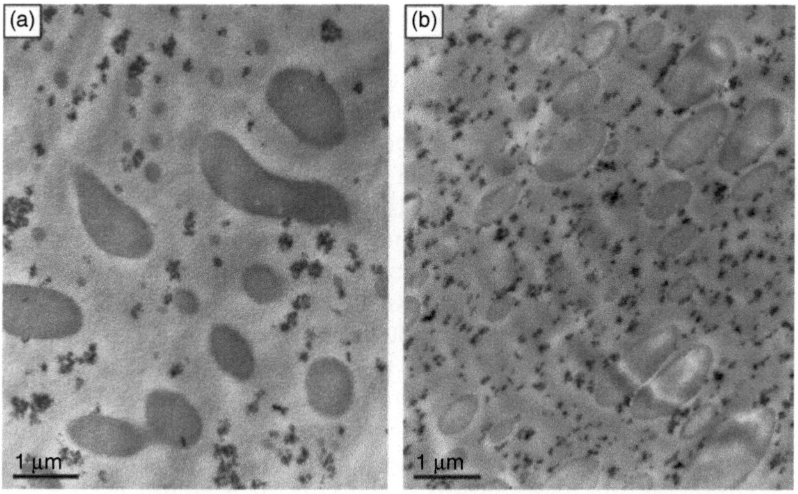

Figure 17.3. TEM micrographs of (PP/PP-*g*-MAn)/POE 80/20 TPO composites. (a) 5 wt.% SiO_2; (b) 5 wt.% $mSiO_2$. Scale bars represent 1 μm. Reprinted from Liu, Y., Kontopoulou, M., 2006. The structure and physical properties of polypropylene and thermoplastic olefin nanocomposites containing nanosilica. Polymer 47, 7731–7739, with permission from Elsevier.

silica without blending compatibilizers with the polymeric matrix. In such an example, precipitated silica having 10 nm average particle size was grafted with several commercial monomers like styrene, methyl methacrylate, ethyl acrylate, and butyl acrylate. This monomer order, which is related also with their rigidity ranking (highest for styrene), induced a similar order of moduli increase for the respective nanocomposites at low filler loadings (Wu et al., 2005). Apart from the addition of silica in order to improve the mechanical properties of a PP matrix, a synergy by polypropylene-methyl polyhedral oligomeric silsesquioxane (PP-methyl POSS) may be exploited when acting at low filler loadings (Lin et al., 2009). Nonetheless, optimized extrusion processing parameters relating to shear forces and residence time should also be considered for better performance of the PP/silica nanocomposites (Lee et al., 2011).

Although the main advantage of using silica nanoparticles in PP was mechanical property enhancement, a side effect of lower oxygen permeability has been also reported. Nevertheless, when nitrogen was served as gas (lower molecules than oxygen) the relevant permeability values were only marginally decreased for the same specimens (Vladimirov et al., 2006).

Boehmite alumina (BA) nanoparticles have been reported to increase the thermo-mechanical properties of a PP matrix.

A blow-molded PP film having a thickness of 0.6mm revealed property improvement already at 2.5 wt.% BA nanoparticles of 40 nm particle size functionalized with octylsilane. The respective nanocomposite films showed enhanced tensile properties, as well as tear resistance along both the transverse and machine direction compared to the neat film. Interesting to note is the existence of microvoids within the blown film, which were even more pronounced when the functionalization of BA involved C10–C13 alkylbenzene sulfonic acid (Pedrazzoli et al., 2014). For LLDPE, BA nanoparticles in 4 wt.% improved significantly the creep compliance at 30°C of a 500 μm thick compression molded sheet. In these experiments, the unmodified nanoparticles delivered also enhanced creep stability as compared to the plain LLDPE matrix (Pedrazzoli et al., 2013). In a similar case, boehmite alumina nanoparticles in 4 wt.% were found to improve the thermooxidative degradation of a LDPE or a HDPE matrix, as this was detected by measuring the oxidation induction time (Khumalo et al., 2010).

As it has been presented in an example of PS filled with silica, there is an analogy between film thickness and interparticle spacing. More specifically, through measuring the glass transition temperature shift of PS samples it was concluded that the thermomechanical properties of the nanocomposites varying the filler content resembling those of nonreinforced planar (nanolayer) PS films for a respective thickness (Bansal et al., 2005). Such a conclusion might project to the design of multilayer food packaging films, too.

Nanoparticles of calcium carbonate ($CaCO_3$) 40 nm in diameter have been found to increase the toughness of PVC, wherein their dispersion in the PVC matrix could be further optimized by using chlorinated polyethylene as interfacial modifier (Wu et al., 2004). For PET, the incorporation of 3 wt.% precipitated calcium carbonate of 70 nm mean particle diameter required compatibilization with anhydride modified PET for property improvement. The compatibilization improved the filler dispersion within the polymeric matrix, reducing at the same time the oxygen permeability compared to the noncompatibilized composite or the neat matrix. It has to be mentioned that the glass transition temperature (T_g) of the amorphous matrix was increased about 8°C compared to the examined compatibilized compression-molded nanocomposite films of 70 μm in thickness, namely from 83 to 91°C, respectively (Avolio et al., 2013). In general, improvement of the thermal properties such as the T_g or the melting temperature is expected to affect the hot-fill or the form-fill-seal (FFS) applications of the respective films.

2.2 1-D Nanomaterials

A 1-D nanomaterial, which has a potential to be used in food packaging is halloysite nanotube (HNT). Raw halloysite is mined from natural deposits and it is abundantly found in nature. The molecular formula of HNT resembles kaolinite, however its morphology is totally different. HNT has a hollow tubular structure with high aspect ratio. Its length and diameter ranges from 0.2 to 2 μm and 40 to 70 nm, respectively. Additionally, its inner diameter ranges from 10 to 40 nm (Liu et al., 2014).

HNTs have been mixed with PP through extrusion process followed by injection molding. It was found that filler modification was not a prerequisite for property improvement of the nanocomposite, although that adequate modification improved further its performance. A halloysite content of 6 wt.% enhanced the properties of PP in terms of tension, bending, and impact, presenting increased strength and modulus without sacrificing ductility. Differential scanning calorimentry (DSC) revealed that HNTs acted as nucleated agents in PP increasing its crystallization temperature along with the overall crystallinity (Prashantha et al., 2011). Improvement of strength, modulus, and toughness at the same time for polymer/HNT nanocomposites has been also reported for a PVC matrix (Liu et al., 2011).

In another example, HNTs, which were melt-mixed with PA 6 through a twin screw extruder, were found to enhance the mechanical performance of the nanocomposite, while the flow properties of the matrix were only moderately affected (Handge et al., 2010). Apart from mechanical property improvement, HNTs delivered also increased thermal stability, as well as reduced water uptake and oxygen permeability when served as filler in an EVA film (Bidsorkhi et al., 2015). High impact polystyrene was prepared via emulsion polymerization in presence of HNTs using sodium dodecyl sulfate (SDS) as emulsifier. The obtained powder containing 5 wt.% or 10 wt.% filler loading was processed by means of injection molding in order to produce PS/HNTs nanocomposites. The thermal stability of the respective specimens, as this was evaluated by thermogravimetric analysis (TGA), presented a notable improvement. Additionally, as shown in Fig. 17.4, the notched Izod impact strength of the nanocomposite containing 5 wt.% HNTs (PS-05) yielded the best performance compared to the neat matrix or the specimen with 10 wt.% filler loading (PS-10). The observed toughening effect yielded by the 5 wt.% HNTs in the PS matrix was related with the optimized filler dispersion in the PS matrix along with the low filler-matrix interfacial strength due to SDS. Moreover, above effect was manifested in the scanning electron

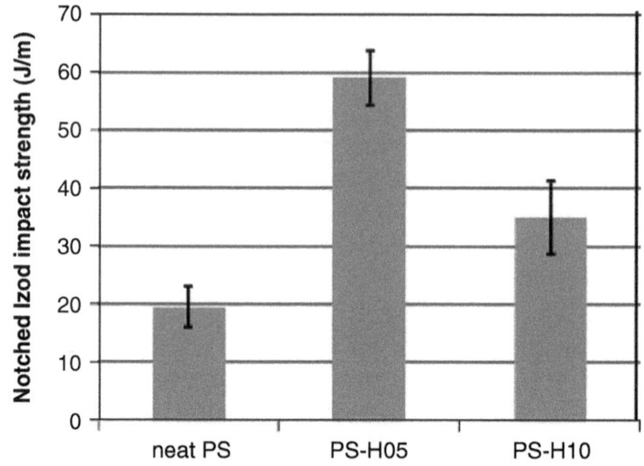

Figure 17.4. Notched Izod impact strength of the neat PS and the PS/HNT nanocomposites. Reprinted from Lin, Y., Ng, K.M., Chan, C-M., Sun, G., Wu, J., 2011. High-impact polystyrene/halloysite nanocomposites prepared by emulsion polymerization using sodium dodecyl sulfate as surfactant. J. Colloid Interface Sci. 358, 423–429, with permission from Elsevier.

microscopy (SEM) images of the fractured surfaces by the existence of a large number of crazes and the dimple-like structures (Lin et al., 2011).

HNT presents a potential in polymer reinforcement. Nevertheless, it has to be mentioned that since halloysite finds already applications in medical and pharmaceutical sector as nanocontainer, for example, controlled release of biological active compounds (Abdullayev and Lvov, 2011), its role as active or intelligent nanomaterial in food packaging in the near future seems rather attractive.

Another nanomaterial abundant in nature that could be used in food packaging is cellulose (de Azeredo 2009). It has been reported to be mixed with polymers imparting improved properties in terms of high transparency, low coefficient of thermal expansion and improved mechanical performance (Khan et al., 2014). However, in many cases its hydrophilic character and the questionable durability may limit extensive industrial applications (Hubbe et al., 2008). Cellulose belongs to carbohydrates and it is obtained by both plant and nonplant sources. Plants containing large amounts of cellulose include hemp, flax, jute, ramie, and cotton (Eichhorn et al., 2010).

PP has been solution mixed with cellulose nanowhiskers yielding a marginal tensile property and thermal stability improvement, as well as a decrease in the contact angle of a water droplet

(Bahar et al., 2012). Surface modification with organic acid chloride aliphatic chains of different sizes was found to reduce the hydrophilicity of cellulose enabling melt mixing with the LDPE in a twin-screw extruder. Nevertheless, the mechanical and thermal properties were only slightly altered while the potential advantage for food packaging applications of low water contact angle was dramatically diminished (de Menezes et al., 2009). Cellulose has been reported to have a profound effect on the crystallization of PLA. Its role as nucleating agent was revealed at 1.5 wt.%, increasing the crystallization rate of PLA and providing smaller spherulite sizes. However, the storage modulus of the relevant nanocomposites was improved by mixing more than 5 wt.% of cellulose nanocrystals within the matrix (Kamal and Khoshkava, 2015).

2.3 2-D Nanomaterials

In several food-packaging applications, the barrier properties of the primary packaging play a significant role in food preservation extending the shelf life (Adur et al., 2002; Bensur 2004). As expected, the 2-D nanomaterials create a tortuous path for the gas molecules within the polymer matrix hindering gas permeability (Gatos and Karger-Kocsis, 2011). This kind of nanoreinforcement has been well established for improving the oxygen barrier properties of food packaging (Sablani, 2015).

A very popular 2-D nanomaterial, which has been extensively investigated in academia finding already applications in industry, is clay mineral (de Azeredo, 2009). Therein, the group of great importance for mixing with polymers relates to layered silicates or phyllosilicates and especially to those that are able to expand their basal spacing under specific circumstances. It is usually advantageous to treat phyllosilicates with appropriate low molecular weight organic chains bearing a polar head. These latter are tethered onto the silicate sheets, which become organophilic (ie, organoclay) and the interlayer spacing may facilitate accommodation of macromolecules. Note that the thickness of each silicate sheet is about 1nm. As such, montmorillonite has received the majority of research due to its exceptional swelling capabilities (Theng, 1979). The goal of dispersing the silicate sheets within a polymer matrix can be achieved by exploiting various processing methods like melt compounding, solution mixing, or latex technique. Nevertheless, nanocomposite is obtained when the polymer chains are intercalated into the gallery spacing, that is, intercalated nanocomposite, or when the silicate sheets are pulled apart within the matrix, that is, exfoliated nanocomposite (Gatos and Karger-Kocsis, 2010). As schematically shown in Fig. 17.5, both

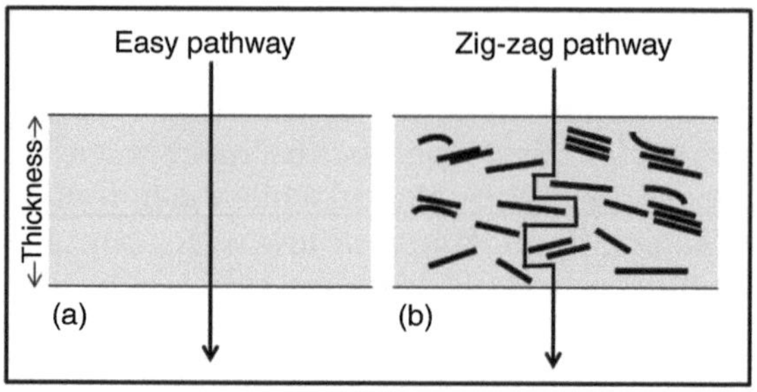

Figure 17.5. Schematic illustration of the pathway of the gas molecules through (a) a neat polymeric matrix and (b) a polymer/clay nanocomposite with exfoliated and intercalated clay structures.

intercalated and exfoliated silicate layers contribute to the zig-zag route of the gas molecules, which induces reduction of the gas permeability of the respective nanocomposites compared to the neat polymeric matrix. In general, it has been presented that using layered silicates of a higher aspect ratio, for example, fluorohectorite instead of montmorillonite, the respective barrier performance may further be improved (Gatos and Karger-Kocsis, 2007).

Polarity has proven to be an important issue for mixing polymers with organoclay promoting nanocomposite formation. PP grafted with maleic anhydrite (PP-g-MA) has been shown to be efficient for producing PP/organoclay nanocomposites with enhanced mechanical performance (Nam et al., 2001). A similar improvement holds also when the articles of PP/organoclay nanocomposites concern polymeric foams (Okamoto et al., 2001). In this respect, food applications related to trays and pads can be envisaged. The property of heat distortion temperature (HDT), which presents industrial applicability, has been significantly improved for PP from 109 to 144°C with the addition of only 3 wt.% organo-montmorillonite (Manias et al., 2001). In general, addition of filler during polymer manufacturing is expected to deteriorate the transparency of the respective articles. In the case of PP mixed with poly(ethylene oxide) (PEO), incorporation of 5 wt.% montmorillonite modified with a quaternary ammonium bearing hydroxyl polar groups changed the initial opaque appearance of the PP/PEO (70/30) blend into a transparent one. The observation was directly noticeable by the bare eye examining a sheet of 1.5 mm in thickness. This performance was attributed mainly to the lowering of the domain size of the dispersed PEO phase within the PP matrix due to presence of the polar organoclay (Kelarakis and Yoon, 2008). Similarly, in another example, the mixing of 5 wt.% organoclay in a PP/EVA (75/25) blend reduced the EVA droplet dispersion within the continuous PP phase. Increasing further the EVA weight content in the blend to reach a 1:1 weight ratio induced a cocontinuous blend morphology wherein the organoclay was found mainly dispersed in the polar EVA phase (Goodarzi et al., 2013). Such a kind of blends may improve sealability, mechanical, and barrier performance in food packaging applications at the same time.

For flowpack and lidding films, biaxially oriented polypropylene maybe proposed. Its practice might be benefited when clay is going to be served as filler (Abu-Zurayk et al., 2009). Accordingly, the degree of clay orientation, which is related with the property improvement, may be conveniently determined with spectroscopic methods (Cole et al., 2005). Nevertheless, the property of PP, which is especially enhanced due to the platy structure of clay, is gas permeability (Villaluenga et al., 2007).

The gas barrier improvement is unveiled also for other polyolefins suitable for food packaging applications like low density polyethylenes. In such an example, blown films of LDPE filled with 7 wt.% of organoclay (compatibilized with maleic anhydride) presented 24% oxygen permeation decrease compared to the neat matrix (Arunvisut et al., 2007). An even more pronounced oxygen permeability reduction has been reported for LLDPE/organoclay nanocomposites compatibilized with maleic anhydride wherein the same barrier performance was exhibited for N_2 and CO_2 gases (Hotta and Paul, 2004). In another case, the use of low molecular weight oxidized polyethylene has also been proposed as an efficient compatibilizer to improve the gas barrier performance of polyethylene/clay films (Durmuş et al., 2007). Moreover, for LLDPE, a zinc-neutralized carboxylate ionomer, has been presented to induce compatibilization comparable with that of MA (Sánchez-Valdes et al., 2006).

Polyethylene and specifically HDPE is often being exploited as a nonflexible food packaging material. Addition of only 2 wt.% of organoclay improved the creep resistance of the matrix provided that maleic anhydrite would act as compatibilizer (Pegoretti et al., 2007). In a rather different application, organoclay has been presented to alter the heat-sealing behavior of an LLDPE/LDPE (80/20) blend mixed with EVA. The combination of 30% EVA copolymer (of 18% vinyl acetate content) with 6 wt.% organomontmorillonite created a synergy that induced on the one hand sealant integrity and on the other hand easy crack propagation upon seal peeling (Manias et al., 2009).

Another advantage for mixing clay within polymer packaging was presented for EVOH/bentonite films, which contain carvacrol as antimicrobial additive. The presence of 2 wt.% bentonite increased the solubility of the carvacrol and the water vapor, while reducing the water vapor permeability and the carvacrol diffusivity. The clay layers controlled the release of the carvacrol, which was triggered by the water content, suggesting extended shelf life for the packed food (Cerisuelo et al., 2012). Although that such a packaging should be classified as active, nevertheless, the role of bentonite was rather passive.

In several semirigid food packaging applications foamed PS trays are used for carrying the food. In meat, poultry, and seafood packaging industry, PS foamed trays and absorbent pads are usually considered. Organoclay has been proposed as potential filler for enhancing the performance of the PS packaging. Injection molded PS/organoclay foamed nanocomposites with 1 wt.% filler loading were found to yield improved strength and wear resistance compared to the neat foamed PS. This performance was related to the lower cell size induced by the presence of the clay (Hwang et al., 2009). It has to be noticed that several side effects such as alteration of the thermal conductivity are expected to be affected by the cell size differentiation (Notario et al., 2015).

PLA has been proposed as a suitable material for the manufacturing of plastic bottles and trays. For such purposes, addition of clay has proved to be beneficial for reducing water transmission rate and oxygen permeability. PLA reinforced with 4 wt.% of organoclay was melt-mixed in a twin screw extruder followed by pelletizing and injection stretch-blow molding to produce bottles. As shown in Fig. 17.6, the transmission electron microscopy (TEM) images depict that the organoclay is nicely dispersed in the PLA matrix. The clay platelet dispersion has provided a tortuous path, which induced a 37% and 42% reduction of water transmission rate and oxygen permeability for the nanocomposite bottles compared to nonreinforced ones, respectively (Jorda-Beneyto et al., 2014). The advantageous effect of organoclay addition in a PLA matrix has been also demonstrated for PLA/organoclay blown films where mixing of 5 wt.% organoclay raised about 50% improvement of the oxygen and water vapor barrier property compared to the plain matrix (Thellen et al., 2005).

Polyamide, due to its relatively high gas barrier performance, good thermomechanical properties, and puncture resistance, finds plethora of applications in food packaging. Nonetheless, it is susceptible to humidity and it is usually met in multilayer structures along with polyolefins, which occupy the outer layers appointed for printing or sealing. Halim et al. examined the effect of pasteurization, high pressure processing and retorting on the barrier properties of nanoreinforced nylon 6 in coextruded multilayer structures. These processes deteriorated the oxygen

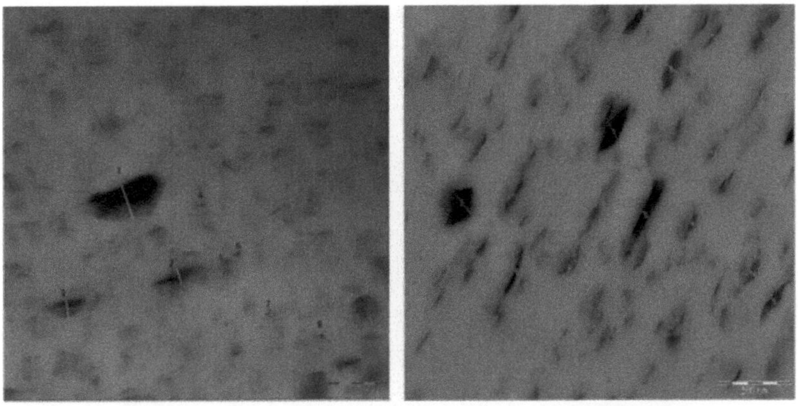

Figure 17.6. Representative transmission electron microscopy (TEM) pictures of organoclay (4% w/w) in PLA matrix (PLA/clay nanocomposite bottles). Jorda-Beneyto, M., Ortuño, N., Devis, A., Aucejo, S., Puerto, M., Gutiérrez-Praena, D., Houtman, J., Pichardo, S., Maisanaba, S., Jos, A., 2014. Use of nanoclay platelets in food packaging materials: technical and cytotoxicity approach. Food Addit. Contam. Part A 31, 354–363, reprinted by permission of the publisher (Taylor & Francis Ltd, http://www.tandfonline.com).

and water vapor permeability of the respective films, however, the nanoreinforcement compensated the property deterioration involved (Halim et al., 2009). Currently, PA is a material that has been extensively examined in academia for its property improvement due to clay reinforcement. A moderate amount of organoclay has been found adequate for improving tensile and flexural strength (Chow and Mohd Ishak, 2007), creep behavior (Seltzer et al., 2012), thermal stability (Liu et al., 2003), flame retardant performance (Kashiwagi et al., 2004), and oxygen barrier properties (Yeh et al., 2009). Fig. 17.7 presents a high magnification TEM image of PA6/organoclay nanocomposite at 5 wt.% filler loading manufactured by injection molding, wherein the clay platelets are indicated with the letter A (Kashiwagi et al., 2004).

Regarding PET, it has been reported an acceleration of its thermal degradation in presence of organoclay. Its molecular weight reduction during processing could be compensated by using a pyromellitic dianhydride during extrusion (Xu et al., 2009). In another case, a multifunctional epoxy-based chain extender has been used for this purpose. Aside from the improvement of the viscoelastic properties of the PET/clay nanocomposites containing the chain extender, the modulus, and the oxygen permeability were also improved (Ghanbari et al., 2013). Besides, PVC has received less attention to be filled with organoclay mainly due to the presence of ammonium-based clay intercalants, which accelerate the degradation of PVC (Awad et al., 2009).

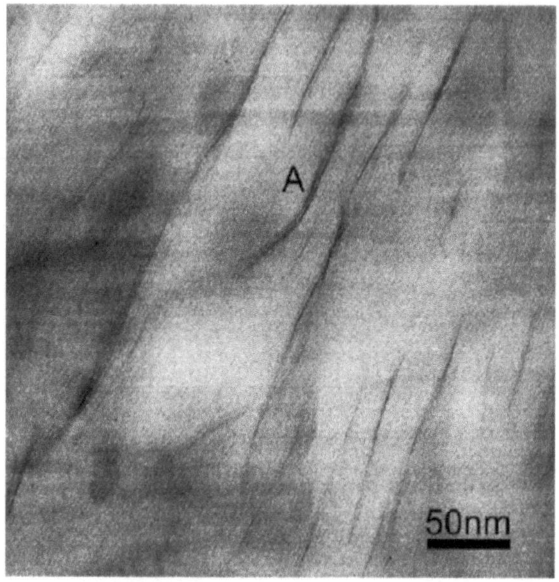

Figure 17.7. TEM image of an injection molded PA6/clay (5%) sample. Reprinted from Kashiwagi, T., Harris Jr, R.H., Zhang, X., Briber, R.M., Cipriano, R.H., Raghavan, S.R., Awad, W.H., Shields, J.R., 2004. Flame retardant mechanism of polyamide 6-clay nanocomposites. Polymer 45, 881–891, with permission from Elsevier.

Apart from the negatively charged layered silicates, which accommodate cations into their interlayer spacing, layered double hydroxides (LDH) are positively charged with anionic interlayer galleries. Likewise, organo-modified LDHs can be dispersed within thermoplastics through melt-mixing expecting the advantages of 2-D nanofillers (Purohit et al., 2014). Another 2-D nanomaterial, which yields mechanical and gas barrier improvement in thermoplastic matrices, is graphene (Kuila et al., 2012), however, it is discussed later for its active and intelligent role in food packaging.

3 Active Role of Nanomaterials in Food Packaging

The nanomaterials that would deliberately be incorporated within a polymeric packaging material to release or absorb substances into or from the packed food and its environment are considered to play an active role in food packaging (Silvestre et al., 2013). Such a role might include alteration of the content of oxygen, ethylene, moisture, carbon dioxide, flavors, odors, or inhibition of oxidation and microbial growth (Vermeiren et al., 1999).

Metals such as silver (Ag), gold (Au), and copper (Cu), as well as metal oxide nanoparticles such as titanium dioxide (TiO_2) and zinc oxide (ZnO) are well known for their antimicrobial action (Moritz and Geszke-Moritz, 2013).

The antimicrobial activity of Ag nanoparticles is mainly governed by Ag^+ ions, which are released from the surface of the particles (de Azeredo, 2012). In order to avoid the agglomeration of the silver nanoparticles especially when they are dispersed in suspensions, it has been proposed to immobilize them on silica (SiO_2) by spray combustion. These nanoparticles presented antimicrobial action against *Escherichia coli* bacteria, while the silver surface area concentration correlates well with the related antimicrobial action (Sotiriou et al., 2011). Such types of nanostructured materials may find extended applications in food packaging (Schiffmann and Schlösser, 2011). On a relevant example, silver nanoparticles of 10 nm in diameter were immobilized on TiO_2 particles in a 5/95 ratio. These nanostructured materials were melt-mixed with LDPE followed by blown extrusion. It was found that even at 1.5 wt.% filler concentration the respective monolayer blown film was able to prolong the shelf life of fresh orange juice for several days. A more or less similar result was also produced by 0.25 wt.% ZnO nanoparticles of 70 nm in diameter (Emamifar et al., 2010).

Apart from the efforts to prepare a fine dispersion of nanoparticles in a polymeric matrix, which would promote antimicrobial action, a nanolayer made by nanoparticles may produce adequate results. In such an example, a PVC film, which was coated with ZnO nanoparticles presented antibacterial activity against both *E. coli* and *Staphylococcus aureus*. UV light irradiation was observed to enhance this action. Besides, the coating reduced the tensile properties of the PVC, as well as its water vapor transmission (Li et al., 2010). Likewise, an oriented polypropylene (OPP) film of 30 μm in thickness, which was coated with TiO_2 nanoparticles, inactivated prominently *E. coli* both in vitro and in actual tests under UVA light irradiation (Chawengkijwanich and Hayata, 2008). As shown in Fig. 17.8, using a colony counting method, the cells of *E. coli* on OPP and on TiO_2-coated OPP films were decreased by 1 log CFU/mL and 3 log CFU/mL after 180 min of black-light illumination, respectively.

Solution mixing has also been exploited for preparing antibacterial films. Polylactic acid was mixed with copper nanoparticles of about 36nm in diameter, which were generated using pulse laser ablation in liquid. The resulting nanocomposites presented antibacterial properties against a mix of *Pseudomonas fluorescens* and *P. putida* (*Pseudomonas* spp.) (Longano et al., 2012). For many packaging applications paper maybe combined with polymers for

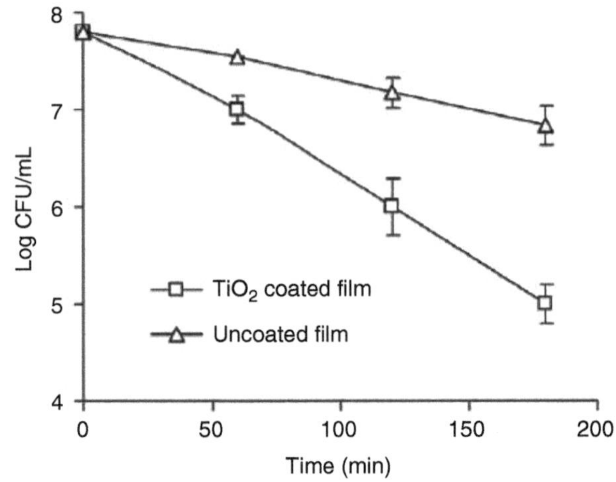

Figure 17.8. Inactivation of *Escherichia coli* in an in vitro test with uncoated OPP film and TiO_2-coated packaging film under black-light illumination (UVA light intensity of 1 mW/cm²). Each value corresponds to the mean of the three replicates, three plates per replicate. The vertical bar indicates the standard error (SE). Reprinted from Chawengkijwanich, C., Hayata, Y., 2008. Development of TiO_2 powder-coated food packaging film and its ability to inactivate *Escherichia coli* in vitro and in actual tests. Int. J. Food Microbiol. 123, 288–292, with permission from Elsevier.

enhanced performance. Paper sheets that were coated by PS/Ag nanocomposite solution presented antibacterial action against *Pseudomonas*, *S. aureus*, *Staphylococcus*, and *Candida*. However, when the PS coating involved TiO_2 nanoparticles instead of silver, the packaging material showed noninhibition against *Pseudomonas* and *Candida* (Youssef et al., 2013).

Nowadays, it is widely accepted that silver nanoparticles at the appropriate weight percentage have the advantage to induce antimicrobial properties. Nevertheless, the size of these nanoparticles affects the efficiency of this effect. Hydroxypropyl methyl cellulose (HPMC) is a film-forming biomaterial that may find applications in food packaging. It has been presented that silver nanoparticles of 41 nm in diameter were more effective than those of 100 nm particle size. This was evaluated in terms of improving the mechanical performance, reducing the water vapor permeability and enhancing the antimicrobial behavior of the neat matrix against *E. coli* and *S. aureus* (de Moura et al., 2012).

Incorporation of silver nanoparticles in a packaging material may be advantageous for other reasons, too. HPMC was reinforced through solution mixing with 4 wt.% cellulose along with 1 wt.% silver nanoparticles. It was found that the obtained films of about 250 μm in thickness yielded reduced moisture sorption and increased tensile modulus and strength compared to the films without reinforcement. It was interesting to measure that the presence of silver nanoparticles improved considerably the ductility of HPMC/cellulose nanocomposite (George et al., 2014).

Presence of titania nanoparticles within a polymer matrix under ultraviolet (UV) illumination has been recognized to induce oxygen scavenging effect. In such nanocomposite films, it has been found a correlation of the optical transparency of the film with the deoxygenation activity. More specifically, by reducing the light scattering it was detected an improvement of the oxygen scavenging efficiency (Xiao-e et al., 2004). Besides, a polyethylene film of 40 μm in thickness containing nanoparticles of silver and titania along with kaolin was measured to delay the ripening of Chinese jujube. It was suggested that this performance was due to decomposition of ethylene into water and carbon dioxide by the TiO_2 nanoparticles (Li et al., 2009).

An oxygen scavenging effect has been also detected for iron particles that were melt-blended with various polymeric matrices in an extruder to produce active films. A multilayer structure of this active layer (30 μm-thick) with a PE layer (20 μm-thick) has been proposed for packaging sausages or cheese. It has been found that reduction of the testing temperature from 23 to 5°C induced a decrease in the oxygen consumption rate of the oxygen scavenger multilayer film within the first 4 days. However, an increase to this rate was observed by the use of a high oxygen permeable matrix and more specific, mixing EVA instead of PP for the active layer. This matrix substitution raised faster oxygen consumption (Gibis and Rieblinger, 2011).

Since cellulose nanofibers have been recognized for their biodegradable behavior in food packaging materials, relevant potential applications have been investigated. Hybrid suspension of poly(lactic) acid (PLA) with 5 wt.% cellulose and 1 wt.% silver nanoparticles was electrospun in air. The produced fibrous mats comprised fibers of about 500 nm in diameter. Due to a synergy effect between cellulose and silver, the hybrid mats presented enhanced dynamic mechanical and tensile performance compared to the neat matrix or the binary nanocomposites (Cacciotti et al., 2014).

ZnO oxide nanorods have been proposed for reinforcing fish gelatin films, which were prepared through solution mixing. Such films have already been introduced in food packaging applications (Rouhi et al., 2013). It has to be reported that the action of ZnO nanoparticles having a diameter of 30–40 nm have been suggested as more efficient in terms of their antibacterial properties than ZnO nanorods of 250–300 nm in length and 40–50 nm in diameter (Ramani et al., 2013). In general, ZnO nanoparticles present antibacterial action against both Gram-positive and Gram-negative bacteria. This emerges due to the release of ions, the interaction of nanoparticles with the microorganisms and the

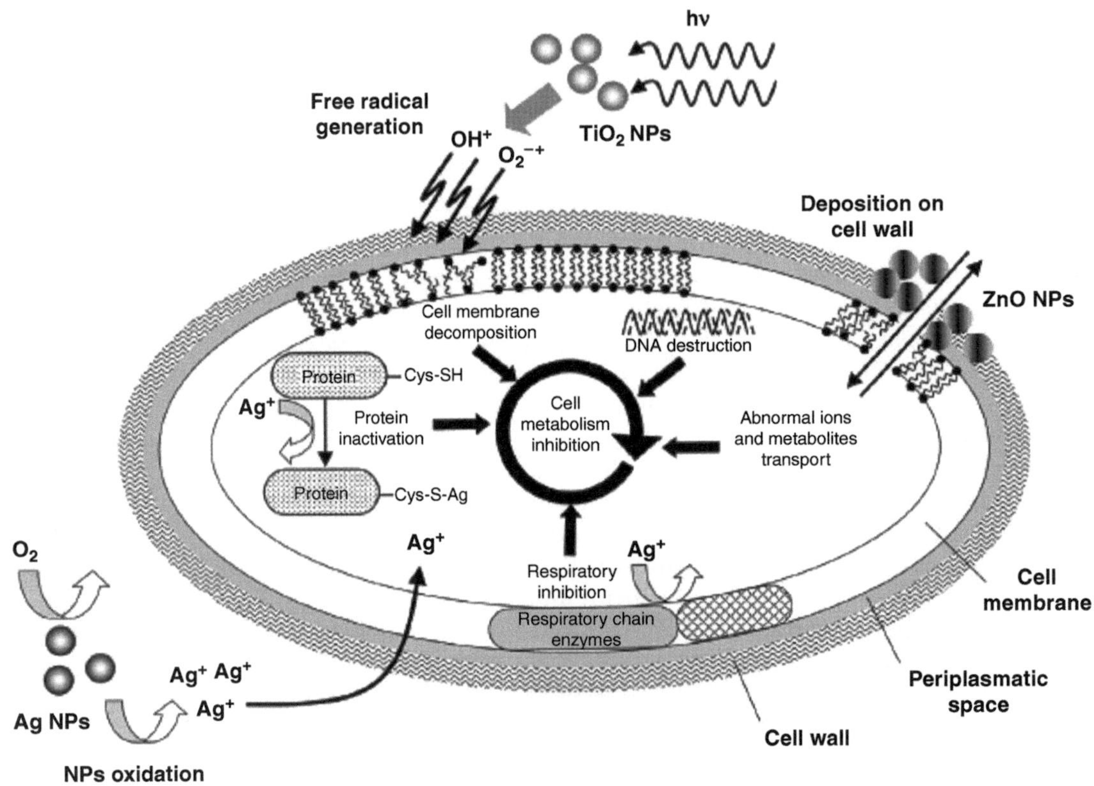

Figure 17.9. Suggested mechanism of antibacterial nanoparticles activity. Reprinted from Moritz, M., Geszke-Moritz, M., 2013. The newest achievements in synthesis, immobilization and practical applications of antibacterial nanoparticles. Chem. Eng. J. 228, 596–613, with permission from Elsevier.

formation of reactive oxygen species (ROS) by the effect of light radiation (Espitia et al., 2012). A general approach of the cytotoxicity mechanism of Ag, TiO_2, and ZnO nanoparticles against bacterial cells is schematically presented in Fig. 17.9.

In order to maintain the advantages of layered silicates within a polymer matrix and at the same time to create nanofiller with active role on packaging, cation exchanged ionic silver has been tethered on the surface of montmorillonite layers. Following, these nanofillers have been mixed with PLA via solution mixing at concentrations up to 10 wt.% and the resulting nanocomposites were evaluated for their antibacterial role against *Salmonella* spp. It was found that at 5 wt.% filler content (and above) the films exhibited antimicrobial action along with reduced water vapor permeability. However, higher filler loadings up to 10 wt.% hindered the dispersion of the stacked silicate sheets, which affected the transparency of the respective films (Busolo et al., 2010).

Graphene is a 2-dimensional arrangement of carbon atoms in a honeycomb structure creating a 1-atom thick layer. Its role in polymers has been extensively investigated imparting significant improvement in mechanical, thermal, electrical, and barrier properties (Kim et al., 2013). It has been demonstrated that its presence in polymers is able to induce some antimicrobial activity (Liu et al., 2011). Nevertheless, it has been found that the presence of graphene oxide is able to assist further the action of silver nanoparticles against *E. coli* and *Salmonella typhi* optimizing, at the same time, the overall amount of silver (Chook et al., 2012).

4 Intelligent Role of Nanomaterials in Food Packaging

As it has been presented, the intelligent role of nanomaterials in food packaging relates to communicating information about the history of the package or the quality of the food. Beyond the conventional and static indications of food quality such as "Best Before," the role of intelligent nanomaterials may involve oxygen and carbon dioxide sensors, time-temperature indicators, and freshness indexes. The nanomaterials may interact with the package internal (eg, food components or headspace species) or external (eg, environment) factors generating a response (eg, visual, electrical signal), which correlates with the state of the food (Mihindukulasuriya et al., 2014).

Exploiting TiO_2 nanoparticles it was proposed the preparation of a solvent-based ink, which would be able to indicate the level of oxygen within a package. Such a printing on a food packaging film could be used either to indicate the period that a package had been opened or for modified atmosphere packaging (MAP) applications (Arvanitoyiannis, 2012). In a relevant example presented by Mills et al. this process was based on the initial UV activation of titania, which played the role of the semiconductor photocatalyst to create the electron-hole pairs. The glycerol used in the ink acted as sacrificial electron donor reacting with the photo-generating holes. Thus, the photo-generating electrons were able to react with the methylene blue, which was used as redox dye in the mix, inducing its bleaching. This reduced form of the redox dye was oxygen sensitive so that in the absence of oxygen remained in the bleached condition while upon exposure to air it was able to restore its initial color within 5 days (at 21°C and RH above 20%). As shown in Fig. 17.10, this oxygen indicator ink was sensitive to the level of humidity, which altered the respective photo-bleaching recovery detected by a UV-visible spectrophotometer. Thus, such inks might

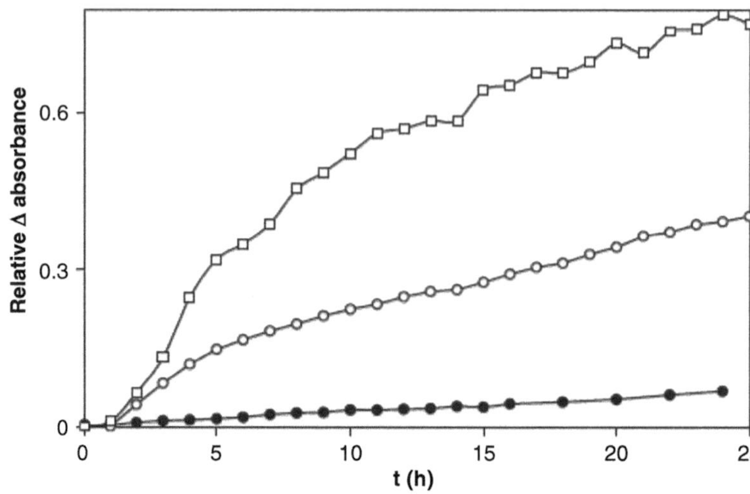

Figure 17.10. Plot of the relative absorbance of a typical, photobleached indicator versus time of recovery, in air at 21°C, as a function of relative humidity (RH). In this work the humidity was maintained at: 18%, 57%, and 96% RH, respectively (bottom to top). Reprinted from Mills, A., Hazafy, D., Lawrie, K., 2011. Novel photocatalyst-based colorimetric indicator for oxygen. Catal. Today 161, 59–63, with permission from Elsevier.

find applications as humidity sensors, too (Mills et al., 2011). Moreover, it has been shown that the rate of bleaching recovery can be suppressed upon reducing the temperature from 21 to 0°C (Lawrie et al., 2013). In similar cases, the use of SnO_2 nanoparticles (instead of titania), which can be activated by UVB light might extend the relevant MAP applications (Mills and Hazafy, 2009).

Nanoparticles in an array designed to fluoresce in different colors upon contact with food pathogens have been proposed to act as a sensor to reduce the time required to detect food spoilage (Sozer and Kokini, 2008). Similarly, an array of sensors based on a polyaniline/carbon black mix within a polyethylene adipate matrix was proved to be suitable for the detection of the odors raised from the food-borne pathogens *Salmonella* spp., *Bacillus cereus*, and *Vibrio parahaemolyticus*. The vapor-induced expansion of the conductive polymer nanocomposite (CPC) caused an increase in the electrical resistance of the CPC due to the reduction of the conductive pathways. The higher sensitivity of the sensor was observed for the mixture with the lower volume fraction of conductive nanoparticles (Arshak et al., 2007). Moreover, various types of sensors based on graphene have been suggested in the literature (Sundramoorthy and Gunasekaran, 2014). These include among others, substrates of graphene/silver nanocomposites for detecting several prohibited colorants (Xie et al., 2012), modified glassy carbon electrodes with graphene/gold-nanoparticle multilayers for detecting the toxic colorant malachite green (Hou et al., 2013), or reduced graphene oxide based quantum resistive vapor sensors for the detection of organic solvents released from food during its degradation (Tung et al., 2012).

Exploiting the chemoresistive properties of a PLA/CNT nanocomposite film, a sensor for the detection of volatile organic compounds has been manufactured. Initially, a PLA/CNT suspension was sprayed onto an electrode followed by solvent evaporation. This process was repeated several times until a film of 1.5-μm-thick was obtained. The sensor was exposed at periodic cycles of 10 min of dry nitrogen followed by 10 min of vapor streams recording simultaneously the electric response. As shown in Fig. 17.11a, the nanocomposite with 3 wt.% CNT presented a higher response

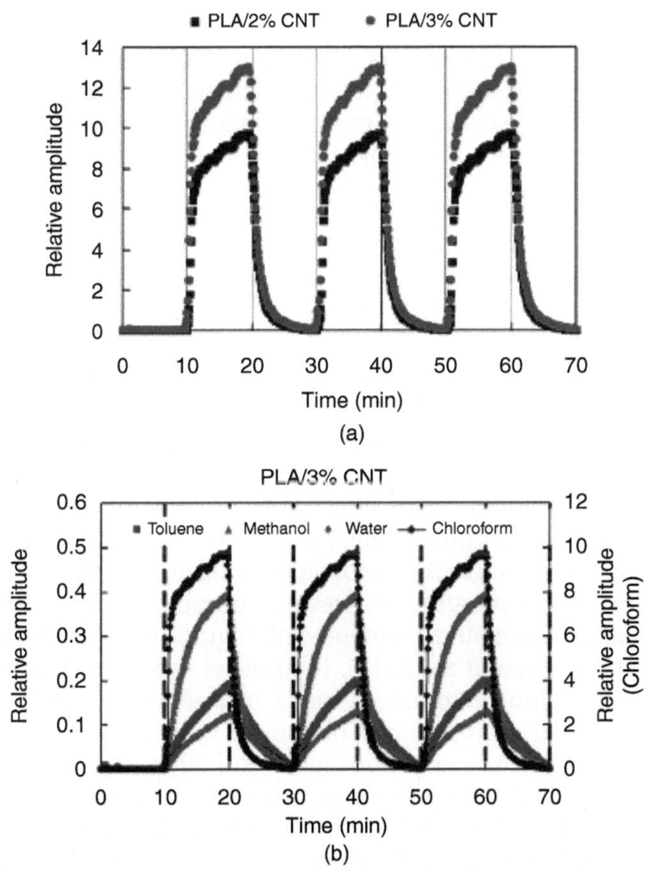

Figure 17.11. Typical positive vapor coefficient behavior of PLA/CNT conductive polymer nanocomposite transducer. (a) toward chloroform vapor and (b) toward all vapors by PLA/3% CNT. Reprinted from Kumar, B., Castro, M., Feller, J.F., 2012. Poly(lactic acid)-multi-wall carbon nanotube conductive biopolymer nanocomposite vapour sensors. Sens. Actuat. B Chem. 161, 621–628, with permission from Elsevier.

than that containing 2 wt.%, indicating that the former filler loading was more suitable for this application. For this filler content, the chemo-resistive properties of the sensor were investigated by its exposition to the vapors of toluene, methanol, water, and chloroform. It was found that the highest transducer's sensitivity (see Fig. 17.11b) was due to chloroform (Kumar et al., 2012).

5 Direct Health Concerns

The effect of nanomaterials on epigenetic processes has already been recognized, as well as their possible health impact (Smolkova et al., 2015). In general, people are influenced to approve or disapprove nanotechnology in food packaging thinking mainly of the benefits rather than any risks involved. Their choice is governed by their level of knowledge on nanotechnology, as well as existing social factors such as public trust to responsible institutions (Stampfli et al., 2010). Besides, a plethora of countries around the world has agreed to follow specific rules on the exploitation of nanomaterials in food packaging. Especially in Europe, product manufacturers and importers should be in accordance with European directives, which relate to authorized materials and the possible migration of the permitted nanomaterials to the food. Any nanomaterial migration above critical limits could raise health concerns and is strictly prohibited. In special cases, end-users and consumers might even require declarations that the packaging materials they use do not contain any nanomaterials.

Silver nanoparticles have been tested for their migration to boneless chicken breast. The packaging material was based on polyethylene and the filler content was set at 0.5 wt.%. It was found that the migration of silver was lower than that described in the European directive limits for various storage scenarios changing time and temperature. However, when copper nanoparticles were used as active filler, the degree of migration was considerably increased (Cushen et al., 2014). In another example, commercial antimicrobial food containers made of polyolefins, which were filled with silver nanoparticles of about 0.33 wt.%, were tested on their migration levels in various food simulants. It was found that the higher migration level was in presence of acetic acid (3%) at 40°C for 10 days, however it appeared to be in accordance with the relevant European legislation. Interesting to note is that the silver migration (although within limits) was higher when the food container was heated in a microwave over at 700 W for 2 min rather than for 2 h at 70°C in a conventional oven (Echegoyen and Nerín, 2013).

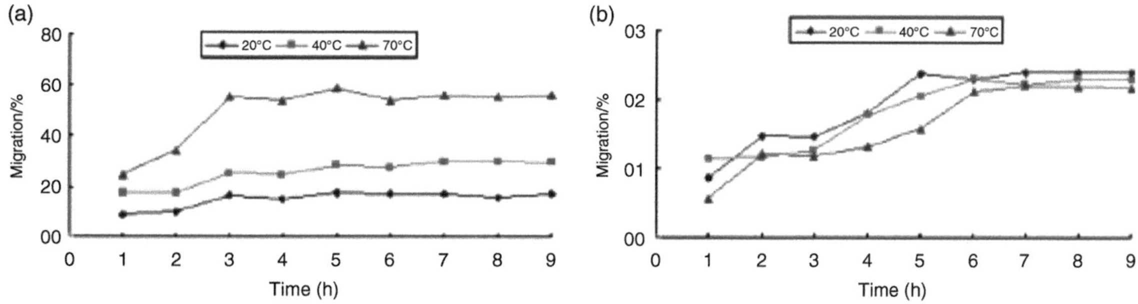

Figure 17.12. Migration of Ag from plastic film into 3% (w/v) aqueous acetic acid (a) and 95% (v/v) aqueous ethanol (b). Song, H., Li, B., Lin, Q.,-B., Wu, H.-J., Chen, Y., 2011. Migration of silver from nanosilver-polyethylene composite packaging into food simulants. Food Addit. Contamin. 28, 1758–1762, reprinted by permission of the publisher (Taylor & Francis Ltd, http://www.tandfonline.com).

In order to investigate the migration levels of silver nanoparticles of 7 nm in diameter from polyethylene/silver nanocomposite films, a high amount of filler content was selected along with two different types of food simulants at various temperatures. The respective film of 55 μm in thickness, which was filled with 3 wt.% silver, was tested in 3% (w/v) aqueous acetic acid and 95% (v/v) aqueous ethanol after being treated in a microwave oven on a suitable digestion program. As shown in Fig. 17.12, the maximum migration ratio in presence of acetic acid was found to be 1.7%, 3.0%, and 5.6% at 20°C, 40°C, and 70°C, respectively. Nevertheless, the maximum migration ratio of silver in case of ethanol was lower than using acetic acid and it was about 0.23% in all three tested temperatures (Song et al., 2011).

The toxicity of Ag nanoparticles is not always size dependant. This has been observed for silver nanoparticles of 50 nm, 80 nm, and 200 nm coated by polyvinyl-pyrrolidone (PVP) in order to ensure absence of silver ions. It was found that silver in 50 nm diameter had strong cytotoxic and genotoxic effect in mammalian cells, however, silver in 200 nm in diameter had the most mutagenic potential (Huk et al., 2014). Nonetheless, the low amount of silver nanoparticles in plastic packaging, as well as their minimum migration in food in accordance with the relevant directives maximizes safety levels for end-users. Silver nanoparticles of 50 nm and 10 nm in diameter have been evaluated for their migration in skinless, boneless chicken breast at 5°C and 20°C. For this purpose, a PVC film of 50 μm in thickness was filled with 5 wt.% and 0.5 wt.% silver content. It was found that the effect of the nanoparticle size on the migration level was marginal, however, for the high filler loading a 10-fold increase in silver levels was detected (Cushen et al., 2013). Albeit, silver migration results from the worst

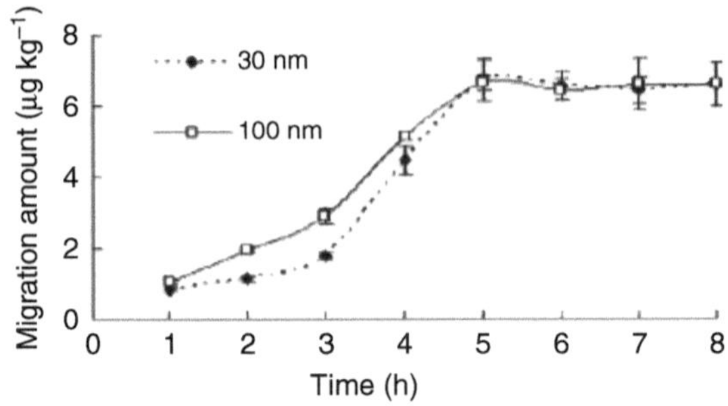

Figure 17.13. Effect of nanoparticles (NP) size on migration of the composite film to 3% (w/v) aqueous acetic acid at 70°C. Lin, Q-B., Li, H., Zhong, H-N., Zhao, Q., Xiao, D-H., Wang, Z-W., 2014. Migration of Ti from nano-TiO_2-polyethylene composite packaging into food simulants. Food Addit. Contamin. Part A 31, 1284–1290, reprinted by permission of the publisher (Taylor & Francis Ltd, http://www.tandfonline.com).

case scenario were found below the critical migration levels of relevant legislation. Similar to these results, when TiO_2 was served as filler in polyethylene, on the one hand the maximum migration levels were detected for the 3% acetic acid food stimulant, and on the other hand there was almost no difference in filler migration amount between the 30 nm and the 100 nm particle size at equilibrium (Lin et al., 2014). This migration behavior for the two nanoparticle sizes as a function of time at 70°C is presented in Fig. 17.13.

Regarding CNTs, in general, pristine versions tend to agglomerate and accumulate in organs like liver and lungs, while their functionalized versions are being tolerated in vivo depending on the type of the functionalization. In this respect, well-dispersed functionalized CNTs may be more easily cleared from the body by excretion compared to the nonmodified CNTs. For the former, the type, level, surface density of functional groups, as well as their purity are issues of great importance for inflammatory responses and cytotoxicity (Marchesan et al., 2015).

Since platy nanofillers have started to supply solutions to existed technological problems in food packaging industry, research on their migration levels and health impact has been emerged (Souza et al., 2013). Initial in vitro tests have shown that the presence of the modifier of the organoclay affected the toxicity of clay (Maisanaba et al., 2013). In the case of polymer/organoclay nanocomposites, the various types of intercalants used for clay modification differ not only in terms of their nanocomposite formation efficiency but also in their temperature stability and migration degree. For applications of PLA in the food packaging industry, surfactants that have been proposed for montmorillonite modification involve the quaternary ammoniums salts of acetylcholine chloride and hexadecyltrimethylammonium bromide (HDTA) (Maisanaba et al., 2014). Apart from in vitro analysis, additional studies on PLA/organoclay nanocomposite containing 4 wt.% organoclay modified with HDTA were performed in vivo. Wistar rats were exposed for 90 days to the relevant migration extract followed by histopathological examination of tissue from their organs through optical and electron microscopy. The obtained results in connection with clinical biochemistry, inflammation, and oxidative stress biomarkers showed no apparent evidence of damage (Maisanaba et al., 2014). On the other hand, graphene, graphene oxide (GO), and reduced graphene oxide (rGO), are considered as cytotoxic and genotoxic. The most recognized mechanism for their toxicity in living systems is ROS (Guo and Mei, 2014).

6 Environmental Issues

The use of nanomaterials in food packaging applications has started to receive notable attention by the industry (Duncan, 2011). In the best case scenario, recycling may provide an initial solution to the waste management. Nevertheless, property deterioration related to yellowness index and mechanical performance for the new product may be observed (Sánchez et al., 2014). As a consequence, in several cases the packaging material may end up in the environment either as solid waste in landfills or as residual from incineration processes (Caballero-Guzman et al., 2015). Through these paths, nanomaterials can be released in the environment affecting terrestrial and aquatic systems. For example, graphene oxide has been reported to decrease soil enzyme activity in the short term (Chung et al., 2015) and to be ecotoxic to phytoplankton (Nogueira et al., 2015).

Besides, the engineered nanomaterials may experience multiple transformations in their primary particle, coating, and conformational structure during their life cycle. These changes may take place from their production and use, up to their final disposal. The transformed nanomaterials may have differences with their initial versions in terms of hazard, transport, fate, and environmental impact (Mitrano et al., 2015).

Nanodimensional silver has been reported to undergo transformations when it is released in the environment. These transformations, which follow its subsequent aggregation and agglomeration, may involve for example formation of silver chloride and silver sulfide. Nonetheless, the mode of action of silver is governed mainly by the release of ionic silver, which is considered to interact with multiple microbial target sites with structural and metabolic disruption (Hartemann et al., 2015). Silver nanoparticles are considered as dissolvable according to their dissolution behavior. A similar performance is due to ZnO in contrast to the nondissolvable pristine nanoparticles like CNTs and graphene.

Moreover, a smaller size of silver nanoparticles is expected to induce faster dissolution time. In general, the dissolution of nanoparticles helps them persist in the environment for a longer time (Dwivedi et al., 2015). The colloidal stability of the nanoparticles differs in freshwater and seawater due to the effect of multivalent cations or sorbing anions. Additionally, nanoparticles exhibiting photocatalytic activity, like for example, TiO_2, present a disproportionally increased ecotoxicity under UV irradiation due to ROS formation (Schaumann et al., 2015).

Besides, aquatic systems like the oceans may receive in the future a great deal of packaging materials containing nanomaterials. It has been shown that particulate nanomaterials may affect gill function while 1-D dimensionalities like carbon nanotubes can be toxic to the respiratory systems of fishes (Maccormack et al., 2014). Nanomaterials tend to be removed from the water columns and they can be found in sediments and organisms. Silver nanoparticles have been shown to bioaccumulate in organisms like snails, worms, and clams. Titania nanoparticles have been reported to peril life within the ecosystem and more specifically algae, zooplankton, bacteria, and fish (Zhang et al., 2012). It has been stated that TiO_2 nanoparticles may be transferred from prey to consumers. In this respect, its bioaccumulation from algea to daphnids can eventuate, while the fish *D. rerio* has presented nanoparticle uptake via the ingestion of contaminated daphnids. Usually, nanoparticle accumulation via the trophic route is greater than the accumulation via water (Bour et al., 2015).

7 Outlook

Nanomaterials are expected to improve functional properties of food packaging bringing value to both manufacturers and end-users. The former should identify profit by operating with nanomaterials, while the latter should perceive that the functions of packaging in terms of containment, protection, convenience, and communication are fulfilled in the best way due to their use.

Going through actual needs and future trends of the food packaging sector, potential applications of nanomaterials could be identified (Torres et al., 2011). In an effort to reduce packaging cost and weight, packs for food packaging have been proposed to employ the technology of film melted with the supporting rigid frame during the injection molding process. Thus, lighter packaging compared to traditional ones can be obtained wherein all sides of the pack can be decorated delivering a better shelf appeal. Such a film structure may further enhance its packaging performance for example, by reducing its oxygen barrier and improving mechanical properties, exploiting platy nanomaterials.

Taking into consideration the property enhancement of packaging materials due to the addition of a small amount of nanofillers, the relative heavy structures used in rigid and semirigid packaging may be avoided. Moreover, use of nanomaterials in films may eliminate ceramic or metallic layers from laminates used in high-barrier films for food packaging applications. Thus, product manufacturing and recycling becomes cost effective. Besides, demanding future applications in food packaging sector may require combination of passive and active (Beckwith et al., 2014) or even intelligent role of nanomaterials on the same packaging.

Modern packages for edible highly viscous products may include bottles or tubes comprising polymers of low modulus for providing better squeezeability. Their susceptibility to scratching may be reduced by introducing the appropriate nanomaterials within their recipes. In addition, the relevant manufacturing blowing process can be optimized providing less wall-thickness variation and eliminating sagging.

In several cases the beverage industry shifts its production from glass bottles to plastic and more specifically to polyolefin or PET ones. Such packages are much lighter than the traditional bottles yielding, at first place, transportation cost optimization. These plastic bottles require excellent gas barrier performance, a property that is benefited by using platy nanomaterials. Furthermore, the titanium nitride (TiN) nanoparticles, which have been proposed to enable better heat-transfer during plastic bottle manufacturing, present very low migration level to food simulants minimizing relevant health concerns (Bott et al., 2014).

The potential of nanomaterials in food packaging sector will be further developed in combination with future findings in materials science, engineering, electronics, and medicine. The synergy of these sciences would be capable to boost market-pull solutions and initiate technology-push differentiated food packaging products fulfilling, at the same time, stringent health and environmental safety standards.

References

Abdullayev, E., Lvov, Y., 2011. Halloysite clay nanotubes for controlled release of protective agents. J. Nanosci. Nanotech. 11, 10007–10026.

Abu-Zurayk, R., Harkin-Jones, E., McNally, T., Menary, G., Martin, P., Armstrong, C., 2009. Biaxial deformation behavior and mechanical properties of a polypropylene/clay nanocomposite. Compos. Sci. Technol. 69, 1644–1652.

Adur, A.M., Volpe, R.A., Shih, K.S., 2002. Clay-filled polymer barrier materials for food packaging applications. US 6,358,576 B1.

Arshak, K., Adley, C., Moore, E., Cunniffe, C., Campion, M., Harris, J., 2007. Characterisation of polymer nanocomposite sensors for quantification of bacterial cultures. Sens.Actuat. B 126, 226–231.

Arunvisut, S., Phummanee, S., Somwangthanaroj, A., 2007. Effect of clay on mechanical and gas barrier properties of blown film LDPE/clay nanocomposites. J. Appl. Polym. Sci. 106, 2210–2217.

Arvanitoyiannis, I.S., 2012. Principles of MAP and definitions of MAP, CA and AP. In: Arvanitoyiannis, I.S. (Ed.), Modified Atmosphere and Active Packaging Technologies. CRC Press/Taylor & Francis, Boca Raton, pp. 3–7.

Avolio, R., Gentile, G., Avella, M., Carfagna, C., Errico, M.E., 2013. Polymer-filler interactions in PET/$CaCO_3$ nanocomposites: chain ordering at the interface and physical properties. Eur. Polym. J. 49, 419–427.

Awad, W.H., Beyer, G., Benderly, D., Ijdo, W.L., Songtipya, P., Jimenez-Gasco, M.D.M., Manias, E., Wilkie, C.A., 2009. Material properties of nanoclay PVC composites. Polymer 50, 1857–1867.

Ayhan, Z., 2013. Potential application of nanomaterials in food packaging and interactions of nanomaterials with food. In: Silvestre, C., Cimmino, S. (Eds.), Ecosustainable Polymer Nanomaterials for Food Packaging. CRC Press/Taylor & Francis, Boca Raton, pp. 253–279.

Bahar, E., Ucar, N., Onen, A., Wang, Y., Oksüz, M., Ayaz, O., Ucar, M., Demir, A., 2012. Thermal and mechanical properties of polypropylene nanocomposite materials reinforced with cellulose nanowhiskers. J. Appl. Polym. Science 125, 2882–2889.

Bansal, A., Yang, H., Li, C., Cho, K., Benicewicz, B.C., Kumar, S.K., Schadler, L.S., 2005. Quantitative equivalence between polymer nanocomposites and thin polymer films. Nat. Mater. 4, 693–698.

Beckwith, S., Edwards, F.B., Rivett, J., Ebner, C.L., Kennedy, T., McDowell, R., Speer, D.V., 2014. Multilayer film having passive and active oxygen barrier layers. US 8,815,360 B2.

Bensur, F.J., 2004. Laminate packaging material. EP 1 232 855 B1.

Bidsorkhi, H.C., Adelnia, H., Pour, R.H., Soheilmoghaddam, M., 2015. Preparation and characterization of ethylene-vinyl acetate/halloysite nanotube nanocomposite. J. Mater. Sci. 50, 3237–3245.

Bott, J., Störmer, A., Franz, R., 2014. A model study into the migration potential of nanoparticles from plastic nanocomposites for food contact. Food Packaging Shelf Life 2, 73–80.

Bour, A., Mouchet, F., Silvestre, J., Gauthier, L., Pinelli, E., 2015. Environmentally relevant approaches to assess nanoparticles ecotoxicity: a review. J. Hazard. Mater. 283, 764–777.

Busolo, M.A., Fernandez, P., Ocio, M.J., Lagaron, J.M., 2010. Novel silver-based nanoclay as an antimicrobial in polylactic acid food packaging coatings. Food Addit. Contam. 27, 1617–1626.

Caballero-Guzman, A., Sun, T., Nowack, B., 2015. Flows of engineered nanomaterials through the recycling process in Switzerland. Waste Manag. 36, 33–43.

Cacciotti, I., Fortunati, E., Puglia, D., Kenny, J.M., Nanni, F., 2014. Effect of silver nanoparticles and cellulose nanocrystals on electrospun poly(lactic) acid mats: morphology, thermal properties and mechanical behavior. Carbohydr. Polym. 103, 22–31.

Cerisuelo, J.P., Alonso, J., Aucejo, S., Gavara, R., Hernández-Muñoz, P., 2012. Modifications induced by the addition of a nanoclay in the functional and active properties of an EVOH film containing carvacrol for food packaging. J. Membr. Sci., 423–424, 247–256.

Chawengkijwanich, C., Hayata, Y., 2008. Development of TiO_2 powder-coated food packaging film and its ability to inactivate *Escherichia coli* in vitro and in actual tests. Int. J. Food Microbiol. 123, 288–292.

Chook, S.W., Chia, C.H., Zakaria, S., Ayob, M.K., Chee, K.L., Huang, N.M., Neoh, H.M., Lim, H.M., Jamal, R., Rahman, R.M.F.R.A, 2012. Antibacterial performance of Ag nanoparticles and AgGO nanocomposites prepared via rapid microwave-assisted synthesis method. Nanoscale Res. Lett. 7 (541), 1–7.

Chow, W.S., Mohd Ishak, Z.A., 2007. Mechanical, morphological and rheological properties of polyamide 6/organo-montmorillonite nanocomposites. Express Polym. Lett. 1, 77–83.

Chung, H., Kim, M.J., Ko, K., Kim, J.H., Kwon, H-a., Hong, I., Park, N., Lee, S.-W., Kim, W., 2015. Effects of graphene oxides on soil enzyme activity and microbial biomass. Sci. Total Environ. 514, 307–313.

Cole, K.C., Perrin-Sarazin, F., Dorval-Douville, G., 2005. Infrared spectroscopic characterization of polymer and clay platelet orientation in blown films based on polypropylene-clay nanocomposites. Macromol. Symp. 230, 1–10.

Cushen, M., Kerry, J., Morris, M., Cruz-Romero, M., Cummins, E., 2013. Migration and exposure assessment of silver from a PVC nanocomposite. Food Chem. 139, 389–397.

Cushen, M., Kerry, J., Morris, M., Cruz-Romero, M., Cummins, E., 2014. Evaluation and simulation of silver and copper nanoparticle migration from polyethylene nanocomposites to food and an associated exposure assessment. J. Agric. Food Chem. 62, 1403–1411.

de Azeredo, H.M.C., 2009. Nanocomposites for food packaging applications. Food Res. Int. 42, 1240–1253.

de Azeredo, H.M.C, 2012. Antimicrobial activity of nanomaterials for food packaging applications. In: Cioffi, N., Rai, M. (Eds.), Nano-Antimicrobials. Springer, Heidelberg, pp. 375–394.

De Menezes, A.J., Siqueira, G., Curvelo, A.A.S., Dufresne, A., 2009. Extrusion and characterization of functionalized cellulose whiskers reinforced polyethylene nanocomposites. Polymer 50, 4552–4563.

De Moura, M.R., Mattoso, L.H.C., Zucolotto, V., 2012. Development of cellulose-based bactericidal nanocomposites containing silver nanoparticles and their use as active food packaging. J. Food Eng. 109, 520–524.

Duncan, T.V., 2011. Applications of nanotechnology in food packaging and food safety: barrier materials, antimicrobials and sensors. J. Colloid Interface Sci. 363, 1–24.

Durmuş, A., Woo, M., Kaşgöz, A., Macosko, C.W., Tsapatsis, M., 2007. Intercalated linear low density polyethylene (LLDPE)/clay nanocomposites prepared with oxidized polyethylene as a new type compatibilizer: structural, mechanical and barrier properties. Eur. Polym. J. 43, 3737–3749.

Dwivedi, A.D., Dubey, S.P., Sillanpää, M., Kwon, Y.-N., Lee, C., Varma, R.S., 2015. Fate of engineered nanomaterials: implications in the environment. Coord. Chem. Rev. 287, 64–78.

Echegoyen, Y., Nerín, C., 2013. Nanoparticle release from nano-silver antimicrobial food containers. Food Chem. Toxicol. 62, 16–22.

Eichhorn, S.J., Dufresne, A., Aranguren, M., Marcovich, N.E., Capadona, J.R., Rowan, S.J., Weder, C., et al.,2010. Review: current international research into cellulose nanofibers and nanocomposites. J. Mater. Sci. 45, 1–33.

Emamifar, A., Kadivar, M., Shahedi, M., Soleimanian-Zad, S., 2010. Evaluation of nanocomposite packaging containing Ag and ZnO on shelf life of fresh orange juice. Innov. Food Sci. Emerg. Technol. 11, 742–748.

Espitia, P.J.P., Soares, N.D.F.F., Coimbra, J.S.D.R., de Andrade, N.J., Cruz, R.S., Medeiros, E.A.A., 2012. Zinc oxide nanoparticles: synthesis, antimicrobial activity and food packaging applications. Food Bioprocess Technol. 5, 1447–1464.

Gatos, K.G., Karger-Kocsis, J., 2007. Effect of the aspect ratio of the silicate platelets on the mechanical and barrier properties of hydrogenated acrylonitrile butadiene rubber (HNBR)/layered silicate nanocomposites. Eur. Polym. J. 43, 1097–1104.

Gatos, K.G., Karger-Kocsis, J., 2010. Rubber clay nanocomposites: preparation properties and applications. In: Thomas, S., Stephen, R. (Eds.), Rubber Nanocomposites. Wiley, Singapore, pp. 169–195.

Gatos, K.G., Karger-Kocsis, J., 2011. Rubber-clay nanocomposites based on apolar diene rubber. In: Galimberti, M. (Ed.), Rubber Clay Nanocomposites. Wiley, Hoboken, pp. 369–407.

Gatos, K.G., Leong, Y.W., 2016. Classification of nanomaterials and nanocomposites. In: Parameswaranpillai, J., Yu, Y., Hammed, N., Kurian, Th. (Eds.), Nanocomposite Materials: Synthesis, Properties and Applications. CRC Press/Taylor & Francis, Boca Raton.

George, J., Kumar, R., Sajeevkumar, V.A., Ramana, K.V., Rajamanickam, R., Abhishek, V., Nadanasabapathy, S., Siddaramaiah, 2014. Hybrid HPMC nanocomposites containing bacterial cellulose nanocrystals and silver nanoparticles. Carbohydr. Polym. 105, 285–292.

Ghanbari, A., Heuzey, M.C., Carreau, P.J., Ton-That, M.T., 2013. A novel approach to control thermal degradation of PET/organoclay nanocomposites and improve clay exfoliation. Polymer 54, 1361–1369.

Gibis, D., Rieblinger, K., 2011. Oxygen scavenging films for food application. Procedia Food Sci. 1, 229–234.

Goodarzi, V., Jafari, S.H., Khonakdar, H.A., Ghalei, B., Mortazavi, B., 2013. Assessment of role of morphology in gas permselectivity of membranes based on polypropylene/ethylene vinyl acetate/clay nanocomposite. J. Membr. Sci. 445, 76–87.

Guo, X., Mei, N., 2014. Assessment of the toxic potential of graphene family nanomaterials. J. Food Drug Anal. 22, 105–115.

Halim, L., Pascall, M.A., Lee, J., Finnigan, B., 2009. Effect of pasteurization, high-pressure processing, and retorting on the barrier properties of nylon 6, nylon 6/ethylene vinyl alcohol, and nylon 6/nanocomposite films. J. Food Sci. 74, 9–15.

Handge, U.A., Hedicke-Höchstötter, K., Altstädt, V., 2010. Composites of polyamide 6 and silicate nanotubes of the mineral halloysite: influence of molecular weight on thermal, mechanical and rheological properties. Polymer 51, 2690–2699.

Hartemann, P., Hoet, P., Proykova, A., Fernandes, T., Baun, A., Jong, W.D., Filser, J., Hensten, A., Kneuer, C., Maillard, J.-Y., Norppa, H., Scheringer, M., Wijnhoven, S., 2015. Nanosilver: safety, health and environmental effects and role in antimicrobial resistance. Mater. Today 18, 122–123.

Hotta, S., Paul, D.R., 2004. Nanocomposites formed from linear low density polyethylene and organoclays. Polymer 45, 7639–7654.

Hou, J., Bei, F., Wang, M., Ai, S., 2013. Electrochemical determination of malachite green at graphene quantum dots-gold nanoparticles multilayers-modified glassy carbon electron. J. Appl. Electrochem. 43, 689–696.

Hubbe, M.A., Rojas, O.J., Lucia, L.A., Sain, M., 2008. Cellulosic nanocomposites: a review. Bioresources 3, 929–980.
Huk, A., Izak-Nau, E., Reidy, B., Boyles, M., Duschl, A., Lynch, I., Dušinska, M., 2014. Is the toxic potential of nanosilver dependent on its size? Part. Fibre Toxicol. 11 (65), 1–16.
Hwang, S-s., Usu, P.P., Yeh, Y-m., Hu, C-h., Chang, K-c, 2009. Effect of organoclay on the mechanical/thermal properties of microcellular injection molded polystyrene-clay nanocomposites. Int. Commun. Heat Mass Trans. 36, 799–805.
Jorda-Beneyto, M., Ortuño, N., Devis, A., Aucejo, S., Puerto, M., Gutiérrez-Praena, D., Houtman, J., Pichardo, S., Maisanaba, S., Jos, A., 2014. Use of nanoclay platelets in food packaging materials: technical and cytotoxicity approach. Food Addit. Contam. Part A 31, 354–363.
Kamal, M.R., Khoshkava, V., 2015. Effect of cellulose nanocrystals (CNC) on rheological and mechanical properties and crystallization behavior of PLA/CNC nanocomposites. Carbohydr. Polym. 123, 105–114.
Kashiwagi, T., Harris, Jr., R.H., Zhang, X., Briber, R.M., Cipriano, R.H., Raghavan, S.R., Awad, W.H., Shields, J.R., 2004. Flame retardant mechanism of polyamide 6-clay nanocomposites. Polymer 45, 881–891.
Kelarakis, A., Yoon, K., 2008. Optical transparency in a polymer blend induced by clay nanofillers. Eur. Polym. J. 44, 3941–3945.
Khan, A., Huq, T., Khan, R.A., Riedl, B., Lacroix, M., 2014. Nanocellulose-based composites and bioactive agents for food packaging. Crit. Rev. Food Sci. Nutr. 54, 163–174.
Khumalo, V.M., Karger-Kocsis, J., Thomann, R., 2010. Polyethylene/synthetic boehmite alumina nanocomposites: structure, thermal and rheological properties. Express Polym. Lett. 4, 264–274.
Kim, H., Abdala, A.A., Macosko, C.W., 2013. Graphene polymer nanocomposites. In: Mukhopadhyay, P., Gupta, R.K. (Eds.), Graphite, Graphene and Their Polymer Nanocomposites. CRC Press/Taylor & Francis, Boca Raton, pp. 513–556.
Kuila, T., Bose, S., Mishra, A.K., Khanra, P., Kim, N.H., Lee, J.H., 2012. Effect of functionalized graphene on the physical properties of linear low density polyethylene nanocomposites. Polym. Test. 31, 31–38.
Kumar, B., Castro, M., Feller, J.F., 2012. Poly(lactic acid)-multi-wall carbon nanotube conductive biopolymer nanocomposite vapour sensors. Sens. Actuat. B Chem. 161, 621–628.
Laoutid, F., Estrada, E., Michell, R.M., Bonnaud, L., Müller, A.J., Dubois, Ph, 2013. The influence of the nanosilica on the nucleation, crystallization and tensile properties of PP-PC and PP-PA blends. Polymer 54, 3982–3993.
Lawrie, K., Mills, A., Hazafy, D., 2013. Simple inkjet-printed UV-activated oxygen indicator. Sens. Actuat. B 176, 1154–1159.
Lee, S.H., Zhang, Y., Kontopoulou, M., Park, C.B., Wong, A., Zhai, W., 2011. Optimization of dispersion of nanosilica particles in a PP matrix and their effect on foaming. Int. Polym. Process. 4, 388–398.
Li, H., Li, F., Wang, L., Sheng, J., Xin, Z., Zhao, L., Xiao, H., Zheng, Y., Hu, Q., 2009. Effect of nano-packing on preservation quality of Chinese jujube *(Ziziphus jujuba Mill. var. inermis (Bunge) Rehd)*. Food Chem. 114, 547–552.
Li, X.H., Xing, Y.G., Li, W.L., Jiang, Y.H., Ding, Y.L., 2010. Antibacterial and physical properties of poly(vinyl chloride)-based film coated with ZnO nanoparticles. Food Sci. Technol. Int. 16, 225–232.
Lin, O., Mohd Ishak, H., Akil, Z.A.H.M., 2009. Preparation and properties of nanosilica-filled polypropylene composites with PP-methyl POSS as compatibiliser. Mater. Design 30, 748–751.

Lin, Y., Ng, K.M., Chan, C.-M., Sun, G., Wu, J., 2011. High-impact polystyrene/halloysite nanocomposites prepared by emulsion polymerization using sodium dodecyl sulfate as surfactant. J. Colloid Interface Sci. 358, 423–429.

Lin, Q.-B., Li, H., Zhong, H.-N., Zhao, Q., Xiao, D.-H., Wang, Z.-W., 2014. Migration of Ti from nano-TiO_2-polyethylene composite packaging into food simulants. Food Addit. Contam. Part A 31, 1284–1290.

Liu, Y., Kontopoulou, M., 2006. The structure and physical properties of polypropylene and thermoplastic olefin nanocomposites containing nanosilica. Polymer 47, 7731–7739.

Liu, T., Lim, K.P., Tjiu, W.C., Pramoda, K.P., Chen, Z.-K., 2003. Preparation and characterization of nylon 11/organoclay nanocomposites. Polymer 44, 3529–3535.

Liu, C., Luo, Y.F., Jia, Z.X., Zhong, B.C., Li, S.Q., Guo, B.C., Jia, D.M., 2011. Enhancement of mechanical properties of poly(vinyl chloride) with polymethyl methacrylate-grafted halloysite nanotube. Express Polym. Lett. 5, 591–603.

Liu, S., Zeng, T.H., Hofmann, M., Burcombe, E., Wei, J., Jiang, R., Kong, J., Chen, Y., 2011. Antibacterial activity of graphite, graphite oxide, graphene oxide, and reduced graphene oxide: membrane and oxidative stress. ACS Nano 5, 6971–6980.

Liu, M., Jia, Z., Jia, D., Zhou, C., 2014. Recent advance in research on halloysite nanotubes-polymer nanocomposites. Progress Polym. Sci. 39, 1498–1525.

Longano, D., Ditaranto, N., Cioffi, N., di Niso, F., Sibillano, T., Ancona, A., Conte, A., del Nobile, M.A., Sabbatini, L., Torsi, L., 2012. Analytical characterization of laser-generated copper nanoparticles for antibacterial composite food packaging. Anal. Bioanal. Chem. 403, 1179–1186.

Maccormack, T.J., Goss, G.G., Handy, R.D., 2014. Emerging threats to fishes: engineered organic nanomaterials. In: Tierney, K.B., Farrell, A.P., Brauner, C.J. (Eds.), Fish Physiology: Organic Chemical Toxicology of Fishes. Elsevier, San Diego, pp. 439–479.

Maisanaba, S., Puerto, M., Pichardo, S., Jordá, M., Moreno, F.J., Aucejo, S., Jos, Á., 2013. In vitro toxicological assessment of clays for their use in food packaging applications. Food Chem. Toxicol. 57, 266–275.

Maisanaba, S., Gutiérrez-Praena, D., Puerto, M., Liana-Ruiz-Cabello, M., Pichardo, S., Mayano, R., Blanco, A., Jordá-Beneyto, M., Jos, Á., 2014. In vivo toxicity evaluation of the migration extract of an organomodified clay-poly(lactic) acid nanocomposite. J. Toxicol. Environ. Health A 77, 731–746.

Maisanaba, S., Pichardo, S., Jordá-Beneyto, M., Aucejo, S., Cameán, A.M., Jos, Á., 2014. Cytotoxicity and mutagenicity studies on migration extracts from nanocomposites with potential use in food packaging. Food Chem. Toxicol. 66, 366–372.

Manias, E., Touny, A., Wu, L., Strawhecker, K., Lu, B., Chung, T.C., 2001. Polypropylene/montmorillonite nanocomposites: review of the synthetic routes and materials properties. Chem. Mater. 13, 3516–3523.

Manias, E., Zhang, J., Huh, J.Y., Manokruang, K., Songtipya, P., Jimenez-Gasco, M.D.M., 2009. Polyethylene nanocomposite heat-sealants with a versatile peelable character. Macromol. Rapid Commun. 30, 17–23.

Marchesan, S., Kostarelos, K., Bianco, A., Prato, M., 2015. The winding road for carbon nantubes in nanomedicine. Mater. Today 18, 12–19.

Mihindukulasuriya, S.D.F., Lim, L.T., 2014. Nanotechnology development in food packaging: a review. Trends Food Sci. Technol. 40, 149–167.

Mills, A., Hazafy, D., 2009. Nanocrystalline SnO_2-based, UVB-activated, colourimetric oxygen indicator. Sens. Actuat. B 136, 344–349.

Mills, A., Hazafy, D., Lawrie, K., 2011. Novel photocatalyst-based colourimetric indicator for oxygen. Catal. Today 161, 59–63.

Mitrano, D.M., Motellier, S., Clavaguera, S., Nowack, B., 2015. Review of nanomaterials aging and transformations through the life-cycle of nano-enhanced products. Environ. Int. 77, 132–147.

Moritz, M., Geszke-Moritz, M., 2013. The newest achievements in synthesis, immobilization and practical applications of antibacterial nanoparticles. Chem. Eng. J. 228, 596–613.

Morris, V.J., 2011. Emerging roles of engineered nanomaterials in the food industry. Trends Biotechnol. 29, 509–516.

Nam, P.H., Maiti, P., Okamoto, M., Kotaka, M., Hasegawa, N., Usuki, A., 2001. A hierarchical structure and properties of intercalated polypropylene/clay nanocomposites. Polymer 42, 9633–9640.

Nogueira, P.F.M., Nakabayashi, D., Zucolotto, V., 2015. The effects of graphene oxide on green algae *Raphidocelis subcapitata*. Aquat. Toxicol. 166, 29–35.

Notario, B., Pinto, J., Solorzano, E., de Saja, J.E., Dumon, M., Rodríguez-Pérez, M.A., 2015. Experimental validation of the Knudsen effect in nanocellular polymeric foams. Polymer 56, 57–67.

Okamoto, M., Nam, P.H., Maiti, P., Kotaka, T., Nakayama, T., Takada, M., Ohshima, M., Usuki, A., Hasegawa, N., Okamoto, H., 2001. Biaxial flow-induced alignment of silicate layers in polypropylene/clay nanocomposite foam. Nanoletters 1, 503–505.

Pedrazzoli, D., Ceccato, R., Karger-Kocsis, J., Pegoretti, A., 2013. Viscoelastic behaviour and fracture toughness of linear-low-density polyethylene reinforced with synthetic boehmite alumina nanoparticles. Express Polym. Lett. 7, 652–666.

Pedrazzoli, D., Khumalo, V.M., Karger-Kocsis, J., Pegoretti, A., 2014. Thermal, viscoelastic and mechanical behavior of polypropylene with synthetic boehmite alumina nanoparticles. Polym. Test. 35, 92–100.

Pegoretti, A., Dorigato, A., Penati, A., 2007. Tensile mechanical response of polyethylene-clay nanocomposites. Express Polym. Lett. 1, 123–131.

Prashantha, K., Lacrampe, M.F., Krawczak, P., 2011. Processing and characterization of halloysite nanotubes filled polypropylene nanocomposites based on a masterbatch route: effects of halloysites treatment on structural and mechanical properties. Express Polym. Lett. 5, 295–307.

Purohit, P.J., Wang, D.-Y., Wurm, A., Schick, C., Schönhals, A., 2014. Comparison of thermal and dielectric spectroscopy for nanocomposites based on polypropylene and layered double hydroxide-proof of interfaces. Eur. Polym. J. 55, 48–56.

Ramani, M., Ponnusamy, S., Muthamizhchelvan, C., 2013. Preliminary investigations on the antibacterial activity of zinc oxide nanostructures. J. Nanopart. Res. 15 (1557), 1–11.

Robertson, G.L., 2013. Food Packaging Principles and Practice, 3rd ed. CRC Press/ Taylor & Francis, Boca Raton.

Rouhi, J., Mahmud, S., Naderi, N., Ooi, C.R., Mahmood, M.R., 2013. Physical properties of fish gelatin-based bio-nanocomposite films incorporated with ZnO nanorods. Nanoscale Res. Lett. 8 (364), 1–6.

Sablani, S.S., 2015. Polymer nanocomposites for food packaging applications. In: Hernández-Sánchez, H., Gutiérrez-López, G.F. (Eds.), Food Nanoscience and Nanotechnology. Springer, New York, pp. 205–212.

Sánchez, C., Hortal, M., Aliaga, C., Devis, A., Cloquell-Ballester, V.A., 2014. Recyclability assesement of nano-reinforced plastic packaging. Waste Manag. 34, 2647–2655.

Sánchez-Valdes, S., López-Quintanilla, M.L., Ramírez-Vargas, E., Medellín-Rodríguez, F.J., Gutierrez-Rodriguez, J.M., 2006. Effect of ionomeric compatibilizer on clay dispersion in polyethylene/clay nanocomposites. Macromol. Mater. Eng. 291, 128–136.

Schaumann, G.E., Philippe, A., Bundschuh, M., Metreveli, C., Klitzke, S., Rakcheev, D., Grün, A., Kumahor, S.K., Kühn, M., Baumann, T., Lang, F., Manz, W., Schulz, R., Vogel, H.-J., 2015. Understanding the fate and biological effects of Ag- and TiO_2-nanoparticles in the environment: the quest for advanced analytics and interdisciplinary concepts. Sci. Total Environ. 535, 3–19.

Schiffmann, J., Schlösser, H., 2011. Antimicrobial food packaging. US 7,915,325 B2.

Seltzer, R., Mai, Y.-M., Frontini, P.M., 2012. Creep behaviour of injection moulded polyamide 6/organoclay nanocomposites by nanoindentation and cantilever bending. Compos. Part B 43, 83–89.

Silvestre, C., Duraccio, D., Cimmino, S., 2011. Food packaging based on polymer nanomaterials. Progress Polym. Sci. 36, 1766–1782.

Silvestre, C., Pezzuto, M., Cimmino, S., Duraccio, D., 2013. Polymer nanomaterials for food packaging: current issues and future trends. In: Silvestre, C., Cimmino, S. (Eds.), Ecosustainable Polymer Nanomaterials for Food Packaging. CRC Press/Taylor Francis, Boca Raton, pp. 1–25.

Smolkova, B., El Yamani, N., Collins, A.R., Gutleb, A.C., Dusinska, M., 2015. Nanoparticles in food Epigenetic changes induced by nanomaterials and possible impact on health. Food Chem. Toxicol. 77, 64–73.

Song, H., Li, B., Lin, Q., Wu, -B., Chen, H.-J.Y., 2011. Migration of silver from nanosilver-polyethylene composite packaging into food simulants. Food Addit. Contam. 28, 1758–1762.

Sotiriou, G.A., Teleki, A., Camenzind, A., Krumeich, F., Meyer, A., Panke, S., Pratsinis, S.E., 2011. Nanosilver on nanostructured silica: antibacterial activity and Ag surface area. Chem. Eng. J. 170, 547–554.

Souza, P.M.S., Morales, A.R., Marin-Morales, M.A., Mei, L.H.I., 2013. PLA and montmorilonite nanocomposites: properties, biodegradation and potential toxicity. J. Polym. Environ. 21, 738–759.

Sozer, N., Kokini, J.L., 2008. Nanotechnology and its applications in the food sector. Trends Biotechnol. 27, 82–89.

Stampfli, N., Siegrist, M., Kastenholz, H., 2010. Acceptance of nanotechnology in food and food packaging: a path model analysis. J. Risk Res. 13, 353–365.

Sundramoorthy, A.K., Gunasekaran, S., 2014. Applications of graphene in quality assurance and safety of food. Trends Anal. Chem. 60, 36–53.

Thellen, C., Orroth, C., Froio, D., Ziegler, D., Lucciarini, J., Farrell, R., D'Souza, N.A., Ratto, J.A., 2005. Influence of montmorillonite layered silicate on plasticized poly(L-lactide) blown films. Polymer 46, 11716–11727.

Theng, B.K.G., 1979. Formation and Properties of Clay-Polymer Complexes, 1st ed. Elsevier, Amsterdam.

Torres, G., Cabello, L., Redon, V., 2011. Polymeric matrix nanocomposite materials having improved mechanical and barrier properties and procedure for preparing the same. EP 2 305 747 A1.

Tung, T.T., Castro, M., Kim, T.Y., Suh, K.S., Feller, J.-F., 2012. Graphene quantum resistive sensing skin for the detection of alteration biomarkers. J. Mater. Chem. 22, 21754–21766.

Vermeiren, L., Devlieghere, F., van Beest, M., de Kruijf, N., Debevere, J., 1999. Developments in the active packaging of foods. Trends Food Sci. Technol. 10, 77–86.

Villaluenga, J.P.G., Khayet, M., López-Manchado, M.A., Valentin, J.L., Seoane, B., Mengual, J.I., 2007. Gas transport properties of polypropylene/clay composite membranes. Eur. Polym. J. 43, 1132–1143.

Vladimirov, V., Betchev, C., Vassiliou, A., Papageorgiou, G., Bikiaris, D., 2006. Dynamic mechanical and morphological studies of isotactic polypropylene/fumed silica nanocomposites with enhanced gas barrier properties. Compos. Sci. Technol. 66, 2935–2944.

Wu, C.L., Zhang, M.Q., Rong, M.Z., Friedrich, K., 2002. Tensile performance improvement of low nanoparticles filled-polypropylene composites. Compos. Sci. Technol. 62, 1327–1340.
Wu, D., Wang, X., Song, Y., Jin, R., 2004. Nanocomposites of poly(vinyl chloride) and nanometric calcium carbonate particles: effects of chlorinated polyethylene on mechanical properties, morphology, and rheology. J. Appl. Polym. Sci. 92, 2714–2723.
Wu, C.L., Zhang, M.Q., Rong, M.Z., Friedrich, K., 2005. Silica nanoparticles filled polypropylene: effects of particle surface treatment, matrix ductility and particle species on mechanical performance of the composites. Compos. Sci. Technol. 65, 635–645.
Xiao-e, L., Green, A.N.M., Haque, S.A., Mills, A., Durrant, J.R., 2004. Light-driven oxygen scavenging by titania/polymer nanocomposite films. J. Photochem. Photobiol. A Chem. 162, 253–259.
Xie, Y., Li, Y., Niu, L., Wang, H., Qian, H., Yao, W., 2012. A novel surface-enhanced Raman scattering sensor to detect prohibited colorants in food by graphene/silver nanocomposite. Talanta 100, 32–37.
Xu, X., Ding, Y., Qian, Z., Wang, F., Wen, B., Zhou, H., 2009. Degradation of poly(ethylene terephthalate)/clay nanocomposites during melt extrusion: Effect of clay catalysis and chain extension. Polym. Degrad. Stabil. 94, 113–123.
Yeh, J.-T., Chang, C.-J., Tsai, F.-C., Chen, K.-N., Huang, K.-S., 2009. Oxygen barrier and blending properties of blown films of blends of modified polyamide and polyamide-6 clay mineral nanocomposites. Appl. Clay Sci. 45, 1–7.
Youssef, A.M., Kamel, S., El-Samahy, M.A., 2013. Morphological and antibacterial properties of modified paper by PS nanocomposites for packaging applications. Carbohydr. Polym. 98, 1166–1172.
Zhang, R., Bai, Y., Zhang, B., Chen, L., Yan, B., 2012. The potential health risk of titania nanoparticles. J. Hazard. Mater., 211–212, 404–413.

18

A NEW APPROACH FOR FLAVOR AND AROMA ENCAPSULATION

Danielle Branta Lopes, Paula Speranza, Gabriela Alves Macedo

University of Campinas (UNICAMP), Department of Food and Nutrition, Food Engineering Faculty, Campinas, São Paulo, Brazil

1 Introduction

Aroma compounds are substances of great importance in food, textile, pharmaceutical, cosmetic, and perfume industries. They are used as additives, accounting for up to 90% of the characteristic odor and flavor of processed foods (Berger, 1995). Flavors are homogeneous organic compounds with a molecular weight of 50–300 Da and high volatility with boiling points of 20–250°C. Flavoring agents include thousands of molecular compounds, and can be complex mixtures of solvents, pure, synthetic, or natural isolates, and flavorists can mix these together to produce many of the common flavors (Weerawatanakorn et al., 2015; Uhlemann and Reiß, 2010).

Flavors play an important role in consumer preferences for many products. However, the key factors that limit flavor industry development are their long residual action and thermostability (Xiao et al., 2011). These highly valuable compounds lose their effect beginning from the production step because of their volatile and oxidation labile nature (Madene et al., 2006). The major purpose of the flavor industry is to provide good flavors to their customers, and these must be stable over time in addition to being economically advantageous. The industry is therefore constantly researching new more efficient and cheaper ways of stabilizing its delicate flavor ingredients (Belingheri et al., 2015).

The world market for flavors and fragrances was $22 billion per year in 2010 and shows up constantly growing (Han et al., 2013), and the 10 largest companies of these categories are responsible for almost 73% of this total (Speziali, 2012). A study conducted by the Freedonia Group in the United States, shows that flavor and fragrance demand is forecast to increase 4.4% per annum to

Novel Approaches of Nanotechnology in Food. http://dx.doi.org/10.1016/B978-0-12-804308-0.00018-2

$26.5 billion in 2016, which is the market corresponding to 30% in North America, 30% in Asia and the Pacific, 24% in Western Europe, and 16% for other regions (Freedonia, 2012).

The food aroma quality is governed by some features, such as chemical reactivity flavor; the environment like availability of light and atmospheric oxygen; and the matrix system and its constituents such as carbohydrate, protein, transition metal, fat, radical, and other polymers. Among the many elements associated with aroma quality, flavor stability is the most important one, and the chemical structure of individual flavor compounds is associated with the chemical reaction that is responsible for its stability (Weerawatanakorn et al., 2015).

Every year, tons of food powders are required with some specific properties that the natural product does not offer. Food technologists are looking at alternatives to help improve food production for human consumption (Tapia-Hernández et al., 2015; Bakhshi et al., 2013). Encapsulation of flavors into different materials is a well-known technique to elucidate this problem because it is a methodology by which one material or a mixture of materials is coated with or entrapped within another material or system (Madene et al., 2006). Furthermore, the encapsulation of food powders can separate the reactive components within a mixture; mask undesirable flavors; protect unstable ingredients from degradation factors, such as heat, moisture, air and light. It can provide controlled or delayed release and reduce hygroscopicity. It also helps in changing the physical characteristics of the original material, such as flowability and compression, dust reduction, and density modification (Dezarn, 1995).

2 Encapsulation of Flavor Compounds

Flavors, as seen so far, are among the most important ingredients that define the quality and acceptability of food products, being able to improve the smell, taste, price, and consumer satisfaction. However, this compound is highly volatile and susceptible to oxidation as well as being sensitive to light, moisture, and heat, which hampers its handling. Microencapsulation becomes a very important technique used in industries to solve these issues, and coat flavors with secondary and protective wall materials for resistance during processing and storage (Wang et al., 2015). This is a method which surrounds sensitive ingredients with wall materials or carriers, or entraps them in homogeneous or heterogeneous matrices, to give small capsules with many useful properties (Wang et al., 2015; Gharsallaoui et al., 2007). The microencapsulation started in the 1950s in the research into pressure-sensitive coatings for the production of carbonless copying paper (Green and Scheicher, 1955). Nowadays, the encapsulation technology is well developed and accepted among many industries, like pharmaceutical, cosmetics, foods, and other related fields (Brisa et al., 2012; Madene et al., 2006).

In its simplest form, a microcapsule is a small sphere with a uniform wall around it (Gharsallaoui et al., 2007). The complete encapsulated system is composed by two main parts: the first one, named as active material, internal phase or core, that could be flavors (essential oils, resins), bioactive ingredients (enzymes, probiotics), food additives (vitamins, antioxidants, colorants), oils and lipids (vegetable oils, fish oils, milk fat, butter); and the second one, named coating material, also called carrier, wall material, or shell (Jafari et al., 2008). One of the main important and initial steps for flavor encapsulation is the assortment of a suitable wall material according to the properties of the internal phase material and demand of final product. Usually, an encapsulation matrix could be from a wide variety of natural and synthetic polymers, providing important features like good emulsifying and drying properties, besides solubility in water and low viscosity at high solid content, low hygroscopicity, providing many advantages in view of protection from losses and environmental factors, improving resistance to undesirable reactions, retaining flavors in the product, being a good film-former, promoting a stability reinforcement and allowing sustained release (Sun et al., 2013; Nori et al., 2011; Zhang et al., 2011). Moreover, they should present low cost, soft taste, and a stable supply (Ma et al., 2013; Milanovic et al., 2010). Since practically no wall material can meet all the properties enumerated, they are employed in combination with each other (Tan et al., 2015; Fernandes et al., 2014; Ma et al., 2013). The most used wall materials and their properties are shown in Table 18.1.

In counterpart, the main objectives of the core material are reducing the reactivity to the outside environments, like oxygen, light, and water; controlling the release at the right time; enhancing handling and flow properties; and masking the unfavorable tastes (Jafari et al., 2008).

The particles formed, depending on the process used, are classified according to their size and can be macro (>5000 μm), micro (1–5000 μm), or nano (<1 μm). Capsules below 1 μm in size are often referred to as nanocapsules, which are often made by a very specialized nanoencapsulation process (Jafari et al., 2008; Couvreur et al., 2002).

There are several techniques used for encapsulating flavors. The methods could be chemical (molecular inclusion and

Table 18.1 Different Wall Materials and Their Properties for Diverse Flavors Encapsulation

Wall Materials	Properties	Products	References
Carbohydrates			
Maltodextrins	Low cost and good retention	Tuna oil; liquid tomato flavor	Kwamman and Klinkesorn (2015); Belingheri et al. (2015)
Porous starches	Effective absorption and sustained release, simple production technology, and low cost	Coffee oil; limonene	Buttery et al. (1999); Yao and Yao (2002)
Hydrolyzed starches	Excellent oxygen-barrier, low viscosity at high solid level	Meat flavor; wasabi flavor	Jeon et al. (2003); Ratanasiriwat et al. (2013)
Proteins			
Milk proteins	Excellent functional and nutraceutical properties	β-Pinene; citronella	Koupantsis et al. (2014); Baranauskiene et al. (2006)
Gelatin	Early formation of surface crust, high emulsifying and stabilizing activities	Lime oil; lavender oil	Kumari et al. (2015); Xiao et al. (2013)
Arabic gum	Excellent emulsifying properties and provides good volatile retention	Rosemary oil; mint oil	Fernandes et al. (2013); Sarkar et al. (2012)
Cyclodextrins	Good resistance to oxidation, excellent inclusion of volatiles	Shiitake flavor; pine flavor	Bhandari et al. (1998); Lee et al. (2005)
Liposomes	Lipid bilayer that increases emulsifying and emollient effects	Resveratrol; sesame	Liu et al. (2013); Nasirullah et al. (2011)
Inulin	Prebiotic effects, dietary fiber action, and good solubility in water	Rosemary oil; olive oil	Fernandes et al. (2014); Turchiuli et al. (2014)

interfacial polymerization), physical (spray drying, fluidized bed coating, freeze-drying, spray cooling and spray chilling, extrusion, and cocrystallization), or physicochemical (organic phase separation, coacervation, and liposome entrapment) (Jafari et al., 2008). Depending on the technique used, two main structures will be formed: a single-core, usually produced by droplet coextrusion, complex coacervation, fluidized bed drying, and molecular inclusion that has high core loading (90% of total capsule weight); and a multiple-core, mostly produced by spray drying, that has a core loading of 20–30%. In this last structure

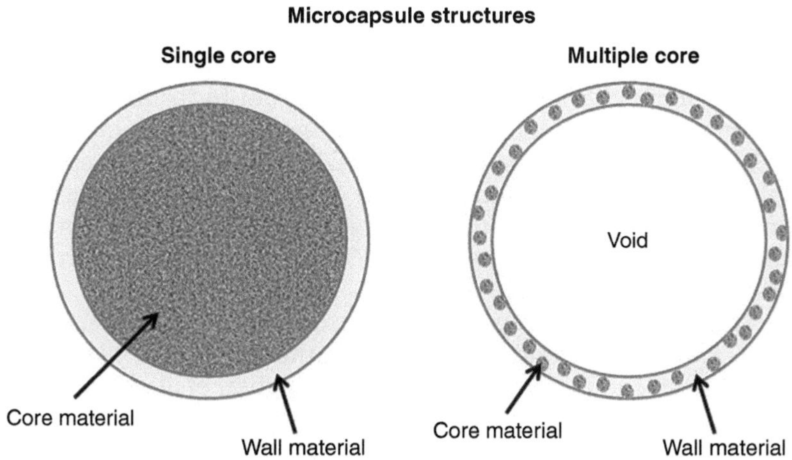

Figure 18.1. Principle microcapsule structures for flavor encapsulation.

(multiple-core), the core material is distributed entirely through the carrier and the central area is filled by the void resulting from expansion of particles throughout the later drying stages (Jafari et al., 2008; Wu et al., 2005). Both microcapsule structures are shown in Fig. 18.1.

The microencapsulation technique can offer numerous benefits to flavors being encapsulated. Several properties of the core can be altered by encapsulation. For example, hygroscopic materials may be safeguarded from moisture and stability of ingredients that are sensitive or volatile to light, heat, or oxidation can be preserved, as well as materials that are on the other hand incompatible may be blended and used securely together (Jafari et al., 2008; Augustin et al., 2001).

An extensive amount of flavor compounds are mainly in the form of liquid at room temperature. This transformation of liquid flavors to dry powders is a relevant usage of microencapsulation for industries. Moreover, one of the key aims for the microencapsulation of flavors is the release control of these active ingredients in the right moment (Jafari et al., 2008; Madene et al., 2006).

The most commonly used techniques for flavor encapsulation are spray drying, freeze-drying, extrusion, spray cooling, or spray chilling, coacervation, fluidized bed coating, and molecular inclusion. The choice of appropriate encapsulation methodologies will depend on the end use of products, the characteristics of core and carrier materials, and the processing conditions related to product manufacturing (Wang et al., 2015; Penbunditkul et al., 2012). The most widely used methodologies are discussed in this chapter.

2.1 Spray Drying

The spray-drying technique is a unit operation by which a liquid product is atomized in a hot gas current to instantaneously obtain a powder. Commonly, the gas used is air or sometimes an inert gas such as nitrogen. The initial liquid feeding the sprayer can be a solution, an emulsion, or a suspension. This method produces, depending on the starting feed material and operating conditions, a very fine powder (10–50 μm) or large-size particles (2–3 mm) (Gharsallaoui et al., 2007).

This is the most common encapsulation technique used in large scale in the food industry and other segments, like pharmaceutical, chemical, and cosmetic (Belingheri et al., 2015; Wang et al., 2015), and is one of the oldest encapsulation processes. It was used in the 1930s to make the first encapsulated flavors with the use of gum arabic as the coating material (Shahidi and Han, 1993). It is a flexible and economical process that uses equipment of easy availability, and produces powder particles of good quality (Jafari et al., 2008). Fig. 18.2 summarizes the main characteristics of this technique.

A great spray-drying encapsulation depends on reaching high retention of the active materials mainly volatiles and tiny

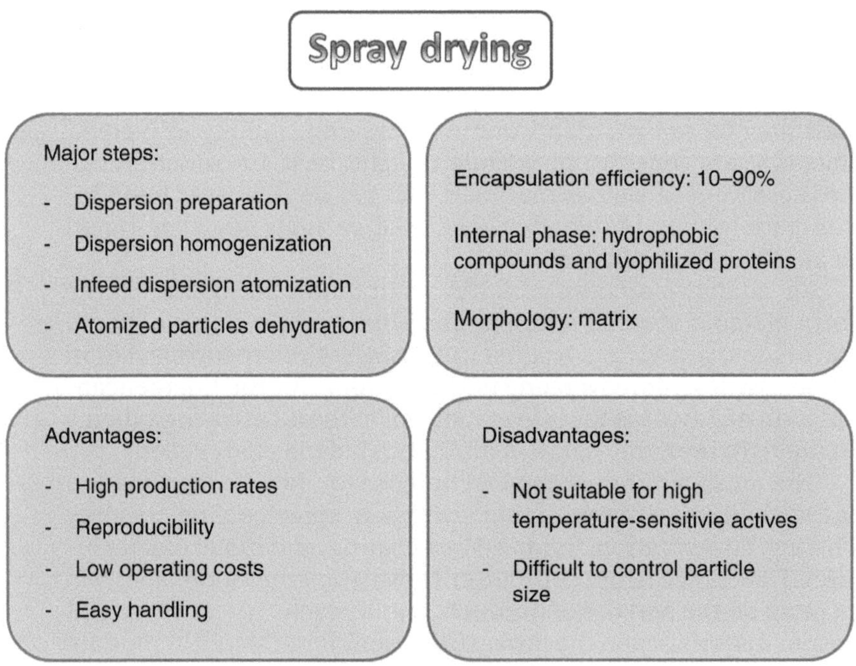

Figure 18.2. Overview of spray-drying encapsulation.

amounts of the surface flavor on the powder particles for both nonvolatiles and volatiles through the process and storage (Jafari et al., 2008). The most common ingredient for spray-drying flavor encapsulation is the gum arabic, natural hydrocolloid exudates from acacia trees, due to its emulsifying properties and excellent retention of volatiles during drying (Jafari et al., 2008; Reineccius, 1994).

The application of this technique in microencapsulation involves four basic stages: preparation of the dispersion or emulsion to be processed; homogenization of the dispersion; atomization of the mass; and dehydration of the atomized particles into the drying chamber (Shahidi and Han, 1993). The first step consists of the formation of a fine and stable emulsion of the core material in the wall solution. This mixture, that will be atomized, is prepared by dispersing the active material or core (usually of hydrophobic nature), into a solution of the coating agent or carrier with which it is immiscible. The dispersion must be heated and homogenized, adding or not an emulsifier depending on the emulsifying properties of the wall materials because some of them have themselves interfacial activities (Gharsallaoui et al., 2007). In the next step, the emulsion obtained in the previous stage is then atomized into a heated air stream supplied to the drying chamber and the solvent evaporation (commonly water) consequently leads to the microcapsules formation, that will take a spherical shape with the oil enclosed in the aqueous phase as the sprayed particles fall through the gaseous medium (Dziezak, 1988).

Some parameters have great influence on microencapsulation by spray drying, like emulsion viscosity and particle size distribution. High viscosities interfere in the atomization step and take to the formation of lengthened and large droplets that adversely affect the drying rate (Rosenberg et al., 1990). The core material retention during microencapsulation by spray drying is affected by the composition and the properties of the emulsion and by the drying conditions (Gharsallaoui et al., 2007).

The selection of the wall material for the microencapsulation is very important for efficiency of the process and for the stability of the microcapsules. The criteria used for selecting a wall material are mainly based on the costs and physicochemical properties such as molecular weight, solubility, crystallinity, diffusibility, glass/melting transition, film forming, and emulsifying properties (Klinkesorn et al., 2006). The wall materials most commonly used in spray drying are carbohydrates such as starches, maltodextrins, and corn syrup solid (Santhalakshmy et al., 2015; Wang and Zhou, 2015); and proteins like sodium caseinate, soy protein isolate, and whey protein concentrates and isolates (Fernandes et al., 2014; Park et al., 2014); and gums, mainly acacia gum, generally call, gum arabic (Lotfy et al., 2015; Cynthia et al., 2015).

The microencapsulation of hydrophobic flavors by spray drying is very important in the flavoring and food industries, once solid or liquid microencapsulated food flavor demonstrate a controlled release and a good chemical stability. This technique is commonly used to produce flavor powders in a short time, and many studies have been carried out to evaluate the influence of wall materials compositions and the operating conditions on the retention and controlled release of encapsulated flavors (Madene et al., 2006).

Fernandes et al. (2014) studied the effects of partial or total replacement of gum arabic by modified starch, maltodextrin and inulin on the microparticles' properties of rosemary essential oil encapsulated by spray-drying process. They studied the oil retention, product stability, retention of the volatiles, and morphology and size of the powders. Although gum arabic is considered to be an excellent wall material, many problems are related with the use of this carrier for microencapsulation, including its high cost and limited supply (Kanakdande et al., 2007), and for this reason the search has been encouraged as modified starch and maltodextrin have become a viable alternative to gum arabic because they are abundant and low-cost materials (Peng et al., 2013). The inulin, a fiber with proven functional activity, improved the particles wettability and decreased the hygroscopicity under high relative humidity while decreasing the encapsulation efficiency. However, the combination of modified starch and inulin was shown to be a viable substitute for gum arabic in foods, because they observed that this mixture retained oil better than the combination of gum arabic and inulin and was similar to the combination of gum arabic and maltodextrin. Moreover, modified starch and inulin presented a high glass transition temperature, which is important for processing and storage. These results indicate an alternative encapsulating material with functional claims, increasing the possibilities of developing new formulations of encapsulants.

Currently, a successful encapsulation of flavors must result in an encapsulated powder with minimum surface flavor content on the powder particles and an utmost retention of the active material, especially volatiles, inside the particles, extending the shelf life of the products. The infeed emulsion size also plays an important role on the retention and the stability of the encapsulated oils and flavors. Emulsions with smaller sizes originate better volatiles retained in the spray dried powder, reaching a straight economic advantage to the producer and user of the product (Jafari et al., 2008).

2.2 Coacervation

Coacervation, often known as phase separation, is generally defined as the division of two liquid phases in a colloidal solution, wherein one phase is rich in polymer and called coacervate phase and the other does not contain polymer and is called equilibrium solution (Kaushik et al., 2014). It was the first microencapsulation technique, developed in 1950 (Srivastava et al., 2013). It is considered a true microencapsulation process because the core material is completely entrapped by the encapsulating material (Kaushik et al., 2014). Fig. 18.3 provides a summary of main features of this technique.

The coacervation technique was classified by Bungenberg de Jong (1949), into simple or complex. Simple coacervation is based on the addition of a poor solvent to a hydrophilic colloidal solution resulting in the creation of two phases: one is rich in colloid molecules (coacervate), and the other is almost coacervate free. One example is the gradual addition of sodium sulfate solution, acetone, or alcohol to a gelatin solution under stirring (Shimokawa et al., 2013). In this type of coacervation just one polymer is used. The complex coacervation is a spontaneous phenomenon that happens between two oppositely charged polymers, and the

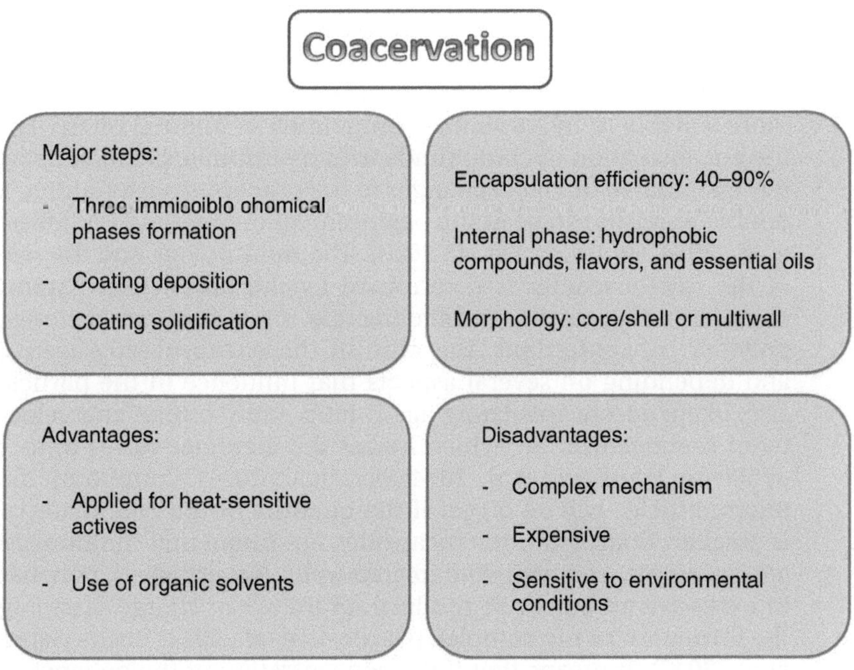

Figure 18.3. Main characteristics for coacervation encapsulation technique.

charges neutralization induces a phase separation (polymer rich phase versus aqueous phase). For this process it is necessary to use two or more kinds of polymer. This method has been applied extensively in microencapsulation (Dordevic et al., 2014; Kaushik et al., 2014). Generally, coacervates are compounds of protein or polysaccharide, however the protein–protein mixtures were also under analysis. Gum arabic and gelatin are the most common and widely used couple in complex coacervation (Dordevic et al., 2014). The complex coacervation technique involves four steps: in the first one, the hydrophobic material is emulsified in an aqueous solution containing two different polymers, generally at a temperature above the gelling point of protein and pH that is above the isoelectric point of protein. In the second step, the emulsion separation occurs in two liquid phases (an insoluble polymer rich phase and an aqueous phase that is depleted in both polymers) as a result of attractive electrostatic interactions between oppositely charged polymers. The third stage comprises the wall formation due to deposition of the polymer rich phase around the droplets of the hydrophobic material-induced by controlled cooling below the gelling temperature. Finally, the fourth step consists in the wall hardening by the addition of a crosslinking agent with the aim of obtaining hard microcapsules (Asbahani et al., 2015). The most common wall materials used in complex coacervation are gelatin and gum arabic (Lemetter et al., 2009).

Among the limitations of this technique is the difficulty to encapsulate the hydrophilic compounds because the coacervation is more suitable to hydrophobic compounds as internal phase. For the encapsulation of compounds with hydrophilic characteristic, it is essential to do many changes to the process, such as adding a double emulsion stage at the beginning of the method (Dordevic et al., 2014; Comunian et al., 2013). The morphology and the size of the microcapsules is determined by the production parameters, including stirring and cooling rate, amount of core material, polymer concentrations, the ratio of the encapsulating agents, and depending on several aspects that influence in the particle size, like core/encapsulating agent ratio and the time and velocity of homogenization, which makes the diameter varies from 1 to 500 μm (Dordevic et al., 2014; Silva et al., 2011). Commonly, the microparticles will be bigger if the quantity of the core material is smaller. Coacervate microcapsules are frequently multinucleate, spherical, and with demarcated walls. Nevertheless, polymer in excessive amount can produce an excess of charge, harming the formation of the complex (Dordevic et al., 2014; Rocha-Selmi et al., 2013). It means that if the concentration of the polymer in the solution is too high, the macromolecules movement will be low as well as the competition extension as the solvent molecules get larger (Dordevic et al., 2014; Silva et al., 2011).

Another challenging issue with complex coacervation is the burst release effect, aggregation, and conglutination problems, which is not desirable in applications of most microcapsules. Moreover, a toxic aldehyde curing agent is used in most of microcapsule preparation process, such as formaldehyde and glutaraldehyde, which cause the toxic material residue problem in the microcapsule products and also restrict its use in a wide-range application (Dong et al., 2011).

Hygroscopicity is another important property of the coacervates. Low moisture content is essential to preserve the powders' stability by preventing caking, which initiates with the accumulation of particles that are wet. The active compound reduces the caking retention, preventing dispersion, and powder flow. The coacervated microcapsules are generally dried in the production final step by freeze-drying, spray drying, or dehydration. The type of drying procedure moreover has an influence in the final properties (Dordevic et al., 2014; Silva et al., 2011).

The microcapsules produced by coacervation method are heat resistant and present an excellent controlled release characteristics based on mechanical stress, shifts, pH, temperature, and sustained release (Dong et al., 2011; Briones and Sato, 2010). Furthermore, this technique provides excellent storage stability, protecting sensitive compounds from oxidation.

Yang et al. (2014) developed a flavor microcapsule with vanilla oil (VO) employing complex coacervation technique, with the objective of a controlled release of vanilla oil with an extended residual effect and aiming an increase in its thermostability for spice application. This oil is employed as a flavor enhancer in the food industry for chocolate, tea, pastry, wine, candy, dairy products, and so on. Chitosan (CS) and gum arabic were used as wall materials and the viscosity of CS was optimized for the production of microcapsules. Genipin was used as a nontoxic curing agent. The microcapsules were then checked by the use of different microscopic techniques (SEM and LSCM), besides infrared spectrometer (FT-IR), controlled-release analysis, particle size analyzer, and thermal analysis. With this study, they achieved 94.2% of encapsulation efficiency in VO/CS ratio of 2:1, when moderate viscosity of CS was used. The SEM analysis showed that the flavor microcapsules had spherical shape and smooth surface. Analysis using FT-IR demonstrated a physical interaction between CS and VO, and a chemical crosslinking reaction between genipin and CS. A core-shell structure of microcapsule was confirmed by LSCM, which was important to increase the thermostability of VO in microcapsule. Furthermore, flavor microcapsules proved to have great potential to serve as a high-quality food spice with long residual action and high thermostability since VO could be remained about 60% in the microcapsules after release for 30 days.

Despite the considerable interest of scientists and the complex coacervation clear benefits, the food industry not yet accepts this method completely for being a complex and expensive technique. If the complex coacervates supply unique properties that are not reachable without them, the encapsulation high cost may be accepted (Dordevic et al., 2014).

2.3 Extrusion

Extrusion is the second most widely used encapsulation technique for the production of encapsulated flavoring powders, and its use in food ingredients is relatively new compared to spray drying (Bhandari et al., 2013). It is also commonly used in the polymer and pharmaceutical industries for affecting product microstructure, product chemistry, or the macroscopic shape of products. This technique is often applied to encapsulate flavors, nutrients, and drugs (Wolf, 2010). Schultz et al. (1956) were pioneers in the extrusion/encapsulation processes, encapsulating orange peel oil and obtaining good stability and flavor retention. With the combination of the basic formulation of Schultz et al. with extrusion, Swisher (1957) created and patented a novel encapsulating process in the 1950s, and since then it has been widely exploited in various industries and is similar to the one currently used today in food industry (Desai and Park, 2005). Fig. 18.4 presents a summary of main characteristics of this technique.

The extrusion method used for encapsulation is not the same as the extrusion used for cooking and texturizing cereal-based products. In fact, this technique, as applied to flavor encapsulation, is a relatively low-temperature entrapping process that comprises to force a core material in a molten carbohydrate mass through a series of dies into a bath of dehydrating liquid, causing hardening of coating material when entering in contact with liquids and forming an encapsulated matrix to entrap the internal phase (Desai and Park, 2005). After that, filaments that were extruded are segregated from the liquid bath, dried, and sized (Shahidi and Han, 1993). The dimension of membrane pore is the most important parameter to take into consideration because it affects the final microcapsule size and size distribution depends mainly of it. However, the pressure applied on the membrane has also an impact (Patil and Jadhav, 2014). The carrier used in this technique may be composed of more than one ingredient,

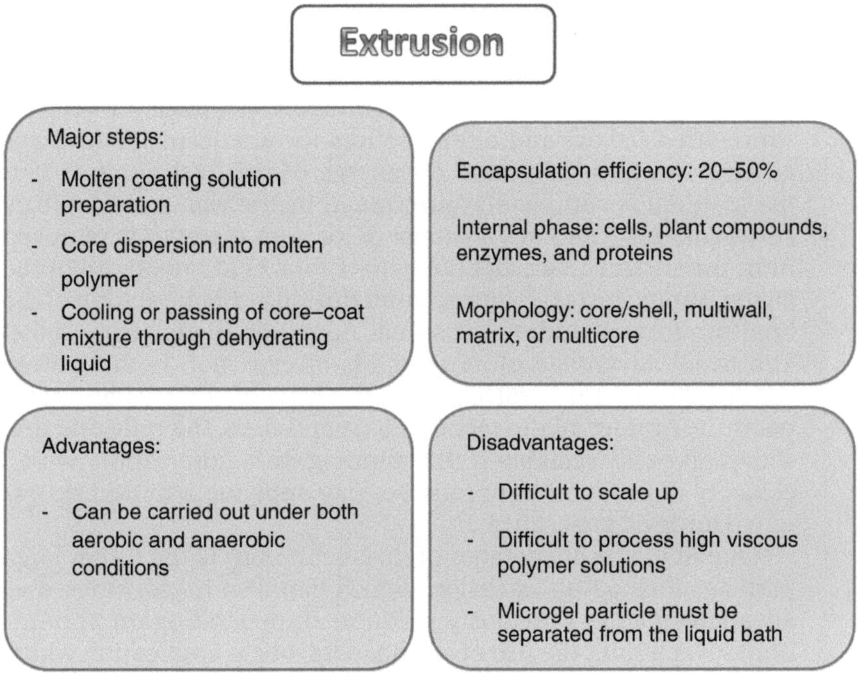

Figure 18.4. Principal features of the extrusion technique.

such as maltodextrin, glucose, sucrose, glucose syrup, and glycerin (Arshady, 1993).

The principal postulate behind all techniques of extrusion comprises the liquid mixture extrusion of a carrier (encapsulating material) and core (bioactive substance) through an aperture and droplet creation at the nozzle discharge point. Several apparatuses of droplet generation appear due to the contact of surface tension, gravitational, frictional forces, and impulse, according to which the methods of extrusion are classified into electrostatic extrusion, vibrating jet/nozzle, jet cutting, simple dripping, coaxial airflow, and spinning disk atomization. The droplets that are generated solidify instantaneously to capsules by chemical (eg, gelation) or physical methods (eg, cooling or heating) (Dordevic et al., 2014). A usual extrusion process for flavors encapsulation is: melting, emulsification, and formation of a matrix with a given polymer (Eraso and Aníbal, 2014).

Encapsulation via extrusion offers advantages due to production of large particles (0.25–2.00 mm); this process provides a strong barrier to oxygen because atmosphere gases diffuse very slowly through the hydrophilic glassy matrix, thus providing an almost impermeable barrier against oxygen, which promotes an extension of the shelf life of flavors and the product can be kept for 1–2 years without any substantial quality degradation (Reineccius, 2005). Shelf lives of up to 5 years have been related for extruded flavor oils, compared to typically 1 year for spray-dried flavors and a few months for unencapsulated citrus oils (Gouin, 2004). Another advantage of this technique is that the material is completely surrounded by the wall material (true encapsulation), and any residual oil or core material is removed from the surface in an alcohol bath (Gibbs et al., 1999). All of the factors should be taken into account during implementation of the final product, including consumer demands (Reineccius, 2005). The usual advantage of all methods of extrusion is that severe conditions are not involved, both in terms of solvents and temperature. Among all the techniques mentioned, the only one that shows an easily scalable is the spinning disk atomization, with a capacity of production of tons per day applying a multidisk system (Dordevic et al., 2014).

One of the disadvantages of this technology is the rather large particles formed by extrusion (which can also be considered as advantageous, as previously mentioned, depending on application), which limit the use of extruded flavors in application where mouth feel is a crucial factor. Moreover, a very limited range of wall material is available for extrusion encapsulation. Most of the applications developed so far make use of carbohydrates of various dextrose equivalents (DE), starch, and mixture of additives (Gouin, 2004).

Extrusion is the most frequently used technique for processing starch-based polymers, because it is an effective and low-cost way to encapsulate substances in starch matrices (Eraso and Aníbal, 2014). It easily handles high-viscosity polymers in the absence of solvents, has good operational flexibility under a wide range of processing conditions (0–500 atm and 70–500°C), and offers the ability to control residence times and degrees of mixing through multiple injections (Liu et al., 2009; Turchiuli et al., 2005). In this case the most common objectives of extrusion are to break down the starch granule to render it digestible and to produce a shaped product (Wolf, 2010).

Levic et al. (2013) applied the electrostatic extrusion to the encapsulation of ethyl vanillin in calcium alginate and calcium alginate/poly(vinyl alcohol) beads. Calcium alginate and poly(vinyl alcohol) (PVA) were used as wall material. Beads loaded with ethyl vanillin were dried in order to obtain a stable spherical encapsulated form. The flavor entrapment in beads was investigated by FT-IR and thermal analysis (thermogravimetry/differential thermal gravimetry analysis; TGA/DTG). According to the DTG curve, the release of ethyl vanillin occurs at about 260°C. During the air-drying process, the microspheres underwent volume contraction at a low extent of 32 mass percentage due to the presence of ethyl vanillin acting as a filler substance. The presence of PVA enhanced the ability of calcium alginate to protect against evaporation. TGA measurements of the stored samples confirmed that formulations were stable for a period of 1 month. FT-IR analysis provides no evidence for chemical interactions between flavor and alginate that would alter the nature of the functional groups in the flavor compound.

2.4 Fluidized Bed Coating

Fluidized bed coating technology is a potent instrument for microparticles manufacture with the objective of improving the functionality of food products (Dordevic et al., 2014; Meiners, 2012). It is a very effective way to apply a uniform layer of wall material onto solid particles (Gouin, 2004). This method was first developed by Wurster in the 1950s, originating the term "Wurster process," which became synonymous with the batch fluid bed apparatus (Arshady, 1993). Fig. 18.5 summarizes the key features of this technique.

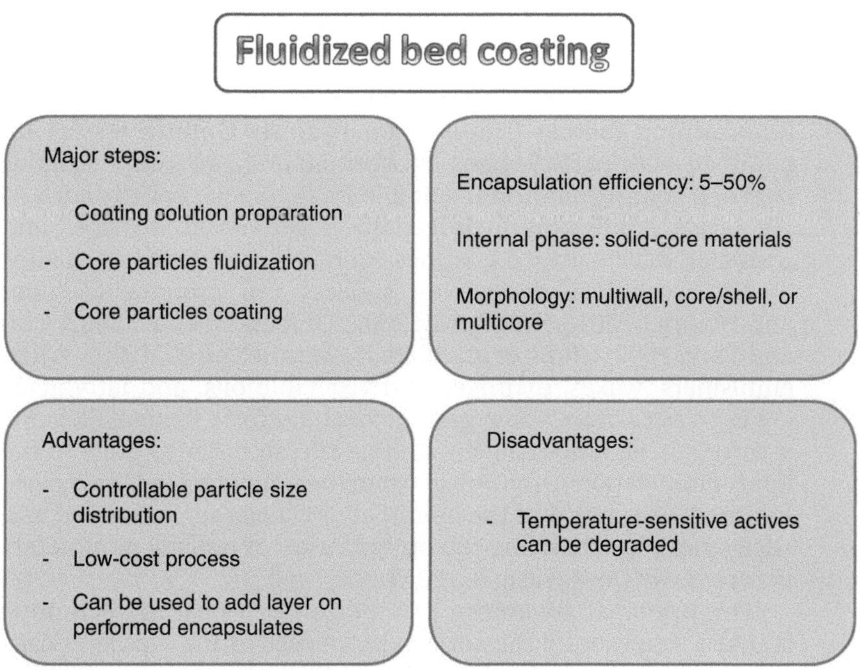

Figure 18.5. Overview of fluidized bed coating technique.

This is an encapsulation methodology in which a coating is administrated onto powder particles, employing a batch processor or a continuous setup. The powder particles are suspended by an air stream at a specific temperature and controlled humidity, and sprayed with an atomized coating material (Zuidam and Heinrich, 2010; Dewettinck and Huyghebaert, 1999; Gibbs et al., 1999). The carrier must have a suitable viscosity to allow pumping and atomizing, must present thermal stability, and should be capable of forming a film over a particle surface. Commonly, 5–50% of wall material is applied, depending on the particle size of the internal phase and application of the encapsulate (Zuidam and Shimoni, 2010). Particle size distribution should also be in a close area, with good flowability and the amount of material that coats the particles is dependent on the length of time that the particles are in the chamber (Gibbs et al., 1999). Particles that are spherical have the lowest possible surface area and require less wall material for the same shell thickness than nonspherical ones. Sharp edges could damage the carrier during handling. Low-dense and thin particles might deal with the risk of accumulating on the filter bags in the top of the machine (Zuidam and Shimoni, 2010). Interestingly, fluidized bed coating technique is one of the few advanced technologies that is able to coat particles with basically any kind of coating material (polysaccharides, proteins, emulsifiers, fats, complex formulations, enteric coating, powder coatings, yeast cell extract, etc.) (Gouin, 2004). The wall materials have to possess an adequate thermal stability, viscosity, and film-forming abilities (Dordevic et al., 2014). Optimal results are obtained with particle sizes between 50 and 500 μm. The fluidized bed coating method is suitable for hot-melt coatings such as fatty acids, emulsifiers, hydrogenated vegetable oil, stearines, and waxes or solvent-based coatings such as gums, starches, maltodextrins, cellulose derivatives, shellac, and proteins (Zuidam and Heinrich, 2010; Trophardy, 2007; Uhlemann et al., 2002; Cho and Park, 2002; Gibbs et al., 1999; Dewettinck et al., 1999). When emulsifiers, waxes, hydrogenated vegetable oils, and fatty acids are used as carriers, some parameters have to be heated, like line, storage vessel, atomizing air, and nozzle, in order to prevent the lipid solidification. In contrast, when the temperature is very close to the melting point of the lipid it could cause agglomerated and sticky particles, therefore the optimization of process parameters is necessary (Dordevic et al., 2014).

The principal advantage of fluidized bed coating technique is that it can modify the surface properties of the aroma encapsulates, like thermal release or additional aroma protection. However, some disadvantages are related to the relatively long coating times (which may easily take 2–20 h in a batch process), the uncontrolled agglomeration of particles and the significant amount of wall materials required to ensure complete covering (depending on the application and particle size of the wall material). These three disadvantages could make fluid bed coated encapsulates reasonably costly (Zuidam and Heinrich, 2010; Saleh et al., 2003). Many studies displays that atomization pressure, coating solution concentration, and inlet temperature are the principal variables, which must be controlled with the purpose of decrease agglomerations (Dordevic et al., 2014; Prata et al., 2012; Hede et al., 2007; Jimenez et al., 2006). In spite of that, this phenomenon is quite difficult to completely explain, meanwhile, besides the operating parameters, a great amount of interrelated parameters hamper the agglomeration prediction (Dordevic et al., 2014; Prata et al., 2012).

There are three different types of fluidized bed coating methods. The first one is the top-spray, which consists of passing the air though a bed of carrier particles to suspend them in air and spray countercurrently the coating solution onto the randomly fluidized particles. The coated particles are carried through the coating zone into the expansion chamber, and then they go down back into the product container and continue cycling throughout the process (Jones, 1988). The second type is the bottom-spray method, known as the "Wurster system," in which particles are recycled through the coating zone at a faster rate and the fluidization model is more controlled than the top-spray process (Mehta and Jones, 1985). Finally, the third type is the tangential-spray, an innovative process for the fluidization and coating of very fine particles. In this technique a rotation perforated drum that contains the particle multiplies the gravitational force and the air flow is then applied tangentially to the rotation of the drum as compensation for the "gravity force" (Matsuda et al., 2001).

The most used process in the food industry is the top-spray method due to its high versatility, relatively high batch size, and relative simplicity. The continuous process is an attractive alternative for food powder coating because producers can adjust the system to their own necessities while maintaining the flexibility required for a large material throughput and wide product ranges, and it provides the coating quality demanded in the food industry (Rumpler and Jacob, 1998). The fluidized bed techniques efficacy is governed by ambient variables, process variables, and thermodynamic factors. Suitable changes or combinations of these variables will yield the desired results (Desai and Park, 2005).

2.5 Spray Cooling and Spray Chilling

Spray cooling and spray chilling are the most affordable encapsulation methods and are habitually employed for aroma encapsulation to delay release in wet environments, to increase heat stability, and/or convert liquid flavor into free-flowing powders (Madene et al., 2006; Gouin, 2004). They are techniques to produce lipid-coated active agents that might be soluble in the lipids, or be present as dry particles or aqueous emulsions (Gouin, 2004; Uhlemann et al., 2002; Kjaergaard, 2001). These technologies are similar to spray drying, however, the carrier/active mixture is cooled instead of evaporated (Madene et al., 2006).

The difference between spray cooling and spray chilling is only the melting point of the carrier material. Spray chilling is in range of 34–42°C and for spray cooling the temperature is higher, achieving 45–122°C. First, the active compound and molten carrier are emulsified, further atomized in small droplets, and subsequently subjected to solidification process. From the aspect of the equipment, the construction characteristics are important concerning the resulting particles size. The rotary of high-speed with a chamber of cocurrent spray cooling is suitable for the production of 50 and 150 μm particles' size; for the production of 150 and 500 μm particles' size it is employed the atomization fountain nozzle and finally, the use of atomizer with low-speed rotary and prilling wheel is convenient for the particles' production of 500 and 2000 μm. The final features of the particles are also affected by various processing parameters such as compressed air pressure, temperature, the atomizer size orifice, air flow rate, feed flow rate, and emulsion viscosity to be powdered (Dordevic et al., 2014; Zuidam and Shimoni, 2010; Gouin, 2004). Particles manufactured by spray-chilling techniques are frequently matrix type, which occurs when the active constituent is dispersed or dissolved through the particle volume and not only in its center. The particles manufactured by this technique are commonly dense and are usually mechanically resistant, remaining intact with agitation because this method does not comprise solvent evaporation (Dordevic et al., 2014; Okuro et al., 2013). Fig. 18.6 gives a summary of main characteristics of these techniques.

Spray cooling is a technique with the possibility to achieve high yields and it can be run in both continuous and batch processing modes. In case of spray chilling, the particles are kept at a low temperature in a setup similar to the fluidized bed spray granulation (Dordevic et al., 2014; Zuidam and Shimoni, 2010; Gouin, 2004). Both techniques are safe, rapid, reproducible, and are considered environmentally friendly compared to other procedures, such as spray drying (Dordevic et al., 2014).

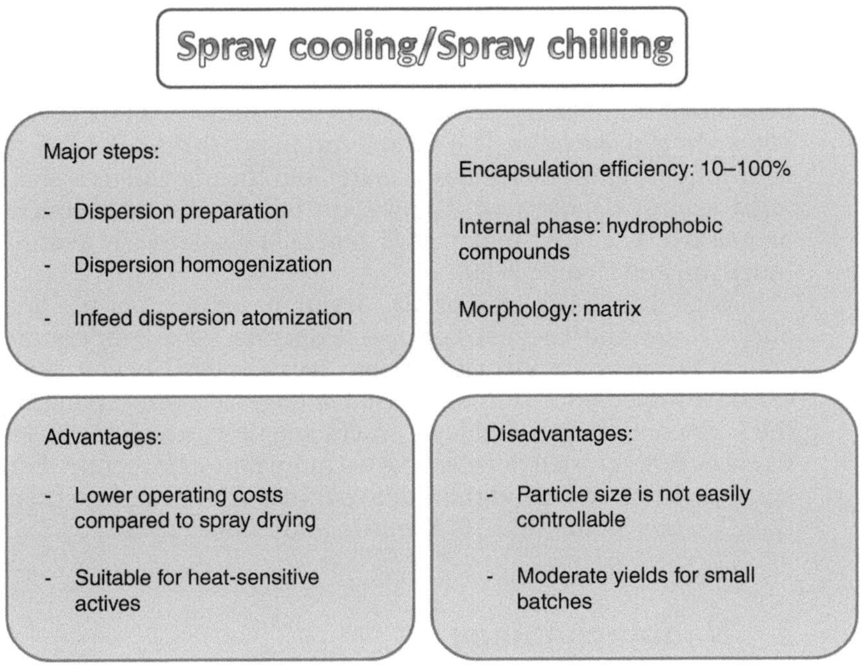

Figure 18.6. Summary of spray-cooling and spray-chilling encapsulation techniques.

Frequently, these methodologies are extensively used for dry products to preserve all thermosensitive actives, like vitamins and minerals, flavors, and proteins (Di Sabatino et al., 2012; Gamboa et al., 2011; Zoet et al., 2011). Approximately 5% of all preparations of flavor are spray chilled; nevertheless, there is minor or no barrier to loose flavors by dispersion if the aroma is soluble in fat. Otherwise, a wax or fat might be utilized as the coating material, if it was immiscible with the flavor, however, it is a challenge to discover such materials with food grade and with a good cost (Dordevic et al., 2014).

2.6 Other Techniques

In the flavor encapsulation field, there are other methodologies used for the creation of these microparticles that are, sometimes, more specific for some groups.

Molecular inclusion is the association of the internal phase in a cavity-based material (Zuidam and Shimoni, 2010). It is one of the most expensive technologies, and therefore less exploited (Nedovic et al., 2011). Cyclodextrin is the best known example because it has a hydrophobic interior and a hydrophilic exterior, being employed in microencapsulation for its ability to form molecular inclusion complexes (Zuidam and Shimoni, 2010; Augustin and Hemar, 2009; Regiert, 2008). The internal size of the hydrophobic cavity that hosts hydrophobic molecules is dependent on the number of glucose units, being composed of six, seven, and eight glucose units. The complexation raises the solubility of hydrophobic guest molecules in water and the ingredient is sheltered against degradation. Unpleasant flavors and odors might be masked. The guest molecule is generally displaced by heating (Augustin and Hemar, 2009).

Freeze-drying is a very similar drying processes, but the first one is faster and cheaper because it operates at a temperature above the freezing point of the solvent. It can be used for very heat-sensitive ingredients. The major disadvantages of freeze-drying are the high energy input and long processing time, which increases its costs. Freeze-dried powders have a more porous structure than spray-dried powders and this adds to transport and storage costs (Zuidam and Shimoni, 2010; Augustin and Hemar, 2009).

3 Nanotechnology

Nanotechnology is the manipulation of materials and structures at sizes in the nanoscale range (Handford et al., 2014). According to National Nanotechnology Initiative, the FDA defines a nanomaterial as a particle with a size below 100 nm. Based on units, a nanomaterial is in the nanometer range, for example, from 1 nm to just below 1000 nm (Keck and Muller, 2013). A nanomaterial is also defined as a discrete entity that has one or more dimensions of the order of 100 nm or less in length (Keck and Muller, 2013; Scenihr, 2007; Tamjidi et al., 2013). According to the Draft EU Commission recommendation, a nanomaterial is a material that meets at least one of the following criteria, which consists of particles with one or more external dimensions of size between 1 and 100 nm; that has internal or surface structures in one or more dimensions between 1 and 100 nm; and having specific surface area per volume of greater than 60 m^2/cm^3, excluding materials consisting of particles smaller than 1 nm (Mihranyan et al., 2012). Nanoparticles apply the modified particles and does not apply to those below 100 nm in that occur naturally or are by-products of other processes (Hoyt and Mason, 2008; Arora et al., 2012).

These nanomaterials exhibit a number of potential advantages over systems containing larger particles. They usually have better stability to particle aggregation and gravitational separation; contain particles that only scatter light waves weakly; and can often be designed to have novel rheological properties (McClements, 2012;

Chen et al., 2011; Mason et al., 2006). The melting temperature of silver is 96°C, whereas silver nanoparticles can be molten at a substantially lower temperature (Asoro et al., 2010); gold is yellow, whereas gold nanoparticles are red or blue (depending on the particle size) (Eustis and El-Sayed, 2006). Nanosizing can also greatly improve the properties of bioactive compounds such as delivery properties, solubility, targetability, and efficient absorption through cells, and prolonged compound residence times in the gastrointestinal tract (Handford et al., 2014; Ezhilarasi et al., 2013).

Nanotechnology is a solution for the development of products with improved performance in very different areas from many consumer products for medicine (Keck and Muller, 2013). In the United States, the market for nanotechnology is projected to grow to $1 trillion by 2015 (Aluya and Garraway, 2010; Aluya, 2015). In the mid- to late 1990s, technological developments in chemistry, engineering, and microscopy have enabled scientists to manipulate tiny structures at the nanoscale level. Such structures include dendrimers, nanoshells, fullerenes, quantum dots, nanotubes, and nanomedicine (McCabe, 2010; Aluya, 2015). As a result, nanotechnology has spawned the development of products like stain-resistant clothing, elasticity of clothing, sports equipment, military uniforms, transparent sunscreens, biological medicines, and many others applications in pharmaceutics, as well as cosmetics, electrical, textile, and food sciences (Aluya, 2015; Raynes et al., 2014; Reade, 2010).

In food science, nanotechnology has potential applications in all aspects of food sectors including food processing, food packaging, food monitoring, production of functional foods, development of foods capable of modifying their color, flavor, or nutritional properties according to a person's dietary needs as well as production of stronger flavorings, colorings, and nutritional food additives (Alfadul and Elneshwy, 2010). Nanocarriers provide more surface area and have the potential to enhance solubility, retain volatile flavors, improve bioavailability, and ameliorate controlled release and targeting of the encapsulated food ingredients, in comparison to microsize carriers (Mozafari, 2006; Weiss et al., 2009). Mohammadi et al. (2016) evaluated the antioxidant activity of olive leave extract (OLE) encapsulated by nanoemulsions in soybean oil. Among different parts of olive tree, the leaf is one of the richest sources of phenolic compounds (Rafiee et al., 2011). The average droplet size 1 day after production was 6.16 nm for primary water in oil nanoemulsion and 675 and 1443 nm for multiple emulsions stabilized by whey protein concentrate alone and complexed with pectin. Nanoencapsulation of bioactive compounds by nonionic emulsifiers and through oil-in-water or water-in-oil nanoemulsions can protect and increase their properties, such as antioxidant, stability, and solubility (Lawrence and Rees, 2012). The antioxidant activity of these emulsions containing three concentrations of 100, 200, and 300 mg OLE during storage was evaluated in soybean oil by peroxide value, TBA value, and thermal stability test Rancimat and compared with the blank (OLE unencapsulated) and the synthetic antioxidant TBHQ. The nanoencapsulated OLE was able to controlling the peroxide value better than the blank. But, due to phenolic compounds retention within the emulsion droplets encapsulated OLE, thermal stability was lower. With increased solubility and because of the controlled release of phenolic compounds from olive leaf nanoencapsulated, a higher antioxidant activity was obtained.

The cosmetics industry was among the first to implement nanotechnological principles in product development. Of more than 1000 registered nanotechnology-based products on the global market in 2009, more than 13% were classified as products for cosmetic use. According to the Woodrow Wilson Project on Emerging Nanotechnologies (PEN), the number of registered nanotechnology products increased by nearly 1900% between 2005 and 2009 from 54 registered products to 1015, respectively. Of these 1015 registered products, 137 are classified as nanotechnology products for cosmetic use and an additional 33 products are separated under the subheading of sunscreens (Mihranyan et al., 2012). Examples of current or intended uses of nanotechnology in cosmetic include packaging, air/moisture barriers to increase the shelf life of the product, and various radio-frequency identification tags, which allow monitoring of temperature and moisture profiles during storage and damage during transportation (Mihranyan et al., 2012). Cosmetic nanocarriers, which do not possess any biological activity of their own, are employed to deliver vitamins, antioxidants, chemical UV filters, antiacne, or antiaging substances in the target layer of skin where they are then dissolved or eroded, after which the effect of the active ingredients is no different from that of the molecular form. Nanocarriers offer several advantages, which are manifested in the improved efficacy, low toxicity, enhanced permeation, chemical stability, and controlled rate of delivery of the active ingredients (Mihranyan et al., 2012; Wu and Guy, 2009).

In pharmaceutical, nanocarriers enhance the in vivo efficiency of many drugs well established over the past decade both in pharmaceutical research and clinical setting. The most important results of nanoencapsulation of drugs includes an increased stability and half-life of nanocarriers in the circulation, required biodistribution, passive or active targeting into the required pathological zone, responsiveness to local physiological stimuli such as pathology-associated changes in local pH and/or temperature, and ability to serve as imaging/contrast agents for various imaging modalities (Torchilin, 2012).

The textile industry is also experiencing the benefits of nanotechnology in its diverse field of applications. Textile-based nanoproducts starting from nanocomposite fibers, nanofibers to intelligent high-performance polymeric nanocoatings are getting their way not only in high-performance advanced applications but nanoparticles are also successfully being used in conventional textiles to impart new functionality and improved performance (Joshi and Bhattacharyya, 2011). The research mainly focuses on using nanosize substances and generating nanostructures during manufacturing and finishing processes in order to impart antibacterial, water-repellency, increase fragrance durability, soil-resistance, antistatic, flame-retardancy, and improved dyeability properties to textiles (Joshi and Bhattacharyya, 2011). Velmurugan et al. (2014) synthesis ecofriendly silver nanoparticles using a natural polymer pine gum solution as the reducing and capping agent. The pine gum solution was combined with silver nitrate ($AgNO_3$) solution to produce silver nanoparticles (AgNPs). The reaction process was simple; formation of the nanoparticles was achieved by autoclaving the silver ions with the pine gum. Application of synthesized AgNPs onto cotton fabrics and leather, in order to evaluate their antibacterial properties against odor- or skin infection-causing bacteria, indicated that the nanoparticles displayed excellent antibacterial activity against *Brevibacterium* linens.

3.1 Flavors Nanoencapsulation

Many of the studies with nanoencapsulation target their use as a delivery vehicle for food nutrition and the controlled release of pharmaceutics and cosmetics; however, it can also be applied to retain volatile flavors in different areas (Reineccius, 1995; Thies, 2007). Maintaining the appropriate intensity level of flavors in commercial products is desirable for both product function and consumer satisfaction. But, due to the delicate nature and high volatility of these compounds, the sustained presence of flavors in consumer products is challenging. The volatility of these molecules is usually due to the low molecular weight and the presence of functional groups, such as hydroxides, aldehydes, and ketones (Sansukcharearnpon et al., 2010; Elia et al., 2015).

For improved stability and delivery, various techniques have been employed to nanoencapsulate flavors in the food, pharmaceutical, and chemical industries. A significantly large part of current literature on the encapsulation of flavors deals with micrometric size capsules, which are used to decrease volatility and for the protection of the flavor against harsh environmental conditions during the processing or storage. Encapsulation within nanometric particle size is an alternative to eliminate these problems, but in addition, because of the size subcellular, may increase the mechanisms of absorption and bioavailability. (Bilia et al., 2014).

The production of nanocapsules with fragrance to be applied to textiles is also very promising. For textiles, encapsulation techniques allow an opportunity for variable fragrant finishing that can favor its durability. It has been proved that the aroma perception is maintained for several days in fabrics if fabrics impregnated with capsules containing perfume and the capsules continue to emit aroma for up to five wash cycles (Liu et al., 2014; Rodrigues et al., 2009; Feczkó et al., 2010; Haefliger et al., 2010). Fragrance encapsulated in small particles (nanometer range) and narrow dispersion are preferred for use in textiles as they can provide a strong adhesive force and interact with the fiber due to their large surface area (Ye et al., 2006). Liu et al. (2014) prepared methylmethacrylate (MMA)-styrene (St) copolymer nanocapsules with cologne essential oil for fragrant fabrics. The physical–chemical composition and thermal properties were characterized by SEM, FT-IR, and TGA. The nanocapsules had relatively small size (c. 136 nm of mean particle diameter), regular sphericity and uniform particle size distribution. The loading capacity and encapsulation efficiency of the produced nanocapsules were 42.7 and 85.4%, respectively. The release experiments showed that 6.8% essential oil still remained after being washed 15 times in the fabrics treated with nanocapsules, indicating the good washing durability of the nanocapsules.

In the food science, the overall perceived flavor is due to sensory inputs to the brain resulting from the interactions of volatile and nonvolatile food components with odor, taste, trigeminal, and tactile receptors in the nose and mouth before, during, and after mastication (McClements, 2005; Smith and Margolskee, 2001). Flavor is one of the most important factors determining the perceived quality of foods (Roberts and Taylor, 2000; Gilbert and Firestein, 2002). Consumers expect that each type of food product will have its own particular characteristic flavor profile (McClements, 2005). Nanoencapsulation of flavors offers many potential advantages in the processing and manufacture of foods: enhanced bioavailability, color, and flavor; novel food textures; new delivery mechanisms; and access to biosensors to enhance food safety (Raynes et al., 2014). Nanoencapsulation helps to protect flavors from degradation, to stabilize and prolong shelf life, and to enable controlled flavor release during consumption. Encapsulated flavors enable a high sensory impact of the final food product. They can also provide natural flavors and/or flavor combinations that are not yet known. Nanoencapsulation of food systems is an engineering technique that involves fabricating nano- and microscaled architectures with food-grade materials such as carbohydrates, proteins, and lipids. Flavor compounds are bound to the architecture with certain interactions that depend on the physicochemical properties of the substrates and the condition of the surrounding environment (Nakagawa, 2014). The flavor retention in a food emulsion is largely dependent on the physicochemical properties of both the aroma compounds and the components of the food matrix (McClements, 2005).

Recently, electrospinning has received great attention in functional food and active food packaging systems (Bhardwaj and Kundu, 2010; Mascheroni et al., 2013). This technique allows the production of nanofibrous polymeric unwoven membranes formed by polymer fibers with diameters ranging from tens to hundreds of nanometers (Frenot and Chronakis, 2003; Uyar and Hacaloglu, 2011; Huang et al., 2003). By their submicron to nanoscale diameter and very large surface area, electrospin fibers are more responsive to changes in the surrounding atmosphere, which enables a tunable release of the entrapped compounds (Lugo and Lim, 2009). Mascheroni et al. (2013) applied electrospinning a single process step in a mixture of edible carbohydrate polymers to encapsulate bioactive aroma compounds and to allow triggered release-humidity. Encapsulation is fast and efficient and an active nanofibrous membrane that can be used in food packaging or active forms the final product. The membrane hosts are small and are dispersed in cyclodextrin-aromatic complexes that are formed during the electrospinning. With this type of structure, the release of aromatic compound is negligible at ambient conditions (23°C and 55% RH) even at high temperatures (up to 230°C) and it occurs beyond a certain humidity threshold relative (90%). Essential oils are complex blends of a variety of volatile molecules such as terpenoids, phenol-derived aromatic components, and aliphatic components having a strong interest in pharmaceutical, sanitary, cosmetic, agricultural, and food industries. Nanoencapsulation in drug delivery systems has been proposed for their capability of decreasing volatility, improving the stability, water solubility, and efficacy of essential oil-based formulations, by maintenance of therapeutic efficacy (Bilia et al., 2014). The advantages of using particulate systems include the possibility to reduce the volatility of the oils, and to formulate these liquid compounds into a carrier form. In addition, particulate delivery systems may also mask the taste and flavor of unpleasant oils, avoid the pharmaceutical/chemical incompatibility, and enhance the bioavailability and stability of the labile essential oils. The organoleptic properties of essential oils are therefore improved, contributing also for a longer shelf life, with the improvement of other properties, for example, stability, compatibility, and in vivo absorption (Severino et al., 2015).

Lv et al. (2014) produced a heat-resistant flavor nanocapsules of jasmine essential oil using gelatin and gum arabic based complex coacervation. Their heat-resistance capability against 80°C was evaluated by both structural characteristics (size, polydispersity index, and zeta potential) and flavor analysis. The results showed that the nanocapsules were stable at 80°C for 7 h, even if the analysis of volatile compounds by gas chromatograph mass spectrometry (GCMS) revealed that jasmine essential oil began to destroy above 5 h. The results suggested that these electrostatic complexes might have potential application as a delivery vehicle for flavors, fragrances or functional ingredients in food, textile, and pharmaceutical industry.

The effect of emulsion size on the retention and shelf life of spray-dried orange oil were investigated. Orange peel oil was spray-dried using either gum arabic or Amiogum 23 as a carrier. Five different emulsion sizes were created by mixing to various extents, homogenization, and microfluidization. The spray-dried powders were tested for oil retention, moisture content, and extractable surface oil. Samples were stored at elevated temperatures and monitored by gas chromatography for the formation of limonene-1,2-epoxide to determine the shelf life. Rehydrated powders were analyzed for particle size to confirm that different emulsion sizes were prepared. A smaller emulsion size yielded a higher percent retention of orange oil with a smaller amount of surface oil but did not give a longer shelf life. Sensory analysis of reconstituted powders from the smallest and largest emulsion size indicated that the emulsion size did not have a significant effect on the perceived intensity of the orange flavor (Risch and Reineccius, 1988).

Rao and McClements (2011) designed conditions where stable microemulsions, nanoemulsions, or emulsions may be produced from a nonionic surfactant and fragrance oil. Colloidal dispersions with different characteristics can be formed depending on the surfactant-to-oil ratio. High-energy homogenization or low-energy homogenization could be used to form nanoemulsions. Systems that only contain small droplets scatter light weakly, accounting for the low turbidity and transparency of the system. There are a number of food and beverage products that should be clear or slightly cloudy, and therefore it is important to utilize a delivery system which does not increase its opacity. This study provides some opacity and provides some ideas for the production and stabilization of colloidal dispersions containing food grade flavor oils. This information may be useful for the formulation of colloidal dispersions for incorporation of flavoring oils into specific products, for example, clear beverages.

Nanoparticles of fish oil encapsulated by spray drying was performed by sonication and microfluidization (Jafari et al., 2008). The encapsulation efficiency of the fish oil was investigated and the droplet size of the emulsion has a greater effect on the efficiency of encapsulation of core materials during spray drying (Soottitantawat et al., 2003, 2005). Reducing the size of nanoemulsions can produce powders encapsulated with greater retention of volatiles and lower nonencapsulated oil content on the surface of the powder. The use of maltodextrin with whey protein as an emulsifier in a ratio of 3:1 was used as the wall material. The results showed that microfluidization is a very efficient emulsification technique, resulting in encapsulated fish oil in powder form with a lower nonencapsulated oil content on the surface of the particles, primarily due to the ability to produce particles of nanometric size.

3.2 Nanotoxicology

The unusual physicochemical properties of engineered nanomaterial are attributable to their small size, chemical composition, surface structure, solubility, shape, and aggregation (Arora et al., 2012). As a result, their properties differ substantially from the bulk material with the same composition as conductivity, reactivity, and optical sensitivity. Possible adverse results of these features are harmful interactions with biological systems and the environment, with the potential to generate toxicity. As the size of a particle decreases, its surface area increases and also allows a greater proportion of its atoms or molecules to be displayed on the surface rather than the interior of the material (Nel et al., 2006). As with any other man-made materials, both in vitro and in vivo studies on biological effects of nanoparticles need to be performed (Arora et al., 2012). Their use in food and pharmaceutics is more diversified and difficult to assess. Nanomaterials made from biological molecules such as lipids, sugars, or proteins, are intended to encapsulate fragile molecules such as flavors or vitamins to protect them during gastrointestinal transit.

It is likely interesting, but the risk may be a modification in the intestinal absorption of the active ingredient transported. Metallic nanoparticles are more problematic; their physicochemical properties and, in particular, their surface reactivity are the causes for their biological reactivity and their potential capacity to overcome barriers and penetrate into the organism. These nanometals are able to generate oxidative stress and induce multiple cellular responses including inflammation and genotoxicity (Marano and Guadagnini, 2013).

Size is a key factor in evaluating the potential toxicity of a particle. However, it is not the only important factor. Other properties of nanomaterials that influence toxicity include: the shape of the particles, the chemical composition, such as the polymers and surfactants used, as well as the organic solvents used during the synthesis process, solubility and presence or absence of functional groups of other chemicals (Chen et al., 2014; Kotsilkova et al., 2001; Balogh et al., 2007). The large number of variables influencing toxicity means that it is difficult to generalize about health risks associated with exposure to nanomaterials, since each factor must be assessed individually and all material properties must be taken into account (Chen et al., 2014).

Tan and Nakajima (2005) prepared nanodispersions from β-carotene to enhance the solubility and bioavailability. The β-carotene in these nanodispersions was chemically unstable and dependent on the average particle diameter. The change in size alters the chemical stability and is required to develop optimal formulations. Huang et al. (2010) reviewed development of nanoemulsions to achieve enhanced water solubility, oral bioavailability, and biological benefits for phytochemicals. The examples included development of nanoemulsion for increasing the bioavailability of hydrophobic compounds for improving the water dispensability. The studies indicated that nanoemulsion droplet size and lipid component are important for emulsion-based delivery systems to increase bioavailability of encapsulated bioactive compounds and may facilitate cellular uptake of micelles. The droplet size of the particles determines the extent of absorption (Jani et al., 1990). Desai et al. (1996) show that the efficiency of absorption of polystyrene nanoparticles was found to be up to 250-folds higher compared to larger sized. Polystyrene nanoparticles of 50 and 100 nm were found to be absorbed for 34 and 26%, respectively (Jani et al., 1990). However, no data are available that link gastrointestinal absorption of these nanoparticles to negative effects in rats.

The excretion of nanoparticles remains poorly understood. The data indicate that not only the chemical composition of the particles, but also the size influences its excretion (Rajagopalan et al., 1996; Qingnuan et al., 2002). Other studies are necessary to check half-lives of nanoparticles. As studies of kinetic data of nanomaterials become available, more reliable extrapolation to other situations and species will become possible (Hagens et al., 2007).

Although we found no studies specifically evaluating the toxic effects of nanoencapsulated flavors, the care of these possible effects should be considered before applying nanoencapsulated flavorings in foods, drugs, and cosmetics.

The encapsulation of flavor compounds is shown as a very attractive alternative to protect flavors from degradation, to stabilize and prolong shelf life, and to enable controlled release flavor during consumption and new developments have been made in this field. However, limitations in some techniques have occurred due to high costs of production and the lack of food-grade available materials.

The choice of which technology and wall material must be employed is frequently based on trial and error and not on a fundamental understanding of the physical and chemical phenomena determining the stability, release, perception, and digestion (Wandrey et al., 2010). Ubbink and Krüger (2006) suggested a strategy focusing on the intended functionality in the food product, analyzing scientific principles including materials science, physical chemistry, and biophysics. It is advisable to accurately determine the functionality and performance of the encapsulate in the final application, and then analyze the physical, chemical, and biological properties of the encapsulate and the prevalent conditions in the matrix/capsule (Wandrey et al., 2010). Interdisciplinary scientific developments at the interfaces between biomaterials science, physical chemistry, biophysics, and encapsulation technology will increasingly be incorporated (Ubbink and Krüger, 2006) and will contribute to the progress of using microencapsulation in the food industry.

4 Conclusions

Flavor compounds exhibit major importance in consumer satisfaction and induce further consumptions of food products. Encapsulations techniques take afford an effective approach to protect these compounds against undesirable reactions and evaporation throughout storage by covering them with a protective wall material. In the past years, there is an inclination to use environmentally friendly coating materials in microencapsulation due to the intensification of the awareness of safeguarding ecology and environment, reaching sustainable development. For this purpose, the carrier employed for encapsulation should have fundamental properties, like emulsification stability, low viscosity, good film-forming, and high solid content. However, the selection of a suitable coating material will be dependent of the flavors properties, and many factors must be considered, comprising encapsulation efficiency and techniques, the sensory quality of the final products, and the cost and source of the raw material that will be used.

The requirement for food products that are healthy (that could incorporate nutrients which frequently cause off-flavors), tasty and appropriate will continue to require better flavor delivery systems. Moreover, this improved system can promote decreased costs and increased stability upon storage. These enhancements could be accomplished by the modification of current encapsulation processes, with the choice of new wall materials and new processing conditions, with the combination of existing encapsulation processes, or with the development of new ones.

Every encapsulation process generally are developed to settle a specific problem which arises during a product development, and presents advantages and disadvantages. Nowadays, the comprehensive technology of encapsulation allows satisfaction of all relevant product necessities, which includes easy product handling, tailoring food properties, controlled release, and improved shelf life. Nevertheless, the most important aspect is a comprehension of how industrial restrictions and demands to create a microencapsulation technology viable can be demonstrated at the laboratory scale in order to make it achievable to accomplish the conversion to full-scale production and marketing of the final product.

Due to the unique characteristics of nanomaterials, the nanoencapsulation can be a further advantageous alternative to encapsulate flavors. However, when using nanoparticles, the care of the possible toxic effects should be checked.

References

Alfadul, S.M., Elneshwy, A.A., 2010. Use of nanotechnology in food processing, packaging and safety: review. Afr. J. Food Agric. Nutr. Dev. 10, 2719–2739.

Aluya, J., 2015. Nanotechnology implications and global leadership perspectives. Energy Sources Part B 10, 31–37.

Aluya, J.U., Garraway, O.L., 2010. Nanotechnology, Solar, Wind, and Hybrid Alternative Energy: Global Leadership Perspectives. AuthorHouse, Bloomington, IN.

Arora, S., Rajwade, J.M., Paknikar, K.M., 2012. Nanotoxicology and in vitro studies: the need of the hour. Toxicol. Appl. Pharmacol. 258, 151–165.

Arshady, R., 1993. Microcapsules for food. J. Microencapsul. 10, 413–435.
Asbahani, A.E., Miladi, K., Badri, W., Sala, M., Addi, E.H.A., Casabianca, H., Mousadik, A.E., Hartmann, D., Jilale, A., Renaud, F.N.R., Elaissari, A., 2015. Essential oils: from extraction to encapsulation. Int. J. Pharm. 483, 220–243.
Asoro, M.A., Kovar, D., Damiano, J., Ferreira, P.J., 2010. Scale effect on the melting behavior of silver nanoparticles. Microsc. Microanal. 16, 1802–1803.
Augustin, M.A., Hemar, Y., 2009. Nano- and micro-structured assemblies for encapsulation of food ingredients. Chem. Soc. Rev. 38, 902–912.
Augustin, M.A., Sanguansri, L., Margetts, C., Young, B., 2001. Microencapsulation of food ingredients. Food Aust. 53, 220–223.
Bakhshi, P.K., Nangrejo, M., Stride, E., Edirisinghe, M., 2013. Application of electrohydrodynamic technology for folic acid encapsulation. Food Bioprocess. Technol. 6, 1837–1846.
Balogh, L., Nigavekar, S.S., Nair, B.M., Lesniak, W., Zhang, C., Sung, L.Y., 2007. Significant effect of size on the in vivo biodistribution of gold composite nanodevices in mouse tumor models. Nanomedicine 3, 281–296.
Baranauskiene, R., Venskutonis, P.R., Dewettinck, K., Verhé, R., 2006. Properties of oregano (*Origanum vulgare* L.), citronella (*Cymbopogon nardus* G) and marjoram (*Majorana hortensis* L.) flavors encapsulated into milk protein-based matrices. Food Res. Int. 39, 413–425.
Belingheri, C., Ferrillo, A., Vittadini, E., 2015. Porous starch for flavor delivery in a tomato-based food application. LWT Food Sci. Technol. 60, 593–597.
Berger, R.G., 1995. Aroma Biotechnology. Springer-Verlag, Berlin, Heidelberg, Germany.
Bhandari, B., Bansal, N., Zhang, M., Schuck, P., 2013. Handbook of Food Powders. Processes and Powders. Woodhead Publishing, Philadelphia, PA.
Bhandari, B.R., D'Arc, B.R., Thi, B., 1998. Lemon oil to β-cyclodextrin ratio effect on the inclusion efficiency of β-cyclodextrin and the retention of oil volatiles in the complex. J. Agric. Food Chem. 46, 1494–1499.
Bhardwaj, N., Kundu, S.C., 2010. Electrospinning: a fascinating fiber fabrication technique. Biotechnol. Adv. 28, 325–347.
Bilia, A.R., Guccione, C., Isacchi, B., Righeschi, C., Firenzuoli, F., Bergonzi, M.C., 2014. Essential oils loaded in nanosystems: a developing strategy for a successful therapeutic approach. Evid. Based Complement. Alternat. Med. 2014, 651593.
Briones, A.V., Sato, T., 2010. Encapsulation of glucose oxidase (GOD) in polyelectrolyte complexes of chitosan-carrageenan. React. Funct. Polym. 70, 19–27.
Brisa, P., Cinta, P., Gemma, A., Ricard, G.V., Tania, G., 2012. Preparation and characterization of polysulfone microcapsules for perfume release. Chem. Eng. J. 179, 394–403.
Bungenberg de Jong, H.G., 1949. Complex colloid systems. Kruyt, H.R. (Ed.), Colloid Science, vol. 2, Elsevier, Amsterdan, The Netherlands, pp. 335–432.
Buttery, R.G., Glenn, G.M., Stern, D.J., 1999. Sorption of volatile flavor compounds by microcellular cereal starch. J. Agric. Food Chem. 47, 5206–5208.
Chen, H.B., Khemtong, C., Yang, X.L., Gao, J.M., 2011. Nanonization strategies for poorly water-soluble drugs. Drug Discov. Today 16, 354–360.
Chen, G., Shim, S.-M., Wen, J., 2014. Potential toxicity of food ingredients loaded in nano and microparticles. In: Kwak, H.-S. (Ed.), Nano- and Microencapsulation for Foods. Wiley Blackwell, UK, pp. 363–381.
Cho, Y.H., Park, J., 2002. Characteristics of double-encapsulated flavor powder prepared by secondary fat coating process. J. Foods Sci. 67, 968–972.

Comunian, T.A., Thomazini, M.A., Alves, A.J.G., de Matos Junior, F.E., de Carvalho Balieiro, J.C., Favaro-Trindade, C.S., 2013. Microencapsulation of ascorbic acid by complex coacervation: protection and controlled release. Food Res. Int. 52, 373–379.

Couvreur, P., Barratt, G., Fattal, E., Legrand, P., Vauthier, C., 2002. Nanocapsule technology: a review. Crit. Rev. Ther. Drug. 19, 99–134.

Cynthia, S.J., Don Bosco, J., Bhol, S., 2015. Physical and structural properties of spray dried tamarind (*Tamarindus indica* L.) pulp extract powder with encapsulating hydrocolloids. Int. J. Food Prop. 18, 1793–1800.

Desai, M.P., Labhasetwar, V., Amidon, G.L., Levy, R.J., 1996. Gastrointestinal uptake of biodegradable microparticles: effect of particle size. Pharm. Res. 12, 1838–1845.

Desai, K.G.H., Park, H.J., 2005. Recent developments in microencapsulation of food ingredients. Dry. Technol. 23, 1361–1394.

Dewettinck, K., Huyghebaert, A., 1999. Fluidized bed coating in food technology. Trend. Food Sci. Technol. 10, 163–168.

Dewettinck, K., Messens, W., Deroo, L., Huyghebaert, A., 1999. Agglomeration tendency during top-spray fluidized bed coating with gelatin and starch hydrolyzate. LWT Food Sci. Technol. 32, 102–106.

Dezarn, T.J., 1995. Food ingredient encapsulation. In: Risch, S.J., Reineccius, G.A. (Eds.), Encapsulation and Controlled Release of Food Ingredients. American Chemical Society, Washington, DC, pp. 74–86.

Di Sabatino, M., Albertini, B., Kett, V.L., Passerini, N., 2012. Spray congealed lipid microparticles with high protein loading: preparation and solid state characterization. Eur. J. Pharm. Sci. 46, 346–356.

Dong, Z., Maa, Y., Hayat, K., Jia, C., Xia, S., Zhang, X., 2011. Morphology and release profile of microcapsules encapsulating peppermint oil by complex coacervation. J. Food Eng. 104, 455–460.

Dordevic, V., Balanc, B., Belscak-Cvitanovic, A., Levic, S., Trifkovic, K., Kalusevic, A., Kostic, I., Komes, D., Bugarski, B., Nedovic, V., 2014. Trends in encapsulation technologies for delivery of food bioactive compounds. Food Eng. Rev. 7, 452–490.

Dziezak, J.D., 1988. Microencapsulation and encapsulated ingredients. Food Technol. 42, 136–151.

Elia, R., Guo, J., Budijono, S., Normand, V., Benczédi, D., Omenetto, F., Kaplan, D.L., 2015. Encapsulation of volatile compounds in silk microparticles. J. Coat. Technol. Res. 12, 793–799.

Eraso, M.O., Aníbal, H., 2014. Use of starches and milk proteins in microencapsulation. Int. J. Veg. Sci. 20, 289–304.

Eustis, S., El-Sayed, M.A., 2006. Why gold nanoparticles are more precious than pretty gold: noble metal surface plasmon resonance and its enhancement of the radiative and nonradiative properties of nanocrystals of different shapes. Chem. Soc. Rev. 35, 209–217.

Ezhilarasi, P.N., Karthik, P., Chhanwal, N., Anandharamakrishnan, C., 2013. Nanoencapsulation techniques for food bioactive components: a review. Food Bioprocess Technol. 6, 628–647.

Feczkó, T., Kokol, V., Voncina, B., 2010. Preparation and characterization of ethylcellulose-based microcapsules for sustaining release of a model fragrance. Macromol. Res. 18, 636–640.

Fernandes, R.V.B., Borges, S.V., Botrel, D.A., Silva, E.K., Costa, J.M.G., Queiroz, F., 2013. Microencapsulation of Rosemary essential oil: characterization of particles. Drying Technol. 31, 1245–1254.

Fernandes, R.B.D.V., Borges, S.V., Botrel, D.A., 2014. Gum arabic/starch/maltodextrin/inulin as wall materials on the microencapsulation of rosemary essential oil. Carbohydr. Polym. 101, 524–532.

Freedonia, 2012. The Freedonia Group. World flavors & fragrances—Industry study with forecasts for 2016 & 2021. Study #2952, pp. 1–388.
Frenot, A., Chronakis, I.S., 2003. Polymer nanofibers assembled by electrospinning. Curr. Opin. Colloid Interface Sci. 8, 64–75.
Gamboa, O.D., Gonçalves, L.G., Grosso, C.F., 2011. Microencapsulation of tocopherols in lipid matrix by spray chilling method. Procedia Food Sci. 1, 1732–1739.
Gharsallaoui, A., Roudant, G., Chambin, O., Voilley, A., Saurel, R., 2007. Applications of spray-drying in microencapsulation of food ingredients: an overview. Food Res. Int. 40, 1107–1121.
Gibbs, B.F., Kermasha, S., Alli, I., Mulligan, C.N., 1999. Encapsulation in food industry: a review. Int. J. Food Sci. Food Nutr. 50, 213–234.
Gilbert, A.N., Firestein, S., 2002. Dollars and scents: commercial opportunities in olfaction and taste. Nat. Neurosci. 5, 1043–1045.
Gouin, S., 2004. Micro-encapsulation: industrial appraisal of existing technologies and trends. Trends Food Sci. Technol. 15, 330–347.
Green, B.K., Scheicher, L., 1955. United States Patent 2,217,507. Pressure sensitive record materials.
Haefliger, O., Jeckelmann, N., Ouali, L., Leon, G., 2010. Real-time monitoring of fragrance release from cotton towels by low thermal mass gas chromatography using a longitudinally modulating cryogenic system for headspace sampling and injection. Anal. Chem. 82, 729–737.
Hagens, W.I., Oomen, A.G., Jong, W.H., Cassee, F.R., Sips, A.J.A.M., 2007. What do we (need to) know about the kinetic properties of nanoparticles in the body? Regul. Toxicol. Pharm. 49, 217–229.
Han, D., Ryu, J.-Y., Lee, H., Hur, H.-G., 2013. Bacterial biotransformation of phenylpropanoid compounds for producing flavor and fragrance compounds. J. Korean Soc. Appl. Biol. Chem. 56, 125–133.
Handford, C.E., Deana, M., Henchionb, M., Spencea, M., Elliotta, C.T., Campbella, K., 2014. Implications of nanotechnology for the agri-food industry: opportunities, benefits and risks. Trends Food Sci. Technol. 40, 226–241.
Hede, P., Bach, P., Jensen, A., 2007. Small-scale top-spray bed coating: granule impact strength, agglomeration tendency and coating layer morphology. Powder Technol. 176, 156–167.
Hoyt, V.W., Mason, E., 2008. Nanotechnology: emerging health issues. J. Chem. Health Saf. 15, 10–15.
Huang, Q., Yu, H., Ru, Q., 2010. Bioavailability and delivery of nutraceuticals using nanotechnology. J. Food Sci. 75, R50–R57.
Huang, Y.Z.-M., Zhang, Z., Kotaki, M., Ramakrishna, S., 2003. A review on polymer nanofibers by electrospinning and their applications in nanocomposites. Compos. Sci. Technol. 63, 2223–2253.
Jafari, S.M., Assadpoor, E., He, Y., Bhandari, B., 2008. Encapsulation efficiency of food flavours and oils during spray drying. Dry. Technol. 26, 816–835.
Jani, P., Halbert, G.W., Langridge, J., Florence, A.T., 1990. Nanoparticle uptake by the rat gastrointestinal mucosa: quantitation and particle size dependency. J. Pharm. Pharmacol. 42, 821–826.
Jeon, Y., Vasanthan, T., Temelli, F., Song, B., 2003. The suitability of barley and corn starches in their native and chemically modified forms for volatile meat flavor encapsulation. Food Res. Int. 36, 349–355.
Jimenez, M., Garcia, H.S., Beristain, C.I., 2006. Spray-dried encapsulation of conjugated linoleic acid (CLA) with polymeric matrices. J. Sci. Food Agric. 86, 2431–2437.
Jones, D.M., 1988. Controlling particle size and release properties. In: Risch, S.J., Reineccius, G.A. (Eds.), Flavor Encapsulation. American Chemical Society, Washington, DC, pp. 158–176.

Joshi, M., Bhattacharyya, A., 2011. Nanotechnology: a new route to high performance and functional textiles. Text. Prog. 43, 155–233.
Kanakdande, D., Bhosale, R., Singhal, R.S., 2007. Stability of cumin oleoresin microencapsulated in different combination of gum arabic, maltodextrin and modified starch. Carbohydr. Polym. 67, 536–541.
Kaushik, P., Dowling, K., Barrow, C.J., Adhikari, B., 2014. Microencapsulation of omega-3 fatty acids: a review of microencapsulation and characterization methods. J. Funct. Foods 19, 1–14.
Keck, C.M., Muller, R.H., 2013. Nanotoxicological classification system (NCS)—a guide for the risk-benefit assessment of nanoparticulate drug delivery systems. Eur. J. Pharm. Biopharm. 84, 445–448.
Kjaergaard, O.G., 2001. Multiple-core encapsulation: prilling. In: Vilstrup, P. (Ed.), Microencapsulation of Food Ingredients. Leatherhead Publishing, Surrey, BC, pp. 197–214.
Klinkesorn, U., Sophanodora, P., Chinachoti, P., Decker, E.A., McClements, D.J., 2006. Characterization of spray-dried tuna oil emulsified in two-layered interfacial membranes prepared using electrostatic layer-by-layer deposition. Food Res. Int. 39, 449–457.
Kotsilkova, R., Petkova, V., Pelovski, Y., 2001. Thermal analysis of polymer-silicate nanocomposites. J. Therm. Anal. Calorim. 64, 591–598.
Koupantsis, T., Pavlidou, E., ParasKevopoulou, A., 2014. Flavour encapsulation in milk proteins—CMC coacervation-type complexes. Food Hydrocoll. 37, 134–142.
Kumari, P., Rose, N.M., Jeet-Singh, S.S., 2015. Microencapsulation of lime essential oil for fragrant textiles. Ann. Agric. Bio. Res. 20, 152–157.
Kwamman, Y., Klinkesorn, U., 2015. Influence of oil load and maltodextrin concentration on properties of tuna oil microcapsules encapsulated in two-layer membrane. Dry. Technol. 33, 854–864.
Lawrence, M.J., Rees, G.D., 2012. Microemulsion-based media as novel drug delivery systems. Adv. Drug Deliv. Rev. 64, 175–193.
Lee, S.J., Lee, Y.B., Hong, J.H., Chung, J.H., Kim, S.S., Lee, W.J., Yoon, J.R., 2005. Optimization of pine flavor microencapsulation by spray drying. Food Sci. Biotechnol. 14, 747–751.
Lemetter, C.Y.G., Meeuse, F.M., Zuidam, N.J., 2009. Control of the morphology and the size of complex coacervates microcapsules during scale-up. AIChE J. 55, 1487–1496.
Levic, S., Djordjevic, V., Rajic, N., Milivojevic, M., Bugarski, B., Nedovic, V., 2013. Entrapment of ethyl vanillin in calcium alginate and calcium alginate/poly(vinyl alcohol) beads. Chem. Pap. 67, 221–228.
Liu, C., Liang, B., Shi, G., Li, Z., Zheng, X., Huang, Y., Lin, L., 2014. Preparation and characteristics of nanocapsules containing essential oil for textile application. Flavour Frag. J. 30, 295–301.
Liu, H., Xie, F., Yu, L., Chen, L., Li, L., 2009. Thermal processing of starch-based polymers. Prog. Polym. Sci. 34, 1348–1368.
Liu, W., Ye, A., Liu, W., Liu, C., Singh, H., 2013. Liposomes as food ingredients and nutraceutical delivery systems. Agro Food Ind. Hi Tec. 24, 68–71.
Lotfy, S.N., Fadel, H.H.N., El-Ghorab, A.H., Shaheen, M.S., 2015. Stability of encapsulated beef-like flavourings prepared from enzymatically hydrolysed mushroom proteins with other precursors under conventional and microwave heating. Food Chem. 187, 7–13.
Lugo, A.C.V., Lim, L.T., 2009. Controlled release of allyl isothiocyanate using soy protein and poly(lactic acid) electrospun fibers. Food Res. Int. 42, 933–940.
Lv, Y., Yang, F., Li, X., Zhang, X., Abbas, S., 2014. Formation of heat resistant nanocapsules of jasmine essential oil via gelatin/gum arabic based complex coacervation. Food Hydrocoll. 35, 305–314.

Ma, M., Tan, L., Dai, Y., Zhou, J., 2013. An investigation of flavor encapsulation comprising of regenerated cellulose as core and carboxymethyl cellulose as wall. Iran. Polym. J. 22, 689–695.
Madene, A., Jacquot, M., Scher, J., Desobry, S., 2006. Flavour encapsulation and controlled release: a review. Int. J. Food Sci. Technol. 41, 1–21.
Marano, F., Guadagnini, R., 2013. Nanoparticles and alimentation: what risks for consumers? Cah. Nutr. Diét. 48, 142–150.
Mascheroni, E., Fuenmayor, C.A., Cosio, M.S., Silvestro, G., Piergiovanni, L., Mannino, S., Schiraldi, A., 2013. Encapsulation of volatiles in nanofibrous polysaccharide membranes for humidity-triggered release. Carbohydr. Polym. 98, 17–25.
Mason, T.G., Wilking, J.N., Meleson, K., Chang, C.B., Graves, S.M., 2006. Nanoemulsions: formation, structure, and physical properties. J. Phys. Condens. Matter. 18, R635–R666.
Matsuda, S., Hatano, H., Kuramoto, K., Tsutsumi, A., 2001. Fluidization of ultrafine particles with high G. J. Chem. Eng. Jpn. 34, 121–125.
McCabe, E.R., 2010. Nanopediatrics: enabling personalized medicine for children. Pediatr. Res. 67, 453–457.
McClements, D.J., 2005. Food emulsions: principles, practices, and techniques, second ed. CRC Press, Boca Raton, FL.
McClements, D.J., 2012. Nanoemulsions versus microemulsions: terminology, differences and similarities. Soft Matter 8, 1719–1729.
Mehta, A.M., Jones, D.M., 1985. Coated pellets under the microscope. Pharm. Technol. 9, 52–60.
Meiners, J.A., 2012. Fluid bed microencapsulation and other coating methods for food ingredient and nutraceutical bioactive compounds. In: Garti, N., McClements, D.J. (Eds.), Encapsulation Technologies and Delivery Systems for Food Ingredients and Nutraceuticals. Woodhead Publishing Limited, Cambridge, UK, pp. 151–176.
Mihranyan, A., Ferraz, N., Strømme, M., 2012. Current status and future prospects of nanotechnology in cosmetics. Prog. Mat. Sci. 57, 875–910.
Milanovic, J., Manojlovic, V., Levic, S., Rajic, N., Nedovic, V., Bugarski, B., 2010. Microencapsulation of flavors in carnauba wax. Sensors 10, 901–912.
Mohammadi, A., Jafari, S.M., Esfanjani, A.F., Akhavan, S., 2016. Application of nano-encapsulated olive leaf extract in controlling the oxidative stability of soybean oil. Food Chem. 190, 513–519.
Mozafari, M.R., 2006. Bioactive entrapment and targeting using nanocarrier technologies: an introduction. In: Mozafari, M.R. (Ed.), Nanocarrier Technologies. Springer, The Netherlands, pp. 1–16.
Nakagawa, K., 2014. Nano- and microencapsulation of flavor in food systems. In: Kwak, H.-S. (Ed.), Nano- and Microencapsulation for Foods. Wiley Blackwell, UK, pp. 249–271.
Nasirullah, Kumar, P., Shariff, R., 2011. Development of nutraceutical carriers for functional food applications. Nutr. Food Sci. 41, 34–43.
Nedovic, V., Kalusevic, A., Manojlovic, V., Levic, S., Bugarski, B., 2011. An overview of encapsulation technologies for food applications. Procedia Food Sci. 1, 1806–1815.
Nel, A., Xia, T., Madler, L., Li, N., 2006. Toxic potential of materials at the nanolevel. Science 311, 622–627.
Nori, M.P., Favaro-Trindade, C.S., Alencar, S.M., Thomazini, M., Camargo Balieiro, J.C., Contreras Castillo, C.J., 2011. Microencapsulation of propolis extract by complex coacervation. LWT Food Sci. Technol. 44, 429–435.
Okuro, P.K., de Matos Junior, F.E., Favaro-Trindade, C.S., 2013. Technological challenges for spray chilling encapsulation of functional food ingredients spray chilling encapsulation. Food Technol. Biotechnol. 51, 171–182.

Park, C.W., Bastian, E., Farkas, B., Drake, M., 2014. The effect of feed solids concentration and inlet temperature on the flavor of spray dried whey protein concentrate. J. Food Sci. 79, C19–C24.
Patil, Y.P., Jadhav, S., 2014. Novel methods for liposome preparation. Chem. Phys. Lipids 177, 8–18.
Penbunditkul, P., Yoshii, H., Ruktanonchai, U., Charinpanitkul, T., Assabumrungrat, S., Soottitantawat, A., 2012. The loss of OSA-modified starch emulsifier property during the high-pressure homogenizer and encapsulation of multi-flavour bergamot oil by spray drying. Int. J. Food Sci. Technol. 47, 2325–2333.
Peng, Z., Li, J., Guan, Y., Zhao, G., 2013. Effect of carriers on physicochemical properties, antioxidant activities and biological components of spray-dried purple sweet potato flours. LWT Food Sci. Technol. 51, 348–355.
Prata, A.S., Maudhuit, A., Boillereaux, L., Poncelet, D., 2012. Development of a control system to anticipate agglomeration in fluidised bed coating. Powder Technol. 224, 168–174.
Qingnuan, L., Yan, X., Xiaodong, Z., Ruili, L., Qieqie, D., Xiaoguang, S., Shaoliang, C., Wenxin, L., 2002. Preparation of (99m) Tc-C (60) (OH) (x) and its biodistribution studies. Nucl. Med. Biol. 29, 707–710.
Rafiee, Z., Jafari, S.M., Alami, M., Khomeiri, M., 2011. Microwave-assisted extraction of phenolic compounds from olive leaves: a comparison with maceration. J. Anim. Plant Sci. 21, 46–51.
Rajagopalan, P., Wudl, F., Schinazi, R.F., Boudinot, F.D., 1996. Pharmacokinetics of a water-soluble fullerene in rats. Antimicrob. Agents Chemother. 40, 2262–2265.
Rao, J., McClements, D.J., 2011. Formation of flavor oil microemulsions, nanoemulsions and emulsions: influence of composition and preparation method. J. Agric. Food Chem. 59, 5026–5035.
Ratanasiriwat, P., Worawattanamateekul, W., Klaypradit, W., 2013. Properties of encapsulated wasabi flavour and its application in canned food. Int. J. Food Sci. Technol. 48, 749–757.
Raynes, J.K., Carver, J.A., Gras, S.L., Gerrard, J.A., 2014. Protein nanostructures in food: should we be worried? Trends Food Sci. Technol. 37, 2–50.
Reade, L., 2010. Putting textiles to work. Chem. Ind. 5, 18–20.
Regiert, M., 2008. Molecular encapsulation in cyclodextrins. Speciality Chemicals Magazine, vol. March, pp. 22–24.
Reineccius, G.A., 1994. Flavor encapsulation. In: Reineccius, G.A. (Ed.), Source Book of Flavors. second ed. Chapman and Hall, New York, NY, pp. 605–625.
Reineccius, G.A., 1995. Liposomes for controlled release in the food industry. In: Risch, S.J., Reineccius, G.A. (Eds.), Encapsulation and Controlled Release of Food Ingredients. ACS Symposium Series, vol. 590. American Chemical Society, Washington, DC, pp. 113–131.
Reineccius, G.A., 2005. Flavor Chemistry and Technology, second ed. CRC Press, New York, NY.
Risch, S.J., Reineccius, G.A., 1988. Spray-dried orange oil: effect of emulsion size on flavor retention and shelf stability. Risch, S.J., Reineccius, G.A. (Eds.), Flavor Encapsulation. ACS Symposium Series, vol. 370, American Chemical Society, Washington, DC, pp. 67–77.
Roberts, D.D., Taylor, A.J., 2000. Flavor release: a rational for its study. In: Roberts, D.D., Taylor, A.J. (Eds.), Flavor Release. ACS Symposium Series, vol. 763. American Chemical Society, Washington, DC, pp. 1–6.
Rocha-Selmi, G.A., Fávaro-Trindade, C.S., Grosso, C.R.F., 2013. Morphology, stability, and application of lycopene microcapsules produced by complex coacervation. J. Chem. Article ID 982603, 7 pages.

Rodrigues, S., Martins, I., Fernandes, I., Gomes, P., Mata, V., Barreiro, M., Rodrigues, A., 2009. Scent fashion: microencapsulation perfumes for textile applications. Chem. Eng. J. 149, 463–472.

Rosenberg, M., Kopelman, I.J., Talmon, Y., 1990. Factors affecting retention in spray-drying microencapsulation of volatile materials. J. Agric. Food Chem. 38, 1288–1294.

Rumpler, K., Jacob, M., 1998. Continuous coating in fluidized bed. Food Market Technol. 12, 41–43.

Saleh, K., Steinmetz, D., Hemati, M., 2003. Experimental study and modelling of fluidized bed coating and agglomeration. Powder Technol. 130, 116–123.

Sansukcharearnpon, A., Wanichwecharungruang, S., Leepipatpaiboon, N., Kerdcharoen, T., Arayachukeat, S., 2010. High loading fragrance encapsulation based on a polymer-blend: preparation and release behavior. Int. J. Pharm. 391, 267–273.

Santhalakshmy, S., Don Bosco, S.J., Francis, S., Sabeena, M., 2015. Effect of inlet temperature on physicochemical properties of spray-dried jamun fruit juice powder. Powder Technol. 274, 37–43.

Sarkar, S., Gupta, S., Variyar, P.S., Sharma, A., Singhal, R.S., 2012. Irradiation depolymerized guar gum as partial replacement of gum arabic for microencapsulation of mint oil. Carbohydr. Polym. 90, 1685–1694.

Scenihr (Scientific Committee on Emerging and Newly-Identified Health Risks), 2007. The existing and proposed definitions relating to products of nanotechnologies: health & consumer protection directorate-general, European Commission.

Schultz, T.H., Dimick, K.P., Makower, B., 1956. Incorporation of natural fruit flavors into fruit juice powders I. Locking of citrus oils in sucrose and dextrose. Food Technol. 10, 57–60.

Severino, P., Andreani, T., Chaud, M.V., Benites, C.I., Pinho, S.C., Souto, E.B., 2015. Essential oils as active ingredients of lipid nanocarriers for chemotherapeutic use. Curr. Pharm. Biotechnol. 16, 365–370.

Shahidi, F., Han, X.Q., 1993. Encapsulation of food ingredients. Crit. Rev. Food Sci. 33, 501–547.

Shimokawa, K., Saegusa, K., Wada, Y., Ishii, F., 2013. Physicochemical properties and controlled drug release of microcapsules prepared by simple coacervation. Colloids Surf. B 104, 1–4.

Silva, D.F., Favaro-Trindade, C.S., Rocha, G.A., Thomazini, M., 2011. Microencapsulation of lycopene by gelatin-pectin complex coacervation. J. Food Process. Preserv. 575, 1–6.

Smith, D.V., Margolskee, R.F., 2001. Making sense of taste. Scientific American, vol. 284, pp. 32–39.

Soottitantawat, A., Bigeard, F., Yoshii, H., Furuta, T., Ohkawara, M., Linko, P., 2005. Influence of emulsion and powder size on the stability of encapsulated D-limonene by spray drying. Innov. Food Sci. Emerg. Technol. 6, 107–114.

Soottitantawat, A., Yoshii, H., Furuta, T., Ohkawara, M., Linko, P., 2003. Microencapsulation by spray drying: influence of emulsion size on the retention of volatile compounds. J. Food Sci. 68, 2256–2262.

Speziali, M.G., 2012. From aromas and perfumes, the market of the "smell" industry. Quim. Nova. 35, 861–864.

Srivastava, Y., Semwal, A.D., Sharma, K., 2013. Application of various chemical and mechanical microencapsulation techniques in food sector: a review. Int. J. Food Ferment. Technol. 3, 1–13.

Sun, P., Zeng, M., He, Z., Qin, F., Chen, J., 2013. Controlled release of fluidized bed-coated menthol powder with a gelatin coating. Dry. Technol. 31, 1619–1626.

Swisher, H.E., 1957. United States Patent 2,809,895. Solid flavouring composition and method of preparing same.
Tamjidi, F., Shahedi, M., Varshosaz, J., Nasirpour, A., 2013. Nanostructured lipid carriers (NLC): a potential delivery system for bioactive food molecules. Innov. Food Sci. Emerg. Technol. 19, 29–43.
Tan, S.P., Kha, T.C., Parks, S.E., Stathopoulos, C.E., Roach, P.D., 2015. Effects of the spray-drying temperatures on the physicochemical properties of an encapsulated bitter melon aqueous extract powder. Powder Technol. 281, 65–75.
Tan, C.P., Nakajima, M., 2005. β-Carotene nanodispersions: preparation, characterization and stability evaluation. Food Chem. 92, 661–671.
Tapia-Hernández, J.A., Torres-Chavez, P.I., Ramírez-Wong, B., Rascon-Chu, A., Plascencia-Jatomea, M., Barreras-Urbina, C.G., Rangel-Vazquez, N.A., Rodríguez-Felix̌, F., 2015. Micro- and nanoparticles by electrospray: advances and applications in foods. J. Agric. Food Chem. 63, 4699–4707.
Thies, C., 2007. Microencapsulation of flavors by complex coacervation. In: Lakkis, J.M. (Ed.), Encapsulation and Controlled Release Technologies in Food Systems. Blackwell Publishing Inc., Ames, IA, pp. 149–170.
Torchilin, V.P., 2012. Multifunctional nanocarriers. Adv. Drug Deliv. Rev. 64, 302–315.
Trophardy, G., 2007. WO Patent 2007/135583. One step spray drying process.
Turchiuli, C., Fuchs, M., Bohin, M., Cuvelier, M.E., Ordonnaud, C., Peyrat-Maillard, M.N., Dumoulin, E., 2005. Oil encapsulation by spray drying and fluidised bed agglomeration. Int. Food Sci. Emerg. Technol. 6, 29–35.
Turchiuli, C., Munguia, M.T.J., Sanches, M.H., Ferre, H.C., Dumoulin, E., 2014. Use of different supports for oil encapsulation in powder by spray drying. Powder Technol. 255, 103–108.
Ubbink, J., Krüger, J., 2006. Physical approaches for the delivery of active ingredients in foods. Trends Food Sci. Technol. 17, 244–254.
Uhlemann, J., Reiß, I., 2010. Product design and process engineering using the example of flavors. Chem. Eng. Technol. 33, 199–212.
Uhlemann, J., Schleifenbaum, B., Bertram, H.J., 2002. Flavor encapsulation technologies: an overview including recent developments. Perfum. Flavor 27, 52–61.
Uyar, J., Hacaloglu, F., Besenbacher, F., 2011. Electrospun polyethylene oxide (PEO) nanofibers containing cyclodextrin inclusion complex. J. Nanosci. Nanotechnol. 11, 3949–3958.
Velmurugan, P., Lee, S.-M., Cho, M., Park, J.-H., Seo, S.-K., Myung, H., Bang, K.-S., Oh, B.-T., 2014. Antibacterial activity of silver nanoparticle-coated fabric and leather against odor and skin infection causing bacteria. Appl. Microbiol. Biotechnol. 98, 8179–8189.
Wandrey, C., Baratkowiak, A., Harding, S.E., 2010. Materials for encapsulation. In: Zuidam, N.J., Nedovic, V.A. (Eds.), Encapsulation Technologies for Active Food Ingredients and Food Processing. Springer, New York, NY, pp. 39–108.
Wang, X., Yuan, Y., Yue, T., 2015. The application of starch-based ingredients in flavor encapsulation. Starch 67, 225–236.
Wang, W., Zhou, W., 2015. Characterization of spray dried soy sauce powders made by adding crystalline carbohydrates to drying carrier. Food Chem. 168, 417–422.
Weerawatanakorn, M., Wu, J.-C., Pan, M.-H., Ho, C.-T., 2015. Reactivity and stability of selected flavor compounds. J. Food Drug Anal. 23, 176–190.
Weiss, J., Gaysinksy, S., Davidson, M., McClements, J., 2009. Nanostructured encapsulation systems: food antimicrobials. In: Barbosa-Cánovas, G.V., Mortimer, A., Lineback, D., Spiess, W., Buckle, K. (Eds.), IUFoST World Congress Book: Global Issues in Food Science and Technology. Elsevier, Amsterdam, The Netherlands, pp. 425–479.

Wolf, B., 2010. Polysaccharide functionality through extrusion processing. Curr. Opin. Colloid Interface Sci. 15, 50–54.
Wu, K.G., Chai, X.H., Chen, Y., 2005. Microencapsulation of fish oil by simple coacervation of hydroxypropyl methylcellulose. Chin. J. Chem. 23, 1569–1572.
Wu, X.J., Guy, R.H., 2009. Applications of nanoparticles in topical drug delivery and in cosmetics. Drug Deliv. Sci. Technol. 19, 371–384.
Xiao, Z., Liu, W., Zhu, G., Zhou, R., Niu, Y., 2013. Production and characterization of multinuclear microcapsules encapsulating lavender oil by complex coacervation. Flavour Frag. J. 29, 166–172.
Xiao, J.X., Yu, H.Y., Yang, J., 2011. Microencapsulation of sweet orange oil by complex coacervation with soybean protein isolate/gum arabic. Food Chem. 125, 1267–1272.
Yang, Z., Peng, Z., Li, J., Li, S., Kong, L., Li, P., Wang, Q., 2014. Development and evaluation of novel flavour microcapsules containing vanilla oil using complex coacervation approach. Food Chem. 145, 272–277.
Yao, W., Yao, H., 2002. Adsorbent characteristics of porous starch. Starch 54, 260–263.
Ye, W., Xin, J., Li, P., Lee, D.K.L., Kwong, T.L., 2006. Durable antibacterial finish on cotton fabric by using chitosan-based polymeric core-shell particles. J. Appl. Polym. Sci. 102, 1787–1793.
Zhang, Z.-Q., Pan, C.-H., Chung, D., 2011. Tannic acid cross-linked gelatine-gum arabic coacervate microspheres for sustained release of allyl isothiocyanate: characterization and in vitro release study. Food Res. Int. 44, 1000–1007.
Zoet, F.D., Grandia, J., Sibeijn, M., 2011. NL Patent 050668. Encapsulated fat soluble vitamin.
Zuidam, N.J., Heinrich, E., 2010. Encapsulation of aroma. In: Zuidam, N.J., Nedovic, V.A. (Eds.), Encapsulation Technologies for Active Food Ingredients and Food Processing. Springer, New York, NY, pp. 127–160.
Zuidam, N.J., Shimoni, E., 2010. Overview of microencapsulates for use in food products or processes and methods to make them. In: Zuidam, N.J., Nedovic, V.A. (Eds.), Encapsulation Technologies for Active Food Ingredients and Food Processing. Springer, New York, NY, pp. 3–30.

19

EMERGING TRENDS IN THE APPLICATION OF NANOBIOSENSORS IN THE FOOD INDUSTRY

Sophiya Devi Lamabam, Robert Thangjam
Mizoram University, Department of Biotechnology, School of Life Sciences, Mizoram, India

1 Introduction

The issues related to food quality have long been a major concern for human health and safety. Most issues in the food industry that pose great challenge in their safety are related to the contamination of food with pharmaceuticals, agrochemicals, food-borne pathogens, and biotoxins. Thus, it is essential for the food and beverage industries to have techniques that are efficient and fast for the detection of phytochemicals like sugars, proteins, vitamins, as well as for the microbial and other chemical contaminants. The conventional methods for detection of food contaminants are based mostly on chromatographic and immunoassay techniques, which are laborious and time consuming. On the other hand, biosensors are increasingly becoming popular as an interesting alternative to the conventional analytical methods with the greater prospect of rapid and sensitive label-free bioelectronic detection of various contaminants such as detection of pathogens, pesticides, and drug residues, heavy metals, and other toxic substances.

2 Biosensors

Biosensors and affinity sensor devices have been shown to have the ability to provide rapid, cost-effective, specific, and reliable quantitative and qualitative analysis (Tothill, 2001). Recent advances and breakthroughs in the field of nanobiotechnology

Novel Approaches of Nanotechnology in Food. http://dx.doi.org/10.1016/B978-0-12-804308-0.00019-4

also led to the development of nanomaterial based biosensors as versatile tools for many applications in different disciplines such as biomedical, pharmaceutical, agricultural, and food industries. A biosensor, often called bioreceptors or transducers, is an integrated receptor-transducer device, which converts the biological-recognition events into a measurable physicochemical signal that are proportional to the target concentration (Pilolli et al., 2013). The classification of biosensors was based on the transducing elements such as the electrochemical, optical, piezoelectric, and electromechanical sensors (Arora et al., 2011) (Fig. 19.1). In biosensors, unlike the conventional chemical sensors, biological structures (cells, enzymes, or nucleic acids) are used as the sensing elements to measure the different biological processes. Biosensors are developed mostly for applications in environmental and bioprocess monitoring, quality control of food, agriculture, and biomedical systems. Many sensors operate through the variation of a surface parameter, like surface conductivity, with analyte concentration. Hence, the effective surface area of the device, that is, the area actually interacting with the analyte, determines the sensitivity. With advent of nanobiotechnology, nanomaterials are playing an increasingly important role in the development of biosensors. The integration of

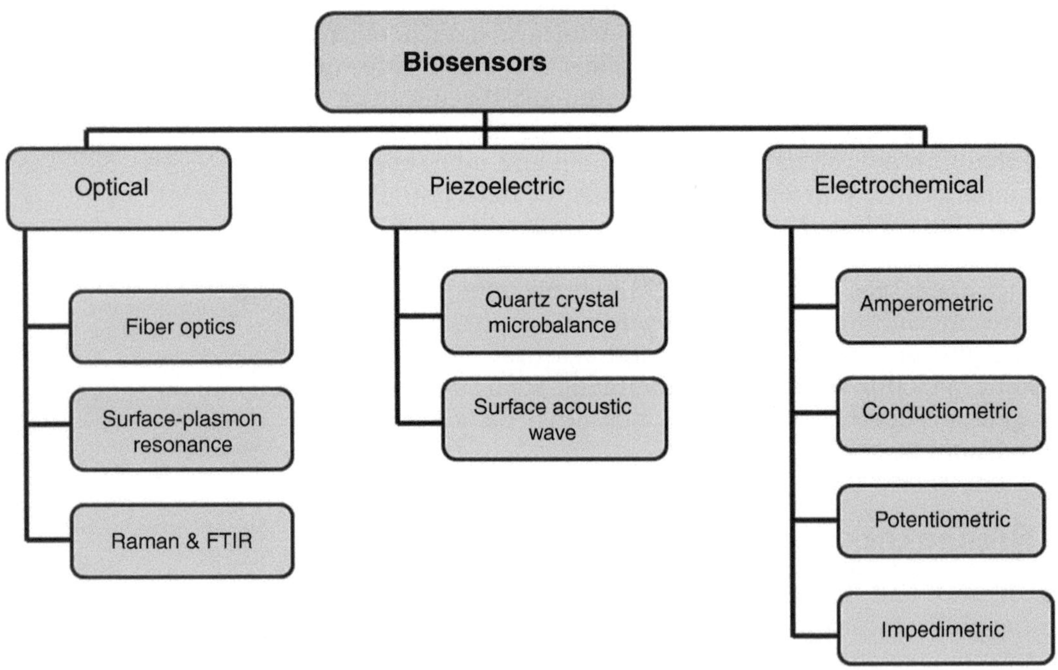

Figure 19.1. Classification of biosensors based on transducers. Adapted from Velusamy et al., 2010.

the high specificity of biological receptors with the unique optical, electrical, and electrochemical properties and large surface-to-volume ratio due to their small size (1–100 nm) of the nanomaterials provides a novel interesting alternative platform for biosensor development to the conventional biosensors. Nanobiosensors are one of the most promising novel-sensing devices that offer higher sensitivity, accuracy, rapid, and convenient for industrial applications. Nanoparticles serve as excellent platforms in development of biosensors such as electric nose, array biosensors, microfluidic devices, and carbon nanotube-based sensors (Viswanathan and Radecki, 2008). Uses of various kinds of nanoparticles have been expanding rapidly in biosensors applications. Recent advancement and improvement of the biosensors has also been achieved by the use of various nanomaterials such as metal nanoparticles, carbon nanotubes, nanorods, quantum dots, and nanowires. According to Perez-Lopez and Merkoci (2011), nanomaterials can also be used as biosensors as labels or transducer modifiers efficiently for lower level detection and higher sensitivity. Nanoparticles, owing to their small size, have physical, electronic, and chemical properties that are different from those of bulk metals. Such properties strongly depend on the number and kind of atoms that make up the particle. The high surface-to-volume ratio and novel electron transport properties of these nanostructures enabled their electronic conductance to be strongly influenced by minor surface perturbations. Therefore, nanomaterial such as colloidal gold (Au) was used to enhance the amount of immobilized biomolecules in construction of a biosensor because of its ultrahigh surface area (Cai et al., 2001).

2.1 Optical Biosensors

Optical biosensors represent powerful detection tools with applications in food industry, health care, and environmental monitoring (Narayanaswamy and Wolfbeis, 2004). The working principle of optical biosensors was based on the measurement of changes in the surface properties the sensor when the analyte reacts with the receptor on surface of the sensing device (Pilolli et al., 2013). Optical transducers are particularly suitable for developing robust devices, which are easy to use, portable, and also inexpensive. Various optical biosensors were developed for rapid detection of pathogenic bacteria (Baeumner et al., 2003), toxins (Bae et al., 2004), and contaminants (Willardson et al., 1998; Tschmelak et al., 2004) in food materials. Various chemical-physical phenomena such as UV-visible absorption, fluorescence, and surface-plasmon resonance (SPR) are being used for the development of different optical biosensors. UV-visible absorption is usually applied in qualitative as well as quantitative analysis of food contaminants. For the detection and determination of the analytes changes in optical density or color is sufficient thus making it easier and simpler than the techniques used in other methods (Pilolli et al., 2013). The fiber-optic biosensor is one of most widely used biosensors in the food sector. It uses tapered fiber to send excited laser light to the detection surface and receive emitted light. Propagation of light through a fiber or waveguide can be very sensitive to the surroundings, which makes the optical fibers excellent detectors for a variety of applications in foods. Among optical biosensors, fiber optic is the first commercially available sensor marketed by Research International (Monroe, WA) (Arora et al., 2011). The fiber-optic biosensors can rapidly detect pathogens from complex matrices but the confirmation tests may take up to several hours to complete. The basic principle of fiber-optic biosensor lies behind fluorescently labeled pathogens or toxins, which when bound to the surface of biosensor gets an excitation by the laser wave (635 nm) generating fluorescent signals (Taitt et al., 2005), which were detected by fluorescent detector in the real-time system. In fluorescent biosensors, a fluorescent compound in conjugation with antibodies aids in the detection of pathogenic bacterium. Most routinely used fluorescent marker is FITC (fluorescein isothiocyanate) (Li et al., 2004), though some lanthanides are also considered for the same purpose (Selvin, 2002). Fluorescence-based optical detection system is also widely used due to its several advantages over other techniques as it is the well-established and highly sensitive nature of its selective labeling techniques. But the techniques are expensive owing to the use of costly fluorescent dyes, limited shelf life, and are often influenced by variable chemical factors that may vary for different samples. SPR is one of the most commonly exploited label-free optical biosensing techniques. The SPR phenomenon due to high surface sensitivity and real-time and label-free response has been widely employed in different transduction configurations to monitor various processes, that is, DNA hybridization and antibody–antigen binding and DNA–protein interactions. SPR-based nanobiosensor is one among the target sources for food-borne pathogen detection. Nanomaterials, such as Au nanoparticles, Au nanorods, Fe_3O_4 metal nanoparticles, and quantum dots have very good optical properties and have been implemented for improving the sensitivity of optical transducer surfaces of biosensors. Antibodies were initially immobilized on thin Au film just over the reflecting surface of waveguide, to capture various pathogens. At certain wavelengths and near IR region, strong resonance was generated through the interaction of light with the electron cloud in the metal. Binding of pathogen to the metal surface causes a shift in resonance to longer wavelengths and the amount of shift accordingly reflects the concentration of bound pathogen or toxins. SPR nanobiosensors can detect molecules even in femtomolar range (Bhunia, 2007). This optical-based SPR system has been used for detection of whole cells of *Escherichia coli* O157:H7, *Salmonella,* and *Listeria* at traceable concentrations (Arora et al., 2011). Raman spectroscopy is another reliable method for specific identification of molecules. Surface enhanced Raman spectroscopy (SERS) have been applied in a number of analytical applications including for the detection of pathogenic microbes (Tripp et al., 2008). Both intrinsic and extrinsic formats of SERS have been has successfully been used to identify the pathogens and also for finger printing of bacteria, viruses, and protein molecules (Huh et al., 2009). SERS-based identification of microbes is less time consuming and simple, which does not require trained personnel (Driskell and Tripp, 2009; Patel et al., 2008). Cheng et al. (2009) have reported use of Au-NP/PVP/Au substrate for the development of SERS-based assay for the detection of dipicolinic acid, a biomarker for bacterial spores including pathogenic bacteria such as *Bacillus anthracis.* Also, silver nanoparticles have been successfully used in SERS-based detection of target analytes.

2.2 Electrochemical Biosensors

The electrochemical biosensors have been used for analyzing food and beverages (Mannino and Wang, 2005; Pellegrini et al., 2004), detection of GMO content in food (Volpe et al., 2006), measuring the freshness of food (Winquist et al., 1997). They possess certain advantages over optical devices such as low cost, simple operation, small size, disposable, and incorporation of multiple sensing elements in a single chip-like device. These biosensors can be classified into potentiometric, amperometric, conductiometric, and impedimetric types (Pilolli et al., 2013). The potentiometric biosensors are used for the detection of ions in a solution. They are portable and cost effective, however, the major disadvantages associated with these biosensors are the poor selectivity toward some environmental samples. In potentiometric-based detection, the biorecognition process is converted into a potential signal. Usually a high-impedance voltmeter is used to measure the electrical potential difference or electromotive force (EMF) between two electrodes at near zero current. Since potentiometry generates a logarithmic concentration response, the technique allows the detection of extremely small concentration changes. Electrochemical biosensors, in particular amperometric biosensors have an important position among biosensors as a class of the most widespread, numerous, and successfully commercialized devices based on biomolecular electronics (Dzyadevych et al., 2008; Goriushkina et al., 2009). Amperometric biosensors are able to detect the electrochemically active analyte, an analyte capable of being oxidized or reduced on the electrode. In amperometric-based detection, the sensor potential is set at a value where the analyte produces current. Thus, the applied potential serves as the driving force for the electron transfer reaction, and the current produced is a direct measure of the rate of electron transfer. Generally, these are prepared with thin-film technology and consist of Au, platinum (Pt), or carbon. Inks in form of thin films in a particular pattern and thickness were deposited on the electrode substrate such as glass, plastic, or ceramic for screen printing. The introduction of nanomaterials into electrochemical sensors and biosensors certainly brings advantages such as decrease over potentials of many analytically important electrochemical reactions, ensure the reversibility of some redox reactions, which are irreversible at unmodified electrodes or bring novel labeling opportunities including multidetection capabilities. Impedimetric biosensor is a powerful technique used for the detection of electrochemical systems. A small amplitude sinusoidal excitation applied to the system measures the changes in electrical impedance of the medium. Analysis is done on the basis of changes in conductance, capacitance, and impedance. The changes in the conductance and capacitance and impedance due to microbial metabolic processes can be monitored to detect the presence of pathogen (Invitski et al., 2000). Nanoparticles have a chemical behavior similar to small molecules, which can be used as specific electrochemical labels and therefore novel applications of nanobased biosensors are being studied for their applications in biomedical researches and food analysis (Perez-Lopez and Merkoci, 2011). An amperometric biosensor design involving coimmobilization of fructose dehydrogenase (FDH) and inulinase (INU) on an Au nanoparticle-cysteamine (Cyst) self-assembled monolayer (SAM)-modified Au electrode (Aucoll-Cyst-AuE), for the determination of the carbohydrate inulin in foodstuffs, has been developed by Manso et al. (2008). In conductometric-based biosensors, there is change in the ionic species concentration, which leads to a change in electrical conductivity or current flow during a biorecognition event. Muhammad-Tahir and Alocilja (2003) developed a conductometric biosensor for detecting food borne pathogens such as *E. coli* O157:H7 and *Salmonella* spp. Nanobased electrochemical biosensors have been fabricated using mostly metallic nanoparticles. Metal nanoparticles can be used to enhance the amount of immobilized biomolecules in construction of a sensor. Because of its ultrahigh surface area, colloidal Au has been used to enhance the DNA immobilization on Au electrode, to ultimately lower the detection limit of the fabricated electrochemical DNA biosensor (Cai et al., 2001). Enzyme-Au colloids have been used on the surface of electrodes to fabricate biosensors for hydrogen peroxide (H_2O_2), glucose, xanthine, and hypoxanthine (Crumbliss et al., 1992; Zhao et al., 1996; Xu et al., 2003). Xu et al. (2003) also studied the electrochemistry of horseradish peroxidase (HRP) immobilized on a colloidal Au modified screen-printed carbon electrode. The immobilized HRP displayed fast amperometric response and an electrocatalytic activity to the reduction of H_2O_2 without the aid of an electron mediator. The biosensor exhibited high sensitivity, good reproducibility, and long-term stability for the determination of H_2O_2 with a linear range from 0.8 mM to 1.0 mM and a detection limit of 0.4 μM. Nanosized semiconductor crystals can also increase efficiency of photochemical reactions and can be effectively coupled to biomolecules such as enzyme, to generate novel photoelectrochemical systems (Curri et al., 2002). Since most biological molecules can be labeled with metal nanoparticles without compromising their biological activities metal nanoparticles are widely used as an electrochemical label. Affinity assays can then be performed by monitoring the electrochemical signal of these metal nanoparticles.

2.3 Piezoelectric Biosensors

Piezoelectric biosensors are based on the piezoelectric effect—alternating current applied on the crystal stimulates high-frequency vibrations. For biosensing, the thickness shear vibrations represented by the quartz crystal microbalance (QCM) systems are mostly used due to robust performance and simple construction of the driving electronic (Abad et al., 1998). Mass-sensitive biosensors are suitable for very sensitive detection, in which the transduction is based on the small changes in mass. The principal means of mass analysis depends on the use of piezoelectric crystals, which can be made to vibrate at a specific frequency with the application of an electrical signal of a specific frequency. The application of QCM requires immobilization of antibodies or DNA fragments on the transducer surface, which is based on the principle that during the biological-recognition process a mass change is generated from the sensing layer, which resulted in the resonant frequency change of the microbalance. In addition, Au nanoparticles can be integrated both as labels in sandwich assays and as surface modifiers onto the quartz crystal, which help in an increase in the amount of immobilized biomolecules (Pilolli et al., 2013). A number of QCM biosensors have been studied for detection of microorganisms such as *Listeria monocytogenes* (Vaughan et al., 2001), *Bacillus cereus* (Vaughan et al., 2003), *B. subtilis* (Lee et al., 2005), *E. coli* O157:H7 (Ngo et al., 2014), *Salmonella typhimurium* (Ye et al., 1997; Babacan et al., 2002), and *Vibrio cholerae* O139 (Carter et al., 1995). Hao et al. (2009) demonstrated the use of oriented immobilization of monoclonal antibodies via protein A on the monolayer-activated QCM for detection of *B. anthracis* (10^3 spores per ml) in less than 30 min. The integration of Au nanoparticles with QCM-based (quartz crystal microbalance) biosensors is a promising methodology of enhancement of the efficiency of the biosensor. The nanoparticle layer provides 3-dimensional architecture and increases the surface area, which can accommodate more ligand-molecules for more target analytes attachment. Zhang et al. (2008a) reported detection of antithrombin III by using a biosensor based on Au nanoparticles integration on the electrode employing QCM technique. Heparin was immobilized on Au nanoparticles by cysteamine self-assembled-monolayer (SAM). A reusable piezoelectric immunosensor using magnetic hydroxyapatite (HAP) nanocrystals embedded with g-Fe_2O_3 and colloidal Au nanoparticles for antibody immobilization was also developed for human IgG detection, which is carried out in a homogeneous bulk solution and the immunocomplexes formed were detected in a vessel by magnetic-separation using QCM measurement (Zhang et al., 2008b). The nanocomposites provided a large loading capacity of biomolecules maintaining the well-preserved bioactivity, also the immunocompound structure exposed possess a good degree of flexibility and accessibility to the analyte. In general, the use of Au nanoparticles in immunosensors as versatile and efficient substrates for antigen or antibodies immobilization showed not only the increased amount of adsorbed biomolecules on the metallic surface, but also the preservation of immunoactivity. Additionally, the Au nanoparticles can be used for enhancing the immunoreaction signals, improving the sensitivity, real-time QCM detection of shrimp allergen in food and detection of gliadin (Xiulan et al., 2010; Chu et al., 2012).

3 Application of Nanobiosensors in Food Analysis

Biosensor- and nanotechnologies are being used together in various applications of food analysis (Fig. 19.2). The integration of nanomaterial with the conventional biosensors has been

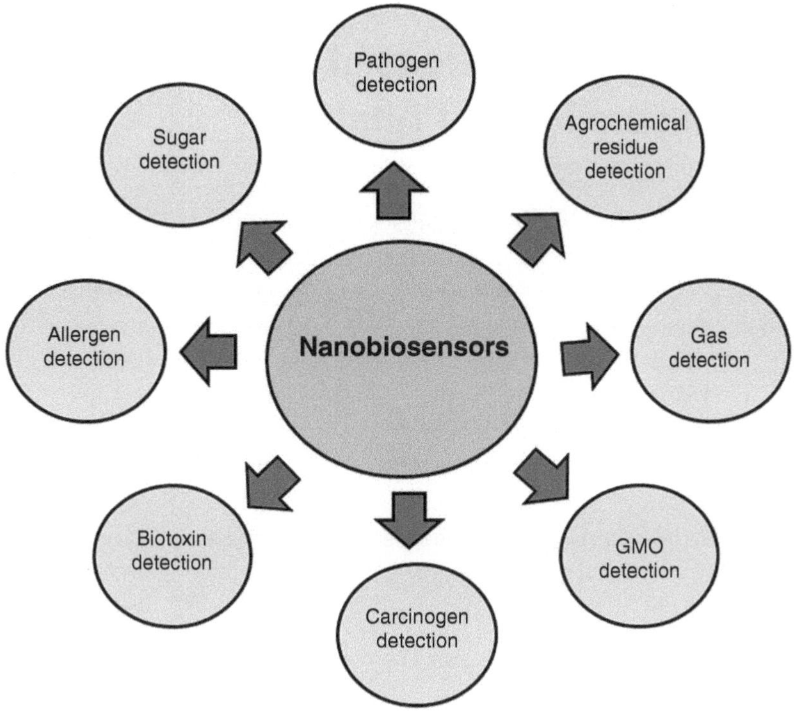

Figure 19.2. Applications of nanobiosensors in food industry.

continuously growing in the past decades, with several applications in food industry. The use of nanobiosensors in various food analyses is summarized in Table 19.1.

3.1 Detection of Harmful and Pathogenic Microbes

Microbiological contamination of food can occur at any point of time during production, processing, transport, retailing, domestic storage, or meal preparation. Several pathogenic microorganisms such as *Campylobacter* spp., *Clostridium botulinum, Staphylococcus aureus, Shigella* spp., *Enterococcus* spp., *E. coli, S. typhimurium, L. monocytogenes,* are frequently found to have contaminated the food materials and products (Borck et al., 2002; Carlin et al., 2004; Ghosh et al., 2007; Geornars et al., 2007; Fredriksson-Ahomaa et al., 2007; Zunabovic et al., 2011; García-Cano et al., 2014; Karabıyıklı et al., 2014). Contaminations of the food products with toxin-producing and infectious microorganisms can lead to enormous public health problems. Therefore, it is necessary to have a rapid and sensitive approach for stringent detection and identification of such food-borne pathogenic

Table 19.1 List of Applications of Nanobased Biosensors in Food Analysis

Applications	Analytes	Nanomaterials Used	Food Sample	Biosensor Type	References
Pathogen detection	*Salmonella typhi*	QDs	Chicken carcass wash water	Optical	Yang and Li (2005)
	Escherichia coli	Silica particles (200 nm) coated with silver shells (c. 20 nm)	Water	Optical	Kalele et al. (2006)
	E. coli	Glyco-NPs	Phosphate buffer solution	Optical	El-Boubbou et al. (2007)
	Salmonella typhi	Au NPs (c. 15 nm)	Phosphate buffer solution	Electrochemical	Dungchaia et al. (2008)
	Mycobacterium avium subsp. *paratuberculosis*	Au NPs	Milk	Optical	Yakes et al. (2008)
	E.coli O157:H7	Au NPs (13 nm diameter)	Milk	Electrochemical	Lin et al. (2008)
	Pseudomonas aeruginosa	Au Nanorods (68 nm–18 nm)	0.85% sodium chloride	Optical	Norman et al. (2008)
	E. coli	Cu@Au NPs (c. 15 nm)	Surface water	Electrochemical	Zhang et al. (2009)
	Salmonella spp.	Au NPs (c. 25 nm)	Pork	Electrochemical	Gong-Jun et al. (2009)
Pesticide detection	Paraoxon	MWCNT	0.05 M phosphate buffer (pH 7.4)	Optical	Prakash-Deo et al. (2005)
	Paraoxon	(CdSe)ZnS coreshell QDs	50% CH3OH/H2O (v/v) solvent	Optical	Ji et al. (2005)
	Paraoxon	Au NPs (c. 25 nm)	20 mM Glycine buffer (pH 9.0)	Optical	Simoniana et al. (2005)
	Parathion	ZrO2 NPs (c. 22 nm)	Buffer solution (pH 2.0e7.0)	Electrochemical	Wang and Li (2008)
	2,4 D (herbicide)	CdTe QD	50 mM phosphate buffer solution (pH 7.4)	Optical	Vinayaka et al. (2009)
	Methyl parathion and chlorpyrifos	SWCNT	0.002 M phosphate buffer solution (pH 7.0)	Electrochemical	Viswanathan et al. (2009)

Table 19.1 List of Applications of Nanobased Biosensors in Food Analysis (*cont.*)

Applications	Analytes	Nanomaterials Used	Food Sample	Biosensor Type	References
Sugar detection	Fructose	SWCNT	Honey	Electrochemical	Antiochia et al. (2004)
	Maltose	CdSe@ZnS NPs (4.5–50 Å diameter)	Buffer solution	Optical	Sandros et al. (2006)
	Glucose	×GnPs (thickness of 10 nm) decorated with Pt and Pd NPs (c. 1–4 nm diameter)	50 mM phosphate buffer solution (pH 7.4)	Electrochemical	Lu et al. (2007)
	Glucose	ZnO:Co Nanoclusters (5 nm)	0.1 M phosphate buffer solution (pH 7.4)	Electrochemical	Zhao et al. (2007)
	Glucose	Fe_3O_4 MNPs	0.2 M acetate buffer solution (pH 4.0)	Optical	Wei and Wang (2008)
	Fructose	MWCNT (c. 5–10 nm diameter)	0.1 M phosphate buffer solution (pH 5.0)	Electrochemical	Tominaga et al. (2009)
	Glucose	CdTe QDs (c. 3.4 nm)	20 mM phosphate buffer solution (pH 7.4)	Optical	Li et al. (2009)
	Glucose	Au NPs	Fruit juice	Electrochemical	Ozdemir et al. (2010)
Biotoxin detection	Aflatoxin B1	Au NPs	Pig feed	—	Delmulle et al. (2005)
	Cholera toxin	Au NPs	Water	Optical	Schofield et al. (2007)
	Bolutinum toxin serotype A	QDs	—	—	Warner et al. (2009)
	Aflatoxin B2	MnGMs	Peanuts, hazelnuts, pistachios and almonds	Fiber optics	Tang et al. (2009)
	Microcystin-LR toxin	CNTs	—	Electrochemical	Wang et al. (2009a)

(*Continued*)

Table 19.1 List of Applications of Nanobased Biosensors in Food Analysis (*cont.*)

Applications	Analytes	Nanomaterials Used	Food Sample	Biosensor Type	References
Gas detection	Acetone	$ZnO–TiO_2$ nanocomposites or SnO nanobelts	—	—	Comini et al. (2005)
	Ethanol	$ZnO–TiO_2$ nanocomposites or SnO nanobelts	—	—	Comini et al. (2006)
	Carbon monoxide	$ZnO–TiO_2$ nanocomposites or SnO nanobelts	—	—	Barreca et al. (2007)
	Ethylene gas	WO_3-SnO_2 nanocomposites	Fruits	—	Pimtong-Ngam et al. (2007)
	Gaseous amines	SnO_2 nanoparticles and TiO_2 microrods	Fish and meat	Fluorescence	Zhang and Zhang (2008)
Carcinogen detection	Lead	Au NPs	—	Optical	Wang et al. (2008)
	Hg^{2+}	Au NPs	Drinking-water samples	QCM sensor	Chen et al. (2011)
	Hg^{2+}	CNT–Ag nanoclusters	—	Optical	Wang et al. (2014)
Genetically modified organisms (GMO) detection	DNA	Au NPs	Soybean powder	Dipstick format	Kalogianni et al. (2006)

Au NPs, gold nanoparticles; CNTS, carbon nanotubes; SWCNT, single-walled carbon nanotubes; MWCNT, multiwalled carbon nanotubes; xGnPs, exfoliated graphite nanoplatelets; QDs, quantum dots; CdTe QD, cadmium telluride quantum dot nanoparticle; MNPs, magnetic nanoparticles; CdSe@ZnS, ZnS coated CdSe; MnGMs, magnetic nanogold microspheres.

microbes in food samples (Sanvicens et al., 2009, 2011). Different detection methods for pathogenic microbes such as culture-based methods, immunologically based biosensors including immunoblotting, enzyme-linked immunosorbent assays, electrochemical assays, flow cytometry, and surface plasmon resonance have been developed by various researchers (Allen et al., 2004; Cheung and Kam, 2012; Cho and Irudayaraj, 2013; Kim et al., 2006; Van et al., 2001). However, these methods are usually time-consuming, labor-intensive, and lack sensitivity and specificity (Agasti et al., 2010). Molecular biology methods such as PCR and real-time PCR are used to overcome these lacunas but most of these methods are designed for the detection of a single species of pathogenic bacteria in one assay and require highly qualified and trained personnel. Use of biosensors provides new strategies for simple, rapid, and sensitive detection of microbial contaminations in a short period of time. Biosensors based on nanomaterials such as metal nanoparticles (Gong-Jun et al., 2009), nanorods (Norman et al., 2008), nanowires (Patolsky et al., 2004; Tok et al., 2006), and carbon nanotubes (So et al., 2008) are being used for the analysis and rapid detection of pathogens. The recognition element in the biosensing device is typically bound to the surface of the nanomaterial, and the interaction of this hybrid with a pathogen is monitored through a signal transduction mechanism, which explicitly detect the interaction between the pathogen and the recognition element improving the sensitivity of the analysis. Nanobased biosensors can be placed directly into the food packaging material, which serves as electronic tongue or noses by detecting chemicals released during food spoilage (Lange et al., 2002; Garcia et al., 2006). Other types of nanosensors are based on microfluidics devices (Baeummer, 2004) and can also be used to detect pathogens efficiently in real time and with high sensitivity. A major advantage of microfluidic sensors is their miniature format and their ability to detect compounds of interest rapidly in only microliters of required sample volumes, which has already led to widespread applications in medical, biological, and chemical analysis (Vo-Dinh et al., 2003; Mabeck and Malliaras, 2006). Au nanoparticles find a significant position among all nanotechnology-based assays for microbial detection. Various physicochemical properties of the microorganisms such as optical, electrochemical, biochemical, and physical properties have been evaluated for the development of detection methods using biosensors. For the detection of pathogens in food materials techniques like surface plasmon resonance (SPR), amperometric, potentiometric, and acoustic wave biosensor have been commonly used (Leonard et al., 2003). Optical techniques generally rely on monitoring the change of the optical signal that occurs between a functionalized nanomaterial and a pathogen (Comparelli et al., 2007). Different biosensors for detecting bacteria based on the measurement of the fluorescence emission of organic dyes, protein-based fluorophores, and so on have been reported (Haes and Van-Duyne, 2002; Ko and Grant, 2006; Curtis et al., 2008; Simpson-Stroot et al., 2008).

Kalele et al. (2006) have demonstrated a rapid and reliable test based on the SPR band associated with the coupling of the antibodies to the silver nanoshells for the extremely specific detection of *E. coli*. Norman et al. (2008) also used Au nanorods covalently or electrostatically linked to anti-*Psuedomonas aeruginosa* primary antibodies specific to selectively destroy *P. aeruginosa*. They demonstrated by irradiating near-infrared (NIR) radiation to conjugated nanorods followed by nanorod attachment to the bacterial cell surface resulted in significant reduction in bacterial cell viability. Nanoparticles modified with antibodies are reported to bring advances in the immunological process for detection of the microorganism. Yakes et al. (2008) studied the detection of *Mycobacterium avium* subsp. *paratuberculosis* in milk samples, using a sonicated immunoassay based on the surface-enhanced Raman scattering (SERS) where Au nanoparticles used as extrinsic Raman labels and an immobilized layer of monoclonal antibodies that target a surface protein on the microorganism for a specific and rapid diagnosis. Gong-Jun et al. (2009) immobilized *Salmonella* spp. monoclonal antibodies (McAbs) on Au nanoparticles and evaluated the interaction of McAbs with *Salmonella* spp. using the electrochemical impedance spectroscopy (EIS) technique as an efficient method for fabricating a capacitive immunosensor for the detection of *Salmonella* spp. in pork samples. The linear relationship between the relative change in capacitance and logarithm of salmonella concentration was obtained in the range of $1.0–10^2$ to $1.0–10^5$ CFU mL^{-1}, having as linear regression equation: DC (%) ¼ 17.18e14.92log C (CFU mL^{-1}), and a correlation coefficient of 0.991. The lowest detection limit of Salmonella antigen concentration was $1.0–10^2$ CFU mL^{-1}. This capacitive immunosensor has the advantage of high nonspecific interactions and the short analysis time (40 min) in comparison with the polymerase chain reaction (PCR) method. A highly sensitive strategy for detecting *S. typhi* based on Au nanoparticles has been studied by Dungchaia et al. (2008). They immobilized monoclonal antibodies on polystyrene microwells and captured *S. typhi* bacteria. After that a polyclonal antibody-colloidal Au conjugate was added to bind to the *S. typhi* bacteria followed by a copper enhancer solution with ascorbic acid and copper (II) sulfate added into the polystyrene microwells. The ascorbic acid reduced the copper (II) ions to copper (0), which was deposited onto Au nanoparticles tags. The released copper ions were detected by using anodic stripping voltammetry and the amount of deposited copper was related to the amount of Au nanoparticles tag. Thus the amount of *S. typhi* attached to the polyclonal antibody-colloidal Au conjugate was controlled. Therefore, the anodic stripping peak current was linearly dependent of the *S. typhi* concentration obtaining a detection limit of 98.9 CFU/mL. Monitoring contamination of water and food by various strain of *E. coli* is crucial for the public health. Several immunoassay-based electrochemical biosensors have been developed for detection of *E. coli*. Lin et al. (2008) reported an amperometric immunosensing strip for detection of food-borne pathogen *E. coli* O157:H7 in milk samples. The screen-printed carbon electrode (SPCE) immunosensor was fabricated by using carbon inks modified with Au nanoparticles and ferrocenedicarboxylic acid (FeDC). The detection method consists of a sensitive detection of horseradish peroxidise (HRP) activity coupled with Au nanoparticles and FeDC to amplify the amperometric effect. It has the potential for further applications in the rapid pathogen detection. One important advantage of these amperometric immunosensing strips is the detection limit that approximately is of 50 CFU of *E. coli* O157:H7 detected in 1 h in milk samples. Zhang et al. (2009) also described a sensitive electrochemical immunoassay for rapid detection of *E. coli* by anodic stripping voltammetry (ASV) based on coreshell Cu@Au nanoparticles as anti-*E. coli* antibody labels.

Magnetic nanoparticles have been integrated with PCR, which is often used for detection of food pathogen (Lee et al., 2009; Li et al., 2011). Since the purity and the presence of a sufficient number of target DNA molecules affect the reliability of PCR for pathogen, evaluation of the preconcentration of target pathogens are required (Yang et al., 2007). Submicron superparamagnetic anion-exchangers were used to concentrate target bacteria for multiplex PCR detection (Yang et al., 2011). The PCR detection limit (10^4–10^5 CFU/mL) was lowered to a level of 10^2 CFU/mL using *E. coli* and *Agrobacterium tumefaciens* as model bacteria. The integration of submicron magnetic ion exchangers and PCR detection was also studied in fresh produce samples and the results showed that the PCR detection limit was slightly impaired to about 10^3 CFU/mL. The submicron superparamagnetic anion exchangers based preconcentration enabled improvement of PCR detection sensitivity by at least 10–100 times even in fresh produce samples. Pal and Alocilja (2009) developed electrically active polyaniline coated magnetic (EAPM) nanoparticle-based biosensor for the detection of *B. anthracis* endospores in contaminated food samples. In their study, the EAPM nanoparticles were biologically modified to act as an immunomagnetic concentrator of *B. anthracis* spores from lettuce, ground beef, and whole milk samples and are directly applied to a direct-charge transfer biosensor. The detection mechanism of the biosensor depends on the capillary flow of the captured spores on the biosensor surface along with direct-charge transfer across the EAPM nanoparticles. The EAPM nanoparticle-based biosensor is able to detect *B. anthracis* spores at concentrations as low as 4.2 × 102 spores/mL from the samples. The EAPM-based biosensor detection system is reported to be fast and reliable with a total detection time of 16 min. Abbaspour et al. (2015) reported a sensitive and highly selective dual-aptamer-based sandwich immunosensor for the detection of *S. aureus*. In this bioassay system, a biotinylated primary anti-*S. aureus* aptamer was immobilized on streptavidin coated magnetic beads (MB), which serves as a capture probe. A secondary anti-*S. aureus* aptamer was conjugated to silver nanoparticles (Apt-AgNP) that reports sensitively detection of the target.

3.2 Detection of Microbial DNA

In the food industry sensing of DNA for the detection of pathogenic contamination plays an important role. Several works have been reported on the development of simple and efficient DNA detection methods, which does not require expensive instrumentation. Au nanoparticles have been integrated with the DNA sensing system to enhance the selectivity and sensitivity for specific target DNA or oligonucleotide detection (Zeng et al, 2007). The principle was based on the electrostatic properties of DNA that only the single-stranded DNA specifically adsorbs and stabilizes the negatively charged Au nanoparticle, which prevents their aggregation and optical properties enabling the design of many fluorescent based methods for DNA detection. Demers et al. (2000) have reported a fluorescence-based method that determined the number of thiol-derivatized single-stranded oligonucleotides bound to Au nanoparticles and their extent of hybridization with complementary oligonucleotides in solution. On the other hand, Li and Rothberg (2004) demonstrated a simple fluorescent assay employing functionalized Au nanoparticles that can easily detect untagged target oligonucleotides in complex mixtures of DNA. Liu et al. (2006) also developed a colorimetric adenosine biosensor based on the aptazyme-directed assembly of Au nanoparticles. A hybrid type surface bioaffinity sensor based on the adsorption of Au nanoparticles onto Au diffraction gratings for sensing DNA and demonstrated that these biosensors can detect target DNA at low concentration was also developed by Wark et al. (2007). Other biosensors were also reported based on the integrated assembly of dehydrogenase enzymes and Au nanoparticle (Jena and Raj, 2006) and quartz crystal microbalance DNA-based sensor using streptavidin-coated Fe_3O_4 nanoparticles as "mass enhancers" for detection of *E. coli* (Mao et al., 2006).

3.3 Detection of Pesticides

The presence of pesticide residues in food represents one of the major issues for the food industry. Among the pesticides, organophosphate (OP), and carbamate compounds are the most widely used due to their high insecticidal activity and relatively low persistence (Mora et al., 1996). These neurotoxins act as inhibitors of acetylcholinesterase, an enzyme that catalyzes in a very efficient way the hydrolysis of the neurotransmitter acetylcholine. The inhibition of this enzyme can disrupt the transmission of nerve impulses (Simonian et al., 1997). Therefore, the presence of residues of such pesticides in natural waters and in foodstuffs is of major concern for public health. OP neurotoxins generally show low environmental persistence, but they have a high acute toxicity and a wide range of biological activities. Therefore, there is a high demand for developing highly sensitive and selective methods for residue analysis of these pollutants. Viswanathan et al. (2009) reported an electrochemical biosensor based on acetylcholinesterase immobilized on polyaniline deposited on vertically assembled single wall carbon nanotubes (SWCNTs) wrapped with thiol terminated single strand oligonucleotide (ssDNA) for the determination of the most commonly used organophosphorous insecticides; methyl parathion and chlorpyrifos in vegetable crops. The thiol terminated ssDNA-SWCNTs has been immobilized onto Au surface via Au-S chemical bonding. The incorporation of SWCNTs with the electrochemical biosensor not only provided the conductive pathways to promote the electron transfer, but also increased the surface area of flexible 3-dimensional conductive supports for acetylcholinesterase enzyme leading to more effective sensing of pesticides. CNTs-modified transducer has been used for the development of biosensors using an organophosphorus hydrolase (OPH) biocatalyst for detection of paraoxon (Prakash-Deo et al., 2005). The CNTs/OPH biosensor has been prepared by using a bilayer approach with the OPH layer atop the CNTs film. The CNTs-modified electrodes greatly improve anodic detection of the p-nitrophenol (product of the OPH reaction) including higher sensitivity and stability compared with the bare electrode. Wang et al. (2003) also reported carbon nanotube modified glassy carbon electrodes for stable and sensitive electrochemical detection of phenolic compounds by decreasing its electrocatalytic activity and minimizing the surface fouling associated with the oxidation process of the phenolic compounds. Another important OP compound to be detected is parathion, which is a nonsystemic organophosphorus insecticide and acaricide with some fumigant action. A Zirconium dioxide/gold (ZrO_2/Au) nanocomposite film electrode for detection of parathion has been described by Wang and Li (2008). The sensing performance of the ZrO_2/Au nanocomposite film electrode toward parathion was studied with square wave voltammetry. The nanofilm showed strong affinity to parathion obtaining a limit of detection (LOD) of 3 ng mL^{-1} in buffer solution pH 2.0–7.0. Llopis et al. (2009) developed an ultrasensitive method to determine pesticides in a lab-on-a-chip system based on enzymatic inhibition of acetylcholinesterase immobilized onto magnetic beads. This method was shown to be useful for carbofuran detection down to the nanomolar level in buffer solution, and it can be extended to other pesticides in connection also to the use of other enzymes and magnetic nanoparticles. Another biosensor for direct detection of OP neurotoxins such as paraoxon has been developed based on the change in fluorescence of a competitive inhibitor of the OPH enzyme when the inhibitor is displaced by the OP substrate (Simoniana et al., 2005). The change in fluorescence intensity was correlated with concentration of paraoxon presented in the solution. The sensitivity to paraoxon was obtained when enzyme inhibitor and OPH-Au nanoparticle conjugates were present at near equimolar levels.

3.4 Detection of Sugar

The real-time monitoring of carbohydrate levels is useful for the control of the fermentation process. Furthermore, sugars sometimes are limiting factors in fermentative processes given that low concentrations reduce the productivity of the bioreactor. In some cases such as fed batch cultures it is desirable to maintain low sugar concentrations. Studies have been carried out on the use of amperometric biosensors and also incorporation of nanoparticles with such biosensors to analyze glucose with glucose oxidase (GOx); lactose with β-galactosidase and GOx; and lactulose with β-galactosidase and fructose dehydrogenase (Serna-Cock et al., 2009). Biosensors based on Au nanoparticles with pyranose oxidase (PyOx), with high affinity for D-glucose (Kmw 1 mM), has been developed by Ozdemir et al. (2010) and applied for detection of sugar level in various fruit (pomegranate, peach, orange, and mixed fruit) juices. Au nanoparticles showed a good biocompatibility and ability for being functionalized, which give to the immobilized enzyme stable operational conditions and fast, stable, and sensitive responses in biosensing due to the retaining of the enzymatic activity of the protein. Anik et al. (2007) developed a microbial biosensor based on a CNT epoxy composite platform, which was used as a supporting electrode for *Pseudomonas fluorescens* immobilization, for glucose detection.

Lu et al. (2007) developed an amperometric biosensor using exfoliated graphite nanoplatelets (xGnanoparticles) decorated with Pt and Pd nanoparticles. These nanomaterials increase the electron active area of the electrode and at the same time decrease the overpotential in the detection of H_2O_2. Zhao et al. (2007) also developed an amperometric biosensor for the detection of glucose, by immobilizing GOx into the ZnO:Co nanocluster-assembled thin film, through Nafion-assisted crosslinking technique. The electrochemical responses of these biosensors have been of high sensitivity for glucose detection with a LOD of 1 and 20 mM(S/N¼3), respectively. Another amperometric biosensor for fructose determination in honey using CNTs as transducer nanomaterial has been reported by Antiochia et al. (2004). The biosensor was constructed by modifying the transducer of CNTs paste with an electropolymerized film of 3.4-dihydroxybenzaldehyde (3.4-DHB) and D-fructose dehydrogenase (FDH) immobilized onto a membrane placed on the top of the electrode surface. The detection of fructose with this biosensor in real samples gives a high current sensitivity, represents a linear calibration curve and its detection limit is of one order of magnitude lower than that obtained with conventional solid and paste electrodes. Tominaga et al. (2009) have studied the direct heterogeneous electron transfer reaction of fructose dehydrogenase (FDH) adsorbed onto MWCNTs placed onto platinum electrode and demonstrated a well-defined catalytic oxidation current in c. −0.15 V (versus Ag/AgCl/sat'd KCl), which was close to the redox potential of heme c as a prosthetic group of FDH. Li et al. (2009) developed a glucose biosensor using the photoluminescent CdTe QDs/GOx multilayer film, which was immobilized on an optically transparent substrate by LBL technique with a good optical response. H_2O_2 produced by the oxidation of glucose with glucose oxidase leads to quenching of QDs photoluminescence, thus the glucose concentration can be determined by monitoring the photoluminescence change. Another optical biosensing device for glucose sensing was developed by Wei and Wang (2008) based on catalytic reaction of Fe_3O_4 magnetic nanoparticles and H_2O_2 produced by the oxidation of glucose with glucose oxidase. The oxidation of a peroxidase substrate 2, 20-azino-bis(3-ethylbenzothiazoline-6-sulfonic acid) diammonium salt (ABTS) was catalyzed by Fe_3O_4 magnetic nanoparticles (MNPs) in presence of H_2O_2. This colorimetric method for glucose detection using GOx and Fe_3O_4 MNPs as artificial enzymes has enabled obtaining a sensitive and selective optical biosensing. Wu et al. (2006) reported silver-DNA hybrid nanoparticles for electrochemical sensing of H_2O_2 and glucose with good sensitivity for rapid detection of H_2O_2 in the absence of oxygen with a wide linear range and acceptable reproducibility. The silver-DNA hybrid nanoparticle was fabricated on an electrode surface by electro-reductive deposition of silver-DNA complex with the help of (sol-gel) ormosil for the immobilization of glucose oxidase. The amperometric glucose biosensor developed by Wang et al. (2009b) was based on the direct electron transfer of glucose oxidase (GOx) immobilized onto the inner wall of highly ordered polyaniline nanotubes (nanoPANi). The biosensor had been reported to have good electrocatalytic activity toward oxidation of glucose and a low detection limit with higher biological affinity. The nonenzymatic electrochemical glucose biosensor of Zhou et al. (2009), which uses a gold nanotube array electrode, showed efficient glucose detection limit upto 10 μM. Whereas, the biosensing system for sensing maltose developed by Sandros et al. (2006) was significant for the detection of sugar in food. It was based on the electron transfer quenching of unimolecular protein CdSe@ZnS nanoparticle assemblies, which is provided by a protein attached Ru^{II} complex.

3.5 Detection of Biotoxin

Analysis for biotoxin in foods is a very important practice to ensure food quality and safety and to eliminate and control the risk of consuming contaminated foods. Mycotoxins are toxic secondary metabolites naturally produced by certain molds. Detection of mycotoxin in foodstuffs is important because it can appear in food as a result of fungal contamination posing a risk for human health through mycotoxin poisoning. Many researchers have studied and reported several efficient biosensors for mycotoxin detection in foodstuffs. The advent of electronic noses has opened tremendous advantages in the food industry. Electronic noses are able to distinguish between moldy and nonmoldy samples. Jonsson et al. (1997) reported the use of electronic nose to classify oat samples as moldy, weakly musty, and strongly musty, and also to predict ergosterol levels and fungal colony-forming units (CFU) in oats, rye, barley, and wheat. Olsson et al. (2002) also used the electronic nose for detection of ochratoxin A, citrinin, and ergosterol production in wheat. Studies have been reported on the application of electronic noses for detection of mycotoxin formation by *Fusarium* strains (Falasconi et al., 2005; Presicce et al., 2006). There are several reports on the developments of the electronic nose based on nanomaterials, however, their application in industrial level is still a challenge (Ramgir, 2013). Maragos and Thompson (1999) developed a fiber-optics sensor for the detection of aflatoxin B1 based on evanescent waves. An optical sensor for detection of aflatoxins, based on fluorescence polarization was also reported by Nasir and Jolley (2002). Integration of nanomaterials with such biosensors can enhance the efficiency and also can save time by detecting multiple target analytes simultaneously (Min and Baeumner, 2004). Delmulle et al. (2005) demonstrated a simple nanobased method for determining aflatoxin B1 in pig feed using immunoassay-based lateral flow dipstick technique. In their study the dipsticks membrane was biofunctionalized with aflatoxin B1-bovine serum albumin (BSA) conjugate and rabbit anti-mouse antibodies as capture reagents and the detector reagent consisted of colloidal Au particles coated with affinity-purified monoclonal antiaflatoxin B1 antibodies, which saturated the conjugate pad (Delmulle et al., 2005). The color intensity decreased with increasing aflatoxin B1 concentration, a pink line appeared for blank pig feed samples.

Tang et al. (2009) reported biosensors based on magnetic nanoAu microspheres (MnGMs) provided the aflatoxin B2 detection in peanuts, hazelnuts, pistachios, and almonds. MnGMs were conjugated with monoclonal antiaflatoxin B2 antibodies as detector reagents by binding to the aflatoxin B2-BSA coating antigen (analyte) at the test line. The assay gave accurate results and moreover it is a simple operation method, which provides a high sensitivity due to the good synergy between the biomolecules and nanomaterials involved in the analysis.

Many researchers have integrated magnetic nanoparticles with ELISA to detect mycotoxin, food allergens, and pesticide residue. Speroni et al. (2010) described a protein A-coated magnetic microparticles-based ELISA method for the detection of Ara h3/4 peanut allergen in foods. This method had a detection limit as low as 0.2 mg/kg and achieved excellent precision and repeatability. Therefore, it has good applicability to detect and quantify peanut allergens.

Several protein-based bacterial toxins (Goldman et al., 2004), including bolutinum toxin serotype A (Warner et al., 2009), have been detected at picomolar (pM) levels using antibody-labeled luminescent quantum dots, which would be useful in food safety. The optical-based SPR system has been used for detection of whole cells of *E. coli* O157:H7, *Salmonella*, and *Listeria* and it also showed strong signals with small toxin molecules, such as staphylococcal or botulinum (Arora et al., 2011). Many electrochemical sensors operate by binding selective antibodies to a conductive nanomaterial such as CNT and then monitoring changes to the material's conductivity when the target analyte binds to the antibodies. For example, conduction changes that occur when microcystin-LR (MCLR), a toxin produced by cyanobacteria, binds to the surface of anti-MCLR-coated single-walled carbon nanotubes are easily detectable down to MCLR concentrations of 0.6 nM, which easily satisfies guidelines set by the World Health Organization for this substance in drinking water (Wang et al., 2009a); this technique improved the sampling time over traditional MCLR measurement methods such as ELISA by an order of magnitude.

It has been reported that Au nanoparticles are very effective detectors of cholera toxin secreted by the bacteria *Vibrio cholera*, which causes acute infection of the intestine (Schofield et al., 2007). Their bioassay comprised of 16 nm Au nanoparticles coated with thiolated lactose where the SH group anchors the lactose to the Au nanoparticles surface. The thiolated lactose derivative plays the role of molecular recognition ligand and mimics the GM1 ganglioside—the receptor to which cholera toxin binds in the small intestine. The lactose nanoparticles solution has a surface plasmon band centered on 520 nm and appears red in color. When the cholera toxin binds to the lactose derivative, aggregation of the nanoparticles is induced. The aggregation causes a broadening and a red shift of the surface plasmon absorption band such that the nanoparticles solution appears deep purple color. The bioassay could detect the toxin within 10 min and provide quantitative toxin concentration data. The lactose stabilized nanoparticles were stable on freeze drying and could be resuspended in water to detect the cholera toxin in an electrolyte solution that mimicked the watery stool that is typical of cholera patients to test the selectivity for the toxin. This bioassay provides a rapid and simple colorimetric tool for the measurement of the infectious bacterial cholera toxin.

3.6 Detection of Gas

Excess moisture and oxygen allows growth of aerobic microorganisms leading to food spoilage. Assays for detection of vapor or gas or O_2 content inside of a package require destruction of the package (Smolander et al., 1997; Arndt, 2008). Therefore, packaged foods are randomly tested for gas content during a production run, that is, one in every 300–400 packages (Mills, 2005). This testing process is expensive and time-consuming; however, it does not ensure that packages meet quality and safety standards (Luechinger et al., 2007). The development of an efficient and easy method for noninvasive leak detection and gas content of food packages, which provide a means to maintain quality and safety of the foodstuffs long after it has left the production facility, is important for food safety for human consumption (Arndt, 2008). Numerous noninvasive gas-sensing methods based on nanotechnology have been developed. Many researchers have developed promising photoactivated indicator inks for in-package oxygen detection based on nanomaterials. Lee et al. (2005) developed a UV-activated colorimetric oxygen indicator using titania (TiO_2) nanoparticles to photosensitize the reduction of methylene blue (MB) by triethanolamine in a polymer encapsulation medium. Mills and Hazafy (2009) also demonstrated a colorimetric O_2 indicator using nanocrystalline tin dioxide (SnO_2) as a photosensitizer to reduce a redox dye (MB) in encapsulating polymer (hydroxyethyl cellulose). In both the studies, the detectors gradually change color in response to even minute quantities of oxygen. Several other gas sensing methods related to food safety have been developed. Von Bultzingslowen et al. (2002) have devised a noninvasive method of measuring carbon dioxide content based on lifetime analysis of luminescent dyes standardized by fluorophore-encapsulated polymer nanobeads. This CO_2 sensor has a detection range of 0.8–100%, a resolution of 1%, and only 0.6% cross-sensitivity with molecular oxygen. Gaseous amines are indicators of fish and meat spoilage (Hernandez-Jover et al., 1996). Studies have reported detection of gaseous amines at the parts-per-trillion (ppt) level (theoretical) using fluorescence quenching of nanofibrils of perylene-based fluorophores (Che et al., 2008; Che and Zang, 2009) or at the ppm level using conductance changes in composites of SnO_2 nanoparticles and TiO_2 microrods (Zhang and Zhang, 2008). Also, a number of electronic sensors have been developed, which utilize ZnO-TiO_2 nanocomposites or SnO_2 nanobelts to detect the presence of volatile organics, including acetone, ethanol, and carbon monoxide (Comini et al., 2005; Comini et al., 2006; Barreca et al., 2007). Tungsten oxide-tin oxide (WO_3-SnO_2) nanocomposites have also been used to detect the presence of ethylene gas, a hormone responsible for fruit ripening (Pimtong-Ngam et al., 2007).

3.7 Detection of Carcinogen

Three approaches involving nanoparticles in the biosensor systems have been developed for the detection of carcinogens: as enzyme carriers, as direct signal source or signal amplification, and as catalysts (peroxidase or oxidase-like activity) (Iranifam, 2013; Yi et al., 2013; Hu et al., 2014; Lin et al., 2014). To achieve high sensitivity for the determination of carcinogens, many amplification strategies have been developed. Au nanoparticles are ideal transducers for biorecognition as it can be easily conjugated with biomolecules and retain the biochemical activity (Cao et al., 2011). An Au@AgNPs film was embedded in graphene oxide (GO), and then Au@AgNPs were immobilized again to form an Au@Ag–GO–Au@Ag sandwich structure (Zhang et al., 2013) and used for SER detection. A label-free fluorescent sensor involving quenching and reappearing processes, based on a CNT–Ag nanoclusters system, was used for determination of Hg^{2+} with exonuclease III activity on DNA hybrids containing thymine–Hg^{2+}–thymine (Wang et al., 2014). An oligonucleotide-functionalized Au nanoparticles-enhanced QCM biosensor was fabricated for Hg^{2+} determination. The sensor has detection limits as low as 4 nmol L^{-1}, and thus able to monitor Hg^{2+} in drinking-water samples (Chen et al., 2011). A highly selective colorimetric detection method, via direct formation of Au nanoparticles was used to detect dissolved Hg^{2+} ions, with a minimum detectable concentration of 35 nmol L^{-1}(Chansuvarn and Imyim, 2012). Au nanoparticles can also be used as a signal amplifier to detect heavy-metal ions (Jin et al., 2013). A colorimetric lead sensor was developed based on the assembly of Au nanoparticles by a Pb^{2+}-dependent DNAzyme by Liu and Lu (2003). Wang et al. (2008) also reported a colorimetric method for lead sensing using combination of both unlabeled DNAzyme and Au nanoparticles with a detection limit of 3 nM.

3.8 Detection of Genetically Modified Organisms

The increasing number of genetically modified organisms (GMO) derived products has led to specific legislation worldwide to regulate and label the presence or absence of GMO in foodstuff. Therefore, there is a growing interest in the development of reliable and rapid analytical methods for detection of GMO in foodstuffs (Erickson, 2000). The need for simple, efficient, and rapid methods has stimulated the development of nanobased biosensors for GMO screening. Detection methods have wide applications because screening is the first step of any procedure for GMO testing of food. For detection of GMO, Mariotti et al. (2002) fabricated a DNA hybridization based SPR affinity biosensor by immobilizing single-stranded DNA probes on the sensor chip of an SPR device. A SERS-barcoded nanosensor was also developed for detection of GMO by Chen et al. (2012), which was designed by encapsulation of Au nanoparticles with silica and conjugation of oligonucleotide strands for targeting DNA strands. Kalogianni et al. (2006) developed a nanobased DNA biosensor in a dry-reagent dipstick configuration for visual detection and confirmation of GMO-related sequences by hybridization within minutes. Oligonucleotide-conjugated Au nanoparticles were used as probes, which constitute an integral part of the biosensor. The sensor is disposable and does not require special instrumentation.

4 Conclusions

This chapter reviews the advances in the development and application of biosensors in the food sector. Improved detection and analysis of microbial and chemical contamination in foods achieved with these techniques ensure food quality and safety. With the advances in nanobiotechnology, biosensors, and innovative diagnostic devices, numerous new food analytical devices have been developed that showed better sensitivity, selectivity, cost effectiveness, and less time. Nanomaterial-based biosensors ensure enhanced efficiency and robust applications in the process of food analyses by providing devices that are highly versatile, easy to use, inexpensive, and portable.

Acknowledgment

The authors gratefully acknowledge the Department of Biotechnology (DBT), Ministry of Science and Technology, Government of India for providing financial assistance through DBT-Research Associateship Program to the first author (SDL) and Department of Biotechnology, Mizoram University, India for the infrastructural support.

References

Abad, J.M., Pariente, F., Hernández, L., Abruña, H.D., Lorenzo, E., 1998. Determination of organophosphorus and carbonate using a piezoelectric biosensor. Anal. Chem. 70, 2848–2855.

Abbaspour, A., Norouz-Sarvestani, F., Noori, A., Soltani, N., 2015. Aptamer-conjugated silvernanoparticles for electrochemical dual-aptamer-based sandwich detection of *Staphylococcus aureus*. Biosens. Bioelectron. 68, 149–155.

Agasti, S.S., Rana, S., Park, M.H., Kim, C.K., You, C.C., Rotello, V.M., 2010. Nanoparticles for detection and diagnosis. Adv. Drug Delivery Rev. 62, 316–328.

Allen, M.J., Edberg, S.C., Reasoner, D.J., 2004. Heterotrophic plate count bacteria-what is their significance in drinking water? Int. J. Food Microbiol. 92, 265–274.

Anik, U., Timur, S., Odaci, D., Perez, B., Alegret, S., Merkoci, A., 2007. Carbon nanotube composite as novel platform for microbial biosensor. Electroanal. 19 (7–8), 893–898.

Antiochia, R., Lavagnini, I., Magno, F., 2004. Amperometric mediated carbon nanotube paste biosensor for fructose determination. Anal. Lett. 37 (8), 1657–1669.

Arndt, G.W., 2008. Leak testing. In: Yam, K.L. (Ed.), The Wiley Encyclopedia of Packaging Technology. third ed. John Wiley and Sons, Inc, New York.

Arora, P., Sindhu, A., Dilbaghi, N., Chaudhury, A., 2011. Biosensors as innovative tools for the detection of food borne pathogens. Biosens. Bioelectron. 28 (1), 1–12.

Babacan, S., Pivarnik, P., Letcher, S., Rand, A., 2002. Piezoelectric flow injection analysis biosensor for detection of *Salmonella typhimurium*. J. Food Sci. 67, 314–320.

Bae, Y.M., Oh, B.K., Lee, W., Lee, W.H., Choi, J.W., 2004. Detection of insulin–antibody binding on a solid surface using imaging ellipsometry. Biosens. Bioelectron. 20 (4), 890–895.

Baeummer, A., 2004. Nanosensors identify pathogens in food. Food Technol. 58, 51–55.

Baeumner, A.J., Cohen, R.N., Miksic, V., Min, J., 2003. RNA biosensor for the rapid detection of viable *Escherichia coli* in drinking water. Biosens. Bioelectron. 18, 405–413.

Barreca, D., Comini, E., Ferrucci, A.P., Gasparatto, A., Maccato, C., Maraguo, C., Sberveglieri, G., Tondello, E., 2007. First example of ZnO-TiO_2 nanocomposites by chemical vapour deposition: structure, morphology, composition and gas sensing performances. Chem. Mater. 19, 5642–5649.

Bhunia, A.K., 2007. Biosensors and bio-based methods for the separation and detection of foodborne pathogens. Adv. Food Nutr. Res. 54, 1–44.

Borck, B., Stryhn, H., Ersboll, A.K., Pedersen, K., 2002. Thermophilic *Campylobacter* spp. in turkey samples: evaluation of two automated enzyme immunoassays and conventional microbiological techniques. J. Appl. Microbiol. 92, 574–582.

Cai, H., Xu, C., He, P., Fang, Y., 2001. Colloid Au-enhanced DNA immobilization for the electrochemical detection of sequence-specific DNA. J. Electroanal. Chem. 510, 78–85.

Cao, X., Ye, Y., Liu, S., 2011. Gold nanoparticle-based signal amplification for biosensing. Anal. Biochem. 417 (1), 1–16.

Carlin, F., Broussolle, V., Perelle, S., Litman, S., Fach, P., 2004. Prevalence of *Clostridium botulinum* in food raw materials used in REPFEDs manufactured in France. Int. J. Food Microbiol. 91, 141–145.

Carter, R.M., Mekalanos, J.J., Jacobs, M.B., Lubrano, G.J., Guilbault, G.G., 1995. Quartz crystal microbalance detection of *Vibrio cholerae* O139 serotype. J. Immunol. Methods 187, 121–125.

Chansuvarn, W., Imyim, A., 2012. Visual and colorimetric detection of mercury (II) ion using gold nanoparticles stabilized with a dithiadiaza ligand. Microchim. Acta 176 (1–2), 57–64.

Che, Y., Yang, X., Loser, S., Zang, L., 2008. Expedient vapor probing of organic amines using fluorescent nanofibers fabricated from an n-type organic semiconductor. Nano Lett. 8, 2219–2223.

Che, Y., Zang, L., 2009. Enhanced fluorescence sensing of amine vapor based on ultrathin nanofibers. Chem. Commun. 34, 5106–5108.

Chen, Q., Wu, X., Wang, D., Tang, W., Li, N., Liu, F., 2011. Oligonucleotide-functionalized gold nanoparticles-enhanced QCM-D sensor for mercury (II) ions with high sensitivity and tunable dynamic range. Analyst 136 (12), 2572–2577.

Chen, K., Han, H., Luo, Z., Wang, Y., Wang, X., 2012. A practicable detection system for genetically modified rice by SERS-barcoded nanosensors. Biosens. Bioelectron. 34, 118–124.

Cheng, H.W., Huan, S.Y., Wu, H.L., Shen, G.L., Yu, R.Q., 2009. Surface-enhanced Raman spectroscopic detection of a bacteria biomarker using gold nanoparticle immobilized substrates. Anal. Chem. 81, 9902–9912.

Cheung, P.Y., Kam, K.M., 2012. *Salmonella* in food surveillance: PCR, immunoassays, and other rapid detection and quantification methods. Food Res. Int. 45, 802–808.

Cho, I.H., Irudayaraj, J., 2013. In-situ immuno-gold nanoparticle network ELISA biosensors for pathogen detection. Int. J. Food Microbiol. 164, 70–75.

Chu, P.-T., Lin, C.-S., Chen, W.-J., Chen, C.-F., Wen, H.-W., 2012. Detection of gliadin in foods using a quartz crystal microbalance biosensor that incorporates gold nanoparticles. J. Agric. Food Chem. 60, 6483–6492.
Comini, E., Faglia, G., Sberveglieri, G., Calestani, D., Zanotti, L., Zha, M., 2005. Comparison of single and binary oxide MoO_3, TiO_2 and WO_3 sol–gel gas *sensors*. Sens. Actuators B111, 2.
Comini, E., Faglia, G., Sberveglieri, G., Zanotti, L., 2006. Oxide nanobelts as conductometric gas sensors. Mater. Manuf. Process 21, 229–232.
Comparelli, R., Curri, M.L., Cozzoli, P.D., Striccoli, M., 2007. Optical biosensing based on metal and semiconductor colloidal nanocrystals. Kumar, C.S.S.R. (Ed.), Nanotechnologies for the Life Sciences: Nanomaterials for Biosensors, vol. 8, Wiley-VCH, Weinheim, pp. 123–174.
Crumbliss, A.L., Perine, S.C., Stonehuerner, J., Tubergen, K.R., Zhao, J., Henkens, R.W., 1992. Colloidal gold as a biocompatible immobilization matrix suitable for the fabrication of enzyme electrodes by electrodeposition. Biotechnol. Bioeng. 40, 483.
Curri, M.L., Agostiano, A., Leo, G., Mallardi, A., Cosma, P., Della, M., 2002. Development of a novel enzyme/semiconductor nanoparticles system for biosensor application. Mater. Sci. Eng. C 22, 449–452.
Curtis, T., Naal, R.M., Batt, C., Tabb, J., Holowka, D., 2008. Development of a mast cell-based biosensor. Biosens. Bioelectron. 23, 1024–1031.
Delmulle, B.S., De Saeger, S.M.D.G., Sibanda, L., Barna-Vetro, I., Van Peteghem, C.H., 2005. Development of an immunoassay-based lateral flow dipstick for the rapid detection of aflatoxin B_1 in pig feed. J. Agri. Food Chem. 53 (9), 3364–3368.
Demers, L.M., Mirkin, C.A., Mucic, R.C., Reynolds, R.A., Letsinger, R.L., Elghanian, R., Viswanadhan, G., 2000. A fluorescence-based method for determining the surface coverage and hybridization efficiency of thiol-capped oligonucleotides bound to gold thin films and nanoparticles. Anal. Chem. 72 (22), 5535–5541.
Driskell, J.D., Tripp, R.A., 2009. Emerging technologies in nanotechnology-based pathogen detection. Clin. Micbrobiol. Newsl. 31, 137–144.
Dungchaia, W., Siangprohb, W., Chaicumpac, W., Tongtawed, P., Chailapakula, O., 2008. *Salmonella typhi* determination using voltammetric amplification of nanoparticles: a highly sensitive strategy for metalloimmunoassay based on a copper-enhanced gold label. Talanta 77, 727–732.
Dzyadevych, S.V., Arkhypova, V.N., Soldatkin, A.P., Elskaya, A.V., Martelet, C., Jaffrezic-Renault, N., 2008. Amperometric enzyme biosensors: past, present and future. IRBM 29 (2–3), 171–180.
El-Boubbou, K., Gruden, C., Huang, X., 2007. Magnetic glyconanoparticles: a unique tool for rapid pathogen detection, decontamination, and strain differentiation. J. Amer. Chem. Soc. 129, 13392–13393.
Erickson, B.E., 2000. Detecting genetically modified products in food. Anal. Chem. 72, 454A–459A.
Falasconi, M., Gobbi, E., Pardo, M., Torre, M.D., Bresciani, A., Sberveglieri, G., 2005. Development of an electronic nose for environmental odour monitoring. Sens. Actuators B 108, 250.
Fredriksson-Ahoma, M., Stolle, A., Stephan, R., 2007. Prevalence of pathogenic in pigs slaughtered *Yersinia enterocolitica* at a Swiss abattoir. Int. J. Food Microbiol. 119, 207–212.
Garcia, M., Aleixandra, M., Gutierrez, J., Horrillo, M.C., 2006. Electronics nose for wine discrimination. Sensors Actuat. B 113, 911–916.
García-Cano, I., Serrano-Maldonado, C.E., Olvera-García, M., Delgado-Arciniega, E., Pena-Montes, C., Mendoza-Hernandez, G., 2014. Antibacterial activity produced by *Enterococcus* spp. isolated from an artisanal Mexican dairy product, Cotija cheese. LWT - Food Sci. Tech. 59, 26–34.

Geornars, I., Yoon, Y., Belk, K.E., Smith, G.C., Sofos, J.N., 2007. Antimicrobial activity of ε-Polylysine against *Escherichia coli* O157:H7, *Salmonella Typhimurium,* and *Listeria monocytogenes* in various food extracts. J. Food Sci. 72 (8), 330–334.

Ghosh, M., Wahi, S., Kumar, M., Ganguli, A., 2007. Prevalence of enterotoxigenic *Staphylococcus aureus* and *Shigella* spp. in some raw street vended Indian foods. Int. J. Environ. Health Res. 17, 151–156.

Goldman, E.R., Clapp, A.R., Anderson, G.P., Uyeda, H.T., Mauro, J.M., Medintz, I.L., Mattoussi, H., 2004. Multiplexed Toxin Analysis using four colors of quantum dot fluororeagents. Anal. Chem. 76, 684–688.

Gong-Jun, Y., Jin-Lin, H., Wen-Jing, M., Ming, S., Xin-An, J., 2009. A reusable capacitive immunosensor for detection of *Salmonella* spp. based on grafted ethylene diamine and self-assembled gold nanoparticle monolayers. Analytica Chimica Acta 647, 159–166.

Goriushkina, T.B., Soldatkin, A.P., Dzyadevych, S.V., 2009. Application of amperometric biosensors for analysis of ethanol, glucose, and lactate in wine. J. Agri. Food Chem. 57 (15), 6528–6535.

Haes, A.J., Van-Duyne, R.P., 2002. A nanoscale optical biosensor: sensitivity and selectivity of an approach based on the localized surface plasmon resonance spectroscopy of triangular silver nanoparticles. J. Amer. Chem. Society 124 (35), 10596–10604.

Hao, R., Wang, D., Zhang, X., Zuo, G., Wei, H., Yang, R., Zhang, Z., Cheng, Z., Guo, Y., Cui, Z., Zhou, Y., 2009. Rapid detection of *Bacillus anthracis* using monoclonal antibody functionalized QCM sensor. Biosens. Bioelectron. 24, 1330–1335.

Hernandez-Jover, T., Izquierdo-Pulido, M., Veciana-Nogues, M.T., Vidal-Carou, M.C., 1996. Ion-pair high-performance liquid chromatographic determination of biogenic amines in meat and meat products. J. Agri. Food Chem. 44, 2710–2715.

Hu, Q., Xu, X., Li, Z., Zhang, Y., Wang, J., Fu, Y., Li, Y., 2014. Detection of acrylamide in potato chips using a fluorescent sensing method based on acrylamide polymerization-induced distance increase between quantum dots. Biosens. Bioelectron. 54, 64–71.

Huh, Y.S., Chung, A.J., Erickson, D., 2009. Surface enhanced Raman spectroscopy and its application to molecular and cellular analysis. Microfluid. Nanofluid. 6, 285–297.

Invitski, D., Abdel-Hamid, I., Atanasov, P., Wilkins, E., Striker, S., 2000. Application of electrochemical biosensors for detection of food pathogenic bacteria. Electroanal. 12 (5), 317–325.

Iranifam, M., 2013. Analytical applications of chemiluminescence detection systems assisted by magnetic microparticles and nanoparticles. TrAC Trend. Anal. Chem. 51, 51–70.

Jena, B.K., Raj, C.R., 2006. Electrochemical biosensor based on integrated assembly of dehydrogenase enzymes and gold nanoparticles. Anal. Chem. 78 (18), 6332–6339.

Ji, X., Zheng, J., Xu, J., Rastogi, V.K., Cheng, T.-C., DeFrank, J.J., 2005. (CdSe)ZnS quantum dots and organophosphorus hydrolase bioconjugate as biosensors for detection of paraoxon. Phy. Chem. 109, 3793–3799.

Jin, Y., Huang, Y., Liu, G., Zhao, R., 2013. Gold nanoparticle sensitized quartz crystal microbalance sensor for rapid and highly selective determination of Cu (II) ions. Analyst 138 (18), 5479–5485.

Jonsson, A., Winquist, F., Schnurer, J., Sundgren, H., Lundstrom, I., 1997. Electronic nose for microbial quality classification of grains. Int. J. Food Microbiol. 35, 187.

Kalele, S.A., Kundu, A.A., Gosavi, S.W., Deobagkar, D.N., Deobagkar, D.D., Kulkarni, S.K., 2006. Rapid detection of *Escherichia coli* by using antibody-conjugated silver nanoshells. Small 2 (3), 335–338.

Kalogianni, D.P., Koraki, T., Christopoulos, T.K., Ioannou, P.C., 2006. Nanoparticle-based DNA biosensor for visual detection of genetically modified organisms. Biosens. Bioelectron. 21 (7), 1069–1076.

Karabıyıklı, S., De_girmenci, H., Karapınar, M., 2014. Inhibitory effect of sour orange (*Citrus aurantium*) juice on *Salmonella typhimurium* and *Listeria monocytogenes*. LWT - Food Sci. Tech. 55, 421–425.

Kim, H.J., Bennetto, H.P., Halablab, M.A., Choi, C., Yoon, S., 2006. Performance of an electrochemical sensor with different types of liposomal mediators for the detection of hemolytic bacteria. Sens. Actuators B Chem. 119, 143–149.

Ko, S., Grant, S.A., 2006. A novel FRET-based optical fiber biosensor for rapid detection of *Salmonella Typhimurium*. Biosens. Bioelectron. 21, 1283–1290.

Lange, D., Hagleitner, C., Hierlemann, A., Brand, O., Baltes, H., 2002. Complementory metal oxide semiconductor cantilever arrays on a single chip: mass-sensitive detection of volatile organic compounds. Anal. Chem. 74, 3084–3095.

Lee, S.-K., Sheridian, M., Mills, A., 2005. Novel UV-activated colorimetric oxygen indicator. Chem. Mater. 17, 2744–2751.

Lee, H.J., Kim, B.C., Kim, K.W., Kim, Y.K., Kim, J., Oh, M.K., 2009. A sensitive method to detect Escherichia coli based on immunomagnetic separation and real-time PCR amplification of aptamers. Biosens. Bioelectron. 24, 3550–3555.

Lee, S.H., Stubbs, D.D., Cairney, J., Hunt, W.D., 2005. Rapid detection of bacterial spores using a quartz crystal microbalance (QCM) immunoassay. IEEE Sens. J. 5, 737–743.

Leonard, P., Hearty, S., Brennan, J., Dunne, L., Quinn, J., Chakraborty, T., 2003. Advances in biosensors for detection of pathogens in food and water. Enzyme Microbial. Tech. 32, 3–13.

Li, A., Zhang, H., Zhang, X., Wang, Q., Tian, J., Li, Y., 2011. Rapidseparation and immunoassay for low levels of *Salmonella* in foods using magnetosomeeantibody complex and real-time fluorescence quantitative PCR. J. Separation Sci. 33, 3437–3443.

Li, H., Rothberg, L.J., 2004. Label-free colorimetric detection of specific sequences in genomic DNA amplified by the polymerase chain reaction. J. Am. Chem. Soc. 126 (35), 10958–10961.

Li, J.H., Hu, J.B., 2004. Functional gold nanoparticle-enhanced electrochemical determination of DNA hybridization and sequence-specific analysis. Acta Chim. Sinica 62, 2081–2088.

Li, X., Zhou, Y., Zheng, Z., Yue, X., Dai, Z., Liu, S., Tang, Z., 2009. Glucose biosensor based on nanocomposite films of CdTe quantum dots and glucose oxidase. Langmuir 25 (11), 6580–6586.

Lin, Y., Ren, J., Qu, X., 2014. Nano-gold as artificial enzymes: hidden talents. Adv. Mater. 26 (25), 4200–4217.

Lin, Y.H., Chen, S.H., Chuang, Y.C., Lu, Y.C., Shen, T.Y., Chang, C.A., 2008. Disposable amperometric immunosensing strips fabricated by Au nanoparticles-modified screen printed carbon electrodes for the detection of foodborne pathogen *Escherichia coli* O157:H7. Biosens. Bioelectron. 23, 1832–1837.

Liu, J.W., Lu, Y., 2003. A colorimetric lead biosensor using DNAzyme-directed assembly of gold nanoparticles. J. Am. Chem. Soc. 125 (22), 6642–6643.

Liu, S., Chen, Y., Liu, Z., Hu, X., Wang, F., 2006. A highly sensitive resonance rayleigh scattering method for the determination of Vitamin B1 with gold nanoparticles probe. Microchimica Acta 154, 87–93.

Llopis, X., Pumera, M., Alegret, S., Merkoci, A., 2009. Lab-on-a chip for ultrasensitive detection of carbofuran by enzymatic inhibition with replacement of enzyme using magnetic beads. Lab on a Chip 9, 213–218.

Lu, J., Do, I., Drzal, L.T., Worden, R.M., Lee, I., 2007. Nanometal decorated exfoliated graphite nanoplatelet based glucose biosensors with high sensitivity and fast response. Biosens. Bioelectron. 23, 1825–1832.

Luechinger, N.A., Loher, S., Athanassiou, E.K., Grass, R.N., Stark, W.J., 2007. Highly sensitive optical detection on polymer/metal nanoparticle hybrid films. Langmiur 23, 3473–3477.

Mabeck, J.T., Malliaras, G.G., 2006. Chemical and biological sensors based on organic thin film transistors. Anal. Bioanal. Chem. 384, 343–353.

Mannino, S., Wang, J., 2005. Electrochemical methods for food and drink analysis. Electroanalysis 4, 835–840.

Manso, J., Mena, M.L., Yanez-Sedeno, P., Pingarron, J.M., 2008. Alcohol dehydrogenase amperometric biosensor based on a colloidal gold–carbon nanotubes composite electrode. Electrochimica Acta 53 (11), 4007–4012.

Mao, X., Yang, L., Su, X.L., Li, Y., 2006. A nanoparticle amplification based quartz crystal microbalance DNA sensor for detection of *Escherichia coli* O157:H7. Biosens. Bioelectron. 21, 1178–1185.

Maragos, C.M., Thompson, V.S., 1999. Fiber-optic immunosensor for mycotoxins. Natural Toxins 7, 371–376.

Mariotti, E., Minunni, M., Mascini, M., 2002. Surface plasmon resonance biosensor for genetically modified organisms detection. Anal. Chem. Acta. 453 (2), 165–172.

Mills, A., Hazafy, D., 2009. Nanocrystalline SnO2-based, UVB-activated, colourimetric oxygen indicator. Sens. Actuators B 136, 344.

Mills, A., 2005. Oxygen indicators and intelligent inks for packaging food. Chem. Soc. Rev. 34, 1003–1011.

Min, J.H., Baeumner, A.J., 2004. Characterization and optimization of interdigitated ultra microelectrode arrays as electrochemical biosensor transducers. Electroanal. 16, 724–729.

Mora, A., Comejo, J., Revilla, E., Hermosin, M.C., 1996. Persistence and degradation of carofuran in Spanish soil suspensions. Chemosphere 32, 1585–1598.

Muhammad-Tahir, Z., Alocilja, E.C., 2003. A conductometric biosensor for biosecurity. Biosens. Bioelectron. 18 (5–6), 813–819.

Narayanaswamy, R., Wolfbeis, O.S. (Eds.), 2004. Optical Sensors: Industrial, environmental and diagnostic applications. Anal. Bioanal. Chem. 381(1), 18--19.

Nasir, M.S., Jolley, M.E., 2002. Development of a fluorescence polarization assay for the determination of aflatoxins in grains. J. Agri. Food Chem. 50, 3116–3121.

Ngo, V.K.T., Nguyen, D.G., Nguyen, H.P.U., Tran, V.M., Nguyen, T.K.M., Huynh, T.P., Lam, Q.V., Huynh, T.D., Truong, T.N.L., 2014. Quartz crystal microbalance (QCM) as biosensor for the detecting of *Escherichia coli* O157:H7. Adv. Nat. Sci.: Nanosci. Nanotechnol. 5, 045004–045011.

Norman, R.S., Stone, J.W., Gole, A., Murphy, C.J., Sabo-Attwood, T.L., 2008. Targeted Photothermal Lysis of the pathogenic bacteria, *Pseudomonas aeruginosa*, with gold nanorods. Nano Lett. 8, 302–306.

Olsson, J., Borjesson, T., Lundstedt, T., Schnurer, J., 2002. Detection and quantification of ochratoxin A and deoxynivalenol in barley grains by GC-MS and electronic nose. Int. J. Food Microbiol. 72, 203–214.

Ozdemir, C., Yeni, F., Odaci, D., Timur, S., 2010. Electrochemical glucose biosensing by pyranose oxidase immobilized in gold nanoparticle-polyaniline/AgCl/gelatin nanocomposite matrix. Food Chem. 119, 380–385.

Pal, S., Alocilja, E.C., 2009. Electrically active polyaniline coated magnetic (EAPM) nanoparticle as novel transducer in biosensor for detection of *Bacillus anthracis* spores in food samples. Biosens. Bioelectron. 24, 1437–1444.

Patel, I.S., Premasiri, W.R., Moir, D.T., Ziegler, L.D., 2008. Barcoding bacterial cells: a SERS based methodology for pathogen identification. J. Raman Spectrosc. 39, 1660–1672.
Patolsky, F., Zheng, G.F., Hayden, O., Lakadamyali, M., Zhuang, X.W., Lieber, C.M., 2004. Electrical detection of single viruses. Proceedings of the National Academy of Sciences of the United States of America, 101, pp. 14017–14022.
Pellegrini, G.E., Carpico, G., Coni, E., 2004. Electrochemical sensor for the detection and presumptive identification of quinolone and tetracycline residues in milk. Anal. Chem. Acta 520, 13–18.
Perez-Lopez, B., Merkoci, A., 2011. Nanomaterials based biosensors for food analysis applications. Trend. Food Sci. Tech. 22, 625–639.
Pilolli, R., Monaci, L., Visconti, A., 2013. Advances inbiosensor development based on integrating nanotechnology and applied to food-allergen management. Trend. Anal. Chem. 47, 12–26.
Pimtong-Ngam, Y., Jiemsirilers, S., Supothina, S., 2007. Preparation of tungsten oxide – tin oxide nanocomposites and their ethylene sensing characteristics. Sens. Actuators A139, 7–11.
Prakash-Deo, R., Wang, J., Block, I., Mulchandani, A., Joshi, K.A., Trojanowicz, M., 2005. Determination of organophosphate pesticides at a carbon nanotube/organophosphorus hydrolase electrochemical biosensor. Anal. Chim. Acta 530, 185–189.
Presicce, D.S., Forleo, A., Taurino, A.M., Zuppa, M., Siciliano, P., Laddomada, B., Logrieco, A., Visconti, A., 2006. Response evaluation of an E-nose towards contaminated wheat by *Fusarium poae* fungi. Sens. Actuators B 118, 433–438.
Ramgir N.S., 2013. Electronic nose based on nanomaterials: issues, challenges, and prospects. ISRN Nanomaterials 2013, 21. <http://dx.doi.org/10.1155/2013/941581>
Sandros, M.G., Shete, V., Benson, D.E., 2006. Selective, reversible, reagentless maltose biosensing with core-shell semiconducting nanoparticles. Analyst 131, 229–235.
Sanvicens, N., Pascual, N., Fernandez-Argüelles, M.T., Adrian, J., Costa-Fernandez, J.M., Sanchez-Baeza, F., 2011. Quantum dot-based array for sensitive detection of *Escherichia coli*. Anal. Bioanal. Chem. 399, 2755–2762.
Sanvicens, N., Pastells, C., Pascual, N., Marco, M., 2009. Nanoparticle-based biosensors for detection of pathogenic bacteria. TrAC Trends Anal. Chem. 28, 1243–1252.
Schofield, C.L., Field, R.A., Russell, D.A., 2007. Glyconanoparticles for the colorimetric detection of cholera toxin. Anal. Chem. 79, 1356–1361.
Selvin, P.R., 2002. Principles and biophysical applications of lanthanide-based probes. Biophysics 31, 275–302.
Serna-Cock, L., Zetty-Arenas, A.M., Ayala-Aponte, A., 2009. Use of enzymatic biosensors as quality indices: a synopsis of present and future trends in the food industry. Chilean J. Agri. Res. 69 (2), 270–280.
Simonian, A.L., Rainina, E.I., Wild, J.R., 1997. A new approach for discrimination detection of organophosphate neurotoxins in the presence of other cholinesterase inhibitors. Anal. Lett. 30, 2453–2468.
Simoniana, A.L., Good, T.A., Wang, S.-S., Wild, J.R., 2005. Nanoparticle-based optical biosensors for the direct detection of organophosphate chemical warfare agents and pesticides. Analytica Chimica Acta 534, 69–77.
Simpson-Stroot, J.M., Kearns, E.A., Stroot, P.G., Magana, S., Lim, D.V., 2008. Monitoring biosensor capture efficiencies: development of a model using GFP-expressing *Escherichia coli* O157:H7. J. Microbiol. Meth. 72 (1), 29–37.
Smolander, M., Hurme, E., Ahvenainen, R., 1997. Leak indicators for modified-atmosphere packages. Trend. Food Sci. Technol. 8, 101–106.

So, H.M., Park, D.W., Jeon, E.K., Kim, Y.H., Kim, B.S., Lee, C.K., 2008. Detection and titer estimation of *Escherichia coli* using aptamer-functionalized single-walled carbon-nanotube field effect transistors. Small 4, 197–201.

Speroni, F., Elviri, L., Careri, M., Mangia, A., 2010. Magnetic particles functionalized with PAMAM-dendrimers and antibodies: a new system for an ELISA method able to detect Ara h3/4 peanut allergen in foods. Anal. Bioanal. Chem. 397, 3035–3042.

Taitt, C.R., Anderson, G.P., Ligler, E.S., 2005. Evanescent wave fluorescence biosensors. Biosens. Bioelectron. 20, 2470–2487.

Tang, D., Sauceda, J.C., Lin, Z., Ott, S., Basova, E., Goryacheva, I., Biselli, S., Lin, J., Niessner, R., Knopp, D., 2009. Magnetic nanogold microspheres-based lateral-flow immunodipstick for rapid detection of aflatoxin B2 in food. Biosens. Bioelectron. 25 (2), 514–518.

Tok, J.B.H., Chuang, F.Y.S., Kao, M.C., Rose, K.A., Pannu, S.S., Sha, M.Y., 2006. Metallic striped nanowires as multiplexed immunoassay platforms for pathogen detection. Angew. Chem. Int. Ed. 45, 6900–6904.

Tominaga, M., Nomura, S., Taniguchi, I., 2009. D-Fructose detection based on the direct heterogeneous electron transfer reaction of fructose dehydrogenase adsorbed onto multi-walled carbon nanotubes synthesized on platinum electrode. Biosens. Bioelectron. 24, 1184–1188.

Tothill, I.E., 2001. Biosensors developments and potential applications in the agricultural diagnosis sector. Comput. Electron Agr. 30, 205–218.

Tripp, R.A., Dluhy, R.A., Zhao, Y., 2008. Novel nanostructures for SERS biosensing. Nanotoday 3, 31–37.

Tschmelak, J., Proll, G., Gauglitz, G., 2004. Verification of performance with the automated direct optical TIRF immunosensor (River Analyser) in single and multi-analyte assays with real water samples. Anal. Chim. Acta. 519 (2), 143–146.

Van, D.E., Ieven, M., Pattyn, S., Van, D.L., Laga, M., 2001. Detection of *chlamydia trachomatis* and *neisseria gonorrhoeae* by enzyme immunoassay, culture, and three nucleic acid amplification tests. J. Clin. Microbiol. 39, 1751–1756.

Vaughan, R.D., O'Sullivan, C.K., Guilbault, G.G., 2001. Development of a quartz crystal microbalance (QCM) immunosensor for the detection of *Listeria monocytogenes*. Enzyme Microb. Technol. 29, 635–638.

Vaughan, R.D., Carter, R.M., O'Sullivan, C.K., Guilbault, G.G., 2003. A quartz crystal microbalance (QCM) sensor for the detection of *Bacillus cereus*. Anal. Lett. 36 (2003), 731.

Velusamy, V., Arshak, K., Korostynska, O., Oliwa, K., Adley, C., 2010. An overview of food borne pathogen detection: in the perspective of biosensors. Biotechnol. Advan. 28, 232–254.

Vinayaka, A.C., Basheer, S., Thakur, M.S., 2009. Bioconjugation of CdTe quantum dot for the detection of 2,4-dichlorophenoxyacetic acid by competitive fluoroimmunoassay based biosensor. Biosens. Bioelectron. 24, 1615–1620.

Viswanathan, S., Radecka, H., Radecki, J., 2009. Electrochemical biosensor for pesticides based on acetylcholinesterase immobilized on polyaniline deposited on vertically assembled carbon nanotubes wrapped with ssDNA. Biosens. Bioelectron. 24, 2772–2777.

Viswanathan, S., Radecki, J., 2008. Nanomaterials in electrochemical biosensors for food analysis-a review. Pol. J. Food Nutr. Sci. 58 (2), 157–164.

Vo-Dinh, T., Alocilja, E.C., Radke, S.M., 2003. Market analysis of biosensors for food safety. Biosens. Bioelectron. 18 (5–6), 841–846.

Volpe, G., Ammid, N., Moscone, D., Occhigrossi, L., Palleschi, G., 2006. Development of an immunomagnetic electrochemical sensor for detection of BT-CRY1AB/CRY1AC proteins in genetically modified corn samples. Anal. Lett. 39, 1599–1609.

Von Bultzingslowen, C., McEvoy, A.K., McDonagh, C., MacCraith, B.D., Klimant, I., Kranse, C., Wolfbeis, O.S., 2002. Sol-gel based optical carbon dioxide sensor employing dual luminophore referencing for application in food packaging technology. Analyst 127, 1478–1483.

Wang, G., Xu, G., Zhu, Y., Zhang, X., 2014. A "turn-on" carbon nanotube-Ag nanoclusters fluorescent sensor for sensitive and selective detection of Hg2+ with cyclic amplification of exonuclease III activity. Chem. Commun. 50 (6), 747–750.

Wang, J., Deo, R.P., Musameh, M., 2003. Stable and sensitive electrochemical detection of phenolic compounds at carbon nanotube modified glassy carbon electrodes. Electro. Anal. 15, 1830–1834.

Wang, L., Chen, W., Xu, D., Shim, B.S., Zhu, Y., Sun, F., Liu, L., Peng, C., Jin, Z., Xu, C., Kotov, N.A., 2009a. Simple, rapid, sensitive and versatile SWNT-paper sensor for environmental toxin detection competitive with ELISA. Nano. Lett. 9, 4147–4152.

Wang, M., Li, Z., 2008. Nano-composite ZrO2/Au film electrode for voltammetric detection of parathion. Sens. Actuat. B 133, 607–612.

Wang, Z.D., Lee, J.H., Lu, Y., 2008. Label-free colorimetric detection of lead ions with a nanomolar detection limit and tunable dynamic range by using gold nanoparticles and DNAzyme. Adv. Mater. 20 (17), 3263–3267.

Wang, Z., Liu, S., Wu, P., Cai, C., 2009b. Detection of glucose based on direct electron transfer reaction of glucose oxidase immobilized on highly ordered polyaniline nanotubes. Anal. Chem. 81 (4), 1638–1645.

Wark, A.W., Lee, H.J., Qavi, A.J., Corn, R.M., 2007. Nanoparticle-enhanced diffraction gratings for ultrasensitive surface plasmon biosensing. Anal. Chem. 79 (17), 6697–6701.

Warner, M.G., Grate, J.W., Tyler, A., Ozanich, R.M., Miller, K.D., Lou, J., Marks, J.D., Bruckner-Lea, C.J., 2009. Quantum dot immunoassays in renewable surface column and 96-well plate formats for the fluorescence detection of botulinum neurotoxin using high-affinity antibodies. Biosens. Bioelectron. 25, 179–184.

Wei, H., Wang, E., 2008. Fe_3O_4 magnetic nanoparticles as peroxidase mimetics and their applications in H_2O_2 and glucose detection. Anal. Chem. 80, 2250–2254.

Willardson, B.M., Wilkins, J.F., Rand, T.A., Schupp, J.M., Hill, K.K., Keim, P., Jackson, P.J., 1998. Development and testing of a bacterial biosensor for toluene-based environmental contaminants. Appl. Environ. Microbiol. 64 (3), 1006–1012.

Winquist, F., Wide, P., Lundstrom, I., 1997. Monitoring of freshness of milk by an electronic tongue on the basis of voltammetry. Anal. Chem. Acta. 357, 21–31.

Wu, S., Zhao, H.T., Ju, H.X., 2006. Electrodeposition of silver-DNA hybrid nanoparticles for electrochemical sensing of hydrogen peroxide and glucose. Electrochem. Commun. 8, 1197–1203.

Xiulan, S., Yinzhi, Z., Jingdong, S., Liyan, S., He, Q., Weijuan, Z., 2010. A quartz crystal microbalance-based immunosensor for shrimp allergen determination in food. Eur. Food Res. Technol. 231, 563–570.

Xu, X., Liu, S., Ju, H., 2003. A novel hydrogen peroxide sensor via the direct electrochemistry of horse radish immobilized on colloidal gold modified screen-printed electrode. Sensors 3, 350–360.

Yakes, B.J., Lipert, R.J., Bannantine, J.P., Porter, M.D., 2008. Detection of *Mycobacterium avium* subsp. *Paratuberculosis* by a sonicate immunoassay based on surface-enhanced Raman scattering. Clin. Vacc. Immuno. 15 (2), 227–234.

Yang, H., Qu, L., Wimbrow, A.N., Jiang, X., Sun, Y., 2007. Rapid detection of *Listeria monocytogenes* by nanoparticle-based immunomagnetic separation and real-time PCR. Int. J. Food Microbiol. 118, 132–138.

Yang, K., Jenkins, D.M., Su, W.W., 2011. Rapid concentration of bacteria using submicron magnetic anion exchangers for improving PCR-based multiplex pathogen detection. J. Microbiol. Meth. 86, 69–77.

Yang, L.J., Li, Y.B., 2005. Quantum dots as fluorescent labels for quantitative detection of *Salmonella typhimurium* in chicken carcass wash water. J. Food Protect. 68 (6), 1241–1245.

Ye, J., Letcher, S.V., Rand, A.G., 1997. Piezoelectric Biosensor for detection of *Salmonella typhimurium*. J. Food Sci. 62 (1997), 1067–1086.

Yi, Y., Zhu, G., Liu, C., Huang, Y., Zhang, Y., Li, H., Zhao, J., Yao, S., 2013. A label-free silicon quantum dots-based photoluminescence sensor for ultrasensitive detection of pesticides. Anal. Chem. 85 (23), 11464–11470.

Zeng, X., Kong, F., Halliday, C., Chen, S., Lau, A., Playford, G., Sorrell, C., 2007. Reverse line blot hybridization assay for identification of medically important fungi from culture and clinical specimens. J. Clin. Micriobiol. 45, 2872–2880.

Zhang, L., Jiang, C., Zhang, Z., 2013. Graphene oxide embedded sandwich nanostructures for enhanced Raman readout and their applications in pesticide monitoring. Nanoscale 5 (9), 3773–3779.

Zhang, Q., Huang, Y., Zhao, R., Liu, G., Chen, Y.J., 2008a. In situ growth of nanogold on quartz crystal microbalance and its application in the interaction between heparin and antithrombin III. J. Colloid Interface Sci. 319 (1), 94–99.

Zhang, W.H., Zhang, W.-D., 2008. Fabrication of SnO_2 – ZnO nanocomposite sensor for selective sensing of trimethylamine and the freshness of fishes. Sens. Actuat. B 134, 403–408.

Zhang, X., Geng, P., Liu, H., Teng, Y., Liu, Y., Wang, Q., 2009. Development of an electrochemical immunoassay for rapid detection of *E. coli* using anodic stripping voltammetry based on Cu@Au nanoparticles as antibody labels. Biosens. Bioelectron. 24, 2155–2159.

Zhang, Y., Wang, H., Yan, B., Zhang, Y., Li, J., Shen, G., Yu, R., 2008b. A reusable piezoelectric immunosensor using antibody-adsorbed magnetic nanocomposite. J. Immunol. Methods 332, 103–111.

Zhao, J., O'Daly, J.P., Henkens, R.W., Stonehuerner, J., Crumbliss, A.L., 1996. A xanthine oxidase/colloidal gold enzyme electrode for amperometric biosensor applications. Biosens. Bioelectron. 11, 493.

Zhao, Z.W., Chen, X.J., Tay, B.K., Chen, J.S., Han, Z.J., Khor, K.A., 2007. A novel amperometric biosensor based on ZnO: Co nanoclusters for biosensing glucose. Biosens. Bioelectron. 23, 135–139.

Zhou, Y., Yang, S., Qian, Q., Xia, Z., 2009. Gold nanoparticles integrated in a nanotube array for electrochemical detection of glucose. Electrochem. Commun. 11, 216–219.

Zunabovic, M., Domig, K.J., Kneifel, W., 2011. Practical relevance of methodologies for detecting and tracing of *Listeria monocytogenes* in ready-to-eat foods and manufacture environments: a review. LWT - Food Sci. Technol. 44, 351–362.

20

PSYCHOLOGICAL AND PHYSIOLOGICAL BASES OF UMAMI TASTE PERCEPTION AS RELATED TO NUTRITION

Nobuyuki Sakai*, Hisauyki Uneyama, Visith Chavasit†**

**Tohoku University, Department of Psychology, Graduate School of Arts and Letters, Sendai, Japan; **Chemosensory Research Group, Frontier Research Labs., Institute for Innovation, Ajinomoto Co., Inc., Kawasaki, Japan; †Mahidol University, Research and Education Unit on Sensory Science for Better Nutrition, Institute of Nutrition, Nakhonpathom, Thailand*

1 Why We Eat What We Eat

In this chapter, the authors indicate that human eating behavior is not based only on physiological but also psychological bases. Of course, eating due to the physiological need is certainly essential for human health; however, appetite from psychological effect can also be equally important especially for people who live in more developed societies. Many times, psychological hunger can be aroused by factors such as traditional meal time, nice odor of food, or looking at people having our favorite food. Meanwhile, physiological fullness may not always lead to psychological satiety because the psychological satiety occurs from our contentment on having our favorite dishes, after finishing our dishes, or getting bored of such food.

Abraham Maslow (1943) classified human motives into five (or more) stages, which is called "Maslow's Motivation Theory." Maslow discussed that we seek for the next level of needs only when we have felt that the current level of needs is satisfied, but not when the needs are satisfied physically. This theory shows that human needs, such as hunger, cannot be simply fulfilled only with physiological but also psychological means. The authors begin this chapter by applying the Maslow's Motivation Theory to human eating behavior (Sakai, 2014) (Fig. 20.1).

Novel Approaches of Nanotechnology in Food. http://dx.doi.org/10.1016/B978-0-12-804308-0.00020-0

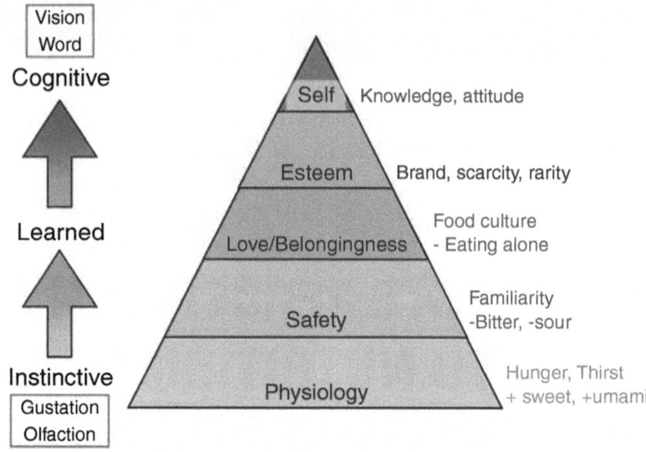

Figure 20.1. A model of human motivation for eating and for the palatability of foods and beverages. This model is modified from the schema of Maslow's Motivation Theory.

1.1 Physiological Motivation Stage

By nature, motivation in human being occurs to satisfy their physiological needs. Babies cry for milk, sleep, and/or caring touch. In our daily life, we eat foods when we feel hunger, and drink beverages when we feel thirst. These are good examples of the physiological motivation.

Regarding to the physiological motivation, humans like foods and beverages that have sweet, salty, and umami tastes. Foods and beverages of these tastes normally contain compounds that satisfy the human body's needs. At this motivation stage, food and beverage serve as the necessary factors for human survival, which is understandable in people of less-developed countries especially during wartime and crisis. Under such conditions, any food choice is consumable and the choices of higher energy are more preferable. In more-developed countries, the liking/disliking for foods and beverages can be quite different, which is corresponded to the phenomenon of Maslow's Motivation Theory. Since the needs of people in more-developed countries for the physiological motivation stage have already been satisfied, they are having foods and beverages based on the upper motivation stages.

1.2 Safety Motivation Stage

The second stage is "Safety." In this stage, humans want to satisfy their needs for safe manner. At the physiological motivation stage, babies like anybody who can provide what they want. As the babies grow up, they prefer to be with their own mothers because they realize that mothers are the safe zone for them.

Similar concepts can be used to explain human attitudes on bitter and sour foods. Bitterness signals for toxin, and sourness signals for unripened or rotten conditions. Humans mostly refuse a novel food due to no confidence on safety. We normally classify safe food from our own experience learned from repeatedly eating such food with no health hazard.

If physiological needs of foods are minimum requirements for basic survival, food safety is the higher need for better survival. In order to fulfill both needs, they require olfaction, gustation, and other chemical senses during eating.

In more-developed countries where both needs have mostly been fulfilled under their efficient management system, people generally have no concern for these needs. However, contamination and mislabeling of foods are occasionally issues of concern for the more-developed countries. Higher safety standards (beliefs) such as organic, no food additives have become the needs of some people in these more-developed countries.

1.3 Love and Belongingness Motivation Stage

The third stage of motivation is named "Love and Belongingness." Children in this stage prioritize their relationships not only with their family members but also with their friends at school. They sometimes belong to a certain group of children.

Through eating experiences with family and friends, children learn to accept and reject foods. At the same time, children also learn the food culture of their own communities and/or family roots. As a social animal, humans also learn pleasantness, enjoyment, and contentment of eating with others. People potentially prefer foods and beverages that are normally consumed by people of similar backgrounds (Robinson and Higgs, 2012). For example, Washoku is a favorite dish among Japanese and others who experience having it since childhood. These phenomena can be explained with the Maslow's Motivation Theory on "Love and Belongingness."

1.4 Esteem Motivation Stage

The fourth stage of motivation is named "Esteem." As people become adolescent and adult, they need to be more recognized and valued by others. These recognitions and values bring the feelings of self-esteem and self-respect to people, which are the core components of next motivation stage, "Self-actualization."

In the Esteem stage, adolescents and adults turn to eat bitter foods and beverages that they used to reject when they were younger. People tend to prioritize on brands of foods and beverages and are willing to pay more for them. This behavior is motivated by the desires to be accepted and valued by the others. These desires cannot be satisfied only by ingesting the foods and/or beverages but by their experiences in exposing to such foods and/or beverages. Thus, people in this motivation stage value the visual and verbal properties of the foods and beverages rather than their chemical components. For example, people who have no experience in eating Washoku come to like Washoku as they learn that it is good for their health. This example can be explained by the Esteem motivation.

1.5 Self-Actualization Motivation Stage

The fifth and the top motivation stage is called "Self-actualization," which is not easily understood. Maslow himself also could not clearly explain this motivation stage. The concept of this motivation stage refers to understanding one's own potential and/or realizing the limit of such potential. People in this motivation stage have already achieved the previous motivation stages and realized the needs.

As applied to eating behavior, people in the Self-actualization motivation stage eat what they think is good. This behavior is triggered by one's own beliefs that come from various factors such as nutrition knowledge, eating habits from daily life, perception on their own body. In more-developed countries, people are suffering from diseases relating to their eating behaviors such as diabetes, obesity, anorexia, and bulimia. These diseases are life-style diseases, which would be rare as people can eat based on their physiologically needs. From this perspective, belief or attitude for eating in this motivation stage is quite important to be understood.

1.6 Human Motivation and Eating Behavior

In Section 1, the authors discuss that human eating behavior is triggered not only by the physiological needs, but also learning and experiences developed through our daily life. From next section, the mechanisms and the phenomena of flavor perception are discussed based on this motivation theory.

2 Physiological Aspects of Umami Taste Perception

When we eat a food, we experience a "taste" sensation. The term of "taste" sensation used in our daily life is different from that used in the aspects of psychology and physiology. The academic term for "taste sensation" is "gustation," which is defined as the sensation caused by activation of taste cells in oral cavity and tongue. The tastants isolated from foods and beverages bind to the taste receptor and activate the taste cells. Based on the definition, there are five basic tastes in humans: sweet, salty, sour, bitter, and umami.

The sensation occurring during eating foods is defined as a flavor perception. The flavor perception consists of several sensations, including gustation, olfaction, and somatosense in the oral cavity. During eating, we also feel (subliminally, in many cases) somatosense and visceral senses evoked in esophagus, stomach, intestine, and so on. In this section, the sensations evoked during eating are reviewed, mainly on umami sensation.

2.1 Physiological Fundamentals Underlying the Feeling of Deliciousness

During meals, we can enjoy foods through simultaneously integrating huge amounts of information inputted from the five sensations (vision, audition, olfaction, gustation, and somatosenses). Among these sensations, people tend to pay more attention on gustatory sensation. Indeed, living organisms start and stop eating based on the gustatory sensation. They can detect the components in foods that are supplied by the gustatory cues. Regarding to physiological and safety needs, the organisms naturally intake sweet-taste foods, and avoid bitter-taste foods.

Sensory information is transferred from the peripheral organs (eye, ear, nose, and mouth) to each sensory cerebral cortex, and integrated to form the total feelings of food liking/disliking at the frontal association cortex after being integrated with the visceral information from gastrointestinal organs during and posteating. This totally integrated "food-related information" is transferred to the limbic system, such as hippocampus and amygdala, to compare with the past ingestion experiences. The hippocampus is involved in processing short-term and declarative memories. Meanwhile, the amygdala is involved in processing emotional memory and making judgment on liking/disliking of the stimuli. As the taste information has been compared with the memory on past ingestion experiences on safety and available essential nutrients, the food eating process is finally triggered. If the food is judged as unsafe and/or unsuitable for ingestion, the eating process is canceled. At this stage, memory and learning experiences are the key factors used for the determination.

In nature, there are millions of substances that can trigger the mentioned human senses. Among them, umami tastants are the well-known substances that can have effects on both taste and visceral senses. Such effects can lead to food appetite and palatability.

2.2 Umami as a Basic Taste: A Sensory Marker Satisfying the Fundamental Physiological Demand

Major macronutrients of high molecular weight (starch, fat, and protein) that are indispensable to our life are normally tasteless, because they are unable to interact with the taste receptors on our tongue. We therefore cannot recognize the tastes of starch, vegetable oil, and casein (milk protein). We are able to know that such macronutrients are included in a food only when there are low molecular weights substances such as sugar, fatty acid, amino acid coexist with those macronutrients. For example, the taste of crab meat is naturally due to glycine, alanine, arginine, and glutamate, therefore, imitated crab meat made from casein needs to be added with these amino acids (Fuke and Konosu, 1991).

Appetite and palatability are strongly influenced by physiological status of the body. Under such influences, living organisms may choose to ingest or avoid certain nutrients by using their taste sensations. For example, in laboratory animals, hypoglycemia induces more preference of sweet foods, and the decrease in plasma sodium concentration by adrenalectomy increases palatability of salty foods (Lucas and Sclafani, 1988; Harriman and Kare, 1964). The mentioned laboratory evidence can be supported by our daily experiences. As we become hungry, we are attracted to sweet foods because they can increase our blood glucose level. Meanwhile, we desire salty foods as we lose minerals from sweating during exercise. Scientific evidences indicate that fasting or exercising affected palatability and pleasantness for sweet or salty taste (Laeng et al., 1993; Passe et al., 2009).

Regarding the preference for umami taste, animal studies showed that the preference for umami taste might link to metabolism of the ingested dietary protein. In this experiment, rats received diets containing stepwise increasing doses of egg protein for certain periods, then the rats were evaluated for their preferences on three solutions, that is, NaCl (salty), glycine (sweet), and monosodium glutamate (MSG) (umami). The rats preferred NaCl and glycine to MSG solutions as the dietary protein was restricted to the deficiency level of 0–5%, Their preferences were gradually changed as the dietary protein increased. At the dietary protein of 20%, rats showed more preference for MSG to NaCl or glycine (Torii et al., 1987).

Based on these phenomena, taste physiologists consider the senses of basic tastes as the significant meanings on nutrition and physiology for our life (Table 20.1). Sweet taste is the signal for energy intake (glucose) to maintain blood glucose, and salty taste is the marker for mineral (NaCl) to maintain plasma osmolality. On the other hand, sour and bitter tastes are the warning signals

Table 20.1 Current Knowledge on Five Basic Tastes

Basic Taste	Typical Substances	Receptor Molecules	Physiological Meaning
Sweetness	Sugars, sweet peptides	GPCR (T1R2 + T1R3)	Energy intake (carbohydrate)
Saltiness	Salts (NaCl, KCl)	ENaC	Mineral intake (sodium, potassium)
Sourness	Proton (acetic acid, citric acid)	TRP channel (PKD1L3 + PKD2L1)	Harmful alarm (decomposition)
Bitterness	Plant alkaloids, pharmacological active agents	GPCR (T2Rs)(~30)	Harmful alarm (toxins)
Umami taste	Monosodium glutamate 5′-nucleotides	GPCR (T1R1 + T1R3)	Protein intake (amino acids)

Recent molecular biology and physiology on basic tastes have revealed the fundamental knowledge on physiological demand mediated by taste sensation.

for rotten foods and toxins, respectively. Umami taste signals for protein intake to recruit amino acids essential for our life. Recent physiological studies using MSG as umami tastant, confirmed that umami taste played important role modulating protein digestion and maintaining homeostasis of blood amino acid.

2.3 Physiological Basis of Umami Taste Perception

As mentioned earlier, the basic principle underlying our eating behavior is to satisfy physiological demand for essential nutrients, which are mainly determined by two sensories (ie, gustatory and visceral sensations).

The mechanism of gustation is relatively simpler as compared to the visceral one. The authors first explain the mechanisms underlying perceiving umami taste in this section.

Binding of the umami tastant, such as free glutamate, in foods to the oral umami taste receptor triggers the umami taste. Recently, it is reported that the metabotropic glutamate receptors (mGluR1/mGluR3) and the amino acid taste receptor (T1R1/T1R3) are the potential umami taste receptor in human (Chandraskekar et al., 2006). Animal studies revealed that the facial (chorda tympani and greater superficial petrosal), the glossopharyngeal, and the vagus (superior laryngeal) nerves, which make synapses with taste cells, convey umami taste information to the first relay nucleus, the rostral part of the nucleus of the tractus solitarius (NTS), and then the taste information are finally transferred to the insular cortex.

Brain mechanisms underlying the perception of oral umami taste has been well addressed. The umami tastants such as MSG,

5′-inosine mono-phosphate (IMP) activate the same regions of the insular cortex that is known as a primary taste cortex in humans, suggesting that both umami tastants could be similarly recognized (de Araujo et al., 2003a).

Oral stimulation of umami taste also triggers series of taste-oral reflexes such as salivation related to food mastication/swallowing. For instance, it was revealed that MSG aqueous solution as synergic enhancement with IMP can induce long-lasting salivation in human subjects (Horio and Kawamura, 1989).

At the same time, the gustation triggers an autonomic nerve reflex, so-called "cephalic phase response," to prepare food digestion in the gut as well as the oral cavity. Regarding to the cephalic phase response, it has been reported that application of MSG into the oral cavity could increase the efferent nerve activities of the abdominal vagus nerve (gastric, celiac, and pancreatic branches) in rats, and induce gastric and pancreatic exocrine secretions in dogs (Niijima, 1991; Ohara et al., 1988). Thus, contributions of the umami taste sensation to health include facilitating food mastication/swallowing by initiating the gastrointestinal exocrine reflexes (saliva, gastric and pancreatic juice) and preparing gastric and pancreatic digestion.

2.4 Physiological Basis of Umami Perception in Guts (Gut-Glutamate Sensation)

The visceral sense is induced when tastants and/or digested nutrients bind to the visceral receptors so-called "visceral nutrient sensors" (ie, nutrient taste receptors and transporters) in gastrointestinal epithelium, and the information activates vagus nerves to release gastrointestinal hormones. Unfortunately, the information at molecular level of the visceral nutrient sensors is not fully understood yet.

After gastric/pancreatic and intestinal digestions of dietary protein, amino acids and peptides are absorbed via mucosal enterocytes. The digestion process is efficiently monitored by neuronal (the vagal and the intrinsic nerves) and hormonal (gastrointestinal and pancreatic hormones) regulators. Recently, some reports indicated possible involvement of oral taste receptors as nutrient-sensing molecules on food digestion and nutrient metabolism processes (Calvo and Egan, 2015; Janssen and Depoortere, 2013).

Both stomach and intestine are found to have specific dietary free glutamate recognizing systems on their epithelial mucosa (Niijima, 1991; Uneyama et al., 2006). For instance, intragastric and intraduodenal injections of free glutamate solution could activate the afferent nerves of abdominal vagus, indicating that dietary free glutamate is sensed by the vagus-mediated gut nutrient monitoring system. In addition to the neuronal recognition, hormonal recognition (somatostatin) of dietary free glutamate had also been preliminarily reported (Nakamura et al., 2011). Regarding to physiological effects of gut glutamate sensation on optimal food digestion, it could involve regulation of gut exocrine and motility (Nakamura et al., 2013; Iwatsuki and Uneyama, 2012; Uneyama, 2011). Meanwhile, the hypothesis on umami taste as marker of protein intake has been proposed based on the finding of gastric action of the visceral sense in the presence of protein from foods (Uneyama, 2011). During gastric food digestion, it has a potential risk that gastric and intestinal walls can be digested because the gut wall is also protein. Therefore, guts need to secret protective factors such as mucin and bicarbonate from the gastric mucosa in order to protect the gut walls from the attacking factors (gastric acid and digestive enzyme). Under this hypothesis, the role of the gut glutamate-sensing system was to regulate secretion of the protective factors to attenuate action of the attacking factors (Uneyama, 2011; Nakamura et al., 2013). Thus, the visceral-sensing system can maintain digestive balance between the attack and the protective factors. These evidences strongly support the hypothesis that gut glutamate-sensing systems can enhance dietary protein digestion by triggering the visceral sense, leading to improved amino acid absorption and utilization.

Recently, a novel finding on the use of spatiotemporal analysis of fMRI data on rat brain response to intragastric administration of MSG or IMP had been reported (Tsurugizawa, 2014). An intragastric load of MSG significantly could activate several brain regions of rats including insular cortex, amygdala, hippocampus as well as hypothalamic including lateral hypothalamus, dorsomedial hypothalamus, and medial preoptic. The largest activation was observed at 5–10 min after the end of the infusion. In contrast, an intragastric infusion of glucose solution at low calorie (0.4 kcal/kg body weight) could activate only nucleus accumbens, insular cortex, somatosensory cortex, and hypothalamic regions in rats (Tsurugizawa et al., 2009). These data indicate that the ingested glucose activates the dopaminergic pathway. A significant increase of the *c-fos* immunoreactive neurons is observed in the hypothalamic area, sensory motor cortex, and the nucleus accumbens following the intragastric infusion of 30% sucrose (Yamamoto and Sawa, 2000) (Fig. 20.2).

2.5 Ingestive Behavior: Beyond Protein Digestion to Appetite Control

Physiologically, appetite is well controlled by the balance between hunger and satiety. As shown in Fig. 20.3a, hunger is

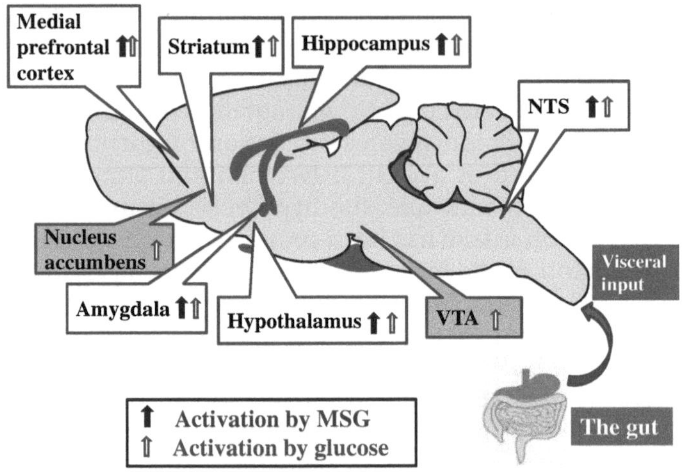

Figure 20.2. Postingestive nutrient specific neural pathways. Animal fMRI studies reveled nutrient (taste) specific neural after swallowing. Please note that sweet perception in the gut might lead to brain activation related to reward system but umami taste perception in the gut failed to activate the system. Modified from Tsurugizawa (2014).

attenuated by satiety developed during and after meal. Hunger ratings lowered with satiety makes a period before the next meal (Blundell and Finlayson, 2004). Satiation during meal and satiety after a meal of living organisms lower their stress during intermeal period and consequently links to their positive emotion such as pleasantness. However, an adequate intermeal period is also quite important because it allows times for food digestion, nutrient absorption, metabolism, and excretion of residue as well as functionally maintains alimentary tract. In general, satiation and the early phase of satiety is mainly developed by the five basic senses, that is, sight, smell, hearing, taste, and touch, especially taste sense during meal. And the late phase of satiety is mainly regulated by the visceral sense (Fig. 20.3b). This cascade indicates that both gustatory and visceral senses affect satiation (which brings a meal to an end) and satiety (which maintains the inhibition of eating) that controls the suppression of appetite via physiological signals. Interestingly, recent evidence suggests that gastrointestinal flora can also affect the satiety formation, which may be a new strategy for preventing overeating (Delzenne et al., 2010).

Since umami receptor has been found to be expressed not only on the tongue but also on the luminal gut wall, it has been unveiled that the vagal afferent from the stomach can respond to umami tastants, such as free glutamate and ribonucleotides, and subsequently activates brain regions associated with

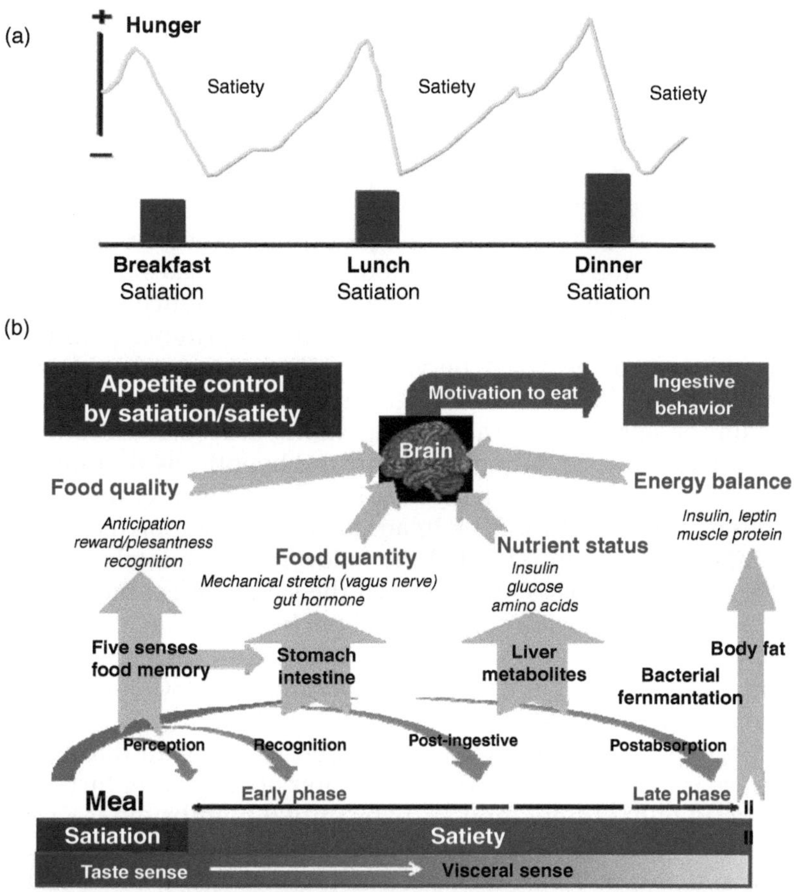

Figure 20.3. Physiological regulation of feeding behavior. (a) Role of satiety on the hunger scale. (b) Physiological basis on appetite control via satiation and satiety. Please note that taste sense is major physiological factor in the early phase during eating, and visceral sensei is major physiological factor in the late phase after the lunch. Partially modified from Blundell and Finlayson (2004).

energy homeostasis (Nakamura et al., 2013). Accumulated evidences indicate that umami tastants influence appetite via their postingestive effects of neural and hormonal mediated pathways in addition to their chemosensory stimulations. Especially, recent evidences with human subjects strongly suggest possible involvement of gustatory and visceral perceptions of umami tastants in appetite control through triggering satiation during the meal and satiety after meals, which can potentially link to enhancement of the gastrointestinal function on protein digestion.

Human studies clearly indicate that three major macronutrients (fat, carbohydrate, and protein) differently affects on satiety after meal. Protein is well known to be the most potent satiety inducer among the nutrients (Westerterp-Plantenga et al., 2012). Potential mechanisms underlying the protein-induced satiety might be from the release of gastrointestinal hormones so-called "satiety hormones," such as neuropeptide Y (NPY), cholecystokinin (CCK), and gastrin-like polypeptide-1 (GLP-1) during protein digestion (Davidenko et al., 2013).

The fact that the umami taste perception enhances gastrointestinal functions related to protein digestion makes the further hypothesis that umami taste enhances protein-induced satiety by enhancing release of these satiety hormones (hormonal pathway), in addition to the vagal nerve activation (neuronal pathway). Activation of the abdominal vagus is well known to regulate the feeding behaviors as well as food digestions. Indeed, recent studies of human subjects strongly support the hypothesis that umami taste substance could enhance the satiety feeling after meal (Masic and Yeomans, 2013, 2014; Imada et al., 2014). Supplementation of MSG to the soup inhibited the following snack intake (Imada et al., 2014). Yoemans reported that the satiety effect induced by MSG supplementation could be differed between macronutrients included in the foods (Masic and Yeomans, 2013). That is, the satiety effect is stronger with the combination of MSG and protein than that of MSG and carbohydrate.

Studies have shown that accelerated weight gain during the first year of life increases the risk of later obesity. After birth, the first and only protein food for baby until 6 months old is casein in breast milk. Therefore, the baby must be able to efficiently digest and absorb breast milk protein. Interestingly, human breast milk contains up to 20 kinds of amino acid, in which free glutamate appears at the highest concentration (Rassin et al., 1978). The free glutamate as an umami tastant potentially promotes breast milk protein digestion and controls the appetite in the baby. Classical observation on taste acceptance of a newborn baby using face response evaluation indicated that umami taste is incorporated into the pleasant feeling similarly to sweet taste (Steiner et al., 2001).

By having breastfed babies as the control group, babies fed with extensive protein hydrolysate formula (ePHF) had normal weight gain, while the group having cow-milk formula (CMF) exhibited the accelerated weight gain. To reach the satiation, the ePHF-fed babies consumed less amount of the formula (Ventura et al., 2012). The level of glutamate in ePHF is markedly higher than in CMF. This finding suggested the role of free glutamate in regulating intake of baby. A study by Ventura et al. had confirmed this assumption. Three groups of babies aged <4 months were fed with isocaloric formulas prepared from (1) CMF, (2) ePHF, or (3) CMF added with free glutamate to similar concentration as in ePHF (CMF + glu). The study indicated that babies fed with formulas ePHF or CMF + glu consumed significantly fewer amounts of formulas with no compensation at the subsequent meal. This study indicated that the babies fed with high glutamate formulas ingested less to reach satiation. These findings suggested a role of free glutamate in breast milk on regulating amount of milk intake in a breast-fed baby.

2.6 Current Hypothesis: Physiological Fundamental on Umami Taste as a Protein Marker

Current hypothesis on how the umami sense can control appetite and palatability as well as act as a protein taste marker is shown in Fig. 20.4 (San Gabriel and Uneyama, 2013; Nakamura et al., 2013). As mentioned earlier, the impacts of umami taste as a gustatory marker of protein can be explained by two aspects of sensory physiology. First, umami taste can be felt consciously and subconsciously during meals. Umami tastant increases palatability of foods containing proteins via the umami taste receptors in the mouth, which leads to consciously feel delicious (taste sense). Subsequently, the taste sense subconsciously triggers taste-vagal reflex (visceral sense) for cephalic phase responses in order to prepare protein digestion and mucosal protective factors.

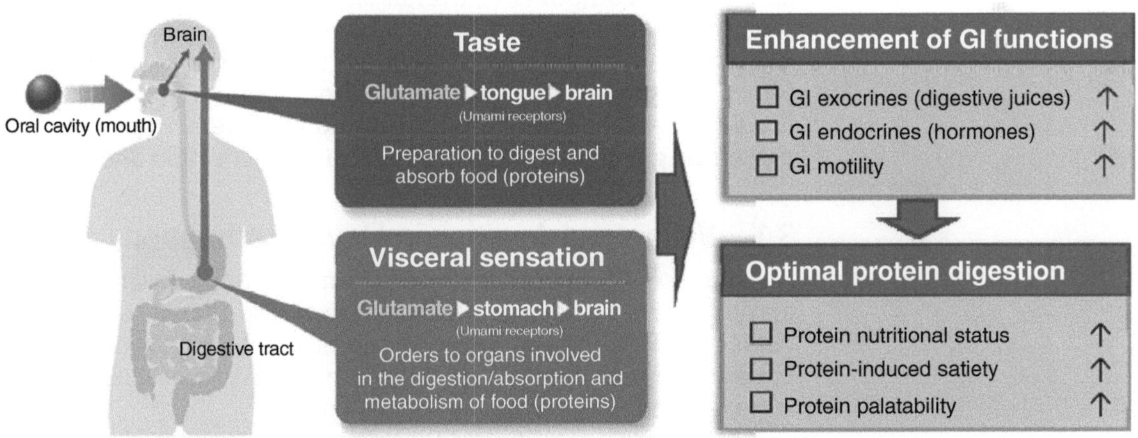

Figure 20.4. Current hypothesis on psychological and physiological role of umami taste as a sensory marked of protein intake. This hypothesis is proposed based on the research using major umami tastant, monosodium glutamate (MSG). Umami taste substance can activate the two sensory systems, the taste and the visceral sensory systems, leading to the appetite and digestive regulations for dietary protein.

The second aspect of sensory physiology happens as the foods entering stomach, which the umami tastant is again sensed by the gut-glutamate sensing system to induce gastric protein digestion from gastric acids and proteolysis enzymes. Meanwhile, the mucosal protective factors, such as mucin and bicarbonate, are induced as well. The enhancing effects of free glutamate on gastrointestinal functions related to protein digestion lead to further enhancement of the postingestive protein-induced satiety. We hypothesize these digestive cascades are related to the food deliciousness, which is a pleasant feeling and increases the protein food acceptance.

3 Flavor Perception and Umami

As mentioned earlier, the sensation evoked during eating foods and/or drinking beverages is not only the gustation but it is the integrated perception of more than one sense, so-called "flavor." As Rozin (1982) noted, we cannot verbally discriminate between the sensation evoked by taste and the perception evoked by flavor (taste plus odor and somatosenses such as irritation, temperature, and touch). This lack of languages leads us to desensitize the discrimination between taste and flavor, and to ignore the importance of odor in flavor perception (Prescott and Stevenson, 2015).

The interactions between gustation and olfaction are in two aspects; one is peripheral, and the other is cognitive. Peripherally, the interactions are caused by chemical interactions between/among food components, overall effects from odorants and tastants released from food, and changes in receptors affinities to odorants and tastants. In this section, the cognitive aspects of interaction, such as the psychological and neural mechanisms underlying the development of flavor and the interactions between flavor and gustation, especially umami taste, are discussed.

3.1 Vision and Audition in Flavor Perception

Japanese cuisine sets its priority on appearance. A typical example is Bento, the lunch box. Mothers of kindergartener make characters in anime or faces of animals with the foods in their children's Bento. Bento available in the market also has colorful appearance with artwork and ornament.

There were a few studies that reported the effect of vision on flavor perception. An experiment with wine flavor indicated that the subjects changed their description of white wine to red wine as the color of the white wine had been changed into red using an odorless and tasteless colorant (Morrot et al., 2001). The authors had also tested visual components of flavor using sushi and chocolate (Miyake et al., 2009; Sakai, 2009, 2011; Sakai and Yamasaki, 2009). For example, when tuna sushi was presented with a photo of red-color tuna meat), the subjects who evaluated the sushi for a stronger fishy smell and less palatable than the same ones that were presented with a photo of white- or light yellow-color tuna meat. Recently, Spence and Piqueras-Fiszman (2014) studied the effect of vision on flavor perception by using colored beverages, color of food plates and spoons. Overall, there are still not enough studies on the effect of vision on flavor perception that is systematically and scientifically performed.

Crispiness of food is detected by the sense of audition (Vickers and Bourne, 1976; Zampini and Spence, 2004). Meanwhile, loud noise can suppress olfaction and all basic taste sensations, but not the umami taste (Spence et al., 2014). Regarding to the mentioned knowledge, an airplane passenger can be advised to have Bloody Mary cocktail drink or tomato juice with meal since they contain umami tastant from tomato that can lift up other taste sensations suppressed by the loud noise of airplane. However, this hypothesis needs to be further investigated.

3.2 Taste–Odor Interaction

If you pinch your nose during eating foods and/or drinking beverages, you cannot experience the taste correctly. This phenomenon becomes more obvious when you taste a canned juice purchased in a market, and cannot differentiate between apple and grape juices without the label. This phenomenon shows that our taste sensation is not only from the gustation but also from the olfaction.

Many studies indicated that taste can be enhanced by odor. Murphy et al. (1977) found that sweetness rating for saccharin flavored with strawberry odor was higher than summation of sweetness of saccharin and of strawberry odor. This phenomenon is called taste-enhancement by the odor, which was found in the studies using sucrose and aspartame as sweetener, and citric acid as acidifier (Stevenson et al., 1999; Sakai et al., 2001). However, not any odor can enhance any taste. Strawberry odor could enhance the sweetness of sucrose in whipped cream but not in peanut butter (Frank and Byram, 1988). Meanwhile, some odors could suppress sourness of citric acid (Stevenson et al., 1999). The authors also showed that vanilla odor could enhance the sweet but not sour taste of juices. On the other hand, lemon odor could enhance the sour but not sweet taste of juices (Sakai et al., 2002). The state of congruity and incongruity between odor and taste plays an important role in this phenomenon. This state of congruity and incongruity between odor and taste are in fact based on food culture that has been learned through eating experiences (Stevenson and Tomiczek, 2007).

The study on taste enhancement by the odor was also performed by using an underthreshold taste solution with underthreshold odor (Dalton et al., 2000). In this study, the sweet taste of underthreshold saccharin solution was detected when the underthreshold almond-like odor was added into it. The mentioned study indicated a clear congruity effect. However, the almond-like odor had no effect on the threshold of umami tastant, MSG. A recent study showed that the reaction time for taste–odor stimulation was faster than the reaction time needed for stimulation of taste or odor only (Veldhuizen et al., 2010). Thus, the gustatory–olfactory integration is not a product of cognitive processing, but a product of information convergence during the early processing stage.

On the contrary, the enhancement of odor by taste has not been so much studied. In an experiment, the mint flavored (added as menthol) chewing gum was evaluated only for sweetness of sucrose, not on the minty odor (Davidson et al., 1999). The product was rated as "no taste, no flavor," which might indicate that taste can hardly enhance odor.

3.3 Learned Synesthesia and Flavor Perception

The gustatory sensation does not appear only when food enters our mouth. Many times, we can detect and judge the characteristics of food that is far away from us by using our olfaction sensation before we decide to let such food entering our mouth. This mechanism helps us to choose delicious and nutritious foods and avoid the poison ones. It is our ability to have synesthetic process of taste and odor.

The integration between olfaction and gustation is the result of the learning experiences in pairing between these two sensations from our daily eating. Prescott and Stevenson (2015) described three types of these learning experiences as follows: redintegration flavor learning, flavor perceptual learning, and flavor recognition learning.

Redintegration flavor learning is a form of learning in which an odor associated with a previously encountered flavor comes to recover the original flavor. In other words, when we smell a sweet odor, we recall the experience of the odor paired with sweet taste, and we feel the sweet taste by smelling such odor. The association between olfaction and gustation is easily acquired by humans with a few pairings of odor and taste (Prescott et al., 2004). This association is resistant to interference (Stevenson et al., 2000) and can be acquired without awareness of the taste-odor contingencies (Stevenson et al., 1998). When the participants experienced a few pairings of the novel odors with sucrose or citric acid, they described the odors as sweet or sour, respectively. Stevenson and coworkers named this learning as a "learned synesthesia," because the association is so stable and highly integrated that the participants are not confused the two components with the odors. In an animal study, rats also acquired the association between olfaction and taste, and they could use odor as a marker for food ingredient (Sakai and Imada, 2003). In the study, rats received pairing of high concentration NaCl solution with odor A, and pairing of plain water with odor B. The rats avoided the plain water flavored with odor A under normal conditions, but they preferred such a sample when they were in salt appetite condition. Thus, the authors concluded that the rats acquired the association of the odor with the taste, and then developed the memory of "salty odor."

Flavor perceptual learning is used to explain the difference between expert and novice. This type of learning consists of two aspects, that is, perceptual expertise and extensive semantic knowledge (Prescott and Stevenson, 2015). Under both aspects, an expert can detect an ingredient at low concentration, describe tastes of foods and beverages, and recognize names of the foods and beverages in a short period of times. The phenomenon can be used to explain why the Japanese can easily detect and recognize the umami taste due to having many descriptions about the umami tastants and Washoku, and to their long-term experiences in eating those in their daily life.

Flavor recognition learning is similar to a flavor template. Normally, humans have a flavor template for each food or beverage category. The template is compared when there is an incoming taste signal. The comparison may stem from detecting food hazards or spoilage based on the food learning experiences. This learning is a basic mechanism underlying the congruity-incongruity or appropriateness-inappropriateness decision.

These phenomena are the bases of flavor perception. During eating foods and/or drinking beverages, flavor perception evoked by olfactory stimulation enhances gustatory sensation, recalls memory related to the foods and/or the beverages, and prepares oneself for ingesting the foods and/or the beverages.

3.4 Brain Mechanisms Involved in Flavor Perception

As described earlier, the association between gustation and olfaction is developed by pairings of gustation and olfaction in our everyday eating experiences. Stevenson and coworkers found that the mechanisms underlying in this association is a type of classical conditioning (Stevenson et al., 2000; Stevenson and Case, 2003).

The authors also studied this association in rats and found that the learning was acquired in the primary gustatory/olfactory cortices because rats with lesions in the primary gustatory cortex could not acquire the learning.

A similar result was also found in human studies. Kettenmann et al. (1996) reported that the human temporal brain, including frontal opercula, showed responses to olfactory stimulus such as vanilla odor. Because a noninvasive tool, that is, magnitoencephalograph (MEG), was used for measuring human brain activities, the brain regions could not accurately identified, which made it hard to identify the activity stem from the primary gustatory area. de Araujo et al. (2003b) found that the human primary gustatory area that responded to banana odor was similar to the sucrose solution by using functional magnetic resonance imaging (fMRI). As compared with both methods, fMRI has better spatial resolution, while MEG has better temporal resolution. Both of fMRI and MEG, however, showed similar results on the activation of primary gustatory cortex by odor stimulus, which indicated that the human primary gustatory area did not respond only to the gustatory stimulation but also to the olfactory stimulation. Through our eating experiences, we have acquired the association between the olfaction and the gustation in our primary gustatory/olfactory cortex, and developed into flavor perception.

Moreover, it was reported that there was secondary gustatory cortex, that is, orbitofrontal cortex in humans and monkeys, and ventral prefrontal cortex in rats, which responded to both gustatory and olfactory stimuli (Rolls, 2009). A human study using fMRI indicated the supralinear additively of umami taste (MSG) and vegetable odor in the secondary cortex, which the participants found the umami taste was more pleasant and full as the MSG was presented with vegetable odor (McCabe and Rolls, 2007). The secondary gustatory cortex therefore has an efferent projection to primary gustatory area and ventral striatum, responses in orbitofrontal cortex reflects the subjective evaluation for the stimulus.

3.5 Umami as Flavor Component

In Japanese traditional cuisine Washoku, Kombu seaweed is used especially with dried- bonito, Katsuo-bushi for Dashi, the base of cuisine. Based on this Japanese food culture, the authors hypothesized that the umami taste in Japanese cuisines was enhanced by the flavor from dried-bonito. To prove this hypothesis, the authors performed a study in a Japanese Woman University (Sakai et al., 2011). The subjects were asked to rate for umami taste intensity in soups that had been flavored with different flavors, on 100 mm visual analogue scale (VAS).

It was found that the umami taste intensity was strongest in the soup flavored with dried-bonito. As the participants were asked to evaluate for the congruity between flavor and taste, we hypothesized that these Japanese participants should find the congruity between Kombu (umami taste) with Katsuo-bushi (flavor of dried-bonito) because both are used in the Washoku cuisine. Unexpectedly, the highest rating on congruity rating was between umami taste and tomato flavor, while very low rating was found in umami taste and dried-bonito flavor. The eating pattern of modern young Japanese people was different from the traditional Japanese one, which Dashi (Kombu + dried bonito) was not always in their regular meals. On the other hand, Westernized cuisines were their preferred meals and regularly consumed. The congruity of umami taste of MSG with tomato flavor therefore had the highest rating, which supported the hypothesis that a congruity of olfaction and gustation sensations was developed based on an individual's eating experiences.

Recently, the authors found that the salty taste of NaCl could be enhanced by MSG (Kondo et al., 2015). In this psychophysical study, the salty taste intensity in corresponding to the stimuli was evaluated by the time-intensity method. As compared to 0.80% NaCl solution, the higher intensity of salty taste was detected in 0.58% NaCl solutions added with 0.20% of MSG. Such evidence could be confirmed in the cognitive neuroscience study with near infrared spectroscopy, which the brain responses to 0.58% NaCl + 0.20% MSG obtained almost the same activations in the channels around temporal regions as those to 1.20% NaCl solution. As considering the sodium content, it can be reduced by 35% with the addition of MSG. These results are still only speculation, and the authors are continuing the experiments.

3.6 Summary

In Section 3 we summarize that the everyday life experiences in pairing a taste with an odor develop the association of gustatory and olfactory information in the primary gustatory area. This association effects the perception of the flavor during food consumption.

4 Perception of Hunger and Umami

4.1 Physiological Hunger and Satiety

As we feel hungry, we start eating. Hunger is a signal for depletion of energy source or other factors related to the homeostasis. The well-known signal is blood glucose level, which triggers the hunger center in brain at lateral hypothalamus. Progress in neuroscience revealed that there are other hunger signals, for example, stored fat as well as other hunger centers, for example, arcuate nucleus of hypothalamus.

During eating, blood glucose and other signals such as insulin, leptin, and melanocortin increase and stimulate some nuclei of hypothalamus, which then activate ventromedial hypothalamus. Then, we are commanded to stop eating. The mentioned hunger-satiety system occurs from physiological mechanism, which basically is found in most animals. In the real living environment, hunger-satiety of animal and human is not totally controlled by such mechanism.

Many times, animals and humans stop their eating long before the ingested nutrients enter their blood circulation. Therefore, there must be other additional mechanisms playing roles in controlling hunger and satiety of humans and animals.

4.2 Psychological Hunger and Satiety

The mentioned feedforward regulation system informs us to eat. Physiologically, the gustatory system, vagus nerve, receptors in gastrointestinal system, play important roles in this feedforward regulation. The gustatory stimulation of umami taste can induce secretions of digestive fluids, for example, saliva (Hayakawa et al., 2008) and pancreatic juice (Ohara et al., 1988).

The feedforward regulation system can also be monitored by eating experiences learned by humans and by animals. For example, when animals anticipate intake of sugar-containing foods (having sweet taste), their bodies release insulin even before the absorption of glucose. This phenomenon is called "cephalic insulin" because it is mediated by a neural connection from brain to pancreas. This cephalic insulin is acquired and easily modified by learning. As animals were presented with sweet taste but no foods, they became hypoglycemic because their bodies release insulin with no glucose to be absorbed. On the other hand, animals that were presented the sugar-containing foods, with no sweet taste, became hyperglycemic because no conditioned cephalic insulin is released. This study indicated that hunger and satiety were not solely controlled by physiological function but also by the experiences learned in everyday life.

4.3 Eating Controlled by the Sensations

"Sensory-specific satiety" is another phenomenon used for explaining satiation in eating behavior. At the beginning of our eating, we normally enjoy and feel palatability of the food. During eating, the food palatability and eating rate tend to reduce. After finishing the food, we can eat other foods but do not want to eat the same food anymore. This phenomenon is sensory-specific satiety, which can occur even before the nutrients in such food have been absorbed, or even after smelling the odor of such food again (Rolls and Rolls, 1997). Thus, the sensory-specific satiety is a part of feedforward regulatory system of satiety (Rolls, 1986). Humans having sensory-specific satiety are not physiologically full, but satiated to the sensory factors of such food. A neuroscientific study showed that the human brain mechanisms underlying sensory-specific satiety involve orbitofrontal cortex, the secondary gustatory area (O'Doherty et al., 2000). This finding suggested that the sensory-specific satiety was evoked by sensory information (gustation and olfaction), which could activate human orbitofrontal cortex and could be an effective ingredient in the diet for regulating body weight. In fact, umami taste was found to activate human orbitofrontal cortex (McCabe and Rolls, 2007).

Human eating behavior is however affected by sensory factors of foods. The results from a survey using Dutch Eating Behavior Questionnaire (DEBQ; van Strien et al., 1986) indicated that humans start eating based on the external cues, such as seeing other people eating and smelling the odor of foods.

Wansink and coworkers (Painter et al., 2002) reported that participants ate more candies when the candies were on their desks (convenient and visible) than those in their desks (convenient but not visible) and on the desks 2 m away (inconvenient and not visible). The other study indicated that the serving size of snacks or popcorn also affected the consumption, that is, the larger serving size, the more amount of consumption (Wansink and Cheney, 2005; Wansink and Kim, 2005). Both studies also support the hypothesis that humans start and stop their eating regarding to the external cues (visibility of foods, serving size) not the internal cues, (hunger, satiety).

Therefore, humans' eating behavior can be manipulated by using external cues of various sensations. To restrict someone's diet, strategies such as strong flavor food, small plate, and visible wrappers of eaten foods may be applied. On the contrary, we can serve foods of nice flavor in a large portion without showing any wrappers of eaten foods to encourage more consumption. Umami tastant is a choice for enhancing and making taste of the foods become more palatable. The mentioned strategy can be beneficial to elderly people who normally lose their sensations from natural senescence and medication, which subsequently, they lose their appetite and become malnourished. The strategy can also be applied to people who need to control their weight (Kawai et al., 2009).

4.4 Eating Controlled by the Psychological Factors

DEBQ also revealed that humans start and stop eating based on their emotional states (van Strien et al., 1986). People can lose their appetite when they are under psychological stresses, such as during exams, competitions, and anxiety about something. Sometimes, these psychological stresses can cause binge eating as well, which is called "emotional eating." Under the psychological stresses, people tend to eagerly eat certain foods called "comfort foods." The typical comfort foods for Americans are chocolates and potato chips (Wansink et al., 2003). On the other hand, Komatsu (2008) found that the comfort foods for Japanese university students under psychological stresses are sushi or other Washoku cuisines. Comfort foods of both American and Japanese have familiar tastes of their own and basically satisfy needs for the safety (described in Section 1). This is the reason why they become their comfort foods. People under psychological stresses try to ease themselves by being in the positive emotional state that is back to the sweet memories of the good old days, evoked by eating. While they are eating their comfort foods, most people feel that they cannot stop eating. Thus, the stress-induced eating is also not based on the physiological hunger and satiety.

Another example of eating behavior that is independent of physiological hunger and satiety was reported by Pliner and Chaiken (1990). In their study, male and female subjects were having a meal in the presence of a male or female partner. Both male and female participants ate less in the presence of a partner of the opposite sex than of the same sex. Furthermore, male participants ate more in the presence of an attractive female partner, while female participants ate less in the presence of attractive male partner. These behaviors can be explained by the social motives leading people to behave in a socially desirable manner that judges the appeal of females and males from eating less and more, respectively. These behaviors are motivated as being esteemed and highly evaluated by the others (described in Section 1). A study with university students indicated that females ingested fewer calories when they were with male than with female companions (Young et al., 2009). In addition, they believed that Washoku is a good menu for females due to low calories and health benefits.

4.5 Eating Controlled by the Beliefs

Modern societies, especially developed countries, are suffering from health problems related to improper eating behaviors. Such a situation can be explained by the knowledge on the motivation stage of needs for self-actualization (described in Section 1). In this stage, people do not eat to satisfy their physiological needs. They eat the foods that reflect their own lifestyle. Or it can be said that their eating behavior is a component of their living style. For example, modern young women eat only small amounts by selecting only low- and noncaloric foods in order to maintain their body weight as light as possible. Such behaviors can lead to eating disorders such as binge eating, bulimia, and anorexia. With certain beliefs, people restrict their eating even when they want to eat more. The tendency for restriction of eating is measured also by DEBQ (van Strien et al., 1986) as a "restraint eating."

The restrained eaters finally tend to fail their dieting behaviors. Herman and Mack (1975) reported that female subjects under high-restraint dieting ate less than low-restraint dieting subjects under normal conditions, however, they ate more than the low-restraint subjects under the preload condition, during which they were asked to drink milk shakes before eating. This disinhibition is explained by the Ironic Process Theory (Wegner, 1994). Based on this theory, the more people try not to think about something, the more they think about those things. The restraint eaters try not to think about foods and eating, ironically they are caught for thinking about foods and eating. Finally, they cannot control their eating, which means they fail their dieting program. The failure can make the restrained eater have low self-esteem and may lead to be bulimia or anorexia.

We hypothesize that food containing umami taste is a good preload food for restrained eaters because it can cause a mouth-filling sense (Green and Nachtigal, 2012; Obrist et al., 2014). Therefore, restrained eaters feel full in spite of eating less and can limit their eating with no suffering. Washoku that consists of many umami tastants, low calories, and low-fat components should be a correct menu for dieting people.

5 Summary and Conclusions

In this chapter, Maslow's Motivational Theory is introduced to explain human eating behavior, especially on starting and stopping of eating. Then physiological and psychological mechanisms underlying umami taste perception are introduced. In each section, applications and implementations of the findings from studies on regulations of hunger, satiety, and body weight are discussed. The authors recommend Washoku as a healthy diet for regulating hunger, satiety, and body weight.

References

Blundell, J.E., Finlayson, G., 2004. Is susceptibility to weight gain characterized by homeostatic or hedonic risk factors for overconsumption? Physiol. Behav. 82, 21–25.

Calvo, S.S., Egan, J.M., 2015. The endocrinology of taste receptors. Nat. Rev. Endocrinol. 11, 213–227.

Chandraskekar, J., Hoon, M.A., Ryba, N.J.P., Zuker, C.S., 2006. The taste receptors and cells for mammalian taste. Nature 444, 289–294.

Dalton, P., Doolittle, N., Nagata, H., Breslin, P.A.S., 2000. The merging of the senses: integration of subthreshold taste and smell. Nat. Neurosci. 3, 431–432.

Davidenko, O., Darcel, N., Fromentin, G., Tomé, D., 2013. Control of protein and energy intake—brain mechanisms. Eur. J. Clin. Nutr. 67, 455–461.

Davidson, J.M., Linforth, R.S.T., Hollowood, T.A., Taylor, A.J., 1999. Effect of sucrose on the perceived flavor intensity of chewing gum. J. Agric. Food Chem. 47, 4336–4340.

de Araujo, I.E., Kringelbach, M.L., Rolls, E.T., Hobden, P., 2003a. Representation of umami taste in the human brain. J. Neurophysiol. 90, 313–319.

de Araujo, I.E.T., Rolls, E.T., Kringelbach, M.L., McGlone, F., Phillips, N., 2003b. Taste-olfactory convergence, and the representation of the pleasantness of flavour, in the human brain. Eur. J. Neurosci. 18, 2374–2390.

Delzenne, N., Blundell, J., Brouns, F., Cunningham, K., De Graaf, K., Erkner, A., Lluch, A., Mars, M., Peters, H.P., Westerterp-Plantenga, M., 2010. Gastrointestinal targets of appetite regulation in humans. Obes. Rev. 11, 234–250.

Frank, R.A., Byram, J., 1988. Taste–smell interactions are tastant and odorant dependent. Chem. Senses 13, 445–455.

Fuke, S., Konosu, S., 1991. Taste-active components in some foods: a review of Japanese research. Physiol. Behav. 49, 863–868.

Green, B.G., Nachtigal, D., 2012. Somatosensory factors in taste perception: effects of active tasting and solution temperature. Physiol. Behav. 107, 488–495.

Harriman, A.E., Kare, M.R., 1964. Preference for sodium chloride over lithium chloride by adrenalectomized rats. Am. J. Physiol. 207, 941–943.

Hayakawa, Y., Kawai, M., Torii, K., Uneyama, H., 2008. The effect of umami taste on saliva secretion. Jpn. J. Taste Smell Res. 15, 367–370, (in Japanese).

Herman, C.P., Mack, D., 1975. Restrained and unrestrained eating. J. Pers. 43, 647–660.

Horio, T., Kawamura, Y., 1989. Salivary secretion induced by umami taste. Jpn. J. Oral. Biol. 31, 107–111.

Imada, T., Hao, S.S., Torii, K., Kimura, E., 2014. Supplementing chicken broth with monosodium glutamate reduces energy intake from high fat and sweet snacks in middle-aged healthy women. Appetite 79, 158–165.

Iwatsuki, K., Uneyama, H., 2012. Sense of taste in the gastrointestinal tract. J. Pharmacol. Sci. 118, 123–128.

Janssen, S., Depoortere, I., 2013. Nutrient sensing in the gut: new roads to therapeutics? Trend. Endocrinol. Metab. 24, 92–100.

Kawai, M., Uneyama, H., Miyano, H., 2009. Taste-active components in foods, with concentration on umami compounds. J. Health Sci. 55, 667–673.

Kettenmann, B., Jousmäki, V., Portin, K., Salmelin, R., Kobal, G., Hari, R., 1996. Odorants activate the human superior temporal sulcus. Neurosci. Lett. 203, 143–145.

Komatsu, S., 2008. Rice and sushi cravings: a preliminary study of food craving among Japanese females. Appetite 50, 353–358.
Kondo, R., Onuma, T., Sakai, N., 2015. Salt enhancement by MSG revealed by the time-intensity scale and measurement of salivary gland activities. Jpn. J. Taste Smell. 22, 325–328.
Laeng, B., Berridge, K.C., Butter, C.M., 1993. Pleasantness of a sweet taste during hunger and satiety: effects of gender and "sweet tooth". Appetite 21, 247–254.
Lucas, F., Sclafani, A., 1988. Polycose and sucrose appetite in rats: influence of food deprivation and insulin treatment. Appetite 11, 201–213.
Masic, U., Yeomans, M.R., 2013. Does monosodium glutamate interact with macronutrient composition to influence subsequent appetite? Physiol. Behav. 27, 23–29.
Masic, U., Yeomans, M.R., 2014. Umami flavor enhances appetite but also increases satiety. Am. J. Clin. Nutr. 100, 532–538.
Maslow, A.H., 1943. A theory of human motivation. Psychol. Rev. 50, 370–396.
McCabe, C., Rolls, E.T., 2007. Umami: a delicious flavor formed by convergence of taste and olfactory pathways in the human brain. Eur. J. Neurosci. 25, 1855–1864.
Miyake, Y., Yoshimatsu, H., Sakai, N., 2009. Color variations of pictures of sushi affect palatability and fishness of sushi of tuna. Jpn. J. Taste Smell Res. 16, 403–406, (in Japanese).
Morrot, G., Brochet, F., Dubourdieu, D., 2001. The color of odors. Brain Lang. 79, 309–320.
Murphy, C., Cain, W.S., Bartoshuk, L.M., 1977. Mutual action of taste and olfaction. Sens. Process. 1, 204–211.
Nakamura, E., Hasumura, M., Uneyama, H., Torii, K., 2011. Luminal amino acid-sensing cells in gastric mucosa. Digestion 83 (Suppl. 1), 13–18.
Nakamura, E., Uneyama, H., Torii, K., 2013. Gastrointestinal nutrient chemosensing and the gut-brain axis: significance of glutamate signaling for normal digestion. J. Gastroenterol. Hepatol. 28 (Suppl. 4), 2–8.
Niijima, A., 1991. Effects of oral and intestinal stimulation with umami substance on gastric vagus activity. Physiol. Behav. 49, 1025–1028.
O'Doherty, J., Rolls, E.T., Francis, S., Botwell, R., McGlone, F., Kobal, G., Renner, B., Ahne, G., 2000. Sensory-specific satiety-related olfactory activation of the human orbitofrontal cortex. Neuroreport 11, 893–897.
Obrist, M., Comber, R., Subramanian, S., Piqueras-Fiszman, B., Velasco, C., Spence, C., 2014. Temporal, affective, and embodied characteristics of taste experiences: a framework for design. CHI 2014 Proceedings. Available from: http://dl.acm.org/citation.cfm?id=2557007&CFID=534855459&CFTOKEN=98946057
Ohara, I., Otsuka, S., Yugari, Y., 1988. Cephalic phase response of pancreatic exocrine secretion in conscious dogs. Am. J. Physiol. 254, G424–G428.
Painter, J.E., Wansink, B., Hieggelke, J.B., 2002. How visibility and convenience influence candy consumption. Appetite 38, 237–238.
Passe, D.H., Stofan, J.R., Rowe, C.L., Horswill, C.A., Murray, R., 2009. Exercise condition affects hedonic responses to sodium in a sport drink. Appetite 52, 561–567.
Pliner, P., Chaiken, S., 1990. Eating, social motives, and self-presentation in women and men. J. Exp. Soc. Psychol. 26, 240–254.
Prescott, J., Johnstone, V., Francis, J., 2004. Odor–taste interactions: effects of attention strategies during exposure. Chem. Senses 18, 389–404.
Prescott, J., Stevenson, R., 2015. Chemosensory integration and the perception of flavor. In: Doty, R.L. (Ed.), Handbook of Olfaction and Gustation. third ed. John Wiley & Sons, Inc., Hoboken, NJ.

Rassin, D.K., Sturman, J.A., Gaull, G.E., 1978. Taurine and other free amino acids in milk of human and other mammals. Early Human Dev. 2, 1–13.

Robinson, E., Higgs, S., 2012. Liking food less: the impact of social influence on food liking evaluations in female students. PLoS One 7, e48858.

Rolls, B.J., 1986. Sensory-specific satiety. Nutr. Rev. 44, 93–101.

Rolls, E.T., 2009. Functional neuroimaging of umami taste: what makes umami pleasant? Am. J. Clin. Nutr. 90 (Suppl.), 804S–813S.

Rolls, E.T., Rolls, J.H., 1997. Olfactory sensory-specific satiety in humans. Physiol. Behav. 61, 461–473.

Rozin, P., 1982. Taste-smell confusions and the duality of the olfactory sense. Percep. Psychophys. 31, 397–401.

Sakai, N., 2009. Cognitive and contextual factors affecting olfactory and gustatory perception and palatability of beverages. ChemoSense 11, 1–6.

Sakai, N., 2011. Tasting with eyes. i-Perception 2, 945, Available from: http://i-perception.perceptionweb.com/journal/I/article/ic945.

Sakai, N., 2014. The psychology of eating from the point of view of experimental, social, and applied psychology. Psychol. Russia State Art 7, 14–22.

Sakai, N., Fujimoto, M., Murata, M., 2011. Flavors which show interaction between umami taste. Jpn. J. Taste Smell Res. 18, 375–378.

Sakai, N., Imada, S., 2003. Bilateral lesions of the insular cortex or the prefrontal block the association between taste and odor in the rat. Neurobiol. Learn. Mem. 80, 24–31.

Sakai, N., Ishihara, Y., Saito, S., 2002. Odor induced taste enhancement is affected by the effect of taste images for odor. Jpn. J. Taste Smell Res. 9, 423–426.

Sakai, N., Kobayakawa, T., Gotow, N., Saito, S., Imada, S., 2001. Enhancement of sweetness ratings aspartame by a vanilla odor presented either by orthonasal or retronasal routes. Percet. Motor Skills 92, 1002–1008.

Sakai, N., Yamasaki, Y., 2009. Packages of chocolate affect taste perceptions and palatability of chocolate. Jpn. J. Taste Smell Res. 16, 407–410, (in Japanese).

San Gabriel, A., Uneyama, H., 2013. Amino acid sensing in the gastrointestinal tract. Amino Acids 45, 451–461.

Spence, C., Michel, C., Smith, B., 2014. Airplane noise and the taste of umami. Flavor 3, 2.

Spence, C., Piqueras-Fiszman, B., 2014. The perfect meal: the multisensory science of food and drinking. John Wiley & Sons, Ltd, Oxford, UK.

Steiner, J.E., Glaser, D., Hawilo, M.E., Berridge, K.C., 2001. Comparative expression of hedonic impact: affective reactions to taste by human infants and other primates. Neurosci. Biobehav. Rev. 25, 53–74.

Stevenson, R.J., Boakes, R.A., Prescott, J., 1998. Changes in odor sweetness resulting from implicit learning of a simultaneous odor-sweetness association: an example of learned synesthesia. Learn. Motiv. 29, 113–132.

Stevenson, R.J., Boakes, R.A., Wilson, J.P., 2000. Resistance to extinction of conditioned odor perceptions: evaluative conditioning is not unique. J. Exp. Psychol. Learn. Mem. Cogn. 26, 423–440.

Stevenson, R.J., Case, T.I., 2003. Preexposure to the stimulus elements, but not training to detect them, retards human odor-taste learning. Behav. Proc. 61, 13–25.

Stevenson, R.J., Prescott, J., Boakes, R.A., 1999. Confusing taste and smells: how odours can influence the perception of sweet and sour tastes. Chem. Senses 24, 627–635.

Stevenson, R.J., Tomiczek, C., 2007. Olfactory-induced synesthesias: a review and model. Psychol. Bull. 133, 294–309.

Torii, K., Mimura, T., Yugari, Y., 1987. Biochemical mechanism of umami taste perception and effect of dietary protein on taste preference for amino acids

and sodium chloride in rats. In: Kawamura, Y., Kare, M.R. (Eds.), Umami: A Basic Taste. Marcel Dekker, New York, NY, pp. 513–563.
Tsurugizawa, T., 2014. Neuroimaging of gut nutrient perception. Curr. Pharm. Des. 20, 2738–2744.
Tsurugizawa, T., Uematsu, A., Kondoh, T., Nakamura, E., Hasumura, M., Hirota, M., Uneyama, H., Torii, K., 2009. Mechanisms of neural response to gastrointestinal nutritive stimuli: the gut-brain axis. Gastroenterology 137, 262–273.
Uneyama, H., 2011. Nutritional and physiological significance of luminal glutamate-sensing in the gastrointestinal functions. Yakugaku Zasshi 131, 1699–1709.
Uneyama, H., Niijima, A., San Gabriel, A., Torii, K., 2006. Luminal amino acid sensing in the rat gastric mucosa. Am. J. Physiol. Gastrointest. Liver Physiol. 291, G1163–G1170.
van Strien, T., Frijiters, J.E.R., Bergers, G.P.A., Defares, P.B., 1986. The Dutch eating behavior questionnaire (DEBQ) for assessment of restrained, emotional, and external eating behavior. Int. J. Eat. Disord. 5, 295–315.
Veldhuizen, M., Shepard, T.G., Wang, M.-F., Marks, L.E., 2010. Coactivation of gustatory and olfactory signals in flavor perception. Chem. Senses 35, 121–133.
Ventura, A.K., Beauchamp, G.K., Mennella, J.A., 2012. Infant regulation of intake: the effect of free glutamate content in infant formulas. Am. J. Clin. Nutr. 95, 875–881.
Vickers, Z.M., Bourne, M.C., 1976. Crispness in foods—a review. J. Food Sci. 41, 1153–1157.
Wansink, B., Cheney, M.M., 2005. Super bowls: serving bowl size and food consumption. JAMA 293, 1727–1728.
Wansink, B., Cheney, M.M., Chan, N., 2003. Exploring comfort food preferences across age and gender. Physiol. Behav. 79, 739–747.
Wansink, B., Kim, J., 2005. Bad popcorn in big buckets: portion size can influence intake as much as taste. J. Nutr. Educ. Behav. 37, 242–245.
Wegner, D.M., 1994. Ironic processes of mental control. Psychol. Rev. 101, 34–52.
Westerterp-Plantenga, M.S., Lemmens, S.G., Westerterp, K.R., 2012. Dietary protein—its role in satiety, energetics, weight loss and health. Br. J. Nutr. 108, S105–S112.
Yamamoto, T., Sawa, K., 2000. c-Fos-like immunoreactivity in the brainstem following gastric loads of various chemical solutions in rats. Brain Res. 866, 135–143.
Young, M.E., Mizzau, M., Mai, N.T., Sirisegaram, A., Wilson, M., 2009. Food for thought. What you eat depends on your sex and eating companions. Appetite 53, 268–271.
Zampini, M., Spence, C., 2004. The role of auditory cues in modulating the perceived crispness and staleness of potato chips. J. Sens. Stud. 19, 347–363.

SUBJECT INDEX

FSC
www.fsc.org
MIX
Papier | Fördert gute Waldnutzung
FSC® C083411

Kolibri 360 GmbH
Ferdinand-Jühlke-Straße 7
99095 Erfurt, Deutschland
produktsicherheit@kolibri360.de